JOURNAL OF CHROMATOGRAPHY LIBRARY — volume 52

capillary electrophoresis

principles, practice and applications

JOURNAL OF CHROMATOGRAPHY LIBRARY — volume 52

capillary electrophoresis

principles, practice and applications

S. F. Y. Li

Department of Chemistry, National University of Singapore,
10 Kent Ridge Crescent, Singapore 0511, Republic of Singapore

ELSEVIER
Amsterdam — London — New York — Tokyo 1993

ELSEVIER SCIENCE PUBLISHERS B.V.
Sara Burgerhartstraat 25
P.O. Box 211, 1000 AE Amsterdam, The Netherlands

ISBN 0-444-81590-2

This book is printed on acid-free paper.

Transferred to digital printing 2006

Preface

Capillary electrophoresis (CE) has developed into an exciting and extremely powerful analytical technique in recent years. Along with advances in instrumentation and separation methodologies, a wide range of applications have been developed in greatly diverse fields, such as chemical, biotechnological, environmental and pharmaceutical analysis.

The main aim of this book is to provide a comprehensive reference on CE. Chapter 1 covers the principles of various modes of operation of the technique. In Chapter 2, methods of sample introduction are discussed. In Chapter 3, detection techniques are dealt with rather extensively. Chapter 4 provides a detailed treatment of contemporary column technology for CE. The uses of coated columns and gel-filled columns are discussed in this chapter. In Chapter 5, a detailed discussion is given on the different types of electrolyte systems utilized to obtain unique separation mechanisms, such as the use of surfactants in micellar electrokinetic chromatography. In Chapter 6, special instrumental features and separation methodologies not found in routine CE operations are discussed. In Chapter 7, selected applications in all the important areas of current interest are described. Finally, in Chapter 8, the latest advances in CE and its prospect for growth are considered.

This book has been designed to be a reference covering all aspects of capillary electrophoretic techniques. It is intended to meet the ever-growing need for a comprehensive and balanced text on an analytical technique which has generated tremendous interests in recent years.

In addition to being a reference work, this book can also serve as a modern textbook for advanced undergraduate and graduate courses in many disciplines, including analytical chemistry, analytical biochemistry, environmental science, pharmaceutical analysis and biotechnology.

The writing of a book on CE is an outgrowth of related research endeavors. My knowledge in this field owes much to many colleagues and students, especially Dr. H.K. Lee, L.H. Kwek, C.L. Ng, C.P. Ong, Y.J. Yao, S.K. Yeo and Y.F. Yik. I would also like to express my appreciation to Dr. M. Chung, Prof. N. Dovichi, Mr. B. Egardo, Prof. Y. Hirata, Prof. K. Jinno, Dr. S.B. Khoo, Dr. T. Leung, Dr. P. Marriott, Prof. H. Nakamura and Prof. E.S. Yeung for their interesting discussions, and the many students who have worked in my laboratory in recent years for their contributions in various ways, including K.T. Chan, K.P. Chin, S.F.S. Chong, J.M.Y. Lee, M.L. Lee, K. Li, L.K. Lim, Y.H. Poh, C.S. Seet, S.M. Tan and K.H. Teo.

I am indebted to authors and publishers of relevant papers for the information and inspiration therein, the reprints and preprints, and above all, their kind permission to reproduce copyrighted material. I have also benefited from the numerous suggestions of reviewers of my publications. The many helpful comments of my critical reader, Dr. U. Tjaden, are gratefully acknowledged. I would also like to thank Prof. S. Terabe for reading and commenting on the draft of Chapter 5, and many other discussions and kind suggestions over the years.

In the course of preparing the manuscript, many people have assisted me and offered their support. I would like to express my appreciation to Miss L.H. Kwek, who furnished technical illustrations for five of the chapters, Miss C.P. Ong, for typing the draft of Chapters 2 and 3, Miss Y.F. Yik and Mr. S.M. Tan for sending out permission requests, Miss C.L. Ng and Miss Y.F. Yik for proofreading Chapter 3, and Mr. K.H. Teo for preparing the index of Chapters 4 and 5. I would also like to thank the National University of Singapore for supporting my research, and several companies for their generous loan of equipment, including Beckman (Singapore) Instruments (P/ACE 2000), Diagnostic Biotechnology and ITS Distributors (ABI 270A), Balmar Marketing (ISCO 3850), Fisons Instruments (Singapore) (Carlo Erba) and Schmidt Scientific (Shimadzu).

I would also like to express my sincere gratitude to Dr. A.E. Hollander, Prof. K.Y. Sim, Prof. W.A. Wakeham, my family and numerous friends for their advice and encouragement. Finally, I would like to thank my wife, Han, and children, Anne, Brian and Conrad for their forbearance during the months spent writing this book.

SAM F.Y. LI

Contents

List of Symbols

The symbols are defined as follows unless stated otherwise in the text:

a_j	activity of ion j
A	cross sectional area
b	bias factor
C	concentration
%C	concentration of crosslinking agent (C = [bis + acryl]/acryl, where bis and acryl are weight of bisacrylamide and acrylamide, respectively)
C_f	concentration of fluorescent molecule
C_s	concentration of sample
cmc	critical micelle concentration
d	intermicellar distance
d_c	diameter of capillary
D	diffusion coefficient
D_{aq}	diffusion coefficient in aqueous phase
D_{mc}	diffusion coefficient in micellar phase
DR	dynamic reserve
e	excess charge in solution per unit area
e_0	electronic charge
e_h	acceptable increase in plate height
E	electric field
E_z^{OH}	Ohm's law contribution of electric field in z direction
Ex	effect of a particular factor x
f	focal length; fraction of plate number lost due to extracolumn effects
F	Faraday's constant; fraction of analyte ion free from polymer ion
F_e	enhancement factor
Δh	injection height
H_{aq}	plate height contribution due to intermicelle mass transfer in aqueous phase
H_c	plate height contribution due to intracolumn mass transfer
H_{col}	overall column plate height
H_l	plate height contribution due to longitudinal diffusion
H_{mc}	plate height contribution due to sorption–desorption kinetics
H_T	plate height contribution due to temperature gradient effects
H_{TH}	height equivalent to theoretical plate
I	current; light intensity
I_0	intensity of incident light
I_F	intensity of fluorescent light
J_i	flux of analyte i
$J_{z,i}^{OH}$	flux of analyte i due to Ohm's law contribution to electric field along z direction
k	Boltzman's constant; molar conductivity

Δk	conductivity difference
k'	capacity factor
k_d	desorption rate constant
k_i	rate of excitation of analyte
k_{-i}	non-radiative decay of excited species
k_{ij}	selectivity coefficient
k'_{opt}	optimum capacity factor
K	distribution coefficient
K'	overall formation constant for a particular metal ion
K_f	overall formation constant for a metal ion at infinite dilution
K_{HA}	protolysis constant
K_i	formation constant of complexes of solute i
l	length
l_{inj}	injection plug length
l_{max}	maximum path length
L	length of capillary from injection point to detector
L_{cap}	length of capillary
M_i	molecule i
N	theoretical plate number
n	number
n_i	amount of sample i
N_{max}	maximum number of theoretical plates
N_{ob}	observed theoretical plate number
N_{true}	true theoretical plate number
P	power
ΔP	pressure difference
P_c	critical pressure
$P_{DOD/W}$	partition coefficient in the dodecane/water system
P_{WM}	partition coefficient of solute between water and micellar phase
q_c	volume of sample introduced into capillary
q_{inj}	volume injected
q_s	volume of sample introduced into split-vent tubing
Q	heat released per unit volume
ΔQ	difference between rates of solvent in and out of reservoir
Q_{in}	rate of solvent into reservoir
Q_{out}	rate of solvent out of reservoir
r	radius
r_c	inner radius of capillary
r_{cry}	crystal radius
r_o	outer radius of capillary
R	universal gas constant; resolution
R_s	retention or migration time ratio
$R_{0.5}$	radius of molecules for which 50% of total gel volume is available
RI	refractive index
s	separation factor
s_α	selectivity

S	micelle size
t	time
t_0	migration time of solute with no interaction with stationary phase or with the micelles
t_{ad}	mean residence time of the adsorbed solute
t_{crit}	critical value of t-test
t_m	dead time obtained by measuring the migration time of neutral marker
t_{mc}	migration time of micelle
t_M	migration time
t_R	retention time
t_T	travel time
T	temperature
T_0	temperature of solution
T_c	critical temperature
%T	total gel concentration (T = [bis + acryl]/V, where bis, acryl and V are weight of bisacrylamide, weight of acrylamide and total volume, respectively)
TR	transfer ratio (number of mobile phase molecules displaced by one analyte molecule)
v	velocity
Δv	difference in migration velocity
v^{α}	migration velocity in α phase
v^{β}	migration velocity in β phase
v_b	velocity of neutral band
v_{diff}	velocity difference
$v_{ep}(p)$	electrophoretic velocity of polymer ion
$v_{tot,i}$	total ionic velocity
V	voltage; volume
V_{ac}	ac voltage
V_{dc}	dc voltage
V_G	gate voltage
V_{inj}	electrokinetic injection voltage
V_{obs}	observed voltage
V_{SDS}	specific partial molar volume of sodium dodecyl sulfate
w	amount injected
w_T	amount injected during travel time
W	peak width
x	mole fraction
X	distance in x direction
ΔX_0	original width of sample zone
ΔX_k	zone broadening due to conductivity difference
$\Delta X'_{k,D}$	zone broadening at the front boundary due to combined effects of conductivity difference and diffusion
$\Delta X''_{k,D}$	zone broadening at the rear boundary due to combined effects of conductivity difference and diffusion
z	distance along length of capillary
Z	charge
ΔZ	charge difference

α	degree of dissociation
α_A	degree of dissociation of acid
α_B	degree of dissociation of base
α_f	absorptivity of fluorescent species
α_s	absorptivity of sample
$[\alpha]_D^{25}$	specific rotation
γ	ratio of buffer concentration in the original sample solution to that in the column
δ	thickness of electrical double layer
ε	dielectric constant
ζ	zeta potential
ζ_{eo}	zeta potential of tube wall
ζ_{ep}	zeta potential of solute
η	viscosity
θ	angle
θ_{core}	temperature excess within core of liquid
θ_{wall}	temperature excess at the capillary wall
ϑ	angle
κ	conductance
κ_Z	local solution conductance
λ	thermal conductivity
μ	mobility
$\overline{\mu}$	average mobility
$\mu_{app,s}^{ps}$	pseudo-apparent mobility of solute
$\mu_{eff,s}^{ps}$	pseudo-effective mobility of solute
μ_{eo}	coefficient of electroosmotic flow
μ_{ep}	electrophoretic mobility
μ_i^c	mobility of complexed solute i
μ_i^f	mobility of free solute i
μ_i^k	mobility of ion i in zone k
μ_L	mobility of leading ion
μ_R	mobility of counter-ion
μ_r	relative mobility
$\mu_{tot,i}$	total ionic mobility
ρ	density
σ	zone width
σ_{con}^2	variance of migrating zone width
σ_{det}^2	band broadening due to detection cell volume
σ_{dif}^2	band broadening due to molecular diffusion
σ_{ec}^2	extracolumn variance due to both injection and detection cell volume
σ_{ic}^2	irreversible contribution to variance
σ_{inj}^2	band broadening due to injection volume
σ_{Joule}^2	band broadening due to Joule heating
σ_{rc}^2	reversible contribution to variance
σ_{total}^2	total band broadening
ϕ	capacitance
ϕ_c	capacitance of capillary tubing

ϕ_i	capacitance of diffuse layer at inner capillary/inner solution interface
ϕ_o	capacitance of diffuse layer at outer capillary/outer solution interface
ϕ_T	total capacitance
φ	electrical potential
χ	retention parameter
ω	frequency of modulation

Abbreviations

2-AP	2-aminopyridine
7-OHMTX	7-hydroxymethotrexate
AA	amino acid
ABA	aminobutyric acid
ac	alternating current
ACES	N-(2-acetamido)-2-aminoethanesulfonic acid
ACN	acetonitrile
ACTH	adrenocorticotrophic hormone
ADP	adenosine 5′-diphosphate
AG	agarose gel
AIBA	2-aminoisobutyric acid
AMMEDIOL	2-amino-2-methyl-1,3-propanediol
AMP	adenosine monophosphate
ANS	arylaminonaphthalene sulfonate
AP	alkaline phosphatase
APF	aryl-pentafluoro
APS	(3-aminopropyl)trimethoxysilane
ATP	adenosine-5′-triphosphate
BALF	bronchoalveolar lavage fluid
BHI	biosynthetic human insulin
Bis	N,N′-methylenebisacrylamide
bST	bovine somatostatin
β-CMCD	2-O-carboxymethyl-β-cyclodextrin
CAc	citric acid
CAE	capillary array electrophoresis
cAMP	adenosine 3′,5′-cyclic monophosphate
CAPS	3-(cyclohexylamino)-1-propanesulfonic acid
CAPSO	3-(cyclohexylamino)-2-hydroxy-1-propanesulfonate
CBI	1-cyano-2-substituted-benz[f]isoindole
CBQCA	3-(4-carboxybenzoyl)-2-quinolinecarboxaldehyde
CBS	N-cyclohexyl-2-benzothiazole sulfenamide
CCD	charge-coupled device
CD	cyclodextrin
CD-EKC	cyclodextrin modified electrokinetic chromatography
CD-MEKC	cyclodextrin modified micellar electrokinetic chromatography
CE	capillary electrophoresis
Cd-Te	cadmium–tellurium
CEC	capillary electrokinetic chromatography
CFAB	continuous-flow fast atom bombardment

CGE	capillary gel electrophoresis
CHAPS	3-[3-(chloroamidepropyl)dimethylammonio]-1-propane-sulfonate
CHAPSO	3-[3-(chloroamidepropyl)dimethylammonio]-2-hydroxy-1-propane-sulfonate
CHES	2-(N-cyclohexylamino)ethanesulfonic acid
CIA	capillary ion analysis
CIEF	capillary isoelectric focusing
CITP	capillary isotachophoresis
CPM	cefpiramide
CTAB	cetyltrimethylammonium bromide
CTAC	cetyltrimethylammonium chloride
CVL	laser capillary vibration
CZE	capillary zone electrophoresis
CZE-CF-FAB-MS	capillary zone electrophoresis–continuous flow–fast atom bombardment-mass spectrometry
CZE-ESI-MS	capillary zone electrophoresis–electrospray ionization–mass spectrometry
DABSYL	4,4-dimethylaminoazobenzene-4′-sulfonyl chloride
dc	direct current
DDAC	diallyldimethylammonium chloride
DEAE	diethylaminoethyl
DET	detection
DHBA	3,4-dihydroxybenzylamine
di-O-Me-β-CD	heptakis(2,6-di-O-methyl)-β-cyclodextrin
DM-β-CD	2,6-di-O-methyl-β-cyclodextrin
DNP	dinitrophenyl
DNS	dansylated
dNTP	deoxyribonucleoside triphosphate
DTAB	dodecyltrimethylammonium bromide
DTAC	dodecyltrimethylammonium chloride
EACA	ϵ-aminocaproic acid
ED	electrically driven
EDGE	ethyleneglycol diglycid ether
EDTA	ethylene diaminetetraacetic acid
ED-OTLC	electrically-driven open tubular liquid chromatography
ELISA	enzyme-linked immunosorbent assay
ESI	electrospray ionization
FAB	fast-atom bombardment
FAM	6-carboxy-fluorescein
FDCD	fluorescence-detected circular dichroism
FEP	fluoroethylene propylene
FITC	fluorescein isothiocyanate
FL	fluorescence
FMOC	9-fluorenylmethylchloroformate
FTH	fluorescein thiohydantoin

FT-IR	Fourier-transformed infrared
GC	gas chromatography
GITC	2,3,4-6-tetra-O-acetyl-β-D-glucopyranosyl isothiocyanate
Gly-Gly	glycylglycine
GMP	guanosine 5′-monophosphate
GPC	gel permeation chromatography
HAc	acetic acid
Hb	hemoglobin
HDL	high density lipoprotein
HEC	hydroxyethylcellulase
He-Cd	helium–cadmium
He-Ne	helium–neon
HEPES	N-2-hydroxyethylpiperazine-N′-2-ethanesulfonic acid
HEPPSO	4-(2-hydroxyethyl)piperazine-1-hydroxypropanesulfonic acid
hGH	human growth hormone
His	histidine
HPC	hydroxypropylcellulose
HPEFS	4-(2-hydroxyethyl)piperazine-1-ethanesulfonic acid
hPI	human proinsulin
HPMC	hydroxypropylmethylcellulose
HQS	8-hydroxyquinoline-5-sulfonic acid
HV	high voltage
ICP	inductively coupled plasma
I.D.	internal diameter
IDL	intermediate density lipoprotein
IEF	isoelectric focusing
IgA	immunoglobulin A
IgG	immunoglobulin G
IM	ionic matrix
IMP	inosine 5′-monophosphate
INJ	injection
IPLC	ion-pairing liquid chromatography
ITP	isotachophoresis
JOE	2′,7′-dimethoxy-4′,5′-dichloro-6-carboxy-fluorescein
LC	liquid chromatography
LCPL	left circular polarized light
LDL	low density lipoprotein
LE	leading electrolyte
LIF	laser-induced fluorescence
LMT	N-lauroyl-N-methyltaurate
MAPS	multiple antigen peptide
MCE	microemulsion capillary electrophoresis
MECC	micellar electrokinetic capillary chromatography

MEKC	micellar electrokinetic chromatography
MES	2-(N-morpholino)ethane sulfonic acid
Met	methionine
MHEC	methylhydroxyethylcellulose
MIEEKFED	metal-insulator-electrokinetic field effect device
MM	molecular mass
MoAb	monoclonal antibody
MOPS	3-(N-morpholino)propanesulfonic acid
MS	mass spectrometry
MTX	methotrexate
m/Z	mass to charge ratio
NA	numerical aperture
NaCh	sodium cholate
NaDCh	sodium deoxycholate
NADH	nicotinamide adenine dinucleotide-reduced form
NBD	4-chloro-7-nitrobenzofuran
NDA	naphthalene-dialdehyde
NE	norepinephrine
NTP	ribonucleoside triphosphate
O.D.	outer diameter
ODS	octadecylsilane
OPA	o-phthaldialdehyde
ORM	overlapping resolution mapping
PAG	polyacrylamide gel
PAGE	polyacrylamide gel electrophoresis
PAH	polyaromatic hydrocarbon
par	4(2-pyridylazo)resorcinol
PD	protein distribution
PEG	polyethylene glycol
PEI	polyethyleneimine
PEP	poly(ethylene-propylene)
PFEP	perfluorinated ethylene-propylene
PFIB	perfluoroisobutylene
PIPES	piperazine-N,N′-bis(2-ethane)sulfonic acid
PL	pyridoxal
PM	pyridoxamine
PMG	polymethylglutamate
PMP	1-phenyl-3-methyl-5-pyrazolone
PMT	photomultiplier
PN	pyridoxine
PNP	p-nitrophenyl
Poly(A)	polyadenylic acid
Poly(dA)	polydeoxyadenylic acid
PSL	porous silica layered
PTFE	polytetrafluoroethylene

PVA	polyvinylalcohol
PVP	polyvinylpyrrolidone
RBC	red blood cell
rCD4	recombinant T4 receptor protein
RCPL	right circular polarized light
rDNA	recombinant DNA
rhGH	recombinant human growth hormone
rhIL-3	recombinant interleukin-3
rhSOD	recombinant superoxide dismutase
rHuEPO	recombinant human erythropoietin
ROX	6-carboxy-X-rhodamine
RP	reversed-phase
rt-PA	recombinant tissue plasminogen
SCE	supercritical capillary electrophoresis
SDBS	sodium dodecyl benzene sulfonate
SDS	sodium dodecyl sulfate
SDS-PAGE	sodium dodecyl sulfate-polyacrylamide gel electrophoresis
SDVal	sodium-N-dodecanoyl-L-valinate
SHS	sodium heptyl sulfate
SIM	selected ion monitoring
S/N	signal-to-noise ratio
SS	stainless steel
SSCE	sodium saturated calomel electrode
STC	sodium taurocholate
STS	sodium tetradecyl sulfate
TAA	tetraalkylammonium
TAMRA	N,N,N′,N′-tetramethyl-6-carboxyl-rhodamine
TBA	tetrabutylammonium
TBAB	tetrabutylammonium bromide
TBBS	N-t-butyl-2-benzothiazole sulfenamide
TBE	Tris-borate-EDTA buffer
TE	terminating electrolyte
TEA	triethanolamine
TEMED	N,N,N′,N′-tetramethylethylenediamine
TES	N-tris(hydroxymethyl)methyl-2-aminoethanesulfonic acid
TG	thermogravimetry
THA	tetrahexylammonium
THAP	tetrahexylammonium perchlorate
THF	tetrahydrofuran
TIM	transient ionic matrix
TlI	thallium iodide
TMAB	tetramethylammonium bromide
TMCS	trimethylchlorosilane
TM-β-CD	2,3,6-tri-O-methyl-β-cyclodextrin
TNS	2-p-toluidonaphthalene sulfonate

Tr	transferrin
Tricine	N-[tris-hydroxymethyl)-methyl]glycine
Tris	tris(hydroxymethyl-aminomethane)
TTAB	tetradecyltrimethylammonium bromide
TTHA	triethylenetetraamine hexaacetic acid
USEPA	United States Environmental Protection Agency
UV	ultraviolet
UV-vis	ultraviolet-visible
VLDL	very low density lipoprotein
ZEST	zone electrophoresis sample treatment

CHAPTER 1

Introduction

Capillary electrophoresis (CE) is a modern analytical technique which permits rapid and efficient separations of charged components present in small sample volumes. Separations are based on the differences in electrophoretic mobilities of ions in electrophoretic media inside small capillaries [1–21]. Chemical, biomedical and pharmaceutical applications of CE are discussed in Chapter 7 of this book. Some examples include the separations of proteins and peptides, tryptic mapping, DNA sequencing, serum analysis, analysis of neurotransmitters in single cells, determination of organic and inorganic ions, and chiral separations. CE offers clear advantages over slab-gel electrophoresis in terms of speed, ease of automation, and quantitation. The technique provides efficiencies up to two orders of magnitude greater than high-performance liquid chromatography (HPLC). Currently CE is increasingly seen as being either an alternative separation method capable of faster analysis and higher efficiency than HPLC or as a complementary technique to HPLC to augment the information obtained from the analysis (see Section 6.9).

The 1980s have been a period of rapid growth for CE, which is evident in terms of the increases in the number of publications, scientific meetings, commercial instruments and separation methodologies related to this technique. There will certainly be further developments in CE. To appreciate the reasons for the tremendous interest in CE, it would be worthwhile to examine its historical background, its current state of development and its future potential. The primary purposes of this chapter are to provide a picture of the evolution of CE, to present an overview of the different modes of contemporary CE techniques, to give an outline of the basic mechanism of separation in CE, and finally to make a comparison of CE with other separation techniques to highlight the areas where there may be important future developments.

1.1 HISTORICAL BACKGROUND

The history of development of capillary electrophoresis has been traced back to more than a century ago by Compton and Brownlee [22]. Table 1.1 presents a historical timetable of contributions to the advance of modern CE technology.

TABLE 1.1

HISTORICAL DEVELOPMENT OF CAPILLARY ELECTROPHORESIS (adapted from [22])

Year	Researchers	Development
1886	Lodge [23]	H^+ migration in a tube of phenolpthalein "jelly".
1892	Smirnow [24]	electro-fractionation of diptheria toxin solution.
1899	Hardy [25]	globulin movement in "U" tube with electric current.
1905	Hardy [26]	detailed study of globulins with various "U" tube designs.
1907	Field and Teague [27]	toxin/antitoxin separations via agar tube bridges between sample and water.
1923	Kendall and Crittenden [28]	preparative separation of isotopes in agar "U" tube.
1930	Tiselius [29]	moving boundary studies of proteins in solution.
1937	Tiselius [30]	improved apparatus for moving boundary studies.
1939	Coolidge [31]	electrophoretic separation of serum proteins in tubes of glass wool.
1946	Consden *et al.* [32]	"ionophoresis" of amino acids and peptides in silica gel slab; first "blotting" experiments.
1950	Haglund and Tiselius [33]	electrophoresis in a glass powder column.
1956	Porath [34]	column electrophoresis using cellulose powder.
1964	Ornstein [35]	design of apparatus for tube "disc" electrophoresis.
1965	Tiselius [36]	"free zone" electrophoresis of virus particles in 3 mm I.D. rotating capillary.
1965	Hjerten *et al.* [37]	"particle seiving" electrophoresis of ribosomes in polyacrylamide tube gels.
1967	Hjerten [38]	free solution electrophoresis in 3 mm tube.
1974	Virtenen [39]	demonstrated advantages of small I.D. columns.
1979	Mikkers *et al.* [40]	electrophoresis in polymer capillaries.
1981	Jorgenson and Lukacs [1]	theoretical and experimental approaches to high resolution electrophoresis in glass capillaries.
1983	Hjerten [41]	adaptation of SDS-PAGE to capillary columns for capillary gel electrophoresis.
1984	Terabe *et al.* [42]	micellar electrokinetic chromatography for separation of neutral compounds.
1987	Cohen and Karger [43]	demonstration of high efficiency using small I.D. tubings in capillary gel electrophoresis.
1989		availability of commercial CE instrument.

It is not surprising to note that the growth of CE can be attributed to fundamental contributions in various separation sciences, particulary electrophoresis and chromatography.

As early as the late 1800s, electrophoretic separations were attempted in free

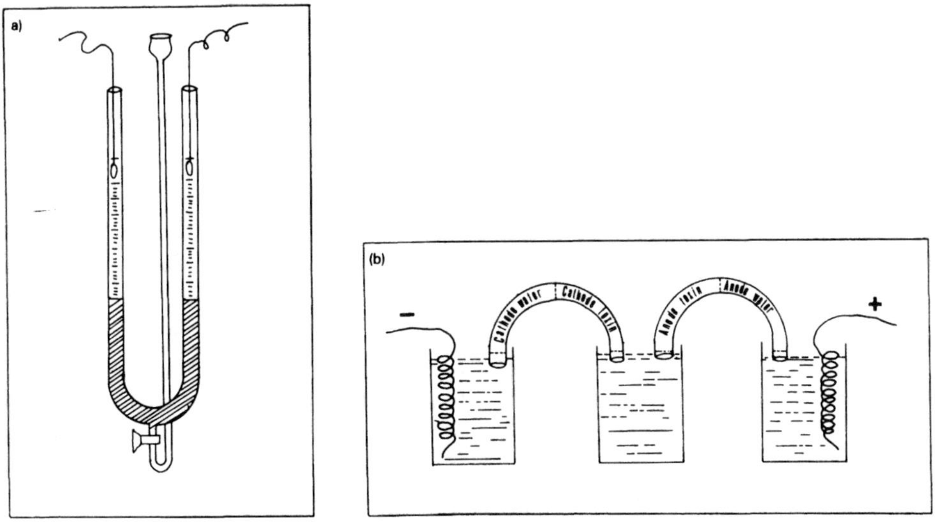

Fig. 1.1. (a) A glass U tube apparatus used in early experiments with free-solution electrophoresis. Electrodes made of platinum foil were immersed in the electrolyte solution. The sample solution with indicator dye was at the bottom of the U tube. (b) An inverted U tube apparatus which consisted of two tubes filled with agar bridging the sample reserviour and reserviours of distilled water. (Adapted from Ref. 22 with permission of Eaton Publishing Co.)

solutions as well as various gels. Many early experiments were performed using glass U tubes with electrodes connected to each of the tubes' arms as shown in Fig 1.1. Figure 1.1a shows a U tube in the upright configuration whereas Fig. 1.1b illustrates an inverted U tube instrument. The experiments were performed using direct current of up to several hundred volts. The separation of various types of samples, such as ions, isotopes, toxins and proteins was investigated. In order to overcome problems of convective mixing which were encountered in electrophoretic separations performed in free solutions, various stabilizing media have been employed, such as agar, cellulose powder, glass wool, paper, silica gel and acrylamide.

An alternative approach to alleviate thermal convection problems in free solution electrophoresis was the use of tubes with small internal diameters. These small tubes or capillaries dissipate heat better and provide a more uniform thermal cross-section of the sample within the tube. Provided ideal conditions can be maintained, samples migrate rapidly as a flat plug with resolution limited only to diffusion [1–6]. Hence, the technique has the potential of achieving extremely high efficiency in separations.

At its early stage of development, capillary electrophoresis was originally described as free solution electrophoresis in capillaries [38]. Hjerten provided the earliest demonstration of the use of high electric field strength in free solution electrophoresis in 3 mm I.D. capillaries in 1967. Virtenen described the advantages of using smaller diameter columns in 1974 [39]. Mikkers *et al.* [40] performed zone electrophoresis in instrumentation adapted from isotachophoresis employing 200 μm I.D. PTFE capillaries. These earlier studies were unable to demonstrate the high separation efficiencies achievable because of sample overloading, a condition induced by poor detector sensitivity and large injection volumes.

The most widely accepted initial demonstration of the power of capillary electrophoresis was that by Jorgenson and Lukacs [1–3]. The pioneering paper on modern CE by these authors included a brief discussion of simple theory of dispersion in CE and provided the first demonstration of high separation efficiency with high field strength in narrow (less than 100 μm I.D.) capillaries [1].

The invention of micellar electrokinetic chromatography, which involved adding a surfactant to the electrophoretic buffer to form micelles to enhance resolution of neutral substances, by Terabe *et al.* [42] represents another significant step in the development of CE. Since then, various type of modifiers for the enhancement of selectivity in CE separation have been investigated (see Chapter 5).

Recent developments in gel-filled capillaries and coated columns have further enhanced the scope and efficiency of capillary electrophoretic techniques [41,43]. Theoretical plate numbers in the multimillion range can now be routinely achieved using gel-filled capillaries in CE separations [43].

At the end of the 1980s, commercial CE instruments have become available. With the rapid advances currently being made, CE is now gaining popularity as an alternative analytical tool for some routine analytical applications.

1.2 DIFFERENT MODES OF CAPILLARY ELECTROPHORESIS

One of the main advantages of CE is that it requires only simple instrumentation. A schematic diagram of the basic CE instrument is shown in Fig. 1.2. It consists of a high-voltage power supply, two buffer reservoirs, a capillary and a detector. This basic setup can be elaborated upon with enhanced features such as autosamplers, multiple injection devices, sample/capillary temperature control, programmable power supply, multiple detectors, fraction collection and computer interfacing.

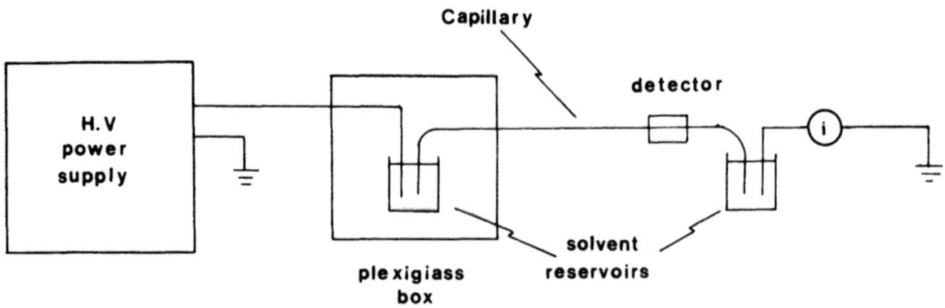

Fig. 1.2. Schematic of a system for capillary electrophoresis. (Reproduced from Ref. 11 with permission of Marcel Dekker, Inc.).

Different modes of capillary electrophoretic separations can be performed using a standard CE instrument. The origins of the different modes of separation may be attributed to the fact that capillary electrophoresis has developed from a combination of many electrophoresis and chromatographic techniques. In general terms, it can be considered as the electrophoretic separation of a number of substances inside of a narrow tube. Even though most applications have been performed using liquids as the separation media, capillary electrophoretic techniques encompass separations in which the capillary contains electrophoretic gels, chromatographic packings or coatings.

The distinct capillary electroseparation methods include:

(A) Capillary zone electrophoresis (CZE) [1–6,44,45]
(B) Capillary gel electrophoresis (CGE) [46–50]
(C) Micellar electrokinetic capillary chromatography (MEKC or MECC) [51–60]
(D) Capillary electrochromatography (CEC) [61–63]
(E) Capillary isoelectric focusing (CIEF) [48,64–66]
(F) Capillary isotachophoresis (CITP) [67,68]

In electrophoresis a mixture of different substances in solution is introduced, usually as a relatively narrow zone, into the separating system, and induced to move under the influence of an applied potential. Due to differences in the effective mobilities (and hence migration velocities) of different substances under the electric field, the mixture then separates into spatially discrete zones of individual substances after a certain time.

Electrophoretic separations may be carried out in continuous or discontinuous electrolyte systems. In the case where a continuous electrolyte is used, the solution of the so-called background electrolyte forms a continuum along the migration path. This continuum does not change with time and provides an electrically conducting medium for the flow of electric current and the formation of an electric field along the migration path. The background electrolyte is usually a buffer which can selectively influence the effective mobilities. By changing the properties of the background electrolyte system along the migration path the separation can be operated either as a kinetic or steady state process.

In a kinetic process, the composition of the background electrolyte is constant along the migration path. The electric potential and the effective mobilities of the separated substances are therefore constant. Consequently, different substances migrate with constant, but different velocities provided constant current is passed through the system. CZE, CGE, MEKC and CEC are examples of this type of separation processes.

In a steady state process, the composition of the background electrolyte is not constant. Both the electric field and the effective mobilities may change along the migration path. The most common practical realization of this type of separation process is to form a pH gradient along the migration path. After a time interval, certain components of the sample, e.g. ampholytes, would stop to migrate and focus at certain characteristic positions corresponding to their isoelectric points.

The result is a steady state where the substances, during the passage of the electric current, are focused in certain places along the migration path. CIEF is an example of this type of separation processes.

In the case of the discontinuous electrolyte system, the samples migrate between two different electrolytes as a distinct individual zone. The discontinuous electrolyte consists of a leading and a terminating electrolyte. The leading electrolyte forms the front zone, and the terminating electrolyte forms the rear zone. CITP is an example of electrophoretic separation in discontinuous electrolyte systems.

In Sections 1.2.1 to 1.2.6, a brief overview of all the six modes of CE is given. The mechanisms by which solutes separate in the six techniques are illustrated in Figs. 1.3 and 1.4. In Fig. 1.3, a diagrammatic representation of kinetic separation processes such as CZE, CGE, MEKC and CEC is shown. The migration of each type of charged species under the influence of the applied voltage is represented by an arrow in the figure. In Fig. 1.4, a schematic representation of CZE, CIEF and CITP is shown. In this figure, the distribution of electrolytes and a two-component sample are shown at three different times. This figure serves to depict the differences in separation mechanisms in CZE, which is a kinetic process in a continuous electrolyte; in CIEF, which is a steady-state process in a continuous electrolyte; and in CITP, which is carried out in a discontinuous electrolyte.

1.2.1 Capillary zone electrophoresis (CZE)

The principles of separation in capillary zone electrophoresis (CZE), or free solution capillary electrophoresis, are discussed in detail in Section 1.3. Currently CZE is the most commonly used technique in CE. Many compounds can be separated rapidly and easily. The separation in CZE is based on the differences in the electrophoretic mobilities resulting in different velocities of migration of ionic species in the electrophoretic buffer contained in the capillary. Separation mechanism is mainly based on differences in solute size and charge at a given pH. Most capillaries used for CE today are made of fused silica, which contains surface silanol groups. These silanol groups may become ionized in the presence of the electrophoretic medium. The interface between the fused silica tube wall and the electrophoretic buffer consists of three layers: the negatively charged silica surface (at $pH > 2$), the immobile layer (Stern layer or inner Helmholtz plane), and the diffuse layer of cations (and their sphere of hydration) adjacent to the surface of the silica, which tend to migrate towards the cathode. The migration of cations results in a concomitant migration of fluids through the capillary. This flow of liquid through the capillary is called electroosmotic (or electroendosmotic) flow.

The electroosmotic flow in uncoated fused silica capillaries is usually significant with most commonly used buffers. It is also significantly greater than the electrophoretic mobility of the individual ions in the injected sample. Consequently, both anions and cations can be separated in the same run. Cations are attracted towards the cathode and their speed is augmented by the electroosmotic flow.

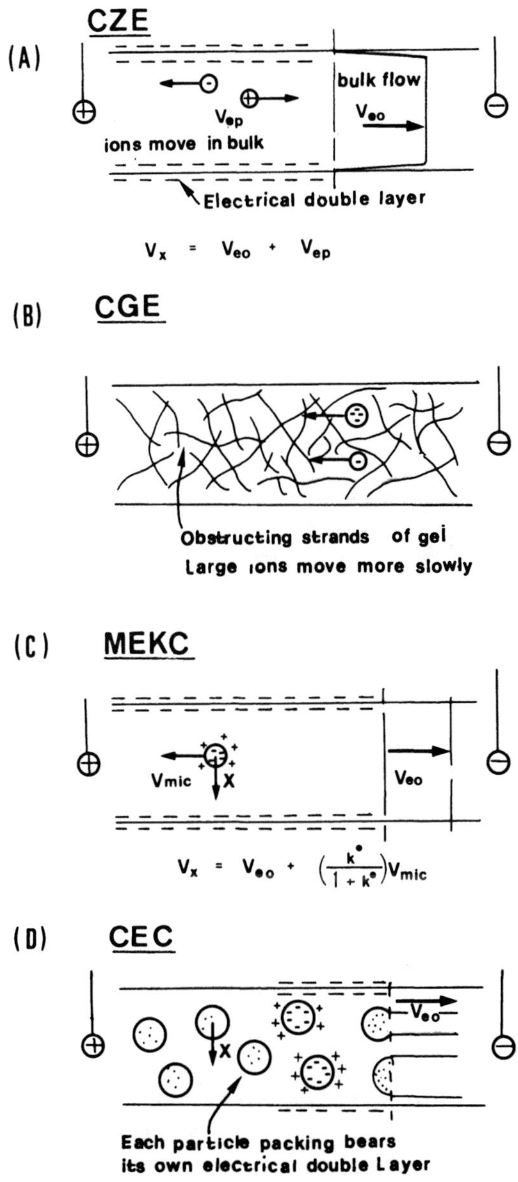

Fig. 1.3. Diagramatic representation of (A) capillary zone electrophoresis (CZE), capillary gel electrophoresis (CGE), (C) micellar electrokinetic chromatography (MEKC), and (D) capillary electrokinetic chromatography. v_x is the linear migration velocity of the analyte X. v_{eo} is the electroosmotic velocity, v_{ep} is the electrophoretic velocity and k^* is the phase capacity ratio. (Adapted from Ref. 61 with permission of Friedr. Vieweg & Sohn, Pergamon Press.)

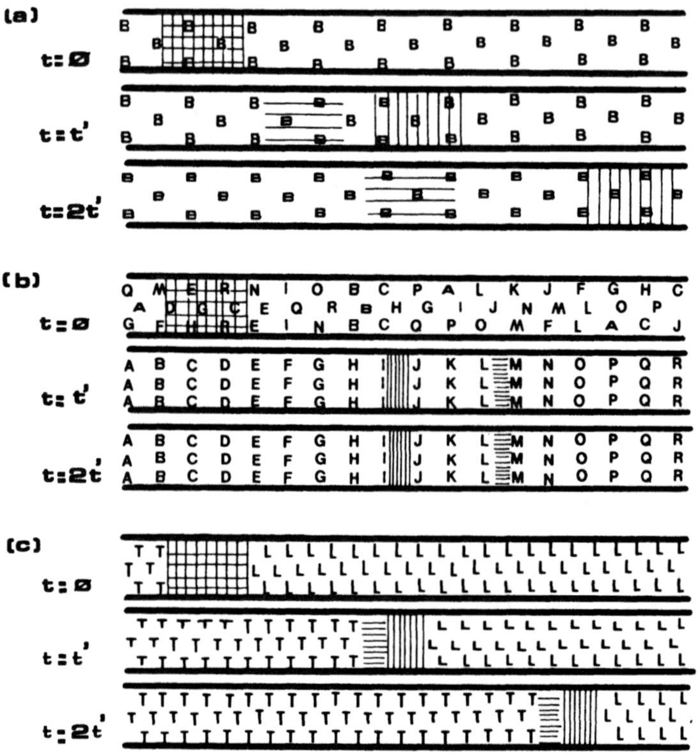

Fig. 1.4. Schemes of electrophoretic techniques: (a) zone electrophoresis, (b) isoelectric focusing, and (c) isotachophoresis. The distribution of electrolytes and a two- component sample are shown at three different times: the start of the analysis ($t = 0$), the time interval t' after the start ($t = t'$), and the double time interval after the start ($t = 2t'$). (Reproduced from Ref. 68 with permission of VCH Verlagsgesellschaft.)

Anions, although electrophoretically attracted toward the anode, are swept towards the cathode with the bulk flow of the electrophoretic medium. Under these conditions, cations with the highest charge/mass ratio migrate first, followed by cations with reduced ratios. All the unresolved neutral components are then migrated as their charge/mass ratio is zero. Finally, the anions migrate. Anions with lower charge/mass ratio migrate earlier than those with greater charge/mass ratio. The anions with the greatest electrophoretic mobilities migrate last. One important point to note is that it is possible to change the charge/mass ratio of many ions by adjusting the pH of the buffer medium to affect their ionization and hence electrophoretic mobility.

As will be discussed in more detail in Section 1.3.3, the electroosmotic velocity, v_{eo}, can also be adjusted by controlling the pH (since more silanol groups are ionized, both the zeta potential and the flow increase), the viscosity (as viscosity increases the velocity decreases), the ionic strength (because of its effect on the zeta potential), the voltage (flow increases proportionally to voltage), and the dielectric constant of the buffer. Rinsing the capillary can alter the ionizable silanol groups on the silica surface and hence the electroosmotic flow.

A significant feature of the electroosmotic flow is that instead of showing parabolic flow profiles as in pressure-driven flows, it tends to flow in a plug shape. This increases the resolution in separations by reducing the band broadening of the analyte peak during its passage along the capillary.

1.2.2 Capillary gel electrophoresis (CGE)

The main separation mechanism in capillary gel electrophoresis (CGE) is based on differences in solute size as analytes migrate through the pores of the gel-filled column. Gel-filled columns used for CGE are discussed in detail in Section 5.2. Gels are potentially useful for electrophoretic separations mainly because they permit separation based on "molecular sieving". Furthermore, they serve as anti-convective media, they minimize solute diffusion, which contributes to zone broadening, they prevent solute adsorption to the capillary walls and they help to eliminate electroosmosis. However, the gel must possess certain characteristics, such as temperature stability and the appropriate range of pore size, for it to be a suitable electrophoretic medium. Furthermore, the technique is subjected to the limitation that neutral molecules would not migrate through the gel, since electroosmotic flow is suppressed in this mode of operation.

Hjerten [41] and Hjerten and Zhu [46,47] employed polyacrylamide-filled and agarose-filled glass capillaries of 150 μm I.D. for electrophoretic separation of both large and small molecules. CGE with fraction collection has also been performed [41,50] for micropreparative purification of macromolecules.

Karger and co-workers [49,50] achieved extremely high separation efficiency (up to 30 million theoretical plates per meter) using gel-filled capillary columns. The capillaries were filled with polyacrylamide gels which contained sodium dodecyl sulfate. This technique is referred to as capillary SDS-PAGE separation and has been used for the separation of proteins, polynucleotides and DNA fragments [48–50]. The remarkable success achieved by the technique could partly be attributed to procedures developed for crosslinking acrylamide and bisacrylamide monomers inside fused silica capillaries. The resulting polyacrylamide has a randomly coiled gel structure which can be bonded to the capillary walls through the addition of a bifunctional reagent. The pore size is determined by the total gel concentration, %T (T = [bis + acryl]/V, where bis, acryl and V are weight of bisacrylamide, weight of acrylamide and total volume, respectively) and the concentration of the cross linking agent, %C (C = [bis + acryl]/acryl). When the gel is bonded to the capillary surface, electroosmosis would be eliminated. Since the protein form complexes with the SDS which are negatively charged, injection and detection are performed at the cathodic and anodic ends of the capillary respectively.

Capillary SDS-PAGE has several advantages over conventional slab gel electrophoresis, including small sample requirement, possibility of automation, and high sensitivity. By exploiting the capability of high throughput and two-dimensional separations of the slab gel format and rapid and efficient molecular mass determination and trace quantitation of the capillary format, rapid advances have been made on the separation and analysis of a wide variety of large biomolecules.

1.2.3 Micellar electrokinetic capillary chromatography (MEKC)

An important development in CE is the introduction of micellar electrokinetic capillary chromatography (MEKC or MECC) by Terabe and co-workers [42,51,52] in 1984. The principles of separation in MEKC are discussed in Section 5.2. In MEKC, the main separation mechanism is based on solute partitioning between the micellar phase and the solution phase. The technique provides a way to resolve neutral molecules as well as charged molecules by CE. Subsequently, investigations on geometrical parameters, column efficiency, wall treatment, and velocity profiles have been performed [53–56]. The power of the technique was demonstrated by the resolution of isotopically substituted compounds by Bushey and Jorgenson [57].

Micelles form in solution when a surfactant is added to water in concentration above its critical micelle concentration (cmc). Micelles consist of aggregates of surfactant molecules with typical lifetimes of less than 10 μs. The most commonly used surfactant in MEKC is sodium dodecyl sulfate (cmc = 0.008 M, aggregate number = 58 at 25°C), which is an anionic surfactant. Other anionic and cationic surfactants have been employed (see Chapter 5). In the case of SDS, the micelles can be considered as small droplets of oil with a highly polar surface which is negatively charged. Even though these anionic micelles are attracted toward the anode, in an uncoated fused silica capillary they still migrate toward the cathode because of electroosmotic flow. However, the micelles move towards the cathode at a slower rate than the bulk of the liquid because of their attraction towards the anode. Neutral molecules partition in and out of the micelles based on the hydrophobicity of each analyte. Consequently the micelles of MEKC are often referred to as a pseudo (or moving) stationary phase. A very hydrophilic neutral molecule, e.g. methanol, will spend almost no time inside the micelle and will therefore migrate essentially at the same rate as the bulk flow and elute earlier. On the other hand, a very hydrophobic neutral molecule, e.g. Sudan III, will spend nearly all the time inside the micelles and will therefore elute later, together with the micelles. All other solutes with intermediate hydrophobicity will migrate within this migration window.

MEKC can be used with ionic substances as well as neutral compounds. A combination of charge/mass ratios, hydrophobicity and charge interactions at the surface of the micelles combine to affect the separation of the analytes.

The use of different surfactants as well as organic modifiers can lead to significant changes in resolution. The micelles of MEKC can also be replaced with any material that reacts differentially with the analytes of separations and affects their velocity through the capillary [58–60]. For instance, soluble ion exchangers, derivatized cyclodextrins and charged colloidal particles can all be added to the buffer to provide selectivity in the separation. In fact, the additives do not necessarily have to be charged. Neutral cyclodextrins can differentially bind aromatic compounds and change the apparent molecular weight and electrophoretic mobility. As will be discussed in detail in Chapter 5, there are numerous ways to enhance selectivity in CE applications. The ability to choose the type of resolution by modification of the buffer is one of the main advantages of CE.

1.2.4 Capillary electrochromatography (CEC)

In capillary electrochromatography (CEC), the separation column is packed with a chromatographic packing which can retain solutes by the normal distribution equilibria upon which chromatography depends [61] and is therefore an exceptional case of electrophoresis. The use of packed column for capillary electrophoresis is discussed in more details in Section 4.4. In CEC the liquid is in contact with the silica wall, as well as the particle surfaces. Consequently, electroosmosis occurs in a similar way as in an open tube due to the presence of the fixed charges on the various surfaces. Whereas in an open tube the flow is strictly plug flow, and there is no variation of flow velocity across the section of the column, the flow in a packed bed is less perfect because of the tortuous nature of the channels. Nevertheless, it approximates closely to plug flow and is substantially more uniform than a pressure-driven system. Therefore, the same column tends to provide higher efficiency when used in electrochromatography than when used in pressure-driven separations [61–63].

1.2.5 Capillary isoelectric focusing (CIEF)

Another separation method which can be conveniently performed using a capillary electrophoresis instrument is isoelectric focusing, in which substances are separated on the basis of their isoelectric points or pI values [64]. The use of capillary isoelectric focusing (CIEF) is discussed in more detail in Section 6.7. Hjerten and co-workers [48,64–66] have described isoelectric focusing of proteins in glass capillaries. In this technique, the protein samples and a solution that forms a pH gradient are placed inside a capillary. The anodic end of the column is placed into an acidic solution (anolyte), and the cathodic end in a basic solution (catholyte). Under the influence of an applied electric field, charged proteins migrate through the medium until they reside in a region of pH where they become electrically neutral and therefore stop migrating. Consequently, zones are focused until a steady state condition is reached. After focusing, the zones can be migrated (mobilized) from the capillary by a pressurized flow, e.g. simply lifting the height of one end of the capillary and permitting the sample to flow through the detection cell. Alternatively, after focusing, salt (e.g. sodium chloride) can be added to the anolyte (acid reservoir) or catholyte. By the principle of electroneutrality, sodium ions can exchange for protons in the tube, generating a pH imbalance gradient which causes the migration of the components [64]. Sharp peaks are obtained with good resolution, and a large peak capacity is observed mainly because the whole tube is simultaneously used for focusing.

The resolving power in isoelectric focusing can be expressed in terms of the difference in pI of the two species for separation [64]. Therefore, high resolutions can be obtained for species with low diffusion coefficients and a high mobility slope at the isoelectric point, a shallow rate of change of pH with tube distance and a high electric field. High fields enable focusing to be performed faster. Cooling of columns can also enhance resolution and separation speed in capillary isoelectric focusing [64]. Another important factor to consider is the coating on the capillary surface. The coating on the walls must be able to minimize electroosmotic flow and remain stable to allow good reproducibility from run to run with the same column.

1.2.6 Capillary isotachorphoresis (CITP)

Another mode of CE operation is capillary isotachorphoresis (CITP). A more detailed discussion on CITP is given in Section 6.8. The main feature of CITP is that it is performed in a discontinuous buffer system. Sample components condense between leading and terminating constituents, producing a steady-state migrating configuration composed of consecutive sample zones [67,68]. This mode of operation is therefore different from other modes of capillary electrophoresis, such as CZE, which are normally carried out in a uniform carrier buffer and is characterized by sample zones which continuously change shape and relative position. In the case of a typical CZE separation, the electropherogram obtained contains sample peaks similar to those obtained in chromatographic separations, whereas in the case of CITP, the isotachopherogram obtained contains a series of steps, with each step representing an analyte zone. Unlike in other CE modes, where the amount of sample present can be determined from the area under the peak as in chromatography, quantitation in CITP is mainly based on the measured zone length which is proportional to the amount of sample present.

1.3 PRINCIPLES OF SEPARATION IN CAPILLARY ZONE ELECTROPHORESIS (CZE)

In this section, the principles of electrophoretic migration in capillaries relative to migration time and efficiency, as well as the physical phenomena that affect the nature of separation are discussed. This discussion will primarily concern aspects of free-zone electrophoresis in capillary tubes. Many of the points addressed are similar for related capillary electrophoretic techniques.

1.3.1 Electrophoretic migration in capillary tubes

As shown in Fig. 1.2 (schematic of a basic CE instrument), the CE system consists of a buffer-filled capillary placed between two buffer reservoirs, and a potential field which is applied across this capillary. In general the flow of electroosmosis is towards the cathode, and hence a detector is placed at this end. Injection of solutes is performed at the anodic end by either electromigration or hydrodynamic flow (see Chapter 2).

One of the main advantages of capillary zone electrophoresis (CZE) is that there is no need for a pressure-driven flow which usually results in a parabolic flow profile and thus band broadening. Since open-tubular capillaries of small I.D. are employed, band broadening due to resistance to mass transfer and heating effects are minimized. Consequently, the only factor contributing to band broadening is logitudinal diffusion [1–3,69]. Under conditions in which electroosmosis does not occur, the migration velocity (v) in electrophoresis is given by [1–3]:

$$v = \mu_{ep} E = \frac{\mu_{ep} V}{L} \tag{1.1}$$

where μ_{ep} is the electrophoretic mobility, E is the field strength (V/L), V is the voltage applied across the capillary, and L is the capillary length. The time taken for a solute to migrate from one end of the capillary to the other is the migration time (t) and is given by;

$$t = \frac{L}{v} = \frac{L^2}{\mu_{ep} V} \tag{1.2}$$

1.3.2 Band broadening due to diffusion

Assuming that the only contribution to band broadening is logitudinal diffusion, the variance of the migrating zone width (σ^2) can be written as [1–8,11,70]:

$$\sigma^2 = 2Dt = \frac{2DL^2}{\mu_{ep} V} \tag{1.3}$$

or

$$\sigma = \left(\frac{2DL}{\mu_{ep} E}\right)^{1/2} \tag{1.4}$$

where D is the diffusion coefficient of the solute. The number of theoretical plates (N) is given by:

$$N = \frac{L^2}{\sigma^2} = \frac{\mu_{ep} V}{2D} \tag{1.5}$$

The efficiency is therefore based on applied voltage but not capillary length. Maximum efficiency and short analysis times are obtained with high voltages and short columns, provided that there is efficient heat dissipation (see Section 1.3.4).

1.3.3 Electroosmosis

An important phenomenon in capillary electrophoresis is electroosmosis, which refers to the flow of solvent in an applied potential field. In Fig. 1.5, a model of the silica–solution interface is shown. Electroosmotic flow originates from the negative charges on the inner wall of the capillary tube, which caused the formation of a double layer at the interface adjacent to the stagnant double layer, a diffuse layer consisting of mobile cations exists in the diffuse region of the double layer shown in Fig. 1.5. The potential across the layers is called the zeta potential, denoted by ζ, which is given by the Helmholtz equation:

$$\zeta = \frac{4\pi\,\eta\,\mu_{eo}}{\varepsilon} \tag{1.6}$$

where η is the viscosity, ε is the dielectric constant of the solution, and μ_{eo} is the coefficient for electroosmotic flow [38,39]. Under the influence of an applied electric field, the mobile cations in the diffuse layer migrate toward the cathode, causing the solvent molecules to migrate in the same direction. The linear velocity, ν, of the electroosmotic flow is given by [38,39]:

$$\nu = \frac{\varepsilon}{4\pi\,\eta} E\zeta \tag{1.7}$$

The double layer is typically a very thin layer (up to several hundred nanometers) relative to the radius of the capillary (typically 50–100 μm). Therefore, the electroosmotic flow may be consider to originate at the walls of the capillary. Consequently, a flat flow profile as shown in Fig. 1.6 is obtained. For comparison, the parabolic flow profile normally observed in pressure-driven systems, such as in HPLC, is also shown in Fig. 1.6. For capillary radius greater than seven times the double layer thickness, a flat flow profile would be expected in CE [71].

Electroosmotic flow should not cause the broadening of solutes zones in the capillary directly. Electroosmotic flow does, however, affect the amount of time a solute would take to migrate through the capillary, and therefore, may affect both

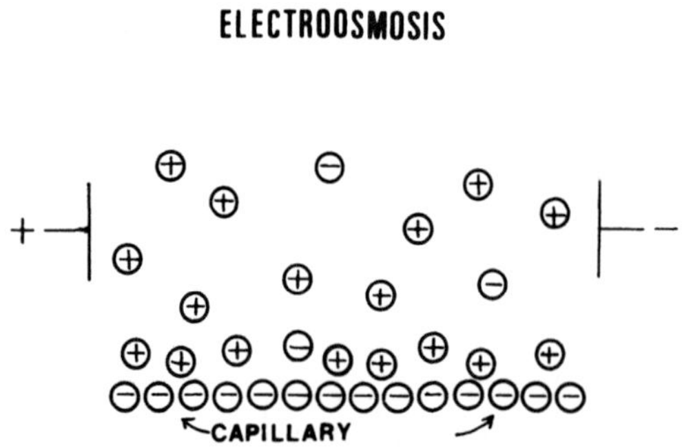

Fig. 1.5. Schematic representation of ions at a silica–solution interface. (Reproduced from Ref. 11 with permission of Marcel Dekker, Inc.)

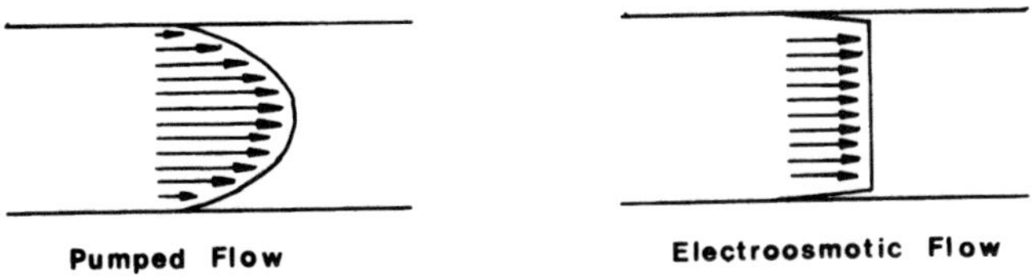

Fig. 1.6. Flow profiles in HPLC (left) and CZE (right).

efficiency and resolution indirectly [11]. In the presence of electroosmotic flow, the migration velocity and time are given by:

$$v = \frac{(\mu_{eo} + \mu_{ep})V}{L} \tag{1.8}$$

and

$$t = \frac{L^2}{(\mu_{eo} + \mu_{ep})V} \tag{1.9}$$

The zone variance and the number of theoretical plates are expressed as:

$$\sigma^2 = \frac{2DL^2}{(\mu_{eo} + \mu_{ep})V} \tag{1.10}$$

and

$$N = \frac{(\mu_{eo} + \mu_{ep})V}{2D} \tag{1.11}$$

According to Eq. (1.8), all ions will migrate in the same direction if the rate of electroosmotic flow is greater in magnitude and opposite in direction to all anions in the buffer. Moreover, non-ionic species will be carried by the electroosmotic flow and migrate at one end of the capillary. The effects of electrophoretic migration and electroosmotic flow on the migration order of cations, neutral species and anions in CZE are shown in Fig. 1.7. Since separation is based on differential electrophoretic migration in CZE, neutral species are not separated.

The resolution of two zones in electrophoresis is given by the equation:

$$R = \frac{1}{4(2)^{1/2}}(\mu_{ep,1} - \mu_{ep,2})\left[\frac{V}{D(\overline{\mu}_{ep} + \mu_{eo})}\right]^{1/2} \tag{1.12}$$

where $\mu_{ep,1}$ and $\mu_{ep,2}$ are the electrophoretic mobilities for the two solutes and $\overline{\mu}_{ep}$ is the average electrophoretic mobility [1–5].

According to Eq. (1.12), the highest resolution is obtained when $\mu_{eo} = -\overline{\mu}_{ep}$. However, the analysis time would approach infinity in this case. It is also noted that electroosmosis toward the cathode should result in better resolution of anions, which migrate against the electroosmotic flow and are carried back toward the cathode, whereas cations will be more poorly resolved under these conditions.

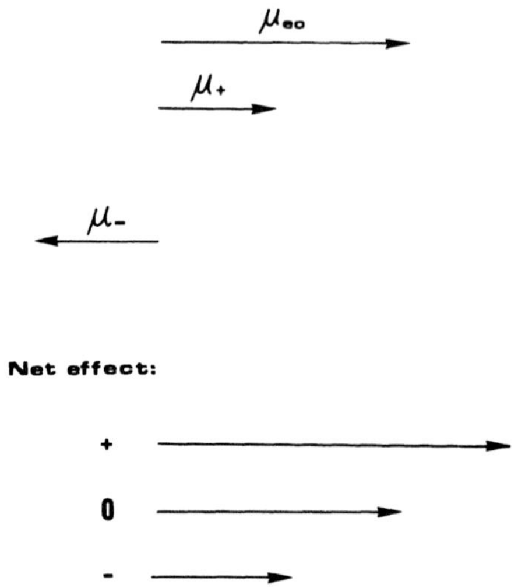

Fig. 1.7. Migration order for cations (+), non-ions (0), and anions (−) based on the cumulative effects of electrophoresis and strong electroosmotic flow toward the cathode. (Reproduced from Ref. 11 with permission of Marcel Dekker, Inc.)

Electroosmotic flow in a CE system can facilitate automation. The buffer is electroosmotically pumped through capillary tubes without the need for a pressure-driven flow as demonstrated in numerous CE applications [1–21].

Jorgenson and Lukacs studied the effect of electroosmosis in small capillaries [3]. The flat flow profile and its effect on net mobility (i.e. migration of positive and negatively charged species in electrophoresis) were demonstrated. They investigated the effect of pH on electroosmosis in pyrex, silica and PTFE capillaries using phenol as a neutral marker [3]. Tsuda et al. also studied the effect of pH and current density on electroosmotic flow in similar types of capillaries [10], using benzene as a neutral marker. The rate of electroosmotic flow was found to be the highest under conditions that increase the zeta potential or double layer thickness or decrease the solution viscosity. The zeta potential was found to depend only on the nature and amount of ions at the capillary surface. If the bulk of these ions are hydroxyl or carboxyl groups, the ionic content will depend on the solution pH. Furthermore, electroosmotic flow is enhanced in the direction of the cathode at elevated pH.

The magnitude of the electroosmotic flow can be measured by several methods. One method involves measuring the rate of electroosmotic flow by measuring the change in weight in one buffer reservoir [44,72]. By weighing the solution emerging from the capillary directly on an analytical microbalance, the problem of possible adsorption of the neutral marker is avoided. It was found that flow rate was inversely proportional to ionic strength, independent of column diameter, and decreased by organic modifiers, e.g. methanol. Huang *et al.* [73] measured electroosmotic flow by measuring the change in electrophoresis current when a buffer with a different ionic strength was introduced. Everaerts and co-workers [74] described several methods for the measurement and control of electroosmotic flow. Van de Goor *et al.* [75] determined the rate of electroosmotic flow by measuring the zeta potential. One of their methods employed the weighing procedure adopted by Altria and Simpson [44,72] and another involved measuring the streaming potential where solvent was pumped through the column. They determined the zeta potential and electroosmotic flow of PTFE capillaries as a function of pH.

A great deal of work has been done to investigate ways of manipulating the electroosmotic flow. In certain cases it is important to totally inhibit electroosmotic flow. There are several approaches to alter electroosmotic flow. The most commonly used methods involve either changing the zeta potential across the solution–solid interfaces or increasing the viscosity at the interface.

In the simplest case, the pH and the ionic composition of the buffer can be adjusted to give the desired electroosmotic flow. An example is the separation of proteins by CZE at buffer pH between 8 to 11. Under these conditions, the capillary wall and many proteins are electronegative. Therefore, they repel one another to minimize surface interaction [76]. Fujiwara and Honda found that addition of sodium chloride reduced electroosmotic flow by decreasing the thickness of the double layer [77].

It is also possible to vary the electroosmotic flow by introducing additives to the buffer to alter the zeta potential developed across the capillary solution interface. By adding a cationic surfactant, such as cetyltrimethylammonium bromide (CTAB) [42] or tetradecyltrimethylammonium bromide (TTAB) [78], the direction of flow could be reversed. Putrescine was used to reduce electroosmotic flow [76]. The addition of 0.02 M S-benzylthiouronium chloride to the electrophoretic buffer at pH 4.5 was found to inhibit electroosmotic flow [72]. Foret *et al.* eliminated electroosmosis at high ionic strength by using Triton X [79]. The addition of organic solvents to the electrophoretic buffer was found to affect electroosmotic flow dramatically [80]. Methanol was found to reduce electroosmotic flow significantly, whereas acetonitrile was found to increase electroosmotic flow, though not as significantly.

Other approaches of varying or eliminating electroosmotic flow include covalently bonding γ-methacryloxylpropyltrimethysilane to the glass surface [48,80] or coating the capillary wall with a polymer such as methylcellulose [37,48]. This will be discussed in detail in Section 4.2.

In summary, electroosmotic flow is advantageous in some systems and deleterious in others. In the cases of capillary gel electrophoresis, capillary isotachophoresis, or isoelectric focusing in capillaries, electroosmosis is not desirable. On the other hand, resolution of zones in free zone electrophoresis is dependent upon the electroosmotic flow rate. In addition, strong electroosmotic flow produces a system that can be readily automated. In micellar electrokinetic chromatography, where anionic micelles are commonly employed, a strong electroosmotic flow towards the cathode occurs and has been found to be beneficial in most MEKC applications.

1.3.4 Power dissipation

The effects of power dissipation on capillary electrophoretic separations have been investigated by many workers [61,82–93]. The capillary tube containing the electrophoretic medium behaves in similar way as a cylindrical ohmic conductor when a voltage is applied across the two ends. When a current is passed along the capillary, ohmic heat is released and the conductor heats up. Figure 1.8 indicates the temperature distribution for an insulated conductor as it would be in the case of a CE system. Over the central region heat is generated homogeneously and the temperature variation across the bore of a cylindrical tube (i.e. conductor) is parabolic. The heat so generated is conducted first through the walls of the tube and then through the surrounding medium, typically air or a cooling liquid.

In order to attain high efficiencies in capillary electrophoresis, it is essential to ensure that efficient heat dissipation can be accomplished in the system. Excess solution heating, leading to a parabolic temperature gradient across the capillary, can increase electrophoretic mobilities by about 2% per degree centigrade [11,61]. Assuming that heat generation as a result of Joule heating by passage of current in the capillary is efficiently dissipated, then the electrical power dissipated per unit length of the capillary is given by:

$$\frac{P}{L} = \frac{\kappa C r^2 V^2}{L^2} \tag{1.13}$$

where P is power, L is the capillary length, κ is the molar conductance of the solution, C is the buffer concentration, r is the column radius and V is the applied voltage [63]. The thermal gradient generated depends on the thermal conductivities of the materials involved. The heat released per unit volume in an electrolyte is

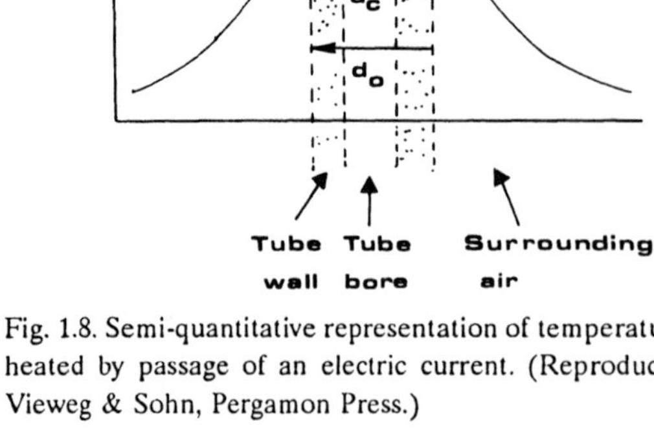

Fig. 1.8. Semi-quantitative representation of temperature profile across a tube containing electrolyte heated by passage of an electric current. (Reproduced from Ref. 61 with permission of Friedr. Vieweg & Sohn, Pergamon Press.)

given by:

$$Q = E^2 k C \psi \tag{1.14}$$

where E is the electric field strength, k is the molar conductivity of the solution, C its concentration and ψ the total porosity of the medium. The value of ψ will be unity for an open tube and ranges from 0.4 to 0.8 for a packed tube.

The temperature excess across the wall of the capillary, θ_{wall}, is given by

$$\theta_{\text{wall}} = \frac{Q r_c^2}{2\lambda_w} \ln\left(\frac{r_o}{r_c}\right) \tag{1.15}$$

where r_c and r_o are the inner and outer radius of the tube and λ_w is the thermal conductivity of the tube wall.

The temperature excess θ_{core}, within the core region (i.e. the difference between the temperature on the axis of the tube and at its inner wall) is given by:

$$\theta_{\text{core}} = \frac{Q r_c^2}{4\lambda} = (E^2 k C \psi)(r_c^2/4\lambda) \tag{1.16}$$

where λ is the thermal conductivity of the solution. Under typical operating conditions, θ_{core} and θ_{wall} would be small (less than 1 K) compared with the temperature excess of the tube wall relative to the surrounding ambient air [92].

Heat loss from a horizontal tube in air is mainly by natural convection or by forced convection rather than by conduction through still air. By using characteristic plots for natural and forced convection given by Roberts [82], Knox estimated the temperature rise, θ, for different tube diameter under natural and forced convection based on typical CE operating conditions [61]. In Fig. 1.9, θ is plotted against tube diameters.

The results show that very large temperature rises may be obtained when only natural convective cooling is employed during capillary electrophoresis, especially

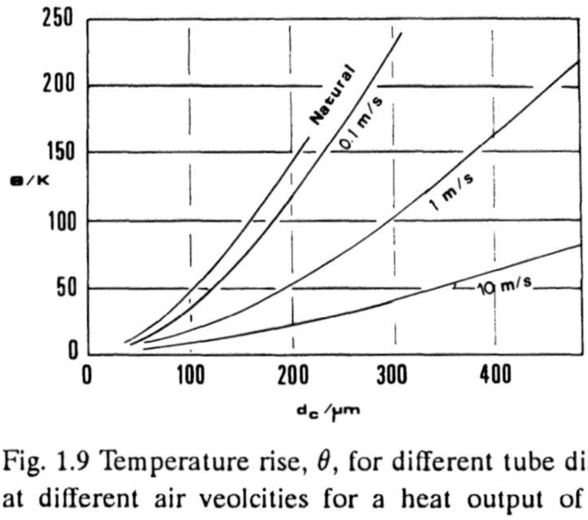

Fig. 1.9 Temperature rise, θ, for different tube diameters (d_c) under natural and forced convection at different air veolcities for a heat output of 300 W cm^{-3}. (Reproduced from Ref. 61 with permission of Friedr. Vieweg & Sohn, Pergamon Press.)

if tubes larger than 200 μm I.D. are used. The use of forced convection was recommended (see Chapter 5) although there may be difficulties in adequately cooling those parts of tubes which pass through the detector and other pieces of equipment [61].

According to Knox [61], the parabolic temperature profile which exists across the capillary tube may cause variations in migration rate due to one or more of three possible effects, which include the changes in the viscosity of the electrophoretic buffer, η, changes in partition ratios between phases, k' and changes in the rates of kinetic process.

A detailed treatment of these temperature effects on efficiency has been presented by Knox [61]. Although the temperature difference between the tube in CE relative to the surrounding air does not affect the plate height or plate count in a direct way, the variation of the temperature within the electrolyte can significantly reduce the efficiency of a CE separation. In the case of CE systems with self heating of the liquid core (Joule heating), there would be a small parabolic disturbance superimposed on the uniform migration velocity of a solute along the tube. The effect is analogous to the dispersion of a solute in the case of chromatography where a pressure-driven flow is employed [83,84]. An expression for the thermal contributions to the plate height, H_{TH}, has been derived [61]:

$$H_{\mathrm{TH}} \cong 6 \times 10^{-6} \frac{\varepsilon_0 \varepsilon_{\mathrm{r}} \delta}{D_{\mathrm{m}} \eta \lambda^2} E^5 r_{\mathrm{c}}^6 k^2 C^2 \tag{1.17}$$

where δ represents the thickness of the electrical double layer, ε_0 is the permittivity in vacuum and ε_{r} is the relative permittivity or dielectric constant. Using typical values of the constants, H_{TH} is estimated to be 0.006 μm and 0.4 μm for capillaries with I.D. of 100 μm and 200 μm, respectively. Thus H_{TH} is generally negligible for narrow bore capillaries, but it becomes significant (compared to H_{TH} = 1.1 μm from axial diffusion) if either E, r_{c} or C are allowed to rise beyond acceptable limits.

Various investigations have been performed to study the effect of power dissipation on separation efficiency. Hjerten minimized the effect of heating by rotating the capillary around its longitudinal axis [38] in an early version of the CE instrument. Mikkers *et al.* [40] adapted an instrument for isotachophoresis for CE, and performed a theoretical evaluation of the effect of electrophoretic migration on concentration distribution in free zone electrophoresis [85]. Zones were found to be unsymmetrical when the concentration gradient induced by differential migration of different solutes produce inhomogeneities in the electric field. Thormann *et al.* [86,87] confirmed these calculations by demonstrating non-symmetrical broadening in overloaded separations due to the variation of electrophoretic velocities caused by variation in electric field strength.

Lukacs and Jorgenson discussed the relative significance of diffusional broadening when other factors (i.e. Joule heating) are eliminated through the use of narrow capillaries [1–4]. They developed the primary relationships governing separation efficiency and relate the separation voltage, column diameter, length, and solute concentration to resolution and separation efficiency [63]. Tsuda *et al.* [10] also demonstrated the current–voltage relationship to efficiency in the separation of several cations and anions. Lauer and McManigill discussed the importance of power dissipation for efficient separation in CE [88]. Small capillaries were also considered by Foret *et al.* [89], who provided both theoretical and experimental evidence for separation efficiency in 125–500 μm I.D. capillaries. Decreased plate height was observed at high field strength (>50 kV/m) due to thermal convection caused by Joule heating.

Due to the high ionic strength used in the separation of large molecules, Joule heating is still a major problem for these separations. Cohen *et al.* demonstrated the importance of efficient dissipation of Joule heat in capillary SDS-PAGE [50]. Nelson *et al.* [90] showed that separation efficiency could be improved by using thermoelectric (Peltier) devices to control the air temperature around the CE capillary.

The effect of heat generation in CE on band broadening and separation efficiency was investigated by Grushka *et al.* [91]. They predicted significant loss of efficiency at high ionic strength and high field in columns larger than 75 μm I.D., due to temperature gradients across the column diameter. Jones and Grushka performed calculations of the radial temperature gradient in polyimide-coated quartz capillaries [92]. Their calculations showed that in typical CE experiments, i.e. capillaries with 50–100 μm I.D., 375 μm O.D., and up to 5 W power input, the temperature profile derived was nearly identical to a parabolic profile. However, at higher power, the parabolic approximation was found to underestimate the temperature at the capillary centre.

1.3.5 Adsorption

Adsorption of solute molecules by the capillary surface results in peak distortion. In certain cases, irreversible adsorption may occur which inhibits migration of the adsorbed species. The tendency of silica surfaces to adsorb analyte molecules presents a potential problem in CE, especially for macromolecules, such as proteins. The approaches which have been adopted to alleviate the problem of adsorption is discussed in more detail in Section 4.2. When adsorption is not completely eliminated, its effect on efficiency can be estimated from the following equation:

$$H_{ad} = 2C^2(1 - C)\left|v_{ep} \pm v_{eo}\right| Et_{ad} \tag{1.18}$$

where H_{ad} is the contribution from adsorption to the plate height, E is the electric field, C is the fractional concentration of the free solute and t_{ad} is the mean residence time of the adsorbed solute [69]. v_{ep} and v_{eo} are the electrophoretic velocity and the electroosmotic velocity, respectively.

1.3.6 Conductivity differences

When the solute ions do not have the same mobility as the buffer ions, the migrating solute may not have the same electrical conductivity as the surrounding buffer. The conductivity difference, Δk, between the migrating zone and the background buffer solution is given by [93]:

$$\Delta k = \frac{C}{v} (\mu_B - \mu)(\mu_c - \mu) \tag{1.19}$$

where C is concentration of the sample ion, v is the electromigration velocity, μ is the mobility of the sample, μ_B is the mobility of the buffer ion which has the same sign as the ion, and μ_c is the mobility of the counter-ion.

Furthermore, the solute molecules are likely to diffuse across the boundary between the injected plug and the carrier buffer solutions. Consequently, zone broadening may occur as a result of the combined effects of conductivity difference (Δk) and diffusion. These effects are illustrated in Fig. 1.10 for a sample which has a mobility lower than that of the buffer ion, i.e. $|\mu| < |\mu_B|$. In Fig. 1.10, α represents the sample whereas β represents the background buffer. M_1, M_2 and M_3 are solute molecules 1, 2 and 3. L is the length of the capillary and k_B is the conductivity of the buffer. v^α and v^β are the electrophoretic migration velocity in the α and β phase, respectively. v_{diff} is the rate of diffusion of the solute molecule, M_1, in the β phase. Δx_0, Δx_k, $\Delta x'_{k,D}$ and $\Delta x''_{d,K}$ represent the original width of the sample zone, the zone broadening due to conductivity difference, the zone broadening at the front boundary and at the rear boundary due to the combined effects of difference in conductivity and diffusion, respectively [70].

On the other hand, it should be noted that since the electric field is inversely proportional to the specific conductivity, the field strength would be higher for a solution of lower conductivity. Consequently, the migration of ions would be faster in this solution. This effect has been exploited to improve efficiency during injection. The technique is called sample stacking and is described in Section 2.2.2.

1.4 COMPARISON WITH OTHER SEPARATION TECHNIQUES

Capillary electrophoresis (CE) has frequently been compared with high-performance liquid chromatography (HPLC) and conventional gel electrophoresis. Bearing in mind the relatively recent development of CE as a separation technique, it should be recognized that there will probably be more rapid growth in CE than in the other more matured techniques. In fact, currently there is such a tremendous interest in CE that rapid progress is constantly being made, both in instrumentation and separation methodologies. There will certainly be an immense scope for further advances in the development of new applications for CE. The comparison made here is therefore partly intended to serve as an indicator of the areas where efforts can be made to achieve significant future developments in CE.

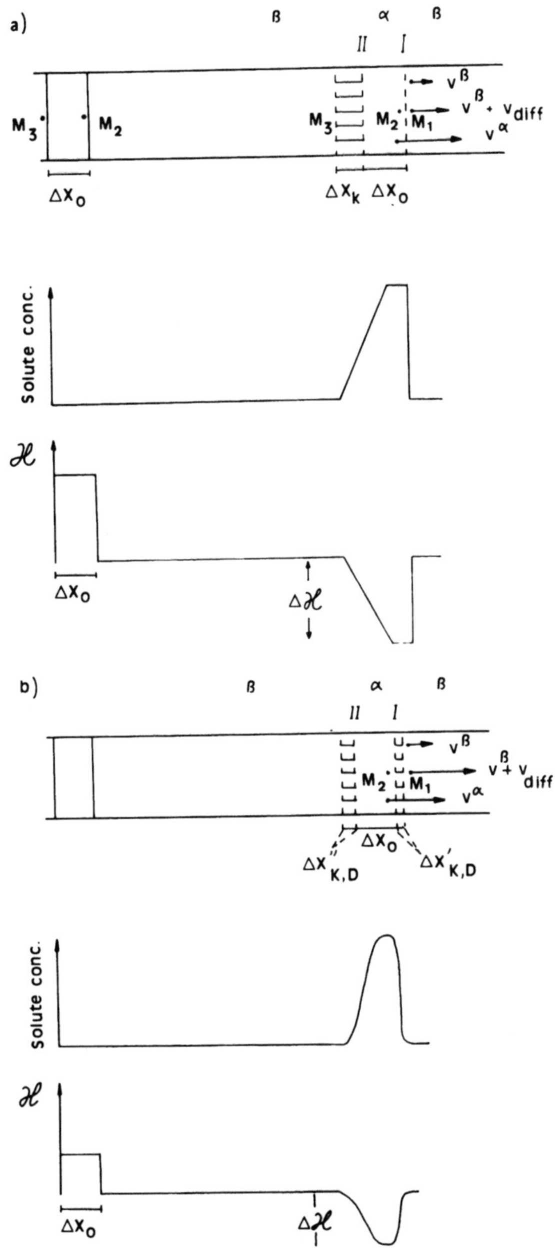

Fig. 1.10. (a) Zone broadening (ΔX_k) caused by a conductivity difference, Δk, between the solute zone and the buffer when $v^{\alpha} \geq v^{\beta} + v_{\mathrm{diff}}$, i.e. boundary I remains sharp and boundary II tails. $\Delta X_k = L\,\Delta k/k$. ΔX_0 is the original width of the sample zone and $\mathcal{H}$ is the peak height. (b) Zone broadening (ΔX_k) caused by the combined effect of conductivity differene (Δk) and diffusion for $v^{\alpha} \leq v^{\beta} + v_{\mathrm{diff}}$. $\Delta X_{k,D} = \Delta X'_{k,D} + \Delta X''_{k,D} = a\sqrt{2Dt}$ where $a \simeq 4$.
(Reproduced from Ref. 70 with permission of VCH Verlagsgesellschaft.)

1.4.1 Comparison with HPLC

CE has attracted considerable attention in recent years because of its potential to achieve very high efficiency. The main reason for the extraordinary high efficiency in CE is attributable to its characteristically flat flow profile. In Fig. 1.6, flow profiles in HPLC and CE are shown. In general, the flow of mobile phase in HPLC is maintained by a pump and therefore under normal operating conditions, a parabolic flow profile is observed. As a result of its contributions to peak broadening, the flow profile inherently limits the separation efficiency theoretically achievable in HPLC separations. In the case of CE, charged species migrate under the influence of an applied voltage, resulting in a practically flat flow profile. This fundamental difference in flow profile is the main reason for the extremely high efficiencies achievable using CE, where narrower peaks and potentially better resolution can be readily obtained, especially when selectivity is also optimized for the separation (see Chapter 5).

Furthermore, peak capacity in CE is much higher than that in HPLC. Column efficiencies in CE of several hundred thousand to millions of theoretical plates have been reported [1–5,49,50], allowing the resolution of closely eluting peaks and the separation of a large number of components in a mixture.

On the other hand, there are a large number of stationary phase and mobile phase systems developed for HPLC to obtain the required selectivity for a particular separation. In this respect, CE is a relatively less developed technique. Nevertheless many interesting approaches have already been demonstrated to enhance selectivity in CE separations [94–97], such as the use of micelles in micellar electrokinetic chromatography, the use of inclusion or complexing agents, e.g. cyclodextrins, the use of chiral additives for enantiomeric seprations, and the use of organic modifiers, such as methanol (see Chapter 5).

In terms of instrumentation, HPLC and CE are similar in some respects but different in others. It is simpler for CE due to the absence of an injector, a pump (or one or more pumps and a solvent mixer for gradient HPLC) and a special detection cell. For CE, injection is usually performed using one end of the separation column, and on-column detection is accomplished with part of the column forming the detection cell.

Sample introduction in HPLC is commonly performed by introducing into the column a known volume of the sample into a fixed-volume sample loop in the injection valve by means of a syringe. The injected sample is than swept into the chromatographic column by the mobile phase. Consequently, the volume of the injected solution can be exactly measured.

In CE, there is no injection valve as in HPLC. The sample solutions are most commonly introduced into the capillary either by the electrokinetic or the hydrodynamic mode. These injection techniques are described in detail in Chapter 2. With the electrokinetic mode, the amount of sampled introduced will be affected by the applied injection voltage and the time of applied voltage, whereas in the hydrodynamic mode, the injection amount is affected by the pressure differential across the column and the injection time. In both injection modes, it is necessary to know the exact measurements of the column dimensions (radius and length) in order to calculate the amount injected.

Under normal operating conditions in CE, the two ends of the capillary need to be immersed in the buffer. Collection of sample fractions can be performed by interrupting the voltage and transferring the outlet (detection) end of the capillary to a small collection vial containing an electrode and a solution which is normally the same as the electrophoretic buffer. In addition, several interesting approaches has been developed to facilitate sample collection in CE [98–101]. Olefirowicz and Ewing [98] utilized a porous glass junction which did not interrupt the flow of current. Huang and Zare [99] used an on-column frit structure that allows the flow of current and would neither interrupt the electrophoretic process nor dilute the zones collected.

HPLC has the advantage that it can be employed as a micro as well as a macro separation technique, since the column diameter can vary considerably. For fraction collection, commercially available fraction collectors can be utilized. In CE, the column diameter is limited by the efficiency of heat dissipation. Since the heat gradient between the center of the capillary and the walls is proportional to the square of the radius, smaller capillaries enhance heat dissipation. So far, capillaries of 2–200 μm diameter and 10–100 cm length are most commonly used in CE separations. The use of multiple capillaries permits larger sample capacity in CE [100,101].

Modes of detection in both HPLC and CE are similar. A wide range of detectors, mostly developed originally for HPLC, such as UV, fluorescence, electrochemical, conductivity, Raman and radioisotope detectors have been successfully adapted for CE detection. Interfaces for mass spectrometric detection have also been developed for both HPLC and CE applications. In the case of optical detection techniques, such as UV detection, the concentration sensitivity for HPLC tends to be better than that of CE [102]. This is due to the fact that the cell path length in CE (capillary width) is usually smaller than that of a conventional HPLC flow cell unless specially designed flow cells are employed (see Chapter 3). On the other hand, extremely sensitive mass detection has been achieved by laser-induced fluorescence [103,104] and electrochemical detection [105–107], with detection limits ranging below attomole (atto = 10^{-18}) level.

Recently Issaq *et al.* [108] performed a comparative study on separations by high-performance liquid chromatography and capillary zone electrophoresis. Their investigation was based mainly on mechanism of separation, instrumentation and fields of applications. They concluded that the two techniques could be complimentary, especially for the separation and analysis of biomolecules. Each technique has its own points of strength and weakness. Capillary zone electrophoresis was found to be superior whenever high peak efficiency was required, such as in the analysis of DNA fragments, while high-performance liquid chromatography was superior for small and neutral molecules and in its quantitative capabilities.

An interesting and extremely useful approach is to couple together the HPLC and CE systems to give an on-line multi-dimensional setup. This aspect is discussed in detail in Section 6.9. The objective of combined analytical separations is to obtain non-redundant information from independent systems [102]. For techniques to be complementary to each other, the acquired data should be orthogonal, so that more information can be obtained from the analysis. Steuer *et al.* [102] defined the retention parameter:

$$\chi_i = \frac{t_i - t_0}{\Delta t} \tag{1.20}$$

where t_i represents the time for the ith component, t_0 the time for the first component and Δt the total range of analysis times. The retention parameters were calculated for several drugs and their by-products and degradation products, which represented a range of substances with vastly different chemical properties. The retention parameter for CZE (χ_{CZE}) were plotted against that of HPLC (χ_{HPLC}) in Fig. 1.11. They demonstrated that CZE and HPLC were highly orthogonal systems. Hence coupling of these techniques would be of considerable benefit.

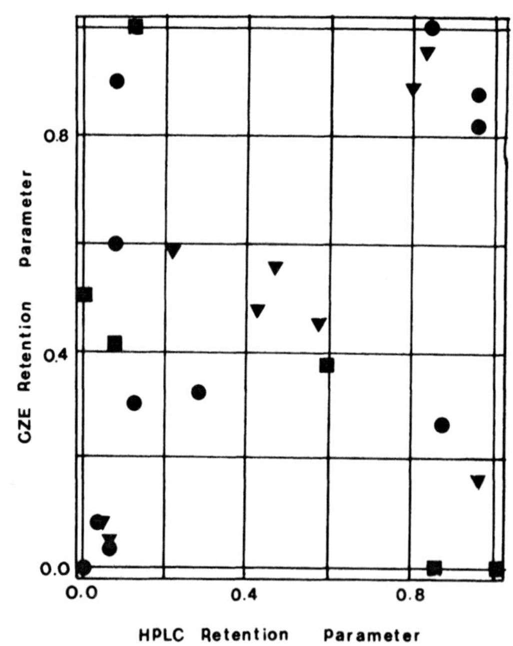

Fig. 1.11. Demonstration of the orthogonality between HPLC and CZE. No obvious correlation is observed between the retention parameters in HPLC and CZE. All HPLC separations were carried out under reversed-phase conditions, with the exception of isradipine (normal-phase). Compounds: ▸, terbinafine; ■, spriapril; •, AH21132. (Reproduced from Ref. 102 with permission of Elsevier Science Publishers.)

1.4.2 Comparison with slab-gel electrophoresis

CE has several advantages over conventional slab-gel electrophoresis. The major limitation in conventional electrophoresis is solution heating owing to the ionic current carried between the electrodes. Joule heating can result in density gradients and subsequent convection and temperature gradients that increase zone broadening, affect electrophoretic mobilities, and can even lead to evaporation of solvent. In large-scale electrophoresis, a supporting medium such as a gel is used to help dissipate heat, thereby minimizing these sources of band broadening. However, the support increases the surface area available for solute adsorption and introduces the band-broadening effect of eddy diffusion. On the other hand, one of the main advantages of capillary tubes is the enhanced heat dissipation relative to the volume of solution in the tube. In the case of CE, dissipation of heat takes place via the capillary wall. Hence, the maximized ratios of inner surface area to volume attained in small-bore capillaries provide more efficient heat dissipation relative to large-scale systems. This permits the use of very high potential fields and free solutions for fast, efficient separations.

In addition, there are several other advantages to the use of capillaries for electrophoresis. One is the possibility to utilize electroosmotic flow in the CE system to facilitate automation. Since electroosmosis is the flow of solvent in a capillary when a tangential potential field is applied, this flow could be deliberately altered so that it is strong enough to cause all solutes to elute at one end of the capillary. Consequently, CE is more readily automated than large-scale electrophoresis, which tends to be rather labour intensive and time consuming. Another advantage of CE is the availability of a wide range of instruments already developed for HPLC which can be easily adapted for CE work. For example, in the area of detection, many types of detection modes for HPLC have already been successfully modified for CE detection. Finally, the ultrasmall volume flow rates typically obtained in CE and the possibility of on-column detection permit analysis to be performed on very small amounts of sample (nanoliters per run). Recently techniques have been developed for sampling from microenvironments [105–107]. In contrast, much larger amounts of sample would be required in conventional gel electrophoresis.

1.5 CONCLUSION

In summary, CE displays an enormous efficiency and possesses inherent advantages over conventional separation techniques. The technique has fundamentally better capability for high-resolution separation as a result of its characteristic flat flow profile. Although currently CE is at an early stage of development and there are still needs to improve column technology, to enhance selectivity in separations, and to refine the instrumentation for CE work at this stage, it can be certain that there will be an immense potential for further developments in the area of CE.

References pp. 28–30

Furthermore, the usefulness of the technique stems from its potential not just in competition or simply as an alternative to HPLC, but as an additional method complementary to HPLC which is capable of augmenting the information that can be obtained from the analysis. Modern CE instruments equipped with automated features which are also capable of achieving faster analysis time would certainly serve as valuable tools to replace some of the time consuming and laborious task involved in conventional electrophoresis.

1.6 REFERENCES

1 J.W. Jorgenson and K.D. Lukacs, Anal. Chem., 53 (1981) 1298
2 J.W. Jorgenson and K.D. Lukacs, J. Chromatogr., 218 (1981) 209
3 J.W. Jorgenson and K.D. Lukacs, J. High Resolut. Chromatogr. Chromatogr. Comm., 4 (1981) 230
4 J.W. Jorgenson and K.D. Lukacs, Clin. Chem., 27 (1981) 1551
5 J.W. Jorgenson and K.D. Lukacs, Science, 222 (1983) 266
6 J.W. Jorgenson, Trends Anal. Chem., 3 (1984) 51
7 J.W. Jorgenson and K.D. Lukacs, in: Microcolumn Separations, Journal of Chromatography Library, Vol. 30, Elsevier, 1985, p. 121
8 J.W. Jorgenson, ACS Symp. Ser., 335 (1987) 182
9 J.W. Jorgenson, D. Rose and R. Kennedy, Amer. Lab., 1988, April.
10 T. Tsuda, K. Nomura and G. Nakagawa, J. Chromatogr., 264 (1983) 385
11 R.A. Wallingford and A.G. Ewing, Adv. Chromatogr., 29 (1989) 1
12 A.G. Ewing, R.A. Wallingford and T.M. Olefirowicz, Anal. Chem., 61 (1989) 292A
13 B.L. Karger, A.S. Cohen and A. Guttman, J. Chromatogr., 492 (1989) 585
14 B.L. Karger, J. Res. Natl. Bur. Stand. (USA), 93 (1988) 406
15 M.J. Gordon, X. Huang, S.L. Pentoney and R.N. Zare, Science, 242 (1988) 224
16 E.S. Yeung, Acc. Chem. Res., 22 (1989) 125
17 P.D. Grossman, J.C. Colburn, H.H. Lauer, R.G. Nielsen, R. Riggin, G.S. Sittampalam and E.C. Rickard, Anal. Chem., 61 (1989) 1186
18 J. Snopek, I. Jelinek and E. Smolkova-Keulemansova, J. Chromatogr., 452 (1988) 571
19 N.A. Guzman, L. Hernandez and B.G. Hoebel, BioPharm, 2 (1989) 22
20 M.V. Pickering, LC-GC, 43 (1989) 134
21 H.M. Widmer, Chimia, 43 (1989) 134
22 S. Compton, R. Brownlee, Biotechniques, 6 (1988) 432
23 A. Lodge, B.A. Thesis, 1886.
24 I. Smirnow, Berl. Klin. Woch. 1892, 32, 645
25 W.B. Hardy, J. Physiol, 1899, 24, 288
26 W.B. Hardy, J. Physiol. 33 (1905) 273
27 C.W. Field and O. Teague, J. Exp. Med. 9 (1907) 86
28 J. Kendall and C. Crittenden, Proc. Nat. Acad. Sci., 9 (1923) 75
29 A. Tiselius, Dissertation, University of Upsala, Sweden, 1930
30 A. Tiselius, Trans. Faraday Soc., 33 (1937) 524
31 T.B. Coolidge, J. Biol. Chem., 127 (1939) 551
32 R.A. Consden, A.H. Gordon and A.J.P. Martin, Biochem. J., 40 (1946) 33.
33 H. Haglund and A. Tiselius, Acta Chem. Scand., 4 (1950) 957

34 J. Porath, Biochim. Biophys. Acta, 22 (1956) 151
35 L. Ornstein, N.Y. Acad. Sci., 121 (1964) 321
36 A. Tiselius, S. Hjerten and S. Jerstedt, Arch. Ges. Virusforsch, 17 (1965) 512
37 S. Hjerten, S. Jerstedt and A. Tiselius, Anal. Chem., 11 (1965) 211
38 S. Hjerten, Chromatogr. Rev., 9 (1967) 122
39 R. Virtenen, Acta Polytech. Scand., 123 (1974) 1
40 F.E.P. Mikkers, F.M. Everaerts and T.P.E.M. Verheggen, 169 (1979) 11
41 S. Hjerten, J. Chromatogr., 270 (1983) 1
42 S. Terabe, K. Otsuka, K. Ichikawa, A. Tsuchiya and T. Ando, Anal. Chem., 56 (1984) 111
43 A. Cohen and B.L. Karger, J. Chromatogr., 397 (1987) 409
44 K. Altria and C. Simpson, Anal. Proc., 23 (1986) 453
45 T. Tsuda, J. High Resolut. Chromatogr. Chromatogr. Comm. 10 (1987) 622
46 S. Hjerten and M.D. Zhu, J. Chromatogr., 327 (1985) 157
47 S. Hjerten and M.D. Zhu, Protides Biol. Fluids, 33 (1985) 537
48 S. Hjerten, K. Elenbring, F. Kilar, J.L. Liao, A.J.C. Chen, C.J. Siebert and M.D. Zhu, J. Chromatogr. 403 (1987) 1987
49 A.S. Cohen and B.L. Karger, J. Chromatogr., 397 (1987) 409
50 A.S. Cohen, A. Paulus and B.L. Karger, Chromatographia, 24 (1987) 14
51 S. Terabe, K. Otsuka and T. Ando, Anal. Chem., 61 (1989) 251
52 K. Otsuka and S. Terabe, J. Microcol. Sep., 1 (1989) 150
53 T. Tsuda, K. Normura and G. Nakagawa, J. Chromatogr. 248 (1982) 241
54 M.J. Sepaniak and R.O. Cole, Anal. Chem., 59 (1987) 472
55 A.T. Balchunas, and M.J. Sepaniak, Anal. Chem., 59 (1988) 1466
56 M. Martin, G. Guiochon, Y. Warbroehl and J. Jorgenson, Anal. Chem. 57 (1985) 559
57 M.M. Bushey and J.W. Jorgenson, J. Microcol. Sep., 1 (1989) 125
58 A. Dobashi, T. Ono, S. Hara and J. Yamaguchi, J. Chromatogr., 480 (1989) 413
59 S. Terabe, H. Utsumi, K. Otsuka, T. Ando, T. Inomata, S. Kuze and Y. Hanaoka, J. High Resolut. Chromatogr. Chromatogr. Commun., 9 (1986) 666
60 D.E. Burton, M.J. Sepaniak and M.P. Maskarinec, J. Chromatogr. Sci., 25 (1987) 514
61 J.H. Knox, Chromatographia, 26 (1988) 329
62 V. Pretorius, B.J. Hopkins and J.D. Schieke, J. Chromatogr., 99 (1974) 23
63 J.W. Jorgenson and K.D. Lukacs, J. High Resolut. Chromatogr. Chromatogr. Commun., 8 (1985) 407
64 S. Hjerten and M.D. Zhu, J. Chromatogr., 346 (1985) 265
65 S. Hjerten, J.L. Liao and K. Yao, J. Chromatogr., 387 (1987) 127
66 J.R. Mazzeo and I.S. Krull, BioTechniques, 10 (1991) 638
67 F.M. Everaerts and P.E.M. Verheggen, in: New Directions in Electrophoretic Methods, J.W. Jorgenson and M. Phillips, Amer. Chem. Soc. Symp. Vol. 335, Washington DC, 1987, Chap. 4.
68 P. Bocek, M. Deml, P. Gebauer and V. Dolnik, Anal. Isotachophoresis, VCH Verlagsgesellschaft, Weinhein, 1988.
69 R.J. Wieme, in Chromatography — A Laboratory Handbook of Chromatographic and Electrophoretic Methods, 3rd ed., (E. Heftman, Ed.), Van Nostrand Reinhold, New York, 1975, Chapter 10.
70 S. Hjerten, Electrophoresis, 11 (1991) 665
71 T.S. Stevens and H.J. Cortes, Anal. Chem., 55 (1983) 1365
72 K. Altria and C. Simpson, Chromatographia, 24 (1987) 527

73 X. Huang, M.J. Gordon, R.N. Zare, Anal. Chem., 60 (1988) 1837
74 B. Wanders, A. Van de Goor and F. Everaerts, J. Chromatogr., 470 (1989) 89
75 A. Van De Goor, B. Wanders and F. Everaerts, J. Chromatogr., 470 (1989) 95
76 H.H. Lauer and D. McManigill, Anal. Chem., 587 (1986) 166
77 S. Fujiwara and S. Honda, Anal. Chem., 58 (1986) 1811
78 X. Huang, J.A. Lucker, M.J. Gordon and R.N. Zare, Anal. Chem., 61 (1989) 766
79 F. Foret, S. Fanali, L. Ossicini and P. Bocek, J. Chromatogr., 470 (1989) 299
80 S. Fujiwara and S. Honda, Anal. Chem., 59 (1987) 487
81 S. Hjerten, J. Chromatogr., 347 (1985) 191
82 J.K. Roberts, Heat and Thermodynamics, 3rd ed., Blackie, London, 1947, 245 pp.
83 G.I. Taylor, Proc. Roy. Soc. (London), A219 (1953) 186
84 R. Aris, Proc. Roy. Soc. (London), A235 (1953) 67
85 F. Mikkers, F. Everaerts and T. Verheggen, J. Chromatogr., 169 (1979) 1
86 W. Thormann, P. Michaud and R.A. Mosher, Electrophor. '86, Proc. Meet. Int. Electrophor. Soc., 5th (1986) 267
87 W. Thormann, Electrophoresis, 4 (1983) 383
88 H. Lauer and D. MaManigill, Trends Anal. Chem, 5 (1986) 11
89 F. Foret, M. Deml and P. Bocek, J. Chromatogr., 452 (1988) 601
90 R. Nelson, A. Paulus, A. Cohen, A. Guttman and B. Karger, J. Chromatogr., 480 (1989) 111
91 E. Grushka, R.M. McCormick and J.J. Kirkland, Anal. Chem., 61 (1989) 241
92 A.E. Jones and E. Grushka, J. Chromatogr., 466 (1989) 219
93 S. Hjerten, in G. Milazzo, (Ed.) Topics in Bioelectrochemistry and Bioenergetics, John Wiley, Vol. 2, 1978, pp. 89–128
94 S. Terabe, K. Otsuka and T. Ando, Anal. Chem., 57 (1985) 834
95 A. Guttman, A. Paulus, A. Cohen, N. Grinberg and B. Karger, J. Chromatogr., 448 (1988) 41
96 S. Terabe, M. Shibata and Y. Miyashita, J. Chromatogr., 480 (1989) 403
97 S.K. Yeo, C.P. Ong and S.F.Y. Li, Anal. Chem., 63 (1991) 2222
98 T.M. Olefirowicz and A.G. Ewing, Anal. Chem., 59 (1987) 1762
99 X. Huang and R.N. Zare, Anal. Chem., 62 (1990) 443
100 C. Fujimoto, Y. Muramatsu, M. Suzuki, Y. Hirata and K. Jinno, Proc. 12th Int. Symp. Cap. Chromatogr., Kobe, Japan, 11–14 Sept. 1990, p. 684
101 N.A. Guzman, M.A. Trebilcock and J.P. Advis, J. Liq. Chromatogr., 14 (1991) 997
102 W. Steuer, I. Grant and F. Erni, J. Chromatogr., 507 (1990) 125
103 S. Wu and N.J. Dovichi, J. Chromatogr., 480 (1989) 141
104 Y.F. Cheng and N.J. Dovichi, Science, 242 (1989) 562
105 R. Wallingford and A.G. Ewing, Anal. Chem., 60 (1988) 1972
106 R. Wallingford and A.G. Ewing, Anal. Chem., 61 (1989) 98
107 R. Wallingford, P.D. Curry, A.G. Ewing, J. Microcol. Sep., 1 (1989) 23
108 H.J. Issaq, G.M. Janini, I.Z. Atamna and G.M. Muschik, J. Liq. Chromatogr., 14 (1991) 817

CHAPTER 2

Sample Injection Methods

2.1 INTRODUCTION

To preserve the high efficiency capabilities of capillary electrophoresis, the injection system must not introduce significant zone broadening [1–42]. It is important to ensure that the sample injection method employed is capable of delivering small volumes of sample (typically several nanoliters) onto the column efficiently and reproducibly [1–3]. Consequently, the most commonly employed injection methods for CE are direct on-column methods such as electromigration [4–6] and hydrodynamic flow [1,6,8]. In both of these types of injection methods, one end of the capillary is used as the sample injector directly, thereby eliminating the zone broadening due to connection with sample injection valves [4–16,42]. In addition, a number of specialized injection systems have also been developed [17–41].

2.1.1 Effect of sample overloading on efficiency

Capillary electrophoresis systems are easily overloaded by large sample volumes. Sample overload can affect system efficiency by two distinct mechanisms [2]. One mechanism relates to the volume of the sample injected (q_{inj}) relative to the total volume of the capillary (q_c). The standard deviation of the injection plug in volume units is [2]:

$$\sigma_{q_{inj}} = q_{inj}/\sqrt{12} \tag{2.1}$$

and

$$N_{max} = 12\left(\frac{q_c}{q_{inj}}\right)^2 \tag{2.2}$$

From Eq. (2.2) it can be seen that the maximum number of theoretical plates (N_{max}) of the overall system is constrained to a value proportional to the square of the ratio of the volume of the injected sample to the volume of the column.

The second mechanism imposes a limit on the concentration of the sample injected and is related to the difference in electrical conductivity of the sample and the electrophoretic medium. At high sample concentration, system efficiency can be degraded due to perturbation in the potential field gradient by the sample within the column. Severely distorted peaks may result. On the other hand, if the sample injected has a slightly lower conductivity than the electrophoretic buffer, sample stacking can be achieved which improves peak shape and hence increases efficiency (see Sections 2.1.2 and 2.2.2).

Grushka and McCormick [3] approximated the maximum allowable injection plug length as

$$l_{\mathrm{inj}} = (24\,De_{\mathrm{h}}\,t)^{1/2} \tag{2.3}$$

$$H = \frac{2D}{v} \tag{2.4}$$

where D is the solute's diffusion coefficient, v is the solute's velocity, t is the migration time, and e_{h} is the acceptable increase in plate height (H) relative to the theoretical minimum HETP of the system. They calculated the allowable injection plug length as a function of diffusion coefficient (or molecular mass) of the solute, and of the migration time from the system. Figure 2.1 shows the allowed injection plug length as a function of the solute's diffusion coefficient at a constant analysis time of 10 min. Three different cases are shown, which correspond to 5, 10 and 20% loss in efficiency. Based on Fig. 2.1, it can be seen that the restriction on the injection plug length could be rather stringent, especially for larger molecules which have small diffusion coefficients.

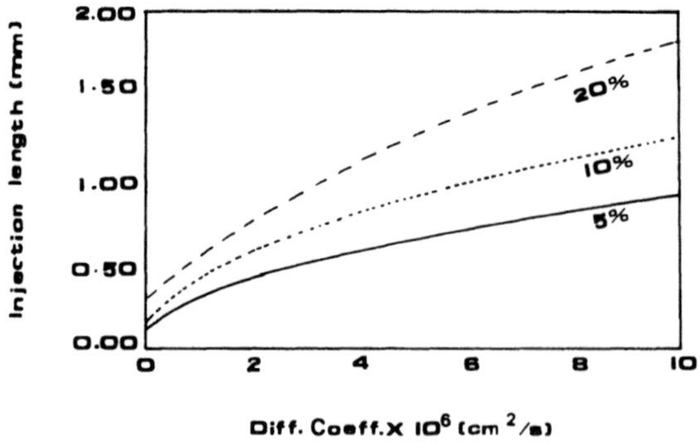

Fig. 2.1. The allowed injection plug length as a function of the solute's diffusion coefficient. The analysis time was assumed to be 10 min. Each line corresponds to a different allowable loss in the efficiency (increasing H). (Reproduced from Ref. 3 with permission of Elsevier Science Publishers.)

2.1.2 Sample stacking

Sample stacking occurs when the conductivity of the injected sample is lower than that of the surrounding buffer, and hence results in concentration of the analyte zone [29–31]. The reason for the narrowing of the analyte zone can be attributed to the fact that the electric field depends inversely on the specific conductivity, i.e. higher field strength at lower conductivity. Therefore, the electric field strength increases in the sample zone of lower conductivity. The electrophoretic velocity increases at the higher field and hence the analyte zone becomes narrower. This effect, described as stacking, can be utilized in both hydronamic and electrokinetic injections to enhance efficiency. An example of its application is in field amplified sample injection, which is described in Section 2.2.2.

The dispersion processes in free solution capillary electrophoresis under both stacking and non-stacking conditions have been investigated by Vinther and Soeberg [29,30]. It was found that moderate stacking conditions should be employed during injections, i.e. the sample solution should have a specific conductivity only slightly lower than that of the buffer solution. This is because radial dispersion increases if there is a large difference between the conductivity of the sample and the buffer, which counteracts the stacking effect (see Section 1.3.6). Furthermore, the applied potential should also be kept at a moderate level during the stacking period, i.e. low enough to minimize the effect of radial dispersion but high enough to exceed the limit imposed by axial diffusion (Eq. 1.3). Optimum condition for sample stacking can be achieved by preparing the sample in a buffer at a concentration about 10 times less than that of the electrophoretic medium and by injecting a sample plug of length about 10 times the diffusion-limited peak width [31].

2.1.3 Extraneous injection

Grushka and McCormick demonstrated that the insertion, withdrawal or both actions could result in sample penetration into the capillary [3]. This constitutes an additional source of peak broadening since sample enters the capillary in an uncontrolled manner. Three mechanisms could be responsible for such extraneous sample injection: (i) displacement of small volume of the sample into the capillary during insertion of the capillary into the sample solution; (ii) convective movements between buffer and sample due to difference in thermophysical properties, such as viscosity, surface tension and/or density and (iii) diffusion of solute into the capillary. Among the three mechanisms the convective movements attributable to density difference between the buffer and the sample have been thought to play the major part.

2.2 ELECTROKINETIC INJECTION

Electrokinetic injection is also called electromigration injection. To perform electrokinetic injection, the electrode is removed from the buffer vial and placed into the sample vial. The buffer reservoir at the high-voltage electrode is replaced with the sample vial such that the capillary and electrode dip into the sample solution. An injection voltage is then applied for a brief period of time, causing sample to enter the end of the capillary by electromigration. Electromigration injection includes contribution from both electrophoretic migration of charged sample ions and electroosmotic flow of the sample solution.

The electrophoretic and electroosmotic velocities can be represented as [40]:

$$v_{\mathrm{ep}} = \mu_{\mathrm{ep}} \frac{V_{\mathrm{i}}}{L} \tag{2.5}$$

$$v_{\mathrm{eo}} = \mu_{\mathrm{eo}} \frac{V_{\mathrm{i}}}{L} \tag{2.6}$$

where μ_{ep} is the electrophoretic mobility of the sample molecule, μ_{eo} is the electroosmotic mobility of the sample solution, V_{i} is the injection voltage and L is the length of the column.

The length of the sample zone is given by

$$l = (v_{\mathrm{ep}} + v_{\mathrm{eo}})\, t_{\mathrm{i}} \tag{2.7}$$

where v_{ep} is the electrophoretic velocity of the sample molecule, v_{eo} is the electroosmotic velocity of the sample solution, and t_{i} is the injection time (i.e. time over which the injection voltage is applied).

By substituting Eqs. (2.5) and (2.6) into Eq. (2.7), the sample zone length becomes:

$$l = \frac{(\mu_{\mathrm{ep}} + \mu_{\mathrm{eo}})\, V_{\mathrm{i}}\, t_{\mathrm{i}}}{L} \tag{2.8}$$

For on-column injection methods, the amount, w, of sample injected (in weight or number of moles, depending on unit for concentration) into the capillary is given by [1]:

$$w = \pi r^2 l\, C \tag{2.9}$$

where r is the radius of the capillary, l is the length of the sample zone, and C is the sample concentration.

Thus we can determine w, the amount of solute injected by electromigration by combining Eqs. (2.8) and (2.9) to give:

$$w = \frac{(\mu_{\mathrm{ep}} + \mu_{\mathrm{eo}})\, \pi r^2 V_{\mathrm{i}}\, t_{\mathrm{i}}}{L}\, C \tag{2.10}$$

From Eq. (2.10), it can be seen that the quantity of sample injected during electrokinetic injection can be controlled through the variation in the injection time, t_{i}, and injection voltage V_{i}.

During electrokinetic injection, two types of bias may occur [12]. One occurs as a result of differences in mobilities of the species in the sample solution. The more mobile components are injected in larger quantities than the less mobile species. Another type of bias is related to the differences in the conductivity between the sample solution and the operating buffer. When injections are made from samples not prepared in the operating buffer, this effect must be considered. The reason is that both the electrophoretic mobilities and electroosmotic flow rate would be different in different solutions and thus changes in the absolute amount injected would occur. Equation (2.10) is applicable only when the conductivity of the sample solution and the operating buffer are approximately equal.

Huang *et al.* [12] performed an analysis on the sampling bias in electrokinetic injection within a single sample and between different samples. For sample solution containing species 1 and 2, the ratio of the amounts electrokinetically injected is given by

$$\frac{w_{(1)}}{w_{(2)}} = \frac{[\mu_{(1)} + \mu_{eo}]}{[\mu_{(2)} + \mu_{eo}]} \frac{C_{(1)}}{C_{(2)}} \tag{2.11}$$

where $w_{(1)}$ and $w_{(2)}$ are the amounts of species 1 and 2 injected into the capillary; $\mu_{(1)}$ and $\mu_{(2)}$ are mobilities of species 1 and 2; $C_{(1)}$ and $C_{(2)}$ are the concentrations of species 1 and 2 and μ_{osm} is the electroosmotic mobility. A bias factor, b, was defined [12]:

$$b = \frac{\mu_{(1)} + \mu_{eo}}{\mu_{(2)} + \mu_{eo}} \tag{2.12}$$

It is noted in Eq. (2.12) that $b = 1$ when μ_{eo} is much greater than $\mu_{(1)}$ and $\mu_{(2)}$. In such cases, $w_{(1)}/w_{(1)}$ is directly proportional to $C_{(1)}/C_{(2)}$, and no sampling bias would occur. For $b \neq 1$, electrokinetic injection introduces a sampling bias which must be taken into account to ensure accurate quantitation. If migration time, i.e. the time for the pieces i to reach the detector located at a distance z from the injection end of the capillary is defined as:

$$t_{m,i} = \frac{z}{v_{tot,i}} \tag{2.13}$$

where $v_{tot,i}$ is the total ionic velocity, which is equal to the total ionic mobility ($\mu_{tot,i} = \mu_{(i)} + \mu_{eo}$), times the electric field strength E.

Then

$$t_{m,i} = \frac{z}{[\mu_{(i)} + \mu_{eo}]E} \tag{2.14}$$

For a two-species system, the bias factor is given by

$$b = \frac{t_{m,2}}{t_{m,1}} \tag{2.15}$$

Therefore, by measuring the ratio of the migration times, the bias factor can be calculated and the apparent ratio of the injection amounts can be related to the concentration in the sample solution.

To use the migration time ratio to correct the bias in electrokinetic injection, it is necessary to ensure that the electroosmotic flow rates of the sample solution and that of the electrophoretic buffer are nearly the same; and that the electroosmotic flow rates during sample injection and that during the capillary electrophoretic run are almost identical. These conditions can be met if the sample ion concentration is low compared to that of the electrolyte and the injection voltage is nearly the same as the applied voltage during the CE run.

When two different sample solutions are considered (S_1 and S_2), the absolute amount of the same species will vary for different solutions because of the variation in $v_{\text{tot,i}}$ from sample solution to sample solution. Equation (2.12) then becomes:

$$\frac{w_{(\text{i};\, \text{s}_1)}}{w_{(\text{i};\, \text{s}_2)}} = \frac{v_{\text{tot}(\text{i};\, \text{s}_1)}}{v_{\text{tot}(\text{i};\, \text{s}_2)}} \times \frac{C_{(\text{i};\, \text{s}_1)}}{C_{(\text{i};\, \text{s}_2)}} \tag{2.16}$$

Huang *et al.* [12] showed for cations (K^+ and Li^+) that electrokinetic injection introduces a linear bias in which more ions are injected for solutions having higher ohmic resistance. This behaviour can be explained by the fact that both the electrophoretic velocity of species i and the electroosmotic flow rate of the solution increase approximately linearly with decreasing electrolyte concentration, and hence $v_{\text{tot,i}}$ varies almost linearly with sample solution resistance.

Rose and Jorgenson [6] designed an automatic sampling system for CZE, utilizing stepping motors controlled by a computer. By performing electrokinetic injection with such an automatic instrument, they were able to achieve an improvement of %RSD in peak area from 13.47% by manual injection to 4.1% by automated injection.

Lux *et al.* [28] considered a modular CE instrument and obtained with it reproducibility in retention time of better than 0.5% RSD and in absolute peak area of about 3.0% RSD. They also found that rinsing of the outside of the capillary inlet after sample introduction helped to prevent peak tailing.

2.2.1 Field amplified sample injection

Recently, field amplified sample injection (FASI) techniques have been investigated [32–35]. Sample stacking was also performed. In FASI, a plug of sample in a buffer of lower ionic strength is injected into the column filled with a buffer of higher ionic strength. Subsequently, sample ions migrate rapidly into the run buffer under the applied voltage, resulting in stacking in front of the water boundary. As the sample concentration increases, the local conductivity of the stacking region becomes higher, resulting in a further drop in the electric field strength. Consequently, the leading edge of the sample region slows down and further enhances

TABLE 2.1

COMPARISION OF AMOUNT INJECTED AND EFFECTIVE PLUG LENGTH FOR CONVENTIONAL ELECTROKINETIC INJECTION AND FASI

	Conventional electrokinetic injection	FASI
Amount of injection	$(\mu_{eo} + \mu_{ep})AC_i Et$	$(\mu + \gamma\mu_{ep})AC_i Et$
Different plug length	$(\mu_{eo} + \mu_{ep})Et$	$\left(\frac{\mu_{eo}}{\gamma} + \mu_{ep}\right)Et$

A is cross-sectional area.

the stacking effect. The peak width will therefore depend on the ratio of buffer concentrations in the original sample solution to that in the column, γ. The larger the differences in the concentrations, the narrower the peak. In Table 2.1, the difference in how the sample ions are injected into the capillary column between conventional electro-injection and FASI are shown. For larger γ, more sample is injected and in a shorter plug length when FASI is used.

On the other hand, it should be noted that the difference in the concentration of two regions inside the capillary column will also cause an electroosmotic pressure to occur at the concentration boundary. The electroosmotic pressure leads to laminar flow which would introduce additional peak broadening. In order to achieve the optimum stacking effect, the difference in conductivity between the sample and the buffer and the applied voltage should be kept at moderate levels (see Section 2.1.2).

To perform FASI, the samples are prepared in a low-conductivity solution, e.g. H_2O, and injected electrokinetically into the column. A field enchancement can be achieved at the injection point. The enchancement factor, F_e, is the ratio of the concentration of the buffer in the column and the concentration of the buffer into which the sample is prepared. Further enchancement can be obtained by injecting a plug of pure water before sample is introduced. Figure 2.2 illustrates the effect of (a) sample stacking and peak narrowing in FASI, (b) FASI without water plug, and (c) FASI with water plug. A comparison of the peak heights obtained from PTH-amino acids (PTH-Arginine, PTH-Histidine, PHT-Aspartic acid and PTH-Glutamic acid), with different injection schemes is given in Table 2.2 [32].

One problem with FASI with either positive or negative polarity alone is that only one type of ions can be injected into the column. Therefore, FASI with polarity switching, i.e. FAPSI, should be performed if both positive and negative ions are to be injected [32].

2.3 HYDRODYNAMIC INJECTION

Hydrodynamic injection, also referred to as hydrostatic injection in some cases, can be performed by gravity flow, pressure or vacuum suction. The main advantage

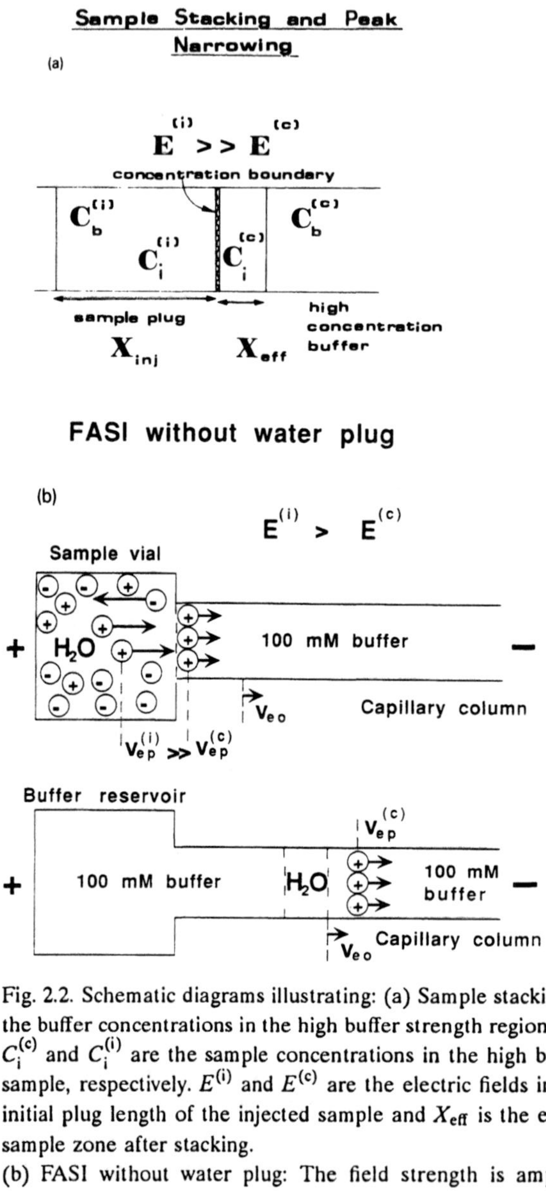

Fig. 2.2. Schematic diagrams illustrating: (a) Sample stacking and peak narrowing: $C_b^{(c)}$ and $C_b^{(i)}$ are the buffer concentrations in the high buffer strength region, and in the injected sample, respectively. $C_i^{(c)}$ and $C_i^{(i)}$ are the sample concentrations in the high buffer strength region and in the injected sample, respectively. $E^{(i)}$ and $E^{(c)}$ are the electric fields in the plug and in the column. X_{inj} is the initial plug length of the injected sample and X_{eff} is the effective plug length, or the length of the sample zone after stacking.
(b) FASI without water plug: The field strength is amplified by injecting a solution of lower conductivity than that of the buffer, resulting in stacking. $v_{ep}^{(i)}$ and $v_{ep}^{(c)}$ are the electrophoretic velocities of the ions at the injection end and in the rest of the column, respectively. v_{eo} is the electroosmotic velocity of the bulk solution.

of this type of injection method is that unlike electrokinetic injection, there is no inherent discrimination of the sample injected. In Fig. 2.3, a comparison of electrokinetic injection and hydrodynamic injection is shown. The peak area is plotted as a function of the sample solution resistance. For hydrodynamic injection,

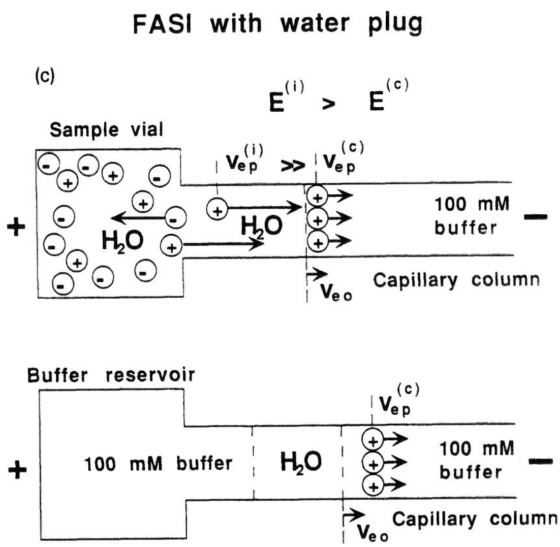

Fig. 2.2 (continued). (c) FASI with water plug (for injection of positive ions only): A short plug of water is introduced into the column before sample injection to further amplified the field strength. (Adapted from Ref. 32 with permission of Elsevier Science Publishers.)

TABLE 2.2

COMPARISION OF PEAK HEIGHTS FOR DIFFERENT INJECTION SCHEMES (NORMALIZED TO GRAVITY INJECTION) (Adapted from Ref. 32)

	PTH-Arginine (+)	PTH-Histidine (+)	PTH-Aspartic (−) acid	PTH-Glutamic (−) acid
Gravity injection	1	1	1	1
Conventional eletro-kinetic injection	0.31	0.23	0.025	0.022
FASI without water plug	17.0	3.7	0	0
FASI with water plug (positive ions only)	28.0	13.4	0	0
FASI with water plug (negative ions only)	0	0	13.7	12.6
FAPSI (positive and negative ions)	32.0	9.3	2.6	2.0

the peak area remained constant, whereas in the case of electrokinetic injection, the peak area increases with solution resistance despite the fact that the injection voltage and duration were kept constant [12].

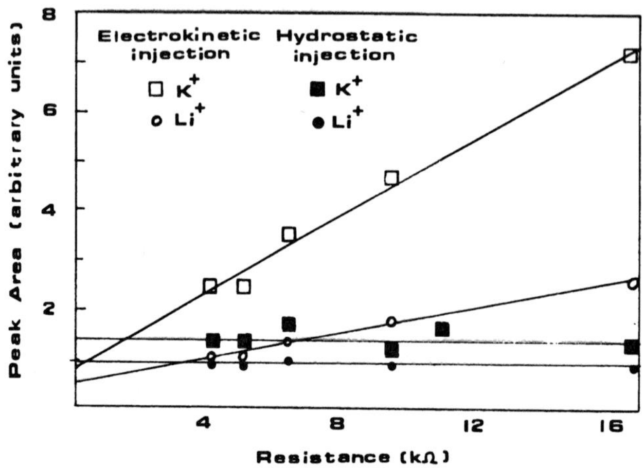

Fig. 2.3. Plot of K^+ and Li^+ peak areas as a function of sample solution resistance for both electrokinetic and hydrostatic injection. Electrokinetic injection causes a bias linear in sample solution resistance (which is inversely proportional to electrolyte concentration). (Reproduced from Ref. 12 with permission of the American Chemical Society.)

2.3.1 Gravity flow injection

Hydrodynamic injection by gravity flow can be achieved by placing the end of the capillary into a sample solution followed by moving the sample container and column end to a certain height, Δh, higher than the opposite end of the capillary for a period of time. The volume of sample injected, q, is given by [40]:

$$q = \frac{\rho g \pi r^4 \Delta h \, t_i}{8 \eta L} \tag{2.17}$$

where ρ is the density of the sample solution, g is the constant for gravitational acceleration, r is the internal radius of the capillary, Δh is the height difference between the liquid levels of the sample vial and the buffer reservoir at the grounded electrode, η is the solution viscosity, and L is the capillary length.

If C represents the sample concentration, the amount of sample injected is then:

$$w = \frac{\rho g \pi r^4 \Delta h \, C t_i}{8 \eta L} \tag{2.18}$$

It can be noted from Eq. (2.18) that the amount injected does not depend on electrophoretic mobility. Furthermore, the composition of the sample solution has no effect on the amounts injected by this method. The quantity introduced during hydrodynamic flow injection can be controlled through variations in the injection time, t_i, and injection height, Δh.

2.3.2 Pressurized and vacuum injection

During pressurized injection, a pressure is applied to the vial containing the sample, pushing it into the capillary, whereas during vacuum injection, a sample is placed at the opposite end of the capillary, drawing the sample into the capillary. Both of these techniques tend to have lower precision than that found in electroinjection although they possess the same advantage of no sample bias as gravity flow injection does. The main problem in pressure injection is the difficulty of controlling the pressure precisely. Atmospheric conditions and elevation can also affect pressure injection. The performance of the system may be further degraded as the equipment ages, when the pressure or vacuum lines become more rigid.

The amount of analyte injected with pressure can be calculated from the Poiseuille law; with an equation similar to Eq. (2.18):

$$w = \frac{\Delta P \pi r^4 C\, t_i}{8\, \eta L} \tag{2.19}$$

where ΔP is the pressure difference across the capillary.

2.3.3 Automated hydrodynamic injection

In the study of Rose and Jorgenson [6], an automated sampling system for hydrostatic injection was developed to minimize operational error. They found that the %RSD in peak area could be improved from 11.8% by manual injection to 2.9% with the automated sampling system. However, to obtain accurate estimate of the amount injected, it is necessary to correct for the time taken to raise and lower the column end (travel time) during the injection process. The reason is that during the raising and lowering process, hydrodynamic pressure is generated, which causes the sample to flow into the column.

This effect is shown in Fig. 2.4. The amount introduced, w_i, during injection time, t_i, is given by,

$$w_i = \Delta h\, t_i C B \tag{2.20}$$

where C is the concentration of the sample and $B = (\rho g \pi r^4)/(8\, \eta L)$. The amount injected, w_T, during travel time, t_T, before or after injection is given by,

$$w_T = 0.5\, \Delta h\, t_T\, C B \tag{2.21}$$

The total amount injected, w_{tot}, is therefore:

$$\begin{aligned} w_{tot} &= w_i + 2\, w_T \\ &= \Delta h\, C(t_i + t_T)\, B \end{aligned} \tag{2.22}$$

The travel time of the sampling system, therefore, extends the hydrodynamic flow introduction by the amount, t_T. This correction factor must be added to the injection time, t_i [6].

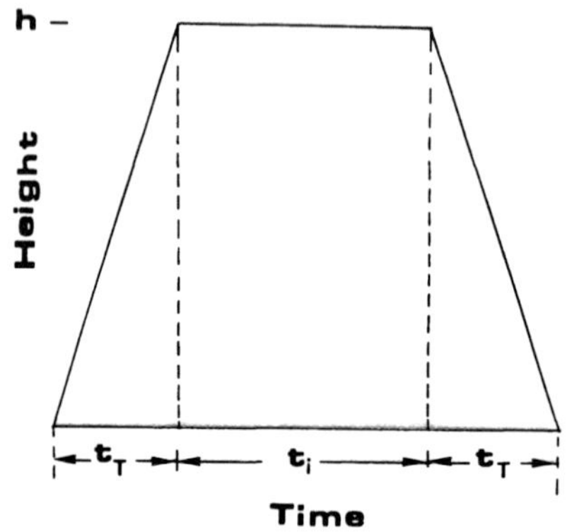

Fig. 2.4. Representation of autosampler travel time effect on quantity introduced during hydrodynamic flow sample introduction. (Reproduced from Ref. 6 with permission of the American Chemical Society.)

Honda *et al.* [8] used an automatic siphonic sampler for CZE. They obtained with a 250 μm tube relative standard deviation (n = 12) of as low as 2.4 and 0.84% for peak height and 1.3 and 0.55% for peak area for 1.5 and 5 s sampling respectively.

2.4 ELECTRIC SAMPLE SPLITTER

Deml *et al.* [36] reported a method based on the principle of the splitter. The splitting ratio is given by the ratio of the electric currents, I_2 to I_3.

The principle of the splitter is shown in Fig. 2.5. The current, I_1, flows through the dosing capillary and drives the original sample, n_1. This current is then split into I_2, which drives part, n_2, of the original sample into the separation capillary and I_3 which drives the rest of the sample, $n_3 = n_1 - n_2$, to the drain.

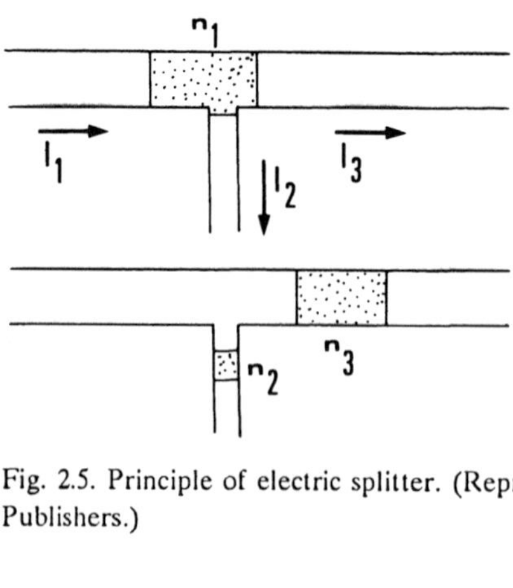

Fig. 2.5. Principle of electric splitter. (Reproduced from Ref. 36 with permission Elsevier Science Publishers.)

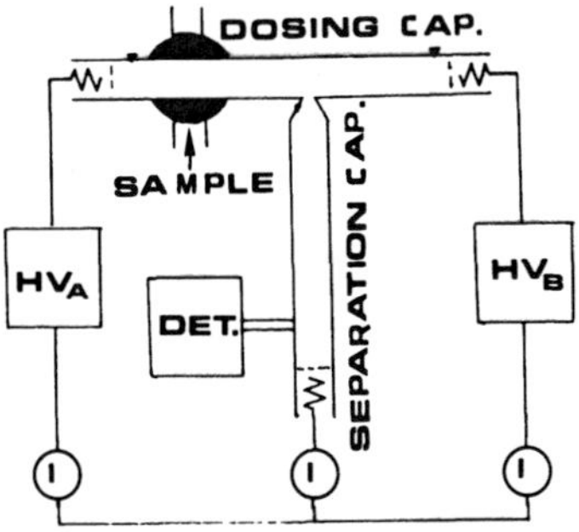

Fig. 2.6. Scheme of the home-made splitter. ●, operating valve; ···, cellophane membrane. (Reproduced from Ref. 36 with permission of Elsevier Science Publishers.)

The amount of component trapped in the separating capillary is [36]:

$$n_2 = \frac{I_2}{I_3} n_3 \tag{2.23}$$

or

$$n_2 = \frac{I_2}{I_1} n_1 \tag{2.24}$$

The quantity sampled by the electric splitter into the separation capillary, n_2, is thus proportional to the total supplied amount and to the ratio of the electric currents passing through the separation and dosing capillaries, I_2 and I_1, respectively, for any component.

A schematic design of the electric sample splitter is shown in Fig. 2.6, which consisted of a monolithic block of polyester resin formed by casting. The system comprises the dosing and separation capillary (500 μm and 200 μm I.D. respectively). A dosing valve (1 μl) was attached to the dosing capillary at a distance of 10 mm from the splitting point. Accuracy of the sample splitter, i.e. the agreement of the splitting ratio with the electric current ratio, was found to be better than 3% RSD.

Use of the splitting system produced significant gains in efficiency over that obtained with a sampling valve assembly, although the overall performance of the system used for their experiments was relatively poor. The main advantage of the splitter is that it allows injection of smaller sample volumes, which decreases the overloading effects seen when the sample valve is used. On the other hand, a major limitation of the electric sample splitting technique is the difficulty in adapting this system to the capillaries typically employed in CE ($<$100 μm). The use of the electric sample splitter has not been reported since the initial development.

2.5 SPLIT FLOW SYRINGE INJECTION SYSTEM

Tehrani *et al.* [37] developed a sample introduction system which can accept a standard HPLC type microlitre syringe. The injection sample is divided proportion-

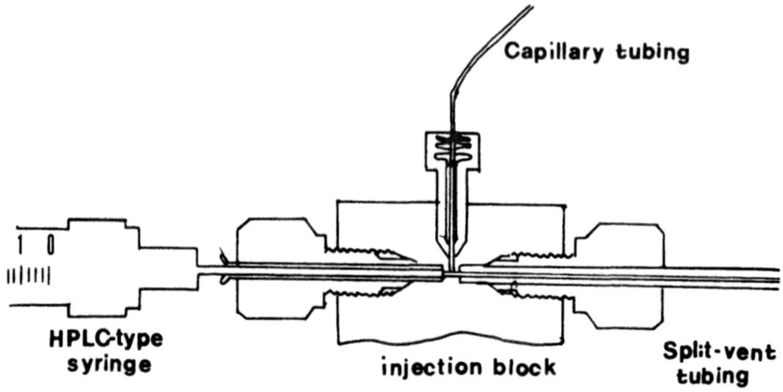

Fig. 2.7. Split-flow injection mechanism. (Reproduced from Ref. 37 with permission of Dr. Alfred Huethig Publishers.)

ally between the separation capillary and an adjustable split-vent. The volume of the sample introduced can be controlled by using split-vent tubing of varying length or I.D., or by changing the volume of the sample injected.

In Fig. 2.7, a schematic diagram of the injection block is shown. The injection block is equipped with three parts: one part is connected to the separation capillary, another part is equipped with fittings to accept the needle of a standard HPLC-type syringe, and a third part is connected to a split-vent tubing which can be of varying length (3–10 cm) and/or I.D. (127–500 μm).

The volume introduced into a capillary tube due to a pressure difference is given by Eq. (2.19). For the split-flow mechanism, Eq. (2.19) can be applied to both the separation capillary and the split-vent tubing.

The following relationship is then obtained by taking the ratio:

$$\frac{q_c}{q_s} = \left(\frac{r_c^4}{L_c}\right) \Big/ \left(\frac{r_s^4}{L_s}\right) \tag{2.25}$$

where q_c and q_s are the volumes of sample introduced into the separation capillary and the split-vent tubing, respectively. r_c and L_c are the radius and length of the separation capillary. r_s and L_s are the radius and the length of the split-vent tubing. Equation (2.25) applies when the viscosities of the sample and the buffer solution are approximately the same. Assuming that the portion of sample introduced into the split-vent tubing (q_s) is much larger than the volume introduced into the separation capillary ($q_s \gg q_c$), the following equation is obtained:

$$\frac{q_c}{q_s} = \frac{q_c}{q_{inj}} \tag{2.26}$$

where q_{inj} is the total volume of sample injected and $q_{inj} = q_c + q_s \cong q_s$. The volume of sample injected into the separation capillary can then be estimated by:

$$q_c = \left(\frac{r_c}{r_s}\right)^4 \left(\frac{L_s}{L_c}\right) q_{inj} \tag{2.27}$$

To inject a sample, the high voltage is turned off and the capillary inlet is moved from the buffer reservoir to the injection block. An HPLC-type syringe containing the sample is inserted into the syringe part and the sample is injected into the capillary. The capillary is then placed into the high-voltage buffer reservoir and the high voltage is switched on to start the run. With this injection system, run to run repeatability obtained was 0.3–0.5% RSD in migration time and 1–3% RSD in peak height and peak area. Day-to-day reproducibility of 1–2% RSD in migration time and 3–4% RSD for peak area was obtained [37].

2.6 ROTARY-TYPE INJECTOR

The rotary injector has several advantages over electromigration and hydrodynamic methods: (i) the amounts of solute introduced are fixed by the injection loop; (ii) it is easy to change sample solution by refilling the injection loop; and (iii) injection is possible while a high electrical field is applied [17].

However, there are also several problems which need to be overcome before the rotary injector can gain widespread acceptance for use in CE: (i) it is difficult to scale down the injection size to permit ultrasmall volume (≅1 nl) injection required to exploit the high efficiency capabilities of CE; (ii) several microlitres of sample is required to fill the injection loop; (iii) it is necessary for the injector to be operated automatically for safety reasons; (iv) the sample solution may decompose if kept in the loop at a high electrical field for long periods of time and hence attention must be paid to the design of the loop and the materials used; and (v) it is difficult to adapt the injector to capillary of small I.D. without introducing significant zone broadening contributing from the connections.

Tsuda *et al.* [17] employed a rotary-type injection for CE which is similar to those employed with conventional HPLC system. However, ceramics and tetrafluorethylene resins were used instead of metals to eliminate electrochemical reactions. The design of the injector is shown in Fig. 2.8. The injector is connected to the capillary at a point near the high-voltage end. The injection loop has a volume of 0.35 μl and is connected to a 200 μm I.D. capillary. Tsuda et al. [17] obtained %RSD in peak area of 2.2% (n = 5) and 1.5% (n = 4) for aniline and dansylated spermidine, respectively with the rotary injector.

2.7 FREEZE PLUG INJECTION

In Yonker and Smith's [38] work on high-pressure and supercritical capillary electrophoresis, a "freeze plug" injection technique was employed for sample injection. The technique involved dipping one end of the analytical column in the

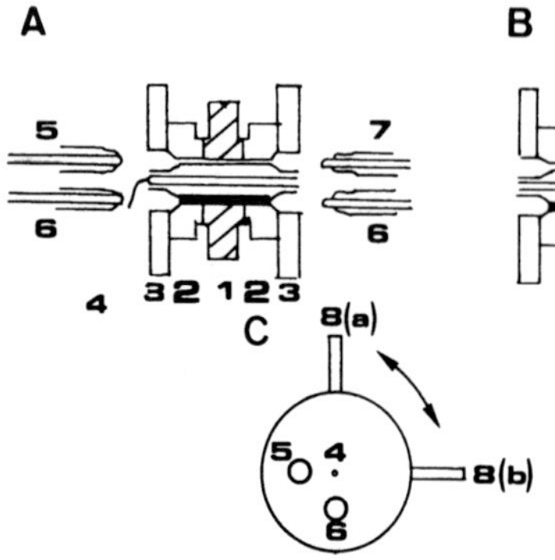

Fig. 2.8. Schematic diagram of rotary injector made of fine ceramics: *1* = rotor; *2* = stator; *3* = plate for setting rotor and stators; *4* = central pin; *5* = supplemental tubing between injector and reservoir; *6* = tubing for sample introduction; *7* = capillary column; *8* = knob made by stainless steel covered with a silicone tube. Positions *a* and *b* are load and injection, respectively. (Reproduced from Ref. 17 with permission of the American Chemical Society.)

solute–electrolyte solution, followed by rapidly cooling a small section (< 1.5 cm) of the column in liquid nitrogen (see Fig. 2.9). The freezing and contraction of the solvent in the capillary drew a sample plug into the column. The column segment was kept frozen while the column was reconnected to the buffer reservoir. The system was then repressurized to the required pressure with a pump. A hydrostatic line was used to equalize the pressure on both side of the frozen column plug upon thawing the sample.

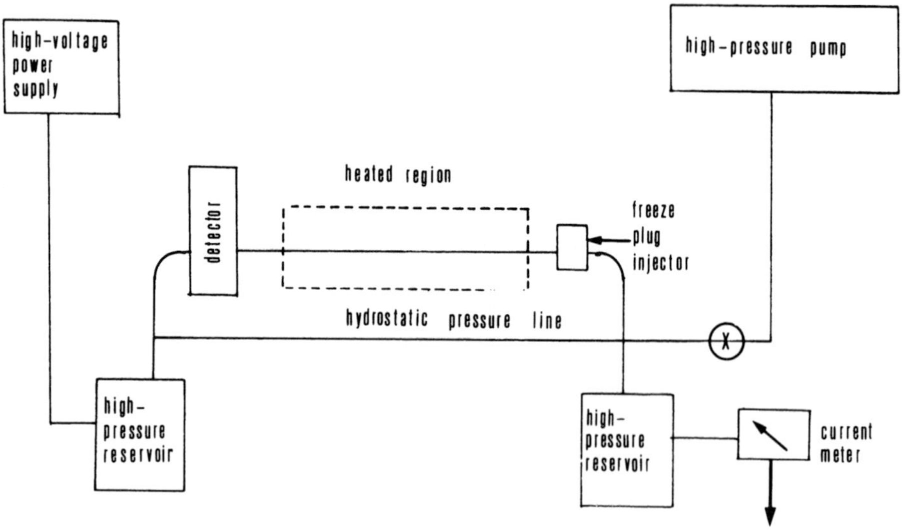

Fig. 2.9. Schematic of supercritical-fluid capillary electrophoresis system. (Reproduced from Ref. 38 with permission of Elsevier Science Publishers.)

Data on the reproducibility of the freeze plug injection technique has not been reported. However, it is unlikely that the reproducibility would be as high as other injection techniques for CE, such as electrokinetic and hydrostatic injection. At the present stage of development, methodology of sample introduction for high-pressure, low-dead volume uses is still at an infant stage.

2.8 SAMPLING DEVICE WITH FEEDER

Verheggen *et al.* [19] constructed a sampling device whereby the sample solution is introduced directly into part of the capillary tube by means of two feeders which are perpendicular to the capillary tube. The sample can be introduced without mixing with the background electrolyte. This device has several advantages. Since there are no moving parts, cleaning can be performed easily. With the use of a valve, the concentration effect of dilute sample can be achieved.

A schematic diagram of the electrophoretic equipment used by Verheggen *et al.* [19] is shown in Fig. 2.10. A cast capillary block (*C*) is connected between the electrode compartment (A_1) and the sampling device (*SD*). In the electrode compartments (A_1 and A_2), Pt-Ir electrode *E* were immersed in the eletrolyte solution. The capillary tube (0.25 mm I.D., separation length 6 cm) contains the measuring electrode *M*. The sampling device *SD* consists of a broadened part of the capillary tube (0.55 mm I.D.) connected with two feeders (0.4 mm diameter) which are perpendicular to the capillary tube. Electrolyte solutions are introduced using valves 1 and 3 and the drain 2. Valves 5 and 6 are used for rinsing of the electrode compartments. The sample can be introduced via valve 4 and the drain 2.

Sample is introduced through one of the feeder capillaries [4] and excess sample is eluted from the drain capillary [2]. The distance between these determines the

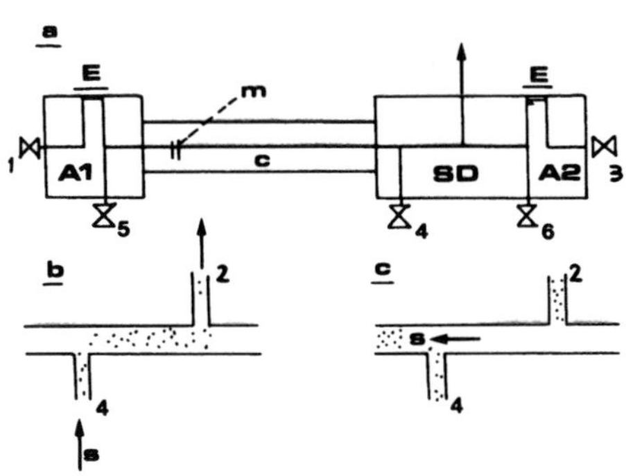

Fig. 2.10. (a) Schematic diagram of electrophoretic equipment with sample device. *A1, A2* = electrode compartments; *C* = capillary block; *E* = Pt-Ir electrodes; *M* = measuring electrodes; *SD* = sampling device. (b) Sample device during sampling (*s* = sample). (c) Sampling device after a certain time. (Reproduced from Ref. 19 with permission of Elsevier Science Publishers.)

sample volume. Sample zone length between 13 and 81 mm, corresponding to samples volume of 1–5 μl were used [19]. Zone lengths for several concentrations of acetate and glutamate were measured five times each on different days. Relative standard deviation in the zone length of up to 2.5% has been reported.

2.9 MICROINJECTORS

Ewing and co-workers [25,26,39,40] developed several versions of microinjection systems for ultra-low-volume sample introduction from biological microenvironment such as single cells and discrete tissue regions. Three such microinjectors are shown in Fig. 2.11. The microinjectors were made from glass capillaries, one end of the glass capillary was drawn to give a very small diameter tip, and the other end was placed over the high-voltage end of the separation capillary to form an extension of the column. For injection, the small tip was inserted into the sample and a voltage is applied for a brief period of time to achieve electrokinetic injection. Microinjectors with two barrels and with one-barrel have been used. The design shown in Fig. 2.11a is a dual-barrel design which employs a 10 μm O.D. carbon fibre electrode. The carbon fibre is aspirated into one barrel of the dual-barrel glass capillary. The carbon fibre electrode is then used to apply an injection voltage at the tip of the assembly to cause electromigration of sample into the other barrel which is filled with buffer and connected to the separation capillary. This earlier design was found to be unreliable [25,26,39] and electrolysis often occurred around the carbon fibre if the glass does not form a tight seal with the fibre.

In Fig. 2.11b, a simplified and improved version of the microinjector is shown. In this design, the carbon fibre was dispensed with and instead a platinum wire was inserted directly into one barrel which was filled with buffer. The separation

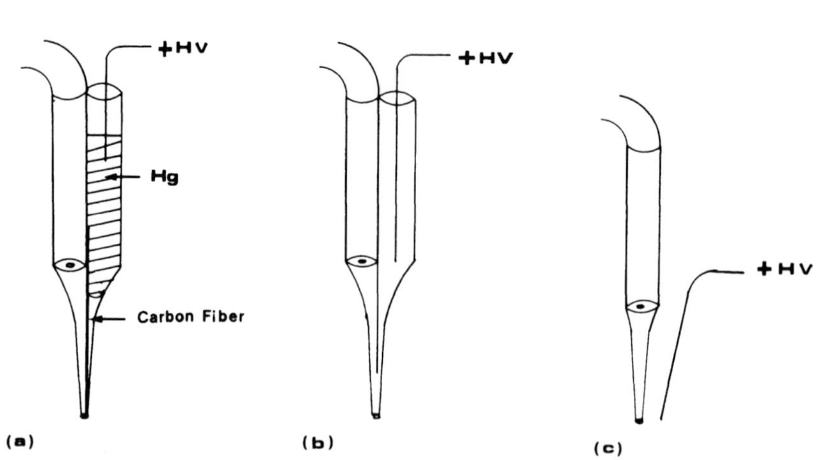

Fig. 2.11. Schematic representations of three microinjector designs. (Reproduced from Ref. 40 with permission of Marcel Dekker, Inc.)

capillary was inserted into the other barrel. In this design, more reliable injections can be achieved and problems associated with electrolysis were eliminated. An advantage is that, small tip diameter can be used since they are no longer limited by the outer diameter of the carbon fibre. Furthermore, since the electrode is contained in a buffer-filled capillary, electroosmotic flow is expected to occur in both barrels. Once placed into a microenvironment, it is possible to perform continuous sampling.

Figure 2.11c shows a single-barrelled design of the microinjector [39]. In this design, the glass capillary were pulled down to sub-micrometer tip diameter and then cut under microscope to the desired tip diameter. For injection, the microinjector was filled with the operating buffer. Then the anodic end of the separation column was placed into the microinjector, the tip of which were placed into the sample reservoir. Injection voltage was applied using a platinum electrode placed outside of the microinjector in the sample reservoir. Since this microinjector employs an anode that is spatially separated from the injection barrel, electrical coupling and electrolysis could be eliminated. Also tip diameter can be made extremely small ($<10\ \mu m$ O.D.) to facilitate penetration into living cells. This system has been used to inject cytoplasmic sample directly from single nerve cells [39].

In order to achieve maximum efficiency when injecting with a microinjector, it is necessary to minimize turbulences that may occur as a result of flow of solutes into and out of channels of different sizes. The problem can be alleviated by matching the inner diameter of the separation capillary with the tip diameter of the microinjector. In other words, when microinjectors with very small tips are used, capillaries with corresponding small inner diameter should be employed. It is also important to minimize the distance from the injector tip to the capillary inlet. The reason is that while in this space, ionic solutes undergoes electrophoretic separation and discriminating effects are magnified. By using a single-barrelled microinjector, a success rate of injection of nearly 90% could be achieved [39].

2.10 OPTICAL GATING

Recently an on-column optical gating injection technique has been developed to exploit the high-speed potential of capillary electrophoresis [41]. Another advantage of this technique is the possibility of making sample introduction while the capillary is maintained at operating voltage. With this technique, the components in the mixture to be determined are first tagged with a fluorescent molecule and then continuously introduced into one end of the column (see Fig. 2.12). Before sample gating, a laser is used to photodegrade the tag so that the molecules are not detected by the fluorescence detector placed at the other end of the column. During operation of optical injection, a sample zone is created by momentarily turning off the laser beam, so that the fluorescent tagged molecules are allowed to pass intact. Separation of the species occurs in the column before they reach

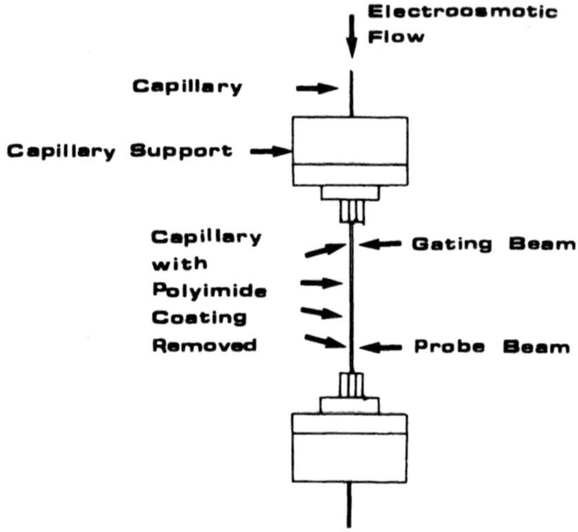

Fig. 2.12. Diagram of capillary mount showing the relative position of the capillary, the "gating" and the "probe" beams. (Reproduced from Ref. 41 with permission of the American Chemical Society.)

the detector. The temporal relationship for this type of on-column optical injection technique is shown in Fig. 2.13. With this system, separation of a mixture of fluorescein isothiocyanate (FITC) labelled amino acids in as short as 1.5 s was demonstrated.

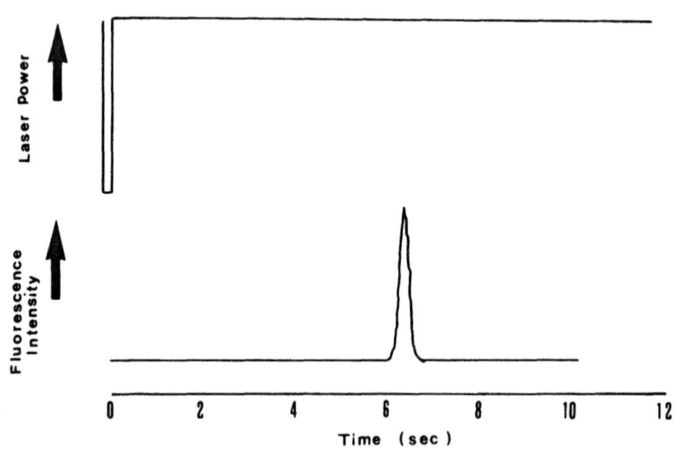

Fig. 2.13. Diagram showing the temporal relationship between the intensity of the "gating" beam and the fluorescence signal generated at the "probe" beam. (Reproduced from Ref. 41 with permission of the American Chemical Society.)

2.11 ON-COLUMN FRACTURE FOR SAMPLE INTRODUCTION

A disadvantage of electrokinetic injection (see Section 2.2) is the sample bias caused be differences in charges of the analyte molecules. However, by utilizing only electroosmotic flow for sample introduction, the problem of sample bias can be alleviated. The elimination of electromigration of the analytes during injection has been achieved by producing an on-column fracture in the separation capillary to create a short section of capillary with no applied electric field [42]. The technique is an adaptation of the porous glass joint used in electrochemical detection (see Chapter 3) developed by Wallingford and Ewing [25,26]. Fig. 2.14 illustrates the design of the fracture assembly. During injection, the positive electrode is connected to the electrode in the fracture assembly and the outlet end reservoir is grounded. When the injection voltage is applied, electroosmotic flow pulls sample into the separation capillary, and hence the amount injected is proportional to the electroosmotic flow. After injection, the positive electrode is placed in the inlet end buffer reservoir before the separation voltage is turned on. Reproducibility in migration times of a neutral marker (mesityl oxide), was within 0.3% (n = 16). For peak areas, %RSD were 2.4% (n = 16) at pH 6.1 and 2.0% at pH 8.8, respectively.

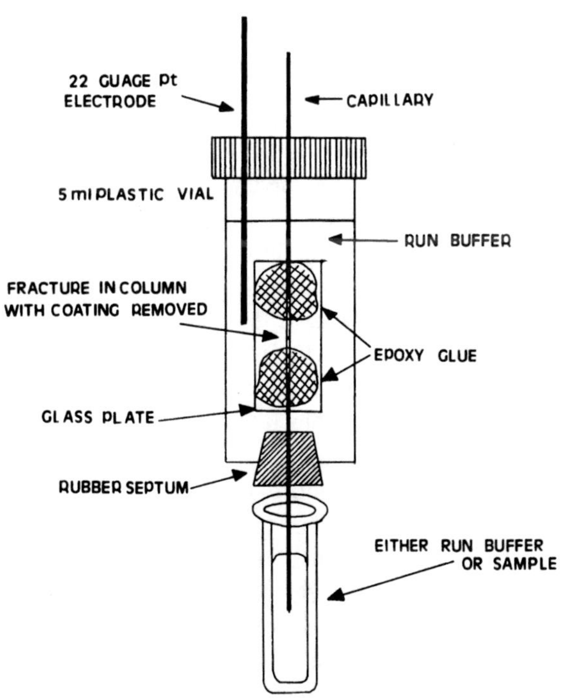

Fig. 2.14. Fracture assembly and buffer reservoir. (Reproduced from Ref. 42 with permission of the American Chemical Society.)

2.12 CONCLUSION

Many ingenious approaches have been developed for sample introduction in CE. Currently, the most commonly used methods are on-column hydrodynamic and electrokinetic injection techniques. With automated systems, reproducibility of within 2–3% RSD in peak area can be readily achieved for these types of methods.

A major limitation of the hydrodynamic methods, including pressure, vacuum and gravity injection, is that they are not suitable for the injection of highly viscous samples or for capillary gel electrophoresis, due to the fact that hydrodynamic flow would be hampered or suppressed. However, hydrodynamic injection systems are easy to operate, not subject to sample bias and can be readily automated. Consequently, they are likely to remain one of the most widely employed sample injection methods in CE.

Electrokinetic injection techniques are also very easy to perform and to automate. Furthermore, they can be utilized even in the cases of viscous samples and capillary gel electrophoresis. The development of field amplified sample injection techniques provides an additional advantage in terms of improving detection sensitivity. Although sample bias has been a major limitation, the development of injection techniques employing only electroosmotic flow, as in the on-column fracture technique, would help to overcome this problem and widen significantly the scope of application of electrokinetic injection methods. In view of these advances and the potential of further refinement of these techniques, it is likely that electrokinetic injection would become the method of choice for sample introduction in CE separations.

Injection techniques based on the use of sampling valves, splitters and syringes suffer from practical limitations, such as the need to design elaborate devices for introducing the sample into the capillary without introducing dead volumes or causing sample overloading. On the other hand, these techniques have the advantages that the amounts injected are usually fixed by the geometry of the sample loop used or the splitter design. Consequently, the absolute amounts of samples introduced into the capillary by these methods are usually determined directly, whereas in hydrodynamic and electrokinetic injection methods, the amounts injected need to be calculated from the injection time, together with either the pressure differential or the injection voltage. Furthermore, the sampling valves and syringes employed are similar to those adopted in other types of chromatographic methods, such as gas chromatography and high-performance liquid chromatography. In view of the ubiquity of syringe-based injection systems for chromatography, it is expected that these injection techniques will continue to be used. The popularity of such off-column techniques would probably increase more rapidly if significant advances in miniaturization of sampling valves and injection devices can be made, since these developments will no doubt help to ensure that the potential of high efficiency separations of CE can be fully exploited with these systems.

Finally, other techniques, such as optical injection, freeze-plug injection and the use of microinjectors, provide a wide range of additional strategies which may be considered for certain special applications due to some of their unique advantages.

2.13 REFERENCES

1 X. Huang, W. Coleman and R. Zare, J. Chromatogr., 480 (1989) 95
2 H. Lauer and D. McManigill, Trends Anal. Chem., 5 (1986) 11
3 E. Grushka and R. McCormick, J. Chromatogr., 471 (1989) 421
4 J.W. Jorgenson and K.D. Lukacs, Anal. Chem., 53 (1981) 1298
5 J.W. Jorgenson and K.D. Lukacs, J. Chromatogr., 218 (1981) 209
6 D.J. Rose and J.W. Jorgenson, Anal. Chem., 60 (1988) 642
7 S. Fujiwara and S. Honda, Anal. Chem., 59 (1987) 487
8 S. Honda, S. Iwase and S. Fujiwara, J. Chromatogr., 404 (1987) 313
9 H. Schwartz, M. Melera and R. Brownlee, J. Chromatogr., 480 (1989) 129
10 D. Burton, M. Sepaniak and M. Maskarinec, Chromatographia, 21 (1988) 583
11 K.H. Row, W.H. Griest and M. Maskarinec, J. Chromatogr., 409 (1987) 193
12 X. Huang, M. Gordon and R.A. Zare, Anal. Chem., 60 (1988) 375
13 W.G. Kuhr and E.S. Yeung, Anal. Chem., 60 (1988) 375
14 K. Otsuka and S. Terabe, J. Chromatogr., 480 (1989) 91
15 Y. Walbroehl and J.W. Jorgenson, J. Microcol. Sep., 1 (1989) 41
16 S. Terabe, K. Otsuka, K. Ichikawa, A. Tsuchiya and T. Ando, Anal. Chem., 56 (1984) 111
17 T. Tsuda, T. Mizuna and J. Akiyama, Anal. Chem., 59 (1987) 799
18 M. Demyl, F. Foret and P. Bocek, J. Chromatogr., 320 (1985) 159
19 T. Verheggen, J. Beckers and F. Everaerts, J. Chromatogr., 452 (1988) 615
20 J. Pospichal, M. Deml, P. Gebauer and P. Bocek, J. Chromatogr., 470 (1989) 43
21 V. Rohlicek and Z. Deyl, J. Chromatogr., 480 (1989) 289
22 A.J. Debets, R. Frei, K. Hupe and W.T. Kok, J. Chromatogr., 465 (1989) 315
23 H. Yamamoto, T. Menabe and T. Okuyama, J. Chromatogr., 480 (1989) 277
24 D. Rose and J. Jorgenson, J. Chromatogr., 438 (1988) 23
25 R.A. Wallingford and A.G. Ewing, Anal. Chem., 60 (1988) 1972
26 R.A. Wallingford and A.G. Ewing, Anal. Chem., 59 (1987) 678
27 R.T. Kennedy, M.D. Oates, B.R. Cooper, B. Nickerson, J.W. Jorgenson, Science (Washington, DC), 246 (1989) 57
28 J.A. Lux, H.F. Yin and G. Schomburg, Chromatographia, 30 (1990) 7
29 A. Vinther and H. Soeberg, J. Chromatogr., 559 (1991) 3
30 A. Vinther and H. Soeberg, J. Chromatogr., 559 (1991) 27
31 D.S. Burgi and R.L. Chien, Anal. Chem., 63 (1991) 2042
32 R.L. Chien and D.S. Burgi, J. Chromatogr., 559 (1991) 141
33 R.L. Chien and D.S. Burgi, J. Chromatogr., 559 (1991) 153
34 D.S. Burgi and R.L. Chien, Anal. Chem., 63 (1991) 2866
35 R.L. Chien and J.C. Helmer, Anal. Chem., 63 (1991) 1354
36 M. Deml, F. Foret and P. Bocek, J. Chromatogr., 320 (1985) 159
37 J. Tehrani, R. Macomber and L. Day, J. High Resolut. Chromatogr., 14 (1991) 10
38 C.R. Yonker and R.D. Smith, J. Chromatogr., in press

39 A.G. Ewing, R.A. Wallingford and T.M. Olefirowicz, Anal. Chem., 61 (1989) 292A
40 R.A. Wallingford and A.G. Ewing, Adv. Chromatogr., 29 (1990) 1
41 C.A. Monning, D.M. Dohmeier and J.W. Jorgenson, Anal. Chem., 63 (1991) 802
42 M.C. Linhares and P.T. Kissinger, Anal. Chem., 63 (1991) 2076

CHAPTER 3

Detection Techniques

3.1 INTRODUCTION

The small capillary dimensions employed in capillary electrophoresis and the small zone volumes produced present a challenge to achieve sensitive detection without introducing zone dispersion. Zone broadening normally caused by joints, fittings and connectors can be eliminated by on-column detection. On-column UV adsorption and fluorescence detection have been the most commonly used detection techniques for CE applications. Many other detection techniques have been explored with varying degrees of success [1].

3.1.1 On-column detection window

To achieve on-column UV or fluorescence detection, a window has to be made on the polyimide coating of the fused silica capillary. The simplest way that can be used to form the window is by burning off a small section of the polyimide [2], although alkaline etching [3] and mechanical scraping [4] can also be used.

More elaborate devices have been designed for this purpose. In Figs. 3.1 and 3.2, two devices designed for the production of detection windows are shown. With the device shown in Fig. 3.1, the removal of polyimide coatings is effected by a 0.13 mm filament which is electrically heated by a low-voltage transformer [4]. The fused silica capillary is positioned above the filaments with the help of two specially machined metal blocks with grooves. The polyimide coating is burnt off at the point of contact between the filament and the capillary. The detection window is then cleaned with acetone. Figure 3.2 shows a device based on mechanical stripping of the polyimide using a chisel-pointed knife [5]. The other components of the device include a variable speed motor and a jig to provide support. The main advantage is that detection windows can be made on the capillary without subjecting it to extreme heat on the surface. Consequently, it is suitable for use on capillary which have undergone chemical derivatization or filled with gel medium.

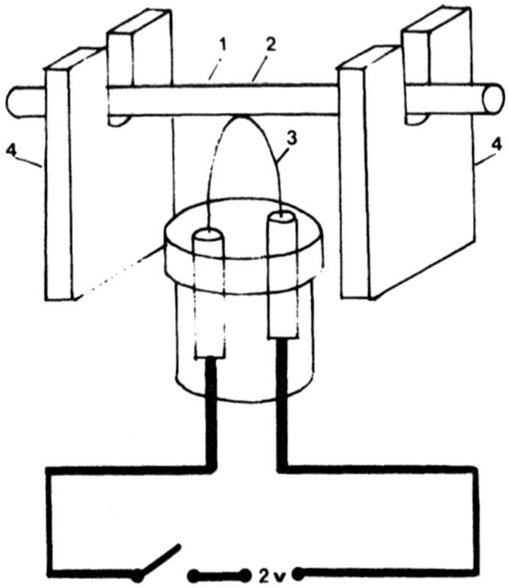

Fig. 13.1. Device for production of detector windows. *1* = Polyimide-coated fused silica capillary; *2* = windows; *3* = filament; *4* = metal blocks. (Reproduced from Ref. 4 with permission of Dr. Alfred Huethig Publishers.)

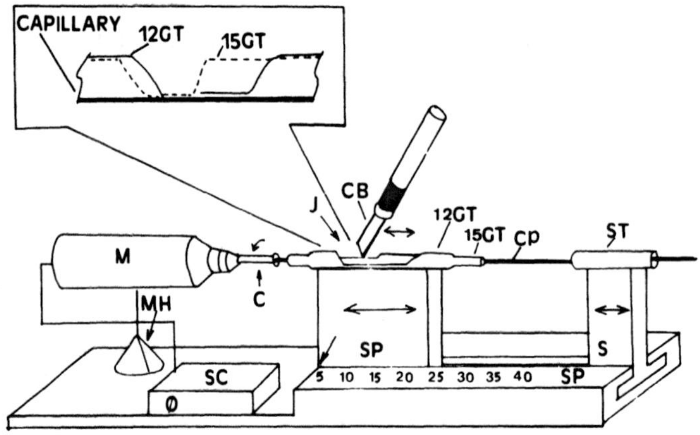

Fig. 3.2. Capillary stripper apparatus for polyimide removal from silica tubing. *J* = jig; *SP* = support; *CP* = capillary; *SC* = speed controller; *C* = chuck; *M* = motor; *12GT* = 12-gauge tubing; *15GT* = 15-gauge tubing; *ST* = support tubing; *MH* = motor-tool holder; *S* = scale; *CB* = chisel blade. (Reproduced from Ref. 5 with permission of American Chemical Society.)

3.2 UV-VISIBLE ABSORBANCE DETECTORS

UV-visible absorption is currently the most popular detection technique for capillary electrophoresis and related techniques [6–39]. The main reasons for its popularity include its relatively universal nature and its widespread availability for HPLC work. Except for a few reports which describe purpose-built UV detectors for CE [6–19], most of the other works employ modified commercial UV detector designed originally for HPLC work with the flow cell replaced by the capillary for on-column detection. Fused silica tubing normally used for CE has a UV cut-off around 170 nm and this is suitable for UV detection. The layer of polyimide coating on the outside can be removed to form a detection window as described in the previous section. In on-column detection the path length is defined by the inner diameter of the capillary. This limits the sensitivity of absorbance detection techniques, since sensitivity is proportional to path length. Another consideration is that with small capillaries, ideally only the capillary is illuminated during detection in order to reduce stray light.

Whether home-built or commercial detectors are used, it would be necessary to consider a few aspects in order to obtain the highest detection sensitivity. These factors include the light source used, the design of the signal amplification system, the background light and the optical path length. In the following sections, the approaches taken to optimize detection sensitivity are considered.

3.2.1 Light source

Because of the difficulty in focusing sufficient amount of light onto small I.D. capillaries, there is usually a need to maximize the intensity of light passing through the sample when UV-absorbance detectors are employed. This can be achieved through a number of techniques, such as the use of high-intensity lamps, the optimum choice of slit-width, the correct and stable positioning of the capillary in the light path, and the use of optical fibres. An alternative approach is to use lasers as the light sources as in laser-induced fluorescence detection (see Section 3.4.2). However, the wavelength obtainable with most lasers are currently not suitable for direct UV absorbance detection.

To investigate the effect of the light intensity on detection sensitivity, Walbroehl and Jorgenson constructed a fixed wavelength UV detector using a Cd "Pen-Ray" source, which was focused onto the capillary [6]. A schematic diagram of the optical layout of this on-column UV absorption detector is shown in Fig. 3.3. The system employed a 7-W cadmium "Pen-Ray" lamp which emits radiation from both sides. Emission from one side of the lamp was used as the reference beam and the other was used as the sample beam. By changing the lamp to zinc or mercury lamps of a similar design, high-intensity UV radiation over narrow band widths could be obtained. The detector employed 100 μm pin-hole as "slits" to cut off stray light to ensure good spatial resolution. With this design, it was possible to ensure that most of the light reaching the photodetector had passed through the capillary. Consequently, the linear dynamic range and the signal-to-noise ratio could be increased. The detector also included features to secure the capillary firmly in the optical path, for maximum signal-to-noise ratio. The capillary occupied the same spot in the optical path during subsequent experiments for high reproducibility and the capillary would not vibrate upon application of high voltages for less noisy

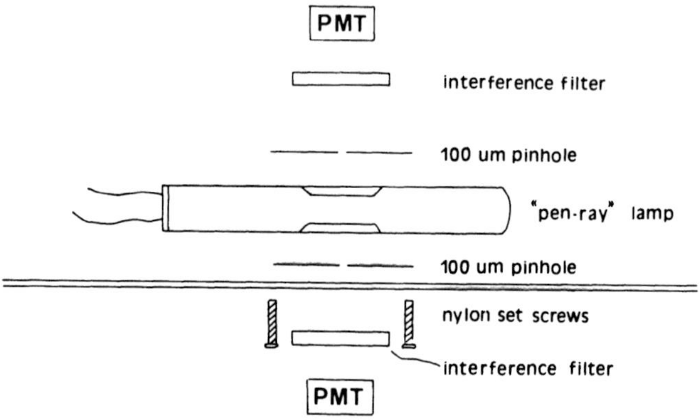

Fig. 3.3. Exploded schematic diagram of optical layout for on-column UV absorption detector. (Reproduced from Ref. 6 with permission of Elsevier Science Publishers.)

baseline. This detector was linear over four orders of magnitude with detection limits as low as 18 fmol ($\simeq 10^{-5}$ M) for lysozyme ($S/N = 2$) [6].

The design, sensitivity and noise characteristic of a deuterium lamp based variable wavelength UV absorption detector for CE was described by Green and Jorgenson [7]. Figure 3.4 shows a schematic diagram of the optical layout of the system. They compared this detector to one used by Walbroehl and Jorgenson which was fitted alternately with a cadmium, zinc or arsenic lamp and found that a

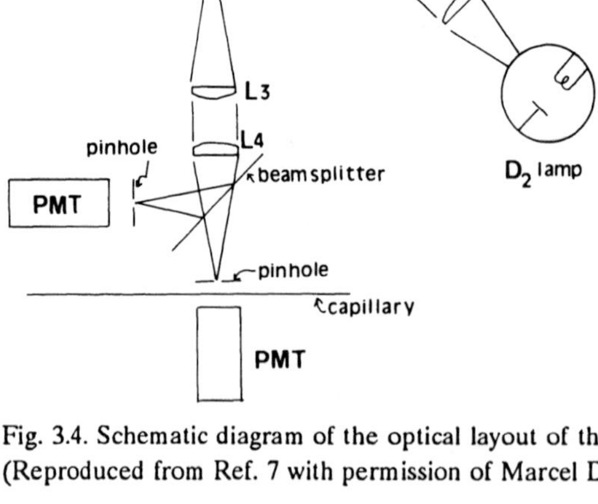

Fig. 3.4. Schematic diagram of the optical layout of the D–2 lamp-based UV absorption detector. (Reproduced from Ref. 7 with permission of Marcel Dekker, Inc.)

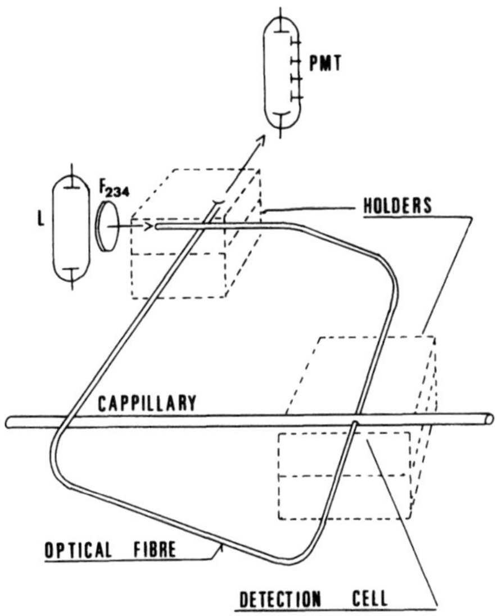

Fig. 3.5. Scheme of the fibre optic UV detection cell. *L* = mercury lamp; *F-254* = interference filter (254 nm); *PMT* = photomultiplier; *CAPILLARY* = separation capillary. (Reproduced from Ref. 8 with permission of VCH Verlagsgesellschaft.)

zinc lamp operating at 214 nm gave the best signal-to-noise and the most uniform response factor.

Optical fibres have been used to enhance detection sensitivity by Foret *et al.* [8]. In their investigation, an on-column detection system was constructed in which two optical fibres (200 μm I.D. fused silica core) were butted against an exposed portion of the fused silica capillary (Fig. 3.5). The optical fibres were positioned on opposite side of the capillary. One fibre was connected to the light source which was a mercury lamp and the other fibre was directed towards a photomultiplier tube for detection. This detector was found to be linear over the range of 10^{-5}–10^{-3} M, with detection limits of 1×10^{-5} M obtained for picric acid ($S/N = 2$).

3.2.2 Signal amplification

Higher detection sensitivity can also be obtained by making improvements in the electronic circuitry and signal processing system used in the detector. Prusik *et al.* [9] developed an on-column fixed-wavelength UV-photometric detector (206 nm), which utilized a high-frequency (100 MHz) excited electrodeless low-pressure iodine discharge lamp (6 W). A schematic diagram of the UV detector is shown in Fig. 3.6 and the scheme of the signal processing in the keying amplifier unit of the UV detector is shown in Fig. 3.7. A mechanical light chopper was used to generate light and dark period of 40 ms duration and a silicon photovoltaic detector used with a high-gain keying amplifier to detect light power on the anode of 10^{-12}

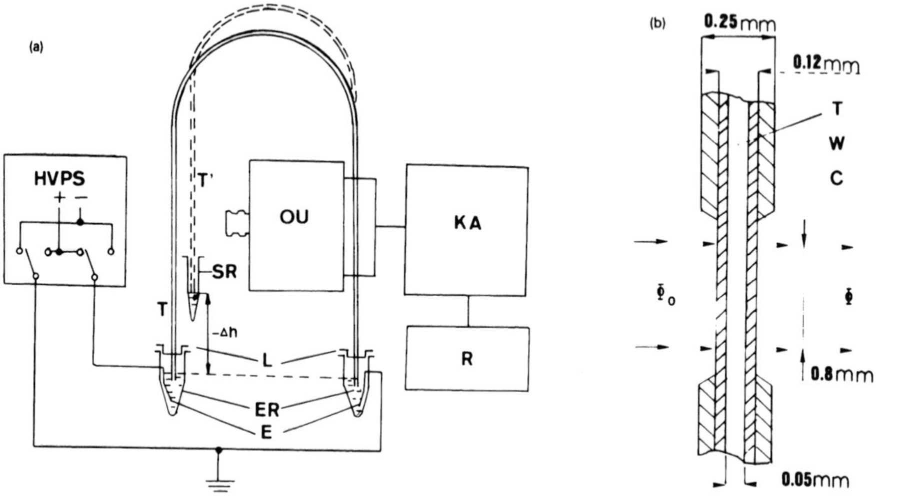

Fig. 3.6. (a) Scheme of experimental device for electromigration chromatography with UV detector. *HVPS* = high-voltage power supply; *T* = quartz capillary tube in the separation position; *T'* = capillary tube in the injection position; *E* = Pt-wire electrodes; *ER* = polypropylene test-tube electrode vessels; Δh = height difference between sample and electrode solution levels; *L* = lids of the electrode vessels; *R* = line recorder; *OU* = optical unit of UV detector; *KA* = keying amplifier unit of UV detector; *SR* = sample reservoir.
(b) Capillary tube T inside the detection cell. *W* = quartz wall of the tube; *C* = dimethylpolysiloxane coating of the tube; ϕ_0 = input light flow; ϕ = output light flow.
(Reproduced from Ref. 9 with permission of Elsevier Science Publishers.)

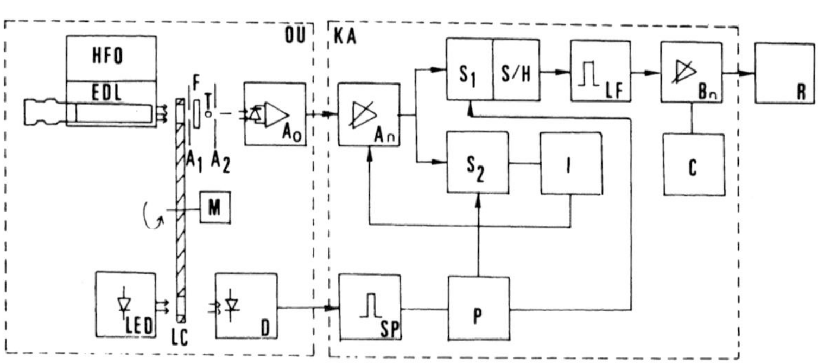

Fig. 3.7. Schematic diagram of the UV detector. *OU* = optical unit; *KA* = keying amplifier unit; *HFO* = high- frequency oscillator for EDL excitation; *EDL* = low-pressure iodine electrodeless discharge lamp; *LC* = light chopper; *M* = light chopper motor; A_1, A_2 = light apertures; *F* = interference filter for 206 nm; *T* = quartz capillary tube; A_0 = silicon photovoltaic detector with pre-amplifier; A_n, B_n = variable gain amplifiers; S_1, S_2 = MOSFET analog switches; *S/H* = sample/hold amplifier; *LF* = low-pass filter; *I* = integrator; *R* = line recorder; *C* = compensation; *LED* = light source for chopper keying amplifier; *D* = light detector for chopper keying amplifier; *SP* = shaper; *P* = phaser. (Reproduced from Ref. 9 with permission of Elsevier Science Publishers.)

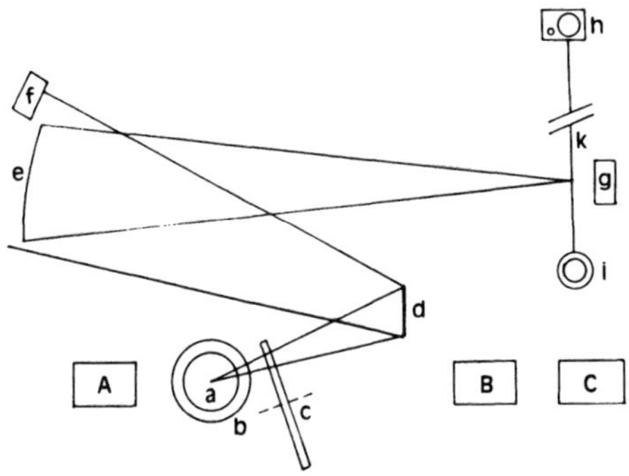

Fig. 3.8. Schematic diagram of the apparatus. *a* = deuterium tube; *b* = thermal shield; *c* = revolving holder with interference filters; *d* = auxiliary mirror; *e* = concave mirror; *f* = reference photodetector; *g* = signal photodetector; *h* = high-potential electrode chamber; *j* = terminal electrode chamber; *k* = capillary; *A* = current source and starting circuit for the deuterium tube; *B* = evaluating electronics; *C* = high-potential source. (Reproduced from Ref. 10 with permission of Elsevier Science Publishers.)

W. A better signal-to-noise ratio was achieved by low-pass filtering, sample/hold integration with dark period integration of noise and drift, followed by feedback integration. With this design, detection limits of 67 fmol (4.3×10^{-5} M) for phenol ($S/N = 2$) were reported.

Rohlicek and Deyl [10] described an apparatus for CE which utilizes an optical system allowing direct absorbance measurement in the capillary in UV light (Fig. 3.8). A deuterium lamp in a thermal shield operated with electronically stabilized current was used as the light source. Light passed through an interface filter selected by rotating a revolving filter holder. The light beam was focused onto a phototube. A current–voltage conversion circuit with an electronic amplifier served as a preamplifier for the photo tube. A schematic diagram of the evaluating electronic module used is shown in Fig. 3.9. The signals from the signal and reference photodetectors are connected with logarithmic amplifiers. In the reference signal path an attenuator was inserted for zero setting at the beginning of the experiment.

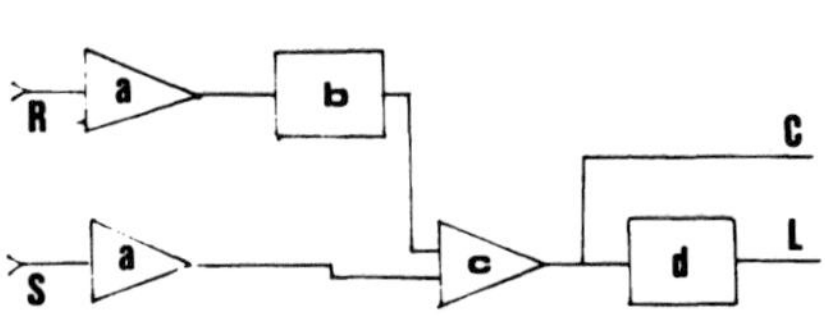

Fig. 3.9. Evaluating electronics. *a* = logarithmic amplifiers; *b* = attenuator; *c* = difference amplifier; *d* = low-pass filter; *R* = input from the reference photodetector; *S* = input from the signal photodetector; *C* = output of the computer; *L* = output to the line recorder. (Reproduced from Ref. 10 with permission of Elsevier Science Publishers.)

The signals were then subtracted in a difference amplifier. The signal then passed through a tunable low-pass filter which suppressed that part of noise which has a higher frequency than the highest frequency expected during the analysis. Detection limits of 10 fmol for collagen polymer was reported.

3.2.3 Background light

In order to reduce background light around the capillary, apertures or slits are often employed. However, apertures which are too small would reduce the light intensity through the capillary excessively whereas apertures which are too large would be ineffective in minimizing background light, and may even lead to losses in the observed column efficiency. The effect of aperture width on sensitivity and efficiency has been investigated by Wang *et al.* [11]. They described an on-column UV-vis detector cell with an adjustable aperture width. The cell increased signal-to-noise ratio almost 6 fold and expanded the linear range of detection about one order of magnitude as compared with a 1 mm diameter aperture cell. Figure 3.10 shows a schematic diagram of an adjustable aperture-width cell. The aperture body was constructed by sandwiching a shim between two pieces of metal which could be either stainless steel or brass. The aperture body was glued on the base. The washer with slits was rotatable. The aperture depends on both the thickness of the shim (25–75 μm) and the diameter of the slit (0.2–2 mm) on the washer. On the aperture body, a fine groove was made for retaining the capillary in the correct position in the light path. A capillary could be installed easily by loosening the capillary retainer, placing the capillary into the groove and tightening the retainer. Computer simulation was also performed to evaluate the effect of aperture width on observed column efficiency. The loss in efficiency was expressed in terms of peak

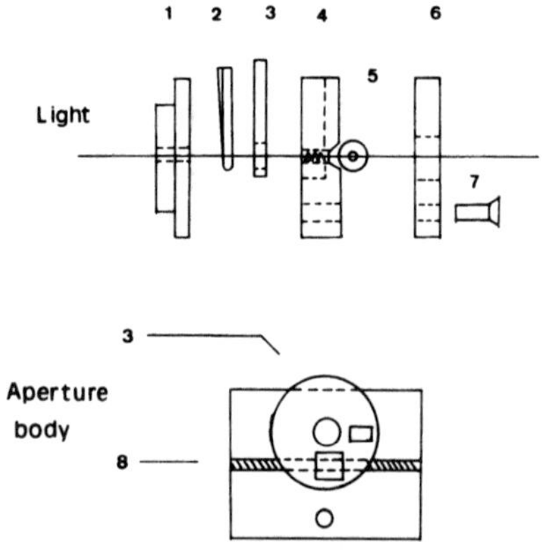

Fig. 3.10. Schematic diagram of the detector cell. *1* = base (Plexiglass); *2* = spring; *3* = washer with slits; *4* = aperture body (stainless steel or brass); *5* = capillary; *6* = capillary retainer (Plexiglass); *7* = screw; *8* = shim. (Reproduced from Ref. 11 with permission of Elsevier Science Publishers.)

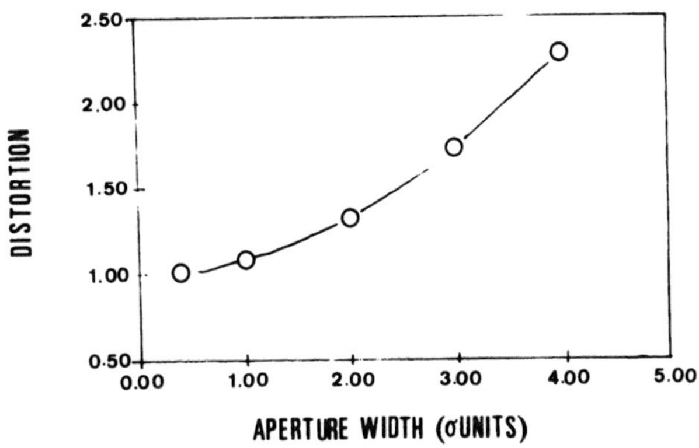

Fig. 3.11. The computer simulation results of the effect of aperture width on observed column efficiency. (Reproduced from Ref. 11 with permission of Elsevier Science Publishers.)

distortion defined as:

$$\text{distortion} = N_{\text{true}}/N_{\text{obs}} \tag{3.1}$$

where N_{true} and N_{obs} are the true and the observed theoretical plate number, respectively.

The results of the computer simulation is shown in Fig. 3.11 where peak distortion is plotted against aperture width in σ units, where σ is peak width calculated with:

$$\sigma = L/(N)^{1/2}. \tag{3.2}$$

From Fig. 3.11, it can be seen that when the aperture width is larger than 1 σ, the distortion becomes significant and increases non-linearly with increasing aperture width. It was concluded that when the aperture width is less than 1 σ of the peak width, the loss in column efficiency resulting from aperture width would be no more than 10% [11].

The response characteristic of an on-column UV-vis absorption detector as a function of wavelength for model solutes was investigated by Moring *et al.* [12]. In Fig. 3.12, the noise and average S/N ratio for dynorphin$_{1-13}$, horse heart myoglobin and β-lactoglobulin A are shown, together with the absorption spectra of bovine serum albumin. They noted that the distinct maximum S/N at 200 nm provided optimum detection performance for peptide and protein. In CE, detection at 200 nm was acceptable despite the short path length used, mainly because of the absence of refractive index changes due to solvent density pulse that result from HPLC pumping. To enhance the sensitivity of the detector, a spherical sapphire lens was used to improve the focusing of the light in the capillary and the aperture was carefully selected to diminish background light around the capillary. The detection sensitivity at 200 nm for different lens and aperture combination are shown in Fig. 3.13. Sensitivity was enhanced by a factor of five if the lens was used and

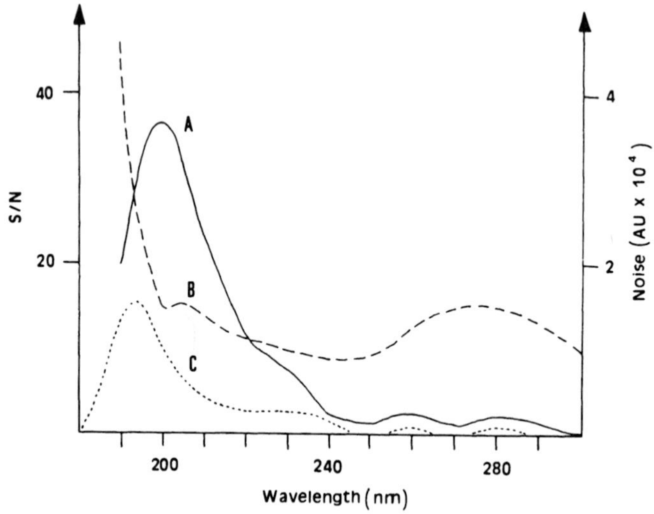

Fig. 3.12. UV detector response for peptides and proteins. Curves: A = average S/N, and B = noise (static measurement) for solutes dynorphin, horse heart myoglobin, and β-lactoglobin A (all 50 μg/ml in deionized water); C = absorption spectrum for BSA. (Reproduced from Ref. 12 with permission of Aster Publishing Corp.)

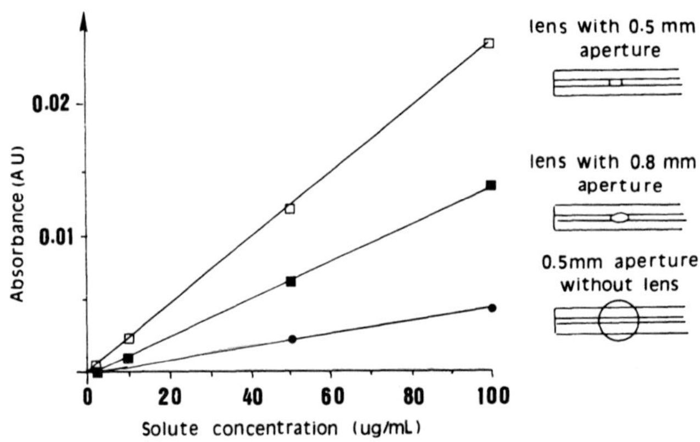

Fig. 3.13. Detector sensitivity at 200 nm for different lens and aperture combinations. Static absorbance measurement of dynorphin$_{1-13}$ in 20 mM sodium citrate, pH 2.5. (Reproduced from Ref. 12 with permission of Aster Publishing Corp.)

the effective cell volume was decreased to about 0.2 nl, which helped to ensure proper detection of nanoliter peak volumes. The detector was linear in peak areas over approximately three orders of magnitude. Minimum detection concentration of <5 ppb could be obtained for the model compounds which correspond to 13.5 pg (2×10^{-7} M) for β-lactoglobulin and 7.5 pg (2×10^{-6} M) for dynorphin$_{1-13}$ at $S/N = 2$.

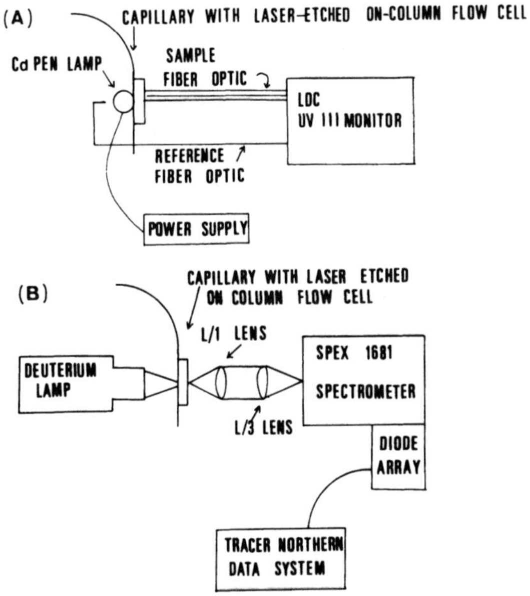

Fig. 3.14. Schematic diagrams of modular flow cell adapted for use with (A) HPLC UV absorbance detector, and (B) multichannel photodiode array detector. (Reproduced from Ref. 13 with permission of Elsevier Science Publishers.)

Sepaniak *et al.* employed a laser-etched flow cell for photometric detection using a modified UV absorbance detector, and spectrophotometric detection using a photodiode array detector [13]. The laser-etched on-column flow cells were produced by focusing the beam of an argon ion laser onto a small section of the capillary column which had its polyimide coating removed and painted black with water-based acrylic paint. The laser was used to create a slit along the flow channel of the capillary. Figure 3.14A and B illustrate diagrams of the flow cell adapted for use with a HPLC UV absorbance detector and a multichannel photodiode array detector respectively. Linear response was observed over three decades in concentration and limits of detection ($S/N = 2$) for isoquinoline as low as 1.6×10^{-5} M could be achieved.

3.2.4 Optical path length

With the use of small capillaries in CE, the optical path length is inherently limited to the I.D. of the capillary when on-column absorbance detection is performed. Several interesting methods have been employed to increase the path length for optical detection in small capillaries. These methods include the use of axial illumination [14], a Z-shaped flow cell [15], and a multireflection cell [16]. An alternative approach is to use tubing with non-circular cross-section. This approach is discussed in more detail in Section 4.1.2.

3.2.4.1 Axial illumination

Several methods have been developed based on the use of axial instead of transcolumn illumination to achieve UV [14,15] and visible [16] absorbance detection. Grant and Steuer [14] described a new on-column UV detection technique which extended the effective path length for UV detection from tens of microns into the millimeter range. A schematic diagram of the axial illumination principle is shown in Fig. 3.15. The amount of excitation energy reaching point A in Fig. 3.15 is governed by both the concentration and the distance between point A and the column outlet, which was chosen to be several millimeters in order to reduce band broadening caused by the increased detection volume. Although it would be difficult to measure directly the amount of light absorbed at point A, the addition of a fluorescent marker to the electrophoresis buffer provided a method for determining indirectly the absorbance with a given length of the capillary (l). This is because the intensity of fluorescence at point A is proportional to the amount of light reaching this point (I), which is given by:

$$I_0 = I \exp(-\alpha_f C_f l) \tag{3.3}$$

where I_0 is the intensity of the incident light source, and C_f and α_f are the concentration and absorptivity of the fluorescent marker, respectively. It was noted that in order to maximize detection signal, a very low concentration of the marker should be used [14]. A He-Cd laser (10 mW, 325 nm) was employed as the excitation source. The laser radiation was coupled to one end of an optical fibre using a quartz lens. The other end of the fibre was positioned close to the outlet of the separation capillary. Fluorescent light was collected with another optical fibre, which was placed perpendicular to the capillary and was connected to the photomultiplier of the detector.

If the sample absorbs at the excitation wavelength, the total fluorescence intensity (F) is given by [14]:

$$I_F = a C_f I_0 \exp[-(\alpha_f C_f + \alpha_s C_s) l] \tag{3.4}$$

where a is a constant which is related to the quantum efficiency of the fluorescent marker and the experimental geometry. α_s and C_s are the absorptivity and concentration of the sample.

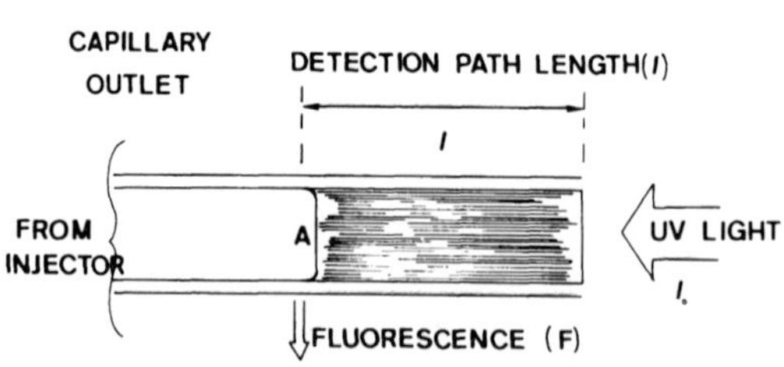

Fig. 3.15. Illustration of the axial illumination principle. The shaded area represents the effective cell volume. (Reproduced from Ref. 14 with permission of Aster Publishing Corp.)

The maximum path length (l_{max}) under typical CZE conditions is give by [14]:

$$l_{max} = \sqrt{12/N}\, f_H L / \sqrt{(1 - f_H)} \qquad (3.5)$$

where f_H is the fraction of plate number lost due to extracolumn effects, N is the number of theoretical plates, and L is the length of the column. The path length is limited to several millimeters, in order to minimize possible losses of efficiency. For a path length of 2 mm, and a capillary of 50 μm I.D., an improvement in sensitivity of 50 times could be obtained compared with the normal transcolumn illumination method, at the expense of 25–30% loss in efficiency.

The method was demonstrated for the CZE separation of dansyl-amino acids. Fluorescein was used as the fluorescent marker. Schematic of the instrument setup and details of the detection cell are shown in Figs. 3.16 and 3.17, respectively.

One of the major limitations of the method is the high background encountered although very stable intense light sources, or an extremely sensitive, low-noise detector should help to alleviate this problem. In addition, the presence of absorbing species in the detection region beyond point A may also affect the amount of excitation energy reaching species present at point A. Another potential problem is the possibility of bleaching of the fluorophore by the laser beam.

Xi and Yeung [15] described an alternative design of the axial-beam absorption detector for capillary electrophoresis. In this system, a ball lens was used to focus the light from a conventional UV source combined with a diffraction grating monochromator into one end of the separation capillary. Light travelled in the column by total internal reflection. An optical fibre was placed at the other end of the capillary to collect the light for detection by a photomultiplier tube. Consequently, the whole column was employed as the detection cell. The total absorbance represents the sum of all the absorbances of the individual compounds

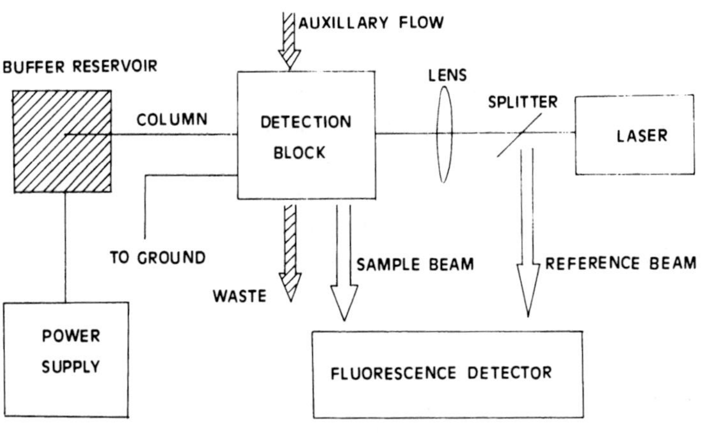

Fig. 3.16. Schematic of the instrumental setup for capillary zone electrophoresis for axial illumination detection. (Reproduced from Ref. 14 with permission of Aster Publishing Corp.)

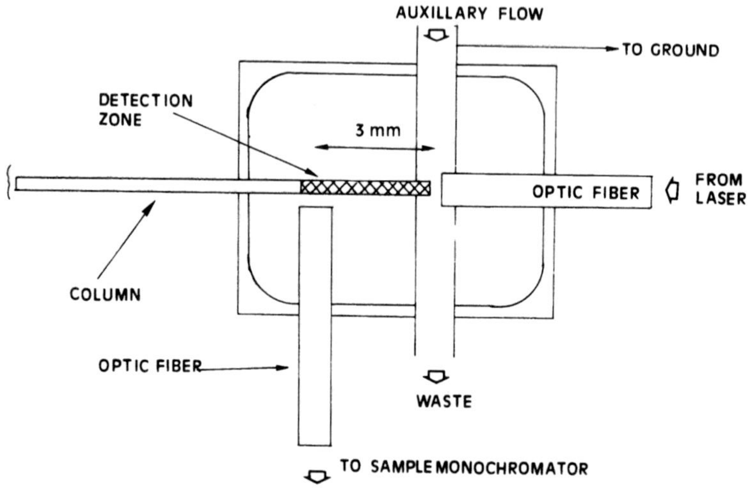

Fig. 3.17. Details of axial illumination detection cell. (Reproduced from Ref. 14 with permission of Aster Publishing Corp.)

within the column. Changes in the light intensity were detected as step changes when absorbing components eluted out from the column. Dimethylsulfoxide was used as the electrophoretic medium in order to ensure total internal reflection of the light in the capillary. With this detection system, a 7-fold improvement was obtained compared with conventional cross-beam methods. The concentration limits of detection (LOD) values were 1.9×10^{-7} M and 1.1×10^{-6} M for acridine and 3-aminoquinolin, respectively. A He-Ne laser has also been employed as the light source for detection in the axial-beam illumination mode by Taylor and Yeung [16]. A 15-fold improvement in detection limits over those for the cross beam arrangement was reported.

Some of the limitations of the axial illumination method have been discussed [15,16]. An important consideration is that aqueous solutions are not suitable media for total internal reflection in fused silica capillaries. This is because the refractive index of water (RI = 1.333) is significantly smaller than that of fused silica (RI = 1.458). Nevertheless, the problem could be alleviated by utilizing PTFE capillaries (RI = 1.35–1.38), although a specially designed jacket needed to be used to prevent damage of the capillary during mounting [15]. Another problem is attributable to electrostatic vibration of the capillary column when the high voltage was on. To alleviate this problem, it would be necessary to secure rigidly the whole column during electrophoresis.

3.2.4.2 Z-shaped flow cell

A Z-shaped flow cell for improved UV detection for capillary electrophoresis was described by Chervet *et al.* [17]. The schematic diagram of the Z-shaped capillary flow cell is shown in Fig. 3.18. The flow cell was prepared by sandwiching a shim

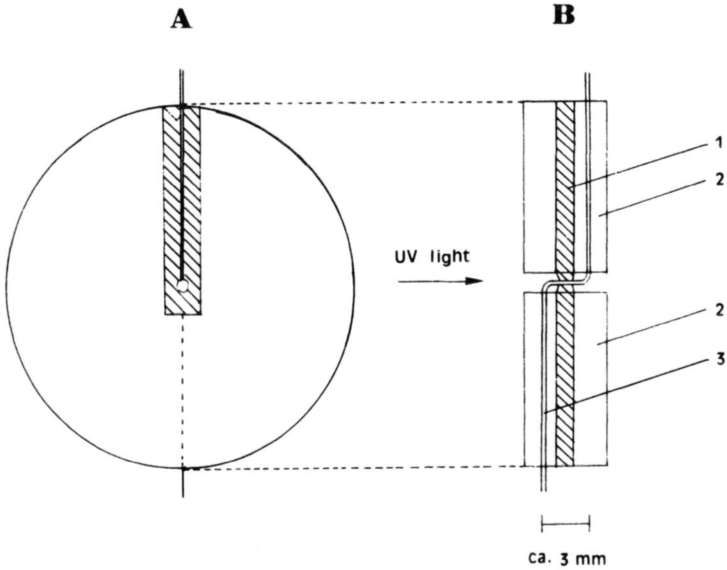

Fig. 3.18. Schematic diagram of the 3-mm Z-shaped capillary flow cell. (A) Front view; (B) cross-sectional view. *1* = shim (alumina) with centered 300 μm I.D. hole; *2* = plastic disks (black polyethylene or Plexiglass); *3* = fused-silica capillary of 50 or 75 μm I.D., 280 μm O.D. (Reproduced from Ref. 17 with permission of Elsevier Science Publishers.)

made of alumina with the bent capillary between two black polyethylene disks. The total path length of the flow cell was determined by the thickness of the shim and the sharpness of the bends. Typically for a shim of 1 mm thickness, and for sharp bends, the path length would be ca 3 mm. Each plastic disk had a groove for fixing the capillary, which was glued onto the disk with epoxy resin. It was shown that enhancement in signal-to-noise ratio of up to 6 fold for 3 mm Z-shaped flow cells could be achieved. The loss in resolution caused by the extended path length was found to be less pronounced than expected [17]. The maximum tolerable path length (l_{max}) of the flow cell can be estimated from Eq. (3.5).

3.2.4.3 Multireflection flow cell

The use of a nanoliter-scale multireflection cell for absorption detection in capillary electrophoresis was investigated by Wang *et al.* [18]. A schematic diagram of the multireflection cell is shown in Fig. 3.19. The capillary was made of fused silica with 75 μm I.D. × 364 μm O.D. and 51 cm in length. The capillary also served as the separation column. An opening of about 1 cm was made on the capillary coating simply by burning off the polyimide layer. A layer of silver was deposited on the opening by redox reaction of $Ag(NH_3)_2^+$ and glucose. Extreme caution is required in preparing this type of silver coating, since it may lead to the generation of explosive compounds. Black paint was applied on the silver layer to protect it from physical damage. The light windows made on the cell were separated by

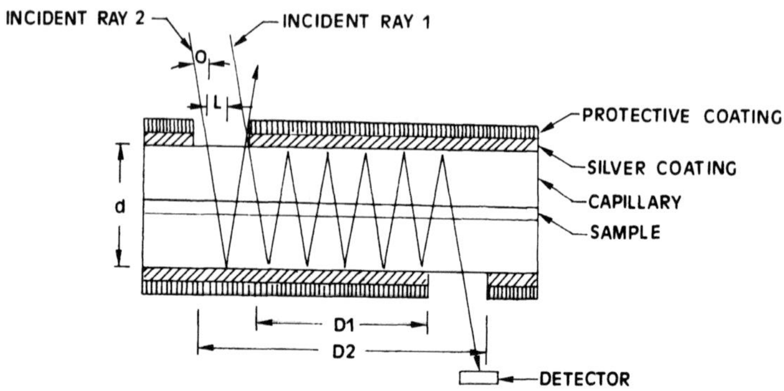

Fig. 3.19. Multireflection cell. (Reproduced from Ref. 18 with permission of the American Chemical Society.)

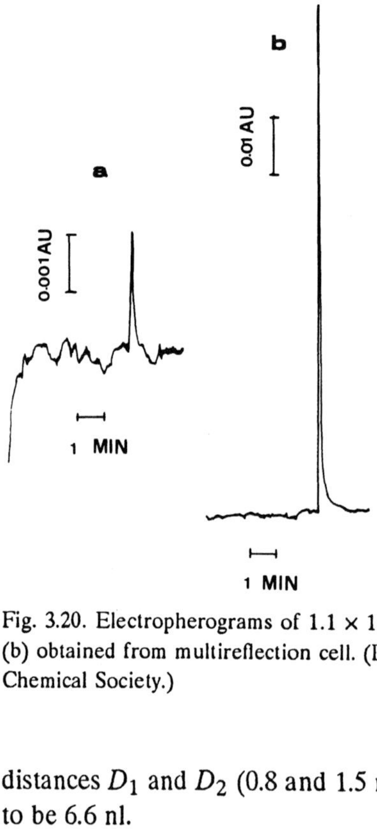

Fig. 3.20. Electropherograms of 1.1×10^{-5} M brilliant green. (a) Obtained from single-pass cell; (b) obtained from multireflection cell. (Reproduced from Ref. 18 with permission of the American Chemical Society.)

distances D_1 and D_2 (0.8 and 1.5 mm, respectively). The cell volume was estimated to be 6.6 nl.

An important consideration in this design is the incident angle, ϑ (see Fig. 3.19). This is because the number of reflections and the light intensity reaching the detector have opposite dependence on this angle. A compromise has to be reached so that the gain in sensitivity due to the increase in path length resulting from multiple reflections is not lost because of decrease in light intensity during reflections and incomplete reflections at the reflecting surface. An additional factor to consider is the effect of increase in detection cell volume on separation efficiency. The increase in cell volume is given by the distance D_2, which is in turn affected by the angle ϑ. Typical distances for D_2 of less than 1 σ (approximately 1 mm for CZE, see Eq. 3.2) should be used. In addition to the angle ϑ, the design of the flow cell, the intensity of the light source and the choice of the reflective coating material are important factors affecting detection sensitivity.

Figure 3.20 shows electropherograms obtained for brilliant green under identical separation conditions for CE with on-column detection (a) and the multireflection cell (b). It was found that the sensitivity of the multireflection cell was over 40 times higher than that of the single-pass cell. The increase in sensitivity was found to be close to the calculated increase of path length (44 times). Noise level were similar for both cells. A detection limit of 6.5×10^{-8} M was reported for brilliant green using the multireflection cell [16].

3.3 PHOTODIODE ARRAY AND MULTIWAVELENGTH UV DETECTION

The main advantage of employing photodiode array (PDA) detection is that multiwavelength spectral information can be obtained. The spectral information can be used to aid in the identification of unknown compounds. Furthermore, peak-purity check, and absorbance ratio at different wavelengths can be performed to confirm whether there is any overlapping of peaks in a single chromatogram.

Kobayashi *et al.* [19] modified a commercial HPLC photodiode array detector for capillary electrophoresis. They demonstrated that CE with PDA permitted accurate characterization of separated components and could be a powerful tool for the investigation of mixed zones. In this investigation, a 512 element diode array was used to provide complete coverage of 200–380 nm UV range. A schematic diagram of the optical path for the modified PDA detection system is shown in Fig. 3.21.

Their data indicated that with a capillary diameter of bigger than 75 μm, spectra of a quality equivalent to or better than those obtained with the standard HPLC

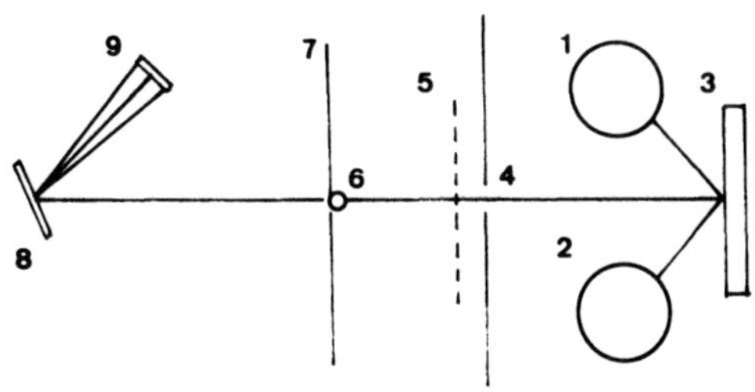

Fig. 3.21. Optical path for the modified photodiode array detection system. *1* = deuterium lamp; *2* = tungsten lamp; *3* = mirror; *4* = aperture; *5* = shutter; *6* = fused silica capillary tubes; *7* = slit; *8* = holographic grating; *9* = photodiode array. (Reproduced from Ref. 19 with permission of Elsevier Science Publishers.)

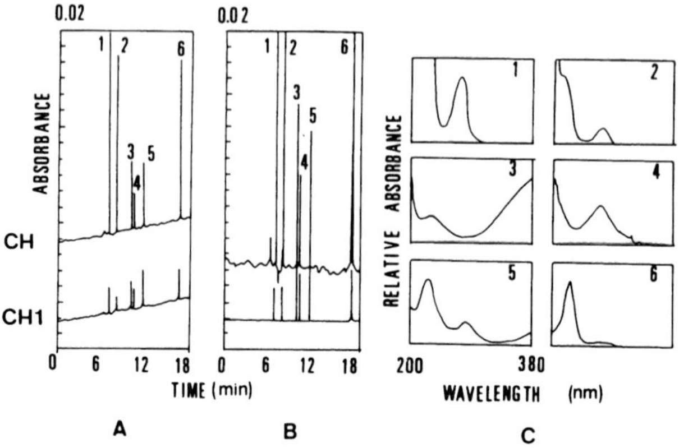

Fig. 3.22. Electropherograms and spectra of aromatics. The electropherograms were obtained by micellar electrokinetic capillary chromatography with 75 μm I.D. (A) and 100 μm I.D. (B) fused silica capillary tubes. Buffer: 50 mM SDS in 10 mM Tris-phosphate, pH 6.9; sample injection: 10 s at 3.5 cm height difference; detection: channel 1 (CH1) at 250–253 nm; channel (CH2) at 207–210 nm. (A) 15 kV, ca. 32 μA, detection at 0.02 a.u.f.s. (B) 15 kV, ca. 38 μA, detection at 0.02 a.u.f.s. (C) Superimposed spectra of components separated: *1* = resorcinol; *2* = phenol; *3* = p-nitroaniline; *4* = nitrobenzene; *5* = o-nitroaniline; *6* = 2-naphtol. Broken line: spectra obtained with 75 μm capillaries; full line: spectra obtained with 100 μm capillaries. (Reproduced from Ref. 19 with permission of Elsevier Science Publishers.)

configuration of the detection could be obtained. Typical electropherogram and spectra of aromatics obtained with the CE-PDA system are shown in Fig. 3.22A and B. In Fig. 3.23, the electropherogram and contour plots for a water-soluble vitamin (calcium panthothenate) and caffeine are shown.

The separation of hop bitter acids by CE and MEKC with UV-diode array detection was investigated by Vindevogel *et al.* [20]. The flow cell of a commercial

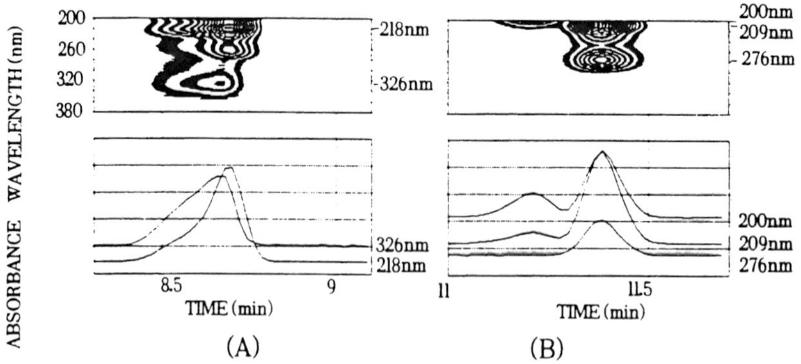

Fig. 3.23. Example of investigation of peaks by the contour method. (A) Extended absorption curves at 218 and 326 nm and the contour plot for peak 1. (B) Extended absorption curves at 200, 209 and 276 nm and the contour plot for peaks 2 and 3. (Reproduced from Ref. 19 with permission of Elsevier Science Publishers.)

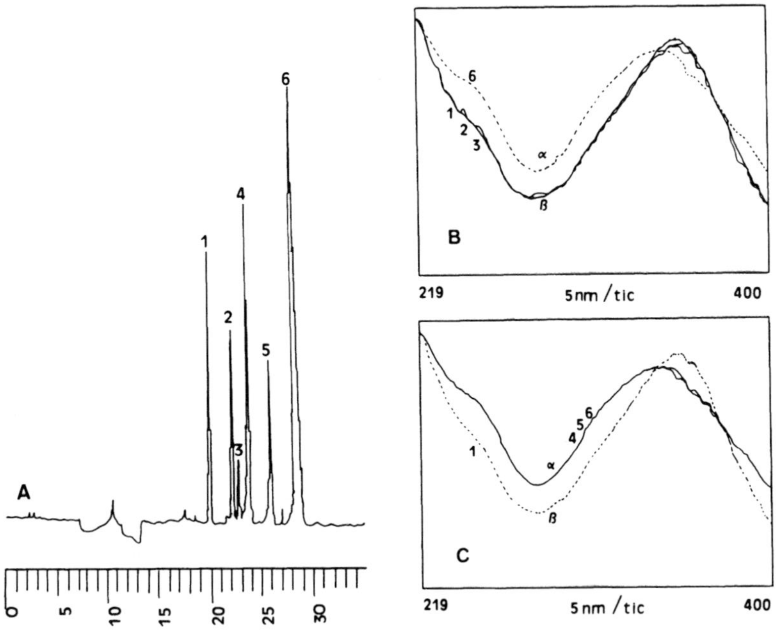

Fig. 3.24. (A) MEKC-analysis of hop extract (25 mM TRIS/HAc + 25 mM SDS, pH 9.0, 10 kV), UV 220 nm. (B) On-line UV-spectra of the α-acids (*4–6*), compared with a β-acid (*1*). (Reproduced from Ref. 20 with permission of Dr. Alfred Huethig Publishers.)

PDA detector was modified for on-column detection. The MEKC electropherogram and spectra obtained for hop extract are shown in Fig. 3.24. These results showed that the spectra obtained using PDA detection can be used to identify the acid type.

Yeo *et al.* [21] investigated the separation of antibiotics using CE with PDA detection. Simultaneous detection at different wavelengths (Fig. 3.25), on-line spectral analysis (Fig. 3.26) and contour plots (Fig. 3.27) were used for the identification and analysis of antibiotics in drug tablets.

3.4 FLUORESCENCE DETECTION

3.4.1 Lamp-based fluorescence detectors

The main advantage of fluorescence detection is that high sensitivity can be readily obtained [40–49]. For analytes which are not normally fluorescent, pre- or post-column derivatization may be employed to introduce fluorophores to the analyte molecules. Different types of derivatizing agents are discussed in Section 3.4.3. To optimize the performance of fluorescence detectors for the detection of non-fluorescent analytes, the design of pre- or post-column reactors for derivatization may need to be considered. In addition, it would be necessary to adopt the same basic approaches as those described in Section 3.2 for absorbance

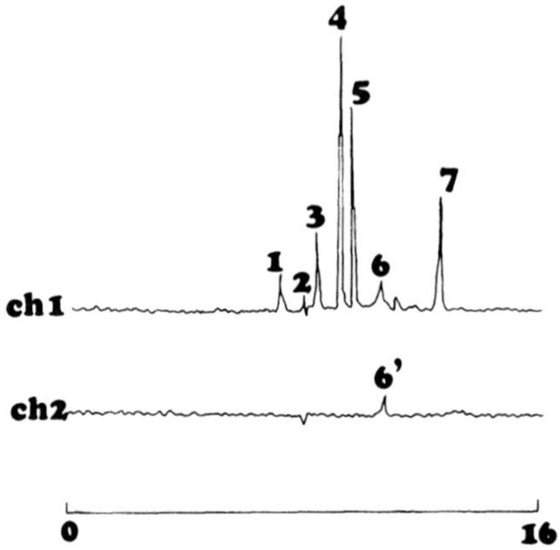

Fig. 3.25. Electropherogram of six antibiotics. *1* = tylosin tartrate, TT; *2* = methanol, MeOH; *3* = nystatin, NYS; *4* = amoxicillin, AMO; *5* = ampicillin, AMP; *6* = chlorotetracycline, CT; *7* = penicillin-G, PEN-G. Electrophoretic solution: 0.05 M phosphate–0.1 M borate at pH 7.06. Separation tube: 50 cm × 50 μm I.D. fused silica capillary. Voltage: 15 kV (36 μA). Detection wavelength: CH1, 198–200 nm; CH2, 380–390 nm. (Reproduced from Ref. 21 with permission of Elsevier Science Publishers.)

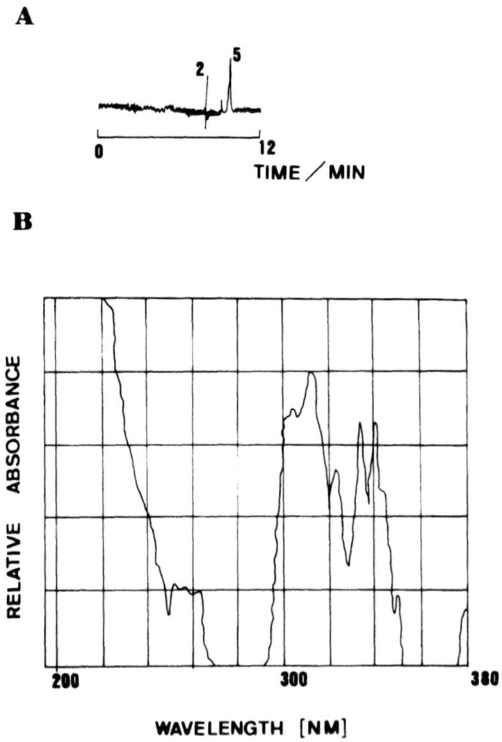

Fig. 3.26. (A) Electropherogram of ampicillin extract. (2) MeOH; (5) AMP. (B) Spectrum of ampicillin. Electrophoretic solution: 0.05 M phosphate - 0.1 M borate at pH 7.6. Separation tube: 50 cm × 50 μm I.D. fused silica capillary. Voltage: 15 kV (36 μA). Detection wavelengths: CH1, 190–200 nm. (Reproduced from Ref. 21 with permission of Elsevier Science Publishers.)

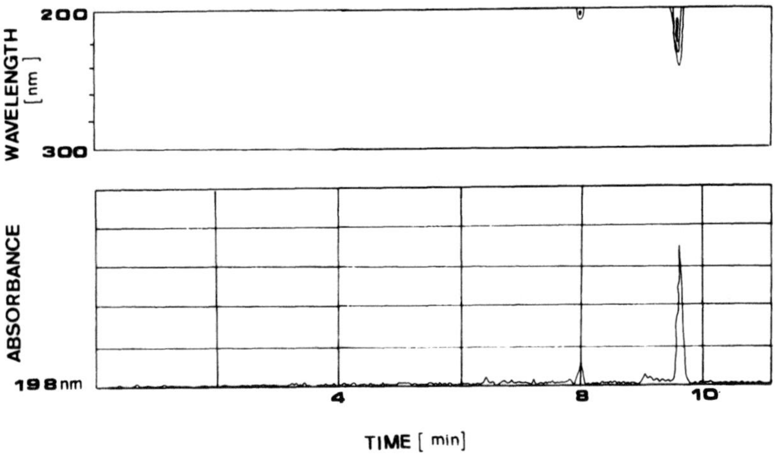

Fig. 3.27. Contour plot of ampicillin extract and its electropherogram at 198 nm. (Reproduced from Ref. 21 with permission of Elsevier Science Publishers.)

detectors in order to optimize the sensitivity of this type of optical detection technique.

In their pioneering work on capillary electrophoresis, Jorgenson and Lukacs [40] employed a fluorescence detector which consisted of a high-pressure mercury arc lamp along with glass filters to isolate the excitation wavelength which was focused onto a portion of the column. Green and Jorgenson [41] modified the design of the detector by replacing the filter system for excitation wavelength isolation with a double monochromator in order to provide better wavelength discrimination, lower stray light levels, and increase excitation wavelength versatility. A schematic diagram of the optical lay out of the on-column fluorescence detector is shown in Fig. 3.28. With this detector, linearity range over 3 orders of magnitude was obtained. Detection limits for this system were in the 10^{-7}–10^{-6} M range. Subsequently, many methods has been developed to further improve the sensitivity of fluorescence detection, such as the use of the pre- and post-column derivatization, epillumination

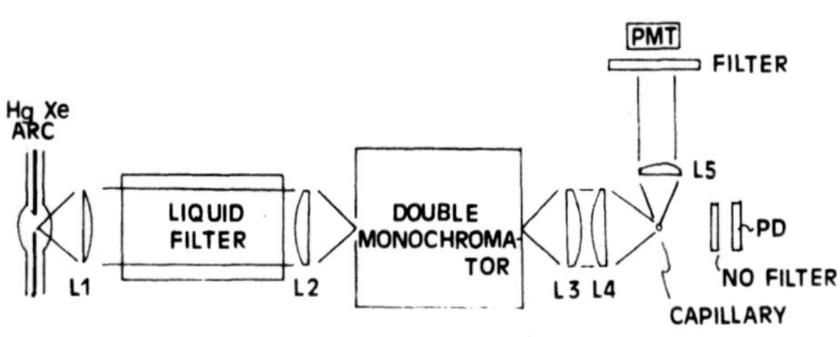

Fig. 3.28. Schematic diagram of the optical layout of the on-column fluorescence detector. (Reproduced from Ref. 41 with permission of Elsevier Science Publishers.)

(Section 3.4.1.3), and the use of laser-induced fluorescence (Section 3.4.2). The sensitivity of fluorescence detection has been extended to below 10^{-12} M level.

3.4.1.1 Post-column derivatization

Rose and Jorgenson utilized a post-capillary fluorescence detection scheme for capillary zone electrophoresis [42]. o-Phthaldialdehyde (OPA) was used as the tagging reagent. A coaxial capillary reactor was used to achieve mixing of the OPA reagent with migrating zones, without causing excessive zone broadening. The experimental setup used is shown in Fig. 3.29. A schematic diagram of the post-capillary reactor is shown in Fig. 3.30. The reaction capillary was held in a stainless steel tee by Vespel ferrules. A detection window was formed by burning off the polyimide coating

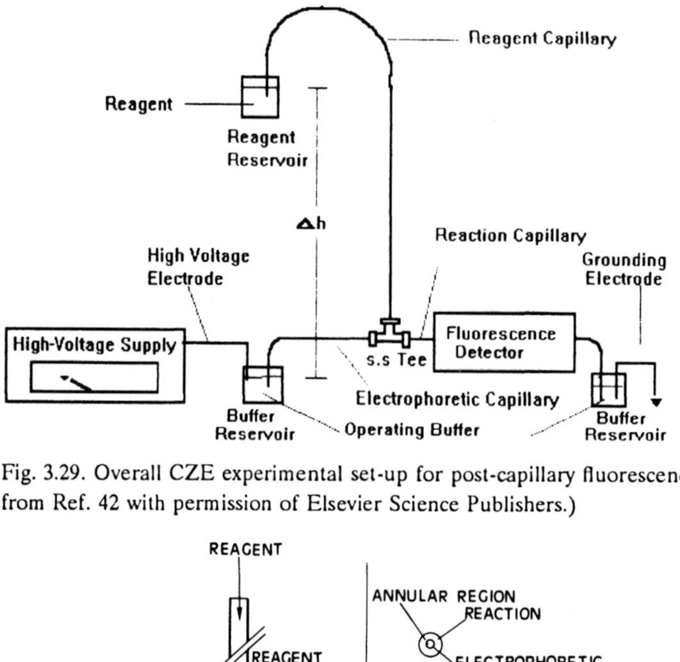

Fig. 3.29. Overall CZE experimental set-up for post-capillary fluorescence detection. (Reproduced from Ref. 42 with permission of Elsevier Science Publishers.)

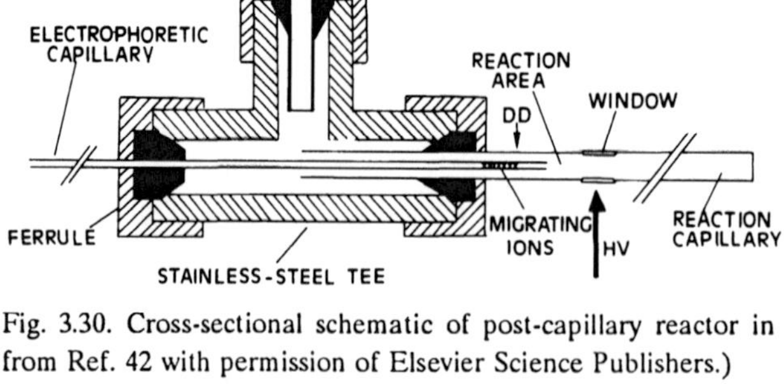

Fig. 3.30. Cross-sectional schematic of post-capillary reactor in stainless steel tee. (Reproduced from Ref. 42 with permission of Elsevier Science Publishers.)

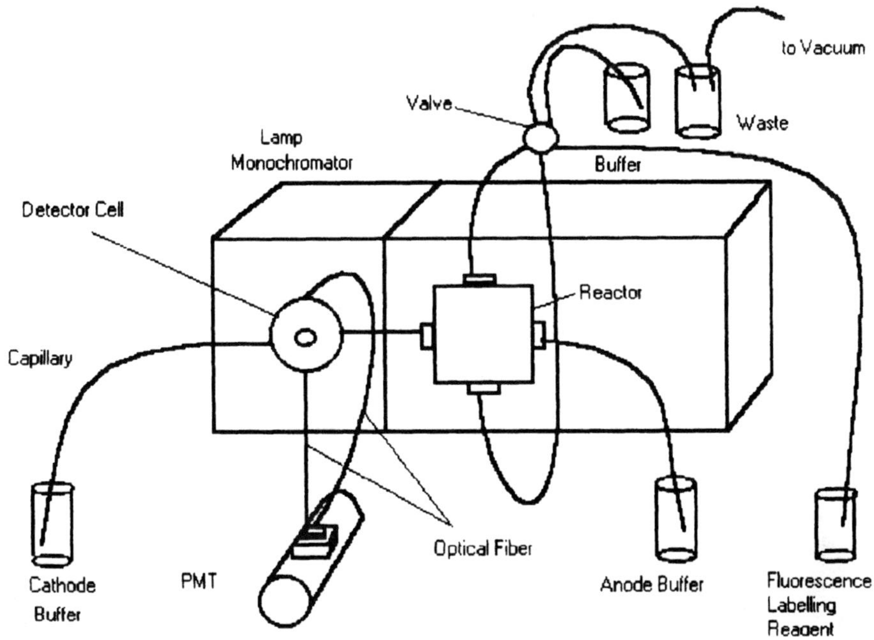

Fig. 3.31. Overview of instrumentation for fluorescence capillary electrophoresis measurements. (Reproduced from Ref. 43 with permission of the American Chemical Society.)

on the reaction capillary. The electrophoretic capillary was passed through the tee and inserted into the reaction capillary to form the coaxial reactor. The inner and outer diameters of electrophoretic and reaction capillaries could be varied according to experimental design. The detector was linear over three orders of magnitude and detection limits in the attomole (10^{-9} M) range were obtained.

A capillary electrophoresis fluorescence detector, which employed either deuterium, tungsten or xenon arc lamps was described by Albin *et al.* [43]. An overview of the instrument is shown in Fig. 3.31 with a schematic representation of the optical path in Fig. 3.32. A grating monochromator was used for the selection of excitation wavelength. Fluorescence emission was monitored by collection of light from the capillary with two optical fibres positioned at right angles to the excitation beam. Filters were used to isolate the emission light. A photomultiplier tube was used to detect the fluorescence light. Separation of amino acids labeled with various fluorescent reagents by using both pre- and post-capillary derivatization was investigated. Postcolumn derivatization was accomplished by using a gap junction reactor in which two capillaries were separated by less than 50 μm, as shown in Fig. 3.33. The reagent buffer was pumped into the capillary by differential electroosmotic flow. Limit of detection obtained for pre-column derivatization using 9-fluorenylmethylchloroformate (FMOC) was 5×10^{-8} M and that for post-column derivatization using OPA was 3×10^{-7} M at $S/N = 2$.

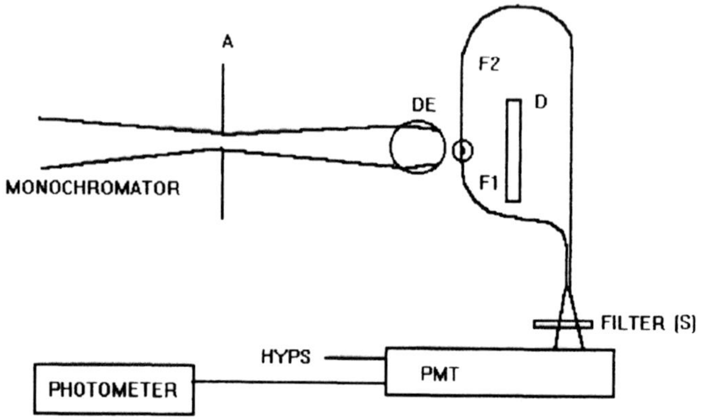

Fig. 3.32. Schematic representation of optical path of monochromatic light through the absorbance/fluorescence cell with dual optical fibres, filter, and photomultiplier tube. Key: aperture controlling image size of light source in center of capillary (*A*), sample lens (*OE*), Optical fibres (*F1* and *F2*), photodiode (*D*), high-voltage power supply (*HVPS*). (Reproduced from Ref. 43 with permission of the American Chemical Society.)

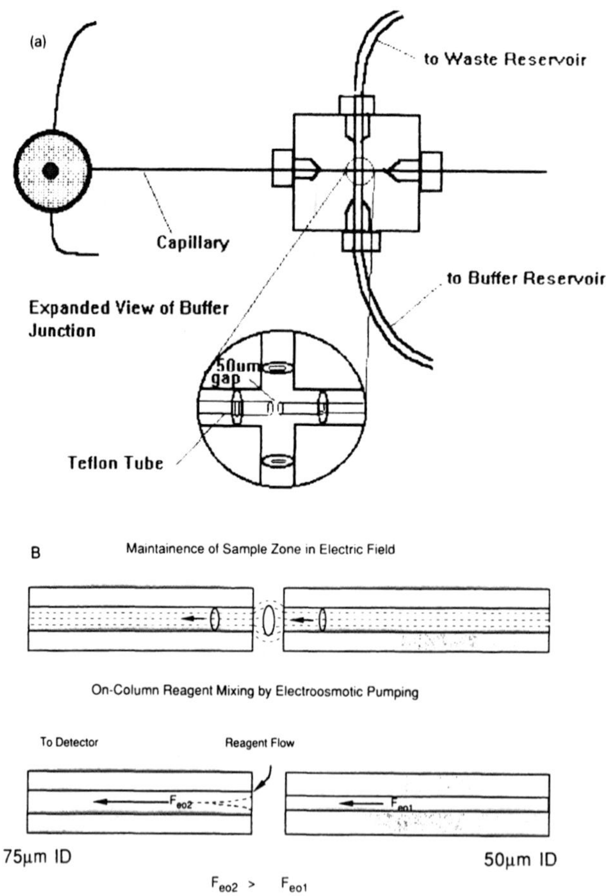

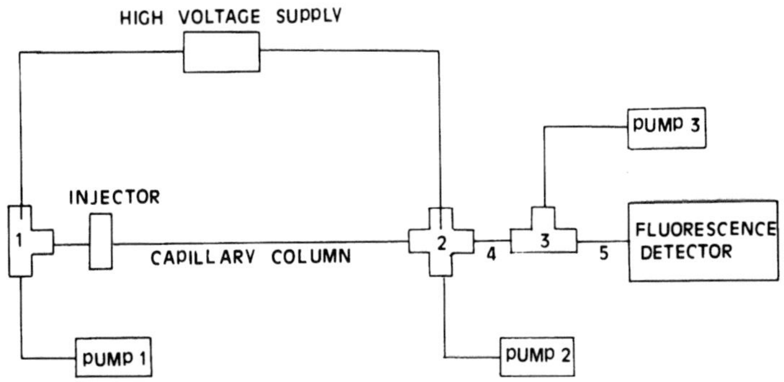

Fig. 3.34. Schematic diagram of post-column detector for detection in capillary electrophoresis. *1* = positive terminal; *2* = four-way connector for earth terminal and mixing of column media and buffer; *3* = three-way connector for mixing with fluorescent reagent; *4*, *5* = PTFE tubes (0.5 mm I.D.) of length 5 and 70 cm, respectively. Column medium, alkaline buffer solution and fluorescent reagent were supplied by pumps 1, 2 and 3, respectively. (Reproduced from Ref. 44 with permission of Elsevier Science Publishers.)

A post-column detection method for capillary zone electrophoresis was described by Tsuda *et al.* [44]. A schematic diagram of the overall setup is shown in Fig. 3.34. The design of the four way polytetrafluoroethylene connector is shown in Fig. 3.35. The end of the capillary column was inserted into the polytetrafluoroethylene tube. The apparatus consisted of three pumps and two mixing parts. Syringe pumps were used to prevent pulsation. The order of starting the three pumps was important to avoid flow backwards into the capillary column. The operational procedure involved first starting pump 1 and than pump 2 and finally pump 3. After starting the three pumps, sample was injected with a rotary injector. Mixing of the fluorescent reagent (fluorescamine) with the buffer medium occurred in a simple three-way connector. Calibration for putrescine was performed in the range of 2–50 pmol although the calibration graph was non-linear and the limits of detection were not reported.

3.4.1.2 Pre-column derivatization

Pre-column derivatization is an alternative method to introduce fluorophores to the analyte molecules in CE [43,45]. This method has the advantage that there are less constraints on the design of the reactor compared with post-column derivatization. Tsuda *et al.* [45] used a modified commercial fluorescence detector for the detection of polyamines by capillary electrophoresis. A microscale procedure

---

Fig. 3.33. (a) Expanded view of the post-column reactor. (b) Principle of function of gap junction reactor. Top: Sample solute dispersion is minimized because of the containment in the electric field bridging the small 10–50 μm gap. Bottom: A secondary buffer containing a fluorescent reagent can be mixed with the run buffer by virtue of the difference in electroosmotic flow in different diameters of capillary tubing. (Reproduced from Ref. 43 with permission of the American Chemical Society.)

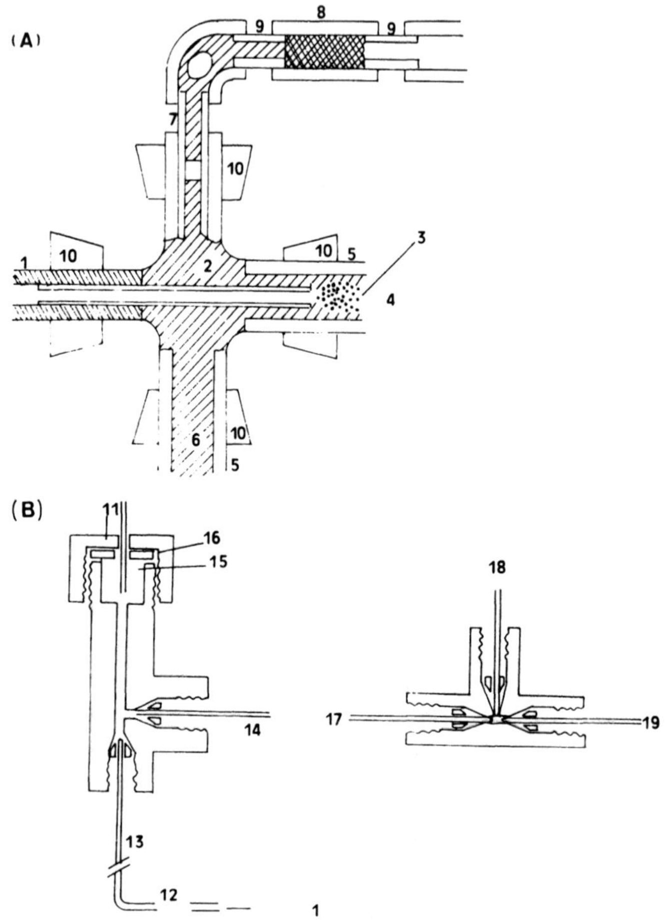

Fig. 3.35. (A) Mixing part for column medium and buffer at four- way connector. This part corresponds to *2* in Fig. 3.34. *1* = FEP capillary column; *2* = fused-silica capillary, inserted into the end of the FEP column; *3* = mixing zone; *4* = to fluorescence detector; *5* = PTFE tube (0.5 mm I.D.); *6* = alkaline buffer; *7* = earth terminal (platinum tubing); *8* = flow resistance, PTFE tubing packed with polymer beads; *9* = stainless steel tubing; *10* = PTFE ferrule. (B) Detailed diagrams of positive terminal (left) and three-way connector (right), corresponding to *1* and *3*, respectively, in Fig. 3.34. *11* = positive terminal, platinum wire; *12* = platinum tubing (3 cm × 0.3 mm I.D. × 0.7 mm O.D.) connected to earth; *13* = PTFE tubing (2 m × 0.1 mm I.D. × 0.2 mm O.D.) as an electric resistance; *14* = FEP tubing connecting positive terminal and injector; *15* = space for bubbles; *16* = silicone-rubber septum. The mixtures of column medium and alkaline solution (*17*) and fluorescent reagent (*18*) were mixed in the three-way connector, and were led into a coil (*19*). (Reproduced from Ref. 44 with permission of Elsevier Science Publishers.)

was used for pre-column derivatization with fluorescamine. To improve detection sensitivity, the fluorescent signal was collected with a cylindrical lens. A schematic diagram illustrating the use of the cylindrical lens in the fluorescence detector

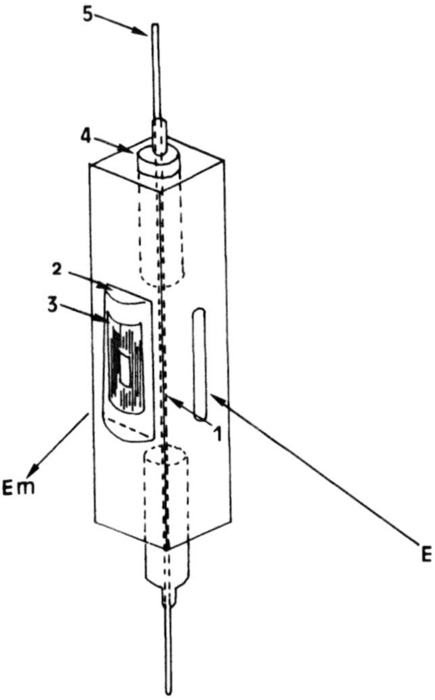

Fig. 3.36. Diagram illustrating the use of a cylindrical lens in the fluorescence detector. *1* = cylindrical flow-through cell, *2* = cylindrical lens, *3* = slit, *4* = ceramic tubing, *5* = fused silica capillary tubing (which was used as both column and cell). *E* = light beam for excitation, and *Em* = light due to fluorescence. (Reproduced from Ref. 45 with permission of Aster Publishing Corp.)

is shown in Fig. 3.36. The cylindrical lens served to concentrate light along the detection window on the capillary without affecting the detection volume, and hence did not introduce additional band broadening. The minimum detectable amount and concentration of the sample containing putrescine, cadaverine and spermidine were around 200 fmol and 10^{-5} M respectively.

3.4.1.3 Epillumination fluorescence microscopy

Hernandez *et al.* [46,47] constructed fluorescence detectors based on fluorescence microscopes for capillary electrophoresis. One design employed an epillumination lamp and a photodiode. The microscope was adapted with a combination of filters: a band pass excitation filter was used to cut off radiation above 405 nm, a chromatic beam splitter was used to reflect radiation above 420 nm. For eye protection, a high-pass filter was used to suppress radiation under 418 nm. The principles of adaptation of fluorescence microscopes are shown in Figs. 3.37 and 3.38. Figure 3.37 shows the excitation of the sample and Fig. 3.38 shows the emission and measurement of fluorescence. The excitation lamp emits a mixed beam of visible and ultraviolet (UV) light. Radiation above 405 nm was suppressed by the excitation filter. The chromatic splitter reflected radiation below 420 nm toward the fluorite

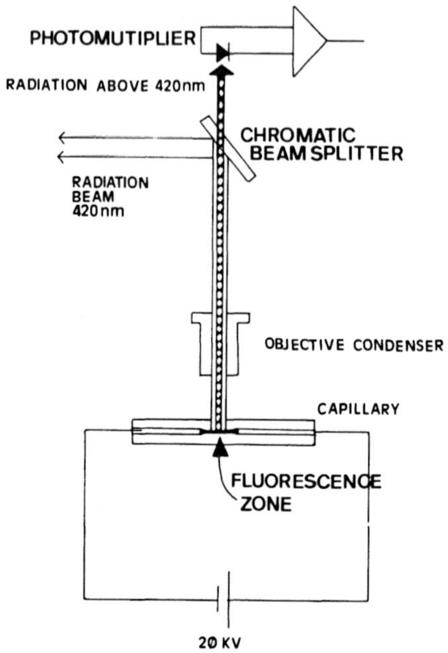

Fig. 3.37. Excitation of the sample with an epillumination fluorescence microscope. The UV beam is shown as thick black line reflected at right-angles by the chromatic beam splitter. The fluorescence zone is indicated in the dark dotted area in the center of the capillary. (Reproduced from Ref. 46 with permission of Elsevier Science Publishers.)

objective of the microscope, which focused the light onto the capillary. The emitted light crossed the objective, the chromatic beam splitter, the high-pass filter, and was then focused onto a photodiode that transformed the light into an electrical signal which was recorded on a strip chart recorder. A carrier made of plexiglass was used to mount the capillary (Fig. 3.39). The objective combined with a 10× ocular was focused to obtain a sharp image of the capillary. By using a 40×, 0.75 NA fluorite objective instead of the conventional 6×, 0.16 NA objective, an improvement in fluorescence detection by a factor of 56 could be achieved. A detection level of 500 amol (5×10^{-7} M) for riboflavin was obtained [46].

Significant improvements in the performance of this type of detection system were obtained in a subsequent design, which employed an air-cooled argon ion laser (488 nm) as the light source [47]. The detection limits ($S/N = 3$) obtained for fluorescein thiocarbamyl-amino acids were around 3.75×10^{-3} amol (10^{-12} M). The linear dynamic range extended over five orders of magnitude.

3.4.2 Laser-induced fluorescence detection

Lasers are superior excitation sources for use with small-diameter capillaries [51–75]. Advantages over arc lamp sources include better focusing capabilities

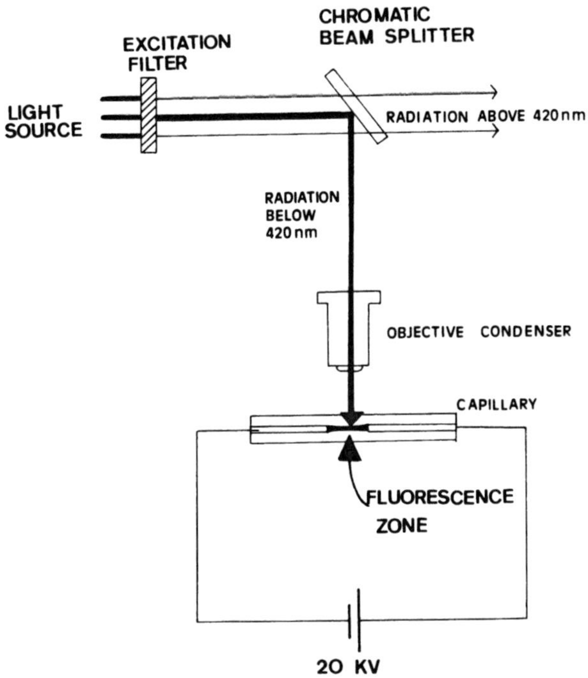

Fig. 3.38. Emission and measurement of fluorescence. The emitted light is indicated by the thick dashed line that crosses the chromatic beam splitter and is detected by the photodiode. (Reproduced from Ref. 46 with permission of Elsevier Science Publishers.)

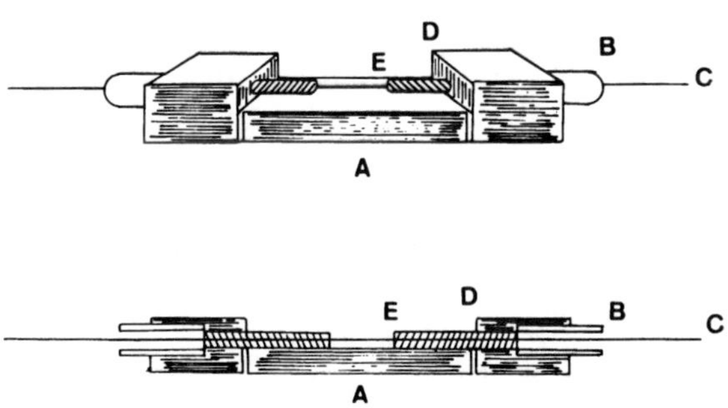

Fig. 3.39. Top: capillary carrier; bottom: cross section. A = 50 × 25 × 5 mm parallelepiped; B = PTFE tubing; C = capillary; D = 20 × 14 × 9 mm parallelepipeds; E = capillary holders. (Reproduced from Ref. 46 with permission of Elsevier Science Publishers.)

which allow the excitation energy to be more effectively applied to very small sample volumes, and better monochromicity which reduces stray light levels. Laser are particularly useful for sensitive detection on capillaries having inside diameter of less than 50 μm because of the ability to be focused into smaller volume than are possible with arc lamp excitation. The disadvantages are that the wavelengths

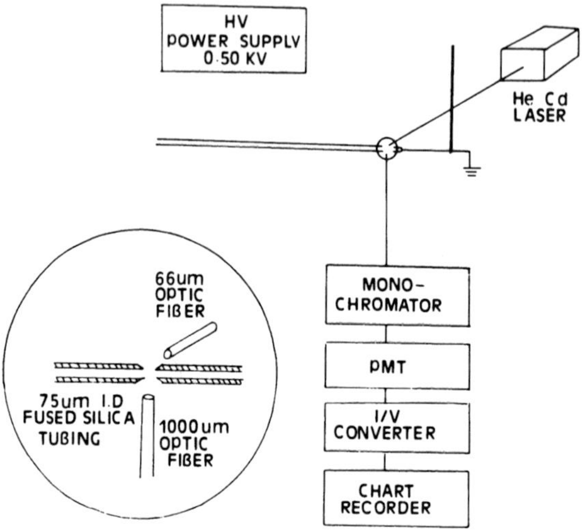

Fig. 3.40. Experimental set-up for capillary zone electrophoresis with laser-induced fluorescence detection. (Reproduced from Ref. 52 with permission of Elsevier Science Publishers.)

available from current types of laser sources are rather limited, and that there are possibilities of photodegradation of the analytes caused by the high light intensity.

Zare and co-workers [51–53] focused the output of a He-Cd laser at 325 nm onto an optical fibre to illuminate the detection region on the capillary. The fluorescence emission was collected by a second optical fibre and detected by a fast monochromator and a photomultiplier. A schematic diagram of the experimental setup is shown in Fig. 3.40. This detection system produced a linear dynamic range of 10^{-3}–10^{-7} M and detection limits on the order of 2 fmol (10^{-7} M) for dansylated amino acids.

An on-column connector for microcolumn was also developed to permit on-column o-phthaldialdehyde derivatization of amino acids separated by CZE [54]. The connector in the form of a cross or tee was fabricated from the fused silica capillary tubing itself. The construction of the on-column capillary cell is shown in Fig. 3.41. The experimental setup of the CZE/LIF detection system is shown in Fig. 3.42. Detection response was found to be linear over more than 3 orders of magnitude with minimum limits of detection in the subfemtomole (10^{-7} M) range.

Burton *et al.* [55] used laser-induced fluorescence to detect vitamin B_6 and its metabolites separated by MEKC in urine samples after extraction. A He-Cd laser (325 nm) was used for fluorescence excitation. Emission was monitored at 430 nm. Fluorescence signals were isolated with a monochromator and detected with a photomultiplier tube. Limit of detection ($S/N = 3$) for pyridoxic acid was less than a picogram injected. This corresponded to approximately 10^{-8} M in terms of concentration.

Jorgenson and co-workers [56,57,59] employed an on-column He-Cd laser-

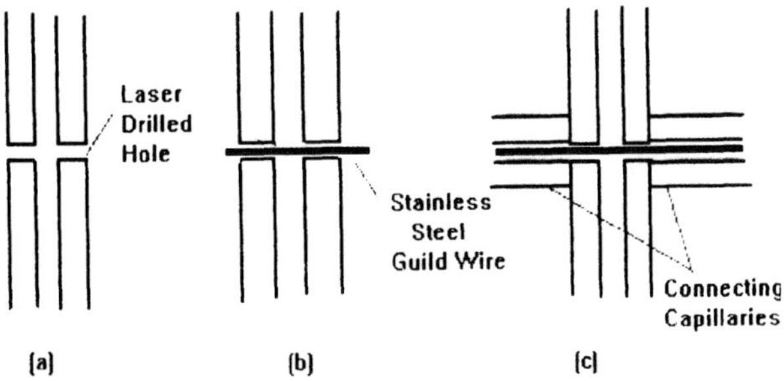

Fig. 3.41. Construction of the on-column capillary connector. (Reproduced from Ref. 54 with permission of the American Chemical Society.)

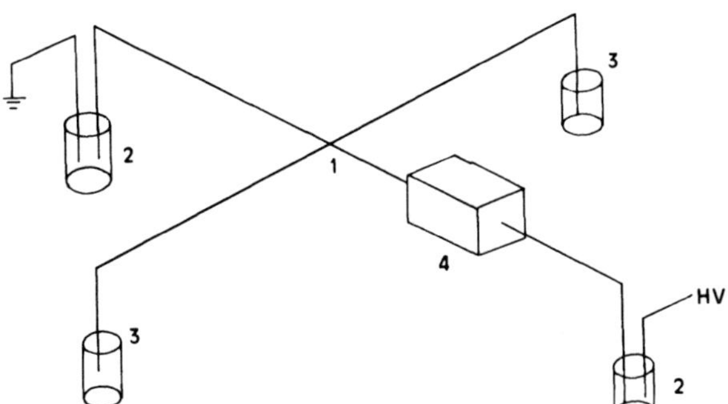

Fig. 3.42. Experimental setup of the CZE/LIF detector system. *1* = on-column connector; *2* = buffer reservoirs; *3* = derivatization reagent reservoirs; *4* = LIF detector housing. (Reproduced from Ref. 54 with permission of the American Chemical Society.)

induced fluorescence detector developed for capillary LC [56] for detection in CZE. The detector was placed out of the perpendicular plane to minimize scattered light. The use of three fluorogenic reagents, o-phthaldialdehyde (OPA), naphthalene-dialdehyde (NDA) and fluorescein isothiocyanate (FITC) for pre-column tagging of amino acids in order to permit LIF detection in CZE was investigated [57]. A He-Cd laser was used as the excitation source and was operated on its 326 nm UV line for detection of OPA derivatives, and on its blue 442 nm line for detection of NDA and FITC derivatives. Detection limits for phenylalanine tagged with OPA, NDA and FITC are given in Table 3.1. In the case of FITC, minimum detectable amount close to 1 amol (0.2×10^{-9} M) was obtained.

CZE/LIF was performed using the post-column reactor developed by Rose and Jorgenson [42] for fluorescence detection. The flow rates of the labelling reagent were accurately controlled and determined in their experiments. A detection limit of 44 amol (1.2×10^{-8} M) horse myoglobin was obtained with a 50 μm I.D. and 100 μm O.D. electrophoretic capillary combination. A disadvantage of this particular post-column reactor is the inconsistency in performance in terms of peak

TABLE 3.1

DETECTION LIMITS FOR PHENYLALANINE TAGGED WITH OPA, NDA AND FITC (adapted from ref. [57])

Reagent	Minimum detection concentration (10^{-9} M)	Minimum detectable amount (attomoles or 10^{-18} moles)
OPA	20.0	80
NDA	1.6	12
FITC	0.22	1.1

efficiencies, from reactor to reactor as a function of reagent flow rate and reactor distance. However, the reactor allowed one to avoid the multiple peaks which would be observed in pre-column protein analysis.

Wright *et al.* [60] applied CE with laser-induced fluorescence detection to the characterization of marine toxins. A schematic diagram of the instrument is shown in Fig. 3.43. On-column detection was achieved by focusing the 325 nm excitation radiation from a 10 mW He-Cd laser on a bare part of the fused silica column a few centimeter from the ground potential end. The fluorescence emission was collected perpendicular to and coplanar with the excitation beam, filtered using appropriate cut-off and band pass filters. The beam was then focused onto a high efficiency end-on photomultiplier tube. A slit of 0.8 mm ×0.3 mm was placed between the capillary holder and the emission collection lens to reduce noise due to scattered

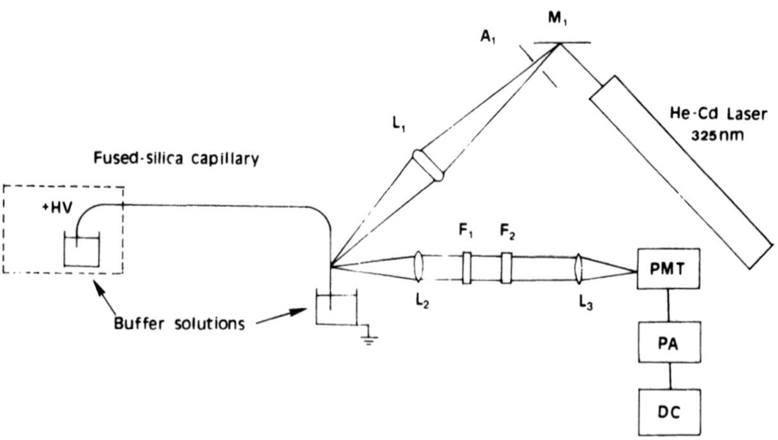

Fig. 3.43. Schematic diagram of the CZE instrument with laser-induced fluorescence detection . *M1* = 325-nm dielectric mirror; 1-mm I.D. iris diaphragm; *L1* = 100-mm focal length, 325-nm laser line lens; *L2* = plano-convex fused-silica lens; *F1* = 400-nm cut-off filter; *F2* = 450-nm cutoff filter or 10 nm narrow bandpass filter at 520 nm; *L3* = fused silica lens; *PMT* = photomultiplier tube; *PA* = picoammeter; *DC* = data collection. (Reproduced from Ref. 60 with permission of Aster Publishing Corp.)

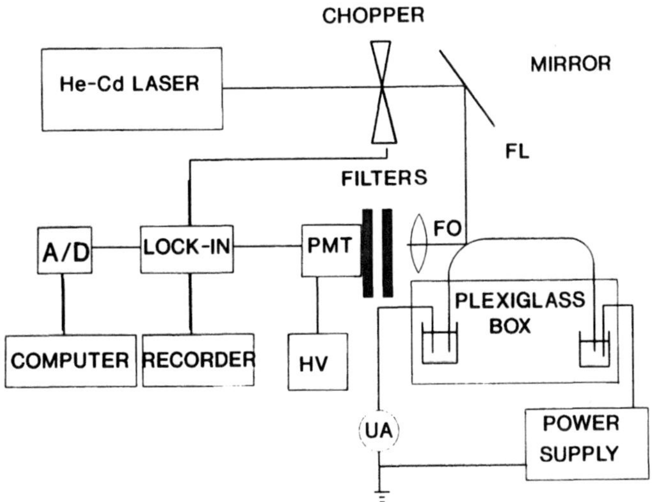

Fig. 3.44 Schematic diagram of the capillary electrophoresis/laser induced fluorescence detection system. *FL* = focal lens; *FO* = fibre optic; *PMT* = photomultiplier tube; *HV* = high-voltage supply. (Reproduced from Ref. 61 with permission of the American Chemical Society.)

light. A picoammeter was used to amplify the photocurrent. Three fluorescent tagging reagents were used to derivatize the primary amine or imine groups of the marine toxins. The three derivatization agents were dansyl, fluorescamine (4-phenylspiro (furan-2(3H)-1′-phthalane)-3,3′-dione and OPA (o-phthaldialdehyde). The detection limit reported for the OPA derivative of saxitoxin was 0.1 amol (3×10^{-11} M).

Novotny and co-workers [61,62] employed a CE/LIF system for the determination of primary amine and amino sugars. A schematic diagram of the instrument is shown in Fig. 3.44. On-column fluorescence measurements were performed with a He-Cd laser as a light source (50 mW power at 442 nm). The on-column optical cell was made by removing the polyimide coating from a short section of the fused silica capillary. Fluorescence emission at 550 nm was collected through a 600 μm fibre optic situated at a right angle to the incident beam. Signal were isolated by a narrow band-pass interference filter. A photomultiplier tube was used to monitor the fluorescence intensity and a lock-in amplifier was used to amplify the signal. 3-(4-carboxybenzoyl)-2-quinolinecarboxaldehyde was used as a derivatizing reagent. Detection limits ($S/N = 3$) in the low attomole range (10^{-9} M) were obtained. The linear dynamic range was over 3 orders of magnitude.

3.4.2.1 Sheath-flow cuvette

Dovichi and co-workers [58,72] reported that stray light can be minimized through the use of sheath-flow cuvette. A schematic diagram of the CE/LIF instrument is shown in Fig. 3.45 and the sheath flow cuvette is shown in Fig. 3.46. Fluorescein

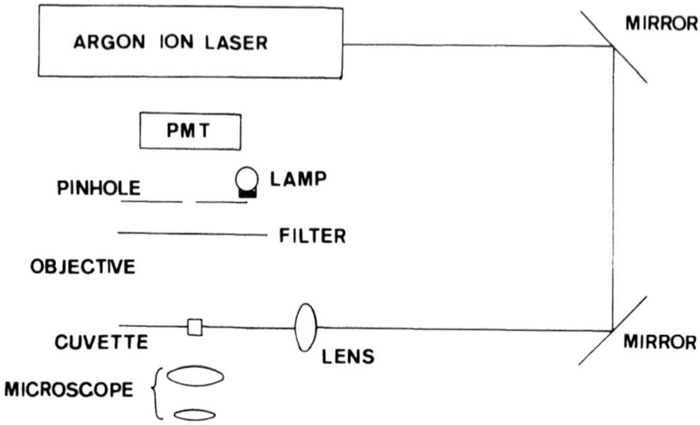

Fig. 3.45. Optical diagram of a laser-induced fluorescence detection system for CE. (Reproduced from Ref. 58 with permission of Elsevier Science Publishers.)

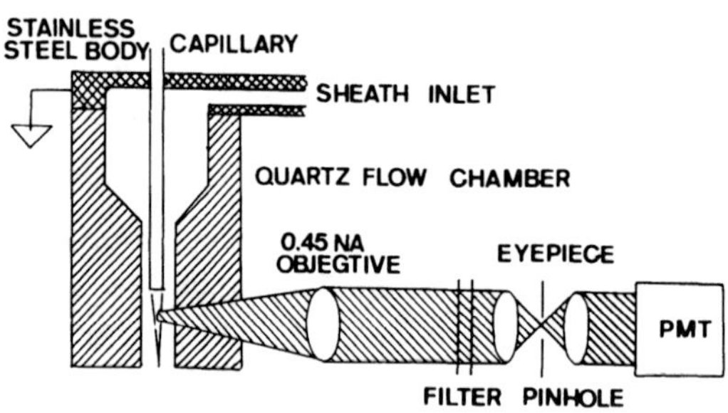

Fig. 3.46. Schematic diagram of a laser-induced fluorescence detection system with sheath flow cuvette. (Reproduced from Ref. 72 with permission of the American Association for the Advancement of Science.)

is excited by a light-regulated argon ion laser beam operating at 488 nm focused to a 10 μm spot about 0.2 mm downstream from the capillary exit. Fluorescence is collected at right angle to both the sample stream and the laser beam with a microscope objective. A long wavelength-pass coloured glass filter (495 nm) was used to block scattered laser light, whereas a short wavelength-pass interference filter (560 nm) was used to block the Raman band of water at 585 nm. A pin hole 200 μm in radius was located in the reticle position of a 20× microscopic objective to restrict the field of view of the photodetector to the illuminated sample stream. Fluorescence is detected with a photomultiplier tube (PMT) that was used for fast response. The end of the capillary was inserted into a square quartz flow chamber 250 μm by 250 μm square. Sheath flow was introduced by a liquid chromatographic pump. The response to fluorescein isothiocyanate (FITC) labelled amino acids was linear over five orders of magnitude and detection limits ($S/N = 3$) of 1.7×10^{-21} mol (1.3×10^{-12} M) FITC-arginine were reported. With this detection system,

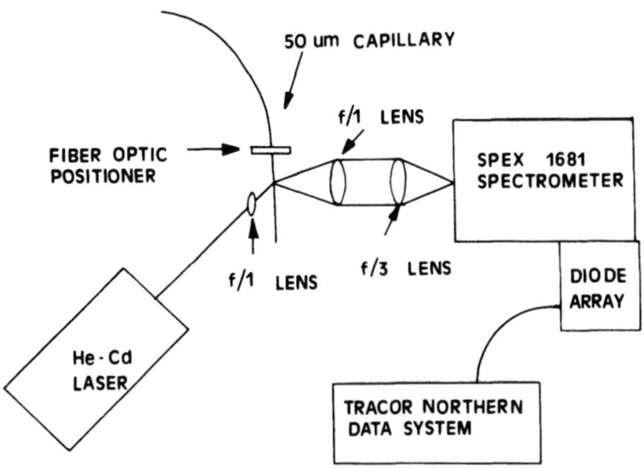

Fig. 3.47. Block diagram of photodiode array detection scheme. (Reproduced from Ref. 49 with permission of Aster Publishing Corp.)

extremely high sensitivity could be achieved. However, it should be noted that the sheath flow cuvette interacts with the electrophoresis, producing back pressure and slowing the separation. Injection should not be performed while the sheath stream is pressurized since the pressure provided by the pump may drive the sample from the capillary tube [72].

3.4.2.2 Fluorometric photodiode array detector

A fluorometric photodiode array detector for CE was developed by Swaile and Sepaniak [49]. A schematic diagram of the instrument is shown in Fig. 3.47. Fluorescence excitation was performed normal to the capillary using a He-Cd laser (30 mW, 442 nm), which is focused onto the on-column flow cell, made by removing a small section of the capillary polyimide coating near its outlet. Emission was collected at a 90° angle from excitation and collimated by a 4 cm diameter, $f/1$ quartz lens, then focused by a 4 cm diameter, $f/3$ quartz lens onto the entrance slits of a spectrometer that dispersed the emission across the diode array. The entrance slit of the spectrometer was set at 1–2 mm, which prevented secular scatter from the sides of the capillary from reaching the photodiode. Detection was accomplished using a photodiode array with 1024 diodes. The diode array was operated at the lowest possible resolution to reduce memory requirement which resulted in a spectral resolution of 4 nm per channel. Calibration data were collected in the histogram mode in order to further reduce memory requirement. Linearity in response over 3 orders of magnitude and detection limits of less than 60 fg ($\simeq 10^{-7}$ M) were obtained for sodium fluorescein.

3.4.2.3 Laser-induced fluorescence detection in capillary gel electrophoresis

The potential of capillary gel electrophoresis with laser-induced fluorescence for rapid, high-resolution DNA sequencing has been demonstrated [63–66]. A

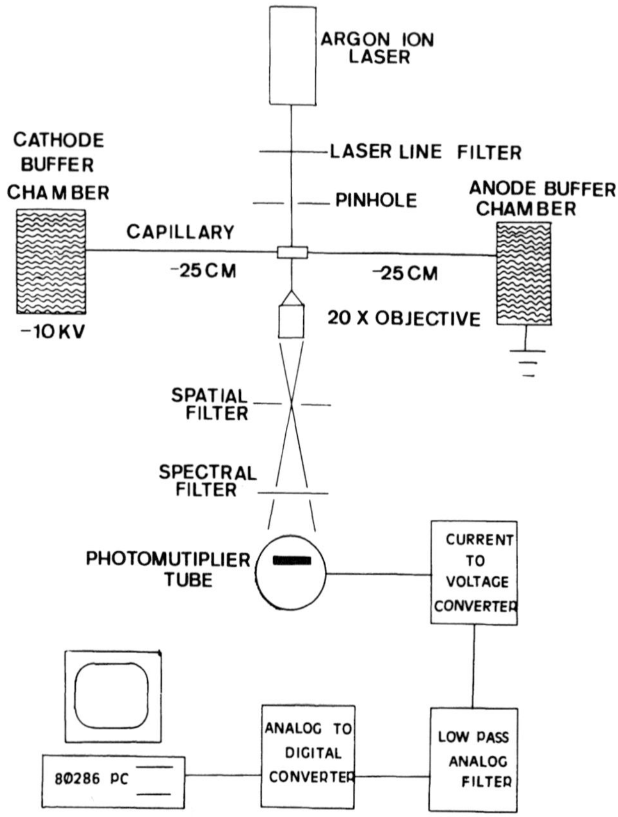

Fig. 3.48. Schematic diagram of the capillary gel electrophoresis apparatus. (Reproduced from Ref. 64 with permission of IRL Press.)

schematic diagram of a typical instrument used for this purpose is shown in Fig. 3.48. An air-cooled argon-ion laser operating at 30 mW single line in the continuous wave mode was used as the excitation source. The 488 nm, 0.8 mm diameter beam was filtered spectrally with a laser line filter (488 nm centre wavelength, 10 nm bandwidth band pass interference filter) and then spatially with a 1.5 nm diameter pinhole. The beam was focused at the centre of the capillary to an approximately 20 μm diameter spot size by a 7.5 mm diameter 18 mm focal length plano-convex lens. This produced an effective detection volume of 30 pl. In the detection region, the polyimide coating on a 1 cm long section of capillary was removed by a fine bunsen-burner flame. The detection region was surrounded by a quartz fluorometer cuvette to reduce scattering due to dust. The laser beam travelled parallel to the surface of the optical table, and the capillary was oriented vertically, but tilted an angle of 25° towards the laser. The tilt angle prevented the majority of scattered laser light from reaching the photomultiplier tube. The light was collected with a 0.4 numerical aperture, 20× microscope objective, oriented horizontally, and at right angle to the laser beam. The objective focused the central illuminated region onto a rectangular aperture which acts as a spatial filter to reduce scattered light which originated from the capillary walls. The light then passed through two colored glass filter and a band pass interference filter (530 nm centre wavelength, 30 nm bandwidth), before entering a light tight box containing the photomultiplier tube. The photomultiplier tube current was converted to a voltage signal by a low-noise operational amplifier circuit. The voltage output was filtered through an analog 1-Hz low-pass filter before entering the analog to digital converter. The acquisition, storage and display of data was accomplished on an 80286-based personal computer using this system. Limit of detection ($S/N = 3$) was 0.01 amol ($\simeq 10^{-11}$ M), which corresponded to approximately 6000 molecules of fluorescently tagged DNA. By using a post-column detector based on the sheath flow cuvette (see Section 3.4.2.1), detection limits ($S/N = 3$) of 2 yoctomoles (ymol) (10^{-21} mol or 10^{-12} M in terms of concentration) were obtained for fluorescently labelled nucleotides [66]. It was shown that the cost of the laser fluorescence detection system could be reduced significantly. For one of the designs, the cost was only several thousand dollars [66].

3.4.2.4 Detection by energy transfer

Garner and Yeung [73] reported the detection of absorbing but non-fluorescing analytes by laser-excited fluorescence. This was possible if the excited analytes transfer their energy to a fluorescent additive in the running buffer to increase the background fluorescence level. Two different fluorophores, F, (fluorescein and riboflavin) and four different absorbing analytes, A, (cresol red, orange G, dabsyl-phenylalanice and dabsyl-glutamine) were tested in this detection scheme. Concentration as low as 6×10^{-9} M and amounts as small as 4 amol were detectable.

A possible energy-transfer scheme is [73]:

$$\mathrm{A} \underset{k_{-1}}{\overset{k_1}{\rightleftharpoons}} \mathrm{A}^* \tag{3.6}$$

$$\mathrm{A}^* + \mathrm{A} \overset{k_2}{\rightarrow} 2\,\mathrm{A} \tag{3.7}$$

$$\mathrm{A}^* + \mathrm{F} \overset{k_3}{\rightarrow} \mathrm{F}^* + \mathrm{A} \tag{3.8}$$

where the asterisks represent the excited forms of the two species. k_1 is the rate of excitation of the analyte, k_{-1} is the non-radiative decay rate of A*, k_2 is the self-quenching rate of A, k_3 is the rate of excitation transfer to the fluorophore.

The rate of formation of the excited analyte is:

$$\mathrm{d}[\mathrm{A}^*]/\mathrm{d}t = k^1[\mathrm{A}] - k_{-1}[\mathrm{A}^*] - k_2[\mathrm{A}^*][\mathrm{A}] - k_3[\mathrm{A}^*][\mathrm{F}] \tag{3.9}$$

The following result is obtained by making use of the steady-state approximation:

$$[\mathrm{A}^*] = \frac{k_1[\mathrm{A}]}{k_{-1} + k_2[\mathrm{A}] + k_3[\mathrm{F}]} \tag{3.10}$$

The relative signal is proportional to [F*]. Hence:

$$\text{relative signal} = k_3[\text{A}^*][\text{F}] = \frac{k_3 k_1[\text{A}][\text{F}]}{k_{-1} + k_2[\text{A}] + k_3[\text{F}]} \tag{3.11}$$

For a fixed fluorophore concentration, one can see that at high analyte concentrations ($k_2[\text{A}] \gg (k_3[\text{F}]+k_{-1})$), the enhanced emission reaches a maximum value. At low analyte concentrations ($(k_3[\text{F}] + k_{-1}) \gg k_2[\text{A}]$), the enhanced emission intensity is proportional to the analyte concentration.

3.4.2.5 Charge-coupled devices

Cheng *et al.* [67] described an instrument combining capillary zone electrophoresis (CZE) separation and charge-coupled device (CCD) fluorescence detection. The experimental configuration of the CZE/CCD system is illustrated in Fig. 3.49. An argon-ion laser producing a 9-mW beam in both 488 and 514 nm lines was used as the excitation source. This laser beam was passed through a 488 nm band-pass filter and then focused with a 20 ×magnification, 10.8 mm focal length lens onto the capillary at a position located 2 cm from the cathodic end of the tube. The fluorescence signal was collected by a lens ($f/1$) and then focused by another lens ($f/2.4$) onto a flat-field polychromator ($f/3$). The polychromator contained a 200 grooves/mm holographically ruled concave grating which formed a

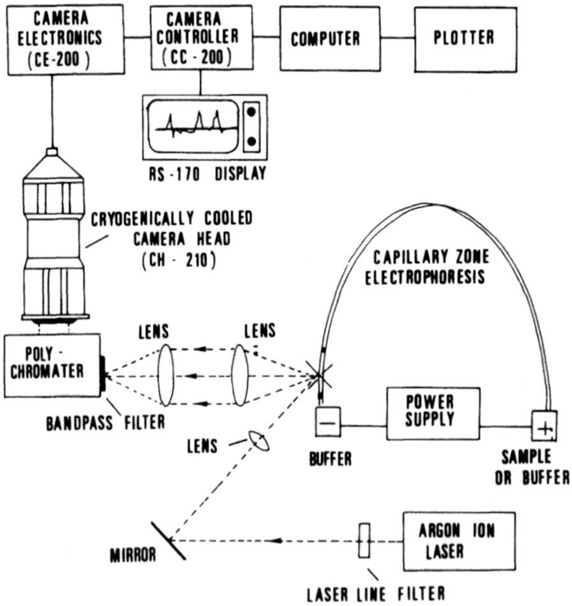

Fig. 3.49. Schematic diagram of system configuration for capillary zone electrophoresis (CZE) with charge-coupled device (CCD) detection. (Reproduced from Ref. 67 with permission of the Society for Applied Spectroscopy.)

planar real image. The reciprocal linear dispersion with this grating was 24 nm/mm dispersion axis within the image plane. The dispersed spectrum was measured with the CCD camera system by locating the camera so that the polychromator plane coincides with the surface of the CCD detector chip. The sample holder, lenses, polychromator and CCD camera head were mounted on micrometer translation stages to allow convenient alignment. The cryostated camera head and CCD chips were cooled to −110°C by liquid nitrogen. A three-dimensional electropherogram was presented [62] to show the spectral and time resolution obtained by injection of 2 fmol of fluorescein into the capillary column. The limit of detection (LOD) for this system was 4 amol of fluorescein ($\simeq 10^{-10}$ M), with a CCD integration time of only 0.2 s. If the fluorescence signal were integrated for 4 s on the CCD, an LOD of less than 1 amol could be achieved for this compound.

Zare and co-workers [68] described a fluorescence detection system for capillary electrophoresis in which a charged-coupled device (CCD) viewed a 2-cm section of an axially illuminated capillary column. The CCD was operated in two readout modes which are illustrated in Figs. 3.50 and 3.51. Figure 3.50 illustrates the snap shot mode which acquired a series of images in wavelength and capillary position. Figure 3.51 illustrates time-delayed integration mode that permitted long exposure times of the moving analyte zones. The ability to differentiate a species based on both its fluorescence emission and migration rate was demonstrated with fluorescein and sulforhodamine 101 [68].

The overall detection system is shown in Fig. 3.52. A detailed schematic of the optical system is shown in Fig. 3.53. The capillary was illuminated end-on to achieve longer exposure and thus high sensitivity with the CCD. The resultant fluorescence from a 2 cm section was imaged onto the CCD. Fluorescence was collected during the entire residence time of the analyte band in this section. Several precautions needed to be taken. The first was to ensure that there is no excessive shadowing effect, which occurs when a leading band absorbs a significant fraction of the

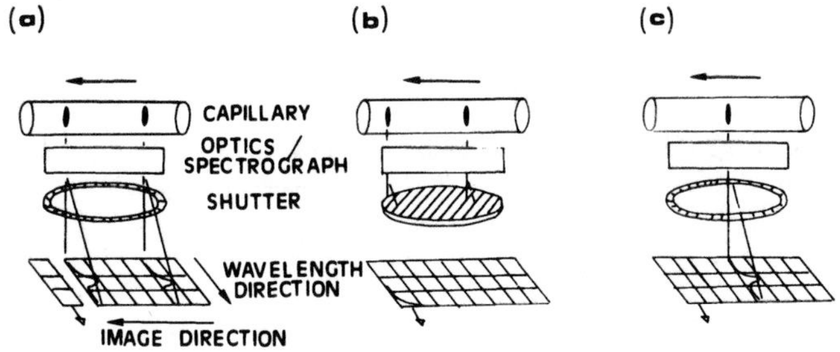

Fig. 3.50. Diagram of the CCD/LIF system illustrating the snapshot mode. (a) The shutter is opened to expose the CCD to the fluorescence of two analyte bands. (b) After exposure, the shutter is closed and the photogenerated charge information is used. (c) When the shutter next opens, the analyte bands have travelled along the capillary, and one band remains in the observation zone. (Reproduced from Ref. 68 with permission of the American Chemical Society.)

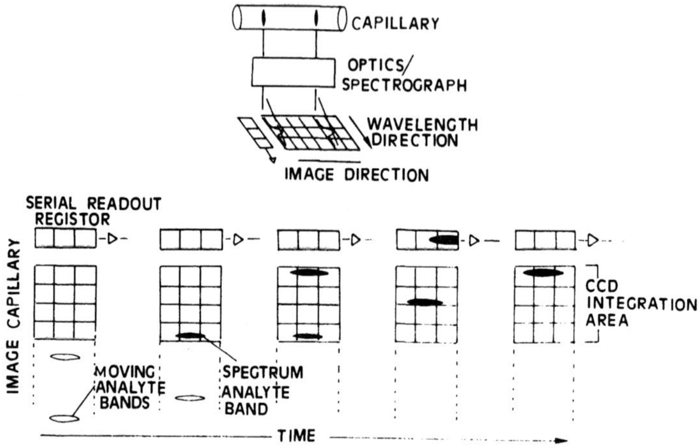

Fig. 3.51. Diagram of the CCD/LIF system illustrating the time-delayed integration (TDI) mode. The CCD is oriented so that the parrel shift direction is the same as the analyte band motion. As the emission of the analyte bands moves across the CCD, the CCD shifts the resultant photogenerated charge information at the same rate. When the photogenerated charge from each band reaches the readout register, the spectrum is read and digitized. (Reproduced from Ref. 68 with permission of the American Chemical Society.)

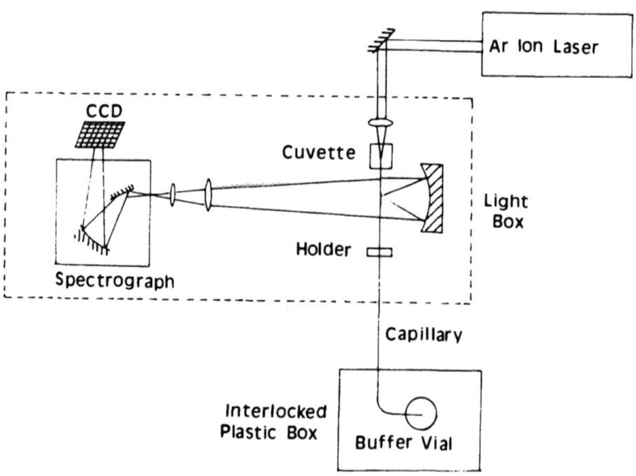

Fig. 3.52. Schematic diagram showing the Ar-ion laser, capillary arrangement, optics, and CCD detector. (Reproduced from Ref. 68 with permission of the American Chemical Society.)

channeled excitation light, thus reducing the fluorescent signal from later bands also resident in the observation zone. Another precaution was to ensure that light travel in the centre of the capillary by the correct choice of laser focusing lens and proper alignment of the laser beam. The optics formed an image of the capillary on the CCD and a spectrograph dispersed the image. Hence the two dimensions of the CCD array contained different information, one contained the image of the

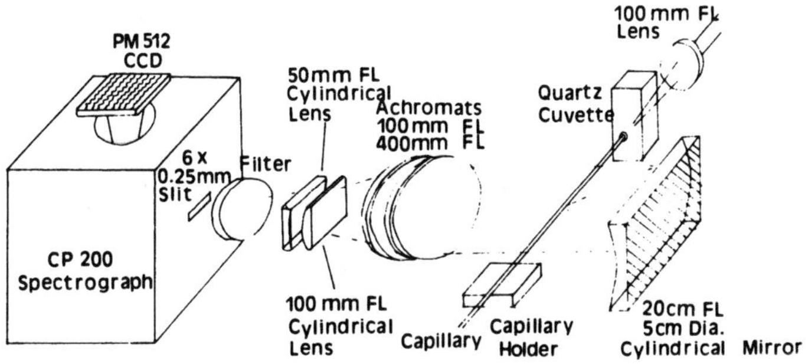

Fig. 3.53. Detailed schematic of the optical system, showing the axially illuminated capillary. The optics from a 250 μm by 6 mm image of a 63 μm section of capillary on the entrance slit of the spectrograph. (Reproduced from Ref. 68 with permission of the American Chemical Society.)

capillary and the other contained wavelength information. The detection system consisted of a 516 by 516 element CCD, controlled with CCD electronics and signal-processing modules. The CCD was contained in a liquid-nitrogen-cooled cryostat to reduce dark current. The detection limit for fluorescein isothiocyanate (FITC) was 1.2×10^{-20} mol, detection limits ($S/N = 2$) for FITC-amino acids were in the $2\text{–}8 \times 10^{-20}$ mol range. In terms of concentration, the detection limits were around 10^{-12} M.

3.4.3 Derivatization

Since most analytes do not fluoresce, pre- or post-column derivatization of the sample with some type of fluorophore allows the extension of fluorescence detection to many analytes. Extensive research has been performed on the derivatization of amino acids [65,73,76–100] for fluorescence detection. Many fluorophores have been investigated including dansyl (DNS) [41,48,51,60,85,88,90], o-phthaldialdehyde (OPA) [59,60,78,86,92], naphthalene-dialdehyde (NDA) [56,69], fluorescein isothiocyanate (FITC) [58,71,72], fluorescamine [60], and phenylthiohydantoin derivatives (PTH) [98], 4-chloro-7-nitrobenzofuran (NBD) [79,99] has been used for the derivatization of *n*-alkylamines. 2-aminopyridine has been used for the derivatization of carbohydrate [100]. The relative sensitivities of OPA, FITC and NDA derivatization procedure have been compared by Nickerson and Jorgenson [57]. The same authors also reported an extremely rapid and efficient separation (75 s) of 8 NDA-labelled amino acids (<0.5 s peak width) in short capillaries (35 cm $\times$10 μm I.D.) with high field (30 kV). Albin *et al.* [43] evaluated various derivatizing agents and techniques for fluorescence detection of amino acids in CE. Pre-capillary derivatization using FITC, OPA and 9-fluorenylmethylchloroformate (FMOC) and post-capillary derivatization using OPA were investigated. Limits of detection (LODs) of 10 and 60 ng/ml ($\simeq 5 \times 10^{-8}$ and 3×10^{-7} M) were obtained for pre-capillary derivatization with FMOC and post-capillary derivatization with OPA respectively.

3.5 LASER-BASED THERMO-OPTICAL AND REFRACTIVE INDEX DETECTORS

3.5.1 Thermo-optical absorbance detectors

By focusing a laser beam onto a capillary containing a solution, the interaction of the laser beam and the capillary results in many dark and light fringes. The position of the fringes changes with the refractive index of the solution. A temperature rise in the sample solution causes its refractive index to change. Typically a temperature change of a few millidegrees results in a change in refractive index (ΔRI) of 10^{-6} units. The change in fringe position due to temperature change can be monitored by a photodetector. In thermo-optical detection techniques, absorbance of the radiation from a high-power pump laser produces a temperature rise within the sample. The temperature rise is proportional to the intensity of the pump laser and the absorbance. A probe laser beam propagating through the heated region can be used to monitor the change in the refractive index of the solution. If a modulated pump laser beam is used, absorbance of the pulsed energy causes periodic rises in temperature and hence refractive index, causing the probe beam to be deflected periodically. A lock-in amplifier driven at the frequency of modulation can be used to demodulate the periodic signal to permit the measurement of sample absorbance. In addition, the unmodulated or dc component of the signal is related to the refractive index of the sample. Therefore, both selective detection based on absorbance measurements and universal detection based on refractive index changes can be performed simultaneously in such a system.

Dovichi and co-workers [81–83,101] exploited the high-intensity and focus ability of lasers to construct a thermo-optical absorbance detector. In this approach, absorption of the pump beam produced a thermal gradient, which deflected the probe beam. The displacement of the probe beam was measured and could be related to the concentration of the analyte. A schematic diagram of the detection system is shown in Fig. 3.54. In the initial design, a He-Cd pump laser (442 nm, 2.8 mW) and a He-Ne (632.8 nm, 1 mW) probe laser crossed on-column in 50 μm I.D. CE capillary. By changing the pump laser to a higher power argon ion laser (458 nm, 130 mW), higher sensitivity could be obtained.

The He-Ne probe laser was focused with a 16-mm-focal-length microscope objective. The pump beam was chopped mechanically at 47 Hz and focused with a 18-mm-focal-length microscope objective into the capillary. The probe beam intensity was monitored with a small-area silicon photocell. The output of the photocell was demodulated with a dual phase lock-in amplifiers operating in the amplitude mode with a 1-s time constant. To perform refractive index measurements, a dc voltmeter was connected to the output of the photocell. Detection limits (S/N = 3) of 40 amol ($\simeq 6 \times 10^{-8}$ M) of methionine were obtained with the 130 mW argon ion pump laser [99].

Bruno *et al.* [102,103] described another thermo-optical absorbance detector which employed UV light at 257 nm from a frequency doubled argon ion laser as

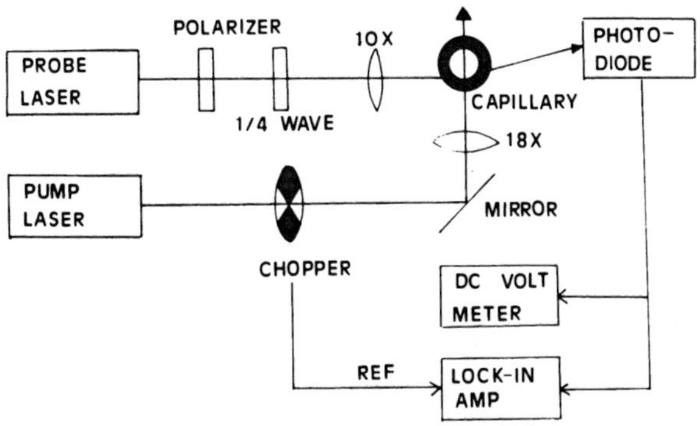

Fig. 3.54. Schematic diagram of thermo-optical detection system for CE. The probe laser is a 1-mW, polarised He-Ne laser. The pump laser is a 2.8-mW, linearly polarised He-Cd laser operating at 442 nm. (Reproduced from Ref. 101 with permission of the American Chemical Society.)

the pump beam to induce refractive index changes inside a capillary tube. Detection limits obtained were around 10^{-5} M ($S/N = 2$) for saccharose and dansylated amino acids.

3.5.2 Refractive index detectors

Morris and co-workers [104,105] applied analyte velocity modulation to improve S/N by modulating the voltage applied across the column. Detection with a laser beam deflection refractive index system may be subject to limitations due to thermal effects, laser beam wander, and other sources of low-frequency drift. To minimize these effects, one approach is to modulate the electrophoretic flow [104]. In this approach, an ac voltage (V_{ac}) is superimposed onto the dc separation voltage (V_{dc}). Consequently, the flux of analyte i (J_i) is given by [104]:

$$J_i = -\frac{\mu_i V_{dc}}{L}\left(1 + \frac{V_{ac}}{V_{dc}} - \cos\omega t\right) C_i\,(x,t) \tag{3.12}$$

where μ_i, C_i, x and t are the mobility and concentration of analyte i, distance from the injection point and time, respectively. L is the length of the capillary and ω is the frequency of modulation.

The time dependence of concentration is:

$$\frac{\partial C_i}{\partial t} = -\frac{\partial J_i}{\partial x} = \frac{\partial C_i\,(x,t)}{\partial x}\,\frac{\mu_i V_{dc}}{L}\left(1 + \frac{V_{ac}}{V_{dc}}\cos\omega t\right) \tag{3.13}$$

By passing the detector output through a lock-in amplifier, operated at the modulation frequency, the component of the output which is synchronous with the modulation voltage can be obtained as follows:

$$V_{obs} = a\mu_i\,\frac{V_{dc}}{L}\,\frac{\partial C_i}{\partial x}\,\frac{V_{ac}}{V_{dc}} \tag{3.14}$$

where V_{obs} is the output of the lock-in amplifier, and a is a constant related to the response factor and the gain of the lock-in amplifier. V_{obs} depends on the concentration gradient $(\partial C_i/\partial x)$, and to the depth of the modulation (V_{ac}/V_{dc}). It was found that the effects of refractive index changes, capillary movement and laser beam wander could be minimized by modulation of the electrophoretic flow at frequencies up to 400 Hz, with a modulation depth of 55% [104]. A schematic diagram of the instrument for capillary zone electrophoresis with an analyte velocity modulation system is shown in Fig. 3.55. The laser beam deflection refractive index detector is illustrated in Fig. 3.56.

The detector consisted of a 3 mW, 750 nm diode laser and a position-sensing photodiode. The laser was focused into the capillary with a 5× microscope objective. The laser beam was focused onto the exit wall of the capillary. Proper focus was obtained when an extended diffraction pattern was observed. The diffraction pattern was centered to focus the laser in the centre of the capillary. A razor blade was inserted into the beam about 1 m beyond the capillary. The blade was adjusted to block half the beam, so that the expansion of the remaining half was observed as a deflection. The half beam was focused on the position-sensing photodiode with a 200 mm focal length lens. Using this system, a detection limit of 10^{-7} M $(S/N = 3)$ for carbonic anhydrase could be obtained.

Pawliszyn [84] described the theory and a practical demonstration of a nanoliter refractive index detector based on a single He-Ne laser employing Schlieren optics which produces a differential signal free from drift. The light beam propagates through the cell and is deflected as a result of refractive index gradient in the cell.

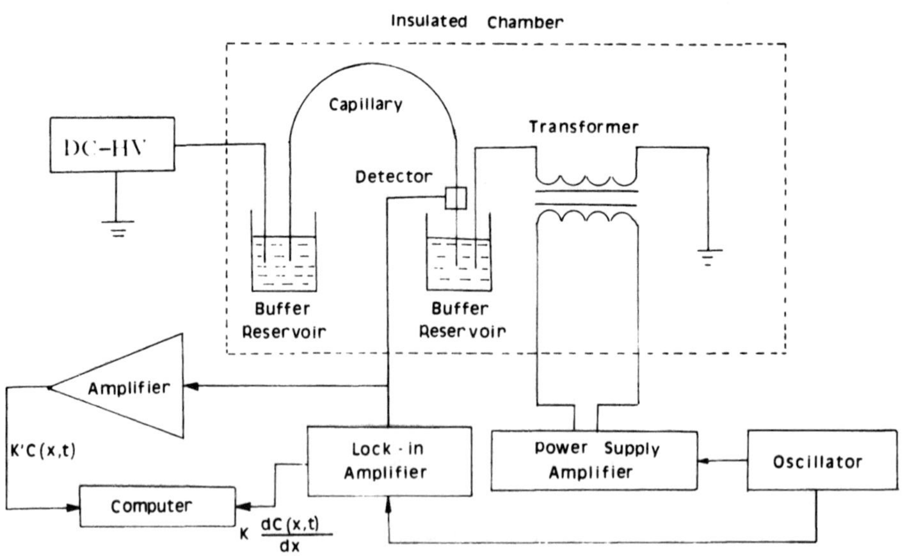

Fig. 3.55. Instrument for capillary zone electrophoresis with an analyte velocity modulation system. (Reproduced from Ref. 104 with permission of the American Chemical Society)

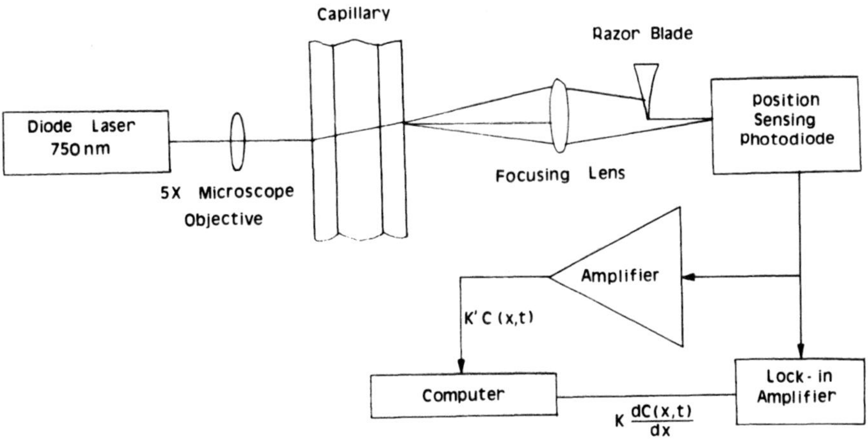

Fig. 3.56. Laser beam deflection refractive index detector. (Reproduced from Ref. 104 with permission of American Chemical Society.)

Since the deflection is proportional to the refractive index gradient, the signal does not depend on concentration directly and hence it is less vulnerable to drift due to thermal fluctuations.

Figure 3.57 shows a schematic diagram of the optical system. Two focused beams propagating close to each other were generated from a single laser beam (a He-Ne laser emitting at 633 nm). A single photodiode was used for the differential detection of the laser beams. Each of the beams was intercepted by a razor blade, so that only half of the beam was allowed to reach the photodiode. The system was capable of compensating for the pointing stability of the laser and mechanical vibrations. No photodiode current change was observed when the laser beam was moved about a micrometer in the vertical direction, since net loss of one beam intensity reaching the detector was compensated by an increase in the light from the other. The electrophoretic apparatus is shown in Fig. 3.58. Separation was performed in 50 cm of 50 μm fused silica capillary. An appropriate length of 10 μm

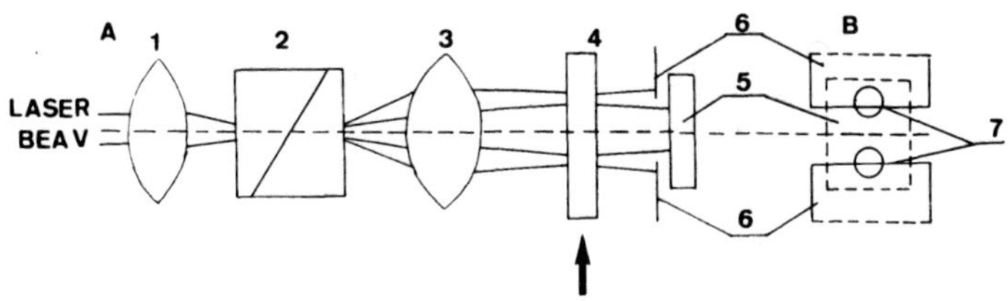

Fig. 3.57. Experimental optical arrangements. (A) Two-beam optical system and sequential differential detector arrangement. (B) Differential detection using single photodiode. *1*, *3* = lenses; *2* = Wallaston prism; *4* = flow detector cell; *5* = razor blade; *7* = laser beams.

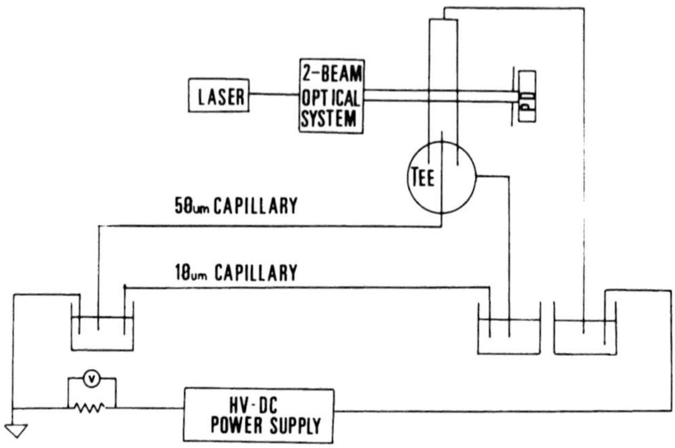

Fig. 3.58. Schematic of capillary zone electrophoresis experimental arrangement with refractive index detection. (Reproduced from Ref. 84 with permission of the American Chemical Society.)

fused silica capillary, which acted as a voltage drop resistorwas used to generate a sheath flow electrokinetically. The signal produced was linear over 4 orders of magnitude with a detection limit of 7×10^{-7} M for sucrose.

3.5.3 Laser-induced fluorescence detected circular dichroism detection

Christensen and Yeung [75] demonstrated laser-induced fluorescence-detected circular dichroism (FDCD) as on-column detection method for CE. This technique takes advantages of the fact that chiral molecules exhibit circular dichroism that varies in relation to molecular conformation. Detection of optical activity was achieved by using alternating left and right polarized light at equal intensity produced from a modulated laser source. Since fluorescence (FL) can be measured simultaneously using this detection, The technique can be used to identify a compound on the basis of its unique FDCD/FL signal ratio in certain cases. A schematic diagram of the detection system is shown in Fig. 3.59. An argon ion laser operating at 488 nm was the excitation source. The linearly polarized laser light was converted into alternating left and right circularly polarized light (LCPL, RCPL) by an electro-optic modulator. The modulated light was focused onto the detection region of the capillary by a short focal length lens, and collection of fluorescence was at 90° to the incident beam with a 10× microscope objective. An enlarged image of fluorescence from within the capillary was focused on the photomultiplier tube (PMT). Two sets of spatial filters and colour filters were mounted in a blackened tube before the PMT to selectively pass fluorescence while rejecting scattered laser light and room light. The column was mounted for normal incidence, and the angle between the column and the collection optics was adjusted to minimize the amount of scattered light collected while retaining normal incidence at the capillary. The fluorescence signal from the PMT was converted to a voltage and sent to the lock-in

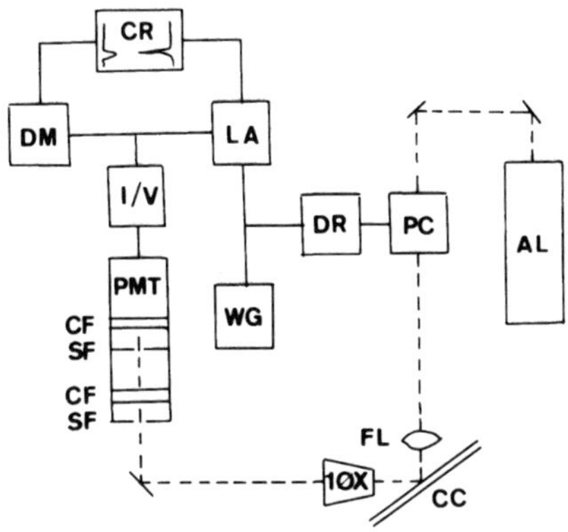

Fig. 3.59. Schematic diagram of FDCD and FL detection system for capillary electrophoresis. *AL* = argon ion laser; *PC* = pockeis cell; *DR* = driver; *CR* = dual-pen chart recorder; *FL* = focusing lens; *CC* = capillary column; 10× = microscope objective; *SF* = spatial filters; *CF* = color filters; *PMT* = photomultiplier tube; *I/V* = current-to-voltage converter. Solid lines indicate electrical connections, and dashed lines show the optical path. (Reproduced from Ref. 75 with permission of the American Chemical Society.)

amplifier and a digital multimeter simultaneously. The demodulated signal from the lock-in amplifier (1-s time constant) gives the FDCD signal, and the digital multimeter displays the average total fluorescence signal. The absolute limit of detection ($S/N = 3$) for riboflavin by CE/FDCD was 0.2 fmol or 0.07 pg. The concentration limit of detection was 1×10^{-6} M.

3.5.4 Laser Raman detection

Chen and Moris [105] demonstrated a Raman spectrometric detector for CE. In this type of detection system, the sample with Raman active vibrational frequencies are excited by an intense, monochromatic light source, and the frequency of the light scattered from the sample is analyzed with a grating monochromator and detected by a photomultiplier or a multichannel detection system. With a multichannel detector, such as a charge-coupled device (CCD), vibrational spectra can be obtained which provide structural information of the analyte. Therefore, the technique has the advantage that specificity of vibrational spectroscopy can be employed for selective detection. Non-resonance Raman spectroscopy can be used to detect molecules which exhibit strong Raman bands, e.g. water-soluble aromatic molecules. Detection sensitivity can be improved by using resonance Raman spectroscopy.

A system equipped with a single stage monochromator and a photomultiplier has been used as detection system for CE [105]. The conventional Raman spectrometer was coupled to the CE capillary via optical fibres, yielding detection limits of 2.5×10^{-6} M methyl red or rhodamine 6-G. A schematic diagram of the instrument

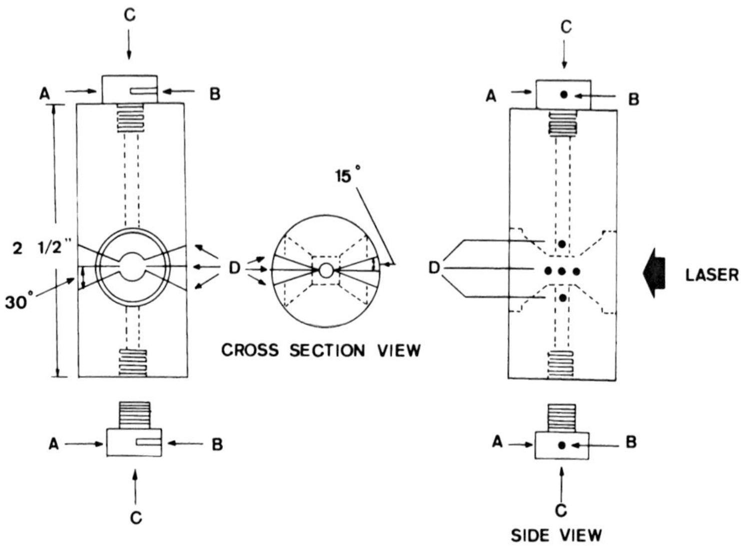

Fig. 3.60. Holder for electrophoresis capillary. Front, cross-section and side views. A = threaded cap; B = set screw; C = bore for capillary; D = holes for optical fibres. (Reproduced from Ref. 105 with permission of American Chemical Society.)

is shown in Fig. 3.60. The electrophoresis capillary was held in a Lucite cylinder. A 5× microscope objective focuses the exciting laser beam directly on the capillary. The signal was gathered by an array of ten 200 μm quartz optical fibres. The fibres are inserted into holes in the Lucite cylinder to form an array of gathering fibres. The fibres were arranged at 15° in the plane normal to the plane polarization of the laser beam, and at 30° above and below the central fibre. The distal ends of the fibres were clamped between two lucite plates and form a vertical line to match the monochromator slit. The Raman spectrometer was equipped with a 0.5 m monochromator, home-built forward-optics and a photomultiplier tube operated in the dc current mode. Interference filters were placed in the exciting laser light path to eliminate spontaneous emission and in the monochromator entrance optics to reduce the effects of light scatter. The excitation source was a 40 mW He-Cd laser, operated at 442 nm. The laser output was chopped at 23 Hz. The photomultiplier current was processed through a lock-in amplifier operated with a 1 s output time constant and 12 dB/octave filter. Measurements were made at 471.2 nm, approximately 1410 cm^{-1}.

It was suggested that further sensitivity increases are possible [105]. Scattering can be reduced by utilizing a double or triple monochromator, and by enclosing the detection zone of the capillary in a refractive index-matching fluid. A cooled photomultiplier operated in the photon-counting-mode would increase the signal-to-noise ratio.

In order to exploit more fully the fingerprinting capability of vibrational spectra, an on-line multichannel Raman spectroscopic detection system for capillary

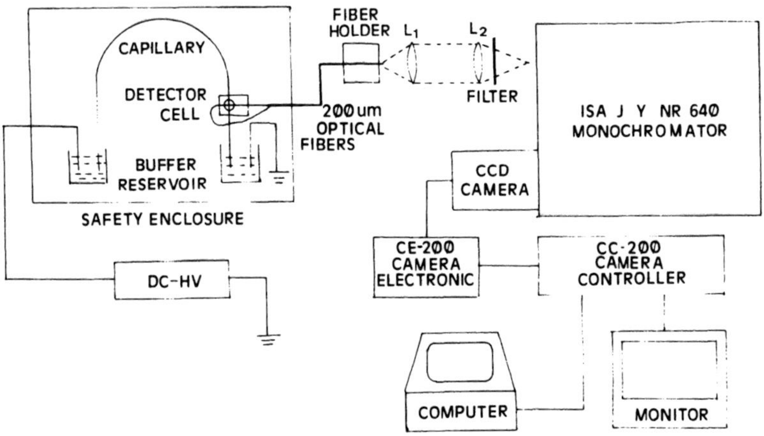

Fig. 3.61. Instrument for capillary zone electrophoresis with a CCD Raman spectroscopic detector. L_1, f = 50 mm; L_2, f = 200 mm; Filter, Schott OG-550. (Reproduced from Ref. 106 with permission of Elsevier Science Publishers.)

electrophoresis using a charge-coupled device as the detector was developed [106]. The detection system is shown in Fig. 3.61. The excitation source was a Nd-YAG laser operated at 532 nm. The laser power at the capillary was about 300 mW. A 5× microscope objective focused the exciting laser beam directly on the capillary. The Raman signal was gathered by two 200-μm quartz optical fibres. These fibres were placed directly against the capillary and normal to the plane of polarization of the exciting laser beam. The distal ends of the fibres were oriented parallel to the spectrograph entrance slit in a Plexiglas holder. A collimating lens, L_1 (f = 50 mm) and focusing lens L_2 (f = 200 mm) were used to couple the fibre output into the spectrograph. The signal collected by the fibres was filtered by a sharp-cut filter to attenuate the laser line scattering.

The Raman spectrometer consisted of a 0.64 m spectrograph with a 1200 grooves/mm holographic grating, operated at $f/5.7$ and a cryogenically cooled charge-coupled device detector . The laser was aligned to the capillary by centering the diffraction pattern obtained by focusing it into the capillary. The collection fibre alignment to the spectrograph input optics were adjusted using intense fluorescence of Rhodamine 6G. For Raman spectroscopy the entrance slit was 300 μm, to give spectral resolution of 12 cm^{-1}. The CCD detector chip had 512 × 512 pixels, each pixel 20 μm square. A Raman spectrum was obtained by binning the CCD vertically to form a 512 × 1 data array. A personal computer was used for the control of the detector, data acquisition and spectra storage. Figure 3.62 shows a time-series of on-line Raman spectra obtained during electrophoresis of 5 × 10^{-4} M methyl orange and methyl red in Trizma buffer at pH 7.2.

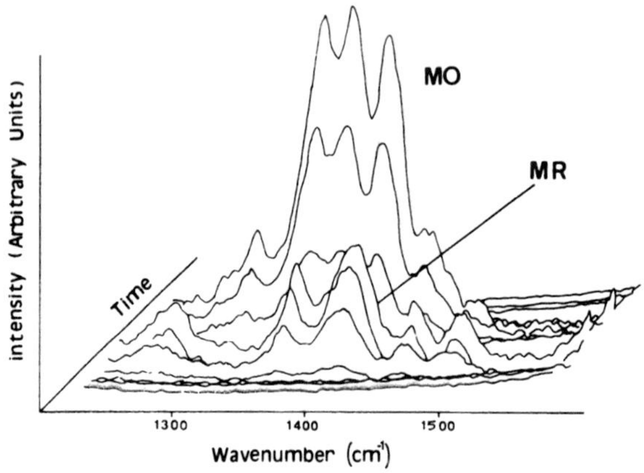

Fig. 3.62. Time series Raman spectra of 5×10^{-4} M methyl orange (MO) and methyl red (MR) in 0.02 M Trizma buffer, pH 7.2. CCD integration time of 5 s for each spectrum. Sample injection: 20 kV, 5 s; CZE at 20 kV; laser power: 250 mW. (Reproduced from Ref. 106 with permission of Elsevier Science Publishers.)

3.5.5 Laser-induced capillary vibration detection

Wu *et al.* [107] described a laser-induced capillary vibration method (CVL) for detection in CE. In this approach, an intensity-modulated laser beam is used to irradiate the midpoint between two fixed points of a capillary, which is supported at these points. The absorption of the excitation radiation results in local temperature expansion at the point of irradiation. Consequently, tension fluctuation occurs in the capillary, which causes the capillary to vibrate like a string. The amplitude of vibration is proportional to sample absorbance. If a probe beam is passed just over the point of irradiation of the excitation beam, the deflection of the probe beam can be utilized for the measurement of sample absorbance. The method has the advantage that the probe beam does not pass through the capillary, and hence the limitation on sensitivity due to a small path length can be overcome.

The experimental arrangement employed is shown in Fig. 3.63. The CE capillary was fixed by two holders, and a tension of 16 g was applied by using a weight before the capillary was fixed in place. An argon ion laser was operated either at 476 nm (40 mW) or at 257.2 nm (5–8 mW) by frequency doubling the 514.5 nm lasing line. The excitation beam was modulated by a mechanical chopper, and the modulation frequency was set at 710 Hz to obtain desirable signal-to-noise ratio. The excitation beam was focused on the capillary by a microscopic objective lens assembly that resulted in a 20-fold magnification for the visible laser beam. For the UV beam, a 30 mm focal length quartz lens was used for focusing. The probe-beam was a He-Ne laser beam. It was focused just above the capillary by a 100 mm focal length lens. Two example electropherograms obtained using this detection system are shown in Fig. 3.64. The sample was riboflavin and the 476-nm argon lasing

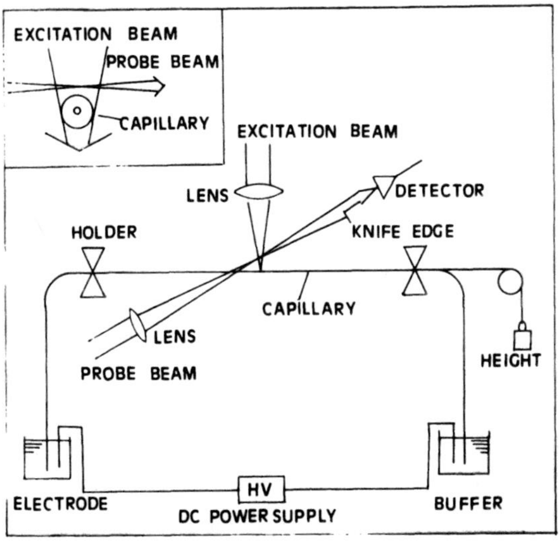

Fig. 3.63. Experimental arrangement of the CZE/CVL system. Insert: vertical arrangement of the capillary, probe beam, and the excitation beam. (Reproduced from Ref. 107 with permission of the American Chemical Society.)

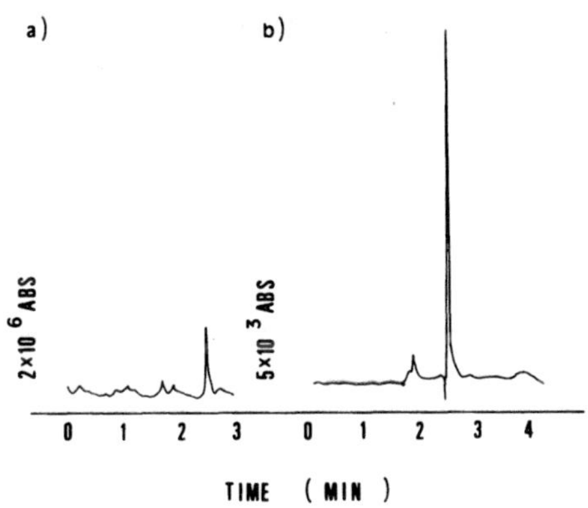

Fig. 3.64. Electropherogram of riboflavin: (a) 2×10^{-4} M; (b) 2×10^{-3} M. (Reproduced from Ref. 107 with permission of the American Chemical Society.)

line was used for visible excitation. The peak heights show good linearity for the sample concentration of 2×10^{-6}–2×10^{-3} M and the lower limit of detection was 1.8×10^{-7} M.

3.6 ELECTROCHEMICAL DETECTION

Electrochemical detection schemes employed in capillary electrophoresis [8,108–126] include methods based on potentiometric measurements [112], conductivity detection [8,108–111,113–117] and amperometry [118–126]. The main challenge in the development of electrochemical detectors for CE lies in the isolation of the high-voltage drop across the separation capillary from the detection system. With suitable designs, relatively universal [8,108–117] or highly sensitive [118–126] detection could be achieved.

3.6.1 Potentiometric detection

In potentiometric detection, the Nernst potential at the surface of an indicator electrode or across an ion-selective barrier, e.g. membrane, is measured.

A potentiometric detector was employed by Virtanen [127] as early as 1974 for detection in electrophoretic separations in relatively large bore tubes. A Ag/AgCl-coated platinum wire was used as the electrode. The electrode was inserted into the end of a capillary tube which was drawn to a diameter of less than 200 μm at one end to serve as a salt bridge. The end of the electrode capillary was inserted into the separation tube. Changes in the electrical potential at the indicating electrode were followed by means of a Wheatstone bridge circuit. This detection system was used to detect eluting zones of small inorganic ions such as K^+, Na^+ and Li^+.

Haber *et al.* [112] described a potentiometric microelectrode as an end-column detector in CE. The microelectrode was placed a few micrometers behind the capillary end. Due to its high internal resistance (10^8–10^{11} Ω), special devices to decouple the potentiometric detector from the electrophoretic current were not necessary. The composition of the liquid membrane of the microelectrode was designed to show a good response for a number of cations except magnesium, which was used as a background electrolyte. With this detector, alkali and alkali earth metals were detected down to concentrations of 10^{-8}–10^{-7} M.

For monovalent ions i and j, the height of peaks in an electrophoretic run arises from the relative difference of the electromotive force (ΔV) between the eluting component and the background electrolyte [121]:

$$\Delta V = 2.303\,\frac{RT}{ZF}\log\left[\frac{(a_i - a_j) + k_{ij}\,a_j}{a_i}\right] \tag{3.15}$$

where Z is the charge of the ions i and j, a_j is the activity of the sample ion j, a_i is the activity of the background ion i, and k_{ij} is the selectivity factor.

Two detector modes could be used: (1) the direct method ($k_{ij} > 1$), by measuring the emf signals directly against a background ion having a low selectivity relative to the sample, and (2) the indirect method ($k_{ij} < 1$) by measuring negative emf signal from a displacement of background ions, having a high selectivity relative to the sample ions.

A schematic diagram of the capillary electrophoresis system with potentiometric detection is shown in Fig. 3.65. The position of the microelectrode and the platinum wire at the column end is shown in Fig. 3.66. The best position of the tip of the microelectrode was found to be several microns beyond the end of the capillary.

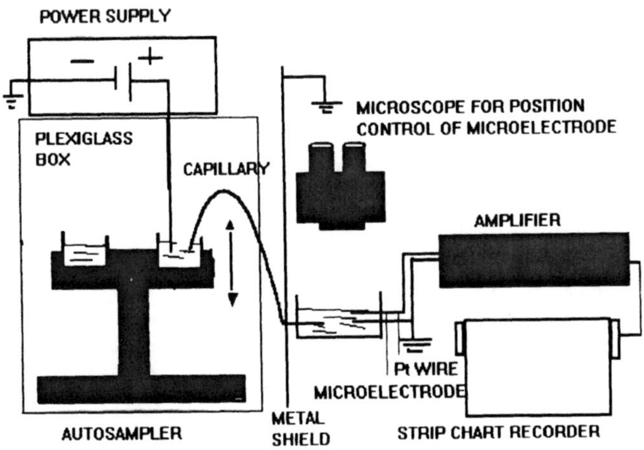

Fig. 3.65. Capillary electrophoresis system with potentiometric detection. (Reproduced from Ref. 110 with permission of Schweiz Chemika Verband.)

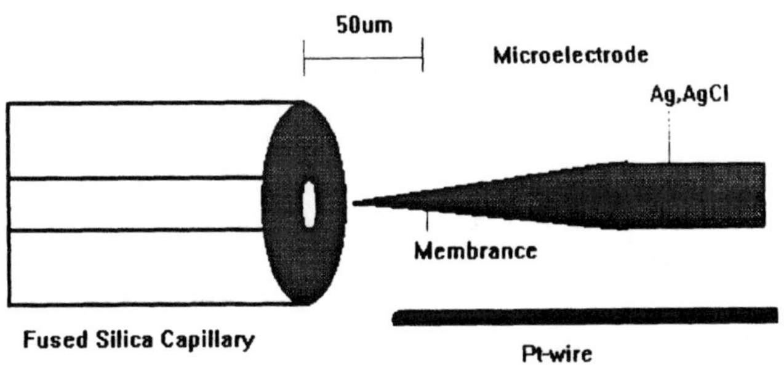

Fig. 3.66. Position of the microelectrode and the Pt wire at the column end. (Reproduced from Ref. 110 with permission of Schweiz Chemika Verband.)

The membrane cocktail consisted of 1% (w/w) solution of the neutral ionophore bis(N,N-diphenyl)-1,2-phenylenebis(oxy-2,1-ethanediyl) bis(oxyacetamide) together with 68.5 mol% (relative to the ionophore, 100%) of potassium tetrakis(4-chlorophenyl)borate in 2-nitrophenyloctyl-ether. With this membrane, by choosing magnesium acetate as buffer electrolyte, the mobile phase showed an extremely low potential and the eluting sample zones of different cations could be detected with the sensitivity according to their relative selectivity ratios to magnesium. The internal filling electrolyte consisted of $MgCl_2$. The concentration of magnesium was chosen to be the same as in the background electrolyte. In this way, diffusion processes across the membrane phase were eliminated and a constant baseline could be obtained. To prepare the ion-selective microelectrode, the micropipette was filled with a 20 mM solution of $MgCl_2$. A slight pressure onto the back-end of the glass body was applied to fill the tip with solution. Then the front of the micropipette was dipped into the membrane cocktail. By applying a short vacuum

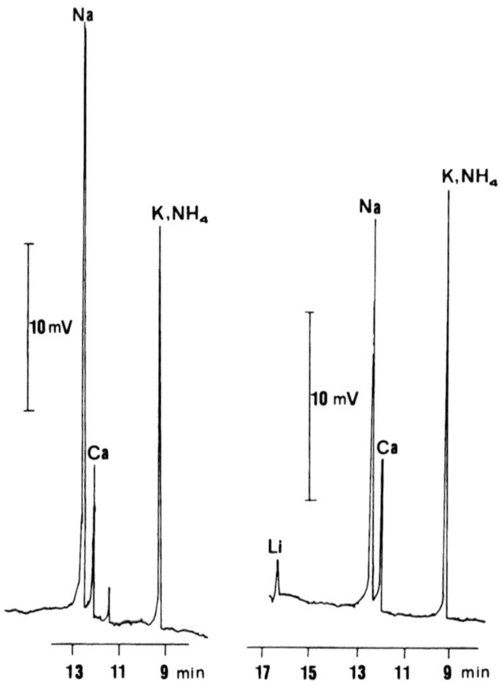

Fig. 3.67. Free-zone electropherogram of deionized (left) and doubly distilled water (right) at pH 5.14. Capillary I.D.: 25 μm; length: 0.99 m; buffer: 20 mM magnesium acetate (HCl). Injection electrokinetically, 5000 V for 5 s; potential 20 kV; detection post-column. (Reproduced from Ref. 110 with permission of Schweiz Chemika Verband.)

pulse onto the back-end of the glass body, the membrane phase was sucked into the tip. The length of the filled zone was between 100 and 300 μm. Figure 3.67 shows a free zone electropherogram of deionized (left) and doubly distilled water (right) obtained with this type of electrode at pH 5.16.

3.6.2 Conductivity detection

Typically in conductivity detection, solution conductivity is measured by placing a pair of electrodes in the capillary and measuring the current passing between the electrodes as a function of potential. One of the main advantages of conductivity detection is its universality. It is particularly useful for species not readily detected by UV absorption. A less obvious advantage of conductivity detection is that by using an internal standard, it is possible to obtain quantification of the components in a mixture on an absolute basis without reference to a detection response curve for each component [108].

Mikkers *et al.* reported the use of a conductivity detector in instrumentation adapted from isotachophoresis for use in large capillaries [109]. The separation of organic and inorganic anions in 200 μm PTFE capillaries was demonstrated. Deml *et al.* [110] investigated some of the problems associated with the use of conductivity detection in CE. Foret *et al.* [8] developed off-column conductivity detector based on a commercial instrument. Detection limits of 10^{-6} M for Cl^-, SO_4^{2-} and NO_3^- were obtained.

Beckers *et al.* [111] described a dual conductivity detection system for the measurement of mobilities in zone electrophoresis. Large capillaries of 250 μm I.D. were used in this study. The ionic species were detected by two detectors mounted at a fixed distance from each other. Since in zone electrophoresis the velocity of an ionic species is proportional to its mobility, the time needed for an ionic species to pass both detectors is inversely proportional to the mobility, provided that the electric field strength is constant and there is no electroosmotic flow. From the ratio of the times required by a sample ionic species and a standard ionic species (with a known mobility) to pass from the first to the second detector, the mobility of the sample ionic species can be calculated.

On-column conductivity detection was first reported by Zare and co-workers. They developed several versions of a conductivity detector for CE [108,113–115]. In an early version, the on-column conductivity cell was constructed by fixing platinum wires through diametrically opposite holes in 50 or 75 μm I.D. capillary tubing. A computer-controlled CO_2 laser was employed for making the 40 μm I.D. holes. A platinum wire of 25 μm O.D. was placed into each of the two holes to serve as electrodes. The wires were aligned under microscope to be exactly opposite to each other, in order to minimize the potential difference between these electrodes when a high electric field strength was applied. The platinum electrodes were then secured in the capillary by first applying poly(ethyleneglycol) as a temporary adhesive and

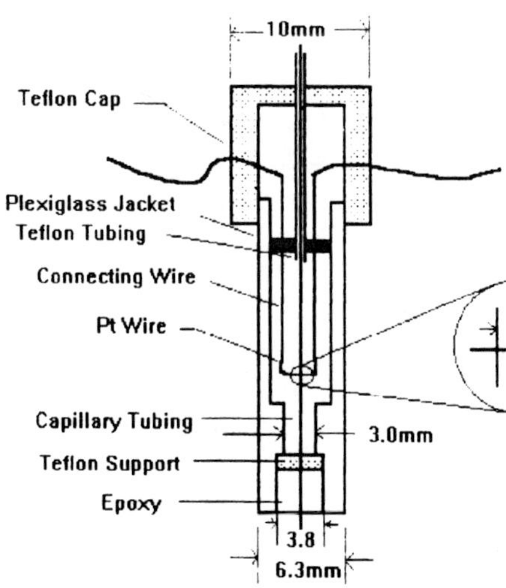

Fig. 3.68. Diagram of the conductivity cell. (Reproduced from Ref. 113 with permission of the American Chemical Society.)

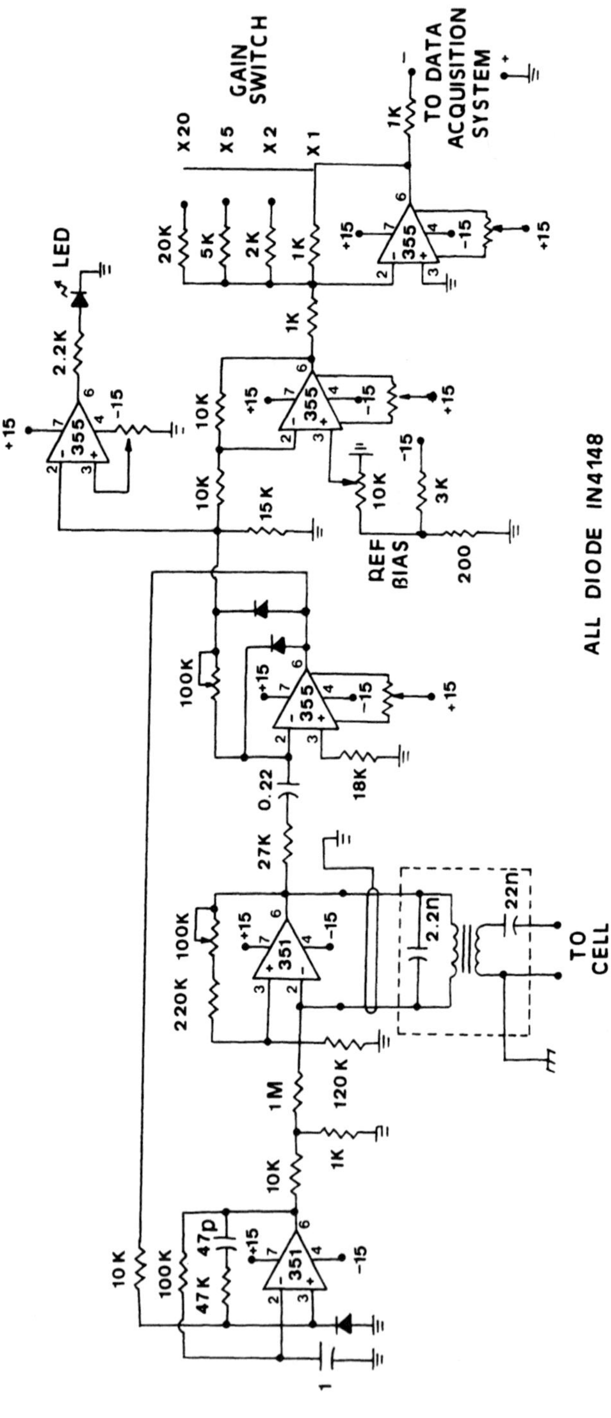

Fig. 3.69. Conductivity detector circuit diagram. (Reproduced from Ref. 113 with permission of the American Chemical Society.)

then epoxy as the permanent seal. A schematic diagram of the conductivity cell is shown in Fig. 3.68, and the circuit employed is shown in Fig. 3.69. Based on a signal-to-noise ratio of 2, the detection limit is found to be about 10^{-7} M (or 10^{-18} mol) for Li^+. The effective detection volume was estimated to be about 30 pl based on the determination of the cell constant (cross-sectional area of the electrodes divided by the distance between them) made by measuring the conductance of a known solution of KCl and using the literature value of the specific conductance for the solution. It was shown that peak area was linearly related to ion concentration over 3 orders of magnitude of concentration from 0.0025 to 2.0 mM for Li^+. Similar results were obtained for Na^+ [113].

The detection system was applied to the quantification of Li^+ in human serum. An electropherogram obtained for a patient on lithium therapy is shown in Fig. 3.70. Since Li^+ was well separated from K^+ and Na^+, there was no interference from these two ions in the CZE analysis procedure.

By maintaining the same ratio of ion analyte concentration to background electrolyte concentration while diluting the latter, it was possible to substantially increase the detection sensitivity without incurring loss of resolution in CZE separations with on-line conductivity detection. Figure 3.71 illustrates the relation

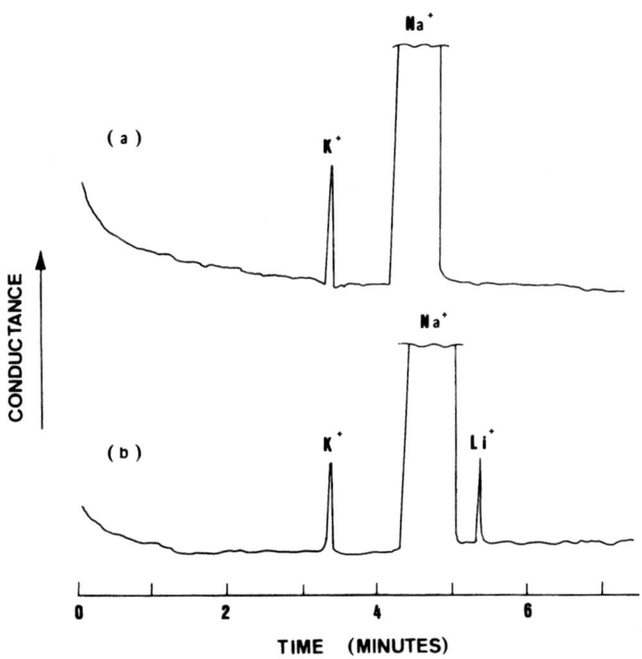

Fig. 3.70. Electropherogram of human serum. (a) Normal subject; (b) patient on lithium therapy. Dilution is 1 : 19 with 20 mM MES–His buffer, pH 6.1. Capillary I.D.: 75 μm; length: 70 cm; gravity injection from 10 cm for 30 s; applied voltage: 25 kV. The Na^+ peak is off scale. (Reproduced from Ref. 114 with permission of Elsevier Science Publishers.)

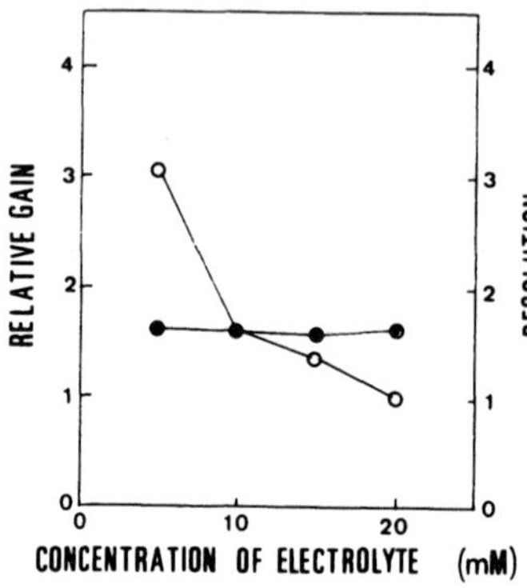

Fig. 3.71. Plot showing the effect of concentration of background electrolyte and sample on the relationship of resolution and sensitivity. Filled data points are resolution; open data points are relative gain. The ratio of sample concentration to background electrolyte concentration is 0.05. (Reproduced from Ref. 115 with permission of Elsevier Science Publishers.)

between resolution and sensitivity at different concentration of background electrolyte and sample while maintaining the same concentration ratio. Since the ratio of sample ion to carrier ion was kept constant, the increase in sensitivity could be caused by a decrease in the background electrolyte conductivity. A fourfold decrease in the electrolyte concentration from 20 mM to 5 mM, while keeping the sample concentration constant, resulted in an increase in absolute sensitivity of more than 12 times. Limits of detection of 10^{-6} M were obtained for a mixture of carboxylic acid. Quantification analysis of low-molecular-weight carboxylic acids by CZE with conductivity detection has also been performed [108]. It was shown that the response of the detector was directly related to the ionic mobility of the species being detected. Since the migration time of a species would also be related to the ionic mobility, the peak area would be expected to correlate with the migration time. By using the response from one species at a known concentration (an internal standard), it would be possible to calibrate the response of all other species present in a mixture on an absolute basis without having to perform calibration for each species detected.

In order to simplify the construction of electrochemical detection systems for CE, Huang *et al.* [116] developed an end-column detector. Both conductivity detection and amperometric detection could be performed with this detector system. For conductivity detection, the end-column detector was placed directly at the outlet of the CZE capillary as shown in Fig. 3.72a. Figure 3.72b shows a diagram of the end-column conductivity detector placed inside a protective plastic jacket. Figure 72c shows an enlarged view of the end-column sensing microelectrode. The sensing microelectrode made of platinum wire of 50 μm diameter was centred in a 1-cm-long fused silica capillary (150 μm I.D., 355 μm O.D.) and held in place by epoxy. The assembly is held in place with a larger fused silica capillary

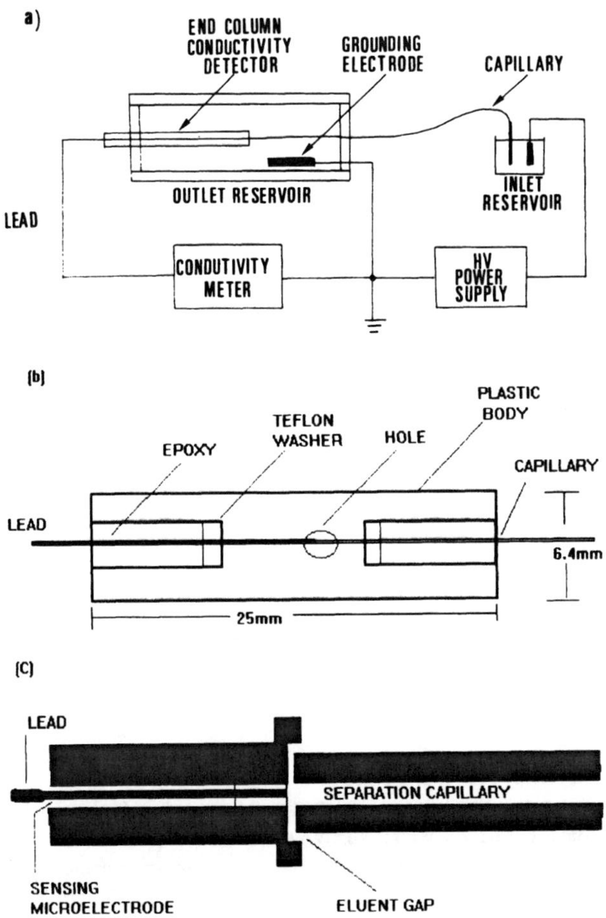

Fig. 3.72. Schematic drawing of (a) the CZE separation device with an end-column conductivity detector, (b) a cross-section view of the plastic jacket assembly, and (c) an enlarged view of the end-column sensing microelectrode. (Reproduced from Ref. 116 with permission of the American Chemical Society.)

(approximately 355 μm I.D.). One end of these larger capillary was sealed with epoxy to the sensing microelectrode holder. The other end extends about 1–2 mm so that the outlet end of the separation capillary could almost be butted against the sensing microelectrode, with a gap of 1–2 μm to form a small path for the eluent. A hole was made on the side of the plastic jacket as shown in Fig. 3.72b to allow eluent to flow through to the buffer reservoir. The conductivity measurement was made between the sensing microelectrode and the grounding electrode, using an ac circuit similar to that shown in Fig. 3.69. The sensitivity of this end-column detector was found to be similar to that obtained for the on-column design [108] for a mixture of

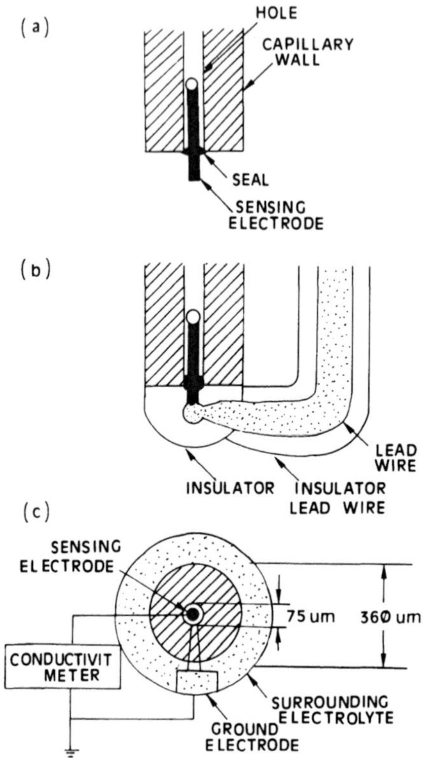

Fig. 3.73. End-column conductivity detector. (a) Alignment of sensing electrode in capillary with eluent hole in capillary wall (not to scale). (b) Same as (a) with electrical connectors. (c) Horizontal cross-section. Not shown is the protective jacket, which surrounds the outlet of the capillary. Conductivity measurements are made between the sensing electrode and the ground electrode, which also acts to complete the electrophoretic circuit. (Reproduced from Ref. 93 with permission of the American Chemical Society.)

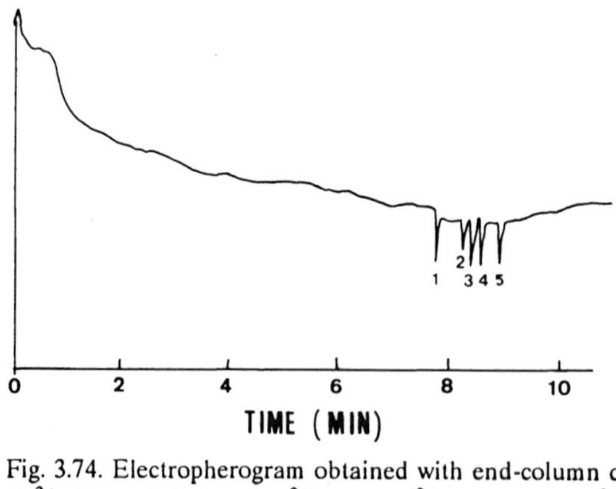

Fig. 3.74. Electropherogram obtained with end-column conductivity detector for a mixture of *1* = Ca^{2+}, *2* = Na^{+}, *3* = Mg^{2+}, *4* = Ni^{2+}, and *5* = Cd^{2+}, about 5×10^{-5} M each. The running buffer is 5 mM potassium acetate, pH 5.0. The applied electric field is 200 V/cm in a 75-μm I.D., 70-cm-long fused silica capillary. (Reproduced from Ref. 93 with permission of the American Chemical Society.)

carboxylic acids. It was estimated that for typical injection volume of 20 nL, and for an 80 μm I.D. capillary, the extra zone broadening caused by the additional dead volume would be less than 25%. However, this detector design has the advantages that it is a simple construction and it can be readily fitted on the outlet of any CE system. An alternative end-column design of the conductivity detector which can be mounted directly to the outlet of the capillary of a commercial instrument has also been described [93]. Using this method, both UV absorbance and conductivity can be recorded during the same run. A schematic diagram of this end-column detector is shown in Fig. 3.73. The application of this detection system is illustrated in Fig. 3.74.

3.6.3 Amperometric detection

Amperometric electrochemical detection with microelectrode is potentially one of the most sensitive detection techniques for CE separations. Wallingford and Ewing were the first to develop an amperometric detector for CE and subsequently there have been several reports on further improvements and applications of this detection technique [87,116,118–123]. The main problem in the use of amperometric detection for CE is due to the fact that the electric current produced in the column upon application of a separation potential of 10–30 kV can be six orders of magnitude greater than the electrochemical currents obtained at a suitable amperometric detector. In order to prevent the detector signal from being overwhelmed by the electrophoretic current, the detection system must incorporate features to electrically isolate the amperometric detector from the applied separation potential.

The main feature of the interfaces developed for amperometric detection is the use of an electrically porous conductive glass joint created near the cathodic end of the capillary. Such a conductive joint between the separation capillary and the detection capillary decoupled the electrochemical detector from the separation capillary. The electroosmotic flow generated in the separation capillary serves as an electroosmotic pump force solute zones and solvents past the joint and through the detection capillary. With this arrangement, the carbon fibre working electrode could be placed directly into the end of the detection capillary in order to minimize detection volume.

A schematic diagram of the CE system with an electrically conductive joint, and the top view of the amperometric detection system is shown in Fig. 3.75. Several procedures have been employed for the construction of the porous joint. In one of the methods, the polyimide coating was removed from the last 2–3 cm of the separation capillary. A 2 cm length of porous glass capillary was placed over the exposed end of the column with approximately 1 cm of the column protruding past the end of the porous capillary. This section of the column was then cemented onto a 3 cm long segment of glass microscope slide with the end of the porous glass kept flush with the end of the slide. Subsequently, the porous capillary was shortened

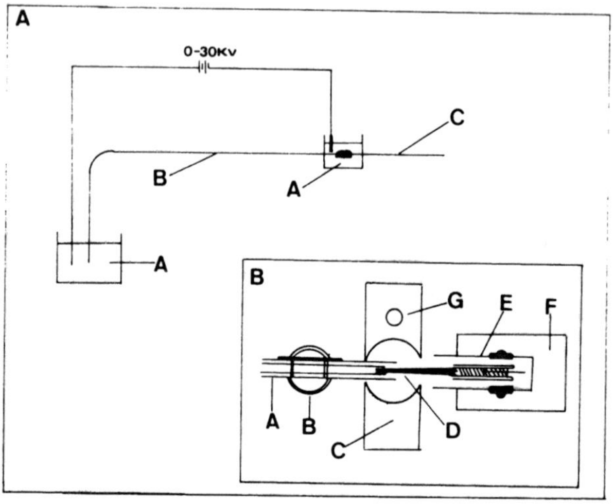

Fig. 3.75. (A) Schematic of CZE system with electrically conductive joint: A = buffer reservoirs; B = separation capillary; C = detection capillary. (B) Top view of amperometric detection system: A = column; B = porous glass joint assembly; C = Plexiglass block; D = carbon-fibre working electrode; E = microscope slide; F = micromanipulator; G = reference electrode port. (Reproduced from Ref. 120 with permission of the American Chemical Society.)

to 1 cm by crushing from each end toward the centre. This section of porous capillary was then moved to the side of the exposed area. The exposed column was scored with a diamond-tipped glass cutter at a location of 1.5–2.5 cm from the end. The column was then fractured at the scored region with the tip of a scalpel blade. The two segments of column were then positioned to form a tight joint. The inner bores of the two segments were aligned as precisely as possible. The porous capillary was then moved into position over the fracture. Subsequently, the ends of the porous capillary were sealed with epoxy and the assembly was mounted in a plastic reservoir. Since buffers and micelle solution may react with certain epoxies and caused them to deteriorate, the choice of epoxy is very important. The epoxy used (e.g. Deucon 2-Ton) should be stable in most types of solution used.

Electrochemical detection was performed with 10 μm diameter carbon fibres protruding 0.2–0.5 mm from drawn glass capillary as the working electrode. The end of the detection capillary was positioned into the center of a 0.635 cm diameter cell in the flexible block. A sodium-saturated calomel reference electrode (SSCE) was placed into the reference electrode part. The cell was then filled with an electrolyte solution of either 0.1 M KCl or 0.01 M phosphate buffer. Potential control between the reference and the working electrode was accomplished with a mercury cell and a voltage divider. The performance of the electrochemical detector was optimized by miniaturizing both the separation column and the electrochemical detector [118]. Figure 3.76 shows an electropherogram of four amines obtained on a 12.7 μm I.D. capillary with a 5 μm O.D. carbon fibre inserted into the end of the detection

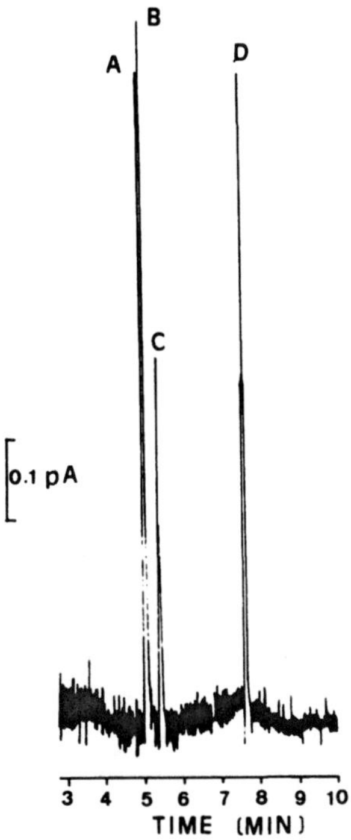

Fig. 3.76. Capillary electrophoretic separation of four amines at low concentration. $A = 5 \times 10^{-8}$ M serotonin (5-HT); $B = 1.1 \times 10^{-7}$ M norepinephrine (NE); $C = 1 \times 10^{-7}$ M isoproterenol (IP); $D = 2 \times 10^{-7}$ M 4-methylcatechol (4-MC). (Reproduced from Ref. 119 with permission of the American Chemical Society.)

capillary. Detection limits ($S/N = 2$) of 6 amol (10^{-8} M) for 5-hydroxytryptamine and 22 amol (3×10^{-8} M) for isoproterenol were obtained. With this detection system, capillaries with small I.D. can be used, since smaller capillaries result in smaller annular flow regions around the electrode, yielding greater coulometric efficiency and therefore increased sensitivity [116]. This is an advantage over optical detection methods, which often lose sensitivity when capillaries with small cross sections are used.

Electrochemical detection has been demonstrated for MEKC in the separation of catecholamines [120] Addition of SDS micelles was found to enhance separation efficiency. Over 400,000 plates could be obtained. A smaller capillary diameter provided better detection limits due to higher conversion efficiencies. Limits of detection of less than 20 fmol were obtained for this system. Separation of Serotonin from catechols by CZE with electrochemical detection was investigated by Wallingford and Ewing [122]. With columns having only 9 μm I.D., an amperometric detection limit of 0.7 amol was obtained for Serotonin. By using a Nafion-coated electrode, sensitivity and selectivity could be further enhanced [123].

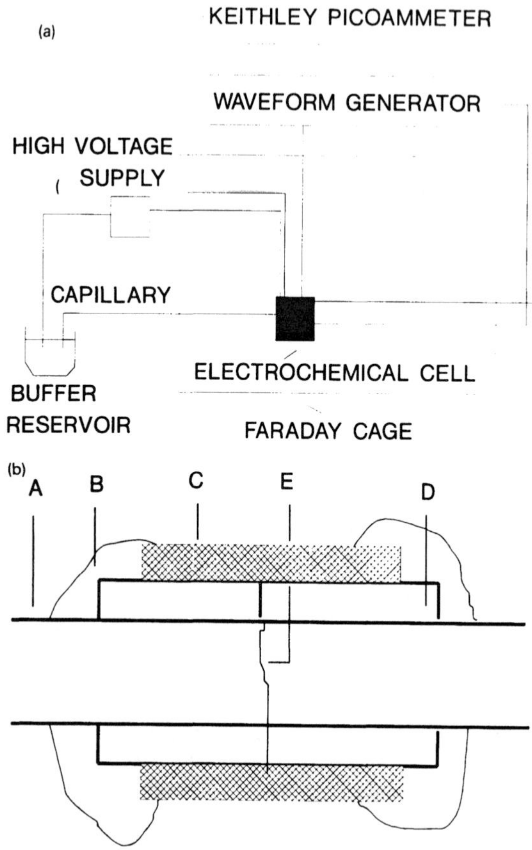

Fig. 3.77. (a) Block diagram of MEKC-ECD system. (b) Detailed schematic of porous graphite joint: A = fused silica capillary; B = epoxy; C = graphite tube; D = PTFE tube; E = joint. (Reproduced from Ref. 125 with permission of Elsevier Science Publishers.)

Yik *et al.* [125] described a simple, rugged yet sensitive electrochemical detection system for CE. This system coupled the separation column to a short length (less than 2.5 cm) of the same column material together with a section of porous graphite tube which formed an electrically conductive joint. The porous joint was constructed easily by pushing the two capillaries through the graphite tube. Since the PTFE capillary formed a tight fit between the graphite tube and the silica capillary, the separation and detection capillaries could be aligned easily within the porous graphite tube (see Fig. 3.77). When this coupler assembly was immersed in the buffer with the cathode (see Fig. 3.78), the separation potential could be applied selectively to the separation column. The strong electroosmotic flow generated in this column served to force the solvent and analyte zones past the joint and through the second section to the detector. This system effectively separated the detector

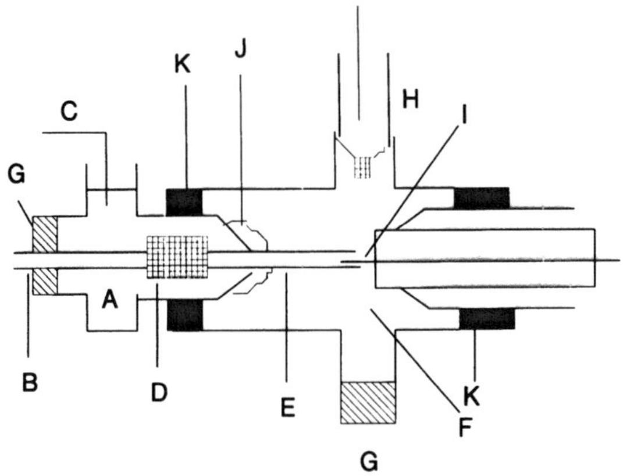

Fig. 3.78. Schematic of the electrochemical cell. *A* = buffer reservoir; *B* = separation capillary; *C* = platinum wire; *D* = graphite joint; *E* = detection capillary; *F* = electrolyte; *G* = stopper; *H* = reference electrode; *I* = carbon fibre working electrode; *J* = epoxy; *K* = ground glass joint. (Reproduced from Ref. 125 with permission of Elsevier Science Publishers.)

from the high separation potential applied. A detection limit ($S/N = 3$) on the order of 0.4 fmol ($\simeq 10^{-7}$ M) was obtained for norepinephrine (NE). Amperometric detection for MEKC separation of mixtures of B_6 vitamers and polycyclic aromatic hydrocarbons was performed. Figure 3.79 shows an electropherogram of the B_6 vitamers, pyridoxine (PN), pyridoxamine (PM) and pyridoxal (PL). A detection limit of 4 fmol ($\simeq 10^{-6}$ M) was obtained.

A new design for an end-column amperometric detector for CE has been reported by Huang *et al.* [116]. This design did not make use of a porous glass joint to decouple the electrophoretic and detection current. Instead, a sensing microelectrode was placed directly at the outlet of the fused silica capillary. A schematic diagram of the end-column amperometric detector is shown in Fig. 3.80. A piece of capillary (5 μm I.D./140 μm O.D.) was cut to the desired length (50–70 cm). It was then positioned in the electrochemical cell with a stainless steel fitting that was epoxied in place. This fitting acted as the cathode for electrophoresis. It was found that with this end-column detector for CZE, there would be no need to isolate the sensing element from the high electric field needed for electrophoresis, because the I.D. of the separation capillary is very small and very little current passes. Also, the carbon fibre microelectrode was not placed directly into the separation capillary. Instead it was aligned with the bore of the capillary and positioned up against but not into the capillary, therefore creating a thin layer cell at the capillary outlet.

Engstrom-Silvermann and Ewing [126] described a copper wire amperometric detector based on a design that was used by Wallingford and Ewing [120] for CE. Detection was accomplished by the use of a porous glass joint, which allows

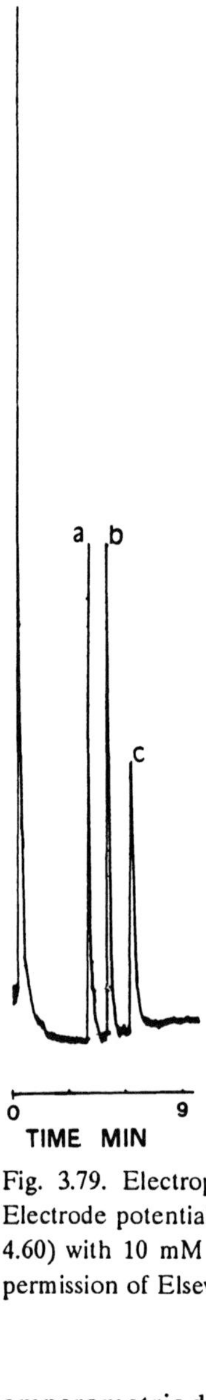

Fig. 3.79. Electropherogram of (a) Pyridoxal, Pl; (b) pyridoxine, PN; (c) pyridoxamine, PM. Electrode potential: 1.2 V vs. Ag/AgCl; column I.D.: 50 μm; buffer: 0.01 M phosphate buffer (pH 4.60) with 10 mM SDS; separation potential: 15 kV (7.4 μA). (Reproduced from Ref. 125 with permission of Elsevier Science Publishers.)

amperometric detection at a copper wire (19 μm diameter) electrode inserted in the end of the capillary. An anodic current is produced by a change in the copper oxide film solubility, resulting from complexation of copper ions with certain analytes at the electrode surface. A schematic diagram of the detection mechanism at the surface of a copper electrode is shown in Fig. 3.81. The copper wire was inserted in the tip of a pulled glass capillary until a length of 200–300 μm protruded from the tip. Electrodes were filled with gallium metal and a 122 μm diameter copper wire was used to make electrical contact. Amperometric detection was performed

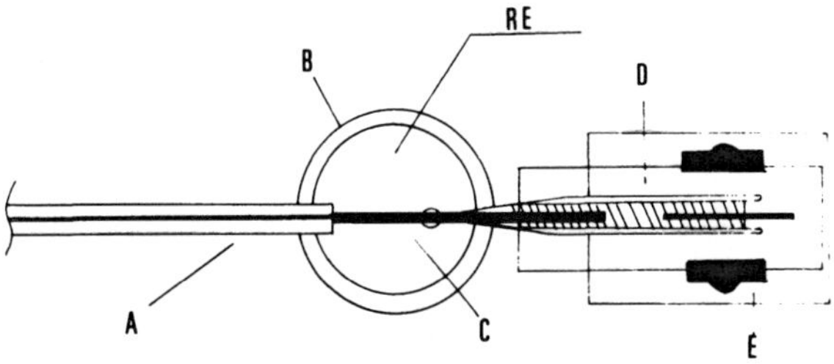

Fig. 3.80. Schematic drawing of CZE with end-column amperometric detection. *A* = capillary; *B* = cathodic buffer reservoir and electrochemical cell; *C* = carbon fibre electrode; *D* = electrode assembly; *E* = micromanipulator; *RE* = reference electrode. (Reproduced from Ref. 116 with permission of the American Chemical Society.)

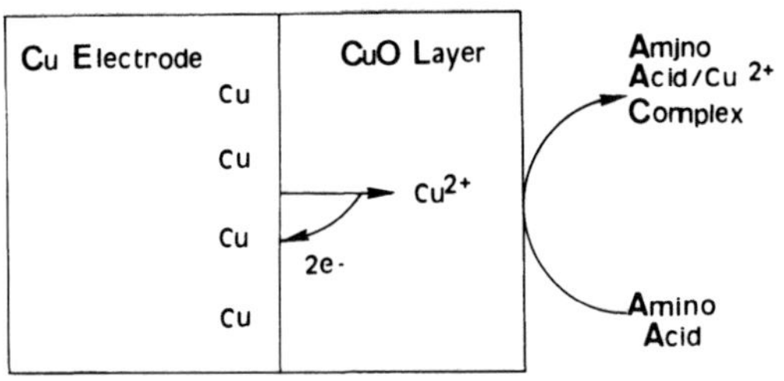

Fig. 3.81. Schematic diagram of the detection mechanism at the surface of a copper electrode. (Reproduced from Ref. 126 with permission of Microseparations Inc.)

with a two-electrode potentiostat. A sodium saturated calomel electrode (SSCE) was used as reference. The detection end of the system was housed in a Faraday cage in order to minimize the effects of external noise sources. This copper/copper oxide electrode has been used to detect non-electroactive native amino acids and dipeptides. Subfemtomole detection limits have been obtained without solute derivatization. In addition, simultaneous analysis of non-electroactive amino acids and electroactive catecholamines has been demonstrated as shown in Fig. 3.82. The response of the non-electroactive amino acids was based on the complexation processes illustrated in Fig. 3.81.

It should be noted that application of CE with amperometric detection has been limited to only several classes of compounds which are easily oxidized at the carbon surface, such as catechols, catecholamines and vitamin B_6. The development of indirect electrochemical detection techniques (see Fig. 3.81 and Section 3.7.3) provides an alternative scheme that promises to be more universal.

3.7 INDIRECT DETECTION

A comprehensive account of indirect detection methods has been given by Yeung [128]. Several reasons for developing indirect detection schemes have been

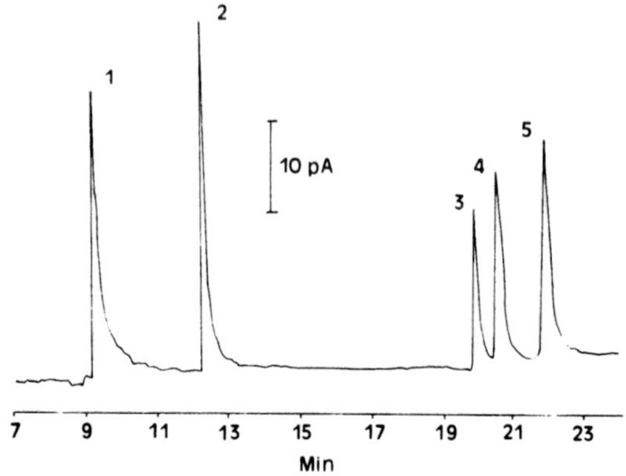

Fig. 3.82. Electropherogram of epinephrine (*1*), catechol (*2*), dihydroxyphenylacetic acid (*3*), glutamic acid (*4*), aspartic acid (*5*) simultaneously detected amperometrically at a copper electrode in a 26 μm I.D. capillary. Buffer: 1 mM potassium dihydrogen phosphate (pH 7.12); 67-cm long separation capillary, 1.35-cm detection capillary. Injection: 3 s at 10 kV; separation voltage: 10 kV; electrode potential: +0.15 V vs. SSCE. (Reproduced from Ref. 126 with permission of Microseparations, Inc.)

mentioned. First, indirect detection is universal, and can be used for compounds which lack chromophores or fluorophores. Second, it is possible to broaden the applicability of highly sensitive but selective detectors by implementing indirect detection. Third, quantification is easier with indirect detection since tedious chemical derivatization procedure can be avoided. Fourth, indirect detection is non-destructive since no chemical manipulation has been introduced. On the other hand, the main disadvantage of indirect detection methods is that the linear dynamic range tends to be rather limited.

In the case of indirect detection, the analyte physically displaces an added component which may be a chromophore, fluorophore or electroactive species. In capillary electrophoresis, a charged added component can be used such that analyte ions of like charge will displace this component, while ions of opposite charge may ion pair with it. This allows visualization of many species which do not contain chromophores. Instrumentations used for indirect detection are similar to those used in direct detection. However, negative peaks are usually observed as the analytes displace the UV-active, fluorescent or electroactive added component from the background electrolyte. A general scheme for indirect detection is shown in Fig. 3.83. While almost all detection schemes can be made to function in the indirect mode as described in Fig. 3.83, it is essential that the mechanism for displacement is clear and unambiguous, and the operating conditions are amenable to optimization at low analyte concentration. An important parameter used in indirect detection is the transfer ratio, TR, which has been defined as the number of mobile phase molecules displaced (or replaced) by one analyte molecules.

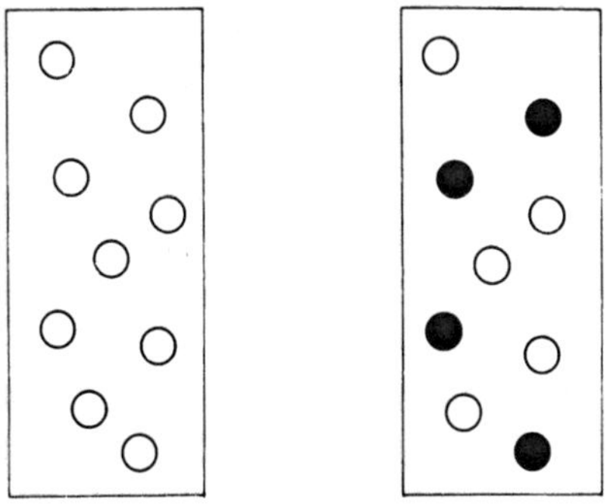

Fig. 3.83. General scheme for indirect detection. Left: The mobile phase contains a component that provides the actual response (open circles). Right: When analytes (solid circles) reach the detector, displacement results in a decrease in the background signal, i.e. a negative peak. (Reproduced from Ref. 128 with permission of the American Chemical Society.)

In the simplest case, displacement is by volume. Another possibility is charge displacement. Displacement can also be due to solubility modification, in which the analyte modifies the partition between stationary and mobile phase components, leading to an increase or a decrease in the background signal when the analyte is eluted out the column and reaches the detector.

Another important parameter is the dynamic reserve, DR, which is defined as the ratio of the background signal to the background noise. In indirect detection, a large background is required. Background noise that normally are negligible can become a serious problem. It can be shown that [128,132]:

$$C_{\text{lim}} = \frac{C_{\text{m}}}{(\text{DR} \times \text{TR})} \tag{3.16}$$

where C_{lim} is the concentration limit of detection and C_{m} is the concentration of the relevant mobile phase added-component. For a given system, the more stable the background signal (large DR), the more effective the displacement process (large TR), and the lower C_{m} is, lower detection limits can be obtained.

3.7.1 Indirect UV detection

Feasibility of indirect detection in CE was first demonstrated by Hjerten *et al.* [129], who employed indirect UV absorbance detection of organic ions. Foret *et al.* [130] used an improved UV detector to perform indirect measurement of benzoic or sorbic acid to visualize metabolic carboxylic acids and obtained a detection limit of 0.5 pmol. Since the sensitivity of indirect detection is dependent on the dynamic reserve (DR), the indirect UV detection was limited to the inherent insensitivity of UV absorbance in on-column capillary detection.

Foret *et al.* described an indirect UV photometric detection method, which was based on the use of an absorbing co-ion as the principal component of the background electrolyte [130]. The zones of non-absorbing ionic species were revealed by changes in light absorption due to charge displacement of the absorbing co-ion. Their results showed that for indirect photometric detection, the highest sensitivity could be achieved for sample ions having an effective mobility close to the mobility of the absorbing co-ion. In such cases, the concentration of the sample components in the migrating zones could be high while electromigration dispersion was still negligible. The useful dynamic range of the detection would then be limited by the linearity and noise of the detector. The best sensitivities were obtained in low concentration background electrolyte containing a co-ion with high absorption at a given detection wavelength.

Bruin *et al.* [131] demonstrated the presence of system peaks in indirect UV detection both theoretically and experimentally. They also found that when the electrophoretic mobility of an ion was close to the mobility of the background ion, the sensitivity of the system increased and the peak shape became Gaussian in case of concentration overload. They suggested that sensitivity could be increased by manipulation of the pH of the background electrolyte (BGE) to effect changes in effective mobilities.

3.7.2 Indirect fluorescence detection

Kuhr and Yeung [132] explored indirect laser-induced fluorescence detection for capillary electrophoresis. The system used is similar to that employed for normal laser fluorescence detection in capillaries. The output power of a He-Cd laser (325 nm, 8 mW) was stabilized to within 0.05% with a laser power stabilizer. The stabilized laser beam at 5 mW was focused into a 15-μm spot in the fused silica capillary (15 μm I.D.). The capillary was mounted at the Brewster angle (angle at which the reflected light is completely polarized) to the incident beam and the illuminated spot was located 10 cm from the cathodic end of the tube. The resulting fluorescence was collected perpendicular to the plane containing the incident beam and the capillary by imaging the illuminated region onto a photomultiplier tube, with a 10× microscope objective after passing through an interference filter and a spatial filter. Salicylate at a concentration of 1.0 mM was added to the buffer to provide a background fluorescence signal. Detection limits in the range of 20 fmol ($\simeq 10^{-5}$ M) have been obtained for negatively charged amino acids.

Indirect fluorescence detection has also been employed for the quantification of inorganic anions and nucleotide, mono, di, and triphosphate [133]. Injections were made at the cathode and detected at the ground-potential anode. Using this system, they found that with injection by electromigration, the amount of sample injected varies significantly with the total ionic concentration of the sample and the response was linear with the resistance of the sample solvation.

The use of indirect fluorescence detection with CZE for the analysis of trace quantities of macromolecular mixtures was also demonstrated [134]. Subfemtomole quantities of tryptic digest mixtures were separated within three minutes with mass limits of detection 3000 times lower than those of commercial HPLC type UV absorbance detection and 180 times lower than those of UV absorbance detection in CZE systems.

3.7.3 Indirect electrochemical detection

A system capable of performing both direct and indirect amperometric detection with CE was demonstrated by Olefirowicz and Ewing [135]. The system consisted of a porous glass coupler which allows amperometric detection at a carbon fibre electrode placed in the end of the capillary.

Indirect electrochemical detection was accomplished by the addition of a cationic electrophore, 3,4-dihydroxybenzylamine (DHBA), to the electrophoresis buffer. When the electrode was held at a constant potential of 0.7 V vs. the sodium saturated calomel electrode (SSCE), the DHBA gave the corresponding orthoquinone by a two electron two proton transfer [135]. A schematic diagram of cation displacement by migrating zone of cationic species is shown in Fig. 3.84. It was noted that continuous oxidation of DHBA produced a constant background current during its passage through the detection region. To maintain neutrality, the cationic DHBA in the buffer was displaced by cationic analyte zones. When these cationic zones passed through the detection region, a lower level of oxidizable species was observed in these zones and the background current decreased. As a result, negative peaks would be expected in the detection of cations. Figure 3.85 shows an electropherogram obtained for the separation of three non-electroactive amino acids and three non-electroactive dipeptides. Detection limits as low as 380 amol ($\simeq 10^{-7}$ M) were obtained for the amino acid arginine using a 9 μm I.D. capillary. Simultaneous direct and indirect amperometric detection was also

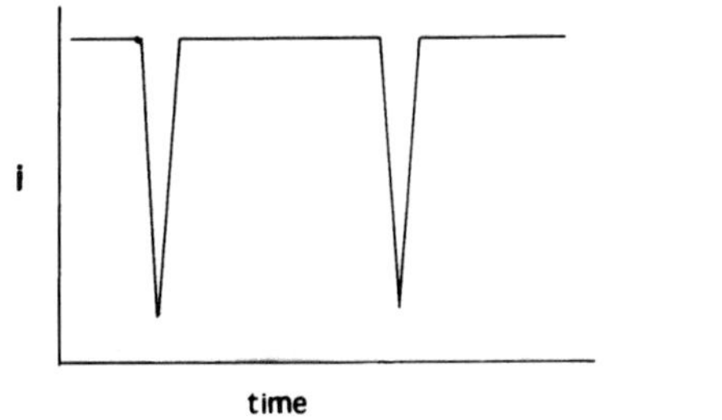

Fig. 3.84. Schematic diagram of cationic displacement by migrating zone of cationic solutes.

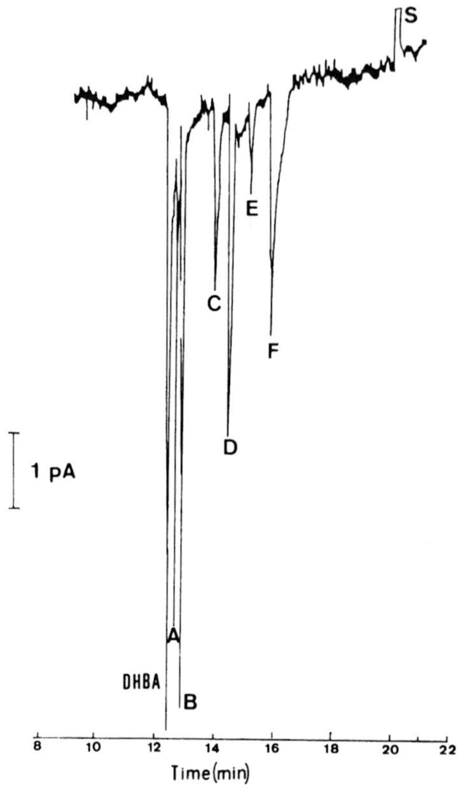

Fig. 3.85. Electropherogram of amino acids and peptides with indirect amperometric detection on a 9-μm I.D. capillary; buffer: 0.01 mM DHBA–0.025 M MES (pH 5.65); 97-cm separation capillary; 1.0-cm detection capillary. Injection: 1 s at 20 kV; separation voltage: 20 kV. Peaks: A = 8.5 fmol Lys; B = 8.2 fmol Arg; C = 7.4 fmol His; D = 7.0 fmol Arg-Leu; E = 6.4 fmol His-Gly; F = 5.8 fmol His-Phe; S = system peak. (Reproduced from Ref. 135 with permission of Elsevier Science Publishers.)

demonstrated using this detector. Figure 3.86 shows the separation of three easily oxidized catecholamines and two non-electroactive dipeptides. In this case, the detector operated in the direct mode for catecholamines and the indirect mode for the dipeptides.

3.8 RADIOISOTOPE DETECTORS

The use of radioisotopic detectors for CE have been investigated. Pentoney *et al.* [136,137] described two simple on-line radioactivity detectors for CE. The first CE/radioisotope detector utilizes a commercially available spectroscopic grade cadmium telloride semiconductor device, which is positioned external to the separation channel and which responds directly to impinging γ- or high-energy

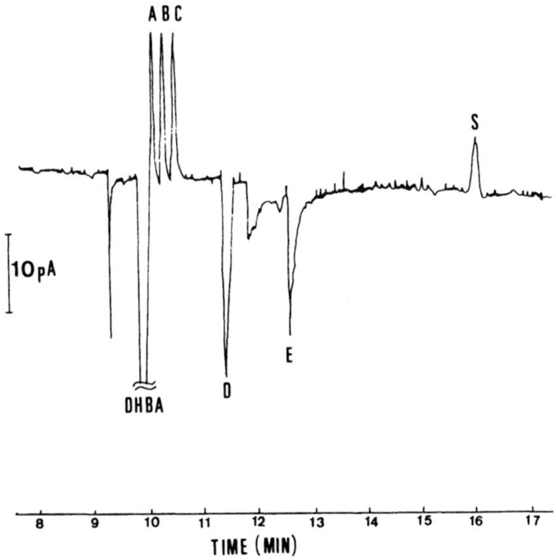

Fig. 3.86. Electropherogram of catecholamines and peptides with combined direct and indirect amperometric detection in a 9-μm I.D. capillary. Conditions are the same as for Fig. 3.85, except as follows. Injection: 5 s at 25 kV; separation voltage: 25 kV. Peaks: A = dopamine; B = norepinephrine; C = epinephrine; D = Lys-Phe; E = His-Phe. The broad peak between peaks D and E is an unknown impurity in the Lys-Phe. (Reproduced from Ref. 135 with permission of Elsevier Science Publishers.)

β-radiation. The second CE/radioisotope detector utilizes a commercially available plastic scintillator material that completely surrounds the separation channel, thereby improving the efficiency of detection. The experimental setup of the CE/semiconductor radioisotope detection system is shown in Fig. 3.87. The cadmium tellurium detector probe consisted of a 2-mm cube of Cd-Te, which was set in a thermoplastic and positioned behind a thin film of aluminized nylon at a distance of approximately 1.5 mm from the face of an aluminum housing. A 2-mm wide Pb-aperture (0.002 cm thickness) was used to shield the Cd-Te detector element from the radiation originating from the region of the capillary adjacent to the detection volume. The Al housing incorporate a BNC-type connector that facilitated both physical and electrical connection to a miniature charge-sensitive preamplifier. the Cd-Te probe and preamplifier assembly were mounted on an $x - y$ translation stage, and the face of the Al housing was brought into direct contact with the polyimide-coated fused silica capillary/Pb-aperture assembly. The Cd-Te detector was operated at a bias voltage of 60 V and the detector signal was amplified by the charge-sensitive amplifier and sent to the counting unit of the semiconductor detector.

The second on-line radioactivity detector consisted of a plastic scintillator material. which was machined from 2.54-cm diameter rod stock into a 1.59 cm

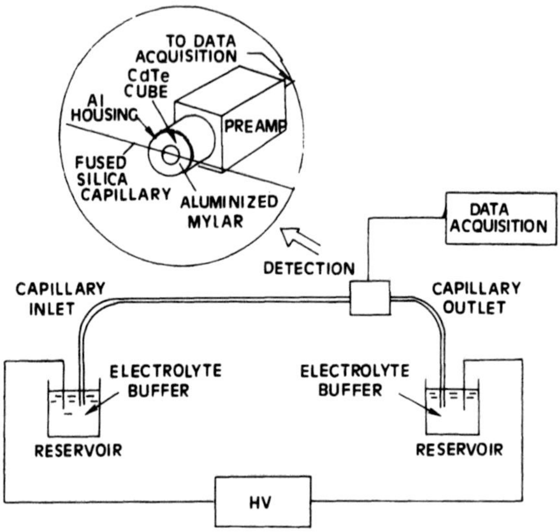

Fig. 3.87. Experimental setup of the capillary electrophoresis/semiconductor radioisotope detector system. The inset shows the positioning of the Cd-Te probe with respect to the capillary tubing. The 2-mm Pb-aperture is not shown in this illustration. (Reproduced from Ref. 136 with permission of the American Chemical Society.)

(⅝ in) diameter (front face) solid parabola. An exploded diagram of the plastic scintillator radioisotope detector is shown in Fig. 3.88. A special rotating holder was constructed for the plastic scintillator and the curved outer surface were coated by vacuum deposition with a thin film of Al in order to reflect the emitted light toward the front face of the scintillator. A 2-mm detection length was defined within the

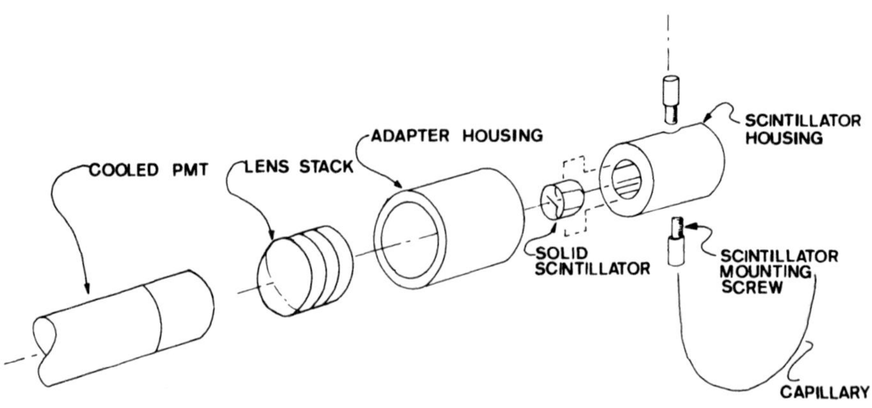

Fig. 3.88. Exploded diagram of the plastic scintillator radioisotope detector. The fused silica capillary is exposed to a 2-mm section of the plastic scintillator that is located in between the press-fit aluminium mounting rods. (Reproduced from Ref. 136 with permission of the American Chemical Society.)

parabola by 6.5-mm outer diameter Al mounting rods, which were press-fitted in the sides of the scintillator. The light emitted by the scintillator as radiolabeled sample passed through the detection region was collected and focused onto the photocathode of the cooled photomultiplier tube by a condenser lens combination. Photon counting was accomplished with a discriminator control unit and a photon counter.

The system performance was evaluated for both detection schemes by using synthetic mixtures of ^{32}P-labeled sample molecules. The efficiency of the semiconductor detector (planar geometry) was determined to be approximately 26%, whereas that of the plastic scintillator was found to be 65%. The detection limits were determined to be in the low-nanocurie range (10^{-7} M) for separations performed under standard conditions. The lower limit of detection was extended to the subnanocurie level by use of flow (voltage) programming to increase the residence time of labeled sample components in the detector.

Altria *et al.* [138] described a gamma-ray detector for CZE. Figures 3.89 and 3.90 show diagrams of the experimental system used and the construction of the detector respectively. The detector as constructed from a Tll doped NaI scintillation crystal ($1.25 \times 2.5 \times 2.5$ cm) placed on the entrance part of a photomultiplier tube whose signal was amplified and the output was recorded on a chart recorder or integrator. The entire detector assembly was shielded with lead to minimize background noise. The precise length of capillary exposed to the scintillator crystal was varied by placing two sheets of lead between the crystal surface and the axis of the capillary to form a slit. The optimum slit width was determined by the signal obtained from duplicate CE runs of a pertechnetate anion solution. Figure 3.91 shows the detector response versus slit width.

A slit width of 20 mm, corresponding to a detection zone volume of 8.84×10^{-2}

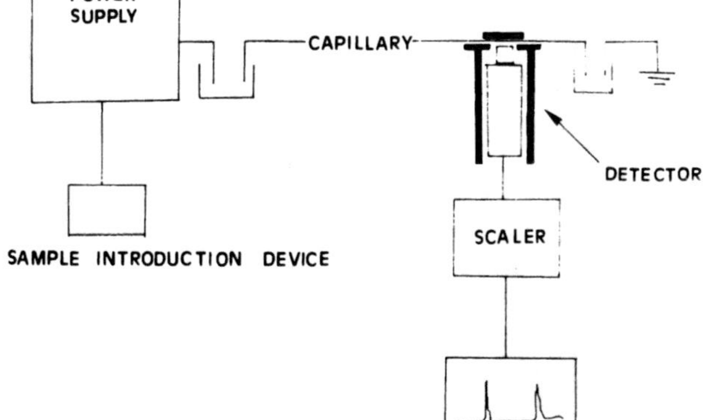

Fig. 3.89. Apparatus used for radiopharmaceutical analysis. (Reproduced from Ref. 138 with permission of VCH Verlagsgesellschaft.)

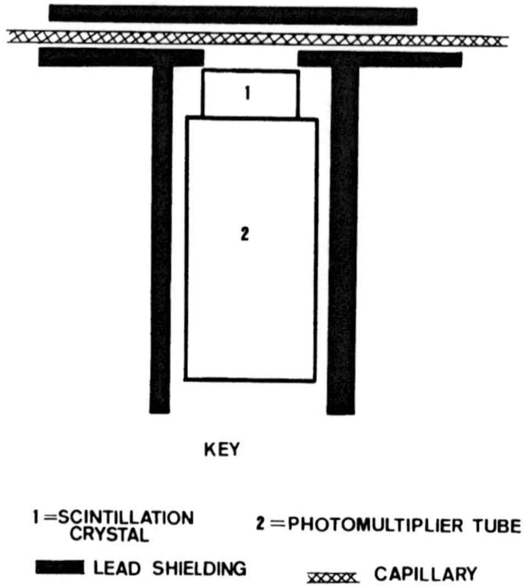

Fig. 3.90. Gamma-ray detector. (Reproduced from Ref. 138 with permission of VCH Verlagsgesellschaft.)

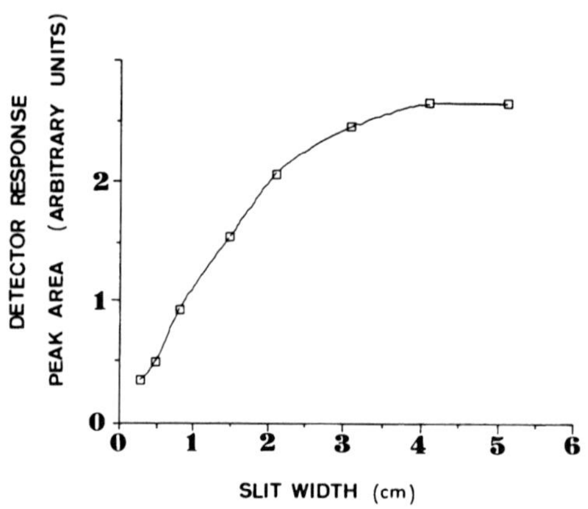

Fig. 3.91. Detector response versus slit width. Operating conditions: −30 kV across a 75 μm × 100 cm capillary filled with 0.002 M cetyltrimethylammonium bromide (CTMAB), pertechnetate solute. Sample introduction, 2 s at −20 kV. (Reproduced from Ref. 138 with permission of VCH Verlagsgesellschaft.)

μl, was found to be a satisfactory compromise between peak efficiency and signal intensity. The reduction in separation efficiency was about 10%. Increasing the slit width further did not markedly increase the response but had a deleterious effect on zone broadening. The detector was shown to have linear response from 10 (the limit of detection) to 5000 Bq cm^{-3} corresponding to 5.1×10^{-17} to 2.55×10^{-15} g cm^{-3} T_{43}^{99m}, or 5×10^{-9} to 2×10^{-7} M. The application of the detector for the analysis of some radiopharmaceuticals was investigated.

3.9 MASS SPECTROMETRIC DETECTION

The use of mass spectrometry (MS) for detection provides the unique capability to identify unknown compounds. The combination of a high-efficiency separation technique, such as CE, with MS detection provides a powerful system for the analysis of complex mixtures/samples [140–156]. The interfacing of CE to MS has been accomplished by two main types of approaches, namely electrospray ionization (ESI) and continuous-flow fast atom bombardment (CF-FAB).

3.9.1 Electrospray ionization (ESI)

Smith and co-worker developed an electrospray interface for CE-MS [140–147]. The electrospray interface makes electrical contact with the electrophoretic buffer via either a small needle [141,147] or a film of metal deposited on the surface of the capillary [140].

A schematic diagram of an apparatus developed for CZE-MS is show in Fig. 3.92. In this design, the low-voltage end of the separation capillaries was terminated in a stainless steel capillary sheath, 300 μm I.D. and 450 μm O.D. The sheath potential was controlled with a 0–5 kV dc power supply and functioned as both the CZE cathode and electrospray needle. The stainless steel capillary ensured immediate electrical contact with the solution flowing out of the fused silica capillary, hence terminating the CZE circuit and initializing the electrospray [141].

Electrospray ionization was performed at atmospheric pressure in a 2.54 cm long by 2.29 cm I.D. stainless steel cylinder. The cylinder terminates in an electrically biased (190 V dc) focusing ring with a 0.475 cm aperture. The sampling nozzle had a 0.5 mm I.D. orifice and was in contact with a copper cylinder at ground potential. The cylinder surrounded the electrospray assembly and was heated to 60°C by a system of cartridge heaters. The electrospray needle, focusing ring, and

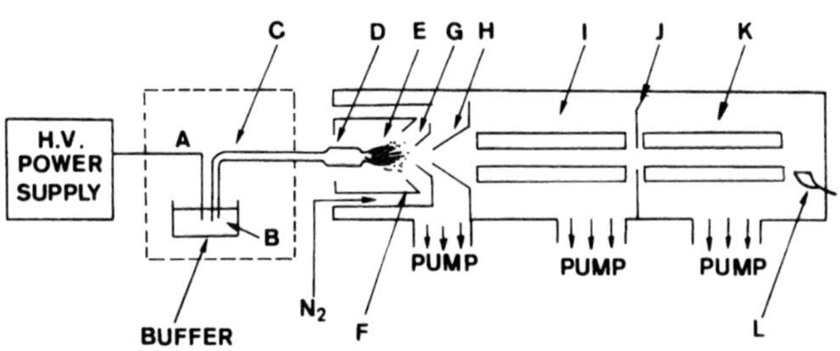

Fig. 3.92. Schematic illustration of the apparatus developed for capillary zone electrophoresis-mass spectrometry (CZE-MS): A = electrically insulated sampling box; B = anode and sample injection reservoir; C = fused silica capillary; D = cathode and electrospray needle; E = electrospray; F = focusing ring; G = nozzle; H = skimmer; I = rf only quadrupole; J = ion entrance aperture; K = quadrupole mass spectrometer; L = channeltron electron multiplier. (Reproduced from Ref. 141 with permission of American Chemical Society.)

ion sampling nozzle were concentric with the mass analyzer. Those components could be positioned independently relative to the fixed skimmer, even when high voltage was on, so as to maximize ion formation and transmission. An electrically isolated stainless steel plate (−28 V dc), with a 0.625 cm orifice, allowed the mass spectrometer chamber to be maintained at 2×10^{-4} Pa. The chamber housed the 2000 amu range quadrupole mass filter and an electron multiplier operated in the analog mode.

The key feature of this system is that the electrospray needle can be used as cathode. Figure 3.93 shows an electrospray ionization mass spectrum of a mixture of five quaternary ammonium salts at 10^{-5} M concentration introduced by continuous electromigration. Figure 3.94 shows a reconstructed total ion chromatogram of the five quaternary ammonium salts at 10^{-6} M (14–17 fmol injection) concentration. Efficiencies ranged between 26,000 to 100,000 theoretical plates based on peak half-widths. Detection limits as low as 10 amol (10^{-9} M) have been reported using single-ion monitoring [141].

A less satisfactory method for making electrical contact between the electrophoretic buffer and the electrospray interface was by depositing a thin film on the surface of the capillary instead of using a needle. The design was demonstrated with enkephalin, other peptides and quarternary ammonium salts [140]. The total ion electropherogram had sensitivities comparable to those obtained with UV detection.

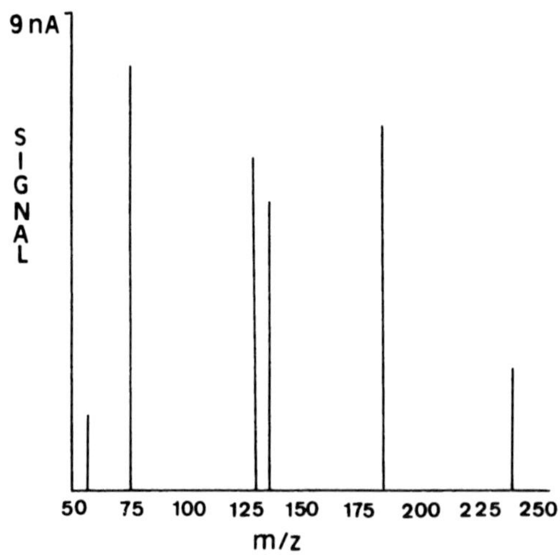

Fig. 3.93. Electrospray ionization mass spectrum of a mixture of five quaternary ammonium salts at 10^{-5} M concentration introduced by continuous electromigration. The dominant peaks are due to the quaternary ammonium cations of tetramethylammonium bromide (*m/Z* 74), tetraethylammonium perchlorate (*m/Z* 130), trimethylphenyl ammonium iodide (*m/Z* 136), tetrapropylammonium hydroxide (*m/Z* 186), and tetrabutylammonium hydroxide (*m/Z* 242) and a background peak due to $Na\text{-}MeOH^{+}$ (*m/Z* 55). (Reproduced from Ref. 141 with permission of the American Chemical Society.)

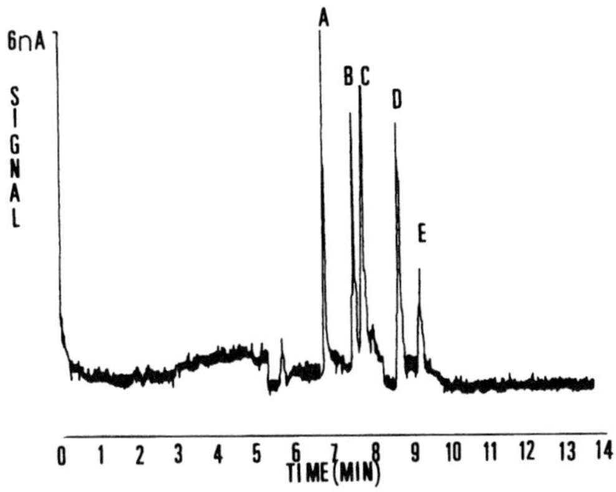

Fig. 3.94. Reconstructed total ion electropherogram of five quaternary ammonium salts at 10^{-6} M (14–17 fmol injection) concentration, obtained by CZE-MS: *A* = tetramethylammonium bormide; *B* = trimethylphenylammonium iodide; *C* = tetraethylammonium perchlorate; *D* = tetrapropylammonium hydroxide; *E* = tetrabutylammonium hydroxide. (Reproduced from Ref. 141 with permission of the American Chemical Society.)

An improved electrospray ionization interface for CZE-MS was described by Smith *et al.* [147]. The interface used a sheath flow of liquid to make the electrical contact at the CZE terminus, thus defining both the CZE and electrospray field gradients. This design allowed the composition of the electrospray liquid to be controlled independently of the CZE buffer. Consequently, operations with aqueous buffer with high ionic strength could be performed. Since the electrospray originated directly from the CZE capillary terminus, additional mixing volumes and metal surfaces could be eliminated and the high separation efficiency of CZE could be preserved.

Figure 3.95 shows a schematic illustration of a version of the liquid sheath electrode ESI interface described by Smith and co-workers [147]. The ESI probe body was machined from a polycarbonate rod and mounted on custom holder which were removed on a small optical bench rail. A 0.16 cm O.D. PTFE tube that contained the CZE fused silica capillary and the two sheath electrode liquid was connected via a PTFE tee outside the probe body. The polycarbonate "tip holder" carried the electrospray electrode fabricated from a 3.3 cm long, 0.25 mm I.D., 0.46 mm O.D. stainless steel (SS) tube soldered into 1.9 cm long, 0.51 mm I.D., 0.81 mm O.D. stainless steel tube. The ESI end of the SS electrode was tapered (~45°) and electropolished. The tip holder was screwed into the probe body. The stainless steel electrode was slided over the protruding CZE capillary, and made contact with the spring-loaded high-voltage connector and fitted into the central PTFE tube. The position of the CZE terminus capillary relative to the SS electrode was adjusted by sliding the capillary in the PTFE tube.

An auxiliary sheath gas flow capability was added to prevent potentially

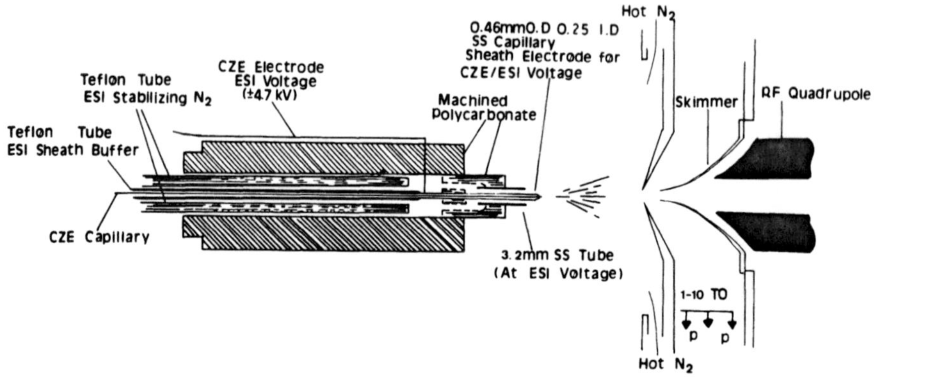

Fig. 3.95. Schematic illustration of the CZE-MS interface utilizing a liquid sheath electrode (not to scale). (Reproduced from Ref. 147 with permission of the American Chemical Society.)

deleterious effects due to heating at either the sheath electrode (due to the CZE current) or the CZE capillary (due to the flow of heated nitrogen). The central axial channel contained six 0.16 cm O.D. PTFE tube that carried nitrogen or oxygen at 0.1–1 l/min for the probe gas sheath. An additional electrode made of 0.5 cm long, 2.4 mm I.D., 3.2 mm O.D. stainless steel tube was mounted around the ESI tip in the central channel of the tip holder. It served to direct the coaxial gas flow over the tip and was held at the ESI potential by a spring connector touching the stainless steel electrode.

Ions created by the ESI process were sampled through a 1 mm nozzle into a region mechanically pumped at 50 l/s. The ions entering this region were sampled through a 2 mm diameter skimmer orifice located 0.5 cm behind the nozzle orifice. Ions passing through the skimmer entered a radio frequency focusing quadrupole lens. The region was differentially pumped with a specially designed cryopump consisting of a standard compressor and cold head with a custom cylindrical second stage baffle cooled to 14 K, which enclosed the quadrupole and provided an effective pumping speed for N_2 of >30,000 l/s. The analyzer quadrupole chamber was pumped at 500 l/s with a turbomolecular pump. A single ion lens with a 0.64 cm aperture separated the ion focusing and analysis quadrupole chamber. The pressure in the focusing and analysis chamber were $\sim 1.3 \times 10^{-9}$ and 2.6×10^{-10} bar, respectively. The counter current flow of N_2 (at ~70°C) for desolvation of the electrospray was in the range of 3–6 l/min. The mass spectrometer used had a range of m/Z 2000.

Figure 3.96 shows a detailed diagram of the ESI interface tip with the liquid sheath electrode. The voltage at the SS capillary (3–6 kV for positive ion production) defined the CZE and ESI field gradient. The actual CZE electrical contact was made with the thin sheath of liquid that flowed over the fused silica capillary. To ensure good performance, the CZE capillary should be extended only a short (but

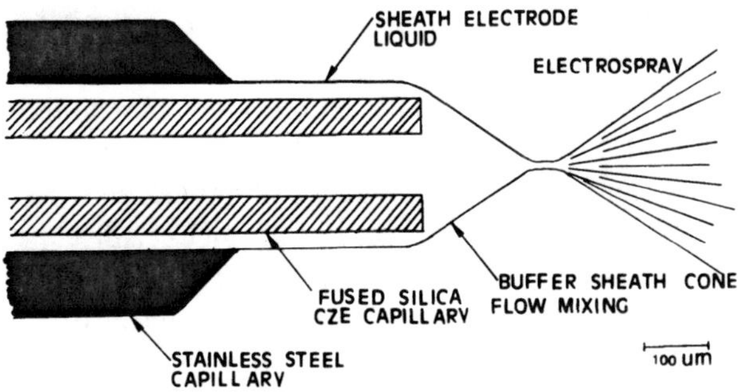

Fig. 3.96. Details of the electrospray ionisation (ESI) interface setup showing the liquid sheath electrode. (Reproduced from Ref. 147 with permission of the American Chemical Society.)

>0.2 mm) distance beyond the metal capillary. If the CZE capillary was retracted into the SS capillary, analyte signal would be lost even though a visibly unperturbed electrospray was still produced. The loss of the analyte ions could be due to an electrochemical process at the SS capillary.

The flow of sheath electrode liquid was maintained by a small syringe pump. Typical sheath electrode flow rates were 5–10 μl/min. The sheath electrode liquid can be the same as the CZE buffer, but other liquids can be used to improve electrospray performance. With acetonitrile, methanol or propanol as the sheath liquid, aqueous CZE buffers having up to at least 0.2 M ionic strength could be used. At high CZE currents (>50 μA), addition of a small amount of electrolyte to the sheath liquid was found to be useful to prevent excessive voltage drop and heat generation which could cause disruption of the electrical contact. Analyte signals were found to be relatively insensitive to the flow rate of the sheath electrode liquid.

CZE-MS separation for mixtures of quaternary phosphonium salts and for epinephine and related amines were reported [147]. The system was also used to demonstrate the operation with high surfactant concentration, which would normally be employed in micellar electrokinetic chromatography (MEKC). The ESI mass spectra of aqueous solution of sodium dodecylsulfate (50 mM) obtained in the negative and positive ion modes are shown.

On-line combination of capillary isotachophoresis (CITP) with electrospray ionization mass spectrometry was demonstrated to produce extremely high sensitivity [142]. The method involved elution of the leading electrolyte to the electrospray followed by a sequence of separated analyte bands and finally, the tailing electrolyte. The CITP-MS system permitted very high resolution separations of quaternary phosphonium ions and other ionic substances having very small differences in electrophoretic mobilities. The potential for application to very dilute sample solution was demonstrated by detection of analytes having 10^{-9} M concentrations, with signal-to-noise ratio of approximately 10^2 for some components. This sensitivity could be attributed to the relatively large sample size which could be introduced into the CITP system without loss of performance.

The extension of CZE-MS to high-molecular-mass biopolymers based upon electrospray ionization was described by Loo *et al.* [143]. Electrospray ionization produced gas-phase intact multiply charged molecular ions of biomolecules from highly charged liquid droplets by a high electric field. For high- molecular-mass substances electrospray ionization resulted in a characteristic bell-shaped distribution of multiply charged ions, with each adjacent major peak in the spectrum differing by one charge. Multiply charged molecular ions of proteins with molecular weight greater than 130,000 was observed with a quadrupole mass spectrometer of limited mass-to-charge range (m/Z 1700). Molecular weights could be readily determined for large proteins with accuracies in the range of ± 0.01 to 0.05%. The electrospray ionization method was highly sensitive and required samples in the 100 fmol to pmol range for proteins. The CZE-MS system was demonstrated for myoglobin and other proteins and polypeptides [143]. CE-MS of very large biomolecules (MM > 100,000) has been further investigated by Smith *et al.* [144,145]. Electrokinetic injection at 15 kV for 10 s resulted in injection of 1–2 pmol/component. Mass spectra of approximately 100 fmol of large proteins (bovine albumin, MM > 100,000) were obtained, where the observed charge was over 100. Selected ion monitoring was used in the detection of CE separations achieving 125,000 plates.

A useful feature of the ESI interface is that solvent clustering with the analyte can be substantially eliminated by a counter current flow of warm nitrogen (80°C) and the high potential (typically 100 to 250 V) relative to the skimmer, which led to collisions in the nozzle-skimmer region that effectively detached weakly bounded solvent molecules. Figure 3.97 shows a plot of the most intense charge state observed vs molecular mass for a set of representative proteins. A crude linear relationship was obtained [145]. In the same figure, lines roughly defining the MS observational window (m/Z 500 to m/Z 2000) are also shown. Proteins observed are expected to fall within this observational window. The maximum charge state for proteins seems to be generally predicted by the number of readily protonated sites. A good linear correlation was also observed for polypeptides and smaller proteins between the maximum number of charges and the number of basic amino acids residues (arginine, lysine, histidine, etc) which are the probable protonation sites.

An additional powerful approach for obtaining structural information by CE-MS is to apply tandem mass spectrometry to collisionally dissociate molecular ions for several of the major charge sites. An example obtained using a triple quadrupole instrument is shown in Fig. 3.98, which gives the MS-MS mass spectra for the +3 to +6 multiply protonated molecular ions of melittin. Extensive fragmentation (singly and multiply charged daughter ions) was observed for each charge state. It has been shown that fragmentation of nearly the complete sequence could be obtained [153]. Therefore, the production of multiply charged molecular ions by

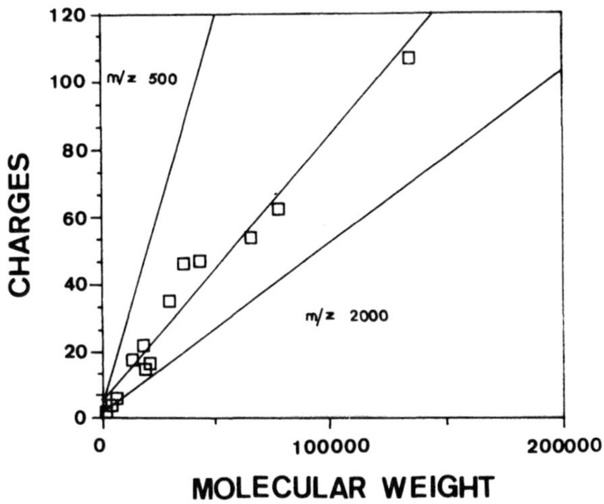

Fig. 3.97. Plot of most abundant molecular ion charge state vs. molecular mass for a representative set of proteins studied by ESI-MS. (Reproduced from Ref. 145 with permission of Elsevier Science Publishers.)

electrospray ionization not only allows for molecular weight determination of larger polypeptides with instruments of limited m/Z range, but a potential method for obtaining sequence information when combined with MS-MS techniques.

Lee *et al.* [148,149] described a variation of electrospray interface, the ion spray interface for CE-MS and CE-MS-MS analysis. An atmospheric pressure source was used to couple the sample ions produced in an ion spray LC-MS interface to a triple quadrupole mass spectrometer. Several sulfonated azo dyes were separated and detected at low ppm levels with full scan MS-MS information.

The design and performance of an ESI interface were discussed by Edmonds *et al.* [155]. The influence of electrophoretic and MS operating parameters were investigated. Examples presented included negative ion electrospray mass spectra of adenosine mono, di and triphosphates and positive ion spectra of biologically important oligopeptides (e.g. bradykinin and gramicidin) and proteins of $M_r > 75$ kDa (e.g. bovine apotransferrin).

Thilbault *et al.* [150] used an ion spray CE-MS interface for the analysis of paralytic shellfish poisons. A schematic diagram of the CE-MS interface is shown in Fig. 3.99. A triple quadrupole MS equipped with an atmospheric pressure ionization (API) source operated in the ion spray mode was used. The interface was based on a co-axial column arrangement, similar in design to that described by Smith *et al.* [144,145]. Detection limit of ca. 100 pg ($\simeq 10^{-4}$ M) was obtained for saxitoxin and neosaxitoxin.

Hallen *et al.* [151] performed preliminary investigation on the interfacing of ion mobility spectrometry to CE. In ion mobility spectrometry, gas-phase ions at atmospheric pressure are separated in time according to their mobilities as

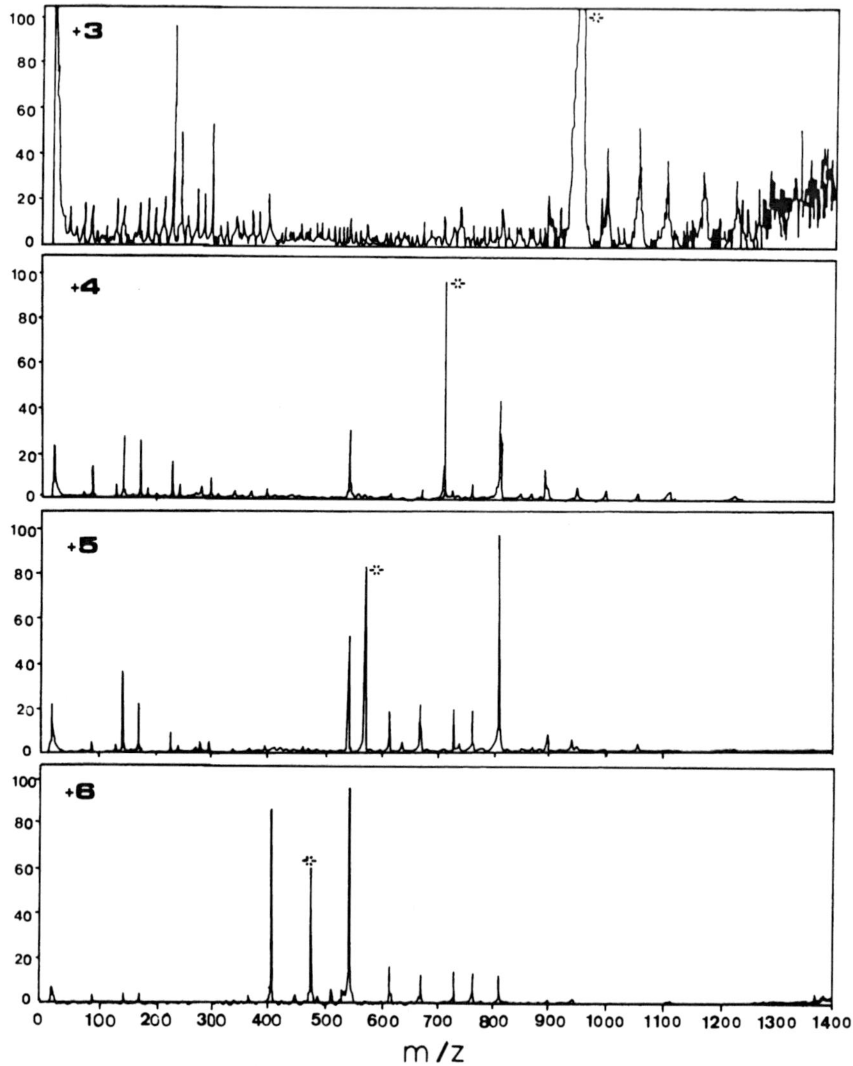

Fig. 3.98. MS-MS spectra of multiply charged protonated molecular ions of melittin (denoted by *). MS-MS of higher charged parent ions (4+, 5+ and 6+) yields higher charged daughter ions, and the information obtained can be combined to yield most of the peptide sequence. (Reproduced from Ref. 145 with permission of Elsevier Science Publishers.)

they travel through an electrical field. Several interfaces were described, including a direct-coupled electrospray interface and a makeup flow-assisted nebulization interface. In the direct-coupled electrospray, the separation column was coupled along the axis with a connector joint to a needle, which is either a metal tube or a fused silica tube containing buffer solution supplied from a reservoir, held

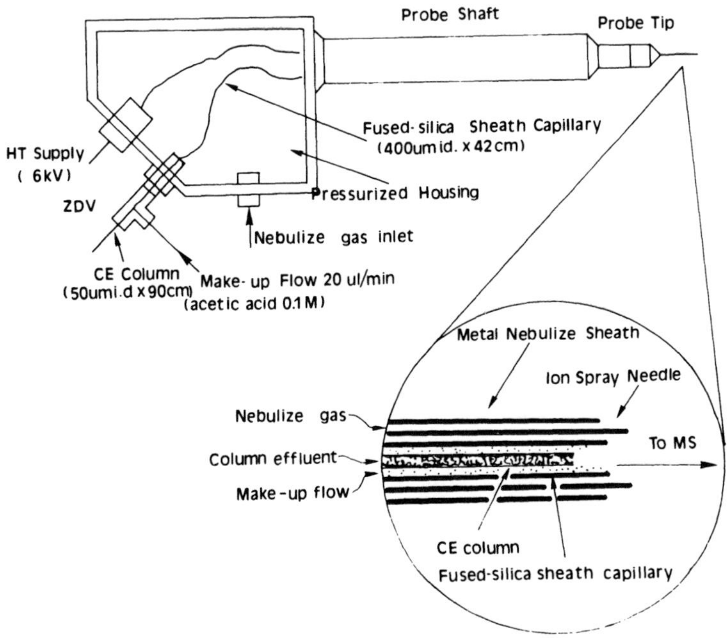

Fig. 3.99. Schematic diagram of the CE-MS interface. (Reproduced from Ref. 150 with permission of Elsevier Science Publishers.)

at the electrospray potential. In the makeup flow-assisted nebulization interface, a T-fitting was employed to couple the separation capillary, the needle and a tubing supplying a flow of a second buffer (sheath flow). Low separation efficiency (3000 plates) was obtained with these interfaces for various amines, although reproducible ion mobility spectra were obtained [151].

3.9.2 Continuous flow fast-atom bombardment (CF-FAB)

An alternative design to electrospray ionization for CE-MS is based on the continuous-flow fast bombardment interface [152–154].

An example of a CZE-coaxial CF-FAB-MS interface is shown schematically in Fig. 3.100. A detailed design of the coaxial CF-FAB probe is shown in Fig. 3.101. The CZE fused silica capillary column (1 m × 13 μm I.D.) was inserted into the sheath fused silica capillary column using a 0.1588 cm stainless steel tee to mate the two columns. The two coaxial capillary columns terminate at the FAB probe tip (Fig. 3.102), which is electrically insulated from the probe shaft with a Vespel insulator. An important feature of this interface is that the +8 kV FAB probe was used as the electrical "ground" of the CZE system, which allowed analytes to travel by electrophoretic transport to the FAB probe tip where ion desorption took place. Therefore, the method eliminates the use of a transfer line from the end of the

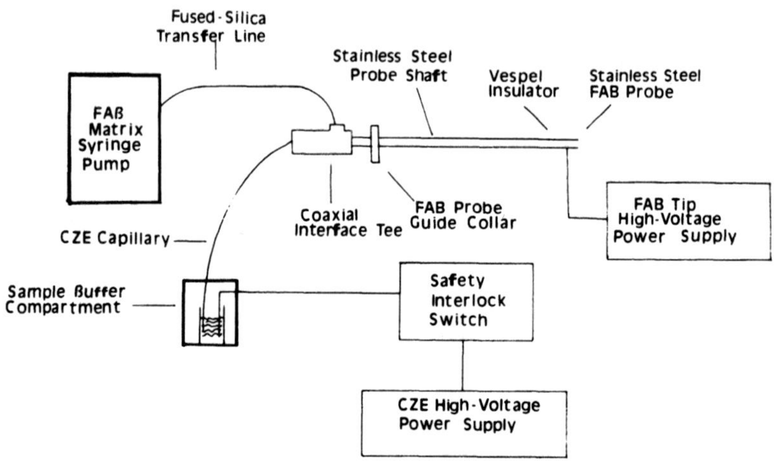

Fig. 3.100. Schematic of on-line CZE-coaxial CF-FAB-MS system. (Reproduced from Ref. 152 with permission of Elsevier Science Publishers.)

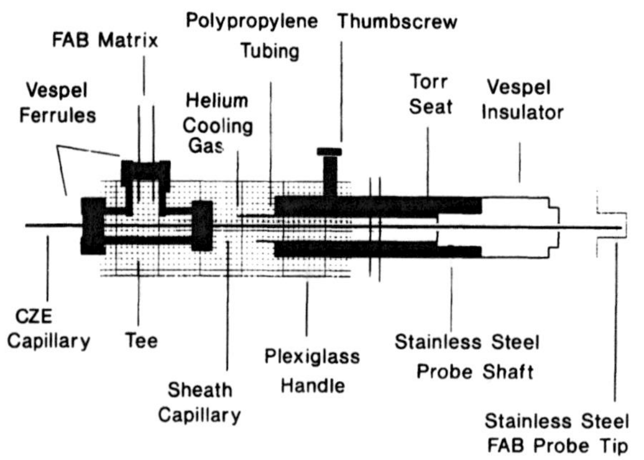

Fig. 3.101. Schematic of on-line capillary zone electrophoresis/coaxial continuous-flow atom bombardment/mass spectrometry probe (not to scale). (Reproduced from Ref. 156 with permission of the American Chemical Society.)

CZE capillary to the FAB tip, although only capillaries with very small I.D. ($\simeq$ 10 μm or less) can be used for the separation.

The FAB matrix used with the CZE-FAB-MS system was glycerol–water (25:75) with 0.0005 M heptafluorobutyric acid. The heptafluorobutyric acid served both to provide ions for electrical contact between the FAB tip and the CZE column effluent, and to acidify the solution of the FAB probe tip, increasing the production of protonated molecular ions. The mass spectrometer used was a tandem four-sector mass spectrometer of B1–E1–E2–B2 geometry. A FAB gun was used with xenon as the FAB gas (8 kV at 1 mA). The desorbed ions were accelerated to 8 kV for analysis. Mass spectra were acquired by scanning MS-I (B1–E1) and detecting the

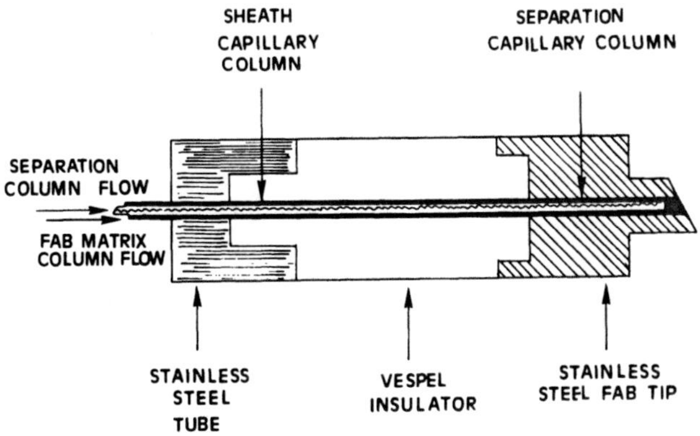

Fig. 3.102. Schematic of coaxial CF-FAB tip. (Reproduced from Ref. 152 with permission of Elsevier Science Publishers.)

ions with a photomultiplier tube based detector. MS-MS spectra were acquired by using MS-I to select the parent ions and direct them into the collision cell located in the third field free region. Helium was used as the collision gas (50% parent ion beam suppression). Daughter ion spectra were acquired by a linked scan of E2–B2 and detection in the fifth field-free region with a photomultiplier tube based detector.

The effect of two classes of buffer used for CZE on the FAB-MS spectrum was investigated [152]. These include the non-volatile buffer such as potassium phosphate buffer and the volatile buffer such as ammonium acetate. A potential problem arising from a non-volatile buffer is the possibility that the ions of the buffer can compete with protons in the formation of adducts with the charged molecular ions of the analyte. The analyte signal may divide into several different ion types, thus lowering the observed signal-to-noise ratio of the protonated molecular ions in the MS spectra, and reducing the signal-to-noise ratio of the daughter ions in MS-MS spectra. The effect of potassium phosphate buffer over a concentration range of 0.05 to 0.01 M at pH 7 was investigated by studying the mass spectra of the tripeptide Met-Leu-Phe. The intensities of the protonated molecular ions $(M+H)^+$ at m/Z 410 and the potassium adduct of the buffer concentration were measured. The results are shown in Fig. 3.103. The data showed that the intensity of the $(M+H)^+$ ion decreased while the ion intensity of $(M+H)^+$ increased as the potassium phosphate buffer concentration increased. The formation of both proton and potassium adducts from the peptide divided the analyte signal into several different signals, reducing the signal-to-noise ratio of the molecular ion data. On the other hand, the use of ammonium acetate buffer was observed to yield only protonated molecular ions with no evidence of the formation of any ammonium adducts with peptide. Therefore, volatile buffer such as ammonium acetate were recommended for use in CZE-FAB-MS operation [152].

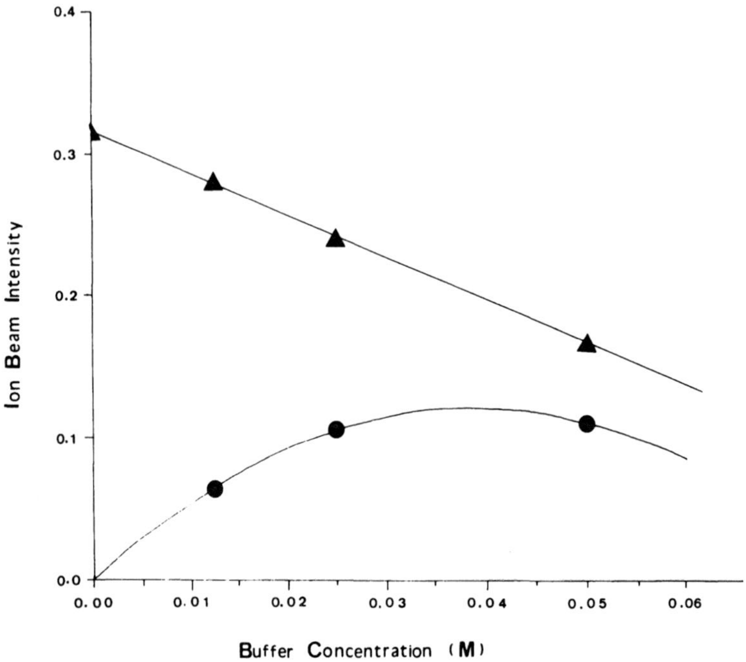

Fig. 3.103. Effects of the concentration of potassium phosphate buffer on the mass spectrum of the tripeptide Met-Leu-Phe. ▸ = (Met-Leu-Phe + H)$^+$, $y = 0.31570 - 2.9863x$, $R^2 = 1.000$; ● = (Met-Leu-Phe + K)$^+$, $y = 2.8460 \times 10^{-19} + 6.2000x - 80.000x^2$, $R^2 = 1.000$. (Reproduced from Ref. 152 with permission of Elsevier Science Publishers.)

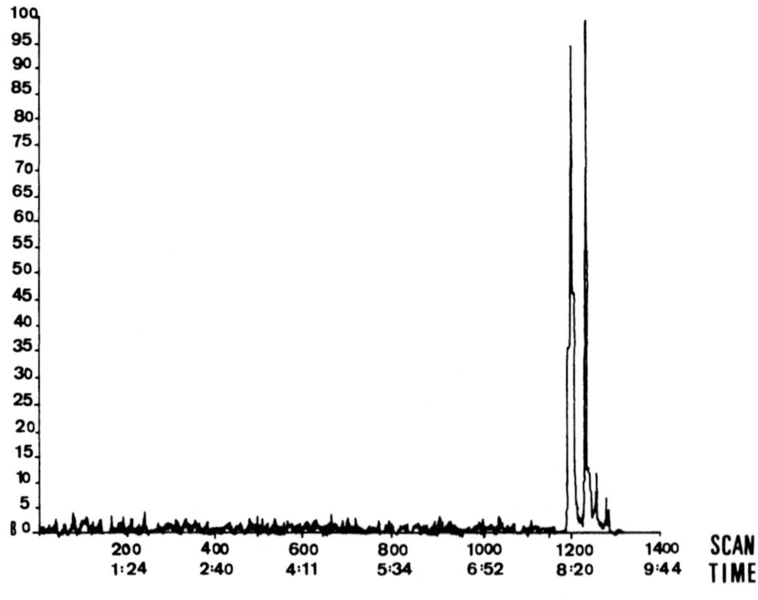

Fig. 3.104. Single-ion electropherogram of the (M + H)$^+$ ion of Met-Leu-Phe resulting from duplicate electromigration sample introductions; 32 fmol, 455,000 plates (average). (Reproduced from Ref. 152 with permission of Elsevier Science Publishers.)

The use of coaxial fused silica capillary columns to independently deliver the microcolumn effluent and the FAB matrix to the tip of the FAB probe offers several advantages: (1) The composition and flow rates of the two liquid streams can be independently optimized; (2) The FAB matrix does not affect the microcolumn separation process; (3) The peak broadening is minimized since the two liquid streams do not mix until they reach the tip of the FAB probe where ion desorption occurs; and (4) With CZE, active electrophoretic transport delivers the analyte directly to the FAB tip. Impressive MS and MS-MS spectra of as little as 32 fmol of several peptides were obtained while maintaining CE separation efficiencies of over 400,000 plates. Figure 3.104 illustrates a single-ion electropherogram of the $(M+H)^+$ ion of Met-Leu-Phe resulting from duplicate electromigration sample introductions. Figure 3.105 shows results of the on-line coaxial CF-FAB-MS-MS spectrum of 130 fmol of Met-Leu-Phe.

Caprioli *et al.* also described a continuous flow FAB interface for CE-MS [154]. A schematic representation of the CZE-CF-FAB instrument analyzer and interface

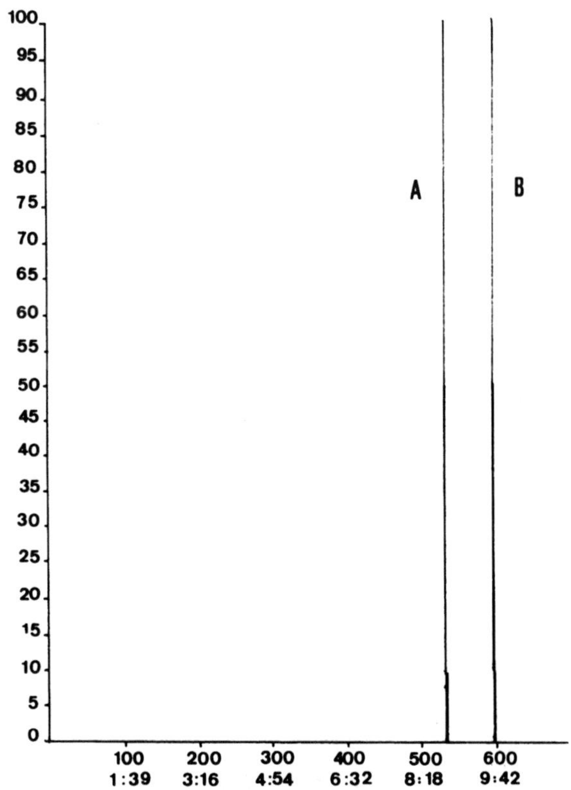

Fig. 3.105. Single-ion electropherograms of the $(M+H)^+$ ions of *N*-acetylangiotensin 1 (peak *A*, 10 fmol) and angiotensin I (peak *B*, 20 fmol) resulting from the analysis of a 10^{-6} M solution of the decapeptides. (Reproduced from Ref. 152 with permission of Elsevier Science Publishers.)

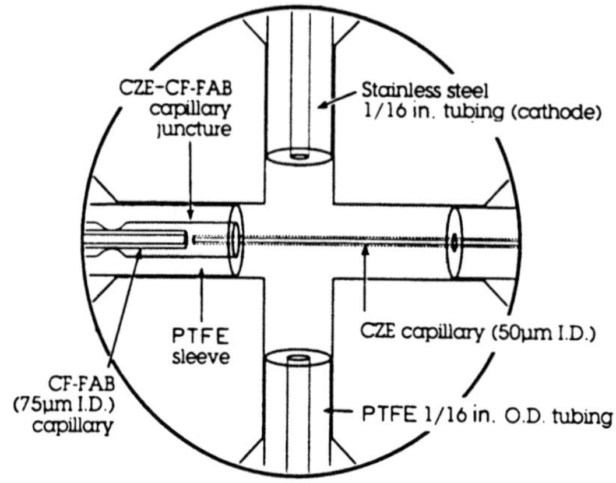

Fig. 3.106. Schematic representation of the CZE-CF-FAB interface. (Reproduced from Ref. 154 with permission of Elsevier Science Publishers.)

is shown in Fig. 3.106. The interface was a 2.54 cm × 2.54 cm × 0.9525 cm plexiglass block consisting of two intersecting passage ways (0.1588 cm I.D.) oriented 90° to each other. The effluent or cathode end of the CZE capillary entered the intake end of the CF-FAB capillary in a short segment of PTFE tubing of 0.5 mm I.D. × 0.1588 cm O.D. placed in the left horizontal passage way of the block. In the upper vertical passage way, a "flow-through" electrode, 4.5 cm × 0.1588 cm O.D. × 0.0762 cm I.D. stainless steel tubing attached to a 10 ml syringe was inserted. In the lower vertical passage, a 0.1588 cm O.D. × 0.0794 cm I.D. PTFE inlet tube was used to allow for the introduction of CF-FAB solvent from a reservoir to the cathode compartment. The "flow-through" electrode permitted periodic flushing of the compartment with CF-FAB solvent to remove bubbles formed in the interface. The 10 ml syringe attached to the interface also provided a means of introducing a sample at the anode end of the CZE capillary. Sample injection volume was estimated to be 30 nl using this injection method.

After introducing a sample, the anodic end of the capillary was transferred from a sample vial to the anode reservoir for the CE run. The flow rate difference between that in the CF-FAB capillary (75 μm I.D. × 280 μm O.D.) created by the pressure gradient (about 5 μl/min) and the electrophoretic flow rate in the CZE capillary (about 0.1 μl/min) permits efficient transfer of CZE eluate to the mass spectrometer. The composition of the CZE buffer and CF-FAB solvent varied depending on the samples analyzed. CF-FAB MS was performed with a high-resolution instrument equipped with a saddle field ion gun using xenon to create energetic atoms, and a CF-FAB probe. The mass spectrometer was operated at a resolving power of 1500 and an accelerating voltage of 4.7 kV. The scanning rate amounted to 10 s/decade. This interface was used to obtain selected ion electropherogram for tryptic digests of cytochrome C and human growth hormone. The MS of a single peptide was shown, which demonstrated the possibility of generating sequence data with CE-MS [154].

3.10 CONCLUSION

In conclusion, a wide range of detection techniques have been employed in CE separations. Currently, the most commonly used detection systems for CE are based on UV and UV-vis absorbance. All the standard commercial instruments are equipped with a UV detector. The majority of home-built CE instruments also employ this type of detection method. UV-vis absorbance detection will continue to play a dominant role in CE. The almost universal detection capability, the simple adaptation for on-column detection, and the relatively low cost are some of the factors which will continue to make UV detection an attractive technique to use. By performing multiwavelength or spectral detection, peak identification can be made easier. Although sensitivity is limited by the path length of the capillary during on-column operation, several approaches have been developed to overcome this limitation, such as the use of axial beam illumination, Z-shape flow cell and multireflection flow cell. Furthermore, it must be realized that the current generation of UV detectors used for CE are predominantly adaptation of the ubiquitous UV detectors for HPLC. There should be tremendous potential in further improving detection sensitivity by completely redesigning UV detectors specifically for CE work. It is likely that further progress in UV detection can be made in several areas, including techniques to enhance stability of the light source, to increase the gain of the photodetector, to improve focusing by suitable choice and arrangement of optical components, and to maximize the path length based on new approaches such as those described in Section 3.2.

Fluorescence is another very popular detection method for CE. As for the UV detectors, the popularity of fluorescence detectors can be partly attributed to the widespread availability of these detectors for HPLC. Additional advantages include high sensitivity and the on-column nature of the technique. Recent advances which would further enhance the capability of the technique include fluorometric diode array detection, which facilitates peak identification by spectral analysis, and programmable excitation and emission wavelengths, which allow maximum sensitivity to be obtained during the analysis. However, it is unlikely that lamp-based fluorescence detection systems will be as popular as the UV absorbance detector. The technique is to some extent hampered by the lack of fluorescent groups in most types of compounds. There is therefore frequently a need to perform derivatization by pre-column or post-column reaction to enhance detection. As a result, additional factors may have to be considered when derivatization needs to be performed, such as the design of the reactors, the stability of the fluorescent derivatives, the effect of the unused derivatizing agents or by-products, the optimum

References pp. 150–154

reaction time, and the dead volume introduced by the connections. On the other hand, for compounds which exhibit intrinsic fluorescence, or are easily and conveniently derivatized, fluorescence detection will still be the method of choice.

Lasers can be used to increase the amount of light focused onto the small detection volume in on-column detection systems. There are potential benefits in their use in both UV absorbance and fluorescence detection systems, despite the fact that only few wavelengths are available from current lasers sources. In particular, laser-induced fluorescence detection has generated tremendous excitement in recent year. The main reason for the growing interest is the impressive sensitivity achievable by this method. The detection of several hundred molecules in CE employing laser-induced fluorescence has already been demonstrated (see Section 3.4), and the ultimate goal of detecting a single molecule will probably be attempted in the near future. More importantly, along with the increase in sensitivity, efforts has also been made to reduce the cost of laser-induced fluorescence detection systems. Commercial CE systems equipped with a laser-induced fluorescence detector have already been developed (e.g. Beckman Instruments, Inc.). It is expected that this type of detectors will be available in more of the future generations of CE systems.

Lasers have also been employed in several other detection methods. Thermo-optical absorbance detection provides an interesting alternative approach, although it is unlikely to be as widely used as UV detectors in the immediate future because of the more sophisticated instrumentation required. As for laser-based refractive index detectors, the promise of universal detection would certainly propel further research efforts on this technique. In fact, sensitivity comparable to UV absorbance detection has already been attained with such a system. It is therefore possible that commercial detectors based on this type of detection technique may soon become available. Another interesting approach is laser-induced fluorescence detected circular dichroism. This technique provides a unique and sensitive method for detecting chiral compounds. Another unique detection system is based on laser-induced capillary vibration. This method overcomes the limitation imposed on sensitivity by path length, and appears to be a very promising approach.

The use of laser Raman detection has been demonstrated (see Section 3.5.4). Combination of CE with other spectroscopic techniques, such as Fourier-transform infrared spectroscopy and inductively coupled plasma spectrometry should provide immense scope for obtaining structural information.

Although detectors for HPLC are more readily adapted for use in CE, there should also be potential and advantages in employing detectors for gas chromatography in CE work. Most notably, the widely used flame ionization detector (FID) is highly sensitive and almost universal. Techniques for the coupling of this type of detection system to CE will certainly be worthwhile investigating.

Electrochemical detection provides an alternative strategy to achieve highly sensitive detection. The main challenge has been in isolating the detection system from the high electric field in the separation capillary. To date, several approaches have already been shown to be successful, and interest in electrochemical detection methods is expected to grow rapidly. Potentiometric and conductivity detection systems are relatively universal and are usually of low costs. A simple conductivity detector can be constructed at a fraction of the cost of other types of detection systems. Amperometric detectors are sensitive and selective. However, the scope of application of this type of technique in the direct mode may be rather limited, since they can only be applied to electroactive compounds. Nevertheless, these methods do not suffer from the limitation of path length as in the case of UV absorbance detection. By constructing ultramicroelectrodes using semiconductor technology, there should also be immense potential in performing multichannel detection. In the near future, the development of inert and rugged microelectrodes will probably be an area which is expected to contribute most significantly to the progress of electrochemical detection methods for CE.

The problem of the lack of chromophore, fluorophore and electrophore in certain groups of compounds can be overcome by indirect detection methods. Although indirect detection methods are usually less sensitive than their direct counterparts, they provide a virtually universal means for detection. Successful implementation of detection in the indirect mode has already been demonstrated for UV, fluorescence and electrochemical detection. Since indirect detection can be performed using similar instruments required by their direct counterparts, they can be utilized whenever necessary using available instruments.

Finally, among all the available detection techniques, mass spectrometry and tandem mass spectrometry would probably be justifiably regarded as the ultimate tools for detection. The hyphenation of CE with MS provides a sensitive method for compound identification and structure determination. CE-MS with the continuous-flow fast atom bombardment interface has been demonstrated. This technique permitted the analysis of moderately high-molecular-mass-compounds while preserving the high efficiency of the CE separation. The remarkably successful combination of CE with MS by the electrospray ionization interface represents another important advance, since the technique permits the analysis of compounds of extremely high molecular mass. The interfacing of commercial CE and MS systems with the electrospary ionization interface has already been demonstrated [157]. Commercial instruments with such an interface have become available (Finnigan Electrospray MS). Furthermore, there should also be potential in utilizing other interfacing techniques. For instance, laser desorption ionization would provide a useful method for the analysis of non-volatile and thermally labile substances [158]. Although CE-MS is a relatively new analytical technique and mass spectrometers are generally much more expensive than other types of detectors, there will certainly be a need for such systems, since they can serve as the definitive method, and they may be indispensable tools in certain applications.

From the viewpoint of improving detection sensitivity, it should be recognized that in addition to instrumental developments, other approaches can also be adopted to enhance detection sensitivity in CE. Sample preconcentration can be performed during injection by stacking and field amplified sample injection (see

Chapter 2). The use of derivatizing agents may be considered in some applications to enhance sensitivity. Alternatively, coupled packed and open-tubular columns may be employed to concentrate the analyte in the capillary (see Chapter 6).

In summary, detection systems for CE is a rapidly developing field. Much progress has already been made. There is great interest and a wide scope for further developments. To help the readers consolidate the information obtained to date, the most important characteristics of the various detection techniques which have been employed for CE are given in Table 3.2.

TABLE 3.2

DETECTION TECHNIQUES FOR CE, TYPICAL SENSITIVITY AND MAIN CHARACTERISTICS

1. *UV absorbance detection*

(a) On-column UV detection ($\simeq 10^{-6}$–10^{-4} M):
- (1) Most commonly used detection method for CE
- (2) Relatively universal
- (3) Instrumentation readily available at low cost
- (4) On-column detection
- (5) Moderately sensitive due to limitation by path length
- (6) Chromophores required

(b) Axial-beam detection (10^{-8}–10^{-6} M):
- (1) Characteristics similar to those for on-column UV detection, except that path length could be increased by about 50 times

(c) Use of Z-shaped flow cell ($\simeq 10^{-7}$–10^{-6} M):
- (1) Z-shaped flow cells are commercially available although special methods may also be used to construct flow cell.
- (2) Characteristics similar to those for on-column UV detection, except that path length could be increased by about 14 times

(d) Multireflection detection (10^{-8}–10^{-6} M):
- (1) Special methods may be used to construct multireflection flow cell
- (2) Characteristics similar to those for on-column UV detection, except that path length could be increased by about 40 times, depending on angle of incidence and reflectivity of coating

(e) Photodiode array or multiwavelength detection (10^{-6}–10^{-4} M):
- (1) Multiwavelength detection capability can be used for peak identification
- (2) Other characteristics similar to those for on-column UV detection, except that sensitivity may be slightly lower

2. *On-column fluorescence detection*

(a) Lamp-based fluorescence detection (10^{-8}–10^{-5} M):
- (1) Available in some of the commercial instruments
- (2) Relatively high sensitivity achievable
- (3) Fluorophore required

(4) Pre- or post-column derivatization may be required
(5) Choice of derivatizing agent and design of post-column reactor may need to be considered

(b) Epillumination fluorescence microscopy (10^{-12}–10^{-10} or 10^{-7}–10^{-5} M with laser or lamp, respectively):
(1) Based on modified fluorescence microscopes
(2) Highly sensitive when laser was used as light source
(3) Other characteristics similar to those for on-column fluorescence detection

3. *Laser-induced fluorescence detection*

(a) On-column laser-induced fluorescence (10^{-9}–10^{-7} M):
(1) Highly sensitive
(2) Wavelengths available from laser sources are limited
(3) Derivatization usually performed as for lamp-based fluorescence detection

(b) Laser-induced fluorescence with sheath flow cuvette (10^{-12}–10^{-9} M):
(1) Extremely sensitive
(2) Sheath flow cuvette needs to be constructed
(3) Other characteristics similar to those for on-column laser-induced fluorescence detection

(c) Fluorometric photodiode array detection (10^{-7}–10^{-5} M):
(1) Characteristics similar to those listed under on-column laser-induced fluorescence, except that multiwavelength detection can be performed

(d) Use of charge-coupled devices (10^{-12}–10^{-9} M):
(1) Two-dimensional detection with imaging capability
(2) Very high sensitivity
(3) Other characteristics similar to those for on-column laser-induced fluorescence

4. *Other laser-based detection techniques*

(a) Thermooptical absorbance detection (10^{-8}–10^{-5} M):
(1) Requires the use of two lasers, a pump laser and a probe laser
(2) Slightly more sensitive than on-column UV absorbance detection

(b) Refractive index detectors (10^{-7}–10^{-5} M):
(1) Universal detection capability
(2) Sensitivity may be improved by minimising drifts in refractive index due to thermal fluctuations

(c) Fluorescence detected circular dichroism detection (10^{-7}–10^{-5} M):
(1) Unique for detection of chiral compounds

(d) Laser Raman detection (10^{-7}–10^{-5} M):
(1) Detection by resonance Raman spectroscopy gives high sensitivity for Raman active compounds
(2) Non-resonance Raman detection exhibits low sensitivity, but may be used to detect certain types of compounds with strong Raman bands

(e) Laser-induced capillary vibration (10^{-7}–10^{-5} M):
(1) Unique approach of detection
(2) Detection of vibration of capillary due to absorption of light
(3) High sensitivity

(4) Independent of path length
(5) Requires the use of two lasers, i.e. an excitation beam and a probe beam

5. *Electrochemical detection*

(a) Potentiometric detection (10^{-8}–10^{-7} M):
(1) Requires the use of ion-selective electrodes
(2) Need to ensure that separation voltage is isolated from detection system
(3) Sensitive provided that interfering ions (e.g. other ions which permeate the ion-slective membrane) are not present

(b) Conductivity detection (10^{-8}–10^{-7} M):
(1) Relatively universal, but lacks sensitivity
(2) Need to ensure that separation voltage is isolated from detection system
(3) Low cost

(c) Amperometric detection (10^{-8}–10^{-6} M):
(1) Detection of electrophore
(2) Highly specific
(3) Highly sensitive for electroactive compounds
(4) Not many compounds are electroactive
(5) Need to ensure that separation voltage is isolated from detection system

6. *Indirect detection* (10^{-7}–10^{-5} M)

(1) Almost universal, does not require chromophore, fluorophore or electrophore
(2) Instrumentation similar to that for direct detection
(3) Slightly less sensitivity than direct detection counterpart

7. *Radioisotope detection* (10^{-9}–10^{-7} M)

(1) Useful for radioactive compounds
(2) Highly sensitive

8. *Capillary electrophoresis–mass spectrometry* (10^{-9}–10^{-5} M)

(1) Excellent for identification and structure determination
(2) High cost
(3) Design of CE-MS interface important
(4) Electrospray ionization interface very promising technique for high molecular mass compounds
(5) Continuous flow fast-atom bombardment interface may be used for moderately high molecular mass compounds

3.11 REFERENCES

1 W.G. Kuhr, Anal. Chem., 62 (1990) 403R
2 M. Yu and N.J. Dovichi, Anal. Chem., 61 (1989) 37
3 A.E. Bruno, E. Gassmann, N. Pericles and K. Anton, Anal. Chem., 61 (1989) 876
4 J.A. Lux, H. Hausig and G. Schomburg, J. High Resolut. Chromatogr., 13 (1990) 373
5 R.M. McCormick and R.J. Zagursky, Anal. Chem., 63 (1991) 750

6 Y. Walbroehl and J.W. Jorgenson, J. Chromatogr., 315 (1984) 135
7 J.S. Green and J.W. Jorgenson, J. Liq. Chromatogr., 12 (1989) 2527
8 F. Foret, M. Deml, V. Kahle and P. Bocek, Electrophoresis, 7 (1986) 430
9 Z. Prusik, V. Kasicka, S. Stanek, G. Kuncova, M. Hayer and J. Vrkoc, J. Chromatogr., 390 (1987) 87
10 V. Rohlicek and Z. Deyl, J. Chromatogr., 494 (1989) 87
11 T. Wang, R. Hartwick and P. Chaplin, J. Chromatogr., 462 (1989) 147
12 S. Moring, J.C. Colburn, P.D. Grossman and H.H. Lauer, LC-GC, 8 (1990) 34
13 M. Sepaniak, D. Swaile and A. Powell, J. Chromatogr., 480 (1989) 185
14 I.H. Grant and W. Steuer, J. Microcol. Sep., 2 (1990) 74
15 X. Xi and E.S. Yeung, Appl. Spectrosc., 45 (1991) 1199
16 J.A. Taylor and E.S. Yeung, J. Chromatogr., 550 (1991) 831
17 J.P. Chevret, R.E.J. van Soest and M. Ursem, J. Chromatogr., 543 (1991) 439
18 T. Wang, J.H. Aiken, C.W. Huie and R.A. Hartwick, Anal. Chem., 14 (1991) 1372
19 S. Kobayashi, T. Ueda and M. Kikumoto, J. Chromatogr., 480 (1989) 179
20 J. Vindevogel, P. Sandra and L.C. Verhagen, J. High Resolu. Chromatogr., 13 (1990) 295
21 S.K. Yeo, H.K. Lee and S.F.Y. Li, J. Chromatogr., 585 (1991) 133
22 S. Hjerten, J. Chromatogr., 270 (1983) 1
23 S. Hjerten and M.D. Zhu, Protides Biol. Fluids, 33 (1985) 537
24 S. Fujiwara and S. Honda, Anal. Chem., 59 (1987) 2773
25 F. Foret and P. Bocek, Chem. Listy, 83 (1989) 191
26 T. Kaspar, M. Melera, P. Gozel and R. Brownlee, J. Chromatogr., 458 (1988) 303
27 S. Terabe, T. Yashima, N. Tanake and M. Araki, Anal. Chem., 60 (1988) 1673
28 V. Dolnik, J. Lui, J.F.J. Banks, M. Novotny and P. Bocek, J. Chromatogr., 480 (1989) 321
29 J. Liu, J.F.J. Banks and M. Novotny, J. Microcol. Sep., 1 (1989) 136
30 M. Zhu, D. Hansen S. Burd and F. Gannon, J. Chromator., 480 (1989) 311
31 Z. Deyl, V. Rohlicek and M. Adam, J. Chromatogr., 480 (1989) 371
32 T. Tsuda, G. Nakagawa, M. Sato and K. Yagi, Appl. Biochem., 5 (1983) 330
33 A. Cohen, D. Nagarian, A. Paulus, A. Guttman, J. Smith and B. Karger, Proc. Natl. Acad. Sci., U.S.A., 85 (1988) 9660
34 V. Sustacek, F. Foret and P. Bocek, J. Chromatogr., 480 (1989) 271
35 H. Yamamoto, T. Manabe and T. Okuyama, J. Chromatogr., 480 (1989) 331
36 R.M. McCormick, J. Liq. Chromatogr., 14 (1991) 939
37 M.E. Swartz, J. Liq. Chromatogr., 14 (1991) 923
38 D.M. Northrop, D.E. Martire and W.A. MacCrehan, Anal. Chem., 63 (1991) 1038
39 C.P. Ong, C.L. Ng, H.K. Lee and S.F.Y. Li, J. Chromatogr., 547 (1991) 419
40 J.W. Jorgenson and K.D. Lukacs, Anal. Chem., 53 (1981) 1298
41 J.S. Green and J.W. Jorgenson, J. Chromatogr., 352 (1986) 337
42 D.R.J. Rose and J.W. Jorgenson, J. Chromatogr., 447 (1988) 117
43 M. Albin, R. Weinberger, E. Sapp and S. Moring, Anal. Chem., 63 (1991) 417
44 T. Tsuda, Y. Kobayashi, A. Hori, T. Matsumoto and O. Suzuki, J. Chromatogr., 456 (1988) 375
45 T. Tsuda, Y. Kobayashi, A. Hori and T. Matsumoto, J. Microcol. Sep., 2 (1990) 21
46 L. Hernandez, R. Marquina, J. Escalona and N.A. Guzman, J. Chromatogr., 502 (1990) 247
47 L. Hernandez, J. Escalona, N. Joshi and N. Guzman, J. Chromatogr., 559 (1991) 183
48 J. Green and J. Jorgenson, J. High Resolut. Chromatogr. Chromatogr. Commun., 7 (1984) 529

49 D.F. Swaile and M.J. Sepaniak, J. Microcol. Sep., 1 (1989) 155
50 D. Rose and J. Jorgenson, J. Chromatogr., 438 (1988) 23
51 E. Gassman, J. Kuo and R. Zare, Science, 230 (1985) 813
52 M. Roach, P. Gozel and R. Zare, J. Chromatogr., 426 (1988) 129
53 P. Gozel, E. Gassmann, H. Michelson and R. N. Zare, Anal. Chem., 59 (1987) 44
54 S. Pentoney, X. Huang, D. Burgi and R. Zare, Anal. Chem., 60 (1988) 2625
55 D.E. Burton, M.J. Sepaniak and M. Maskarinec, J. Chromatogr., Sci., 24 (1986) 347
56 E. Gurthrie, J. Jorgenson and P. Dluzneski, J. Chromatogr., 22 (1984) 171
57 B. Nickerson and J. Jorgenson, J. High Resolut. Chromatogr. Chromatogr. Commun., 11 (1988) 878
58 S. Wu and N.J. Dovichi, J. Chromatogr., 480 (1989) 141
59 B. Nickerson and J. Jorgenson, J. Chromatogr., 480 (1989) 157
60 B. Wright, G. A. Ross, R.D. Smith, J. Microcol. Sep. 1 (1989) 85
61 J. Liu, O. Shirotu and M. Novotny, Anal. Chem., 63 (1991) 413
62 J. Liu, Y.-Z. Hsieh, D. Wiesler and M: Novotny, Anal. Chem., 63 (1991) 408
63 H. Drossman, J.A. Lackus, A.J. Kostichka, J. D'Cunha and L.M. Smith, Anal. Chem., 62 (1990) 900
64 H. Swedlow and R. Gesteland, Nucl. Acid Res., 18 (1991) 1415
65 A.S. Cohen, D.R. Najarian and B.L. Karger, J. Chromatogr., 516 (1990) 49
66 H. Swerdlow, J.Z. Zhang, D.Y. Chen, H. R. Harke, R. Grey, S. Wu and N.J. Dovichi, Anal. Chem., 63 (1991) 2835
67 Y.-F. Cheng, R.D. Piccard and T. Vo-Dinh, Appl. Spectrosc., 44 (1990) 755
68 J.V. Sweedler, J.B. Sheer, H.A. Fishman and R.N. Zare, Anal. Chem., 63 (1991) 496
69 B. Nickerson and J. Jorgenson, J. High Resolut. Chromatogr. Chromatogr. Commun., 11 (1988) 533
70 A. T. Balchunas and M.J. Sepaniak, Anal. Chem., 60 (1988) 617
71 N.J. Dovichi and Y.F. Cheng, Am. Biotechnol. Lab., 7 (1989) 10
72 Y.F. Cheng and N.J. Dovichi, Science, 242 (1989) 562
73 T.W. Garner and E.S. Yeung, Anal. Chem., 62 (1990) 2195
74 D. Swaile, D. Burton, A. Balchunas and M. Sepaniak, J. Chromatogr. Sci., 26 (1988) 406
75 P. Christensen and E. Yeung, Anal. Chem., 61 (1989) 1344
76 J. W. Jorgenson and K.D. Lukacs, Science, 222 (1983) 266
77 R. McCormick, Anal. Chem., 60 (1988) 2322
78 D.J. Rose and J.W. Jorgenson, Anal. Chem., 60 (1988) 642
79 M. Sepaniak, D. Swaile and A. Powell, J. Chromatogr., 480 (1989) 185
80 K. Cobb and M. Novotny, Anal. Chem., 61 (1989) 2226
81 M. Yu and N.J. Dovichi, Anal. Chem., 61 (1989) 37
82 M. Yu and N.J. Dovichi, Mikrochim. Acta, III (1988) 27
83 M. Yu and N.J. Dovichi, J. Appl. Spectrosc., 43 (1989) 196
84 J. Pawliszyn, Anal. Chem. 60 (1988) 2796
85 J.W. Jorgenson and K. D. Lukacs, J. High Resolut. Chromatogr. Chromatogr. Commun., 4 (1981) 230
86 S. Pentoney, X. Huang, D. Burge and R. Zare, Anal. Chem., 60 (1988) 2625
87 R.A. Wallingford and A.G. Ewing, Anal. Chem., 59 (1987) 1762
88 A. Cohen, P. Paulus and B.L. Karger, Chromatographia, 24 (1987), 15
89 P. Grossman, K. Wilson, G. Petrie and H. Lauer, Anal. Biochem., 173 (1988) 265
90 A. Guttman, A. Paulus, A. Cohen, N. Grinberg and B. Karger, J. Chromatogr., 448 (1988) 41

91 P. Grossman, J. Colburn and H. Lauer, Anal. Biochem., 179 (1989) 28
92 J. Liu, K.A. Cobb and M. Novotny, J. Chromatogr., 468 (1989) 55
93 X. Huang and R.N. Zare, Anal. Chem., 63 (1991) 2193
94 S. Terabe, M. Shibata and Y. Miyashita, J. Chromatogr., 480 (1989) 403
95 A. Dobashi, T. Ono, S. Hara and J. Yamagychi, J. Chromatogr., 480 (1989) 413
96 S. Fanali, L. Ossicini, F Foret and P. Bocek, J. Microcol. Sep., 1 (1989) 190
97 P. Gozel, R.N. Zare, ASTM Spec. Tech. Publ., 1009 (1988) (Prog. Anal. Lumin.), 41
98 K. Otsuka, S. Terabe and T. Ando, J. Chromatogr., 332 (1985) 219
99 A.T. Balchunas and M.J. Sepaniak, Anal. Chem., 59 (1987) 1466
100 S. Honda, S. Iwase, A. Makino and S. Fujiwara, Anal. Biochem., 176 (1989) 72
101 D. Bornhop and N. Dovichi, Anal. Chem., 59 (1987) 1632
102 A.E. Bruno, A. Paulus and D.J. Bornhop, Appl. Spectrosc., 45 (1991) 462
103 A.E. Bruno, B. Krattiger, F. Maystre and H.M. Widmer, Anal. Chem., 63 (1991) 2689
104 C.Y. Chen, T. Demana, S.D. Huang and M.D. Morris, Anal. Chem., 61 (1989) 1590
105 C. Chen and M.D. Morris, Apppl. Spec., 42 (1988) 515
106 C.-Y. Chen and M.D. Morris, J. Chromatogr., 540, (1991) 355
107 J. Wu, T. Odake, T. Kitamori and T. Sawada, Anal. Chem., 63 (1991) 2216
108 X. Huang, J.A. Luckey, M.J. Gordon and R.N. Zare, Anal. Chem., 61 (1989) 766
109 F. Mikkers, F. Everaerts and T. Verheggen, J. Chromatogr., 169 (1979) 11
110 M. Deml, F. Foret and P. Bocek, J. Chromatogr., 320 (1985) 159
111 J. Beckers, T. Verheggen and F. Everaerts, J. Chromatogr., 452 (1988) 591
112 C. Haber, I. Silvstri, S. Roosli and W. Siman, Chimia, 45 (1991) 117
113 X. Huang, T.K. Pang, M.J. Gordon and R.N. Zare, Anal. Chem., 59 (1987) 2747
114 X. Huang, M.J. Gordon and R.N. Zare, J. Chromatogr., 425 (1988) 385
115 X. Huang, M.J. Gordon and R.N. Zare, J. Chromatogr., 480 (1989) 285
116 X. Huang, R.N. Zare, S. Sloss and A.G. Ewing, Anal. Chem., 63 (1991) 189
117 P. Gebauer, M. Deml, P. Bocek and K. Janak, J. Chromatogr., 267 (1983) 455
118 R.A. Wallingford and A.G. Ewing, Anal. Chem., 59 (1987) 678
119 R.A. Wallingford and A.G. Ewing, Anal. Chem., 60 (1988) 1972
120 R.A. Wallingford and A.G. Ewing, Anal. Chem., 60 (1988) 258
121 R.A. Wallingford and A.G. Ewing, J. Chromatogr., 441 (1988) 299
122 R.A. Wallingford and A.G. Ewing, Anal. Chem., 61 (1989) 98
123 R.A. Wallingford, P.D. Curry, Jr. and A.G. Ewing, J. Microcol. Sep., 1 (1989) 23
124 C.D. Gaitonde and P.V. Pathak, J. Chromatogr., 514 (1990) 389
125 Y.F. Yik, H.K. Lee, S.F.Y. Li and S.B. Khoo, J. Chromatogr., 585 (1991) 139
126 C.E. Engstrom-Silvermann and A.G. Ewing, J. Microcol. Sep., 3 (1991) 141
127 R. Virtanen, Acta Polytech. Scand., 123 (1974) 1
128 E.S. Yeung, Acc. Chem. Res., 22 (1989) 125
129 S. Hjerten, K. Elenbring, F. Kilar, J. Liao, A.J. Chen, C.J. Siebert and M. Zhu, J. Chromatogr., 403 (1987) 47
130 F. Foret, S. Fanali, L. Ossicini and P. Bocek, J. Chromatogr., 470 (1989) 299
131 G.J.M. Bruin, G. Stegeman, A.C. cvan Asten, X. Xu, J.C. Kraak and H. Poppe, J. Chromatogr., 559 (1991) 163
132 W.G. Kuhr and E.S. Yeung, Anal. Chem., 60 (1988) 1832
133 L. Gross and E.S. Yeung, J. Chromatogr., 480 (1989) 169
134 B.L. Hogan and E.S. Yeung, J. Chromatogr. Sci., 28 (1990) 15
135 T.M. Olefirowicz and A.W. Ewing, J. Chromatogr., 499 (1990) 713

136 S. Pentoney, R. Zare and J. Quint, Anal. Chem., 61 (1989) 1642
137 S. Pentoney, R. Zare and J. Quint, J. Chromatogr., 480 (1989) 259
138 K.D. Altria, C.F. Simpson, A.K. Bharij and A.E. Theobald, Electrophoresis, 11 (1990) 732
139 D. Kaniansky, P. Rajec, A. Svec, J. Marak, M. Koval, M. Lucka, S. Franko and G. Sabanos, J. Radioanal. Nucl. Chem., 129 (1989) 305
140 R.D. Smith, J.A. Olivares, N.T. Nguyen and H.R. Udseth, Anal. Chem., 60 (1988) 436
141 J.A. Olivares, N.T. Nguyen, C.R. Yonker, R.D. Smith, Anal. Chem., 59 (1987) 1230
142 H.R. Udseth, J.A. Loo and R.D. Smith, Anal. Chem., 61 (1989) 228
143 J.A. Loo, H.R. Udseth and R.D. Smith, Anal. Biochem., 179 (1989) 404
144 R.D. Smith, H.R. Udseth, J.A. Loo, B.W. Wright, G.A. Ross, Talanta, 36 (1989) 161
145 R.D. Smith, J. A. Loo, C. Barinaga, C. Edmonds and H. Udseth, J. Chromatogr., 480 (1989) 211
146 J.A. Loo, H.K. Jones H.R. Udseth and R.D. Smith, J. Microcol. Sep., 1 (1989) 223
147 R. Smith, C. Barinaga and H. Udseth, Anal. Chem., 60 (1988) 1948
148 E.D. Lee, W. Mueck, J.D. Henion and T.R. Covey, Biomed. Environ. Mass Spectrom., 18 (1989) 844
149 E.D. Lee, W. Mueck, J.D. Henion and T.R. Covey, Biomed. Environ. Mass Spectrom., 18 (1989) 253
150 P. Thibault, S. Pleasance and M.V. Laycook, J. Chromatogr., 542 (1991) 483
151 R. Hallen, C. Shumate, W. Siems, T. Tsuda and H. Hill, J. Chromatogr., 480 (1989) 233
152 M. Moseley, L. Deterding, K. Tomer and J. Jorgenson, J. Chromatogr., 480 (1989) 197
153 M. Moseley, L. Deterding, K. Tomer and J. Jorgenson, Rapid Commun. Mass Spectrom. 3 (1989) 87
154 R Caprioli, W. Moore, M. Martin, B. DaGue, K. Wilson and S. Moring, 480 (1989) 247
155 C.G. Edmonds, J.A. Loo, C.J. Barinaga H.R. Udseth and R.D. Smith, J. Chromatogr., 474 (1989) 21
156 M.A. Moseley, L.J. Deterding, K.B. Tomer and J.W. Jorgenson, Anal. Chem., 63 (1991) 109
157 R.D. Smith, H.R. Udseth, C.J. Barinaga and C.G. Edmonds, J. Chromatogr., 559 (1991) 197
158 R.N. Zare, J.H. Hahn and R. Zenobi, Bull. Chem. Soc. Jpn., 61 (1988) 87

CHAPTER 4

Column Technology

4.1 INTRODUCTION

Rapid progress in capillary electrophoresis in recent years may be attributed mainly to the availability of high-quality fused silica microcapillaries and advances in column technology. In capillary electrophoresis, the major aim of using capillaries is the achievement of efficient heat dissipation necessary for high efficiencies requiring high separation voltages [1–4]. With the use of microcapillaries, extremely high separation efficiencies can be achieved. However, to further improve the performance of the technique, several other factors need to be considered. First, there are currently very limited types of high-quality tubing materials which would have the necessary thermal, chemical and physical properties, and are available in very small dimensions (less than 100 μm in I.D.). Techniques to improve the properties of the tubing used would probably open up new opportunities for further development of CE. Secondly, with the use of small I.D. capillaries with circular cross-section, detection sensitivity may be compromised, especially in the case of optical detection where sensitivity is path length dependent. Consequently, non-circular cross-section tubings may have to be considered in certain cases. Thirdly, the interaction of analytes with the inner surface of the capillary (e.g. adsorption) may have an effect on the migration of certain species. By applying a suitable surface coating, the properties of the surface can be manipulated to some extent. Fourthly, by filling the capillary with a suitable type of porous gel, the capillary can be used to perform size sieving separations. Last but not least, by introducing packing materials into the capillary, enhanced selectivity due to the use of mixed mechanisms may be obtained in certain separations.

In this chapter, the most important developments in column technology are described which include the use of new types of uncoated open-tubular columns, non-circular cross-section tubings, columns with coatings on the inner surface, gel-filled columns, packed capillaries, as well as a novel chip-like device for capillary electrophoresis.

4.1.1 Uncoated columns

In recent years, nearly all capillary electrophoresis separations have been performed in polyimide-coated fused silica capillaries. The main reasons for the popularity of fused silica capillaries include their flexibility, good thermal and optical properties in the UV range, and most importantly, the availability of high-quality fused silica tubings with internal diameters below 100 μm.

However, there are several potential drawbacks with the use of fused silica capillaries. The first is that the silica surface possesses hydroxyl groups which can interact with charged molecules. The second is that to minimize heating effects, capillaries with small internal diameters are employed. Consequently, limitation on detection sensitivity is imposed by the dimension of the capillary, especially when optical detection is used.

Schomburg *et al.* [5,6] gave accounts of the problems and achievements in column technology for chromatography and capillary electrophoresis. Turner [7] highlighted the new developments in capillary electrophoresis columns. Recent developments include the use of rectangular tubings, optically transparent outer coatings, new types of wall-coated capillary columns and gel columns. In view of the tremendous interests shown in recent years, significant advances are likely to be made in the area of capillary-surface chemistry and column technology for capillary electrophoresis.

4.1.2 Use of rectangular tubings

One of the main problems associated with the use of small bore cylindrical capillaries is the limitation on detection sensitivity when on-column optical detection is employed. The causes of the problem include short path length, and distortion and scatter of light caused by the rounded capillary walls. One method to alleviate this problem is to use rectangular capillaries. Tsuda *et al.* [8] investigated the use of rectangular borosilicate glass capillaries as an alternative to cylindrical capillaries. Typical dimensions ranged from 16 μm $\times$ 195 μm to 50 μm $\times$ 1 mm. Detection across the long cross-sectional axis provides a significant increase in the sensitivity of detection techniques which depend on path length, such as UV-vis absorbance. The enhancement in sensitivity is illustrated in Fig. 4.1, which shows the capillary electropherograms obtained using UV-vis absorbance detection across the short and long axis respectively.

Another advantage of using rectangular capillaries is that, due to their higher surface area–volume ratio which is favourable to heat dissipation, larger volume rectangular capillaries can be used when compared with cylindrical capillaries. For instance, square capillaries (50 μm $\times$ 50 μm) were found to provide slightly higher separation efficiencies than round capillaries, demonstrating that corners do not significantly degrade the separation.

Rectangular tubings are only available in borosilicate glass with no protective coating. Therefore they tend to be much more fragile than polyimide-coated

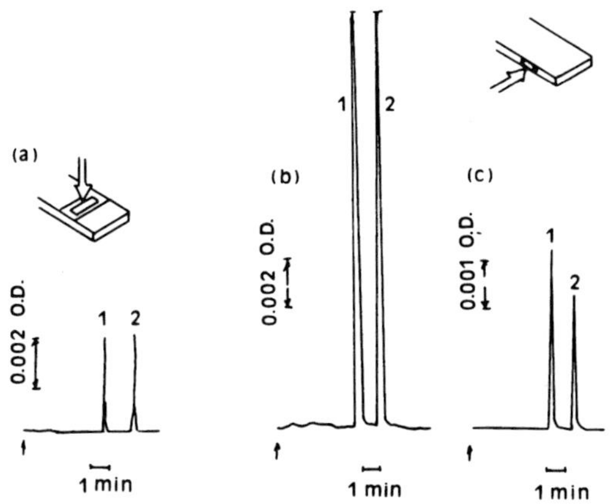

Fig. 4.1. CZE electropherogram of (*1*) pyridoxine and (*2*) dansyl-L- serine, each at 4.2×10^{-5} M, using two different detector arrangements of the rectangular capillary in the UV-vis absorbance detector. In (a), the capillary was positioned so that detection was across the 50 μm axis of the capillary, and in parts (b) and (c), across the 1000 μm axis of the capillary. In (b), the electropherogram is recorded by using the same detector sensitivity as in part a and the peaks are off-scale, while in (c), the sensitivity has been reduced by a factor of 5. (Reproduced from Ref. 8 with permission of the American Chemical Society.)

fused silica tubings. Effective protective coatings may need to be developed for rectangular tubings before they would gain wider acceptance.

Izumi *et al.* [9] used flattened poly(ethylene-propylene) tubing of 0.5 mm I.D., 1.0 mm O.D., 0.2 mm^2 cross-sectional area for CE. The tubing was flattened simply by pressing with a glass bottle, except for 5 cm at each end. The final dimensions of the tubing were 0.2 × 0. 8 mm I.D., 0.7 × 1.2 mm O.D., 0.1 mm^2 cross-sectional area. The tubing was subsequently coated twice by passage of a 1% hydroxypropyl methyl cellulose (HPMC) solution and heating at 120°C for 3 h. Figure 4.2 shows a comparison of the separation of immunoglobulin G (IgG) obtained for flat tubing and round tubing. IgG was resolved into at least 25 peaks in each case. The migration times in the analysis with flat tubing was significantly shorter. This is because a higher voltage could be used without overheating owing to a better surface area/volume ratio.

4.1.3 Capillaries with optically transparent outer coatings

To achieve on-column optical detection it is necessary to remove the polyimide coating in a small section of the separation capillary to form the detection windows. An alternative solution is to replace the polyimide with an optically transparent capillary coating for silica. Since the detection window is usually the most fragile part after the removal of the protective coating, the advantage of an optically transparent coating is that it would help to make CE columns much easier to handle

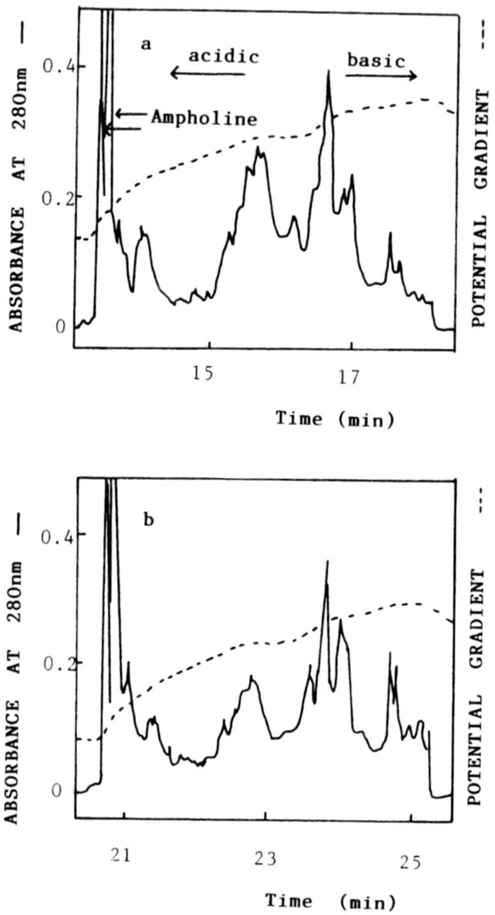

Fig. 4.2. Comparison of the IgG separation patterns obtained from (a) flat tubing and (b) round tubing. (Reproduced from Ref. 9 with permission of Dr. Alfred Huethig Publishers.)

during change of column and everyday use. Recently, fused silica tubings with UV transparent coating has become available commercially (e.g. from Polymicro Technologies, Inc.). As the flexibility and chemical inertness of this type of tubing continue to improve, they will probably become the preferred type of tubing for use in CE.

4.2 COATED COLUMNS

Some of the problems encountered when using fused silica tubing with an uncoated inner surface in CE separations are the possibility of irreproducibility of electroosmotic flow and the adsorption of charged molecules on the capillary surface. The electrostatic wall-analyte interactions cause peak tailing and thereby reduce separation efficiency. One solution involves deactivation of the silica surface by chemical modification. A theoretical explanation for the use of a polymer coating to eliminate electroosmosis has been provided [10]. It is expected that electrophoresis and electroosmosis are roughly governed by the following equations:

$$\mu_{ep} = \frac{\varepsilon\, \zeta_{ep}}{4\pi\, \eta} \tag{4.1}$$

$$\mu_{eo} = \frac{\varepsilon\, \zeta_{eo}}{4\pi\, \eta} \tag{4.2}$$

where μ_{ep} is the electrophoretic mobility, μ_{eo} the electroosmotic mobility, ζ_{ep} the zeta potential of the solute, ζ_{eo} the zeta potential of the tube wall, ε the dielectric constant and η the bulk viscosity. According to these equations, there is no net gain in suppressing electroosmosis by increasing the viscosity of the buffer. The reason is that the electrophoretic mobility will decrease by the same extent as the electroosmotic mobility. However, it is possible to suppress electroosmosis by operating under conditions such that $|\zeta_{eo}| \ll |\zeta_{ep}|$, and therefore $|\mu_{eo}| \ll |\mu_{ep}|$. One way to achieve this is to use materials which are sufficiently inert to prevent electroosmosis. Unfortunately, even the most inert plastic tubes give considerable electroosmosis. It is therefore necessary to adopt another approach by considering the following formula:

$$\mu_{eo} = \frac{\varepsilon}{4\pi} \int_0^{\zeta_{eo}} \frac{1}{\eta}\, d\varphi \tag{4.3}$$

where φ is the electrical potential. It is noted that the value of the integral approaches zero when the viscosity, η, in the double layer close to the tube wall approaches infinity. Consequently, by coating the inner surface of an electrophoresis tube with a polymer solution of high viscosity, electroosmosis can be virtually eliminated. Any neutral polymer that is soluble or swells in water can be used, such as methylcellulose or non-cross-linked polyacrylamide. Moreover, if these polymers are dissolved in the buffer, they will also suppress electroosmosis, probably because the polymers tend to adhere to the tube wall and thereby create a thin surface layer of high viscosity. On the other hand, the electrophoretic mobility is higher in the buffer alone than in a polymer containing buffer which tends to be more viscous. Shorter analysis times can therefore be obtained in the absence of the polymer when capillaries of the same length are used. However, it should be noted that for separations in which electroosmotic flow is not required, a shorter capillary with polymer coating can be used and hence the analysis time can also be reduced.

4.2.1 Techniques for coating CE capillaries

In this section, the techniques for the coating of capillaries for use in CE are considered [10–27]. Some examples of the type of coatings that have been used to modify fused silica surface for CE are give in Table 4.1.

Chemical derivatization of the capillary wall is a widely used technique for changing the properties of the silica surface in the preparation of coated columns for gas chromatography applications. In their early attempts to reduce electroosmotic flow, Jorgenson and Lukacs [11] used trimethylchlorosilane (TMCS) to silylate the silica surface. Subsequently, many approaches have been adopted to improve the effectiveness and stability of the coatings for CE columns. Polyacrylamide [10,12] and polyethylene glycol [13,14] coatings are two of the most commonly employed types of coatings. An aryl-pentafluoro coating [15–17] was also found to be effective in preventing protein adsorption. Recently two types of coatings have been developed which have shown greater hydrolytic stability at both acidic and basic pHs. One is a polyacrylamide coating with Si–C bonding to the silica wall [18].

TABLE 4.1

EXAMPLES OF CAPILLARY COATINGS FOR CE AND THEIR TYPICAL APPLICATIONS

Coating	Typical applications	Reference
Polyacrylamide coating with siloxane bond	Aromatic carboxylic acids; Proteins in pH range 2–10.5	10,12
Polyethylene glycol	Proteins such as lysozyme, trypsin and chymotrysinogen with efficiencies in the range 8×10^4–1.5×10^5 theoretical plates	13,14
Aryl pentafluoro	Mixture of protein markers with efficiences in the range 3–7 $\times 10^5$ theoretical plates	15–17
Polyacrylamide with Si–C bond to silica	Proteins in the pH range 2–10.5	18
Polyethyleneimine	Protein separations optimized using the wide pH range (2–12) of the positively charged coating	19
Non-ionic surfactants	Mixture of proteins in the pH range 4–11	20
LC stationary phases	Proteins	21,22,27
GC stationary phases	DNA restriction fragments	23
Charged-reversal coating	Basic proteins	24
Poly(vinylpyrrolidinone)	Proteins with molecular masses between 12 and 77 kDa	25
Epoxydiol	Mixture of lysozyme and cytochrome C	26
Maltose	Proteins	26

Another is a polyethyleneimine coating [19]. Nonionic surfactant coatings have been shown to be effective in preventing adsorption [20]. Coatings based on LC [21,22,27] and GC [23] types of stationary phases have also been employed. A charge-reversal surface modification technique has been developed [24]. In addition, a number of other miscellaneous coating techniques have been described [13,14,16]. Various coatings have been used with varying degrees of success. Currently column coatings for CE is an area of active research. Undoubtedly further progress will be made which will further enhance the scope of application of CE.

4.2.1.1 Polyacrylamide coating with siloxane bond

Hjerten [10] found that (γ-methylacryloxypropyl)trimethoxysilane was an effective reagent for silylation of the capillary surface. By subsequently cross-linking the surface-bound methylacryl groups with polyacrylamide, electroosmotic flow could be eliminated and adsorption of analyte onto the surface could be minimized. The method was based on the use of a bifunctional compound in which one group reacted specifically with the glass wall and the other with a monomer taking part in a polymerization process. Besides (γ-methyl-acryloxypropyl)trimethoxysilane, other examples of such bifunctional compounds are vinyltriacetoxysilane, vinyltri(β-methoxy-ethoxy) silane, vinyltrichlorosilane and methylvinyldichlorosilane, where one or two of the methoxy, acetoxy, methoxyethoxy or chloro groups react with the silanol groups in the glass wall, whereas the aryl or vinyl groups with acryl or vinyl monomers to form a polymer, e.g. non-cross-linked polyacrylamide, poly(vinylpyrrolidone), poly(vinyl alcohol). It was found that this procedure gave a thin, well defined monomolecular layer of a polymer covalently bound to the glass wall.

The detailed experimental procedure involved first mixing about 80 μl of (γ-methylacryloxypropyl)trimethoxysilane with 20 ml of water, which had been adjusted to pH 3.5 by acetic acid. The capillary was then filled with the silane solution. After reaction at room temperature for about 1 h, the silane solution was withdrawn. After flushing with water, the capillary tube was filled with a de-aerated 3 or 4% (w/v) acrylamide solution containing 1 μl of N,N,N$'$,N$'$-tetramethylethylenediamine (TEMED) and 1 mg potassium persulfate per ml solution. After about 30 min, excess polyacryamide was sucked away and the tubes were rinsed with water. The effectiveness of this type of coating was demonstrated by performing capillary electrophoresis of aromatic carboxylic acids [10]. Subsequently, a similar coating procedure with minor modifications was also employed for the analysis of human serum samples and nucleotides [12].

4.2.1.2 Polyethylene glycol coating

Herren *et al.* [13] evaluated the effectiveness of various coatings for control of electroosmotic flow. Figure 4.3 shows the types of glass coatings investigated, which include methylcellulose, polyethylene glycol (PEG), diol and dextran. It was found that coatings of polyethylene glycol of above 5000 molecular weight could greatly

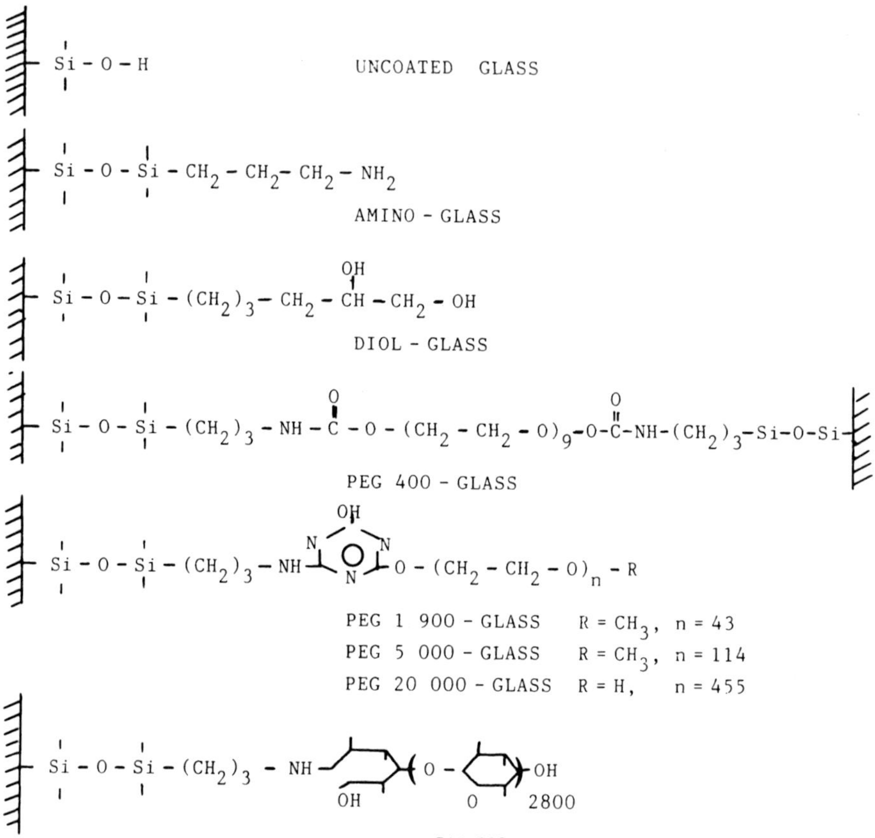

Fig. 4.3. Several different types of glass coatings. (Reproduced from Ref. 13 with permission of Academic Press, Inc.)

reduce electroosmosis. Furthermore, they were stable for long periods of time and were more effective than dextran, methylcellulose or silane coatings.

Before coating was applied, the quartz glass capillaries were cleaned by sonication in 1% (w/w) PEG 8000 solution, rinsed three times with distilled water, and then soaked sequentially for 1 h each in alcoholic NaOH, distilled water, and aqua regia. After being rinsed overnight in distilled water, the items were dried for 4 h at 65 ± 5°C and 666 Pa. Capillaries were further cleaned in a radio frequency glow discharge apparatus.

In the case of the PEG-coated columns, polyethylene glycol coatings of average molecular mass 400 was covalently bonded to glass in one step whereas those of higher molecular masses (1900, 5000 or 20,000) involved two steps. In the first step an aminopropyl sublayer was applied. To achieve this, clean glass in a glass pressure vessel was covered with a 20% (w/v) solution of 3-aminopropyltriethoxysilane and a vacuum of approximately 133 Pa was applied to remove air trapped on the glass surface. The vessel was then sealed and heated in an oil bath at 100°C for 24 h, with occasional stirring. This amino glass (see Fig. 4.3) was washed several times with distilled water. The whole process was repeated once more, and the glass was then washed several times with acetone and dried under vacuum. Except for PEG 400, the second step of PEG coating was then performed. Dry aminopropyl glass was placed in a pressure vessel and covered with a 20% (w/v) solution of cyanuric chloride activated PEG. A vacuum of 133 Pa was then applied, and the vessel was heated at 100°C for 24 h. The glass was washed several times with distilled water, and the entire second-step process was then repeated.

Bruin *et al.* [14] modified fused silica capillaries with (γ-methyl-acryloxypropyl)trimethoxysilane and polyethylene glycol 600 in order to decrease the influence of wall adsorption in CE separation of proteins. The coating procedure is shown schematically in Fig. 4.4. The capillary (50 μm or 100 μm I.D.) was first etched with 1 M potassium hydroxide solution for 3 h at room temperature and rinsed with water for 10 min. The capillary was then flushed with 0.1 M hydrochloric acid to remove K^+ ions from the wall and to produce free silanol groups at the surface of the wall. The capillary was dried at 200°C for 3 h by gently flushing with helium. A solution of (γ-methylacryloxypropyl)trimethoxysilane in dried toluene (10%, v/v) was pumped through the capillary at 110°C for 3 h at an inlet pressure of 0.5 MPa. The unbound reagent was flushed from the column with toluene. Subsequently the epoxide group was opened by a reaction carried out in the same manner with a

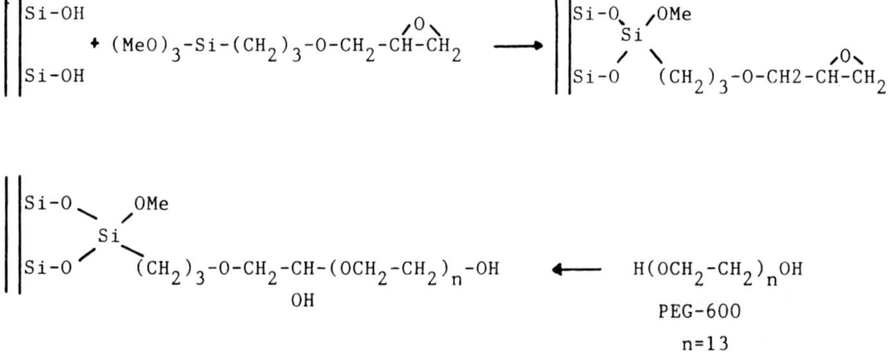

Fig. 4.4. Scheme of the procedure for the deactivation of the silica wall. Me = methyl. (Reproduced from Ref. 14 with permission of Elsevier Science Publishers.)

solution of 20% polyethylene glycol and 2% boron trifluoride etherate in dioxane for 1 h at 100°C. Finally, the capillary was rinsed with distilled water.

Using this type of column, a significant decrease in adsorption was obtained and electro-osmotic flow was also diminished [15]. Symmetrical peaks were obtained for the proteins studied in the pH range 3–5, although some adsorption still occur as the plate numbers were below theoretical expectations. At higher pH values appreciable peak deformations and drastic decreases in resolving power were observed. Nevertheless, the procedure permitted rapid and efficient separation of protein mixtures, which were suitable in the indicated pH range, and the coating showed a good stability.

4.2.1.3 Aryl-pentafluoro coating

The use of an aryl-pentafluoro (APF)-coated column for CE was investigated by Swedberg [15,16]. In this coating procedure, the capillary was first silylated using 0.1% (γ-methylacryloxypropyl)trimethoxysilane solution. The solution was pumped through the capillary using a syringe pump at a flow rate of 1–2 column volume per minute for 30 min. The capillary was then flushed with helium overnight. Dry toluene was used to rinse the columns before a solution of 0.2 M pentafluorobenzoyl chloride in toluene was pumped through the capillaries. To re-equilibrate the column back to aqueous condition, toluene, methanol and finally water were used for washing the column. Finally the column was equilibrated with the appropriate running buffer before use.

The capillary system minimized protein–surface interaction, resulting in high efficiencies ($\geq$300,000 theoretical plates). It allowed the analysis of a set of protein standards over a wide pI range at neutral pH and moderate ionic strength. By using non-ionic and zwitterionic surfactants together with APF-coated capillaries [16], enhanced selectivity was achieved in the analysis of a tricyclic antidepressant (desipramine) and six peptides.

Maa *et al.* [17] investigated the impact of wall modifications on protein elution in CE. Same silylation procedures were used for the preparation of alpha-lactalbumin bonded column as those described for the preparation of APF columns. Average efficiencies of over 250,000 plates were obtained for the separation of a protein mixture.

4.2.1.4 Polyacrylamide coating with Si–C bond to silica

Cobb *et al.* [18] reported the use of a polyacrylamide-coated capillary similar to that described by Hjerten [11], except that the coating was bonded to the silica wall through a Si–C bond, rather than a Si–O–Si bond. The Si–C bond is hydrolytically more stable than the siloxane bond, and hence results in improved coating stability.

The reaction scheme for the preparation of vinyl-bound polyacrylamide-coated capillaries is shown in Fig. 4.5. The scheme consists of three separate stages. First, the silica surface is chlorinated through a reaction of thionyl chloride with the surface silanol groups. Secondly, the chlorinated silica is reacted with a Grignand

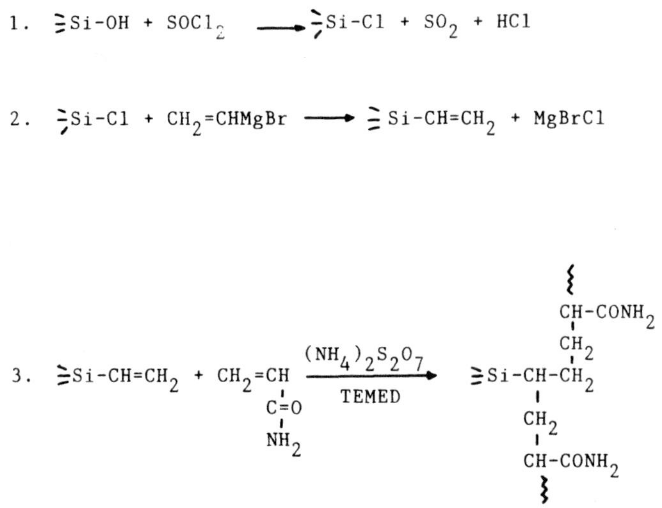

Fig. 4.5. Reaction scheme for the preparation of vinyl-bound polyacrylamide-coated capillaries . (Reproduced from Ref. 18 with permission of the American Chemical Society.)

reagent containing a terminal double bond which provides a site for subsequent bonding with the acrylamide, e.g. vinyl magnesium, to form a direct attachment of the vinyl group with the silica surface. Thirdly, the vinyl group is reacted with acrylamide monomer and polymerising agents ammonium persulfate and TEMED, resulting in a linear non-cross-linked polyacrylamide coating.

The coating was found to be stable over a wide range of pH conditions (2–10.5). It also reduced the electrostatic adsorption of proteins to the silica capillary walls, hence improving the separation efficiencies. Electroosmotic flow was practically eliminated, resulting in reproducible migration times. Although some interactions of the proteins with the capillary walls seemed to still exist, the adsorption appears to be reversible and equilibrium could be reached rapidly. Consequently, there was very little peak tailing. Figure 4.6 shows the capillary electrophoretic separation of model protein mixture at pH 2.7, using coated (A) and uncoated (B) fused silica capillaries. Separation of the protein mixture is clearly improved by the use of the coated capillary.

4.2.1.5 Polyethyleneimine coating

Another type of durable, positively charged, hydrophilic coating was developed by Towns and Regnier [19]. The synthetic route for the coating is shown in Fig. 4.7. High-molecular-mass polyethyleneimines (PEI) is first adsorbed to the capillary wall and then cross-linked with a cross-linking agent, ethyleneglycol diglycid ether (EDGE), to form a relatively thick (≃3 nm), stable layer. This layer is positively charged and causes a reversal of the electroosmotic flow. With a high-molecular-mass polymer, PEI-200 (molecular mass 20,000), the coating was

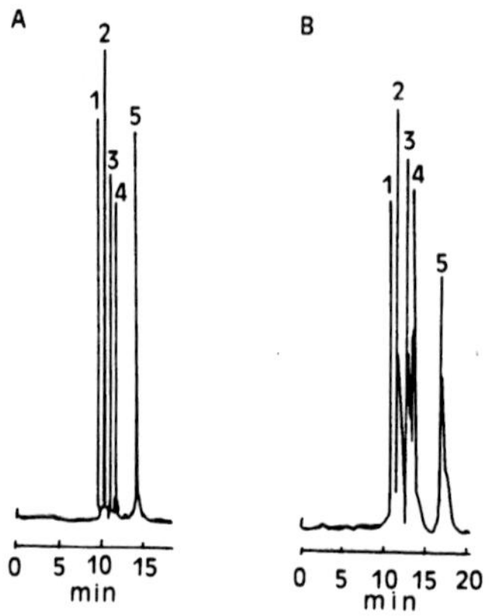

Fig. 4.6. Capillary electrophoretic separations of model protein mixture at pH 2.7, using (A) coated and (B) uncoated fused silica capillaries. Separation conditions for each electropherogram include the following: buffer, 0.03 M citric acid (pH 2.7, adjusted with 1 M NaOH); capillary, 50 μm × 60 cm (45 cm to detector); hydrodynamic injection, 5 s with 20 cm height differential; applied field. (A) 22 kV, 10 μA, and (B) 12 kV, 5 μA. Peaks: *1* = cytochrome C (horse heart); *2* = lysozyme (chicken egg white); *3* = trypsin (bovine pancreas); *4* = trypsinogen (bovine pancreas); *5* = trypsin inhibitor (soybean). (Reproduced from Ref. 18 with permission of the American Chemical Society.)

stable from pH 2–12 and could be used over a wide pH range without substantial change in electroosmotic flow. Figure 4.8 shows the separation of model proteins on a PEI-200–EDGE-coated capillary, with the polarity of the power supply reversed from that used on uncoated fused silica capillaries.

4.2.1.6 Non-ionic surfactant coating

Towns and Regnier [20] also used non-ionic surfactant-coated capillaries for the separation of proteins. Table 4.2 contains a list of the water-soluble TWEEN series surfactants and BRIJ series surfactants which were investigated.

Figure 4.9 shows the structure of the two types of surfactants. Performance parameters of the coatings for the five selected surfactants are shown in Table 4.3.

TABLE 4.2

SELECTED WATER-SOLUBLE NON-IONIC SURFACTANTS

Surfactant	*m* (oxyethylene units)	*n* (alkyl chain length)
TWEEN-20	20	12
TWEEN-40	20	16
TWEEN-80	20	18
BRIJ-35	23	12
BRIJ-78	20	18

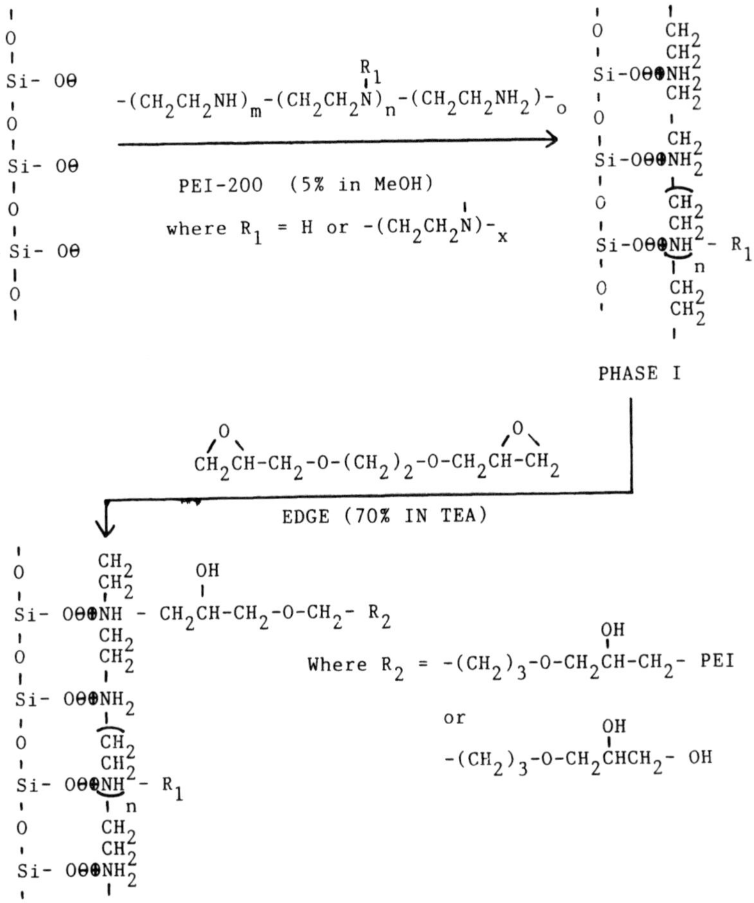

Fig. 4.7. Synthetic route to an adsorbed PEI-bonded phase. The coating process is two steps: (1) adsorption of PEI-200 onto the fused silica capillary, and (2) cross-linking of the PEI-200 polymer in order to stabilize the coating further. (Reproduced from Ref. 19 with permission of Elsevier Science Publishers.)

A comparison of the performance data for the TWEEN series surfactant with those of the BRIJ surfactants reveals that the performance of the coating is not significantly affected by the alkyl chain length (i.e. for TWEEN-20, 40 and 80), but depends strongly on the size and structure of the head-group of the surfactant. The increase in efficiency and decrease in electroosmotic flow for BRIJ-35 may be due to the more compact BRIJ-35 surfactant head-group which is able to cover the alkylsilane surface more effectively and the silanol groups. In the case of BRIJ-78, which has an even smaller head group than BRIJ-35 but longer alkyl chain length, the head group is probably too small to mask the surface, thus allowing proteins to interact with the C_{18}.

Figure 4.10 shows electropherograms of five basic proteins obtained using

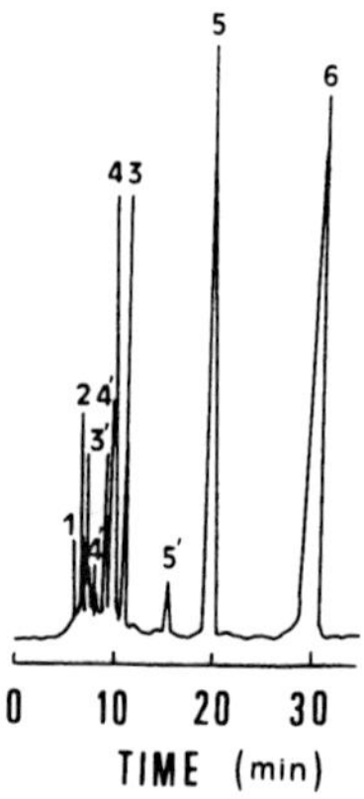

Fig. 4.8. Capillary electrophoretic separation of proteins on a PEI-200–EDGE-coated capillary. Conditions: capillary length, 50 cm; separation length, 35 cm; I.D., 75 μm; buffer, 0.02 M hydroxylamine–HCl at pH 7.0; separation potential, 12.5 kV. Peaks: *1* = mesityl oxide (neutral marker); *2* = horse-heart myoglobin; 3 = bovine ribonuclease A; *4* = bovine chymotrypsinogen A; *5* = horse heart cytochrome C; *6* = hen-egg lysozyme. Peaks *3'*, *4'* and *5'* are impurities in *3*, *4* and *5*, respectively. (Reproduced from Ref. 19 with permission of Elsevier Science Publishers.)

(a) $HO(CH_2CH_2O)_m-(CH_2)_{n-1}CH_3$

(b)

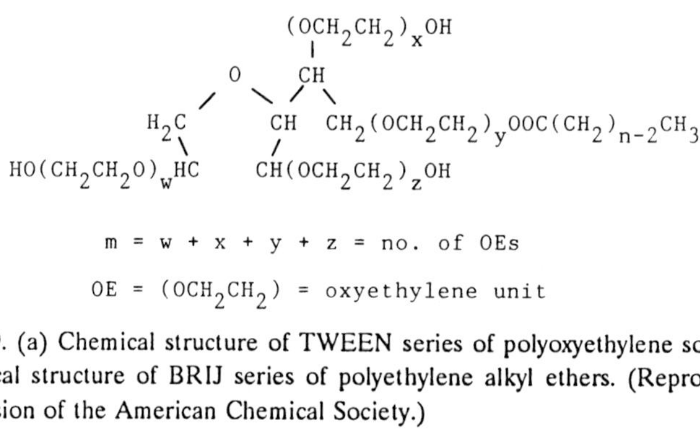

$m = w + x + y + z$ = no. of OEs

OE = (OCH_2CH_2) = oxyethylene unit

Fig. 4.9. (a) Chemical structure of TWEEN series of polyoxyethylene sorbitan monoalkylates. (b) Chemical structure of BRIJ series of polyethylene alkyl ethers. (Reproduced from Ref. 20 with permission of the American Chemical Society.)

capillaries with TWEEN-20 and BRIJ-35 coatings. The surfactant-coated capillaries were found to give high recoveries of proteins, and were stable over the pH range 4 to 11.

4.2.1.7 LC type of coatings

Dougherty *et al.* [21,27] bonded various LC type of stationary phase coatings to the inner wall of fused silica capillaries. The bonded phases investigated included moderately hydrophobic (C_8), highly hydrophobic (C_{18}), hydrophilic (polar) and

TABLE 4.3

ELECTROOSMOTIC FLOW AND PLATE NUMBER FOR THE FIVE SELECTED SURFACTANT MOLECULES ADSORBED ONTO AKYLSILANE-COATED CAPILLARIES (Adapted from Ref. 20)

Surfactant	Electroosmotic flow $\times 10^8$, $m^2/V\ s$	Plate number
TWEEN-20	2.03	170,000
TWEEN-40	2.48	135,000
TWEEN-80	2.27	150,000
BRIJ-35	1.50	240,000
BRIJ-78	1.26	115,000

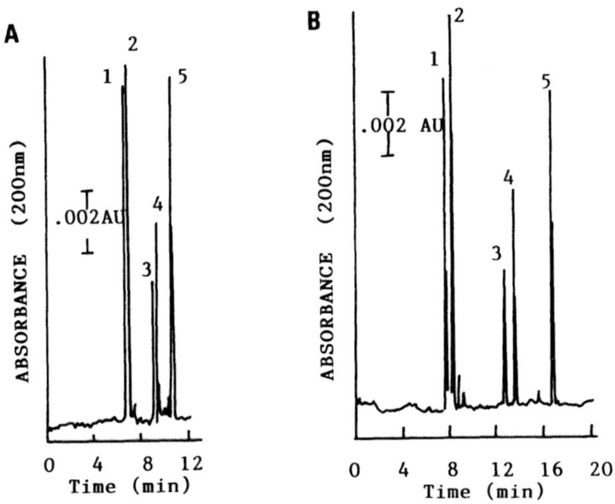

Fig. 4.10. Electropherograms showing the separation of five basic proteins: *1* = lysozyme; *2* = cytochrome C; *3* = ribonuclease A; *4* = α-chymotrysinogen; and *5* = myoglobin. Buffer: 0.01 M phosphate buffer at pH 7.0. Capillary: 75 μm × 50 cm. (A) TWEEN-20/alkylsilane capillary, and (B) BRIJ-35/alkylsilane capillary at 300 V/cm. (Reproduced from Ref. 20 with permission of the American Chemical Society)

low hydrophobic (C_2). The stability of the coatings are compared in Fig. 4.11. The electropherograms obtained for tryptic digest of casein using a bare silica tubing and a C_2-coated tubing at pH 4 and 9 are shown in Fig. 4.12. The bonded CE columns have been reported to separate proteins at near neutral pH, under low ionic strength, and with no buffer additives [21,27].

Bruin *et al.* [22] used octadecylsilica (ODS)-coated column to demonstrate electrically driven (ED) open-tubular liquid chromatography. In their coating procedure, etched and porous silica layered (PSL) fused silica capillaries were prepared according to Tock and co-workers [29,30].

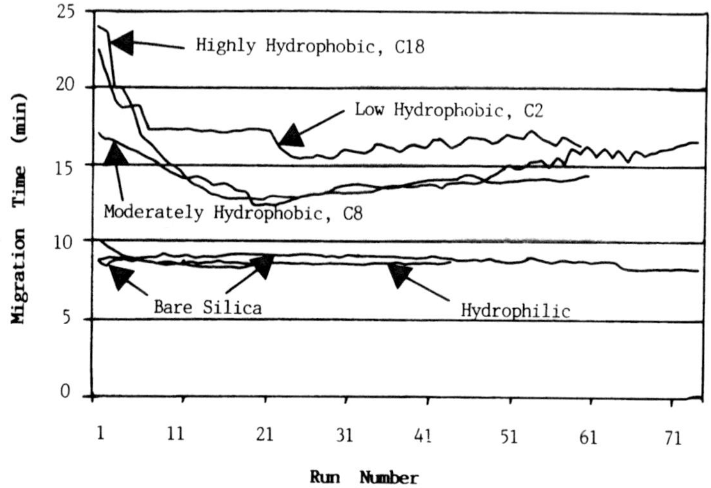

Fig. 4.11. Stability of several different types of coated columns for CE. (Reproduced from Ref. 21 with permission of Supelco Co.)

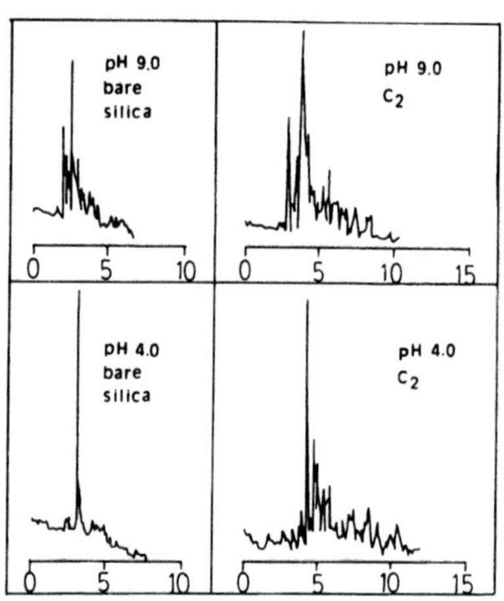

Fig. 4.12. Separation of tryptic digest of casein using bare silica and bonded C_2 columns. Conditions: 10 mM sodium phosphate–6 mM sodium borate; 200 V /cm. (Reproduced from Ref. 21 with permission of Supelco Co.)

The capillaries were dried at 200°C for at least 2 h while being purged with helium. The dried capillary was filled with a 5% (w/v) solution of ODS in toluene. Both ends were sealed in a flame and the capillary was heated at 140°C for 6 h. Finally the capillary was rinsed with toluene and acetonitrile or methanol before use.

Capillaries with inner diameters in the range 5–25 μm were investigated for both electrically driven and pressure-driven chromatography. The efficiency of the former was found to be better by a factor of 2. Injection of the sample in the electrically driven case was much simpler. The results also showed that electroosmotic mobility depended little on the application of an ODS coating.

4.2.1.8 GC type of coatings

The use of capillary gas chromatography (GC) columns for CE is another approach adopted to eliminate wall effects. Ulfelder *et al.* [23] used a microbore DB-17 (50% phenylmethyl silicone stationary phase) for the separation of DNA restriction fragments. A buffer containing hydroxypropylmethylcellulose (HPMC) was used as the electrolyte.

4.2.1.9 Charge-reversal coating

A charge-reversal surface modification technique has also been used to modify the capillary surface in order to improve the separation of proteins [24]. In this method, a positively charged polymeric coating agent is ionically adsorbed to the negatively charged silica surface, forming a neutral surface initially. Subsequently, hydrophobic interactions between surface-bound polymers and free polymers produces a positively charged, second polymer layer. Upon reversal of the surface charge, the electroosmotic flow is reversed. Hence, the polarity of the system must be reversed in order to ensure that all analytes travel past the detector. The procedure required to form the coating involved simply rinsing the additive, which is presumably a cationic surfactant, through the capillary prior to analysis. Separation of several basic proteins were demonstrated.

4.2.1.10 Miscellaneous coatings

Many other types of coated columns have been employed for CE separations. Several of these coatings are described below, including methylcellulose, poly(vinylpyrrolidinone), glycero-glycidoxypropyl, epoxydiol and maltose coatings.

For the preparation of clean capillary surfaces were activated by filling for two 10-min periods with a solution containing (γ-methylacryloxypropyl)trimethoxysilane. This solution was prepared by adding 3.1 g of the reagent to 125 ml of 80% (v/v) methanol in water, acidified with one drop of glacial acetic acid. The capillaries were dried at 65 $\pm$ 5°C and 666 Pa for 1 h, soaked for 10 min in a solution of 1.0 g methylcellulose (MM 110,000) per liter water, and vacuum dried for 1 h, and drained. This rinsing procedure was repeated 25 times over a 72 h period.

For the dextran coating, clean aminopropyl glass capillaries in a glass pressure vessel were covered with an aqueous solution of 10% (w/v) dextran T500 (MM 500,000) in 10% (w/v) NaCl, and were exposed to a vacuum of 133 Pa. The vessel was then opened and 100-fold molar excess of sodium cyanohydride was added to provide coupling via reductive amination of the amino-glass at the dextran-carbonyl end. The reaction was conducted for 24 h at 100°C. The polymer-coated glass was then washed with distilled water, and recoated once.

McCormick [25] studied the CE separation of peptides and proteins using low-pH buffers in modified silica capillaries. Capillaries were modified with phosphate moieties from the separation buffer as well as with conventional silanes.

The procedure for capillary modification involved first flushing the capillaries for 30 min with 1 M KOH and de-ionized water prior to filling with separation buffer or bonding with silane. In the case of poly(vinylpyrrolidinone)-modified (PVP) capillaries, the capillary was flushed with several volumes of aqueous acetic acid (pH 3.5). The silane reagent mixture, 80 μl of (γ-methylacryloxy propyl)trimethoxysilane in 20 ml of pH 3.5 aqueous acetic acid, was introduced by vacuum suction continuously for 3 h. The capillary was then washed with distilled water for 3 h. Subsequently, the reagent mixture consisting of 3% aqueous 1-vinyl-2-pyrrolidine adjusted to pH 6.2 and containing 1 mg/ml ammonium persulfate and 1 μl/ml N,N,N′,N′-tetramethylethlenediamine was introduced into the silylated column by vacuum continuously for 90 min. A polymeric layer on the capillary wall with a hypothetical structure $-[Si(O)\,(CH_2)_3\,OCOCH\,(CH_3)\,CH_2\,(CH_2CH\,C_4H_6NO)_n\,]_m$ was formed. Unbound reagent was flushed from the capillary with water. Other capillaries were modified according to the same procedure, except that acrylamide and acrylic acid were used instead of the 1-vinyl-2-pyrrolidinone reagent. Capillaries prepared in this manner were found to maintain the performance for several weeks at low pH.

In the case of glycero-glycidoxypropyl-derivatized capillaries, (3-glycidoxyopropyl)diisopropylethoxysilane was used. The reagent mixture consisting of 100 μl of silane, 50 μl of N,N-diisopropylethoxysilane and 5 ml of dry toluene was pumped through the heated capillary (100°C) at 50 μl/h for 4 days with a syringe pump. The capillary was then flushed with toluene and dioxane to remove residue reagent. The epoxide functional group on the silane was opened by reacting at 90°C with 5 ml of 1.6 mM glycerol in dry dioxane containing 80 μl of boron trifluoride etherate, which was pumped through the capillary at 140 μl/h. A monomeric hydrophilic bonded phase with the possible structure $-Si(i\text{-}Pr)_2\,[(CH_2)_3\,OCH_2CH\,(OH)\,CH_2\,OCH_2CH\,(OH)\,CH_2\,(OH)]$ was formed. This type of bonded phases exhibit superior stability relative to bonded phases prepared with silanes containing methyl or methoxy group mainly because of the presence of the bulky diisopropyl protecting groups.

The interaction of phosphate with the capillary surface was studied and it was found to bind strongly to the silica surface. Modification reduced electroosmotic flow and shielded the surface from protein adsorption. Synthetic octapeptides with single amino acid substitutions were separated, as were larger proteins up to 77 kDa.

Bruin *et al.* [26] also investigated the use of epoxydiol coatings for the separation of lysozyme, trypsin and chymotrypsinogen and maltose coating for the separation of lysozyme and cytochrome C.

The diol coating of Regnier and Noel [28] has been investigated by Herren *et al.* [13]. After hydrolysis, the glass with this coating contained $-Si-CH_2-CH_2-$

CH_2–O–CH_2 CHOH–CH_2OH groups. However, this coating was found to be less effective in reducing electroosmosis compared with the polyethylene glycol coating (Section 4.2.1.2).

4.3 COLUMNS FOR CAPILLARY GEL ELECTROPHORESIS (CGE)

The use of gel-filled capillaries for CE constitutes an important area of application which has demonstrated great potential in separation science [32–58]. Capillary gel electrophoresis (CGE) has attained the highest separation efficiencies ever achieved by any analytical technique to date. Theoretical plates in the 10–20 million range can be routinely achieved in a gel-filled column less than 1 m in length [32,33]. Gels are potentially useful for electrophoretic separations because they are an anticonvective medium, they minimize solute diffusion, which contributes to zone broadening, they help to prevent adsorption of analytes to the capillary walls, and they eliminate electroosmosis, thus allowing maximum resolution in short lengths' of column.

Capillary gel electrophoresis has several potential advantages over more conventional electrophoresis formats, including nanogram sample capability, ease of automation, accurate quantitation, and high sensitivity, especially when more sensitive detectors are employed (e.g. laser-induced fluorescence).

Gel-filled capillaries provides CE with the capability to perform separation based on differences in size. For a gel with a particular range of pore size, charged species of different sizes within this range migrate through the pores of the gel matrix at different rates. Since separation efficiency is inversely proportional to the diffusion coefficient of solutes in CE separations, the technique is especially useful for macromolecules which tend to have smaller diffusion coefficients in the gel medium and hence potentionally higher separation efficiency.

Despite the immense potential of the CGE technique, currently there remain practical limitations, such as the lack of truly preparative capability and the vulnerability of the gel-filled columns to damage. Nevertheless the technique is gaining in popularity. There is no doubt that improvements in instrumentation and advances in techniques for the preparation of gel columns will continue to be made. These improvements will perhaps help CGE fulfil its promise as potentially the most powerful separation technique ever developed.

4.3.1 Techniques for the preparation of gel-filled columns

Hjerten [36] reported the use of both sieving and non-sieving gels in tubes having inside diameter of 50–300 μm, and wall thickness of 100–200 μm. Separations of monomers, dimers, trimers, tetramers and pentamers of bovine serum albumin in a polyacrylamide gel-filled column, and human serum on agarose gel-filled columns were performed. In this first demonstration of capillary gel electrophoresis (CGE), relative low field strength (150 V/cm) and a high current (0.66 mA) were used, resulting in relatively low separation efficiency. However, the approach taken has encouraged further developments in the field which have subsequently made CGE currently one of the most powerful analytical tools ever developed.

Karger and Cohen [37] have made significant contributions in demonstrating the extremely high separation efficiency of CGE. By developing methods for the preparation of reliable and stable gels, they achieved theoretical plates of as high as 30 million plates in a single run using a gel-filled capillary column less than 1 m long for CE.

The main advantage of the procedures described by Karger and Cohen for the preparation of improved capillary gel electrophoresis columns is the enhanced stability of the gel resulting from the use of a bifunctional reagent, which provides linkages to both the capillary wall and the polymer gel matrix. On the other hand, a potential problem associated with the use of the bifunctional reagent is that there is a tendency for gel shrinkage to occur, which causes bubble formation [41]. In order to further improve the performance of capillary gel electrophoresis, many other methods for the preparation of gel-filled capillaries have been developed in recent years [39–41].

In this section, different techniques for the preparation of gel-filled columns are described. Sufficient experimental details are given to equip a newcomer to the field with the necessary background information for preparing such gel-filled columns. Currently, polyacrylamide is the most commonly used type of gel for CGE, mainly due to its ubiquity and proven success in conventional slab-gel electrophoretic separations. It is worth pointing out that other more suitable gels may need to be developed specially for CGE to fully explore the potential of this technique in the future.

4.3.1.1 Gel preparation with bifunctional reagent

In general, the preparation of the gel-filled column involves first the creation of a window for detection, then the pretreatment and activation of the capillary inner surface, followed by polymerization of the gel (either cross-linked or non-cross-linked) in the tubes. The gel columns are then preconditioned before use. The most important precaution would be the elimination of bubble formation during the whole process.

For the formation of the detection window, the polyimide coating of the capillary is removed from a 1 cm section of one end of the tubing by burning. The column is then activated by heating at a temperature between 110 and 200°C for several hours. This is followed by contact of the inner surface with a solution of a base, e.g. 0.1 N NaOH, for approximately 1–2 h at a temperature in the range 20–35°C. The capillary is then flushed with water.

The activated capillary is then flushed with at least 100 tubing volumes of a solution of the bifunctional reagent to be employed in bonding the column gel to the tubing wall. One end of the bifunctional reagent carries a reactive functional group which can bind chemically to silanol groups or other reactive functionalities on the inner surface of the capillary wall. The opposite end of the bifunctional reagent contains a second reactive group capable of forming a covalent bond with the polymeric gel material. Examples of bifunctional reagents are 3-methyacryloxypropyltrimethoxysilane and 3-methacryloxypropyl-dimethyoxysilane [32,33,37,38].

The bifunctional reagent is left to react with the capillary wall for at least one hour at around room temperature 20–35°C. The solution of bifunctional reagent may be prepared in a non-aqueous solvent such as an alcohol, an ether, or a moderately polar halogenated solvent containing typically between 4 and 60% bifunctional reagent by volume.

After the bifunctional reagent has been allowed to react with the inner wall of the capillary, excess of the unreacted reagent is removed by rinsing the column with a suitable solvent, such as methanol, followed by a further rinse with water.

Separate stock solutions of the monomer, cross-linkers, initiators and free radical sources for the polymerization reaction are then prepared. An example of the monomer used is acrylamide, which is widely used in conventional gel electrophoresis. Possible cross-linking agents are N,N′-methylenebisacrylamine, N,N′-(1,2-dihydroxyethylene)-bisacrylamine, N,N′-cystamine-bisacrylamide, and N,N′-acryloxytris (hydroxymethyl) aminomethane. Examples of the initiators which can be used include ammonium persulfate and N,N,N′,N′-tetramethylethylenediamine (TEMED).

The stock solutions are either prepared in aqueous solution or in aqueous solution containing a denaturing additive (such as urea). Typically the concentration required is in the 7 to 8 molar range. Aliquots of the stock solutions are then taken and mixed together to form a polymerization mixture having predetermined concentrations of monomer, cross-linker and polymerization catalysts. The solutions are then separately degassed for at least an hour. The concentration of the monomer and cross-linking agent are predetermined according to the porosity of the polymeric matrix desired.

In the case that sodium dodecylsulfate-polyacrylamide gel electrophoresis (SDS-PAGE) is to be performed, sodium dodecylsulfate (SDS) is also included in the reaction mixture in the required amount. In the case that a gel containing a hydrophilic polymer, e.g. polyethylene glycol, is to be prepared, the polyethylene glycol is combined with degassed triply distilled water which has been cooled to about 10°C, then stirred while the temperature is slowly raised to room temperature, making sure that a clear, transparent solution is obtained and no precipitation occurs.

The total concentration of monomer and the concentration of cross-linking agent in the gel are generally expressed as %T and %C, respectively [36]. For the acrylamide/N′,N′-methylenebisacryl-amide,

$$\%T = \frac{\text{grams of acrylamide + grams of bisacrylamide}}{\text{100 ml of solvent}} \tag{4.4}$$

$$\%C = \frac{\text{grams of bisacrylamide} \times 100}{\text{grams of bisacrylamide} + \text{grams of acrylamide}} \tag{4.5}$$

The concentrations of the initiator and polymerization catalyst are best determined experimentally. This is done by preparing test solutions containing the desired %T and %C, but varying the amount of initiator and polymerization catalyst employed. The test solutions are allowed to polymerize at the temperature at which electrophoresis is to be performed. The progress of the polymerization reactions can be monitored by UV absorbance levels and initiator and polymerization catalyst are selected to allowed the polymerization reaction to complete in a reasonable time, e.g. 60 min. Once the correct reagent concentrations have been determined, a fresh mixture of the polymerization reagents is prepared and injected into the capillary, taking care not to create any bubbles. A small I.D. PTFE tube is used to connect the mirocapillary to the syringe employed to fill the capillary. When the microcapillary has been filled with the polymerization mixture, the syringe is removed and both ends of the microcapillary are dipped into the running buffer, i.e. the buffer to be used in subsequent electrophoresis, until the polymerization reaction is completed. Preferably the reaction is allowed to proceed for another two hours in addition to the time required for polymerization predetermined by using the test solutions.

After the polymerization reaction in the capillary has completed, the ends of the capillary are removed from the buffer. At least one of the ends must be cut off cleanly. A simple method of cutting is to score it carefully at right angle to its axis by means of a sapphire cleaver, then breaking it cleanly by bending.

Figure 4.13 shows an electropherogram of a mixture of standard protein obtained using a gel-filled capillary column (10% T, 3.3% C).

4.3.1.2 Gel preparation without bifunctional reagent

For applications requiring low electric fields (e.g. less than 300 V/cm), the bifunctional reagent is usually not required and a simpler method for the preparation of the gel columns can be adopted [39]. The procedures are briefly described as follows.

After forming the detection window by burning off the polyimide, the capillary is rinsed with distilled water for 10 min. A 50 ml stock solutions containing 19 g of acrylamide and 1 g of N′,N′-methylene bisacrylamide is prepared. 1 ml of the stock solution is diluted with 7 ml of the buffer solution (0.1 M tris (hydroxymethyl)-aminomethane (Tris) and 0.2 M boric acid with 7 M urea at pH 8.3) This diluted acrylamide solution (5% T, 5% C) is carefully degassed in an ultrasonic bath. 10 μl of 10% N,N,N′,N′-tetramethylethylenediamine (TEMED) solution and 10 μl of 10% ammonium persulfate solution are added into 5 ml of the degassed diluted acrylamide solution to initiate the cross-linking reaction. The polymerization solution is quickly introduced into the capillary by using vacuum suction. Polymerization in the capillary is completed in about 2 h at room temperature. The gel-filled capillary is then preconditioned by running with buffer

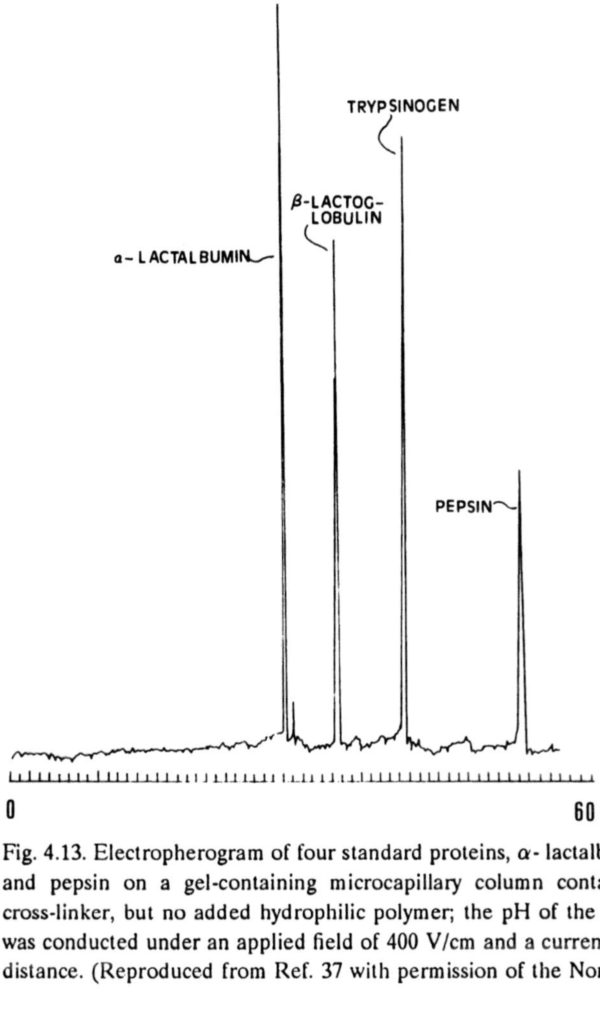

Fig. 4.13. Electropherogram of four standard proteins, α- lactalbumin, β-lactoglobulin, trypsinogen, and pepsin on a gel-containing microcapillary column containing 10% total monomer, 3.3% cross-linker, but no added hydrophilic polymer; the pH of the buffer was 8.6, and electrophoresis was conducted under an applied field of 400 V/cm and a current of 24 mA, over a 20-cm migration distance. (Reproduced from Ref. 37 with permission of the Northeastern University.)

at 10 kV for 20–30 min. Figure 4.14 shows the CGE separation of poly (A) digested by nuclease P1 using a gel-filled column prepared by this method.

4.3.1.3 Gel preparation with γ-radiation initiation

Since gel columns produced by the above methods employing ammonium persulfate and TEMED may contain highly charged by-products and residue amines, and hence may be susceptible to possible deterioration during storage, an alternative procedure is based on γ-radiation initiated formation of the poly-acryamide gel has been proposed [40]. The steps in the production of this type of gel filled columns are summarized as follows.

After forming the detection window (see Section 3.1.1), the capillary is filled with degassed solution of the acrylamide/bisacrylamide mixture (19 g acrylamide and 1 g

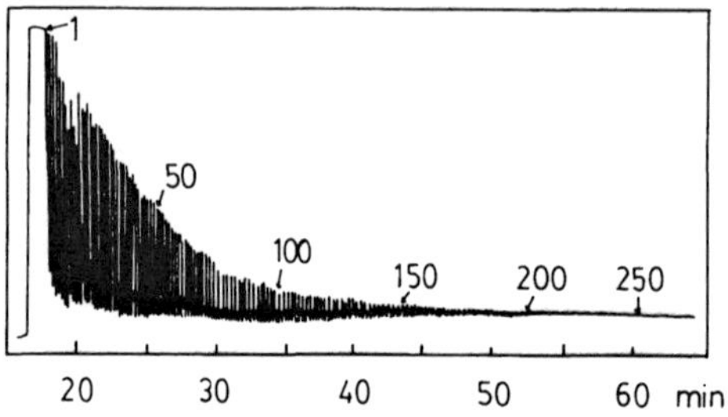

Fig. 4.14. CGE separation of poly(A) digested by nuclease P1. Capillary: 100 μm I.D., 375 μm O.D., length: 50 cm, effective length, 30 cm. Running buffer: 0.1 M Tris, 0.25 M boric acid, and 7 M urea. pH 8.3. Gel composition: 5% T and 5% C. Field: 200 V/cm; current, 10 μA. Injection: 5 kV for 1 s. Detection: 260 nm. (Reproduced from Ref. 39 with permission of the Chemical Society of Japan)

bisacrylamide in 50 ml of triply distilled water) in the buffer solution (1.211 g Tris base, 1.546 g boric acid and 4.204 g urea in 100 ml of triply distilled and degassed water). Both ends of the capillary are closed by silicone rubber septa in order to prevent evaporation of solvent during subsequent polymerization. γ-radiation from a Co source at a dose from 20 krad to 400 krad is used to initiate the polymerization and cross-linking of the acrylamide. Figure 4.15 shows the CGE separation of polydeoxycytidine pd(c)$_{24-36}$ using a capillary prepared by the γ-radiation initiation method.

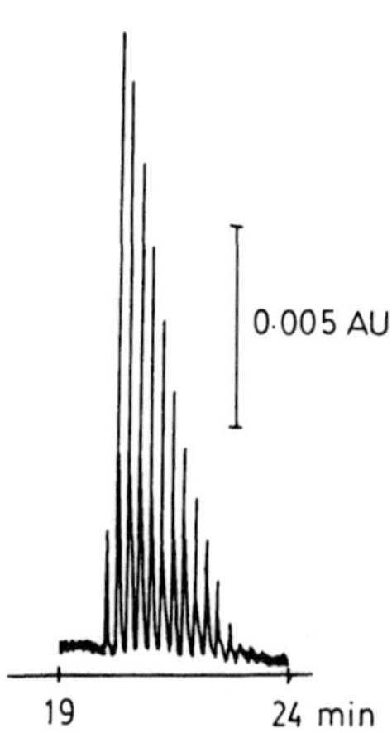

Fig. 4.15. Capillary gel electrophoretic separations of polydeoxycytidine pd(C)$_{24-36}$. Sample: 0.1 mg/ml pd(C)$_{24-36}$; capillary: 45 cm effective, 60 cm total length; 100 μm I.D. polyacrylamide gel filled (6% T, 3% C); buffer: 0.1 M Tris, 0.25 M borate buffer, 7 M urea, pH 7.5. Injection: electrokinetic, 5000 V for 6 s; separation voltage: 300 V/cm; detection: UV/260 nm. (Reproduced from Ref. 40 with permission of Dr. Alfred Huethig Publishers.)

4.3.1.4 Pressurized polymerization

High pressure has been used to reduce bubble formation during gel formation inside the capillary [41,42]. In one of the pressurized polymerization procedure, a fused silica capillary is filled with a de-aerated mixture containing 5.8% acrylamide (6% T), 0.18% N,N′-methylenebisacrylamide (Bis) (3% C), 10 mM triethanolamine, 0.01% ammonium persulfate, 100 mM tris (hydroxymethyl)aminomethane (Tris), 200 mM morpholinoethane sulfonic acid (MES), and 0.1% sodium dodecyl sulfate (SDS). The filled capillary is sealed at one end. A syringe containing the polymerization mixture is connected to the open end of the capillary. Pressure is applied by means of a stiff spring clamp on the syringe until polymerization is completed. Another patented method for pressurized polymerization involves compressing the monomer solution to 7.0×10^7 Pa and maintaining the high pressure until gel formation is completed [42].

4.3.1.5 Gel preparation by sequential polymerization

Dolnik *et al.* [41] proposed that if polymerization of the gel is allowed to proceed sequentially, i.e. gradually from one end of the capillary to the other, the volumetric losses resulting from polymer shrinkage can be compensated from the monomer solution. Figures 4.16 and 4.17 illustrate the conventional simultaneous polymerization process and the sequential polymerization process, respectively. In addition to a pressurized polymerization method, four methods employing sequential polymerization for the preparation of polyacrylamide gel-filled capillaries for CE have been investigated [41]. Although not all these methods have been proven successful in producing stable and bubble-free gel-filled columns, they

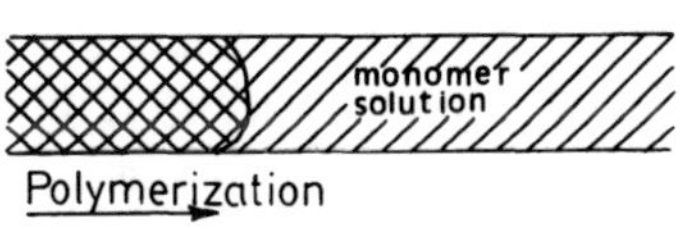

Fig. 4.16. Illustration of the sequential polymerization process, in which the polymer is formed gradually from one end of the capillary to the other. The newly formed polymer is continually in contact with the monomer solution, which fills any voids left by gel shrinkage. (Reproduced from Ref. 41 with permission of Microseparations, Inc.)

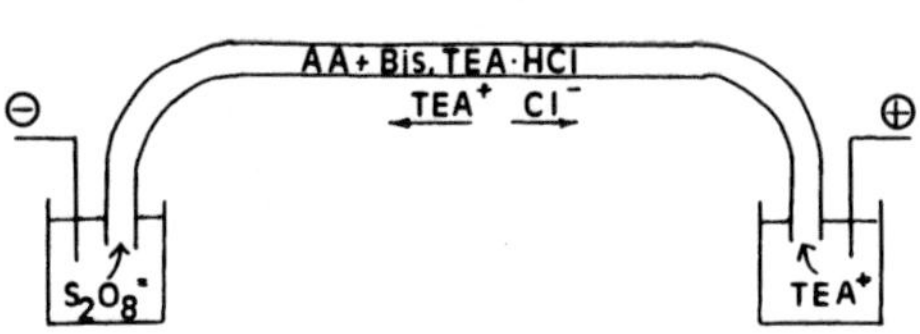

Fig. 4.17. Schematic representation of the isotachophoretic polymerization process. *AA* = acrylamide monomer; *Bis* = bisacrylamide (cross-linking agent); *TEA* = triethanolamine (catalyst); persulfate ion acts as the initiator of polymerization. Applied voltage: 3–6 V/cm. (Reproduced from Ref. 41 with permission of Microseparations, Inc.)

are described briefly below to illustrate the many approaches which can be adopted and the ingenuity of some of these approaches.

Method (A) — Programmed temperature polymerization: A fused silica capillary is filled with the same polymerization mixture as above. The programmed temperature polymerization is achieved either by first immersing completely the capillary in a 4°C water bath and withdrawing at a rate of 0.2 cm/min from the bath, or by immersing the capillary into a 60 or 100°C water bath at a rate of 1 cm/min.

Method (B) — Gradual photopolymerization: A fused silica capillary is filled with a de-aerated solution containing 5.8% acrylamide (6% T), 0.18% Bis (3% C), 10 mM triethanolamine, 0.005% riboflavin (or acriflavin or methylene blue), 100 mM Tris, 200 mM MES, and 0.1% SDS. The filled capillary is placed inside two sections of black opaque tubings separated by a gap of approximately 1 cm. The section of capillary in the gap is exposed to light (100 W light bulb) from a distance of 20 cm for 30 min. The capillary is moved in 1-cm steps section by section at about 30 min intervals until the entire length of the capillary has undergone photopolymerization for at least 30 min.

Method (C) — Laser-induced photopolymerization: A fused silica capillary is filled with a de-aerated photopolymerization mixture as described above in Method (B). One end of the capillary is positioned perpendicular to an argon ion laser beam. The capillary is pulled through the laser beam at a rate of 1 mm/min until the entire length of capillary has been exposed to the beam.

Method (D) — Isotachophoretic polymerization: In isotachophoretic polymerization, the inner surface of the capillary needs to be modified in order to eliminate electroosmosis during the process. To perform this type of modification, a length of capillary is coated with linear polyacrylamide according to the procedure developed by Hjerten [10]. The capillary is then filled with a solution containing 4 μl of γ-methacryloxypropyltrimethoxysilane in 1 ml of 6 mM acetic acid. The solution is rinsed off from the capillary after 1 h or more with distilled water for several minutes. The capillary is then filled with a de-aerated coating solution containing 2.5% acrylamide, 0.1% ammonium persulfate and 0.1% TEMED. The capillary is then rinsed after 30 min with distilled water for 5 min and emptied. For isotachophoretic polymerization, the coated capillary is filled with a mixture containing 5.8% acrylamide (6% T), 0.18% Bis (3% C) and 100 mM triethanolamine-hydrochloride. One end of the capillary is placed in a vial containing 10% ammonium persulfate and a platinum electrode which forms the cathode. The other end of the capillary was placed in another vial (anode) containing 25% triethanolamine and a platinum electrode which forms the anode. An electric field strength of 4 V/cm was applied for 8–12 h. A schematic representation of the isotachophoretic polymerization process is shown in Fig. 4.18. When voltage is applied, persulfate ion enters the capillary isotachophoretically (behind Cl as the leading ion) and initiates the polymerization gradually. The catalyst (triethanolamine) migrates in the opposite direction. The speed of polymerization is mainly determined by the applied voltage. After the polymerization reaction is completed, both vials are replaced by vials

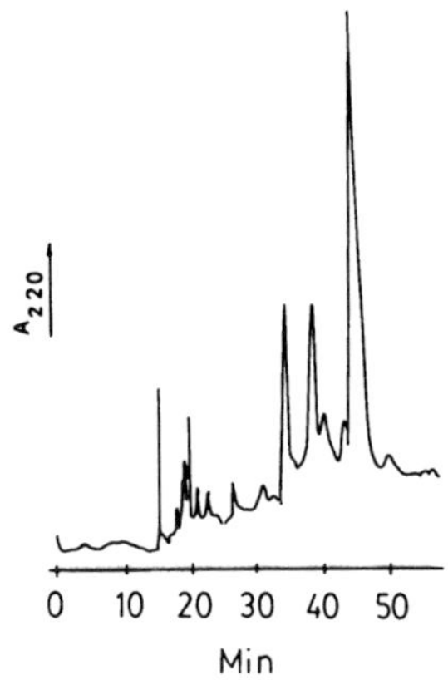

Fig. 4.18. Capillary gel electrophoretic separation of untreated mouse urine. Polyacrylamide gel, T = 6%, C = 3%; capillary, 50 μm I.D. × 35 cm (20 cm to detector); applied voltage, 3.5 kV; background electrolyte, 100 mM Tris, 200 mM MES, 0.1% SDS; electrokinetic injection, 60 s at 3.5 kV; UV absorbance detection, 220 nm. (Reproduced from Ref. 41 with permission of Microseparations, Inc.)

containing the background electrolyte for subsequent electrophoretic separations. Stepwise voltage increase is used to equilibrate the gel-filled capillary (to remove the initiator and catalyst to prevent further polymerization and subsequent shrinkage) until the current stabilizes at a given voltage, or until the baseline of a UV detector output becomes stable.

Among Methods (A) to (D), it has been found that the isotachophoretic polymerization, Method (D), is superior because of reduced bubble formation during polymerization. Preliminary results obtained using a gel-filled column prepared by the isotachophoretic polymerization method is shown in Fig. 4.18. A relatively low voltage (3.5 kV) was used in order to avoid the breakdown of the gel due to Joule heat, resulting in relatively low efficiencies.

4.3.1.6 Non-cross-linked polyacrylamide gel

The use of non-cross-linked polyacrylamide gels have also been reported [45,46]. The procedure for the preparation of a column containing a linear non-cross-linked gel is relatively simple. For instance, to prepare an acrylamide gel of medium chain length, 1.88 g of acrylamide are dissolved in 15 ml of water, 19 mg of TEMED are added and the solution is degassed with a water-jet pump for 10 min. Polymerization is initiated with 23 mg of ammonium persulfate dissolved in 10 ml water. The resulting polyacrylamide gel solution (25 ml) is degassed for 10–25 min using a water-jet pump. A phosphate buffer (0.05 mol/l) containing 0.5% (w/w) SDS is prepared. 2 g of the gel is mixed with the SDS buffer and diluted to 20 ml, giving a buffer solution with 10% (w/v) gel. This buffer is then used to fill the capillary and the electrolyte vessels. Figure 4.19 shows the electropherogram of the

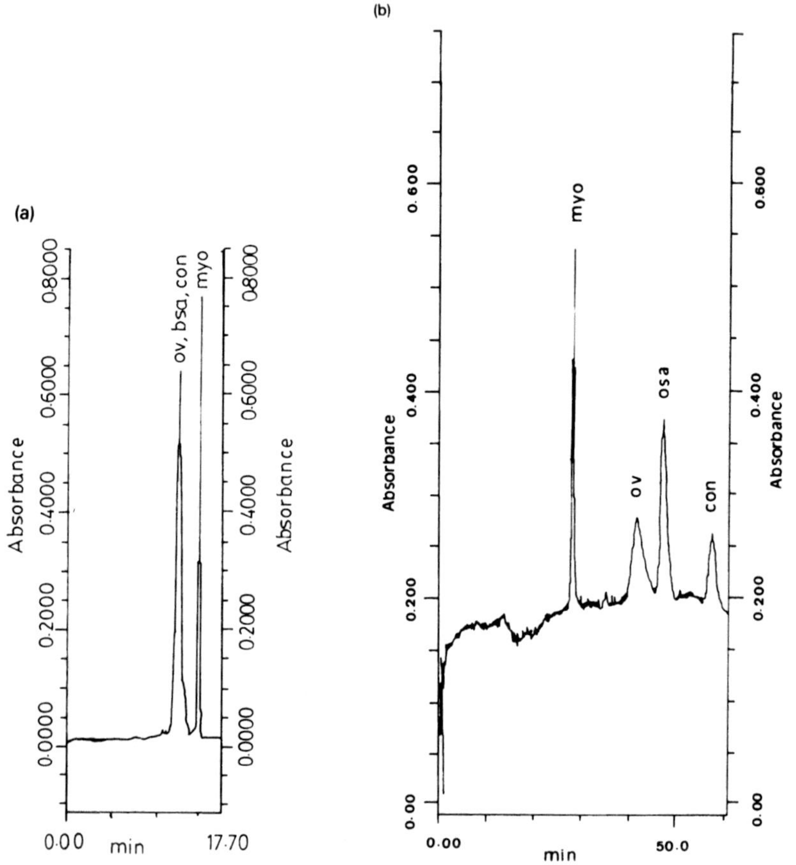

Fig. 4.19. (A) Electropherogram of the protein–SDS complexes by free zone electrophoresis (without polyacrylamide gel). Buffer: phosphate, pH 7, 0.05 mol/l; 0.5% SDS. Proteins recorded at the side of the cathode. Symbols: *ov* = ovalbumin; *bsa* = bovine serum albumin; *con* = conalbumin; *myo* = myoglobin. (B) Electropherogram of protein–SDS complexes obtained in a capillary filled with linear polyacrylamide. The capillary was filled with liquid, non-cross-linked polyacrylamide gel (10%) in a pH 5.5 running buffer (phosphate, 0.05 mol/l; 0.5% SDS). The proteins are separated according to their molecular mass. (Reproduced from Ref. 46 with permission of Elsevier Science Publishers.)

protein–SDS complexes by CZE and CGE using a gel column filled with linear polyacrylamide. Compared with cross-linked gels, linear polyacrylamide gel offers a number of potential advantages. First the preparation of the non-cross-linked gel-filled column is relatively more straightforward. Secondly, the capillary can be more easily emptied and refilled, and therefore replaced after each run if necessary.

4.3.1.7 Agarose gels

Although polyacrylamide gels have been the most commonly used gels in CGE, many other types of gels are available which can enhance selectivity in the separation of large charged molecules. The use of agarose gels has recently been explored [45]. Agarose gels (AGs) are used extensively in slab-gel electrophoresis of biomolecules, mainly because their pore size are relatively large, they are mechanical strong and they are biologically inert.

For the preparation of agarose gels, the required amount of agarose is mixed with buffers in a reaction vessel. Since the melting point of agarose is around 65°C, and the gelling point is around 35°C, the control of the gel formation process can be accomplished relatively easily. To improve the stability of the gel, a small amount of polyalcohol, e.g. sorbitol, may be added to the reaction mixture. The reaction vessel containing the mixture of agarose and buffer should be tightly closed to prevent changes in composition due to evaporation. The reaction vessel is then heated in a water bath for about 15 min. Subsequently, the gel was degassed at 60–70°C in an ultrasonic bath. The molten gel can be transferred directly into the capillary under pressure ($\simeq$20 bar) with the aid of a filling device, which is maintained at an elevated temperature ($\simeq$90°C). Gel formation requires 1–2 h at ambient temperature.

The use of AG-filled capillaries for CGE separations of DNA restriction fragment and unsaturated sulfonate disaccharide were demonstrated. For the separation of the DNA restriction fragments (ϕX-174-RF DNA Hae III digest), the resolution obtained with AG was not as good as that obtained with polyacrylamide gel (PAG), although faster separations were observed with capillaries filled with AG. For the separation of the unsaturated disaccharides, no significant difference in resolution was observed although PAG gave faster separation. An advantage in the use of AG is that UV detection at 232 nm could be performed, whereas for PAG, a longer wavelength had to be used. However, a problem encountered with the use of the AG-filled capillaries was that the capillaries could only be utilized continuously for several days. Therefore, further improvements would need to be make before this type of gels can be of wider use in CGE.

4.3.1.8 Miscellaneous techniques for preparing gel columns

Currently the development of improved gel-filled columns is an active area of research. In addition to the methods above, several others have been proposed. One of the methods involves surface pretreatment with non-cross-linked polyacrylamide for the elimination of electroosmotic flow [43]. Photopolymerization of acrylamide with riboflavin as the initiator has also been employed [44]. The procedure is similar to Method (C) described in Section 4.3.1.6, except that the whole column is illuminated simultaneously.

4.3.2 Effect of gel composition in CGE

The ability of molecules to pass through the gel depends on the size and shape of the holes in the gel, the size and shape of the molecules being sieved and interactions such as adsorption or ion exchange which may occur between the molecules and the matrix of the gel. A knowledge of the range of sizes of the holes present in gels of different concentrations and compositions is therefore useful for selecting the right type of media for CGE.

Figure 4.20 shows the relationship of the average pore size of acrylamide gels to the percentage of bisacrylamide in gels of total acrylamide concentrations of 6.5–20% [60]. As shown in Fig. 4.20, these contain pores of average diameters from 0.6 to 4 nm. The usefulness of this range of pore size is immediately apparent as the molecule diameter of protein likely to require separation range from 1.6 to 8 nm.

A gel of suitable pore size may be selected by consideration of the relative mobilities of the components of the mixture and the range of sizes of the molecules present. Figure 4.21 shows the relationship of gel concentration and mobility for a number of different proteins. It can be seen that in general with increased total concentration of acrylamide plus bisacrylamide (%T), the pore size decreases. For all the proteins examined, log (mobility) is found to be proportional to pore size and hence %T.

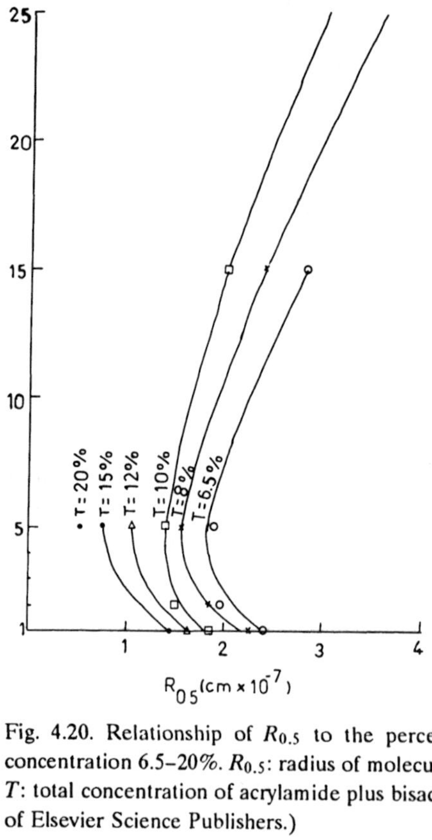

Fig. 4.20. Relationship of $R_{0.5}$ to the percentage of bisacrylamide in gels of total acrylamide concentration 6.5–20%. $R_{0.5}$: radius of molecules for which 50% of the total gel volume is available. T: total concentration of acrylamide plus bisacrylamide. (Reproduced from Ref. 59 with permission of Elsevier Science Publishers.)

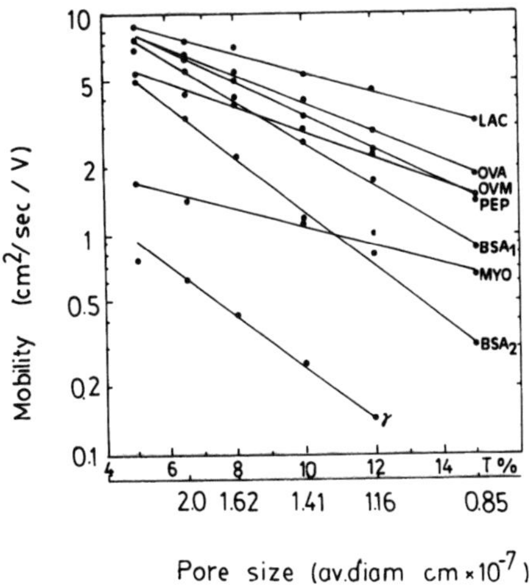

Fig. 4.21. Relationship of gel concentration and mobility for a number of different proteins. With increased total concentration of acrylamide plus bisacrylamide (T) the pore size decreases as shown. For all the proteins examined log mobility is found to be proportional to pore size. All mobilities were measured in gels with 5% of bisacrylamide at pH 8.83. *LAC* = β- lactoglobulin; *OVA* = ovalbumin; *OVM* = ovomucoid; *PEP* = pepsin; *MYO* = myoglobin; γ = bovine γ-globulin; BSA_1, BSA_2 = bovine serum albumin monomer and dimer. (Reproduced from Ref. 59 with permission of Elsevier Science Publishers.)

In addition, gels containing solubilizers may be utilized in certain applications. The use of acrylamide gels containing substances capable of solubilising certain classes of proteins has permitted a number of separations which are difficult to achieve by other methods. Examples of solubilizers include urea and sodium dodecyl sulfate. In most cases, the mixture to be separated has been dissolved in presence of the appropriate solubilizer and because the same substance has been present in the gel, electrophoresis can be carried out without loss by precipitation. Urea at concentrations from 3 to 12 M has been found to be particularly useful for rendering soluble certain classes of proteins. Since H bonds no longer exist in solutions containing urea at high concentrations, protein complexes or aggregates, whose structures are solely maintained by this kind of bonds, are readily dissociated in concentrated urea.

Solubilization of certain proteins and protein aggregates by addition of sodium dodecyl sulfate (SDS) is brought about because the treated proteins are rendered more hydrophilic. Although in such cases no reduction in size of the proteins has occurred, the molecules or aggregates thus formed may still be successfully separated in acrylamide gels. This is possible because gels are available with average pore size of sufficient magnitude to accommodate the rather large molecules thus formed.

4.3.3 Resolution and efficiency of gel-filled columns

In order to evaluate the performance of the gel-filled columns, a quantitative comparison of CGE with conventional slab-gel electrophoresis on the same sample has been performed [47]. It was found that for the electrophoresis of nucleotides, capillary gel electrophoresis was more than 3 times faster than an automated conventional slab-gel electrophoresis instrument. CGE required only 120 min for the analysis of nucleotide C304 whereas conventional gel required 370 min [47].

The following formula was used for the calculation of resolution:

$$R = \Delta t / 4\,\sigma_t \tag{4.6}$$

where Δt is the difference in time of elution between two consecutive peaks (differing by one nucleotide) and σ_t is the standard width of a single peak [61]. The values of resolution obtained using Eq. (4.6) for CGE ranged from 2.3 for the C33–C34 doublet, 1.8 for C156–C157, and 0.8 for C215–C216, compared with 0.85, 0.6 and 0.4, respectively for the same pairs for the slab-gel system.

The number of theoretical plates, given by:

$$N = (t/\sigma_t)^2 \tag{4.7}$$

where t is the total time for a given species to elute. Using this equation, a value of 2.9 million theoretical plates was obtained for residue C100 in a gel-filled column of 21 cm long (10 million theoretical plate per meter). The efficiency measured for the automated slab-gel instrument is lower, typically 460,000 plates (1.8 million plates/m) for C100. It was found that capillary gel electrophoresis was capable of providing superior performance not only in terms of faster analysis time (3×, but also better resolution (2.4×), as well as higher separation efficiency (5.4×) than the conventional automated slab-gel instrument [47].

Despite its impressive performance, CGE is still at an early stage of development with tremendous scope for further developments. In contrast, slab-gel electrophoresis is a relatively more matured technique. At the current stage of development, CGE can be used advantageously in applications involving very small amounts of samples, or requiring very high separation efficiency. There should also be benefits in terms of speed of analysis, automation and low running cost. However, to fully exploit the potential of CGE, there are still needs to improve the stability and lifetime of gel-filled columns, to design systems to handle larger amount of samples, and to develop techniques to perform multidimensional separation.

4.3.4 Use of size-sieving solutions instead of gel-filled columns

Soluble, linear polymers, with or without built-in functional groups tend to entangle in solution. The degree of such entanglement depends on the total polymer concentration and bulk solution properties. By introducing these polymers into the electrophoretic solution, a dynamic sieving system is created. Variable

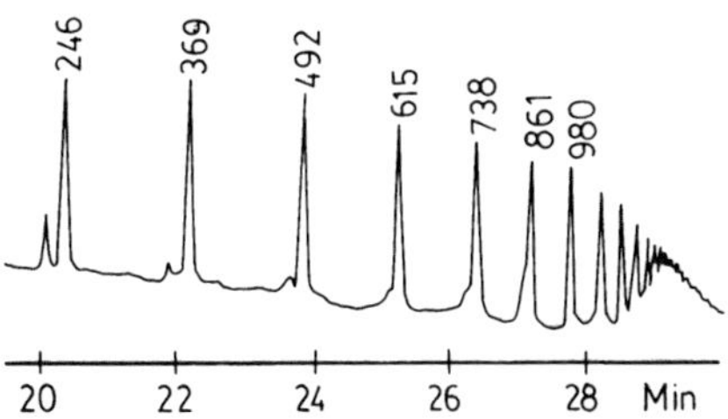

Fig. 4.22. Molecular sieving of 123 base pair ladder, 123 to 4182 DNA base pairs, concentration: 0.5 μg/μl. Separation buffer: 0.089 M Tris–borate–EDTA (TBE), pH 8.0, 0.5% methylcellulose (MC). Electrophoresis at 8 kV in 50 cm $\times$ 50 μm coated capillary. UV detection at 260 nm. (Reproduced from Ref. 49 with permission of Elsevier Science Publishers.)

"pore" size and additional interaction sites may therefore be present in the separation medium. This approach has several advantages over gel-filled capillaries. The most important advantages are their flexibility and ease of use. Specially prepared gel-filled capillaries are not required. There will also be no problem of the degradation of the gels in the capillary. Furthermore, sample introduction can be accomplished by both hydrostatic injection and electromigration, whereas electromigration is the only method of choice for introducing analytes into gel filled capillaries.

The major limitation is that currently polymer solutions have not demonstrated the ultrahigh efficiencies achievable by the use of gel-filled columns. Nevertheless, the many advantages of the technique justify its further development.

Several investigations have been reported on the use of entangled polymer solutions for size sieving capillary electrophoretic separation [48–50]. Typical types of polymers used include methylcellulose and polyethylene glycol. Figure 4.22 shows a typical electropherogram of 123 base pair ladder, obtained in a separation buffer containing 0.089 M Tris–borate–EDTA (TBE) and 0.5% methylcellulose (MC) in a capillary coated with a linear polymer.

4.3.5 Gel containing complexing agent

The incorporation of a complexing agents, e.g. cyclodextrins (CDs), within a polyacrylamide gel column provides an additional method to enhance selectivity of capillary electrophoretic separations. Chiral resolution of dansylated amino acids by CGE with gel columns containing cyclodextrins has been demonstrated [51]. Figure 4.23 shows a possible complex of a dansylated amino acids (Dns-AA) with β-CD. The non-polar dansyl portion of the molecule is found inside the cavity and the amino group forms hydrogen bonds with hydroxyl groups at the rim of the toroid. The differences in the size of the hydrophobic group with respect to the ability of the solute to penetrate the cavity results in differential complexation of individual Dns-AA with CD and hence selectivity in separation.

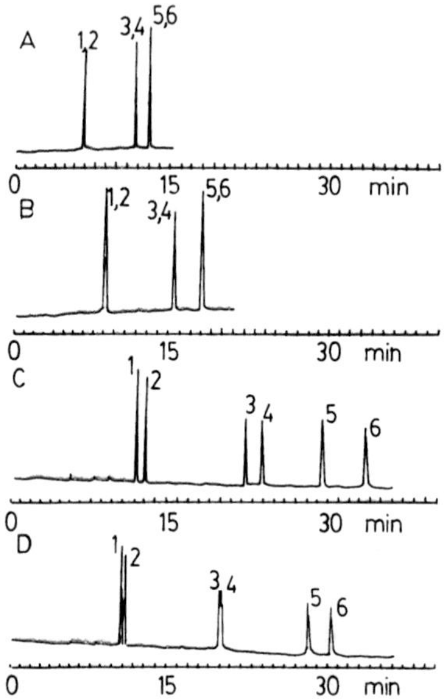

Fig. 4.23. Separation of Dns-DL-AAs. *1* = Dns-L-Glu; *2* = Dns-D-Glu; *3* = Dns-L-Ser; *4* = Dns-D-Ser; *5* = Dns-L-Leu; *6* = Dns-D-Leu. (A) Buffer: 0.1 M Tris–0.25 M boric acid (pH 8.3), 7 M urea. Gel: T = 5%, C = 3.3%, 0.1 M Tris–0.2 M boric acid (pH 8.3), 7 M urea. Capillary: 150 mm × 0.0 75 mm I.D., 400 V/cm, 8 μA. Electrokinetic: 250 V/cm, 5 μA, 30 s, detection wavelength, 254 nm. (B) Addition of 75 mM α-CD to the buffer and the gel mixture. (C) Addition of 75 mM β-CD to the buffer and the gel mixture. (D) Addition of 75 mM γ-CD to the buffer and the gel mixture. (Reproduced from Ref. 51 with permission of Elsevier Science Publishers.)

Guttman *et al.* [51] derived a general expression for relative retention with complexing agents:

$$R_s = \frac{\mu_1^f}{\mu_2^f}\left\{\left(\frac{1 + K_1[C]\,(\mu_1^c/\mu_1^f)}{1 + K_2[C]\,(\mu_2^c/\mu_2^f)}\right)\left(\frac{1 + K_2[C]}{1 + K_1[C]}\right)\right\} \tag{4.8}$$

where μ_1^f and μ_2^f are the mobility of solutes 1 and 2, μ_1^c and μ_2^c are the mobility of the complexed solutes 1 and 2, K_1 and K_2 and the formation constants of complexes 1 and 2 and $[C]$ is the concentration of the complexing agent.

If the mobility of the uncomplexed solute is much greater than that of the complex, i.e. $\mu^f \gg \mu^c$, then Eq. (4.8) simplifies to:

$$R_s = \frac{\mu_1^f}{\mu_2^f}\left(\frac{1 + K_2[C]}{1 + K_1[C]}\right) \tag{4.9}$$

when $K[C] \gg 1$ and for chiral pairs, for which μ_1^f and μ_2^f may be assumed to be equal (for L and D isomers of uncomplexed species), then

$$R_s = \frac{K_2}{K_1} \tag{4.10}$$

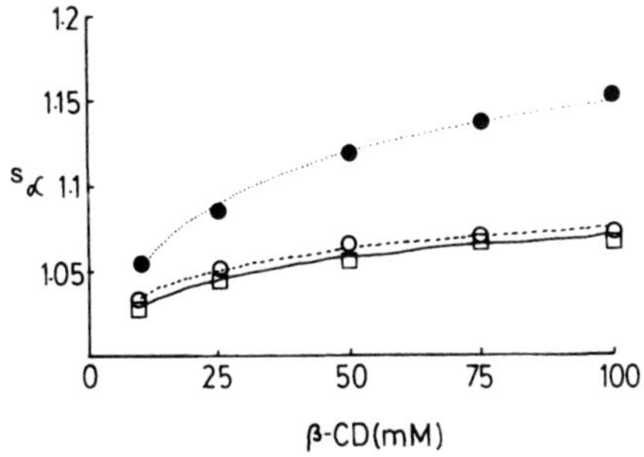

Fig. 4.24. Dependence of chiral selectivity, s_α, on β-CD concentration in the gel. Test mixture: • = Dns-DL-Leu; o = Dns-DL-Ser; □ = Dns-DL-Glu. Buffer: 0.1 M Tris–0.25 M boric acid (pH 8.3), 7 M urea, 75 mM β-CD. Gel: 5% T, 3% C, 0.1 M Tris–0.2 M boric acid (pH 8.3), 7 M urea, 75 mM β-CD. Capillary: 150 mm × 0.075 mm I.D., 400 V/cm, 8 μA. Electroinjection: 250 V/cm, 5 μA, 30 s. Detection wavelength: 254 nm. (Reproduced from Ref. 51 with permission of Elsevier Science Publishers.)

If on the other hand, the complexes moves much faster than the free solute, i.e., $\mu^f \ll \mu^c$, then the first term in Eq. (4.8) dominates and the elution order will be opposite to that observed in the case of Eq. (4.9).

If the complex and the free solute move at the same rate, R_s approaches unity and there will be no separation. Therefore, in order to achieve separation, one of the extreme cases, either $\mu^f \ll \mu^c$ or $\mu^f \gg \mu^c$, would be required.

Figure 4.23 shows the chiral separation of Dns-DL-AAs by addition of α-, β- and γ-CD to the buffer and the gel matrix. β-CD was found to provide the best selectivity in this separation. The reason for this observation is that β-CD provides the optimum fit for the inclusion complexes with Dns-AA.

Figure 4.24 illustrate the dependence of chiral selectivity, s_α, on β-CD concentration and temperature. From Fig. 4.24, it can be seen that the s_α value tends to a plateau at a high concentration of β-CD. The slope of the curve increases with the binding constant and is the greatest for Dns-DL-Leu. Figure 4.25 shows that s_α values decrease with temperature. Therefore, it is expected that temperature may be used as an additional parameter for the enhancement of selectivity.

4.3.6 Field programming CGE

Pulsed field gel electrophoresis is a powerful separation technique especially suitable for the analysis of larger proteins and long DNA molecules which are difficult or impossible to separate by any other methods [62,63]. The enhancement in selectivity observed in pulsed field gel electrophoresis may be attributed to the fact that the time required for a large molecule to orient in an electric field increases with the length of the molecule. For molecules with lengths greater than the pore size of a gel, no net migration would be observed before the molecules become oriented. By periodically altering the magnitude and/or the direction of the electric

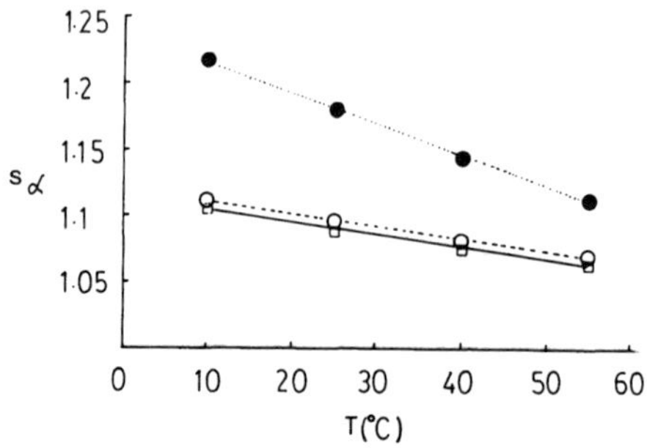

Fig. 4.25. Dependence of chiral selectivity, s_{α}, on column temperature, T. Test mixture and conditions are same as in Fig. 4.24, except that both the buffer and the gel contained 10% (v/v) methanol. (Reproduced from Ref. 51 with permission of Elsevier Science Publishers.)

field, the difference in the rate of reorientation provides a mechanism to enhance separation. The most common methods to effect pulsed field gel electrophoresis are unidirectional [64] and field-inversion [63] methods. In unidirectional methods, the direction of the field is not altered, while the magnitude of the field is varied as pulses or ramps in different waveforms. In field-inversion methods, the direction of the applied electric field is switched periodically, and the magnitude of the electric field may be kept constant, or changed during field inversion. Both methods have been shown to provide enhancement in selectivity for the separation of large species in slab-gel electrophoresis [63,64]. In the case of capillary gel electrophoresis, field programming techniques can also be employed to improve the performance of the system, since the velocity of the solute can be manipulated by varying the electric field.

Field programming has been used to enhance resolution in CGE separation of DNA fragments [53]. By means of electronic circuitry, the direction and magnitude of the applied field could be easily programmed. In Fig. 4.26, the peak separation of 4363- and 7253-base pair fragments is plotted against the frequency of a unidirectional pulse waveform, which was varied between 0.1 and 1000 Hz. Optimum separation is observed at 100 Hz, which is expected for the molecular sizes of the analytes [64]. At 100 Hz, a 20% increase in peak separation is obtained in relation to the continuous field operation. It is expected that pulsed field CGE would be particulary useful for the separation of larger species.

Another important application of field programming CGE is in performing micropreparative collection [52]. The approach involves performing separation at high field to maximize speed and resolution and then collecting at low field where the band would broaden in time without significant loss in resolution. Figure 4.27 shows the results of the use of field programming in CGE separation of polydeoxyadenylic acid mixture. A 6% T, 5% C gel column was used for the separation. It is noted that the narrow sharp peaks in Fig. 4.27A and B would create difficulties in collection. However, in Fig. 4.27C, where a high field is first

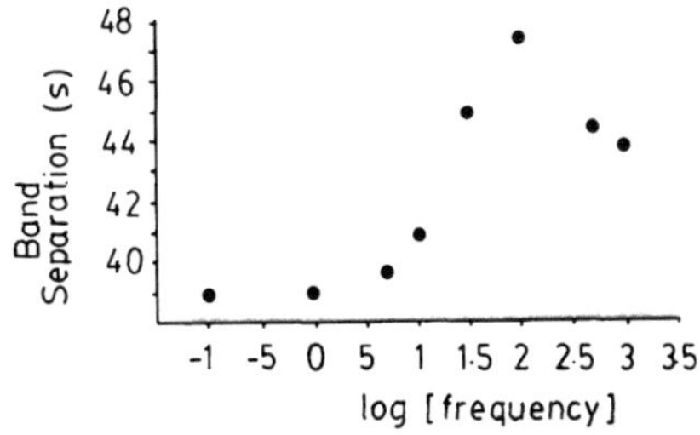

Fig. 4.26. Pulsed field CE of 4363- and 7253-base pair fragments. Plot of peak separation as a function of frequency of unidirectional pulse waveform. Maximum represents optimum frequency for separation of these species. Conditions: symmetric square wave of amplitude 0–300 V/cm; 6% T linear polyacrylamide; capillary length, 30 cm, effective length, 50 cm. (Reproduced from Ref. 53 with permission of Elsevier Science Publishers.)

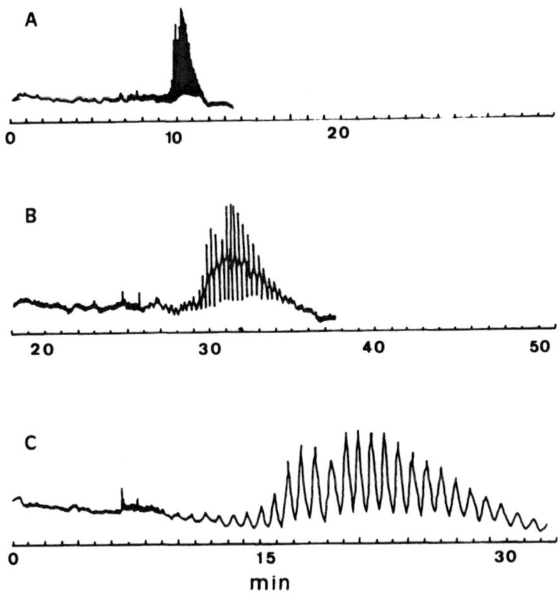

Fig. 4.27. Field programming in the CE separation of polydeoxyadenylic acid mixture, $p(dA)_{40-60}$: capillary dimensions, 300 × 0.075 mm I.D. (effective length, 150 mm). Applied voltages were as follows: (A) 300 V/cm, 17.7 μA; (B) 100 V/cm, 5.6 μA; (C) 300 V/cm, 17.7 μA, 0–10 min and 30 V/cm, 1.9 μA, 10–30 min. Buffer, 0.1 M Tris–0.25 M borate–7 M urea (pH 7.6). Polyacrylamide gel: 6% T and 5% C. (Reproduced from Ref. 52 with permission of the American Chemical Society.)

used for separation, and a low field is used for collection, high separation efficiency is maintained and sample collection can be more easily accomplished.

By manipulating parameters such as frequency, field amplitude, type of waveform and separation medium, pulsed field CGE is potentially a versatile and extremely powerful separation technique.

4.3.7 Typical applications of CGE

Detailed description of the applications of CGE are given in Chapter 7. In this section, the general areas of current interests are briefly described.

To date CGE has been mainly used for the analysis of oligonucleotides and polynucleotides, as well as peptides and proteins. The majority of these applications developed for CGE are based on size sieving separation. An example of this type of application is in the determination of molecular masses of polypeptides and proteins by means of sodium dodecyl sulfate polyacrylamide gel electrophoresis (SDS-PAGE). The protein mixture is denatured and the disulfide bonds are cleaved by subjecting to heat in the presence of SDS and a reducing agent (e.g. β-mercaptoethanol). The polypeptides and SDS reacts to produce complexes of constant charge density and hence similar electrophoretic mobility, as well as similar shapes. As a result, separation based on size or molecular mass (MM) differences via sieving through the polyacrylamide gel matrix can be achieved. A plot of log MM vs. mobility for several proteins (α-lactalbumin, β-lactoalbumin, trypsinogen and pepsin) is shown in Fig. 4.28. Linear plots are obtained for different percentage monomer composition (%T). It is noted that under identical field strength, the proteins were eluted faster as the pore size increased, i.e. lower %T [65]. The size separation mechanism is further validated by using the Ferguson plots [66] shown in Fig. 4.29, where the log mobility is plotted against percent monomer composition. Extrapolation of the plots for the proteins to 0% T (intercept on log mobility axis) gives the mobility of the SDS–protein complex in free solution, i.e. without any gel present. The coincidence of the intercepts confirms that the separation is purely based on MM or size, since no separation occurs when the gel is not present. Furthermore, the slopes of the lines in Fig. 4.29 are expected to be directly proportional to the MM.

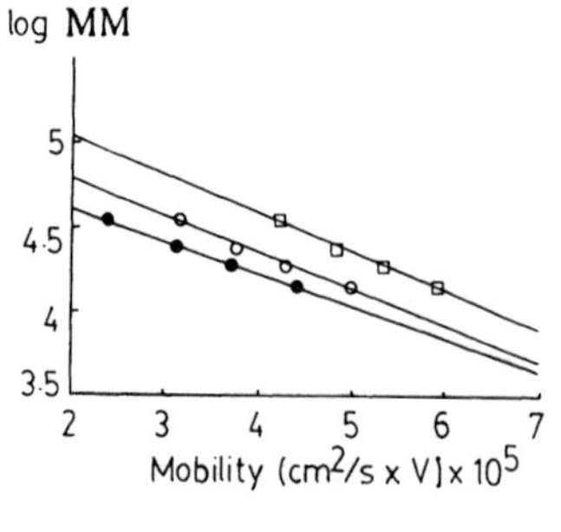

Fig. 4.28. Plot of log MM vs. mobility for proteins, as a function of polyacrylamide composition. • = 10% T, 3.3% C; o = 7.5% T, 3.3% C; □ = 5% T, 3.3% C. (Reproduced from Ref. 65 with permission of Elsevier Science Publishers.)

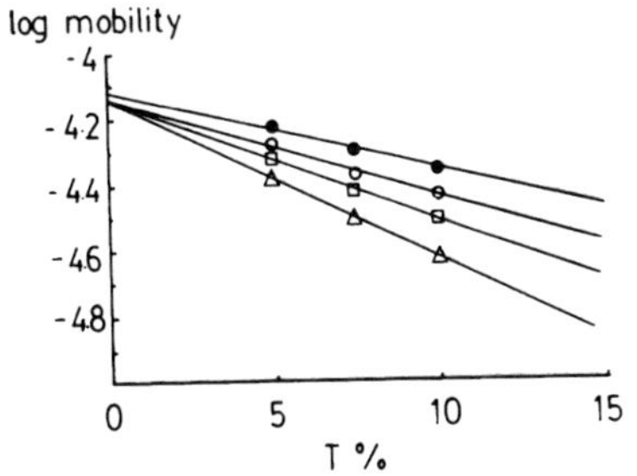

Fig. 4.29. Ferguson plot of log mobility vs. %T for four proteins. ● = α-lactalbumin; ○ = β-lactoglobulin; □ = trypsinogen; △ = pepsin. (Reproduced from Ref. 65 with permission of Elsevier Science Publishers.)

4.3.7.1 DNA sequencing by CGE

Recently an exciting area of development in CE is the use of capillary gel electrophoresis for DNA sequencing [47,53–55]. The tremendous interest stems form the important implications in the progress of the Human Genome Project. An efficient, high-speed and cost-effective automated DNA sequencing technique is required to map and sequence the three billion bases of DNA encoded within the human genome. A main limitation with the conventional sequencers based on slab-gel techniques is the speed of analysis. Typically 14 h of electrophoresis is needed in order to obtain 400 bases of sequence information from each of up to 16 sets of DNA fragments.

The application of CGE to DNA sequencing offers three advantages compared with conventional slab-gel electrophoresis. First, extremely high resolution can be achieved due to the use of a high electric field. Secondly, longer segments of DNA fragments can be sequenced. Thirdly, the speed of separation and scope for automation are potentially higher in CGE, especially if multiple capillaries can be used.

For the detection of the small amounts of materials present, laser-induced fluorescence (LIF) is usually employed (see Section 3.4.2). By using LIF with a sheath flow cuvette, an estimated mass detection limit of 10^{-20} mol of fluorescein labelled DNA fragments was obtained [54].

In Fig. 4.30, the extremely rapid separation that may be achieved with CGE is demonstrated. By employing an electric field of 400 V/cm, peaks up to base 213 were eluted within 17.5 min. The time required by a conventional sequencer would typically be 25 times longer [55].

To fully realize the potential of CGE for DNA sequencing, developments in two areas would be required. The first is to develop gel matrices which can withstand higher applied fields to permit further increase in speed. The second is to develop the capability to analyze many samples concurrently at high speed. Advances made in these areas will certainly help to make the sequencing of the human genome a reality in the near future.

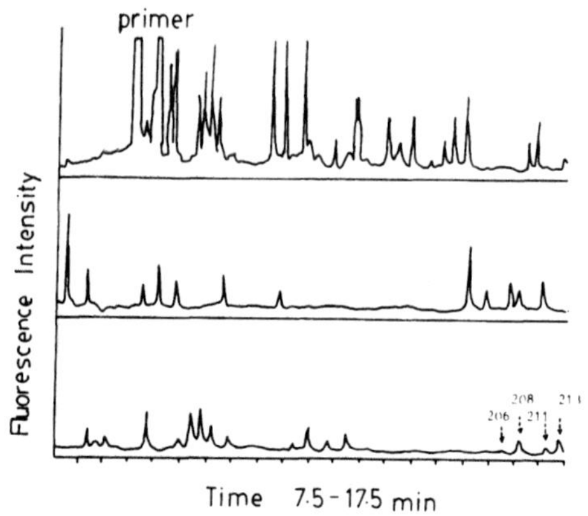

Fig. 4.30. Separation of fluorescein-labeled G reaction of M13mp19 DNA by capillary electrophoresis using a larger electric field. The applied voltage was 20 kV across the 50 cm capillary tube (400 V/cm). The total time between elution of the primer and the peak 213 bases in length is 10 min. (Reproduced from Ref. 55 with permission of the American Chemical Society.)

4.4 PACKED COLUMNS

With packed columns, electroosmotic flow occurs between the particles but not within them. The velocity of the electroosmotic flow is not expected to decline significantly from that achievable with much larger particles, provided that the particles are not smaller than 0.5 μm [8,67–70].

In the case of packed columns, since the tube walls represent only a small proportion of the total surface area of the particles, their condition is relatively less critical than in the case of open tubular columns. It is also expected that uniformity of packing is less important than in pressure-driven chromatography in packed beds. The reason is that the velocity of the electroosmotic flow, to a first approximation, is not dependent on the channel diameter between particles of packing, while the linear flow rate in a pressure-driven flow is proportional to the square of the channel diameter. The effect of slight variation in tube diameter is also expected to be negligible and the introduction of a small pressure-driven element into the flow will hardly affect the general flow profile [68].

Currently the use of packed columns is much less popular than open-tubular, coated or gel-filled columns. However, packed columns can be potentially more robust than the open tubular methods, as have been demonstrated in the case of high-performance liquid chromatography (HPLC). It is expected that if technological breakthrough permits the fabrication and utilization of packed columns for CE to be more easily accomplished, these types of columns may one day even surpass the open-tubular columns in popularity.

4.4.1 Capillary electrochromatography (CEC) in packed capillary

In one of Jorgenson and Lukacs pioneering papers on CE [69], a packed capillary containing 10 μm reversed-phase packing was employed. A pyrex glass tubing of 170 μm I.D. was used as the capillary. To fabricate the packed column, a porous plug was first formed at one end of the column by sintering 30 μm diameter ODS pellicular packing materials into a 5 mm section at the end of the capillary. Subsequently 10 μm reversed-phase packing was pumped at 700 kPa as an acetonitrile slurry into the column in a "down-flow" manner until the column was fully packed. The acetonitrile was removed by gas pressure, and the remaining end was then sealed by sintering another 5 mm plug of pellicular packing at this end. Once sealed, the column was filled with acetonitrile before use. The separation of 9-methylanthracene and perylene was demonstrated, with acetonitrile as the mobile phase and a separation voltage of 30 kV was applied across the column producing a current of 30 nA. The efficiencies obtained for 9-methylanthracene and perylene were 53,000 and 39,000 theoretical plates/m, respectively.

Knox and Grant [67,70] also performed capillary electrochromatography (or electrokinetic chromatography) on polynuclear aromatic hydrocarbon in a 50 μm I.D. capillary packed with 5 μm ODS-bonded stationary phase. To prevent the packing from migrating out, frits were fabricated by sintering spherical silica particles (typically 4–10 μm) at both ends of the capillary. Reduced plate height of below 2 was obtained, which corresponded to 200,000 plates/m. The efficiency obtained in electrochromatography was slightly better than that obtainable in pressure-driven HPLC using a similar column. By utilizing capillaries packed with 1.5 μm particles, efficiencies of 500,000 plates/m were obtainable [70].

Erni and co-workers [71] investigated electrochromatography using fused silica capillaries packed with reversed-phase materials of 3 μm and 1.6 μm diameter. Frits were made by sintering at both ends. Examples of chromatograms obtained with 1.6 μm Monosphere ODS packing is shown in Fig. 4.31. With different capillary of the same size packing (3 μm), reproducibility were 3% and 9% for electroosmotic flow (ν_{eo}) and capacity factor (k'), respectively. The poor reproducibility in capacity factor was attributed to differences in the packing conditions of the capillary [71].

4.5 CAPILLARY ELECTROPHORESIS ON A CHIP

Rapid advances in semiconductor technology has been made in recent years. Currently design, manufacturing and testing of miniaturized devices with features of μm size are standard procedures in the semiconductor industry. The technology required to produce a microchannel on a chip-like structure with dimensions similar to those provided by fused silica tubings is readily available. The use of a planar glass device (a chip) fabricated by photolithography for capillary electrophoresis has been demonstrated [72]. A schematic diagram of the glass structure is shown in

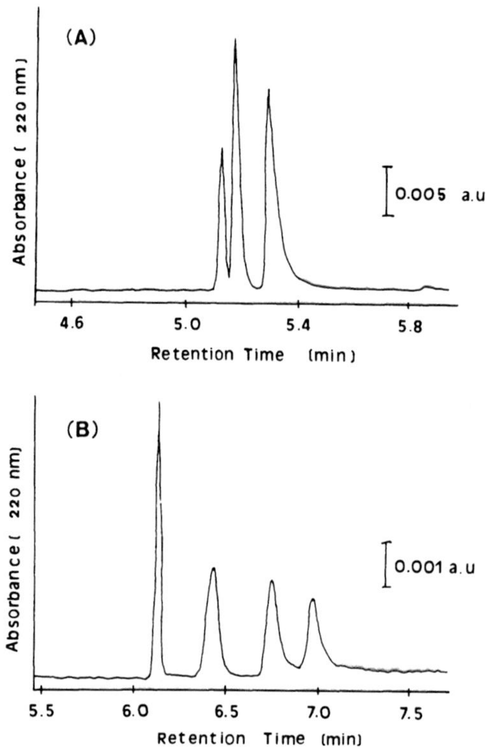

Fig. 4.31. Two examples of chromatograms obtained with 1.6 μm Monospher ODS. (A) Capillary, 680 mm × 50 μm I.D.; mobile phase, 4 mM sodium tetraborate (pH 9.2). ν_{eo} = 2.2 mm/s. Peaks, from left to right: thiourea (N = 243,000, N/s = 790), benzyl alcohol (N = 220,000), N/s = 710), benzaldehyde (N = 108,000, N/s = 340). (B) Capillary, 675 mm, mobile phase, 4 mM sodium tetraborate–20% acetonitrile. ν_{eo} = 1.8 mm/s. Peaks, from left to right: thiourea (N = 248,000, N/s = 670), toluene (N = 47,000, N/s = 120), 2-naphthol (N = 52,000, N/s = 130), 1-naphthol (N = 62,000, N/s = 150). Applied voltage, 35 kV (current 1.1 μA); sampling, 5 kV for 5 s. (Reproduced from Ref. 71 with permission of Elsevier Science Publishers.)

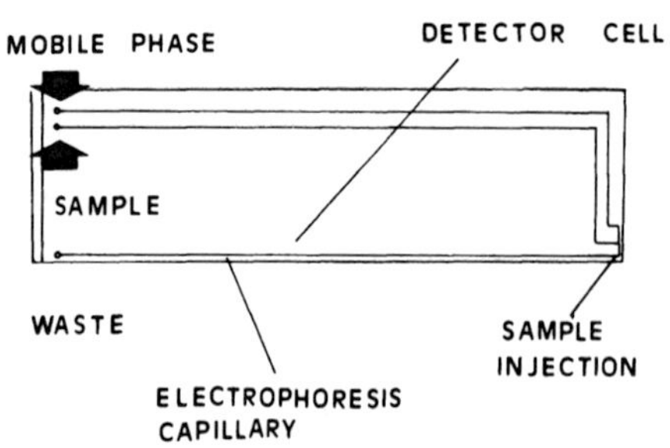

Fig. 4.32. Glass microstructure for injection and CE. Size is 15 × 4 × 1 cm. Electrophoresis channel, 30 × 10 μm. The external laser fluorescence detector was positioned 6.5 cm from the point of injection. (Reproduced from Ref. 72 with permission of Elsevier Science Publishers)

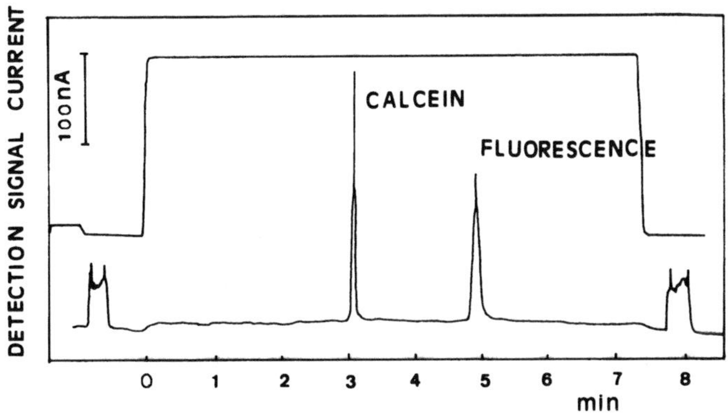

Fig. 4.33. Separation of two fluorescent dyes. Sample: 20 μM calcein, 20 μM fluorescein. Background electrolyte: 50 mM borate, 50 mM Tris, pH 8.5; 3000 V on 13 cm. Detection at 6.5 cm, fluorescence, excitation 490 nm, emission 520 nm, injection through side channel, 500 V for 30 s. (Reproduced from Ref. 72 with permission of Elsevier Science Publishers.)

Fig. 4.32. Three channels were etched onto a glass plate. One of the channels (10 μm deep and 30 μm wide) served as the electrophoretic capillary. Another glass plate was used as cover. Pipette tips were inserted into holes drilled at the ends of the channels to serve as buffer reservoirs. Samples introduction was performed at the injection end (9 pl) electrokinetically at 500 V for 30 s. Detection was by laser-induced fluorescence at a point 6.5 cm from the point of injection. Separation was performed at 3000 V (200 V/cm). The electropherogram obtained for two fluorescent dyes (calcein and fluorescein) is shown in Fig. 4.33. The number of theoretical plates obtained for calcein was approximately 18,000. The efficiency was limited by the voltage breakdown characteristics of the device, despite the use of insulating films (e.g. SiO_2 and SiN_4). Nevertheless, it seems promising that further improvements in the design and in the choice of the materials used for the chip may eventually provide a method for performing CE in an integrated device with an elaborate architecture and features entirely different from those employed in the present state-of-the-art in CE technology.

4.6 CONCLUSION

Advances in column technology are probably the most important factors which have contributed to the progress of CE. With appropriate choices and modifications of the type of column used, different modes of CE can be performed, and the efficiency and selectivity can also be controlled to some extent. A wide variety of compounds have been separated with high efficiency by capillary zone electrophoresis and micellar electrokinetic chromatography in uncoated fused silica capillaries. Recent advances in coating techniques for fused silica have been exploited in CZE separations to achieve reproducible and highly efficient separation of proteins, by effectively preventing their adsorption on the column surface. Coatings have also been used to suppress electroosmotic flow in other modes of CE separations, such as capillary isoelectric focusing (see Section 6.7) and capillary isotachophoresis (see Section 6.8). By utilizing gel-filled capillaries, size sieving separations, as in the analysis and sequencing of protein fragments and polynucleotides can be accomplished with remarkably high efficiency in CGE. Packed capillaries used in electrochromatography have also been shown to demonstrate higher efficiency than that obtained in pressure-driven liquid chromatography. Despite their potential, currently packed capillaries have not become as popular as open-tubular columns in CE applications, partly because of the practical difficulties involved in making and using micropacked columns. Nevertheless, packed columns are widely used in high-performance liquid chromatography, and there are numerous types of packing materials which can be exploited to provide enhancement in selectivity in CE separations.

Fused silica capillaries have been the most commonly used columns for CE. The applications of CE based on this type of columns have been extensively investigated (see Chapter 7). Commercial CE instruments available today have all been designed to utilize fused silica capillaries as the separation column. It is reasonable to expect that significant efforts based on the use of fused silica columns will continue to be made in order to develop new separation methodologies, to explore new applications, and to refine other instrumental features, such as injection and detection for CE. For some of these aspects, there is still tremendous scope and potential for further progress, particulary in the areas of separation chemistry, coating technology, techniques for preparing gel-filled capillaries and the use of multiple capillaries.

Nevertheless, it is important to realize that the potential of CE to achieve high efficiency is not limited to the capillary format. The possibility to perform CE in microchannels fabricated in planar devices has already been demonstrated. Further developments in this type of technology should present interesting new scenarios for advances in capillary electrophoresis.

4.7 REFERENCES

1 J.W. Jorgenson and K.D. Lukacs, Anal. Chem. 53 (1981) 1298
2 J.W. Jorgenson and K.D. Lukacs, Clin. Chem., 27 (1981) 1551
3 E. Grushka, R. M. McCormick and J.J. Kirkland, Anal. Chem., 61 (1989) 241
4 J. Knox, Chromatographia, 26 (1989) 329
5 G. Schomburg, Chromatographia, 30 (1990) 500
6 G. Schomburg, Trends Anal. Chem., 10 (1991) 163
7 K. Turner, LC-GC, 9 (1991) 350
8 T. Tsuda, J.V. Sweedler and R.N. Zare, Anal. Chem., 62 (1990) 2149
9 T. Izumi, T. Nagahori and T. Okugama, J. High Resolut. Chromatogr., Chromatogr. Commun., 14 (1991) 352

10 S. Hjerten, J. Chromatogr., 347 (1985) 191
11 J. W. Jorgenson and K.D. Lukacs, Science, 222 (1983) 266
12 S. Hjerten and M. Kieslling-Johansson, J. Chromatogr., 550 (1991) 811
13 B.J. Herren, S.G. Shafer, S.V. Alstine, J.M. Harris and R.S. Snyder, J. Colloid Interfac. Sci., 115 (1987) 46
14 G. Bruin, J. Chang, R. Kuhlman, K. Zegers, J. Kraak and H. Poppe, J. Chromatogr., 471 (1989) 429
15 S. A. Swedberg, Anal. Biochem., 185 (1990) 51
16 S. A. Swedberg, J. Chromatogr., 503 (1990) 449
17 Y. Maa, K.J. Hyer and S.A. Swedberg, J. High Resolut. Chromatogr., 14 (1991) 65
18 K.A. Cobb, V. Dolnik and M. Novotny, Anal. Chem., 62 (1990) 2478
19 J.K. Towns and F.E. Regnier, J. Chromatogr., 516 (1990) 69
20 J.K. Towns and F.E. Regnier, Anal. Chem., 63 (1991) 1126
21 A.M. Dougherty, C. Wooley and M. Swan, HPCE '91, San Diego, Calif., USA, Feb 3–5, 1991
22 G. Bruin, P. Tock, J. Kraak and H. Poppe, J. Chromatogr., 517 (1990) 557
23 K. Ulfelder *et al.*, Poster PT-5, HPCE '91, San Diego, Calif., Feb 3–6, 1991
24 J.E. Wiktorowicz and J.C. Colburn, Electrophoresis, 11 (1990) 769
25 R.M. McCormick, Anal. Chem., 60 (1988) 2322
26 G Bruin, R. Huisden, J. Kraak and H. Poppe, J. Chromatogr., 480 (1989) 339
27 A.M. Dougherty, C.L. Wolley, D.L. Williams, D.F. Swaile, R.D. Cole and M.J. Sepaniak, J. Liq. Chromatogr., 14 (1991) 907
28 F.E. Regnier and R.J. Noel, J. Chromatogr. Sci., 14 (1976) 316
29 P.P.H. Tock, G. Stegeman, R. Peerboom, H. Poppe, J. C. Kraak and K.K.Unger, Chromatographia, 24 (1987) 617
30 P.P.H. Tock, G. Stegeman, R. Peerboom, H. Poppe, J. C. Kraak and K.K.Unger, J. Chromatogr., 477 (1989) 95
31 A. Paulus and J.I. Ohms, J. Chromatogr., 507 (1990), 113
32 B. Karger, A. Cohen and A. Guttman, J. Chromatogr., 492 (1989) 585
33 A. Cohen, A. Paulus and B. Karger, Chromatographia, 24 (1987) 15
34 R.A. Wallingford and A.E. Ewing, in Advances in Chromatography, J.C. Giddings, E. Grushka and P.R. Brown (editors), Marcel Dekker, N.Y. USA, 29 (1989) 1
35 A. Paulus, E. Gassman and M.J. Field, Electrophoresis, 11 (1990) 702
36 S. Hjerten, J. Chromatogr., 270 (1983) 1
37 B.L. Karger and A.S. Cohen, Eur. Pap. EP 0324 539 A2, (Northeastern University) Jan. 1989
38 A. Cohen, D. Nagarian, J. Smith and B. Karger, J. Chromatogr., 458 (1988) 323
39 Y. Baba, T. Matsuura, K. Wakamoto and M. Tsuhako, Chemistry Letters, (1991) 371
40 J.A. Lux, H. Yin and G. Schomburg, J. High Resolut. Chromatogr., 13 (1990) 436
41 V. Dolnik, K.A. Cobb and M. Novotny, J. Microcol. Sep., 3 (1991) 155
42 P.F. Bente and J. Myerson, Eur. Pat. EP 0272 925 A2, (Hewlett Packard Co.), June 1988
43 H.F. Yin and G. Schomburg, J. High Resolut. Chromatogr., 13 (1990) 625
44 T. Wang, G.J. Bruin, J.C. Kraak and H. Poppe, Anal. Chem., 63 (1991) 2207
45 D.N. Heiger, A.S. Cohen and B.L. Karger, J. Chromatogr., 576 (1991) 33
46 A. Widhalm, C. Schwer, D. Blaas and E. Kenndler, J. Chromatogr., 549 (1991) 446
47 H. Swerdlow and R. Gesteland, Nuclei Acids Res., 18 (1990) 1415
48 A.J.C. Chen, M. Zhu, D.L. Hansen and S. Burd, J. Cell Biol., 107 (1988) 241a
49 M. Zhu, D.L. Hansen, S. Burd and F. Gannon, J. Chromatogr., 480 (1989) 311
50 A.M. Chin and J.C. Colburn, Am. Biotech. Lab., Dec. (1989) 16

51 A. Guttman, A. Paulus, A. Cohen, N. Grinberg and B. Karger, J. Chromatogr., 448 (1988) 41
52 A. Guttman, A. Cohen, D.N. Heiger and B.L. Karger, Anal. Chem., 62 (1990) 137
53 A.S. Cohen, D.R. Najarian and B.L. Karger, J. Chromatogr., 516 (1990) 49
54 H. Swerdlow, S. Wu, H. Harke and N.J. Dovichi, J. Chromatogr., 516 (1990) 61
55 H. Drossman, J.A. Luckacs, A.J. Kostichka, J. D'Cunha and L.M. Smith, Anal. Chem., 62 (1990) 900
56 M. Strege and A. Lagu, Anal. Chem., 63 (1991) 1233
57 S.R. Weinberger and T. Schlabach, HPCE '91, San Diego, Calif., USA, Feb. 3–5, 1991
58 S.R. Motsch, M. Kleem and G. Schomburg , J. High Resolut. Chromatogr., 14 (1990) 629
59 A.H. Gordon, Electrophoresis of Proteins in Polyacrylamide and Starch Gel, T.S. Work and E. Work (Eds), Elsevier Science Publishers, Amsterdam, 1975
60 J.S. Fawcett and C.J.U.R. Morris, Separation Studies, 1 (1966) 9
61 J.C. Giddings, Sep. Sci., 4 (1969) 181
62 D.C. Schwartz and C.R. Canton, Cell, 37 (1984) 67
63 C.F. Carle, M. Fank and M.U. Olson, Science (Washington D.C.), 232 (1986) 65
64 J.C. Sutherland, D.C. Monteleone and J. Trunk, Anal. Biochem., 162 (1987) 511
65 A. Cohen and B. Karger, J. Chromatogr., 397 (1987) 409
66 B.D. Hanes and D. Rickwood, (ed), Gel Electrophoresis of Proteins, IRL, Washington DC, 1983
67 J.H. Knox, and I.H. Grant, Chromatographia, 24 (1987) 135
68 J.H. Knox and K.A. McCormack, J. Liq. Chromatogr., 12 (1989) 2435
69 J.W. Jorgenson and K.D. Lukacs, J. Chromatogr., 218 (1981) 209
70 J.H. Knox and I.H. Grant, Chromatographia, 32 (1991) 317
71 H. Yamamoto, J. Bauman and F. Erni, J. Chromatogr., 593 (1992) 313
72 A. Manz, D.J. Harrison, E.M.J. Verpoorte, J.C. Fettinger, A. Paulus, H. Ludi and H.M. Widmer, J. Chromatogr., 593 (1992) 253

CHAPTER 5

Electrolyte Systems

5.1 INTRODUCTION

The electrophoresis buffer is of key importance in CE because its composition fundamentally determines the migration behaviour of the analytes. A suitable electrolyte system must ensure the correct electrophoretic behaviour of all individual solutes, the overall stability of the system and satisfactory separation of the analytes.

In this chapter, the effects of the electrophoresis buffer on CE separations are first discussed. The use of surfactants and other modifiers to permit the separation of neutral analytes and to enhance selectivity in micellar electrokinetic chromatography is then described. This is followed by discussions on methods to enhance selectivity by using inclusion and complexing additives. Finally, other unconventional types of electrophoretic media are described.

5.1.1 Electrophoresis buffer

A wide variety of electrolyte systems have been used in CE to effect the required separations. The majority of these are aqueous buffers. Some are taken directly from corresponding traditional gel electrophoretic separations and others evolved empirically as the best for a specific separation. Examples of the most commonly used buffers for CE and their appropriate pH ranges are: phosphate (pH 1.1–3.1 and pH 6.2 to 8.2), acetate (pH 3.8 to 5.8), borate (pH 8.1 to 10.1), and zwitterionic buffers, such as MES (pH 5.5–6.7) and Tris (pH 7.3–9.3), which are widely employed for protein and peptide separations. The choice of the electrolyte system in a CE separation involves consideration of many factors, such as the solubility and stability of the analytes in the electrolyte, the degree of ionization of the analytes, the influence of the anions and cations present in the electrolyte on the electromigration of the solutes, the effect of pH, the effects of organic modifiers and other additives, and the dissipation of heat generated in the electrolyte during the passage of the current.

5.1.2 Solubility and stability of substances

In order to perform the electrophoretic run, the analytes must be soluble in the electrophoretic solution used. Most of the electrophoretic solutions used to date are aqueous buffers. For the majority of substances, solubility does not present any extraordinary difficulties. However, in the case of very hydrophobic compounds, such as higher fatty acids, the solubility of the analytes in water may not be sufficient and mixed solvents (e.g. water–methanol, water–acetone) may have to be used for the analysis (see Section 5.1.8).

Another aspect to consider is the chemical stability of the analytes in the presence of the substances forming the electrolyte system and vice versa. Possible oxidation and/orreduction may occur between the electrolyte system and the analytes. Precipitation of ions from solution may also occur.

5.1.3 Ionization of analytes

In the selection of an electrophoretic solution, it is important to consider the degree of ionization of the analytes in the solution.

In the case of weak acids or bases, their degree of ionization depends on the pH of the solution. For a monovalent weak acid HA, which has a protolysis constant (or dissociation constant), K_{HA}, and the degree of dissociation is α_{A}, then:

$$K_{\mathrm{HA}} = \frac{C_{\mathrm{H^+}} \cdot C_{\mathrm{A^-}}}{C_{\mathrm{HA}}} \tag{5.1}$$

$$\alpha_{\mathrm{A}} = \frac{C_{\mathrm{A^-}}}{\overline{C}_{\mathrm{A}}} = \frac{K_{\mathrm{HA}}}{K_{\mathrm{HA}} + C_{\mathrm{H^+}}} \tag{5.2}$$

where $C_{\mathrm{A^-}}$ is the concentration of the ion $\mathrm{A^-}$, C_{HA} is the concentration of the non-ionized weak acid, $\overline{C}_{\mathrm{A}}$ is the total concentration of the weak acid and $C_{\mathrm{H^+}}$ is the concentration of the proton.

In weak electrolytes, both the non-ionized molecules and the corresponding ions may be present. Although each type of particle may have its own value of mobility (non-ionized particles having zero electrophoretic mobility), they behave as a uniform substance since both types of particles mutually interchange by rapid reversible acid–base equilibrium. The mobility of the substance is given by [1]:

$$\overline{\mu}_{\mathrm{A}} = \mu_{\mathrm{A}} \alpha_{\mathrm{A}} \tag{5.3}$$

where μ_{A} is the mobility of the ion $\mathrm{A^-}$. The dependence of α_{A} on pH is similar to the dependence of $\overline{\mu}_{\mathrm{A}}$ on pH. This dependence is represented by the mobility curve (Fig. 5.1). It can be seen, for example, that the weak acid is dissociated to 50% at pH = $\mathrm{p}K_{\mathrm{HA}}$; i.e. $\alpha_{\mathrm{A}} = \overline{\mu}_{\mathrm{A}}/\mu_{\mathrm{A}} = 0.5$.

Similarly for the protonation of a weak base, the dissociation constant of its protonated form ($K_{\mathrm{BH^+}}$), the degree of protonation (α_{B}) and the effective mobility of the base B ($\overline{\mu}_{\mathrm{B}}$) are given by:

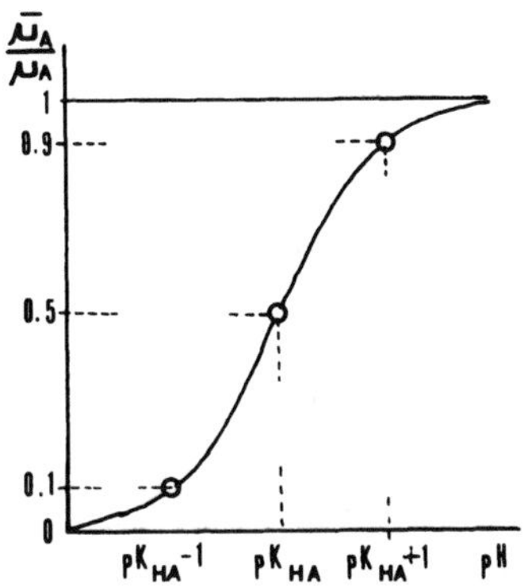

Fig. 5.1. Dependence of the relative effective mobility and the dissociation degree ($\alpha = \bar{\mu}_A/\mu_A$) on pH, for a weak monovalent acid HA. (Reproduced from Ref. 1 with permission of VCH Publishers.)

$$K_{BH^+} = \frac{C_{H^+} \cdot C_B}{C_{BH^+}} \tag{5.4}$$

$$\alpha_B = \frac{C_{BH^+}}{\overline{C}_B} = \frac{C_{H^+}}{K_{BH^+} + C_{H^+}} \tag{5.5}$$

$$\overline{\mu}_B = \mu_{BH} \frac{C_{H^+}}{K_{BH^+} + C_{H^+}} \tag{5.6}$$

or

$$\overline{\mu}_B = \mu_{BH}\alpha_B \tag{5.7}$$

where C_B is the concentration of the non-ionized base, C_{BH^+} is the concentration of the protonated form of the base, $\overline{C}_B$ is the total concentration of the base, C_{H^+} is the concentration of the proton and μ_{BH} is the mobility of the ion BH^+.

The electrophoretic mobility of an ion (μ_{ep}) is related to its charge-to-mass ratio as follows [2]:

$$\mu_{ep} = \frac{Z}{6\pi\,\eta r} \tag{5.8}$$

where Z is the effective charge of the ion, η is the viscosity of the solution, and r is hydrodynamic radius of the ion. According to Huckel, μ_{ep} is given by [3]:

$$\mu_{ep} = \frac{\varepsilon\,\zeta}{6\pi\,\eta} \tag{5.9}$$

where ε is the dielectric constant and ζ is the zeta potential, which is directly proportional to the thickness of the double layer [4]:

$$\zeta = \frac{4\pi\,\delta\,e}{\varepsilon} \tag{5.10}$$

where e is the excess charge in solution per unit area and δ is the thickness of the double layer, which is proportional to the inverse of the square root of the buffer concentration [4].

It is expected from Eq. (5.8) that electrophoretic mobility of a solute depends mainly on the charge-to-radius ratio. Nevertheless, the effective charge of the ions may be affected by the type and concentration of the buffer, as a result of the effect of the buffer on the double layer. This is because μ_{ep} depends on the zeta potential (Eq. 5.9), which is in turn dependent upon the total excess charge in the buffer solution (Eq. 5.10).

In the case of analytes which are ions, the efficiency in CZE separation without electroosmotic flow depends on the charge number of the individual analytes [6]:

$$N = \frac{ZFV}{RT} \tag{5.11}$$

where N, Z, F, V, R and T are the number of theoretical plates, charge of the ion, Faraday's constant, applied voltage, universal gas constant and absolute temperature, respectively.

In the case of weak acids, a theoretical expression for the dependence of efficiency on α has been given [6] as follows:

$$N = \alpha e_0 V/(2kT) \tag{5.12}$$

where e_0 is the electronic charge, k is the Boltzman constant, T is the absolute temperature, V is the applied voltage, and α is the degree of dissociation, given by:

$$\alpha = 1/(1 + 10^{\mathrm{p}K-\mathrm{pH}}) \tag{5.13}$$

Experimental values of plate number obtained for anions, including benzene-1,3-disulfonate, 1-naphthol-3,6-disulfonate, toluene-4-sulfonate, naphthalene-2,sulfonate and pyrocatechol violet agree with those predicted by Eq. (5.12), except that lower efficiencies were obtained experimentally due to peak distortion caused by conductivity differences between the sample and the buffer and Joule heating.

The concentration distribution, i.e. peak (or zone) shape, can be related to the relative mobility (μ_r) of the sample constituent, defined as the ratio of the mobility of the solute to that of the carrier [7,8]. The concentration distribution in zone electrophoresis as a function of the relative mobility of the sample constituent is shown in Fig. 5.2. When the sample constituent has a higher mobility than that of the carrier constituent, i.e. $\mu_r > 1$, the leading side of the sample zone will be diffuse, whereas the rear will be sharp. When the mobility of the sample constituent is equal to that of the carrier constituent, i.e. $\mu_r = 1$, the sample constituent is only diluted or concentrated over the zone boundary. When the sample constituent has a smaller mobility than that of the carrier constituent, i.e. $\mu_r < 1$, the leading side of the zone will be sharp, whereas the rear will be diffuse.

Differences in the degree of ionization of the analytes result in differences in their mobilities which are important for achieving separation in the CE system.

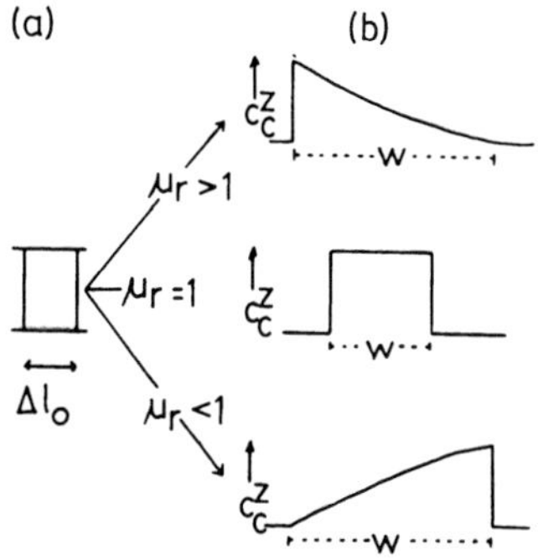

Fig. 5.2. Concentration distributions (C_C^Z) in zone electrophoresis as a function of the relative sample constituent mobility. (a) $t = 0$; (b) $t = t_r$. Δl_0 is the initial width of the sample pulse and w is the peak width at time t.

In differential migration methods, for two ionogenic constituents i and j to be separated, their mobilities (or migration rates) should be substantially different:

$$\frac{\overline{\mu}_i}{\overline{\mu}_j} \neq 1 \tag{5.14}$$

The lower limit of resolution represents the case in which the constituents to be separated forms a mixed zone with each other, and do not separate at all. The criterion for separation can be related to the pH of the mixed state, pH^s [9]:

$$pH^s = pK_j + \log\left[\left(1 - \frac{\mu_{rj}K_j}{\mu_{ri}K_i}\right) \Big/ \left(\frac{\mu_{rj}}{\mu_{ri} - 1}\right)\right] \tag{5.15}$$

where μ_{ri} and μ_{rj} are the ionic mobility of species i and j relative to that of the carrier constituent, whereas K_i and K_j are the protolysis constants of i and j, respectively. In Fig. 5.3, the possible migration configurations for anionic constituents is shown. When the constituent with higher mobility has a higher

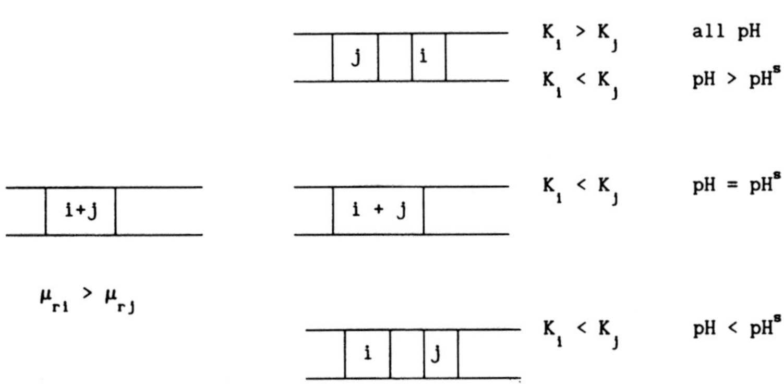

Fig. 5.3. Migration configurations for anionic solutes.

protolysis constant, separation can be obtained at all pH. When the more mobile constituent has a lower protolysis constant, the separation configuration is a function of the pH of the carrier buffer. For cation constituents, similar configurations to those shown in Fig. 5.3 are expected.

5.1.4 Buffer anions

Anions present in the electrolyte system may affect the current and hence the electroosmotic flow, the heat generated, the interaction of analytes with the wall of the capillary, as well as the mobilities of the ions. The relative importance of each of these effects would depend on the system under investigation.

The effect of buffer composition on electroosmotic flow in CE has been studied by VanOrman *et al.* [10]. Figure 5.4 contains plots of the coefficient of electroosmotic flow, μ_{eo}, versus the natural logarithm of buffer concentration for a series of inorganic phosphate, borate and carbonate buffers. Ferguson and Good [11] described some non-interacting buffers, such as N-(2-acetamido)-2-aminoethanesulfonic acid (ACES), 4-(2-hydroxyethyl)piperazine-1-ethanesulfonic acid (HEPES), and 4-(2-hydroxyethyl)piperazine-1-hydroxypropanesulfonic acid (HEPPSO). They found that the coefficient of electroosmotic flow showed a linear relationship with the logarithm of the concentration.

Another important observation made is that if the buffer concentrations were expressed in terms of ionic strength, the coefficient of electroosmotic flow, μ_{eo}, followed a common line when plotted against the logarithm of the ionic strength

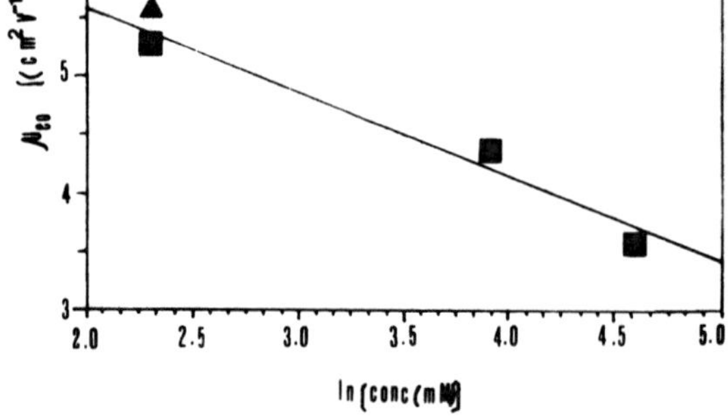

Fig. 5.4. The coefficient of electroosmotic flow (μ_{eo}) at pH = 8.0 vs. the logarithm of the concentration for inorganic buffers. Legend: ■ = phosphate; ● = borate; ▲ = carbonate. (Reproduced from Ref. 10 with permission of Aster Publishing Corp.)

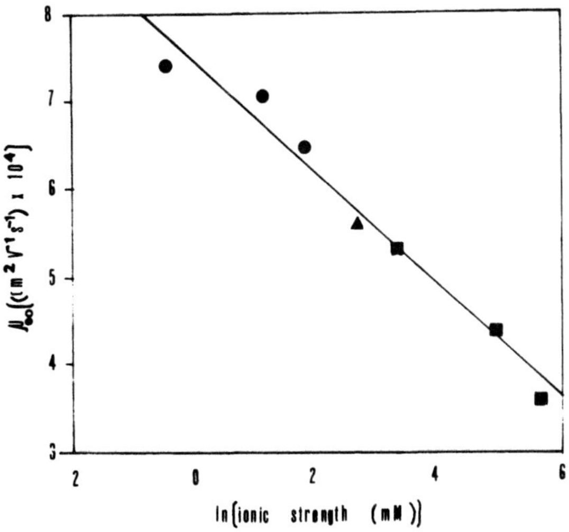

Fig. 5.5. The coefficient of electroosmotic flow (μ_{eo}) at pH = 8.0 vs. the logarithm of the ionic strength for inorganic buffers. Legend: ■ = phosphate; ● = borate; ▶ = carbonate. (Reproduced from Ref. 10 with permission of Aster Publishing Corp.)

(Fig. 5.5). It was therefore suggested that in the absence of a specific buffer-analyte interaction, a number of buffer types can provide the same electroosmotic flow if the buffers are normalized to the same ionic strength.

A study on the role of the buffer's anion on the current, the electroosmotic flow, migration times, resolution and selectivity was performed by Atamna *et al.* [12]. The anions investigated include acetate, borate, phosphate, citrate, carbonate, nitrate and nitrite. Their results showed that at the same applied voltage (20 kV), the current varies widely depending on the anion used. For instance, sodium citrate produced a current (243 μA) which is 3.6 times larger than that produced by sodium hydrogen carbonate having the same concentration. If the voltage is kept constant, the heat generated inside the capillary is directly proportional to the current ($P = VI$, where P is the power, I is the current and V is the applied voltage). It is therefore an important consideration to select the type of buffer which would give the lowest current at the same applied voltage under constant experimental conditions [11]. In practice, this can be achieved by selecting buffers which have low conductance, since the power dissipated per unit length (P) of the capillary is given by (see Section 1.3.3):

$$P = \frac{\kappa C r^2 V^2}{L}$$

where κ, C, r and L are the conductance, concentration of the buffer, radius and length of capillary, respectively.

By using mesityl oxide as a marker to measure μ_{eo} in different buffer systems, Atamna *et al.* found that in most cases μ_{eo} changed by less than 10% in the buffers investigated [12]. This observation is consistent with earlier investigations [13,14]. However, in the case of sodium tetraborate, μ_{eo} differed by over 20% from the other buffers having the same concentration and pH.

Eight dansylated amino acids were selected to investigate the effect of buffer anion on migration (t_M), resolution (R_s) and selectivity[12]. The results showed that large differences in t_M (e.g. over 50% for dans-cysteric acid) may be obtained for different buffers. It was also found that some pairs of peaks were resolved better in one buffer, and others were resolved equally well in all of the buffers. Therefore, difference in selectivity exists which means that the choice of the type of buffer anion can be considered as a method to enhance resolution in CE separations.

The effects of anion on solute migration in CE have been studied by selecting a series of four potassium salts KCl, KNO_3, KBr and K_2SO_4, as buffer additives [13]. The addition of these salts to the electrophoretic buffer helps to prevent adsorption of proteins by the fused silica capillary wall. As long as these salts are used at sufficiently high concentrations (above 0.05 M for K_2SO_4 and 0.1 M for the other

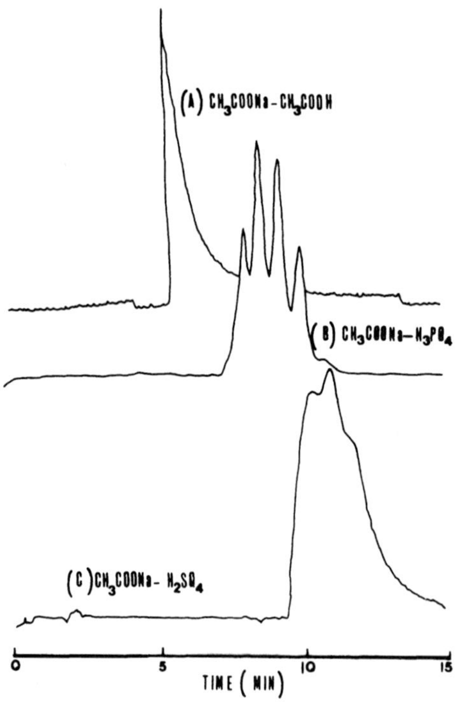

Fig. 5.6. Capillary electrophoresis separation of r-HuEPO in free solution. The electropherograms were obtained using a fused silica capillary tube 20 cm × 75 μm i.d. (A) pH 4.0 (100 mM acetate buffer, 10 kV, 30 μA. (B) pH 4.0 (100 mM acetate–phosphate buffer, 10 kV, 120 μA). (C) pH 4.0 (100 mM acetate–sulfate buffer, 10 kV, 200 μA). (Reproduced from Ref. 14 with permission of Elsevier Science Publishers.)

three salts), the anions do not have a measurable effect on the migration behaviour of the proteins investigated [13].

The interaction of the anions with silica has been explored as a means to improve separation. Tran *et al.* [14] evaluated different types of ions, including acetate, phosphate and sulfate ions in the separation of recombinant human erythropoietin (r-HuEPO) by CE. Their results show that the presence of phosphate results in better separation (Fig. 5.6) due to its stronger interaction with the silica surface of the capillary wall [14].

In view of the significant effects of the buffer's anion on the electroosmotic flow, the current, migration times, resolution and selectivity, it is therefore important to pay attention to the selection of the buffer in order to produce optimum results and generate minimum heat in a CE system.

5.1.5 Buffer cations

Cations present in the electrolyte may also affect the migration of analytes in CE systems. Relatively few studies have been performed so far although the limited amount of information available seems to indicate that the effects can be rather significant.

Migration behaviour has been observed to depend on the size of the cations [5]. Dansyl-alanine and mesityl oxide were used as probes to measure electrophoretic mobility and electroosmotic mobility, respectively. A linear relationship is observed for a plot of electrophoretic mobility vs. reciprocal of crystal radius ($1/r_{cry}$) of buffer cations such as Li^+, Na^+, K^+, Rb^+ and Cs^+. For the electroosmotic vs. $1/r_{cry}$ plot, the cations also obey a linear relationship, except for rubidium and cesium. The deviation is attributed to adsorption of these large cations by the capillary wall [5].

The effectiveness of different alkaline metal cations in preventing adsorption of proteins at the surface of the capillary wall have been demonstrated by Green and Jorgenson [13]. In Table 5.1, the effect of different concentrations of LiCl, NaCl, KCl and CsCl added to 0.1 M 2-(N-cyclo-hexylamino) ethanesulfonic acid (CHES)

TABLE 5.1

CAPACITY FACTORS FOR 0.2% LYSOZYME BY CZE IN 0.1 M CHES BUFFER OF pH 9.0 (Adapted from Ref. 13)

Salt concentration (M)	Capacity factor (k')			
	LiCl	NaCl	KCl	CsCl
0.1	*	0.35	0.49	0.39
0.3	0.10	0.04	0.02	0.03
1.0	0.00	0.00	−0.003	0.00

* k' unmeasurable owing to chemisorption.

buffer of pH 9.0 on the capacity factor in the CE analysis of lysozyme is shown. The capacity factor was defined as: $k' = (t_M - t_0)/t_0$, where t_M is the migration time of the protein and t_0 is the dead time obtained by measuring the migration time of a neutral marker (acetone).

At a concentration of 1.0 M, all four salts show similar effectiveness in preventing adsorption. At 0.3 M, some differences are observed. LiCl is the least effective for preventing adsorption, which gives a k' for lysozyme about 3 times higher than those given by other salts. The other three salts yield k' values that are significant but are approximately equal to each other. At a concentration of 0.1 M, all four salts show high levels of adsorption. For LiCl, the solute peak is not observed. The reason for this is that Li^+ is the most highly hydrated of the four alkali metal ions. With its sphere of hydration, it is effectively the largest of the alkali metals ions. Hence it is the most weakly bound. The effectiveness of the four salts in preventing adsorption is expected to be in the order: $Cs^+ > K^+ > Na^+ > Li^+$. However, Cs suffers from high optical absorbance at short wavelengths. K^+ and Na^+ are therefore more appropriate choices than Li^+ and Cs^+ as buffer cations in preventing the adsorption of proteins, since they can be used effectively at moderately high concentration without causing excessive Joule heating.

The effect of the buffer cations on the CZE separation of aminobenzoic acids is shown in Table 5.2. The buffer cations have an appreciable effect on the electrophoretic mobility of the analytes: μ_{ep} of p- and m-aminobenzoic acids increase in the order $Li^+ < Na^+ < K^+$. Similar trends are observed for p-toluenesulfonic acid (Table 5.3). A possible explanation for this behaviour is that there is an increase in μ_{ep} due to the increase in current and increased Joule heating. However, this explanation is not supported by the observed trend in μ_{eo} values [15].

TABLE 5.2

EFFECT OF THE COUNTER ION ON THE CZE SEPARATION OF AMINOBENZOIC ACID ISOMERS (Adapted from Ref. 15)

Parameter	Lithium (μ = 0.40) *	Sodium (μ = 0.52) *	Potassium (μ = 0.76) *
$\mu_{ep,P}$ (10^{-3} cm^2 V^{-1} s^{-1})	−0.23	−0.24	−0.25
$\mu_{ep,M}$ (10^{-3} cm^2 V^{-1} s^{-1})	−0.25	−0.26	−0.27
μ_{eo} (10^{-3} cm^2 V^{-1} s^{-1})	+0.49	+0.46	+0.44
Analysis time (min)	10	12	14

Conditions: 40 mM acetate buffer (pH 5.4); 25 kV constant voltage; samples 1.5 nl, 10^{-4} M; the mobility data are mean values of duplicate experiments; capillary thermostated at 30°C; detection by UV absorbance at 200 nm. Analytes: P = p-aminobenzoic acid; M = m-aminobenzoic acid.

* Cationic mobility at infinite dilution (10^{-3} cm^2 V^{-1} s^{-1}).

TABLE 5.3

EFFECT OF THE COUNTER ION ON THE CZE DATA FOR p-TOLUENESULFONIC ACID (Adapted from Ref. 15)

Parameter	Lithium	Sodium	Potassium
$\mu_{ep,P}$ (10^{-3} cm^2 V^{-1} s^{-1})	−0.31	−0.32	−0.33
μ_{eo} (10^{-3} cm^2 V^{-1} s^{-1})	+0.49	+0.45	+0.43
Analysis time (min)	14	19	24

Conditions as in Table 5.2.

The effect of the counter ions is an important consideration in separations employing buffer additives. It is advantageous to ensure that the counter ion of the additive used and that of the buffer is the same, in order to avoid adverse effects due to possible exchange of the counterions. An example is in micellar electrokinetic chromatography (see Section 5.2). When an anionic micellar solution is employed, the buffer cations and the counterions of the surfactants should be the same, whereas when a cationic micellar solution is used, the buffer anion and the counterion of the cationic surfactant should be identical.

5.1.6 Ionic strength

Ionic strength or concentration of the buffer has significant effects on solute mobilities and separation efficiency. The dependence of mobility on buffer concentration has been studied by several workers [5,16–19]. Variation of ionic strength has also been used as a method to improve separation of proteins [13,20] and aminobenzoic acid isomers [15].

In general, it has been observed that mobility depends inversely on buffer concentration [16–19]. An expression has been given for the dependence of electrophoretic mobility on concentration [5]:

$$\mu_{ep} \simeq \frac{e}{3 \times 10^7 |Z| \eta \sqrt{C}} \tag{5.16}$$

where e, Z, η and C are defined as in Eqs. (5.9) to (5.11). The reciprocal dependence of μ_{ep} on the square root of concentration (Eq. 5.16) has been demonstrated for both electroosmotic mobility using mesityl oxide, and electrophoretic mobility using dansylalanine in an acetate and phosphate buffer systems [5].

High ionic strength buffers have been used to enhance efficiencies in protein separations. Green and Jorgenson [13] devised a method to minimize the adsorption of proteins on fused silica capillaries in CE by using K^+ concentrations of 0.3 M and above in the operating buffer. The increased ionic strength resulted in a competition between K^+ and proteins for cation-exchange sites on the silica surface. The success of this approach was demonstrated for the separation of five proteins with high efficiencies (140,000 theoretical plates for bovine pancreatic trypsinogen). However, a drawback of this method is that due to the increase in ionic strength and the subsequent increase in conductivity, it would be necessary to use lower voltages and capillaries of small diameter to allow adequate heat dissipation. Consequently relatively long analysis time may be required. Moreover, detection sensitivity would be reduced in the case of optical detection methods due to a decrease in path length.

The influence of KCl concentration on efficiency of protein separation by CZE in 25 μm and 75 μm I.D. capillaries was studied by Sepaniak *et al.* [20]. In order to overcome the problem with decrease in sensitivity for small I.D. capillaries, laser-induced fluorescence (LIF) detection was used. The results (Table 5.4) show that the advantages of high salt concentration can be exploited by LIF with very small diameter capillaries. However, at 45 mM KCl the effects of thermal dispersion becomes significant even with the use of small diameter capillaries.

Similar observations were made by Rasmussen and McNair [21] in their study on the effects of buffer concentration, capillary internal diameter and electric field strength on the coefficient of electroosmotic flow. With a capillary of 50 μm I.D., μ_{eo} decreased when more concentrated Na_2HPO_4 buffers were used (Fig. 5.7). In a 100 μm capillary, the same relationship between buffer concentration was observed at lower electric field strengths. However, Joule heating became more pronounced in the larger capillary, especially when concentrated buffers were employed. μ_{eo} was found to increase markedly with electric field strength (Fig. 5.8), since the viscosity decreased when Joule heating increased. Based on the consideration that the resolution of two species depended on their migration velocities, Rasmussen and McNair calculated the relative velocity difference, $\nu_{rel} = (\mu_{ep,1} - \mu_{ep,2})/(\mu_{ep,av} + \mu_{eo})$, where $\mu_{ep,1}$, $\mu_{ep,2}$ correspond to the electrophoretic mobilities of species 1 (phenol) and 2 (sodium tolutensulfonate), and $\mu_{ep,av}$ is the average of $\mu_{ep,1}$ and $\mu_{ep,2}$. These values are shown in Fig. 5.9. It can be seen that ν_{rel} remains constant for a given buffer concentration, regardless of field strength, since both μ_{eo} and μ_{ep} have the same viscosity dependence [21,22]

TABLE 5.4

INFLUENCE OF SALT (KCl) CONCENTRATION AND COLUMN DIAMETER ON PLATE NUMBER, *N* (plates/m), AND POWER DISSIPATION, *P* (W), FOR CONALBUMIN (Adapted from Ref. 20)

Capillary diameter (μm)	KCl concentration (mM)					
	15		30		45	
	N	*P*	*N*	*P*	*N*	*P*
25	78,000	0.10	390,000	0.20	280,000	0.40
75	80,000	0.24	12,000	0.42	–	–

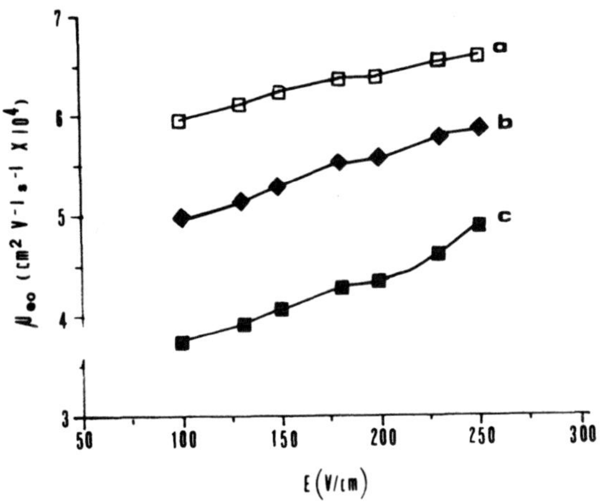

Fig. 5.7. Influence of buffer concentration and electric field strength (E) on the coefficient of electroosmotic flow (μ_{eo}) in 50 μm capillaries. Na_2HPO_4 concentrations: a = 0.01 M; b = 0.02 M; c = 0.05 M. (Reproduced from Ref. 21 with permission of Elsevier Science Publishers.)

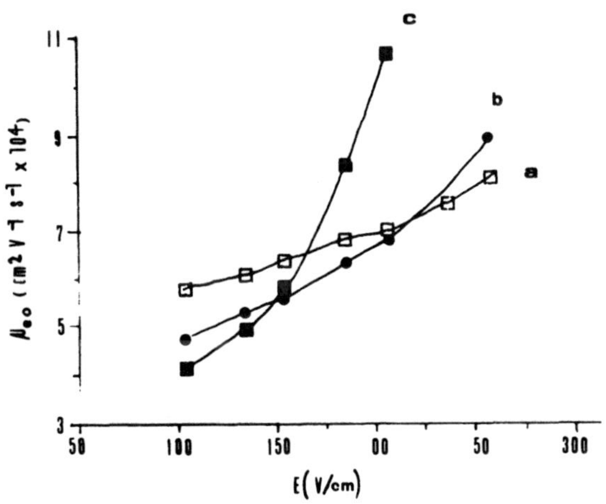

Fig. 5.8. Influence of buffer concentration and electric field strength (E) on the coefficient of electroosmotic flow (μ_{eo}) in 100 μm capillaries. Na_2HPO_4 concentrations: a = 0.01 M; b = 0.02 M; c = 0.05 M. (Reproduced from Ref. 21 with permission of Elsevier Science Publishers.)

The effect of ionic strength on the separation of aminobenzoic acid positional isomers by CE has been investigated by Nielen [15]. The ionic strength of the morpholinolethanesulfonic acid (MES) buffer was varied from 10 to 100 mM at pH 6.0. At lower buffer concentrations, both μ_{eo} and μ_{ep} were found to increase due to an increase in the zeta potential. However, the plate number was found to decrease dramatically (from 210,000 to 40,000) and the resolution decreased to zero at 10 mM (Fig. 5.10). Similar results were obtained for phosphate buffers.

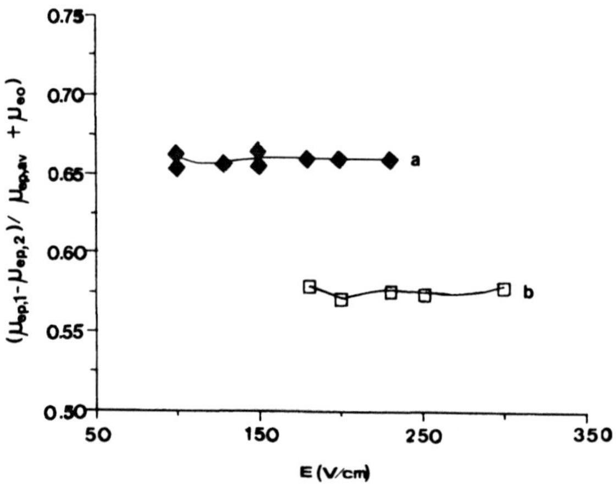

Fig. 5.9. Influence of buffer concentration and electric field strength (E) on the relative velocity difference $(\mu_{ep,1} - \mu_{ep,2})/(\mu_{ep,av} + \mu_{eo})$. Conditions: a = 0.02 M buffer/100 μm capillary; b = 0.01 M buffer/50 μm capillary. (Reproduced from Ref. 21 with permission of Elsevier Science Publishers.)

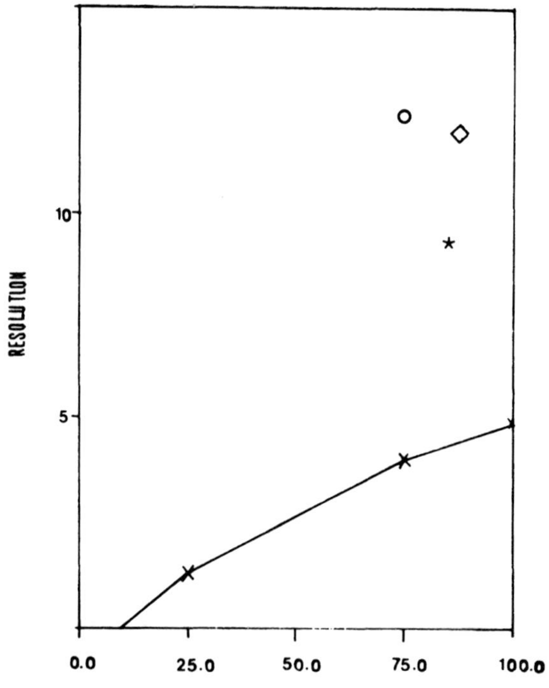

Fig. 5.10. Influence of buffer concentration on the resolution of p- and m-aminobenzoic acid. × = MES buffer (pH 6.0); o = MES buffer-methanol (75 : 25); * = MES buffer-2-propanol (87.5 : 12.5). Capillary thermostated at 30.0°C, except for (◇) MES buffer-2-propanol (87.5 : 12.5) at 60.0dgrC. (Reproduced from Ref. 15 with permission of Elsevier Science Publishers.)

Nielen attributed the observed band broadening to the sample volume, migrational dispersion [8], inhomogeneous electroosmosis or time constants, etc.

5.1.7 Buffer pH

As discussed in Section 5.1.3, the degree of ionization of species present in the electrolyte system depends on the pH of the solution. Differences in the degree of ionization give rise to differences in electrophoretic and electroosmotic mobilities. Consequently, both the separation efficiency and flow velocities may be affected by the buffer pH.

The variation of the electroosmotic flow coefficient with pH in 75 μm I.D. pyrex, 75 μm I.D. fused silica, and 120 μm I.D. PTFE capillaries is shown in Fig. 5.11 [23]. A negative zeta potential is observed for each of the capillaries. The electroosmotic flow is in the direction of the cathode. In the case of pyrex and fused silica, the negative zeta potential can be attributed to the hydroxyl groups at the capillary surface. The electroosmotic flow for the PTFE is lower than that of the other types of columns. This can be explained by the fact that there are no intrinsic ionic groups at the surface of PTFE, and the charges caused by adsorption of hydroxyl and other anions from the solution are relatively much less than those for pyrex and fused silica. In all three cases, the electroosmotic flow is found to decrease with a decrease in pH, since hydrogen ions could neutralize anions at the surface and hence lower the zeta potential.

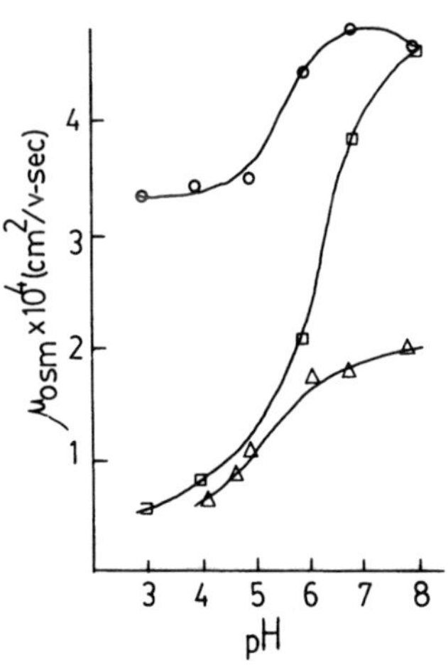

Fig. 5.11. The effect of pH on electroosmosis at constant ionic strength (I = 0.06). Length of capillary: 50 cm. ○ = pyrex; □ = silica; △ = PTFE. Phenol (0.4 M) was employed as a neutral marker for electroosmosis with UV detection at 280 nm. Buffers: pH 3, chloroacetic acid; pH 4, formic acid; pH 5, acetic acid; pH 6, succinic acid; pH 7, phosphoric acid; pH 8, Tris-HCl. Each buffer solution contained 5-fold excess of KCl. Applied power was kept constant (0.33 W) through adjustment of the applied voltage. (Reproduced from Ref. 23 with permission of Dr. Alfred Heuthig Publishers.)

Optimum conditions for CZE separation of oxygen isotopic benzoic acids ($C_6H_5C^{16}O_2H$, $C_6H_5C^{16}O^{18}OH$ and $C_6H_5C^{18}O_2H$) have been considered on the basis of the resolution equation (see Eq. 1.12) by Terabe *et al.* [24]. For the separation of the monobasic acids ($C_6H_5C^{16}O_2H$ and $C_6H_5C^{18}O_2H$), whose dissociation constants are very close, an approximate equation for the optimum pH has been given [24]:

$$\text{pH (optimum)} = \text{p}K_\text{a} - \log 2 \tag{5.17}$$

The optimum pH is calculated to be 3.89. Experiments have been performed which show that maximum resolution is obtained at pH 3.9, which is in good agreement with the calculated value.

The effect of pH on the resolution of p- and m-aminobenzoic acid is illustrated in Fig. 5.12. As expected on the basis of Eq. (5.17), the resolution is found to be optimum at a pH value close to the p*K* values of the analytes. In addition, a minimum is observed at pH 4.2. Below this pH, the migration order of p- and m-aminobenzoic acids reverses. This can be explained in terms of the p*K* values:

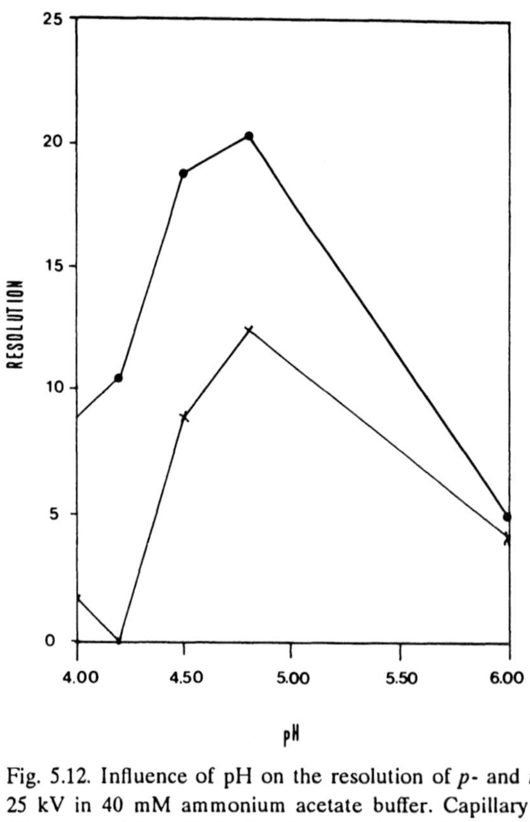

Fig. 5.12. Influence of pH on the resolution of *p*- and *m*-aminobenzoic acid. Conditions: CZE at 25 kV in 40 mM ammonium acetate buffer. Capillary thermostated at 30°C (×) and 60°C (o). (Reproduced from Ref. 15 with permission of Elsevier Science Publishers.)

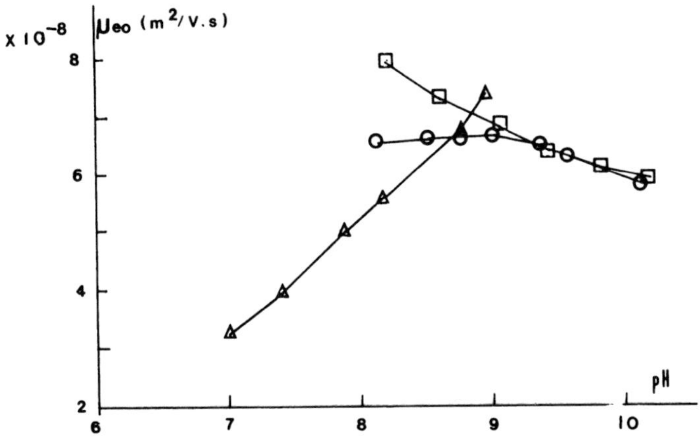

Fig. 5.13. Influence of buffer pH on electroosmotic mobility. △ = Tris–HCl; □ = boric acid–NaOH; ○ = borax–HCl. Buffer concentration: 50 mM. (Reproduced from Ref. 25 with permission of Elsevier Science Publishers.)

pK_1 = 2.41 and pK_2 = 4.85 for p-aminobenzoic acid and pK_1 = 3.12 and pK_2 = 4.74 for m-aminobenzoic acid). Both μ_{eo} and μ_{ep} decrease with pH from pH 6 to 4 because of the increased neutralization of the capillary wall and the decreased ionization of the analytes [15].

Vindevogel and Sandra [25] studied the effect of buffer pH on electroosmotic mobility in the following systems:

(a) Tris + HCl = Tris H^+ + Cl^-
(b) H_3BO_3 + NaOH = Na^+ + $B(OH)_4^-$
(c) $2Na^+$ + $2B(OH)_4^-$ + HCl = $2Na^+$ + $B(OH)_4^-$ + Cl^- + H_3BO_3

The results of this study are shown in Fig. 5.13. For the Tris buffer, case (a), a decrease in pH corresponds to an increase in the number of ions, and hence an increase in the ionic strength. μ_{eo} decreases rapidly with a decrease in pH. For the boric acid–NaOH and borate–NaOH buffers, case (b), an increase in pH results in an increase in the ionic strength. μ_{eo} is seen to decrease with pH. For the borax–HCl buffer, case (c), the amounts of ions remains the same at different pH, and μ_{eo} remains relatively constant although it tends to decrease at higher pH. By adding a suitable amount of salt to keep the ionic strength constant, the decrease of μ_{eo} with increase in pH in cases (b) and (c) is no longer observed. These results show that when constant-concentration buffers are used, μ_{eo} increases with pH for a weak base–strong acid-type buffer, and decreases with pH for a strong base–weak acid type buffer [25]. For buffers at constant ionic strength rather than at constant concentration, μ_{eo} remains relatively constant over the pH range investigated. Nevertheless, it should be noted that the effect of ionic strength can also be observed through changes in pH. For a borate which is 5% neutralized with NaOH, the ionic strength is doubled by doubling the concentration, whereas

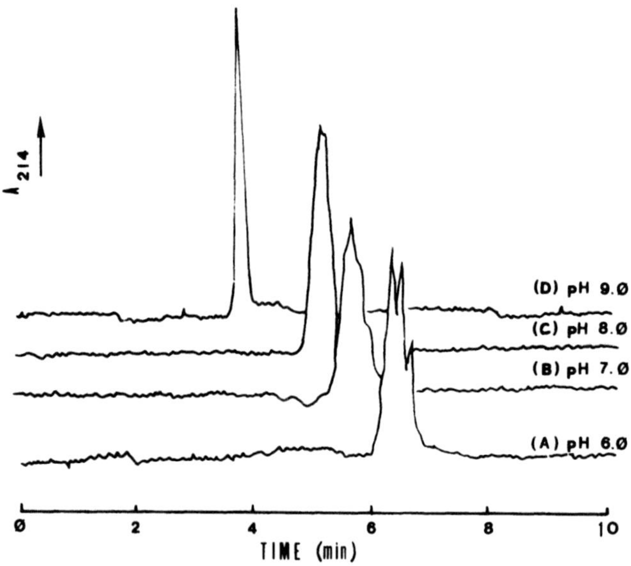

Fig. 5.14. High-performance capillary electrophoresis of r-HuEPO in free solution. The electropherograms were obtained using a fused silica capillary tube 50 cm × 75 μm ID. (A) pH 6.0 (50 mM MES, 25 kV, 44 μA). (B) pH 7.0 (50 mM Bis-Tris, 25 kV, 15 μA). (C) pH 8.0 (50 mM tricine, 25 kV, 70 μA). (D) pH 9.0 (50 mM tricine, 25 kV, 85 μA. (Reproduced from Ref. 14 with permission of Elsevier Science Publishers.)

in order to achieve a pH change of 1 unit, the ionic strength needs to change by five-fold. Therefore, the effect of pH on μ_{eo} is generally observed to be much stronger than that of concentration.

Tran *et al.* [14] attempted to improve the resolution of recombinant human erythropoietin (r-HuEPO) by increasing the differences in electrophoretic mobility and reducing the electroosmotic flow of the running buffer. Separations at pH 6.0, 7.0, 8.0 and 9.0 were investigated (Fig. 5.14). At these pH values, r-HuEPO (pI 4.5–5.0) exists as a negative species and the interaction between solute and the capillary wall is minimized. Consequently, resolution improves with a decrease in pH due to the increase in difference in charge between glycoforms of r-HuEPO and a reduction in electroosmotic flow.

Vinther and Soeberg [26] presented calculations of the radial pH gradient as a function of electroosmotic flow and temperature. Due to the presence of silanol groups and the occurrence of electroosmotic flow, pH value at the capillary wall is expected to be lower than that of the bulk liquid by as much as 2 pH units. Peak tailing or adsorption of analytes may occur as a result, especially when the analysis is performed at a bulk pH slightly above the isoelectric point of the analytes. By increasing the ionic strength of the buffer and hence reducing electroosmotic flow, the undesirable effect of the pH gradient can be reduced.

5.1.8 Effects of organic modifiers

Addition of organic solvents to the electrophoretic buffer permits the analysis of some analytes which are not normally aqueous soluble by improving their solubility in the buffer. Organic solvents are also known to reduce the electroosmotic flow, which may result in better resolution at the expense of a longer analysis time.

The effects on μ_{eo} resulting from the addition of 1% (v/v) ethanol, 2-propanol, butanol, and pentanol to an aqueous McIlvaine buffer is shown in Fig. 5.15. These data suggest that μ_{eo} depends on alcohol chain length. A possible explanation for the dependence is related to the viscosity of the pure alcohols (see Fig. 5.16). In practice, however, the measured viscosities of 1% alcohol solutions do not differ significantly from that of pure water. Therefore, a more likely cause is that the alcohols interact strongly with the capillary wall, resulting in a higher apparent concentration of alcohol within the double layer [10]. These interactions then result in higher apparent viscosities within the double layer which account for the trend observed in Fig. 5.15.

Figure 5.17 illustrates the dependence of μ_{eo} on the concentration of ethanol, 2-propanol and acetonitrile. The change in μ_{eo} generally increases with the concentration of organic solvent. The change in μ_{eo} is the smallest in the case of acetonitrile, due to a weaker interaction between acetonitrile and the capillary wall. The data shown in Fig. 5.17 indicate that acetonitrile can be used as a modifier without causing a significant decrease in μ_{eo}.

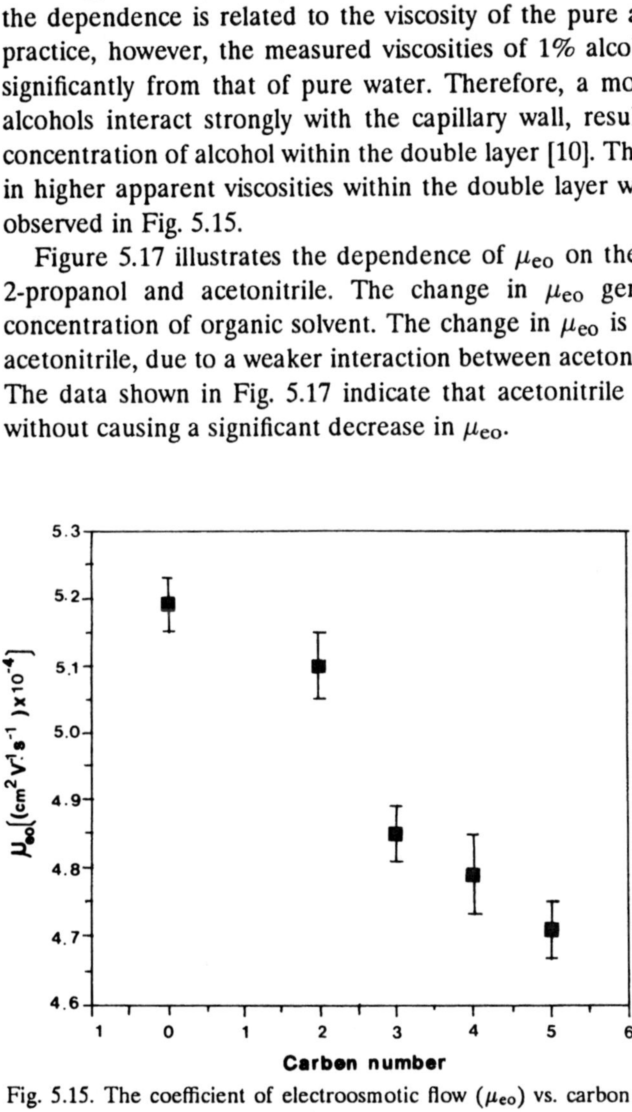

Fig. 5.15. The coefficient of electroosmotic flow (μ_{eo}) vs. carbon number for a series of alcohols (1% v/v). Buffer system: 60 mM McIlvaine, pH 8.0. (Reproduced from Ref. 10 with permission of Aster Publishing Corp.)

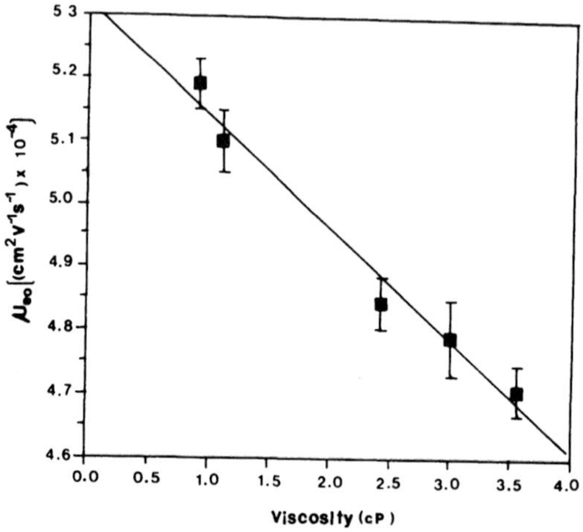

Fig. 5.16. The coefficient of electroosmotic flow (μ_{eo}) vs. viscosity of the neat alcohols (ethanol, 2-propanol, butanol, and pentanol). Buffer system: 60 mM McIlvaine, pH 8.0. (Reproduced from Ref. 10 with permission of Aster Publishing Corp.)

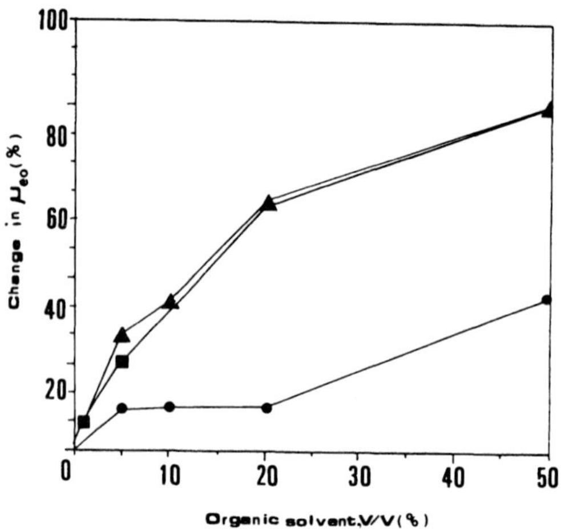

Fig. 5.17. Percent change in the coefficient of electroosmotic flow (μ_{eo}) vs. percent organic solvent (v/v). Legend: ■ = ethanol (in 60 mM McIlvaine, pH 8.0); ● = acetonitrile (in 25 mM MES, pH 5.65); ▲ = 2-propanol (in 25 mM MES, pH 5.65). (Reproduced from Ref. 10 with permission of Aster Publishing Corp.)

The effect of organic modifiers on the electroosmotic flow of the tricine buffer is shown in Fig. 5.18. The dependence of the electroosmotic flow and the chain length of the alcohol is expected to follow the order: methanol < ethanol < propanol < butanol [14]. Diol compounds are more efficient than alcohol in increasing the viscosity and hence decreasing the electroosmotic flow of the running buffer. Typically the efficiencies of diol compounds increase from ethylene glycol <

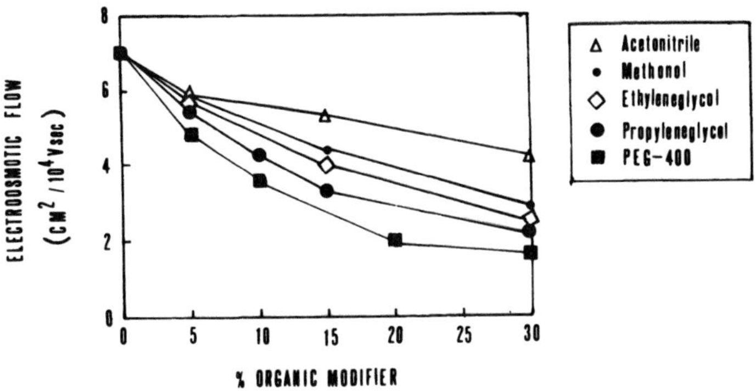

Fig. 5.18. Effect of organic modifier on the electroosmotic flow, μ_{eo} of the running buffer. Conditions: fused silica capillary column 50 cm × 75 μm I.D.; voltage: 25 kV. (Reproduced from Ref. 14 with permission of Elsevier Science Publishers.)

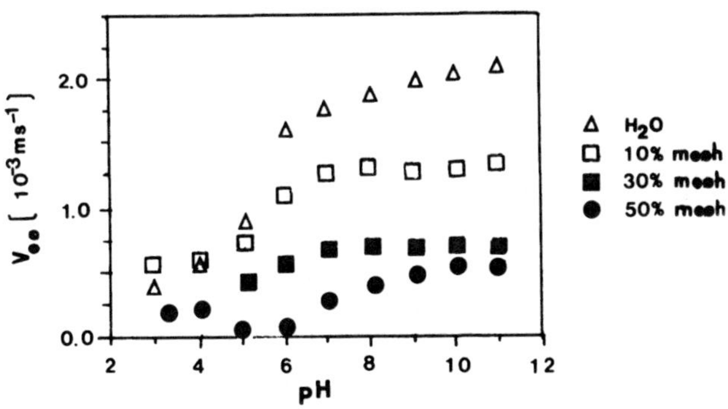

Fig. 5.19. Dependence of the electroosmotic velocity, ν_{eo} on the pH (or pH′) of the buffer electrolyte for pure aqueous and aqueous-methanolic solutions. Percentage given v/v. (Reproduced from Ref. 27 with permission of the American Chemical Society.)

glycerol, 1,2-butanediol, 1,3 or 1,4-butanediol < trans or cis-cyclohexane-1,2-diol < PEG-200 < PEG-400 and are generally more efficient than methanol.

The dependence of the electroosmotic velocity on the pH of the buffer electrolyte for pure aqueous and aqueous-methanolic solutions is shown in Fig. 5.19. It can be observed that μ_{eo} decreases with increasing concentration of methanol, and increases with increase in pH. It can also be observed that the point of inflection corresponding to the pK' of the silanol groups is shifted towards higher values when an increasing amount of the organic solvent is used.

The influence of different organic solvents on pK' of the silanol group is shown in Fig. 5.20 for a constant percentage of 50% of the organic solvents. Results for three protic solvents — methanol, ethanol and 2-propanol — and three aprotic dipolar solvents — acetonitrile, dimethyl sulfoxide and acetone — are shown [27]. An increase in μ_{eo} with increasing pH is observed for all the three solvents. This

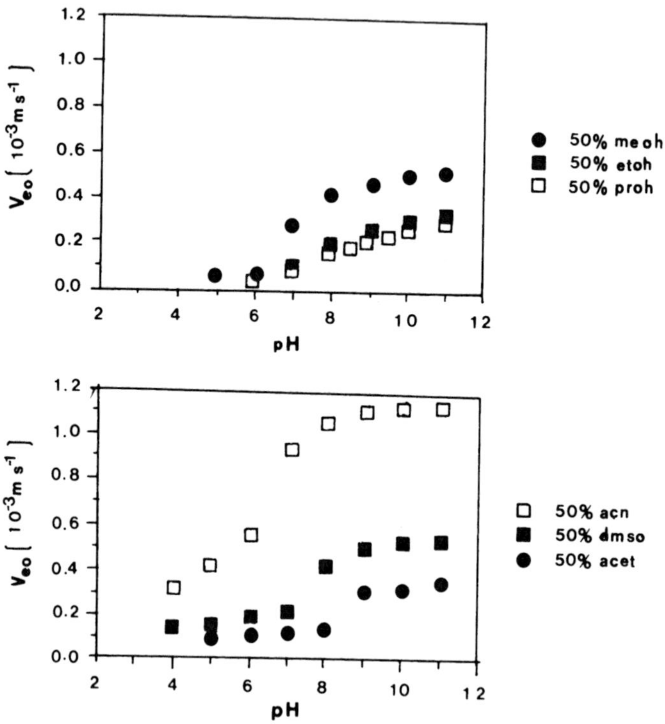

Fig. 5.20. Dependence of the electroosmotic velocity, ν_{eo} on the pH (pH′) of the buffer electrolyte for solvent systems with a constant volume fraction of the organic solvent. Solvent code: *meoh* = methanol; *etoh* = ethanol; *pron* = 2-propanol; *acn* = acetonitrile; *dmso* = dimethylsulfoxide; *acet* = acetone. (Reproduced from Ref. 27 with permission of the American Chemical Society.)

is expected since the silanol groups dissociate more at higher pH, causing the zeta potential to increase. Another observation is that the inflection points shift towards higher pH upon addition of the solvents. As in the case of methanol, the shifts correspond to changes in pK'. Dimethyl sulfoxide affects the pK' of the silanol groups most significantly (3 pH units). This is because of poor ability of this solvent to solvate anions. Acetonitrile, on the other hand, increase the pK' by only 1 pH unit. Hence it acts almost as an inert compound in mixtures with water [27].

The variations of μ_{eo} with the content of the organic solvents is shown in Fig. 5.21. In general, it is observed that μ_{eo} decreases with an increasing fraction of the organic solvent. The decreases for most organic solvents are very steep, e.g. on addition of 30–40% organic liquid, μ_{eo} decreases by a factor of 3 to 4. The only exception is acetonitrile, which reduces μ_{eo} by only about 1/3 on addition of 30% of it into the electrophoretic solution.

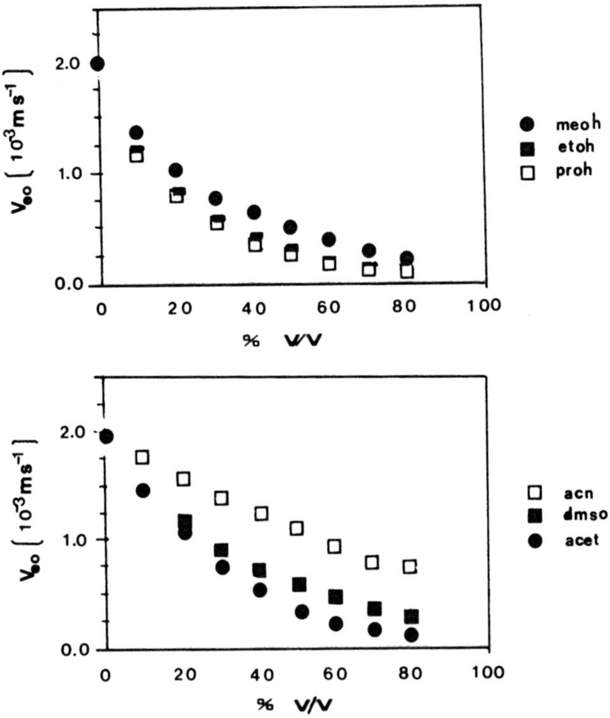

Fig. 5.21. Variation of the electroosmotic velocity, v_{eo} with solvent composition. Solvent code according to Fig 5.20. (Reproduced from Ref. 27 with permission of the American Chemical Society.)

5.1.9 Other modifiers

Other additives can be used to affect electromigration of solutes in the electrophoretic system. The addition of surfactant molecules into the electrolyte to form micelles is considered a separate mode of CE called micellar electrokinetic chromatography (MEKC). The use of micelle-forming modifiers for CE is discussed in detail in Section 5.2. Another important group of modifiers, such as cyclodextrins and crown ethers, which form inclusion complexes are discussed in Section 5.3. In addition, modifiers which form complexes with the analytes are discussed in Section 5.4. In this section, some examples of novel modifiers which do not belong to the above groups are discussed.

The magnitude and sign of the zeta potential of PTFE and fused silica capillary wall in a pH 6.0 buffer containing different additives has been measured by Reijenga *et al.* [28]. Due to their higher viscosities, poly(vinyl alcohol) (PVA), hydroxyethylcellulose (HEC) and hydroxypropylmethylcellulose (HPMC) decrease the zeta potential more significantly than polyvinylpyrolidine (PVP) and Triton X-100. Cationic surfactants, such as cetyltrimethylammonium bromide (CTAB) and Priminox, reverse the sign of the zeta potential and hence the direction of electroosmosis (see Section 1.3.3).

Another approach involves the addition of putrescine to pH 6 buffer to minimize ion exchange of the tris-His peptide in CZE separation [29]. Adsorption of cationic species in CE can be viewed as simple ion exchange with buffer cations at the negative sites on the capillary wall. Ion exchange could be minimized by using buffer cations with strong affinity for surface exchange sites. Putrescine are polyvalent organic cations which has been suggested for sharpening protein peaks in CZE [30]. As can be seen in Fig. 5.22, putrescine at 2 mM concentration gives improved separation of the three His-containing cationic heptapeptides: hydrocinnamyl-Gly-Gly-Gly-His-Gly-Gly-Gly-NH-CH_3 (4-His), hydrocinnamyl-Gly-Gly-His-Gly-His-Gly-Gly-NH-CH_3 (3,5-di His), and p-methoxyhydrocinnamyl-Gly-His-Gly-His-Gly-His-Gly-NH-CH_3 (2,4,6-Tri His).

A metal containing buffer can also be used for obtaining improved separation of peptides. Figure 5.23 shows separations obtained in pH 7.5 tricine buffer with and without added $Zn(ClO_4)_2$. At pH 7.5, His protonation would be minimized and Zn–His interactions maximized. Also, operation at a pH substantially less than the buffer pK_a minimizes $Zn(tricine)_2$ formation, and possible Zn deactivation. With 1 mM Zn^{2+} added into the electrolyte improved separation of the cationic heptapeptides can be obtained [29].

The use of a Tris–borate buffer containing a hydrophilic cellulose derivative and ethidium bromide improve the separations of double-stranded DNA fragments by CE [31]. The effect of ethidium bromide on the separation of double-stranded DNA

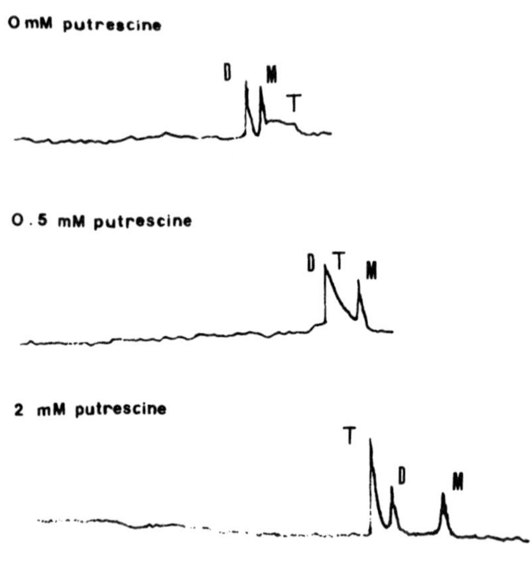

Fig. 5.22. Separation of cationic heptapeptides in pH 6 buffer with varying amounts of putrescine added. M = 4-His, D = 3,5-diHis, T =2,4,6-triHis, 0.5 mg/ml each peptide; 40 cm capillary; 15 kV. (Reproduced from Ref. 29 with permission of Elsevier Science Publishers.)

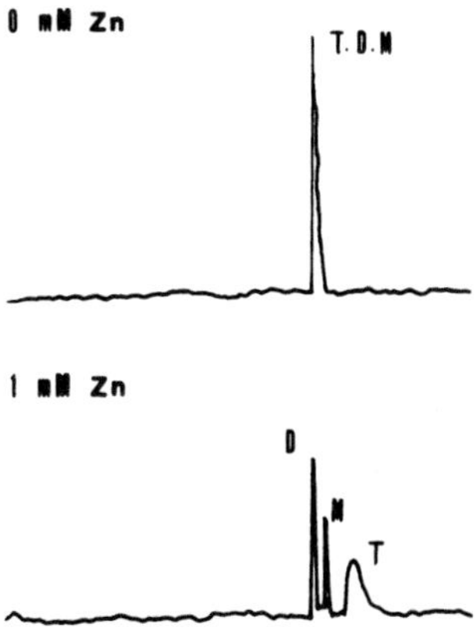

Fig. 5.23. Separation of cationic heptapeptides in pH 7.5 tricine buffer with and without addition of 1 mM $Zn(ClO_4)_2$. Same conditions as Fig 5.22. (Reproduced from Ref. 23 with permission of Elsevier Science Publishers.)

fragments is shown in Fig. 5.24. It can be seen that the capacity factor decreases with increases in the concentration of ethidium bromide. The phenanthidine ring of ethidium bromide is thought to intercalate between two G-C base pairs of the DNA fragments, causing the DNA duplex to derotate [31]. The change in conformation enhances the interaction with the hydrophillic cellulose derivatives, and hence improves resolution of the DNA fragments. The cationic charged nitrogens interact electrostatically with the phosphate group, resulting in a decrease in the electrophoretic velocity of the DNA [32]. As a result, shorter migration times are expected. The impact of ethidium bromide on the number of theoretical plate is shown in Fig. 5.25. It can be observed that high numbers of theoretical plate

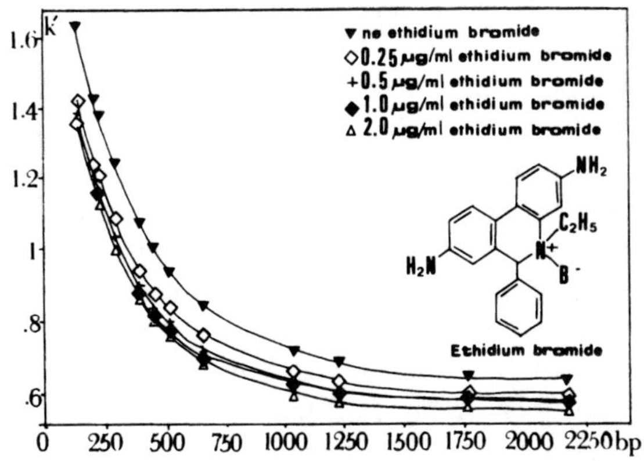

Fig. 5.24. Effect of ethidium bromide on the separation of double-stranded DNA fragments. (Reproduced from Ref. 31 with permission of Applied Biosystems, Inc.)

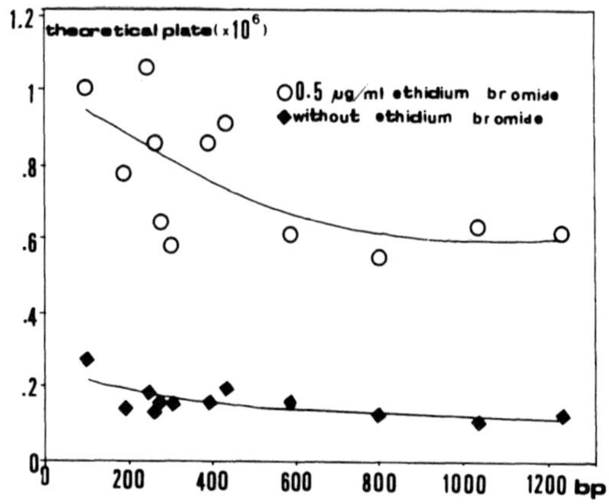

Fig. 5.25. Impact of ethidium bromide on the number of theoretical plates. (Reproduced from Ref. 31 with permission of Applied Biosystems, Inc.)

are obtained with the use of ethidium bromide as a modifier in the electrophoretic solution.

As can be seen from the examples given above, many novel modifiers can be employed to achieve improvements in the separation efficiencies, to prevent adsorption of analytes by the capillary wall, or to influence the electroosmotic flow. One of the unique advantages of CE is that the use of such modifiers can be easily accomplished by simply adding the additives to the electrophoretic solution. The solution can also be easily flushed out by rinsing when it is no longer required.

5.1.10 Effect of temperature

Thermal effects (Joule heating) in CE have been discussed in detail in Section 1.3.3. For certain biological samples, without the use of an efficient cooling system, the species of interest in the sample may thermally degrade, e.g. some proteins may denaturate.

For larger-bore capillaries, Joule heating can be responsible for an increase in the height equivalent to a theoretical plate at high fields due to a temperature difference between the centre of the column and the wall. It has been shown that the overall column temperature can rise to over 70°C even with narrow-bore columns without proper cooling [33]. CE separations are usually performed in narrow-bore columns to reduce the radial temperature effect within the column bore. Forced convection and thermoelectric cooling are commonly used methods to remove Joule heating [34]. At constant field, the method of cooling can have a significant effect on the stability of the system. In Fig. 5.26, the stability of current is plotted against time for natural air convection, forced air convection and solid-state thermoelectric cooling in a CE system with a 100 μm column at 500 V/cm with the 100 mM tris, 250 mM boric acid, 7 M urea buffer. It can be seen from Fig. 5.26 that

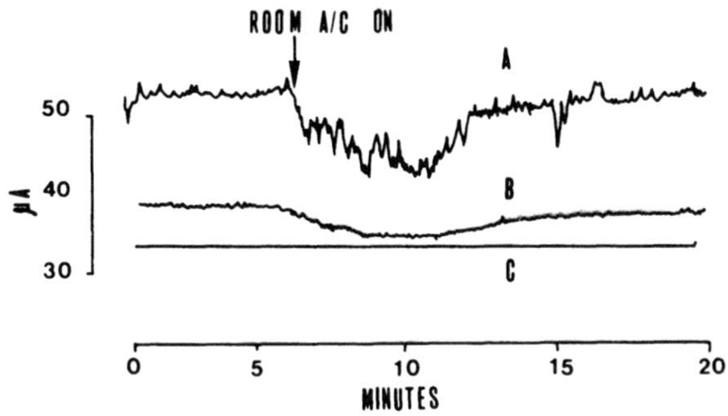

Fig. 5.26. The stability of current with time. The current is monitored at 500 V/cm for natural air convection (*A*), forced air convection (*B*) and solid-state thermoelectric cooling (*C*). Buffer: 0.1 M Tris, 0.25 M boric acid, 7 M urea, pH 7.6. Column: 180 mm × 100 μm. The difference in current levels are the result of the variances in the efficiencies of heat removal. *A*/*C* = air conditioning. (Reproduced from Ref. 34 with permission of Elsevier Science Publishers.)

the current is most unstable when heat dissipation is merely by natural convection, case A. With forced air convection, case B, the peak to peak deviation in the current level is smaller than that in A. In the case of the solid-state cooled column, case C, current fluctuation is the smallest among the three methods. For fan cooling and natural convection with an applied power of 1 W/m, the column temperature can be 5°C or 12°C above ambient [34], respectively. The corresponding differences in electrophoretic mobility from ambient would be estimated to be 10% and 25%, assuming 2% per °C variation in mobility [35]. In the case of the Peltier solid-state cooling device and at the same applied voltage, the temperature difference between the inside wall of the capillary and the environment around the capillary (ΔT_e) is less than 1°C, which corresponds to a mobility difference of approximately 1% [34].

By plotting the current against the applied voltage (*E*-*I* plot or Ohm's law plot), it is possible to determine the maximum applied voltage for a particular electrophoretic system. Figure 5.27 shows the Ohm's law plot for 200, 100 and 50 μm column diameters with natural air convection. In each case, a significant curvature occurs which is due to the decrease in column resistance with an increase in the electric field. At the highest voltages, the column can not dissipate the heat generated by the applied power efficiently, and thermal breakdown may occur. The electrolyte increases in temperature as the resistance decreases until the circuit is broken due to the formation of air or vapour bubbles.

The Ohm's law plot for forced air cooled column of different diameters is shown in Fig. 5.28. As a result of improved heat dissipation by forced convection, higher voltages can be used as thermal breakdown does not occur so readily as in the case of natural cooling. However, the effectiveness of forced air cooling is rather limited. Figure 5.29 shows a plot of μ_{eo} vs. applied electric field (a) for a 100 μm I.D. capillary without cooling, (b) when forced air convection at 3.5 l/min is applied to 50 cm of the 100 μm I.D. capillary and (c) for a 50 μm I.D. capillary without

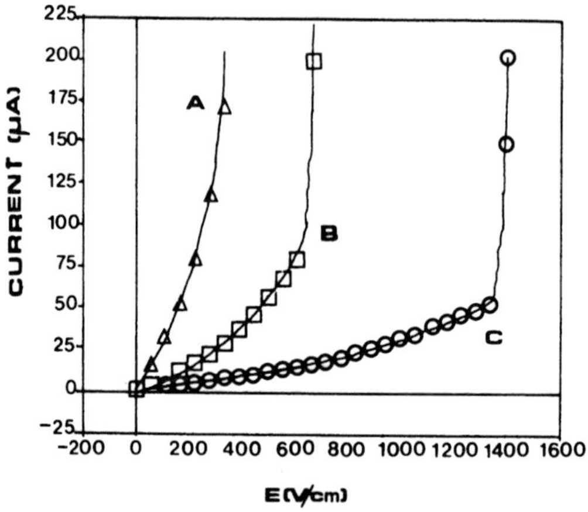

Fig. 5.27. Ohm's law plot of current vs electric field for natural convection on a 180 mm capillary column length with internal diameters of 200 μm (*A*); 100 μm (*B*); 50 μm (*C*). The buffer was the same as in Fig 5.26. The rapid rise in current at high voltage indicates the point of thermal breakdown where the rate of heat generation is greater than heat removal. (Reproduced from Ref. 34 with permission of Elsevier Science Publishers.)

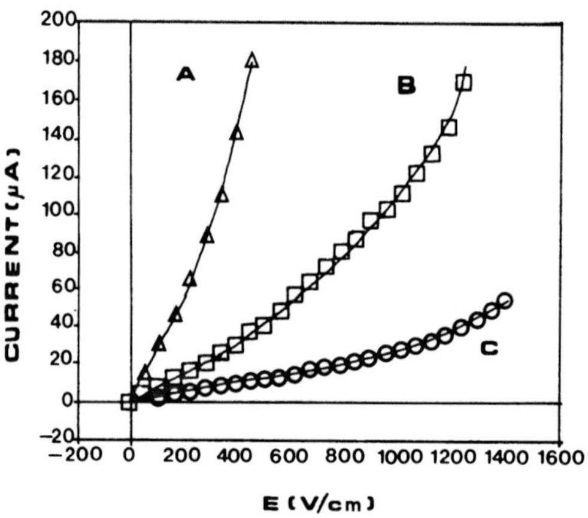

Fig. 5.28. Ohm's law plot for forced air cooling on a 180 mm capillary column length with internal diameter of 200 μm (*A*); 100μm (*B*); 50 μm (*C*). The buffer was the same as in Fig 5.26. (Reproduced from Ref. 34 with permission of Elsevier Science Publishers.)

cooling. It can be seen from Fig. 5.29 when larger capillaries are used, forced convection may not be capable of reducing μ_{eo} to approach the level observed in 50 μm capillaries [21].

Instead, more efficient cooling methods, such as thermoelectric or liquid cooling, can be employed to improve heat dissipation. Fig. 5.30a shows the Ohm's law plot for thermoelectrically cooled columns of different diameter at constant temperature. The linearity can be clearly observed. Furthermore, thermal breakdown is less

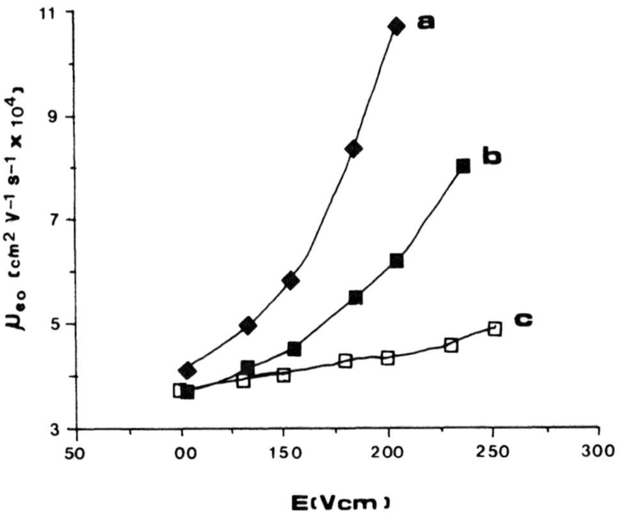

Fig. 5.29. Influence of forced convection on the coefficient of electroosmotic flow (μ_{eo}) in 0.05 M Na_2HPO_4. Capillaries: a = 100 μm; b = 100 μm with convection; c = 50 μm. (Reproduced from Ref. 21 with permission of Elsevier Science Publishers.)

likely than in the case of natural and forced convection. The Ohm's law plot for thermoelectric temperature control on a 180 mm × 50 μm capillary column at several temperature is shown in Fig. 5.30b. The operation of the thermoelectric device at lower temperature results in lower power dissipation for a given field and buffer concentration, and therefore, a lower ΔT_e [34]. Figure 5.30c shows a plot of μ_{eo} vs. applied electric field for the thermoelectric cooling system at 20°C. The plot is linear and passes through the origin. Therefore the temperature of the inside wall of the capillary should be constant over the range of applied voltage.

A consequence of inefficient column temperature control is that the overall temperature of the column increases as a result of Joule heating which may lead to peak broadening. The dependence of peak shape on column temperature is shown in Fig. 5.31 for horse heart myoglobin. The observed decrease in peak height and the increase in peak width may be due to sample denaturation at high temperature [34].

Despite its negative effects in terms of Joule heating, buffer temperature can be exploited as a selectivity parameter. An increase in temperature tends to shorten the analysis time, owing to a decrease in viscosity. Temperature can also influence the chemical equilibrium [15,36,37]. An example of the use of temperature dependence of chemical equilibrium as a selectivity tool is shown in Fig. 5.32, which shows electropherograms obtained for p- and m-aminobenzoic acids at different temperatures. It is noted that the migration order reverses and the separation of the peaks improves when temperature changes from 30°C to 60°C. The reason for this behaviour is that both positional isomers show increased mobilities, μ_{ep},

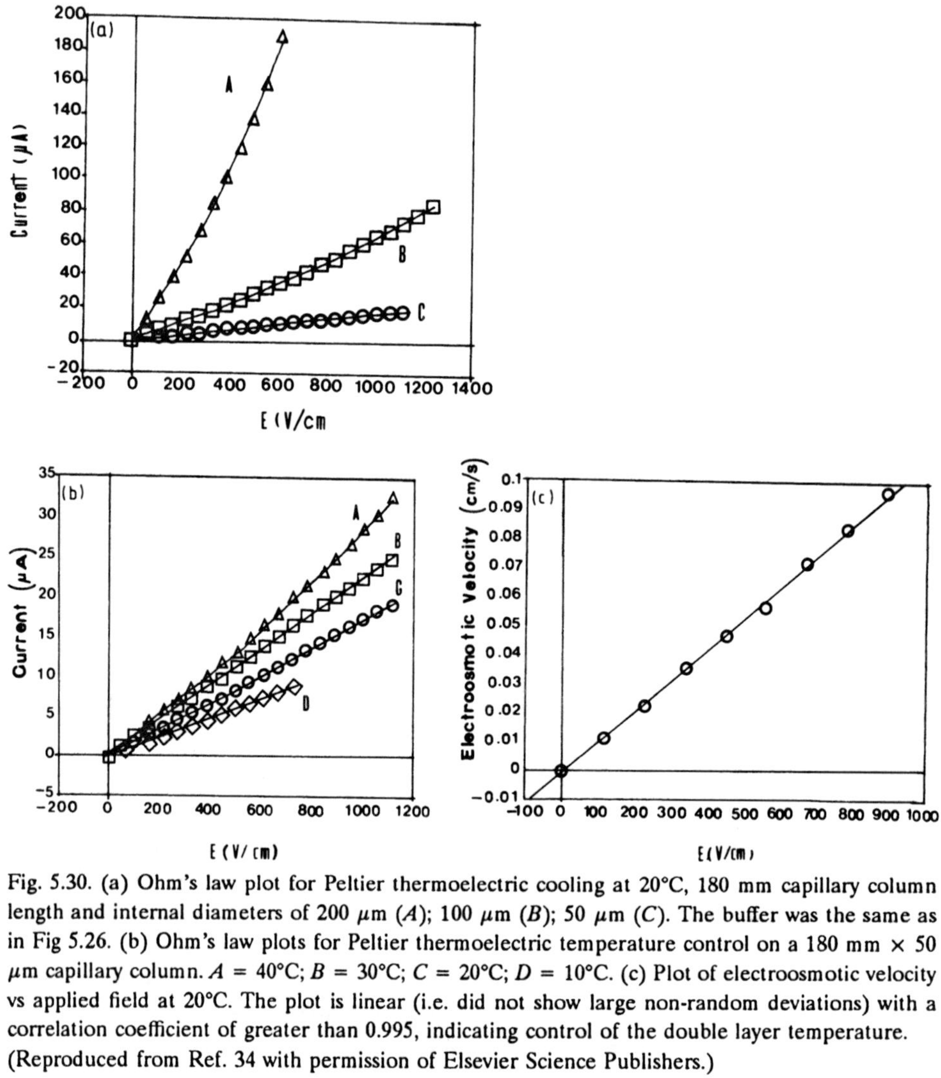

Fig. 5.30. (a) Ohm's law plot for Peltier thermoelectric cooling at 20°C, 180 mm capillary column length and internal diameters of 200 μm (A); 100 μm (B); 50 μm (C). The buffer was the same as in Fig 5.26. (b) Ohm's law plots for Peltier thermoelectric temperature control on a 180 mm × 50 μm capillary column. A = 40°C; B = 30°C; C = 20°C; D = 10°C. (c) Plot of electroosmotic velocity vs applied field at 20°C. The plot is linear (i.e. did not show large non-random deviations) with a correlation coefficient of greater than 0.995, indicating control of the double layer temperature. (Reproduced from Ref. 34 with permission of Elsevier Science Publishers.)

owing to the lower viscosity and the increase in their dissociation constant (lower pK values) at a higher temperature. However, only the meta isomer undergoes a second chemical equilibrium. Two molecules of the meta isomer will be able to form a dimer because of the electrostatic interaction and/or hydrogen bonding ability between each other's amino and carboxyl groups. The additional equilibrium results in a decrease in the effective mobility of m-aminobenzoic acid at 30°C. At 60°C, this chemical equilibrium is less pronounced, causing the μ_{ep} to increase more than expected from the changes in viscosity and acid–base equilibrium alone. Thus, the change in migration order and increase in selectivity are obtained [15].

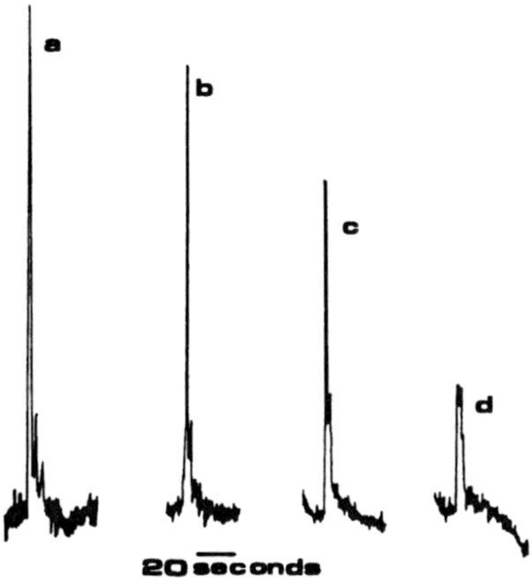

Fig. 5.31. The dependence of horse heart myoglobin peak shape on column temperature. Column conditions were total length 340 mm, effective length 320 mm, 75 μm I.D. capillary column. Operation was under constant current at 9.5 μA with applied fields of 500, 430, 390, and 320 V/cm at 20, 30, 40 and 50°C respectively. The buffer was 0.1 M Tris and 0.025 M boric acid, pH 8.6, with detection at 220 nm. (Reproduced from Ref. 34 with permission of Elsevier Science Publishers.)

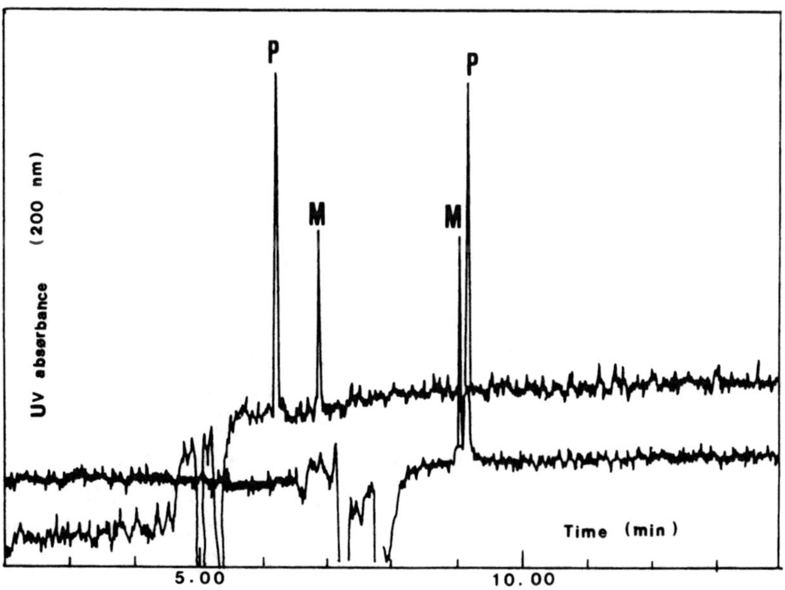

Fig. 5.32. CZE separation of *p*-aminobenzoic acid (*P*) and *m*-aminobenzoic acid (*M*) at 25 kV in 40 mM ammonium acetate buffer (pH 4.0). Solid line, 30°C; dashed line, 60°C. (Reproduced from Ref. 15 with permission of Elsevier Science Publishers.)

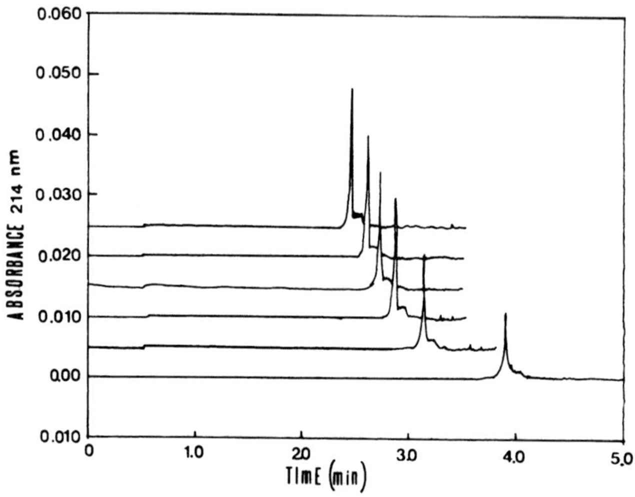

Fig. 5.33. Electropherograms of horse heart myoglobin (0.2 mg/ml) at 20, 30, 35, 40, 45 and 50°C (ascending direction) at a field of 350 V/cm. Sample was dissolved in water and electrophoresis conducted in 0.1 M tetrasodium borate buffer, pH 8.3 and monitored at 214 nm. The 75 μm I.D. capillary had a effective length of 50 cm and a total length of 57 cm. The electropherograms have been corrected for electroosmotic flow. (Reproduced from Ref. 36 with permission of the American Chemical Society.)

Other examples of the influence of temperature on CE are shown in Figs. 5.33 and 5.34. Figure 5.33 depicts a constant field (350 V/cm) electropherogram generated for horse heart myoglobin at six column temperatures [36]. With increasing temperature, the migration time decreases at constant field. Figure 5.34 illustrates constant current (9.8 μA) electropherogram at two wavelengths (214 nm and 410 nm) for the same protein. At 20°C a single peak is observed, whereas a second slower migrating peak appears at 30°C. The second peak becomes larger at the expense of the first at higher temperature. The same behaviour is observed for both wavelengths, which means that the change is unlikely to be due to conformational changes. At 50°C, essentially only the second peak is obtained. A possible explanation for the behaviour is the reduction of Fe^{3+} to Fe^{2+} metal ion coordinated to the heme group. The faster migrating peak corresponds to the more positive ferric form and the slower moving peak to the ferrous form. This order is expected on the basis of charge difference. Temperature dependent migration behaviour is also observed for α-lactalbumin [36], which exhibited temperature-induced conformational changes [36–38].

5.2 MICELLAR ELECTROKINETIC CHROMATOGRAPHY (MEKC)

Terabe *et al.* [39,40] described the fundamental principles of electrokinetic chromatography (MEKC) in 1984. In MEKC, a micellar solution of an ionic

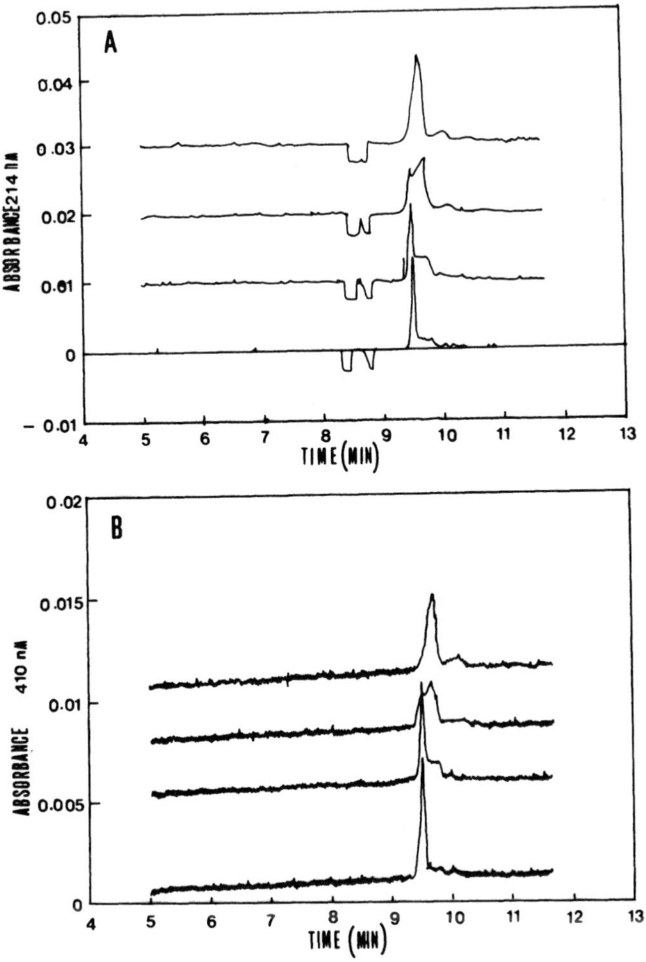

Fig. 5.34. (A) Influence of temperature on the electrophoretic behaviour of horse heart myoglobin (0.2 mg/ml) under constant current conditions (9.8 μA) in 0.1 M Tris and 0.025 mM boric acid, pH 8.6. Electrophoresis was conducted at 20, 30, 40 and 50°C (ascending direction), at fields of 262, 213, 178, and 153 V/cm, respectively, and monitored at 214 nm. The 75 μm I.D. capillary had a effective length of 50 cm and a total length of 57 cm. (B) All conditions as in A except monitored at 410 nm. (Reproduced from Ref. 36 with permission of the American Chemical Society.)

surfactant is used to provide a phase for a chromatographic separation. The micellar phase is sometimes referred to as a pseudostationary phase since it serves similar functions as a chromatographic stationary phase. Neutral species can be separated by micellar electrokinetic capillary chromatography (MEKC or MECC). When no micelles are present, neutral solutes will migrate with the electroosmotic flow, and no separation will result. In MEKC, ionic surfactants are added to the operating buffer at concentrations above the critical micelle concentration (cmc). Under these conditions, surfactant monomers tend to form roughly spherical aggregates, or micelles. Micelles are dynamic structures in equilibrium with the surfactant monomer, which is present at a concentration approximately equal to the cmc. The use of different types of surfactants, mixtures of surfactants, and surfactant solutions containing additives provides a wide variety of methods to enhance selectivity in MEKC separations. Due to their capability for high efficiency and versatility, MEKC and related techniques have been widely explored in recent years [39–88].

5.2.1 Principles of separation in MEKC

MEKC is most commonly performed with anionic surfactants, especially sodium dodecyl sulfate (SDS). In the case of anionic surfactants, the hydrophobic tail groups tend to orientate toward the centre and the charged head groups along the surface. A schematic of the MEKC process with anionic surfactant is illustrated in Fig. 5.35. The MEKC system contains two phases: an aqueous phase and a micellar phase. The surfaces of SDS micelles have a large net negative charge. The micelles therefore exhibit a large electrophoretic mobility (μ_{ep}) toward the anode, which is in opposite direction to the electroosmotic mobility (μ_{eo}) towards the cathode present in most commonly used buffer systems used in CE. The magnitude of μ_{eo} is slightly greater than that of μ_{ep}, resulting in a fast-moving aqueous phase and a slow-moving micellar phase. Solutes can partition between the two phases, resulting in retention based on differential solubilization by the micelles. Consequently, the MEKC technique provides a means of obtaining selective separations of neutral and ionic compounds while retaining the advantages of the capillary electrophoresis format.

Migration behaviour in MEKC is generally governed by hydrophobicity. More hydrophobic solutes interact more strongly with the micellar phase and thus migrate slower than hydrophilic compounds. The migration time (t_M) of a solute that interacts with the micelles will fall in a "migration time window" between the

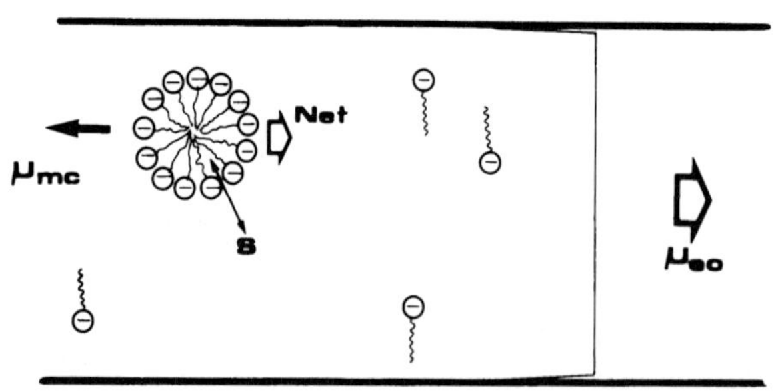

Fig. 5.35. Schematic representation of a system for micellar electrokinetic capillary chromatography showing amphiphilic monomers, micelle, and flow profile. S = solute; μ_{eo} = coefficient of electroosmotic flow; μ_{mc} = electrophoretic mobility of micelle.

migration time (t_0) of a solute that has little or no interaction with the micelles (i.e. H_2O) and the migration time of a solute that is 100% solubilized by the micelles (t_{mc}). The fact that neutral solutes must elute between t_0 and t_{mc} is the most significant difference between MEKC and conventional chromatography.

In conventional chromatography, the capacity factor can be described by the equation:

$$k' = \frac{t_r - t_0}{t_0} \tag{5.18}$$

where k' is the capacity factor describing the ratio of the total moles of solute in the stationary phase to those in the mobile phase. In MEKC, neutral molecules partition between two moving phases, thus requiring modification of the equation for the capacity factor. For electrically neutral solutes eluting between t_0 and t_{mc} with a migration time t_M, Terabe *et al.* [39] have derived the following equation for the capacity factor:

$$k' = \frac{t_M - t_0}{t_0(1 - t_M/t_{mc})} \tag{5.19}$$

As t_{mc} becomes infinite (micellar phase becomes stationary), Eq. (5.19) reduces to the analogous equations for conventional chromatography. The parameters t_0 and t_{mc} are experimentally determined by injecting methanol, which is assumed to not interact with the micelles, and Sudan III, which is assumed to be fully solubilized [39,40].

Resolution in MEKC is given by [40]:

$$R = \frac{N^{1/2}}{4}\left(\frac{s-1}{s}\right)\left(\frac{k_2'}{k_2'+1}\right)\left(\frac{1 - t_0/t_{mc}}{1 + (t_0/t_{mc})k_1'}\right) \tag{5.20}$$

where s is the separation factor given by k_2'/k_1'. The differences between MEKC and conventional chromatography are accounted for in the last term of the equation. As t_{mc} becomes infinite, the latter term equates to unity and results in an expression for resolution that is identical to that of conventional chromatography.

In MEKC, fast, efficient separations can be obtained because of three important phenomena [39,40]. First, the flat flow profile of electroosmotic flow does not require mass transfer in the mobile phase across the capillary diameter. Secondly, fused silica capillaries dissipate heat efficiently, thereby minimizing thermal effects. Thirdly, micelles are dynamic structures, allowing for fast solute entrance/exit kinetics.

5.2.2 Causes of band broadening in electrokinetic chromatography

There are several kinetic processes which serve to disperse solute bands as they are transported through the capillary column. The influences of these processes on column plate height are additive. Extracolumn effects caused by instrumental factors, such as injection, detection and detector time constants can also cause observed peak broadening. For CE separations, on-column injection and detection are usually performed. Hence, extracolumn dispersion is expected to be negligible, except for injection which can potentially cause severe peak broadening (see Chapter 2).

Sepaniak and Cole [41] used 7-chloro-4-nitrobenz-1,1,3-oxadiazole (NBD)-derivatized amines to examine the factors that determine column efficiency in MEKC. By assuming that extracolumn effects and eddy diffusion are negligible, the total plate height (H) for MEKC was expressed in terms of several additive components:

$$H_{col} = H_l + H_{mc} + H_{aq} + H_c + H_T \tag{5.21}$$

where H_{col} is overall column plate height, and H_l, H_{mc}, H_{aq}, H_c, H_T are plate heights generated by longitudinal diffusion, sorption–desorption kinetics in micellar solubilization, intermicelle mass transfer in the aqueous phase, intracolumn mass transfer and temperature gradient effect on electrophoretic velocity, respectively. Van Deemter type plots of plate height versus applied voltage were constructed to describe the effects of experimental parameters on efficiency. A typical plot showing plate height versus applied potential for 4-chloro-7-nitrobenzofurazen-derivatized cyclohexylamine and ethylamine is shown in Fig. 5.36. By inferring from theoretical descriptions of solute band broadening in chromatography and electrophoresis, they made the predictions given in Table 5.5 concerning the qualitative effects

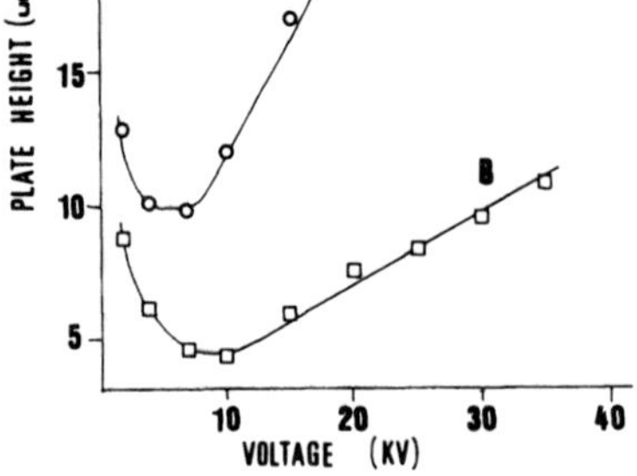

Fig. 5.36. Plate height vs. applied potential for 4-chloro-7-nitrobenzofurazan-derivatized cyclohexylamine (A) and ethylamine (B). Conditions: buffer, 0.01 M SDS, 0.003 M Na_2HPO_4; capillary, 75 cm long, 75 μm I.D., fused silica. (Reproduced from Ref. 41 with permission of the American Chemical Society.)

TABLE 5.5

COLUMN FACTORS CONTRIBUTING TO OBSERVED PLATE HEIGHT IN MEKC (Adapted from Ref. 41)

Factor	Influence of experimental parameters* (effect of increasing parameter)						
	V	D	R	S	d_c	$[m]$	P/L
H_l	↓	↑	↑				
H_{mc}	↑		↑	↑			
H_{aq}	↑	↓	↓			↓	
H_{ep}	↑	↓	↓		↑		
H_T					↑		↑

* Experimental parameters: V = applied voltage; D = diffusion coefficient; R = retention ratio defined as: the fraction of solution which is not solubilized by micelles; S = micelle size; d_c = column diameter; $[m]$ = micelle concentration; P/L = power/length.

(increase ↑ or decrease ↓) that experimental parameters will have on the individual contributions to H_{col}.

Terabe *et al.* [42] examined in detail the causes of band broadening in electrokinetic chromatography with micellar solutions and open-tubular capillaries. By breaking down the causes of band broadening into five mechanisms in a similar way as in Eq. (5.21), expressions for each individual contribution to the plate height have been derived, which are listed in Table 5.6. With the expressions given in Table 5.6, numerical estimates can be made to demonstrate the relative importance of each contribution.

Equation 5.22 indicates that H_l, which represents the plate height due to longitudinal diffusion, is directly proportional to the reciprocal of electroosmotic velocity. The slope of the plot of H_l against $v_{eo}{}^{-1}$ depends on the capacity factor as well as the diffusion coefficient of the solute, provided D_{mc} and t_0/t_{mc} are constant. Approximately linear relationships are obtained between H_l and $v_{eo}{}^{-1}$ as shown in Fig. 5.37, except for one data point on 2-naphthol and Sudan III which does not give linear plots. The results shown in Fig. 5.37 led Terabe *et al.* to conclude that: (1) longitudinal diffusion is operative as a main cause of band broadening throughout the experimental range of electroosmotic velocities, (2) except for Sudan III, a constant plate height needs to be added to each H_l in order to explain the observed plate heights as the linear plots do not pass through the origin, and (3) other electroosmotic velocity-dependent mechanisms should be considered at least for 2-naphthol and Sudan III.

Kinetic or non-equilibrium processes in MEKC are attributed mainly to sorption and desorption of the solute between the aqueous phase and the micellar phase. The prediction according to Eq. (5.23) is that the plate height caused by the kinetic process increases with an increase in velocity v_{eo} and with a decrease in the rate constant k_d. The capacity factor term in the right-hand side of Eq. (5.23) has the

TABLE 5.6

EXPRESSIONS FOR PLATE HEIGHT CONTRIBUTIONS IN MICELLAR ELECTROKINETIC CHROMATOGRAPHY

1. Longitudinal diffusion

$$H_l = \frac{2(D_{aq} + k'D_{mc})}{1 + (t_0/t_{mc})k'} \frac{1}{v_{eo}} \tag{5.22}$$

2. Sorption–desorption kinetics

$$H_{mc} = \frac{2(1 - t_0/t_{mc})^2 k'}{[1 + (t_0/t_{mc})k'] (1 + k')} \frac{v_{eo}}{k_d} \tag{5.23}$$

3. Intermicelle diffusion

$$H_{aq} = \left(\frac{k'}{1 + k'}\right)^2 \frac{(1 - t_0/t_{mc})^2}{1 + (t_0/t_{mc})k'} \frac{d^2 v_{eo}}{4 D_{aq}} \tag{5.24}$$

4. Electrophoretic dispersion

$$H_{ep} = \frac{(1 - t_0/t_{mc})^2 k'}{1 + (t_0/t_{mc})k'} \left(\frac{\sigma(v_{eo})}{v_{ep}}\right)^2 \frac{v_{eo}}{k_d} \tag{5.25}$$

5. Temperature gradient effects

$$H_T = \frac{(1 - t_0/t_{mc})k'}{24 (D_{aq} + k'D_{mc}} \frac{B^2 I^4}{64 \kappa_0^2 \pi^4 r_c^2 \lambda^2 T_0^4} v_{eo} \tag{5.26}$$

k' = capacity factor; D_{aq} = diffusion coefficient of solute in aqueous phase; D_{mc} = diffusion coefficient of solute in micellar phase; t_0 = migration time of solute which does not interact with the micelle; t_{mc} = migration time of solute which is fully solubilized by the micelle; v_{eo} = electroosmotic velocity; k_d = desorption rate constant by the micelle; d = intermicelle distance, $\sigma(v_{ep})$ = standard deviation of the electrophoretic micellar velocity distribution; v_{ep} = electrophoretic velecity; B = constant related to temperature dependence of viscosity; I = electric current; κ_0 = electrical conductance of solution; r_c = radius of column; and T_0 = temperature of solution.

maximum value of 0.26 when k' = 0.75 for t_0/t_{mc} = 0.22, and becomes zero when k' = 0 or k' = ∞. The term H_{mc} must be zero for Sudan III, because the capacity factor of Sudan III is assumed to be infinity. Terabe *et al.* deduced from Eq. (5.23) that the kinetic process should not contribute significantly to band broadening if the desorption rate is not extremely slow. However, very slow desorption kinetics may cause significant band broadening in some solute–micelle systems, in which the solute incorporates with the micelle by electrostatic force or ionic interaction [43,44]. Even if k' is infinite, k_d is equal to the reciprocal of the average lifetime of the micelle [42].

Plate height contributions from intermicelle diffusion are generally small. Although intermicelle distance fluctuates as a result of the Brownian motion of the micelle, the relative positions of micelles may be considered fixed during a short

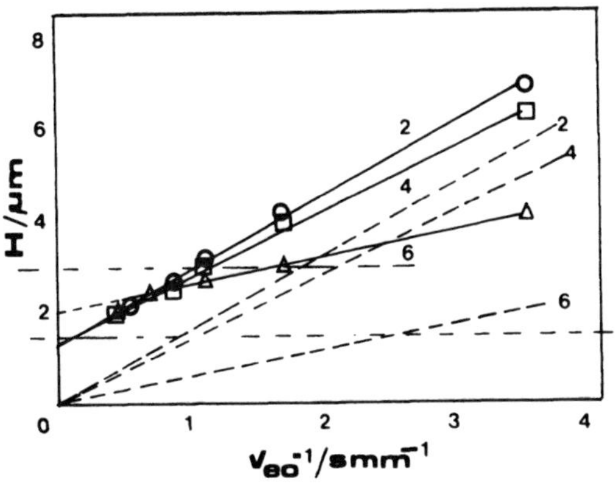

Fig. 5.37. Dependence of plate height on the reciprocal of electroosmotic velocity. Dashed lines are calculated. *1* = resorcinol; *2* = phenol; *3* = p-nitroaniline; *4* = nitrobenzene; *5* = toluene; *6* = naphthol; *7* = Sudan III. (Reproduced from Ref. 42 with permission of the American Chemical Society.)

period of time when solutes diffuse between micelles. This is because the diffusion coefficient of the solute is substantially larger than that of the micelle. Typically the average intermicellar distance is estimated to be ca. 10 nm for a 0.05 M SDS solution [89]. According to Eq. (5.24), this small value of d causes H_{aq} to be negligibly small (1.0×10^{-5} μm), since D_{aq} is about 10^{-3} mm^2/s and v_{eo} is 2 mm/s.

Electrophoretic dispersion arises mainly from microheterogeneity, which is related to randomness in size, shape and/or charge of the migrating species. The micelle fills the whole tube in MEKC. The concentration of the micelle is always kept constant throughout the tube, even when voltage is applied. The solute itself is not directly subject of the electrophoretic force but is indirectly transported by the electrophoretic movement of the micelle. Therefore, electrophoretic dispersion of the micelle containing the solute is a potential cause of band broadening in MEKC. On the right-hand side of Eq. (5.25), the value of the first term containing k' ranges from zero to $(1 - t_0/t_{mc})^2/(t_0/t_{mc})$, while the second term remains approximately constant. Thus, H_{ep} increases linearly with increases of v_{eo} and k_d^{-1}. Since polydispersity of the micelle size seems dominant as the origin of microheterogeneity, the relationship between the distribution of electrophoretic velocity and that of the aggregation number or the micelle size would be needed to evaluate the effect of the polydispersity [42].

The plate height contribution of the temperature gradient, H_T, can be considered in a similar manner as described in Chapter 1 (Section 1.5). Essentially a temperature increase is expected to cause a parabolic flow profile similar to that observed inlaminar flow, except that the velocity is not zero at the wall. The parabolic flow contributes to the plate height. Based on Eq. (5.26), plate height contribution of the temperature gradient, H_T, is proportional to I^4 and inversely proportional to r_c^2. Since current increases in proportion to the square of the radius, the second term of the right-hand side in Eq. (5.36) is independent of current and the tube radius provided the voltage is kept constant. Thus H_T is proportional to v_{eo}, hence proportional to the electrical field strength regardless of the tube radius. Based on Eq. (5.26), Terabe *et al.* [40] deduced that the effect of a temperature gradient on plate height would be negligible, although the solution temperature may rise considerably.

5.2.3 Types of surfactant systems used

Manipulation of the micellar phase in MEKC is analogous to changing the stationary phase in HPLC. MEKC has the advantage that changing the micellar phase is very easy, requiring only that the capillary be rinsed and filled with the micellar solution. Anionic or cationic micelle systems are the most commonly employed micellar phases. Some examples of the typical surfactant systems employed in MEKC are listed in Table 5.7. To date, only very few of the potentially

TABLE 5.7

TYPICAL SURFACTANT SYSTEMS USED FOR MEKC

Surfactant		Ref.
Anionic		
Sodium decyl sulfate	$CH_3(CH_2)OSO_3^- Na^+$	75
Sodium dodecyl sulfate (SDS)	$CH_3(CH_2)_{11}OSO_3^- Na^+$	39–64,86–88
Sodium tetradecyl sulfate (STS)	$CH_3(CH_2)_{13}OSO_3^- Na^+$	40
Sodium dodecyl sulfonate	$CH_3(CH_2)_{11}SO_3^- Na^+$	40
Cationic		
Dodecyltrimethylammonium chloride (DTAC)	$CH_3(CH_2)_{11}N^+(CH_3)_3Cl^-$	75
Dodecyltrimethylammonium bromide (DTAB)	$CH_3(CH_2)_{11}N^+(CH_3)_3Br^-$	46
Cetyltrimethylammonium chloride (CTAC)	$CH_3(CH_2)_{15}N^+(CH_3)_3Cl^-$	75
Cetyltrimethylammonium bromide (CTAB)	$CH_3(CH_2)_{15}N^+(CH_3)_3Br^-$	65
Cationic fluorosurfactant (Fluorad FC 134)	$CF_3(CF_2)_7SO_2NH(CH_2)_3$ $-N(CH_3)_3I$	90
Non-ionic and zwitterionic		
Octyl glucoside		54
3-[3-(chloroamidopropyl) dimethylammonio]-1-propanesulfonate (CHAPS)		54
Chiral surfactants		
Sodium-N-dodecanoyl-L-valinate (SDVal)		91–94
Digitonin		94
Bile salt surfactants (e.g. sodium cholate, sodium taurocholate)		69–73

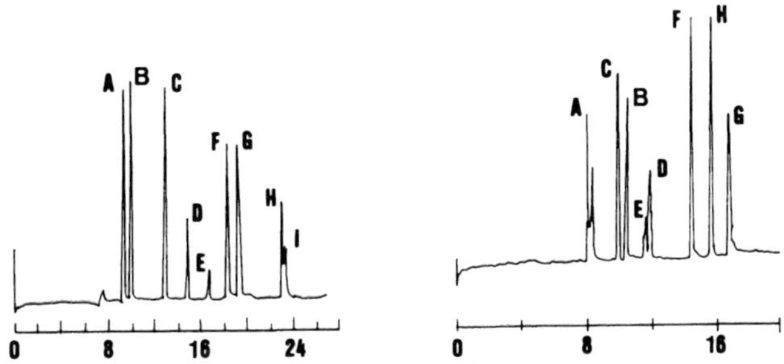

[a] [b]

Fig. 5.38. Electrokinetic separation of test solutes in (a) 0.50 M SDS and (b) 0.50 M DTAC solutions. Solutes: *A* =aniline, *B* = phenol, *C* = nitrobenzene, *D* = benzoic acid, *E* = toluene, *F* = 1-aminonaphthalene, *G* = 1-naphthol, *H* = 1-nitronaphthalene, *I* = 2-nitronaphthalene. Conditions: 100 cm long, 60 μm I.D., fused silica capillary; buffer, 0.01 M Na_2HPO_4, 0.06 M $Na_2B_4O_7$; applied potential, 30 kV. (Reproduced from Ref. 75 with permission of Preston Publications.)

applicable surfactant systems have been explored for use as the micellar phase. In practice, the applicability of a surfactant to MEKC would depend on its solubility and its cmc. Surfactants which have large cmc values are unsuitable because of the high conductivity of such micellar solutions, which may lead to undesirable thermal effects. In addition to the micellar systems listed in Table 5.7, the use of mixed micelles and reversed micelles, and the addition of modifiers would further enlarge the scope and variety of the micellar phase available for use in MEKC.

With different systems, the interactions that can occur between the micelles and the solutes may be different. Consequently, selectivity would change owing to the difference in the solubilization behaviour. Changes in micellar composition can also produce micelles of different sizes, aggregation numbers, and geometries [95]. These factors can all affect selectivity. To illustrate the different selectivity available through the use of different micelles, Burton *et al.* [75] compared the separations of nine test solutes with both SDS (anionic) and DTAC (cationic) micellar phases. The chromatograms obtained are shown in Fig. 5.38. The use of cationic surfactants such as DTAC changes the sign of the zeta potential through adsorption of the positively charged monomers, resulting in a reversal of electroosmotic flow [54,65]. Changes in the micellar phase can also lead to lengthened migration time windows, thus increasing the peak capacity of the system [56].

5.2.4 Electroosmotic flow in electrokinetic chromatography

Factors which affect electroosmotic flow in electrokinetic chromatography include pH, SDS concentration, additives, and polymer coating of the capillary wall. The velocity of the electroosmotic flow is usually measured by elution of an unsolubilized solute or methanol, and the velocity of the micelle with Sudan III.

5.2.4.1 *Dependence of electrokinetic migration on pH*

The effects of pH on the velocity of electroosmotic flow was investigated by Otsuka and Terabe [58]. Figures 5.39 and 5.40 illustrate the dependence of the electroosmotic velocity, v_{eo}, and the electrophoretic velocity of the micelle, v_{ep}, on pH. In Fig. 5.39, the effect of pH in the range 5.5 to 9.0 is shown. Except for the slight increase of v_{eo} observed for the 0.2 M sodium dodecyl sulfate (SDS) solution at pH values from below 7, the electrokinetic velocities remain relatively constant. In Fig 5.40, the dependence of electrokinetic velocities on pH in the pH range of 3 to 7 is shown. The electroosmotic velocity decreased with a decrease in pH below 5.5, while the electrophoretic velocity of the SDS micelle is almost constant throughout the range 3.0 to 7.0. Consequently, the migration velocity of the micelle decrease with a decrease in pH, and its direction changes at pH 5.0, i.e., the direction is the same as that of the electroosmotic velocity above pH 5.0 and the reverse below 5.0.

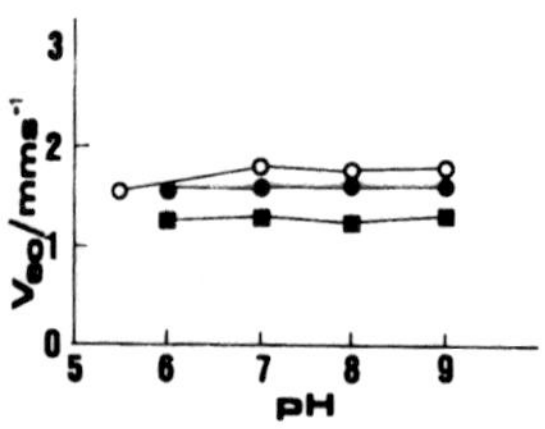

Fig. 5.39. Effect of pH on the electrokinetic velocity. Applied voltage: 15 kV. Capillary tube: 50 μm I.D., 190 μm O.D., 650 mm long. Velocity and solution: v_{eo} in 0.1 M SDS in 0.05 M phosphate-0.1 M borate buffer (•), v_{eo} in 0.2 M SDS in the same buffer as above (○), v_{ep} in 0.1 M SDS in the same buffer as above (■). (Reproduced from Ref. 57 with permission of Dr. Alfred Heuthig Publishers.)

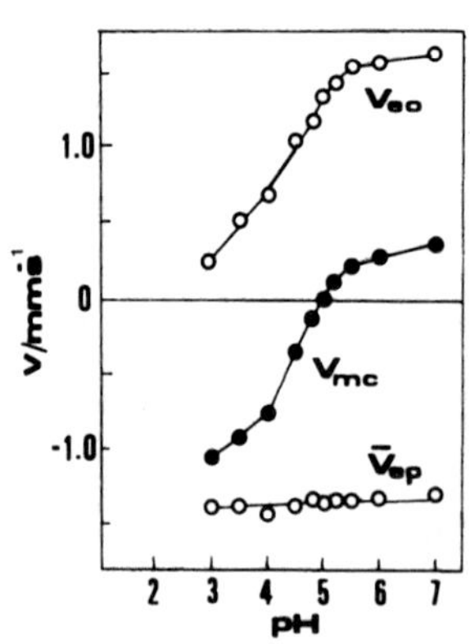

Fig. 5.40. Dependence of electrokinetic velocities on pH: V_{eo} = electroosmotic velocity; V_{mc} = migration velocity of the micelle; V_{ep} = electrophoretic velocity in the micelle. Conditions are the same as in Fig 5.39. (Reproduced from Ref. 58 with permission of Aster Publishing Corp.)

5.2.4.2 *Dependence of electrokinetic migration on surfactant concentration*

The concentration of micelles employed is another factor which affects migration in MEKC. Above the cmc, addition of surfactant serves only to increase the micelle concentration, with the monomer concentration remaining constant [95]. For electrically neutral solutes, the migration time generally increases with increased micelle content as shown in Fig. 5.41. The analogous situation in HPLC is the increase of stationary-phase volume. With anionic surfactants and fused silica capillaries, it has been shown that surfactant content does not appreciably change the electroosmotic flow velocity [40, 65]. This is a result of electrostatic repulsion preventing adsorption and consequent changes in zeta potential, ζ.

Terabe *et al.* [57] plotted the reciprocal of the electrokinetic velocities versus SDS concentration for C_{SDS} up to 0.3 M. It is observed that $1/v_{eo}$ decreases slightly with an increase of C_{SDS} above 0.2 M, and $1/v_{mc}$ increases non-linearly with an increase of C_{SDS} as shown in Fig. 5.42. The reciprocal of the electrophoretic velocity of the micelle, $1/v_{ep}$, decreases linearly as C_{SDS} increases.

5.2.4.3 *Effect of additives on electrokinetic migration*

Various types of additives have been employed in micellar electrokinetic chromatography to enhance efficiency and selectivity, The main aims are to control electroosmotic flow, to prevent adsorption or to introduce an additional separation mechanism.

The effect of adding of 0.1% of various additives, such as hydroxypropylcellulose (HPC), hydroxypropylmethylcellulose (HPMC), BRIJ-35 and TWEEN-20, to 0.05

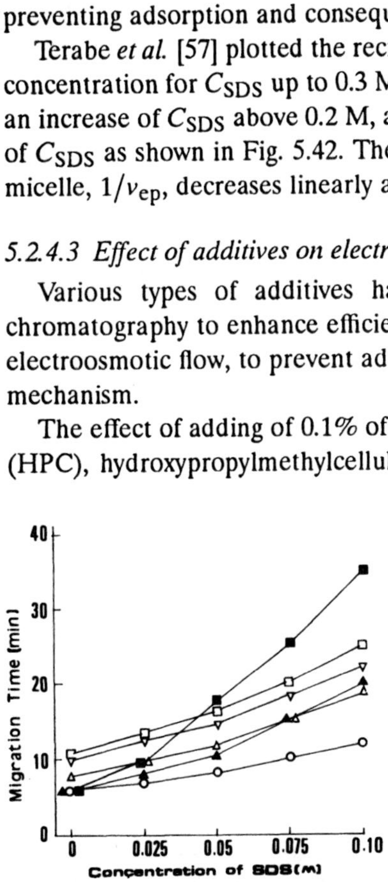

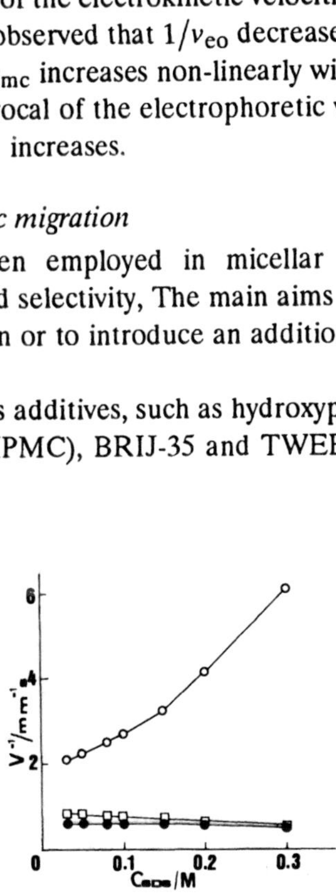

Fig. 5.41 (left). Effect of SDS concentration on migration time. Conditions: 80 cm long, 100 μm I.D. fused silica capillary; buffer, 0.02 M phosphate, pH 11; applied current, 100 μA. Solutes: ○ = methanol; △ = p-acetamidophenol; ▽ = acetylsalicylic acid; ▸ = anhydrous caffeine; ■ = o-ethoxybenzamide; □ = salicylamide. (Reproduced from Ref. 50 with permission of the American Chemical Society.)

Fig. 5.42 (right). Effect of SDS concentration on the reciprocal of electrokinetic velocity under a constant applied voltage (15 kV) and pH 7: $1/v_{eo}$ (●), $1/v_{mc}$ (○), and $1/v_{ep}$ (□). (Reproduced from Ref. 57 with permission of Dr. Alfred Huethig Publishers.)

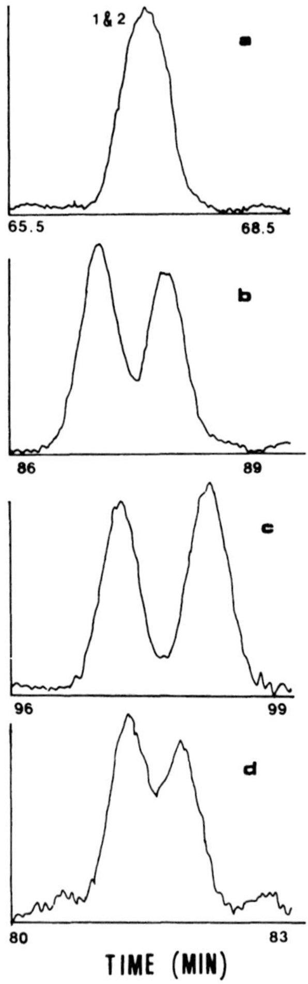

Fig. 5.43. (a) 0% MeOH, 33 μA current. (b) 10% MeOH, 25 μA current. (c) 20% MeOH, 23 μA current. (d) 30% MeOH, 20 μA current. Peak 1 in each case is DNS-$NHCD_3$ and peak 2 in each case is DNS-$NHCH_3$. 10%, 20% and 30% data have all undergone a 9-point Savitzky-Golay smooth. (Reproduced from Ref. 59 with permission of Aster Publishing Corp.)

M SDS solution has been examined [57]. No substantial effect is observed on the electrokinetic velocity, probably because the effect of the additives is diminished due to the presence of the SDS molecule. In other words, two counteracting effects, i.e. a possible decrease in μ_{eo} due to increase in viscosity near the tube wall, and the adsorption of the SDS molecule on the inner surface which diminishes this effect, are observed in the system [57].

Bushey and Jorgenson have demonstrated the excellent separation power of MEKC in the resolution of dansylated methylamine (DNS-$NHCH_3$) from dansylated methyl-d_3-amine (DNS-$NHCH_3$) [59]. The addition of an organic

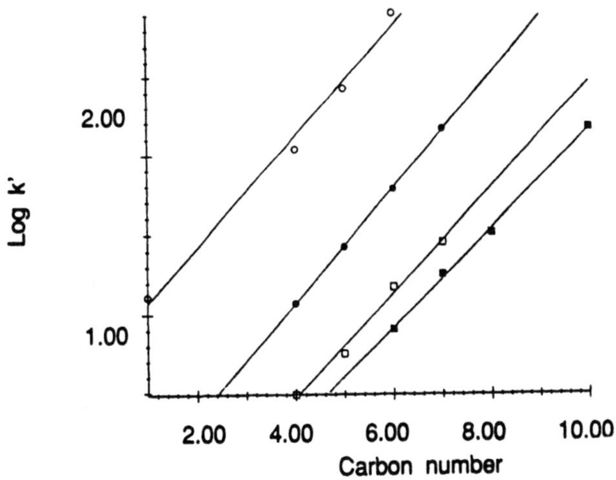

Fig. 5.44. Graph of log of calculated k' versus amine carbon number for each of the four buffer systems: ○ = 0% MeOH, ● = 10% MeOH, □ = 20% MeOH, and ■ = 30% MeOH. (Reproduced from Ref. 59 with permission of Aster Publishing Corp.)

modifier (methanol) decreases the electroosmotic velocity and partitioning with SDS micelles resolves these isomers in 100 min. The electropherogram obtained is shown in Fig. 5.43. Because of the high organic content of the mobile phase, the determination of t_{mc} based on the migration time of a single species (typically Sudan III) may be unreliable. A method for determining t_{mc} based on a homologous series of standards has been proposed. As shown in Fig. 5.44, plots of $\log k'$ versus carbon number yielded straight lines with similar slopes for four buffer systems with different concentrations of methanol. These results indicate that the procedure can be a good method for the determination of t_{mc} [59].

Nishi *et al.* [60] added tetraalkylammonium (TAA) salts to the buffer in MEKC. This improved the resolution of a wide variety of ionic substances. Addition of TAA salts decreases the migration time of cations and increases that of anions, indicating that they may form ion pairs with anions, leading to greater incorporation of the complex into the micellar phase. A schematic diagram illustrating the interaction between the SDS micelle and anionic solute (A) or cationic and zwitterionic solutes (B) in the presence of TAA salts is shown in Fig. 5.45. In Fig. 5.46, the effect of TAA salts on micellar EKC of four carboxylic acids is illustrated. Enhanced selectivity is observed compared with that obtained by pure CZE.

The separation of maltooligosaccharides derivatized with 2-aminopyridine has been studied by Nashabeh and El Rassi [19] using CZE in the pH range 3.0–4.5 with 0.1 M phosphate solutions as the running electrolyte. The addition of small amounts (50 mM) of tetrabutylammonium (TBA) bromide in the electrolyte solution enhances the separation efficiency.

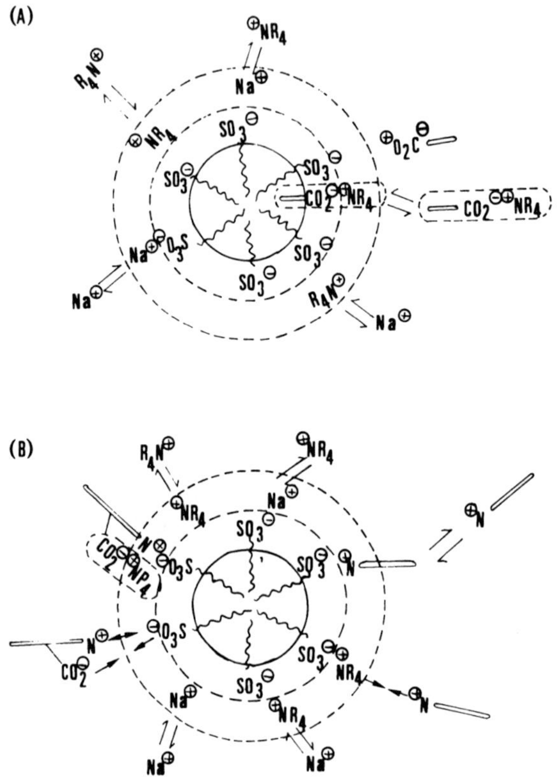

Fig. 5.45. Schematic illustration of interaction between the SDS micelle and anionic solute (A) or cationic and zwitterionic solutes (B) in the presence of the tetraalkylammonium salt. Solid circle shows hydrophobic core and broken circle the Stern layer of the micelle. (Reproduced from Ref. 60 with permission of the American Chemical Society.)

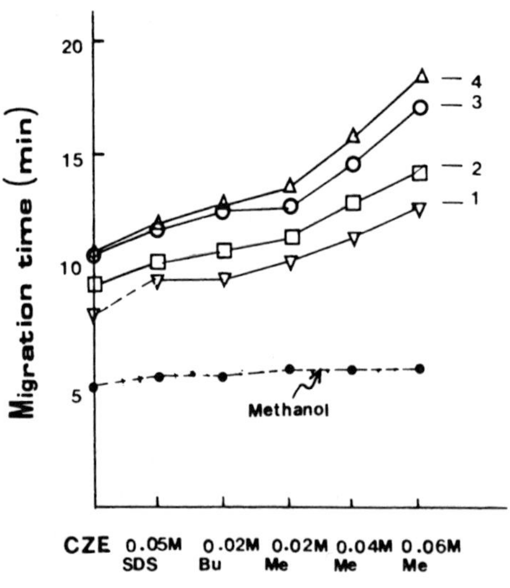

Fig. 5.46. Effect of TAA salts (Bu, TBAB; Me, TMAB) on micellar EKC of four carboxylic acids: *1* = naproxen; *2* = cinnamic acid; *3* = m-hydroxybenzoic acid; *4* = salicylic acid. (Reproduced from Ref. 60 with permission of the American Chemical Society.)

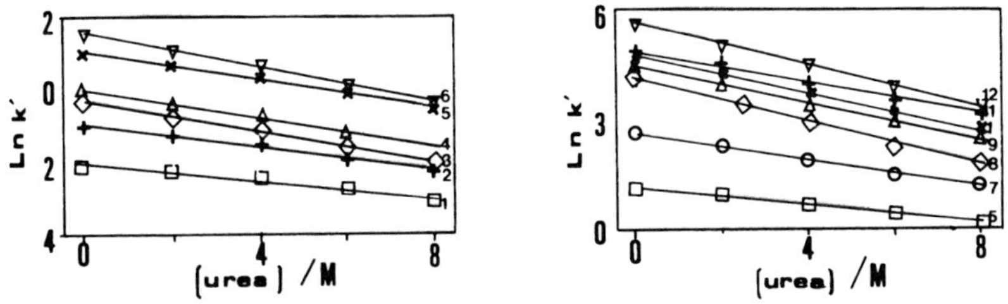

Fig. 5.47. (a) Dependence of $\ln k'$ on the concentration of urea. *1* = Resorcinol; *2* = phenol; *3* = p-nitroaniline; *4* = nitrobenzene; *5* = toluene; *6* = 2-naphthol. Capillary, 70 cm × 52 μm I.D. (50 cm to the detector); separation solution, 50 mM SDS in 100 mM borate–50 mM phosphate buffer; applied voltage, 20 kV.
(b) Dependence of $\ln k'$ on the concentration of urea. *6* = toluene; *7* = naphthalene; *8* = 9-fluorenone; *9* = fluorene; *10* = xanthene; *11* = dibenzyl; *12* = phenanthrene; *13* = stilbene; *14* = fluoranthene. Capillary, 65 cm × 50 μm I.D. (50 cm to the detector); separation solution, 50 mM SDS in 20 mM borate–20 mM phosphate buffer (pH 9.0); applied voltage, 20 kV.
(Reproduced from Ref. 61 with permission of Elsevier Science Publishers.)

Urea has been used as an additive in micellar electrokinetic chromatography by Terabe *et al.* [61]. Micelle formation is affected by urea which exerts an influence on the water structure [96,97]. The addition of urea to micellar solutions results in significant decreases in the capacity factors for most solutes in MEKC. The technique is useful for the separation of hydrophobic compounds from the standpoint of extending the applicability of MEKC.

Figure 5.47 shows the dependence of $\ln k'$ of test samples on the concentration of urea in MEKC with 50 mM SDS. All the $\ln k'$ values decreased linearly with increase in the concentration of urea. These linear relationships suggest that the free energy of transfer from the aqueous phase to the micelle decreases linearly with increase in urea concentration, provided urea does not cause a substantial change on the critical micelle concentration.

The migration time of SDS micelle, t_{mc}, is found to increase with increasing concentration of urea, whereas electroosmotic velocity does not alter significantly. The ratio of t_0/t_{mc} is reduced considerably and hence the migration time window between t_0 and t_{mc} is expanded. A wide migration time window is favourable for high resolution, although long analysis time may be required [39,40].

Figure 5.48 shows the separation of 23 phenylthiohydantoin (PTH)-amino acids using SDS solution with and without urea. In Fig. 5.48a, where urea is not used, five pairs of PTH-amino acids are poorly or not resolved. The use of 100 mM SDS solution containing 4.3 M urea allows a complete separation of 23 PTH-amino acids in one run. Figure 5.49 shows the MEKC separation of eight corticosteroids. These solutes are poorly soluble in water and tend to be mostly incorporated into the micelle. The use of 6 M urea dramatically improves the resolution, as observed by comparing Fig. 5.49a and 5.49b.

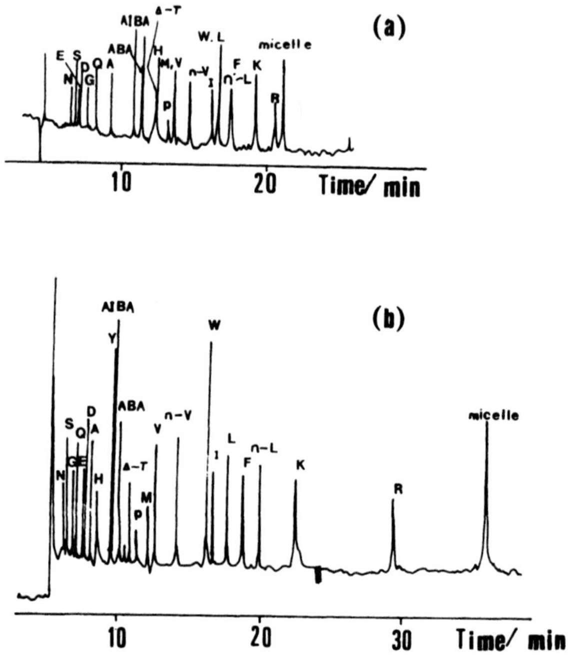

Fig. 5.48. Electrokinetic chromatogram of a mixture of 23 PTH-amino acids. The peaks are identified with one letter abbreviations for the amino acids. *AIBA* = 2-aminoisobutyric acid; *ABA* = 2-aminobutyric acid; Δ-*T* = PTH-dehydrothreonine. The micelle is traced with timepidium bromide. (a) 50 mM SDS in 100 mM borate–50 mM phosphate. Capillary length 50 cm (30 cm to the detector); applied voltage, 10.5 kV. (b) Separation solution, 100 mM SDS and 4.3 M urea in the same buffer as in (a); other conditions as in (a). (Reproduced from Ref. 61 with permission of Elsevier Science Publishers.)

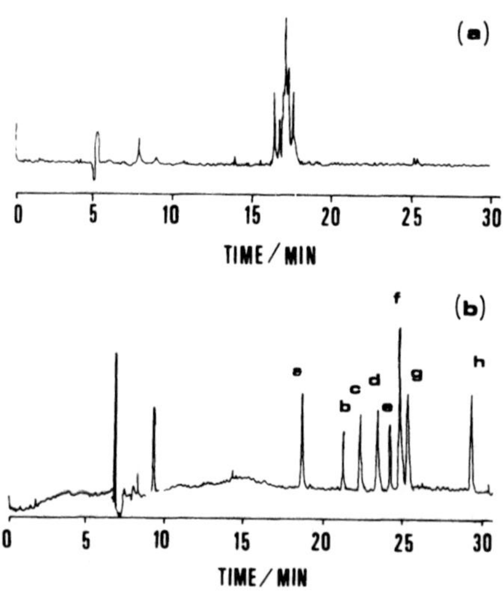

Fig. 5.49. Separation of eight corticosteroids: *a* = hydrocortisone, *b* = hydrocortisone acetate, *c* = betamethasone, *d* = cortisone acetate, *e* = triamcinolone acetonide, *f* = fluorinolone acetonide, *g* = dexamethasone acetate, *h* = fluocinonide. (a) Conditions as in Fig. 5.48a. (b) Conditions as in (a) except that electrophoresis buffer contains 6 M urea. (Reproduced from Ref. 61 with permission of Elsevier Science Publishers.)

5.2.4.4 *Effect of cationic surfactants on electrokinetic migration*

As mentioned in Section 1.3.2, the direction and rate of electroosmotic flow depend on the polarity and magnitude, respectively, of the zeta potential. The direction of electroosmotic flow is normally towards the cathode in aqueous solutions. By introducing cation surfactants into the buffer, which tend to adsorb to the capillary wall and reverse the zeta potential [65,66], the direction of the electroosmotic flow can be reversed (see Fig. 5.50). The effect of cetyltrimethylammonium bromide (CTAB) on coefficient of electroosmotic flow is shown in Table 5.8. The concentration of CTAB for switching of the surface charge of fluoroethylenepropylene (FEP) tubing is found to lie between 3.5×10^{-5} and 3.5×10^{-4} M [65].

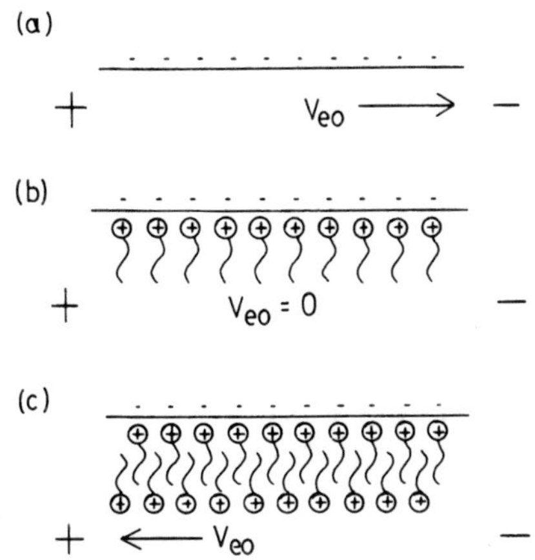

Fig. 5.50. (a) No surfactant added. Electroosmotic flow in normal direction. (b) Electrostatic adsorption of the positively charged surfactant head groups to the negative silanol groups on the silica surface of the capillary inner wall. (c) Admicellar bilayer formation by hydrophobic interaction between apolar chains, resulting in a reversal of the electroosmotic flow.

TABLE 5.8

EFFECT OF CTAB ON COEFFICIENT OF ELECTROOSMOTIC FLOW* IN FLUOROETHYLENEPROPYLENE (FEP) TUBING (Adapted from Ref. 65)

Concentration of CTAB (M)	Coefficient of electroosmotic flow** (10^{-4} cm^2 V^{-1} s^{-1})
0	3.39
3.5×10^{-6}	3.38
3.5×10^{-5}	2.96
3.5×10^{-4}	−1.19

* Experimental conditions: 0.02 M 2-(N-morpholine)ethane-sulphonic acid-L-histidine aqueous solution (pH 6.2) containing ethylene glycol (0.5%) and CTAB. Applied voltage: 10 kV; detection: potential gradient detector; column: FEP tubing (0.2 or 0.3 mm I.D.).

** The minus sign means that electroosmotic flow is towards the positive terminal.

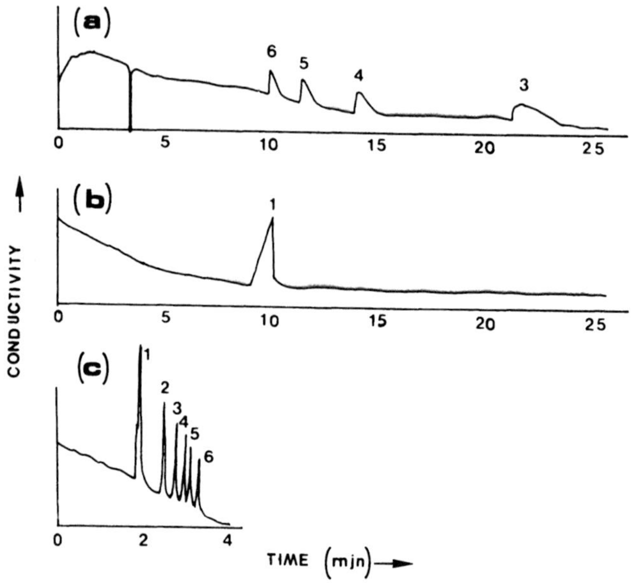

Fig. 5.51. Electropherograms of a six-component mixture of carboxylic acids: *1* = formate; *2* = acetate; *3* = propanoate; *4* = butanoate; *5* = pentanoate; *6* = hexanoate. Conditions: 29 kV was applied to a 75 μm I.D. capillary, 42 cm long; 40 cm to the detector. The electrolyte was 10 mM MES/His at pH 5.9. (a) "Standard" CZE separation. (b) CZE separation with reversed electric field. (c) CZE separation as in (b) with 0.5 mM TTAB added to the electrolyte. The concentration of all sample components was 5×10^{-4} M. (Reproduced from Ref. 52 with permission of the American Chemical Society.)

Huang *et al.* [52] reported the separation of low-molecular-mass carboxylic acids by CZE. The addition of 0.2–0.5 mM tetradecyltrimethylammonium bromide (TTAB) permitted the control of electroosmotic flow and improve the resolution of the carboxylic acid mixture. Figure 5.51 shows the electropherogram of a six-component mixture of carboxylic acids by (a) standard CZE, (b) CZE with reversed electric field and (c) CZE with reversed electric field and 0.5 mM TTAB added to the electrolyte. In case (c), all the six carboxylic acids are fully resolved and sharp peaks are obtained.

Improved CZE separation of basic proteins can be obtained using a fluorosurfactant buffer additive [90]. The cationic fluorosurfactant forms a strong admicellar bilayer (see Fig. 5.50) which results in surface charge reversal. Proteins at a pH below their pI values are repelled from the capillary wall, and their adsorption can be minimized. The hydrophobic nature of the fluorocarbon chain of the surface helps to enhance the stability of the bilayer. High efficiency and symmetrical peaks can be obtained even when low ionic strength buffers are used for the separation of proteins [90].

5.2.4.5 MEKC with mixed micelle systems

The use of mixed micelle system in MEKC to resolve dansylated D- and L-amino acids was reported by Cohen *et al.* [45]. In this method, a chiral chelate detergent (didecyl-L-alanine) is solubilized in a solution of sodium dodecyl sulfate (SDS) in order to form mixed micelles. The chiral reagent incorporates into the SDS micelles with the hydrophilic chiral L-alanine regions oriented at the surface, and interaction with the amino acids can occur. The micelles are transported through the capillary at a different rate than the aqueous phase. The different rate of partitioning into the micelles provides the mechanism to resolve the amino acids.

5.2.4.6 MEKC with non-ionic and zwitterionic surfactants

Swedberg [54] observed enhancement of separation of closely related species by the use of non-ionic and zwitterionic surfactants in CE. These surfactants are only effective in achieving separations at or above the critical micelle concentration (cmc). Capillaries used are aryl pentafluoro (APF) deactivated fused silica tubings. The surfactants used include octyl glucoside and CHAPS (3-[3-(chloroamidopropyl) dimethylammonio]-1-propanesulfonate). With the use of non- or zwitterionic surfactants, there is less impact on the magnitude of the electroosmotic flow than charged surfactants. Unlike CTAB, they do not reverse the electroosmotic flow. Moreover, there is expected to be less impact on protein structure. Figure 5.52 shows electropherograms comparing separations of six peptides with and without addition of octyl glucoside. The addition of 80 mM octyl glucoside to the electrophoresis buffer notably affects selectivity.

5.2.5 Ion-exchange electrokinetic chromatography

Terabe and Isemura [67,68] investigated the effect of polymer ion concentrations on migration velocities in ion-exchange electrokinetic chromatography. The separation principle is based on the differential ion-pair formation of the analyte ion with a polymer ion. Polybrene and poly(diallydimethylammonium chloride) were employed as polymer ions. The technique has been applied to the separation of naphthalenesulfonates and naphthalenedisulfonates, and some isomeric acids.

Figure 5.53 shows schematically the separation principle of ion-exchange EKC, in which a polymer cation and analyte anions, and their electrophoretic migrations are illustrated. The ion-pair formed between the analyte anion and the polymer cation will migrate with the same velocity as other polymer cations, since only a partial neutralization of the positive charge occurs on the polymer, and the electrophoretic velocity of the polymer cation is not expected to be affected significantly by a small increase in molecular size. As shown in Fig. 5.53, the free analyte anion and the bound analytes migrate in opposite directions. Therefore, the apparent velocity of the analyte anion is determined by the extent of ion pair formation between the analyte and the polymer.

The velocity of the analyte solute, v_s, is given by:

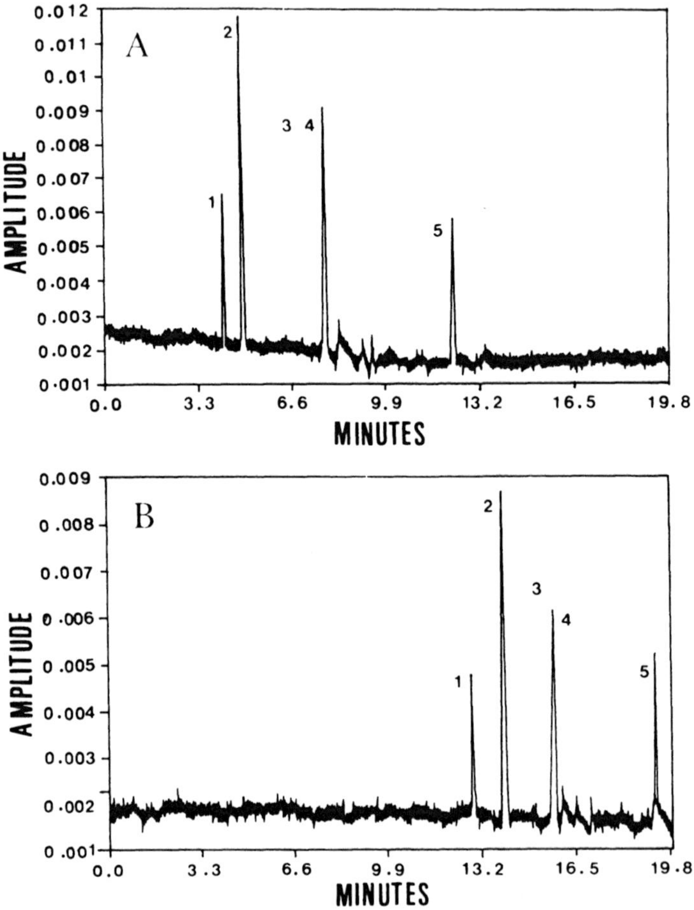

Fig. 5.52. Enhancement of selectivity is demonstrated in (B) by the addition of 80 mM octyl glucoside to 250 mM phosphate (pH 7.0) electrophoresis buffer. Peaks: *1* = bradykinin; *2* = releasing luteinizing hormone; *3* = [Val^4]-angiotensin III; *4* = angiotensin III; *5* = angiotensin II. Conditions: 17 μm I.D. × 375 μm O.D. fused silica APF capillary; 70 cm to detection; detection, 210 nm on-column; field strength, 250 V/cm. (Reproduced from Ref. 54 with permission of Elsevier Science Publishers.)

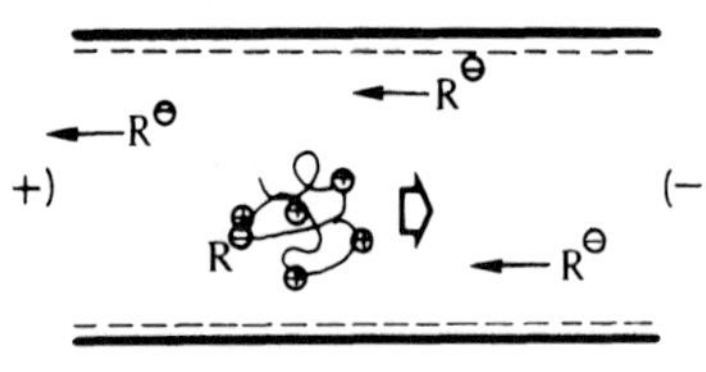

Fig. 5.53. Schematic diagram of the principle of ion-exchange electrokinetic chromatography: a polymer cation is shown; R$^-$, analyte anion; arrows show the electrophoretic migrations. For clarity, other ions, such as buffer constituents and counterions of both analyte anions and the polymer cations have been omitted and the electroosmotic flow is not shown. (Reproduced from Ref. 67 with permission of the American Chemical Society.)

$$\nu_s = \nu_{eo} + F\nu_{ep} + (1 - F)\,\nu_{ep}(p) \tag{5.27}$$

where ν_{eo}, ν_{ep}, and $\nu_{ep}(p)$ are electroosmotic velocity, electrophoretic velocity of the free analyte ion, and the electrophoretic velocity of the polymer ion, respectively. The electrophoretic velocity of the bound ion is supposed to be equal to that of the polymer ion. The quantity F is the fraction of the analyte ion free from the polymer ion and is given by:

$$F = \frac{[S^-]}{[S^-] + [S^- \cdot P^+]} = \frac{1}{(1 + K_{ip}[P^+])} \tag{5.28}$$

and

$$K_{ip} = \frac{[S^- \cdot P^+]}{[S^-][P^+]} \tag{5.29}$$

where K_{ip} is the ion-pair formation constant, [S], [P] and $[S^- \cdot P^+]$ are the concentrations of the solute, the polymer ion, and the ion pair, respectively. The concentration of free polymer ion, $[P^+]$, is considered to be approximately equal to the total concentration of the polymer ion, $[P^+]_0$, provided that the concentration of the analyte ion is low compared with $[P^+]_0$. The difference in migration velocity of two solutes is given by [67]:

$$\Delta\nu = \frac{(K_{ip2} - K_{ip1})\,[P^+]_0\,\{\nu_{ep} - \nu_{ep}(p)\}}{(1 + K_{ip1}\,[P^+]_0)\,(1 + K_{ip2}\,[P^+]_0)} \tag{5.30}$$

where K_{ip1} and K_{ip2} are the ion-pair formation constants of the solutes 1 and 2, respectively. According to Eq. (5.30), $\Delta\nu$ is proportional to the difference between the two formation constants, $(K_{ip1} - K_{ip2})$, and also to the difference between the two electrophoretic velocities, $[\nu_{ep} - \nu_{ep}(p)]$. One point to note is that ν_{ep} and $\nu_{ep}(p)$ have different signs, because they have opposite charges. Typical electropherograms obtained using ion-exchange electrokinetic chromatography with polymer ions are shown in Fig. 5.54.

5.2.6 Effect of polymer coating

The use of coated columns for CE is discussed in detail in Section 4.2. Fused silica tubes coated with two types of polymers were employed to study the effect of coating on MEKC [57]. Chromatograms of the test mixture obtained with these tubes are shown in Fig. 5.55. It is observed that the methyl silicone coating (non-polar) results in an increase in the electroosmotic flow and a decrease in migration times, whereas the polyethylene glycol 20 M (polar) coating resulted in a decrease in the electroosmotic flow and an increase in migration times. The adsorption of more SDS on the methyl silicone coating than on the untreated surface probably leads to the increase in the zeta potential and hence in ν_{eo}. In the case of the polyethylene glycol coating, the effective shielding of the ionized silanol group results in the decrease in ν_{eo}.

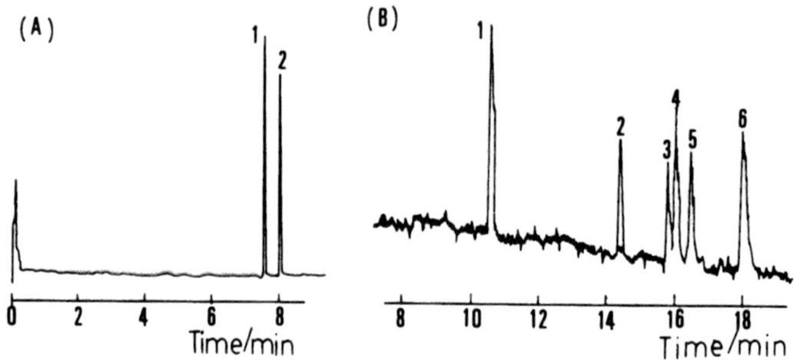

Fig. 5.54. Ion-exchange electrokinetic chromatograms of the analytes. (A) *1, 2* = 2,1-naphthalenesulfonate; (B) *1* = 1- and 2-naphthalenesulfonate; *2* = 2,6-, *3* = 2,7-, *4* = 1,6-, *5* = 1,5-, *6* = 1,7-naphthalenedisulfonates. Separation solution: (A) 0.3% DDAC in 50 mM phosphate buffer (pH 7); (B) 2% DEAE-dextran in the same buffer as in (A); capillary, 75 cm in total length, 50 cm to the detection point; applied voltage, 20 kV; current, (A) 48 μA, (B) 58 μA. (Reproduced from Ref. 67 with permission of the American Chemical Society.)

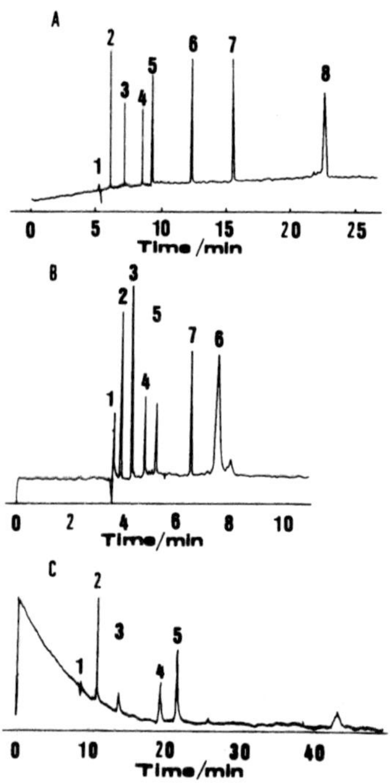

Fig. 5.55. Effect of the coating on the separation of a test mixture: *1* = methanol, *2* = resorcinol, *3* = phenol, *4* = p-nitroaniline, *5* = nitrobenzene, *6* = toluene, *7* = 2-naphthol, *8* = Sudan III. (A) Untreated fused silica tube, 43 μm I.D., 360 μm O.D. (B) DB-1 (methyl silicone), 50 μm I.D., 370 μm O.D. (C) DB-WAX (polyethylene glycol 20 M) 52 μm I.D., 360 μm O.D. Micellar solution, 0.05 M SDS, pH 7; total length of the separation tube, 650 mm; effective length for separation, 500 mm; detection wavelength, 210 nm; temperature of oven, 35°C; applied voltage and current, (A) 16 kV, 28.8 μA; (B) 15.2 kV, 37.4 μA; (C) 14.8 kV, 38.4 μA. (Reproduced from Ref. 57 with permission of Dr. Alfred Huethig Publishers.)

5.2.7 Biological surfactants

The most common examples of biological surfactants are bile salts as an alternative surfactant system in MEKC [69–73]. Bile salts are ionic and possess both hydrophobic and hydrophilic groups which can form chiral micelles. Some of the bile salts which have been employed in MEKC separations include sodium cholate, sodium taurocholate, sodium dehydrocholate, sodium taurolithocholate, sodium deoxycholate and sodium taurodeoxycholate.

An example of the use of bile salts for chiral separations is shown in Fig. 5.56, where MEKC was performed with a 50 mM sodium taurodeoxycholate solution in 50 mM phosphate buffer (pH 3.0) at 40°C. As can be seen in Fig. 5.56, Dns-DL-Nle, Dns-DL-Leu, Dns-DL-Nva and Dns-DL-Met are optically resolved.

Bile salts can also be used in MEKC to enhance resolution of hydrophobic substances. Figure 5.57 shows the separation of a mixture of 7-chloro-4-nitrobenzo-1,1,3-oxadiazole (NBD) derivatized n-alkylamines with sodium cholate (NaCh) and sodium dodecyl sulfate (SDS). It is noted that the late eluting peaks overlap in the case of SDS but are partially resolved with the use of NaCh. It is suggested that the more polar NaCh micelle allows for k' values more near the optimum range.

Another example (Fig. 5.58) of the capability of MEKC using bile salt is the separation of the aflatoxins G_1, G_2 and B_2 with a sodium deoxycholate (NaDCh) micellar phase [74]. When NaDCh is used, baseline separation is obtained for all three compounds without the use of organic modifiers.

Further improvements in resolution can also be attained as in the case of SDS micelles when methanol is added to the electrophoretic solution. Figure 5.59 compares the separation of NBD-derivatized amines using a buffer containing

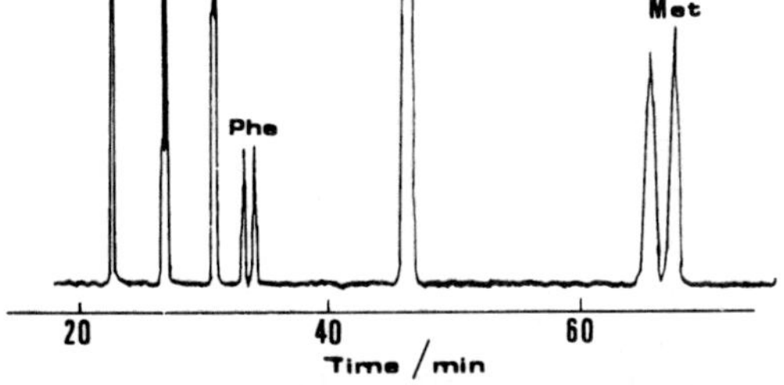

Fig. 5.56. Micellar EKC separation of Dns-D-amino acids with 50 mM taurodeoxycholate in 50 mM phosphate buffer (pH 3.0) in a 500 mm (total length, 700 mm) × 0.05 mm I.D. tube at 40°C. Current, 50 μA. (Reproduced from Ref. 69 with permission of Elsevier Science Publishers.)

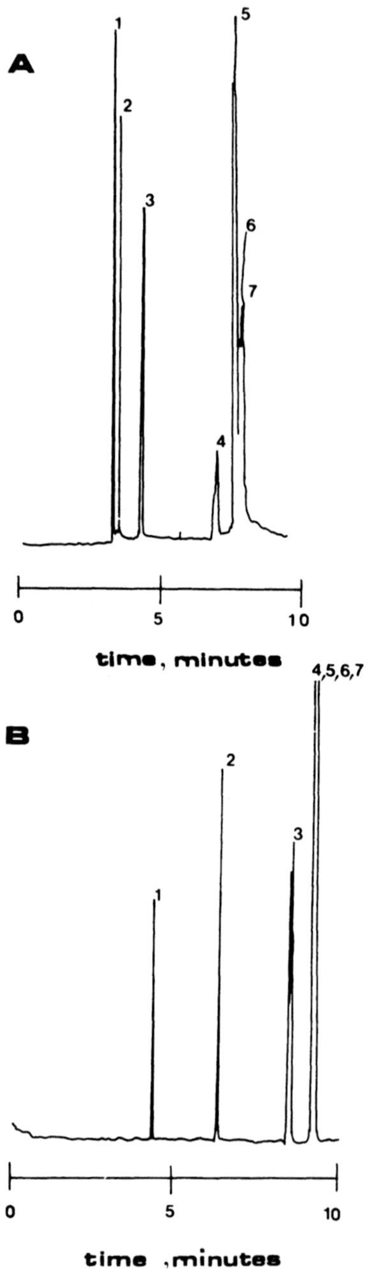

Fig. 5.57. Separations of NBD-derivatized amines in (A) NaCh and (B) SDS. Mobile phase, 0.05 M surfactant–0.01 M Na_2HPO_4–0.006 M $Na_2B_4O_7$; applied voltage, 15 kV; column length, 50 cm, 45 cm to window; detection, Ar-ion laser fluorescence, λ_{ex} 488 nm, λ_{em} 540 nm. Peaks: *1* = methylamine-NBD; *2* = propylamine-NBD; *3* = pentylamine-NBD; *4* = octylamine-NBD; *5* = decylamine-NBD; *6* = deodecylamine-NBD; *7* = quadradecylamine-NBD. (Reproduced from Ref. 73 with permission of Elsevier Science Publishers.)

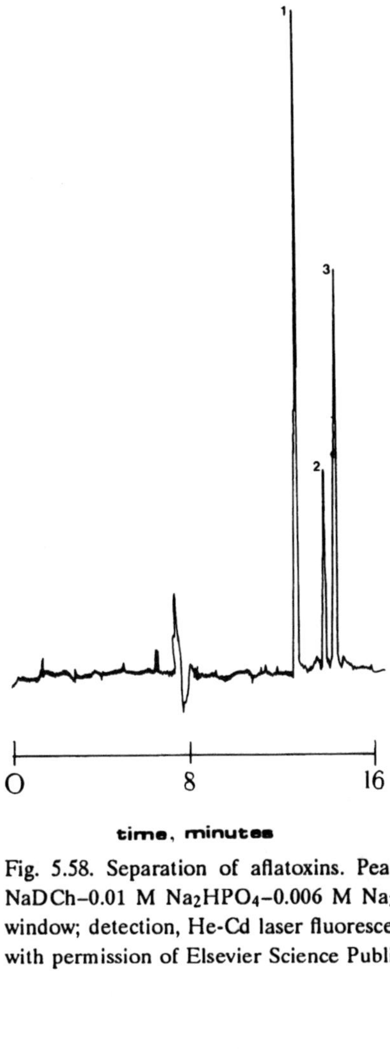

Fig. 5.58. Separation of aflatoxins. Peaks: *1* = G_2; *2* = G_1; *3* = B_2. Mobile phase, 0.05 M NaDCh–0.01 M Na_2HPO_4–0.006 M $Na_2B_4O_7$; applied voltage, 16 kV; column length, 53 cm to window; detection, He-Cd laser fluorescence, λ_{ex} 325 nm, λ_{em} 450 nm. (Reproduced from Ref. 73 with permission of Elsevier Science Publishers.)

NaCh micelles and 20% methanol with that obtained using SDS. With the NaCh system, better resolution is observed and all of the more hydrophobic (C_8–C_{14}) amine derivatives are fully resolved.

5.2.8 Chiral surfactants

Chiral surfactants, such as sodium-N-dodecanoyl-L-valinate (SDVal), sodium N-dodecanoyl-L-alaninate (SDAla) and digitonin have been used for the separation of enantiomers by MEKC [91–94]. Interaction based on hydrophobic entanglement of a solute with chiral functionalized micelles of SDVal permits resolution of N-3,5-dinitrobenzoylated racemic amino acid isopropyl esters [91,92].

For the separation of phenylthiohydantoin derivatives of DL-amino acids (PTH-

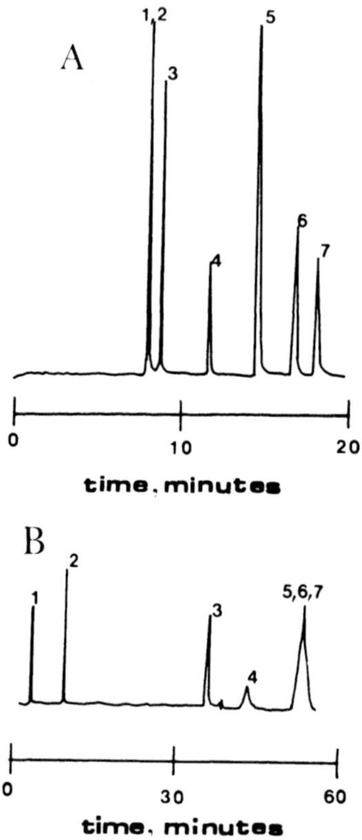

Fig. 5.59. Separation of NBD-derivatized amines. Conditions identical to Fig 5.58 except mobile phase contains 20% methanol. (A) NaCh; (B) SDS. *1* = methylamine-NBD; *2* = propylamine-NBD; *3* = pentylamine-NBD; *4* = octylamine-NBD; *5* = decylamine-NBD; *6* = deodecylamine-NBD; *7* = quadradecylamine-NBD. (Reproduced from Ref. 73 with permission of Elsevier Science Publishers.)

AAs), peak tailing is observed when SDVal is used alone. Although improvement in peak shape can be obtained by adding methanol or urea to the SDVal solution, separation is not satisfactory because of the small capacity factors and narrow migration window [93]. Improved resolution is obtained by using SDVal-SDS co-micellar solutions containing methanol and urea. The enhancement in separation is attributed to increased capacity factors and extended migration windows. The applications of this type of electrophoretic medium for the separation of six PTH-AAs and other chiral compounds, such as benzoin and warfarin, have been demonstrated [93].

A neutral chiral surfactant, digitonin, can be used as an additive to SDS to form mixed micelles having negative charges for the separation of PTH-DL-AAs [94]. Typically, an acidic micellar solution (pH 3.0) containing 25 mM digitonin and 50 mM SDS may be employed. The direction of the mixed micelle is opposite to the electroosmotic flow (i.e. towards the positive electrode). Satisfactory resolution of the six PTH-DL-AAs has been obtained, although a long separation time ($\simeq$100 min) is required [94].

5.2.9 Typical MEKC separations

The applications of MEKC are discussed in detail in Chapter 7. Some examples are mentioned briefly in this section to illustrate the general areas of interest. Nishi *et al.* performed MEKC of B vitamins [76] and β-lactum antibiotics [77]. Their results show that the pH and buffer composition affected peak shape without SDS present. Improved resolution is obtained by MEKC with SDS and sodium N-lauroyl-N-methyltaurate (LMT), although an increase in migration time is observed. In addition to the improvement in selectivity, the peak shapes are also improved by addition of the surfactants. Griest *et al.* [78] utilized MEKC for the separation and detection of normal and modified deoxyribonucleosides and deoxyribonucleotides. Fujiwara *et al.* [79] separated water-soluble vitamins by MEKC. Otsuka *et al.* employed SDS and DTAB in the MEKC separation of 22 PTH-amino acids [80] and aromatic sulfides [81]. Nakagawa et al. [82,83] analyzed cefpiramide in human plasma by MEKC. Otsuka *et al.* [85] performed quantitation of chlorinated phenols and some neutral solutes by MEKC using the internal standard method for peak area and obtained a relative error of less than 1.5%.

5.3 USE OF INCLUSION COMPLEXES

Inclusion complexes are molecular compounds of characteristic structural arrangements, in which one compound (the host molecule) spatially encloses another (the guest molecule) or at least part of it. The inclusion phenomena have been widely exploited in separation methods [99–109]. The most commonly used additives for the formation of inclusion compounds in CE are cyclodextrins. Another group of compounds often used in isotachophoresis are crown ethers, although their use in other modes of CE has rarely been explored to date.

5.3.1 Cyclodextrins

Cyclodextrins (CDs) are cyclic oligosaccharide molecules built of D$-(+)-$glucopyranose units bonded via α-(1,4) linkages. They are also known as Schardinger dextrins, cycloglucopyranoses, cycloamyloses or cycloglucans. In general, CD are soluble in water to some extent. However, their cavity has a non-polar nature. The most commonly used cyclodextrins are α-, β- and γ-cyclodextrins which consist of six, seven and eight glucopyranose units respectively [110]. The chemical structure and a schematic model of a cyclodextrin molecule is shown in Fig. 5.60. In Table 5.9, the physical data of these cyclodextrins are given.

Enhancement of selectivity by the use of cyclodextrins is usually attributable to their ability to selectively include a wide variety of guest organic and inorganic molecules or ions into their hydrophobic cavity.

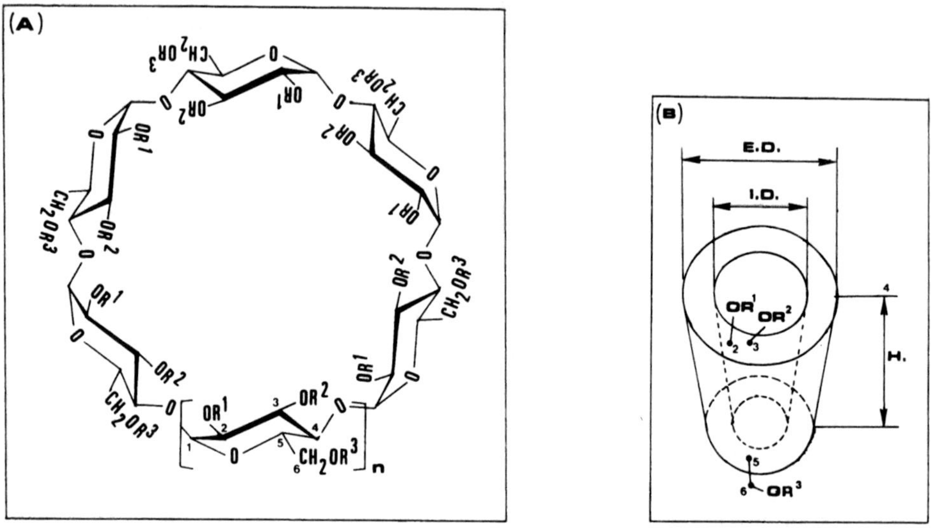

Fig. 5.60. Chemical structure (A) and schematic model (B) of cyclodextrin and/or its derivatives. For n, H, I.D., E.D. values and R^{1-3} substitution see Table 5.9. (Reproduced from Ref. 110 with permission of Elsevier Science Publishers.)

TABLE 5.9

CHARACTERISTICS OF CYCLODEXTRINS[110–114] (Reprinted from Ref. 110)

	Cyclodextrin *		
	α	β	γ
$R^1 = R^2 = R^3$	H	H	H
Number of glucose units	6	7	8
n	1	2	3
Molecular mass	972.86	1135.01	1297.15
Diameter of cavity, I.D. (nm)	0.47–0.52	0.60–0.64	0.75–0.83
Diameter of outer periphery, E.D. (nm)	1.46 ± 0.04	1.54 ± 0.04	1.75 ± 0.04
Height of cavity, H (nm)	0.79–0.80	0.79–0.80	0.79–0.80
Volume of cavity (nm^3)	0.176	0.346	0.510
pK_a range of hydroxyl groups	12.1–12.6	12.1–12.6	12.1–12.6
Solubility in water at 25°C (g/100 ml)	14.50	1.85	23.20
Number of water molecules taken up by cavity	6	11	17
Melting point (K)	551	572	540
Specific rotation, $[\alpha]_D^{25}$	150.5 ± 0.5	162.5 ± 0.5	177.4 ± 0.5

* For meaning of abbreviations see Fig. 5.60.

The relative stabilities of the CD inclusion compounds may be influenced by factors such as hydrogen bonding, hydrophobic interactions, solvent effects and size and shape of the molecules. The effectiveness of the use of CDs to enhance selectivity depends on the size and geometry of the guest molecule, with respect to the dimensions of the CD cavity. Differences in the stability of the inclusion complexes for series of structurally related solutes and optical isomers provide the mechanism to improve resolution of both geometrical and optical isomers.

Although CDs are non-ionic, they can be modified to have ionic groups which are required for the electrophoretic migration in CE. An example of such a derivative is 2-O-carboxymethyl-β-cyclodextrin, which has a carboxyl group [86].

5.3.1.1 Use of cyclodextrins in CZE

Fanali reported the chiral separation of racemic mixtures using cyclodextrins for inclusion complexation of catecholamines [115], terbutaline and propranolol [116] and the separation of diaminobenzene isomers in isotachophoresis [117]. The electropherograms for the resolution of terbutaline into its enantiomers by using an aqueous phosphate buffer containing 5 mM di-OMe-β-CD (R = 2.8) and 15 mM β-CD (R = 2) are shown in Figs. 5.61 and 5.62. For the enantiomeric separation of propanolol enantiomers, satisfactory separation can be obtained by using a background electrolyte containing 4 M urea and 40 mM β-CD, with methanol added as an organic modifier. The resolution of racemic propranolol depends on the amount of organic modifier; the resolution is improved by increasing the methanol content and best resolution is obtained when 30% methanol is added to the background electrolyte (Fig. 5.63).

Enhancement in selectivity and separation efficiency can be obtained by using electrophoretic buffers containing both cyclodextrins and small amounts (0.1%) of alkylhydroxyalkylcellulose derivatives [118]. The derivatives which can be

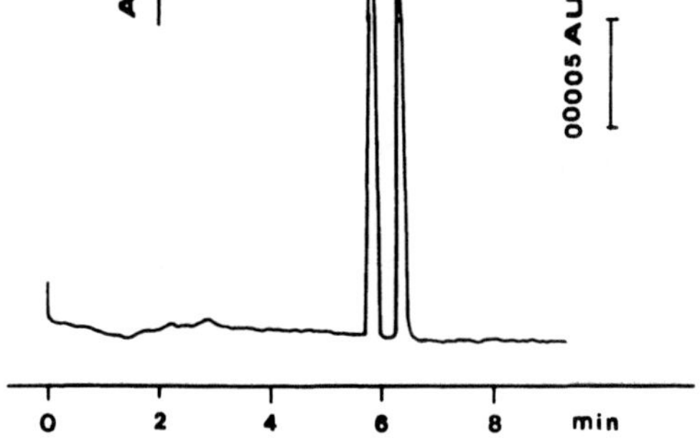

Fig. 5.61. Electropherogram of the enantiomeric resolution of terbutaline. BGE: 0.1 M NaH_2PO_4–H_3PO_3 (pH 2.5) containing 5 mM di-OMe-β-CD. *a* = (−)-terbutaline; *b* = (+)-terbutaline. (Reproduced from Ref. 116 with permission of Elsevier Science Publishers.)

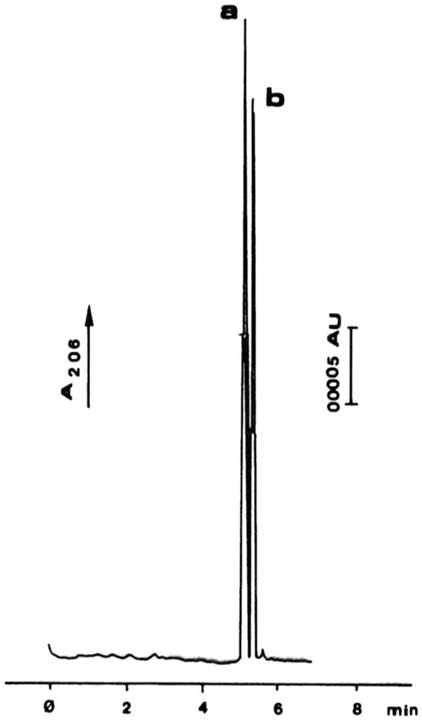

Fig. 5.62. Electropherogram of the separation of racemic terbutaline: *a* =(−)-terbutaline; *b* = (+)-terbutaline. BGE: 0.1 M phosphate buffer (pH 2.5) containing 15 mM β-CD. (Reproduced from Ref. 116 with permission of Elsevier Science Publishers.)

used include methylhydroxyethylcellulose (MHEC), hydroxypropylmethylcellulose (HPMC) and hydroxyethylcellulose (HEC). The separations of benzene- and naphthalene-based isomers, enantiomers of chloramphenical, thioridazine and ketotifen drugs have been demonstrated. For the separation of isomeric compounds, capillary electrophoresis with cyclodextrins and additives has been suggested as one of the best methods available [118].

The behaviour of model peptides in buffer systems containing different types of additives, including α- and β-cyclodextrins has been studied by Liu *et al.* [119]. Two beneficial effects are observed when cyclodextrins are used. First, very narrow peaks are obtained and secondly, enhancement of sensitivity by fluorescence detection is observed. Figure 5.64 demonstrates electropherograms of fluorescamine-derivatized peptides using β-cyclodextrin as a buffer additive. A favourable match between the solute's size or shape and the cyclodextrin hydrophobic cavity is strongly suggested. The excellent peak shapes and high efficiencies can be attributed to decreased diffusion coefficients due to the formation of aggregates or by protection adsorption to the capillary wall.

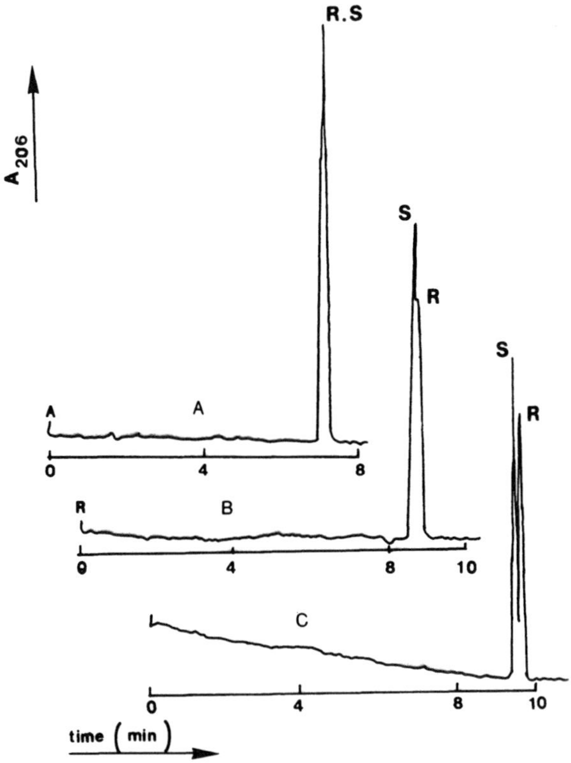

Fig. 5.63. Electropherogram of racemic popranolol separation. BGE: 50 mM phosphate buffer (pH 2.5) containing 4 M urea, 40 mM β-CD and increasing amounts of methanol (*A*, 0%; *B*, 10%; and *C*, 30%). Electrophoresis, 8 kV (constant), 4.6 μA; sampling, 4 kV, 5 s. (Reproduced from Ref. 116 with permission of Elsevier Science Publishers.)

The role of the cyclodextrin cavity in peptide separations is further investigated with 3-(4-carboxybenzoyl)-2-quinolinecarboxaldehyde (or CBQCA) fluorescent derivatives. Using α-cyclodextrin as a buffer additive, good resolution and peak shapes are also achieved for model peptides derivatized with CBQCA (see Fig. 5.65).

5.3.1.2 Cyclodextrin-modified EKC (CD-EKC) and MEKC (CD-MEKC)

By introducing charged groups onto cyclodextrin molecules, electrokinetic chromatography (CD-EKC) can be performed. Terabe *et al.* [86] explored the use of 2-O-carboxymethyl-β-cyclodextrin (β-CMCD) as a phase in CD-EKC. Only the host–guest interaction between β-CMCD and the solute operated as the distribution process. Electroosmosis and electrophoresis in an open-tubular capillary filled with β-CMCD solution permit differential migration between the host and guest molecules. Typical electropherograms obtained for cresol isomers with β-CMCD as a buffer modifier are shown in Figs. 5.66 and 5.67.

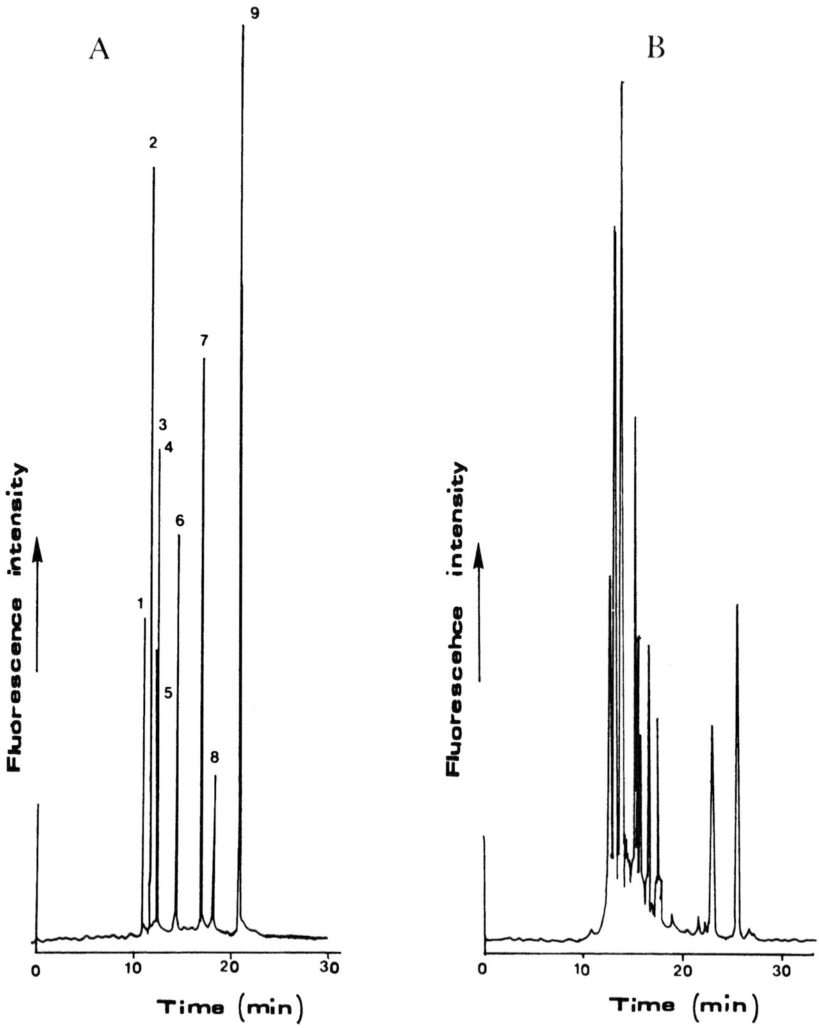

Fig. 5.64. Electropherograms of fluorescamine-derivatized peptides, using β-cyclodextrin as a buffer additive. (A) Model mixture of 9 peptides. Peaks: *1* = angiotensin III; *2* = Gly-Gly-Tyr-Arg; *3* = angiotensin I; *4* = [Val^3]-angiotensin II; *5* = Met-Leu-Phe; *6* = Gly-Leu-Tyr; *7* = Val-Gly-Ser-Glu; *8* = Ala-Gly-Ser-Glu; *9* = Val-Gly-Asp-Glu. Capillary: 80 cm in length (50 cm to detector), 50 μm I.D. Separation buffer: 0.05 M borate buffer (pH 9.50)–20 mM M β-cyclodextrin–15% methanol–1% THF. Operating voltage: 20 kV. (B) Tryptic digest of cytochrome C. Operating voltage: 15 kV. All other conditions are the same as in (A). (Reproduced from Ref. 119 with permission of Elsevier Science Publishers.)

Due to the low electrophoretic mobility of β-CMCD, which results from the small ratio of the charge number to the molecular mass, separation efficiencies are found to be lower than those obtained with micellar solutions. Plate heights of about 4 μm are obtained compared with 2–3 μm observed for micellar electrokinetic

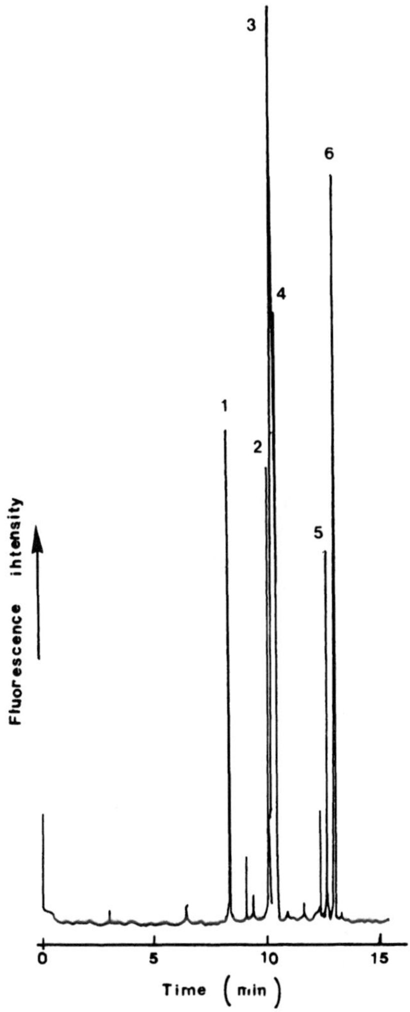

Fig. 5.65. Electropherogram of CBQCA-derivatized peptides using α-cyclodextrin as a buffer additive. Capillary: 90 cm in length (60 cm to detector), 50 μm I.D. Separation buffer: 0.05 M borate buffer (pH 9.50)–20 mM α-cyclodextrin. Operating voltage: 20 kV. Peaks: *1* = Gly-Gly-Tyr-Arg; *2* = Gly-Leu-Tyr; *3* = Met-Leu-Tyr; *4* = Val-Ala-Ala-Phe; *5* = Glu-Gly-Phe; *6* = Glu-Val-Phe. (Reproduced from Ref. 119 with permission of Elsevier Science Publishers.)

chromatography. However, selectivity is found to be higher for the cresol isomers [86].

CD has also been used as an additive in micellar solutions for CD-MEKC separations. Since CD itself is electrically neutral and its outside surface is hydrophilic, it is not expected to interact with the micelle. Therefore, CD in the micellar solution exists as another phase, which migrates with an identical velocity with the bulk solution, and is capable of solubilizing selectively certain

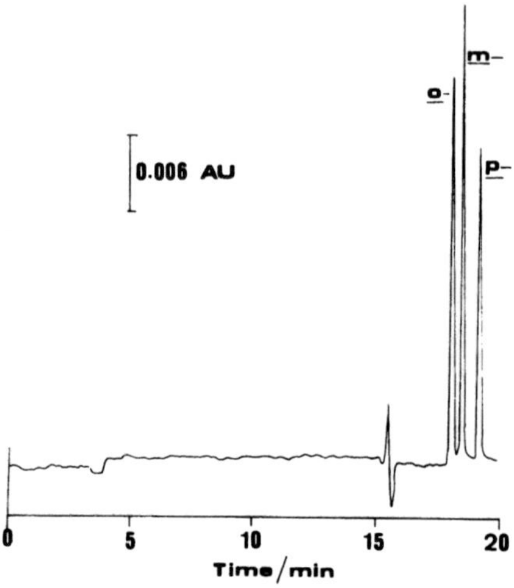

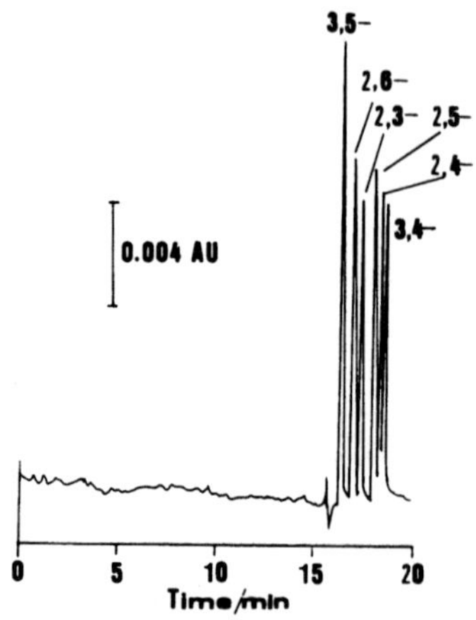

Fig. 5.66 (left). Electrokinetic chromatogram of cresol isomers. Chromatographic solution, 0.025 M β-CMCD in 0.1 M phosphate buffer (pH 7.0); separation tube, 650 $\times$ 0.05 mm I.D.; length of the tube used for separation, 500 mm; current, 50 μA; total applied voltage, ca. 12 kV; detection wavelength, 210 nm. (Reproduced from Ref. 86 with permission of Elsevier Science Publishers.)

Fig. 5.67 (right). Electrokinetic chromatogram of xylidine isomers. Conditions as in Fig. 5.66. (Reproduced from Ref. 86 with permission of Elsevier Science Publishers.)

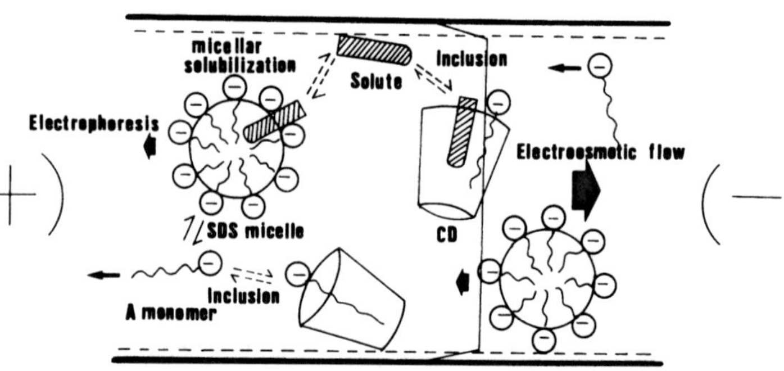

Fig. 5.68. Schematic illustration of the solute interaction in MEKC using SDS and CD. (Reproduced from Ref. 121 with permission of Elsevier Science Publishers.)

types of solutes depending on their size, shape and hydrophobicity. When a highly hydrophobic substance, which is insoluble in water, is injected into the CD-MEKC system, it will distribute itself between the micelle and the CD. A schematic illustration of the separation principle of CD-MEKC is shown in Fig. 5.68. Such a hydrophobic solute is incorporated by either the micelle or CD, but it does not exist in the aqueous medium.

The capacity factor, k', of a highly hydrophobic solute in CD-MEKC is given by:

$$k' = \frac{n_{mc}}{n_{CD}} = K \cdot \frac{V_{mc}}{V_{CD}} \tag{5.31}$$

where n_{CD} and n_{mc} are the total amounts of the solute included by CD and those of the solutes incorporated by the micelle, respectively, V_{CD} and V_{mc} are the volumes of CD and the micelle, respectively, and K is the distribution coefficient. The capacity factor is proportional to the phase (volume) ratio of the micelle to CD. The distribution coefficient means the relative affinity of the solute between CD and the micelle. The ratio of the solute incorporated in the micelle depends on its hydrophobicity but the inclusion-complex formation of the solute with CD depends on the matching of the solute molecular size with the cavity diameter of CD in addition to the hydrophobicity. Consequently, the selectivity in CD-MEKC is mostly determined by the tendency of the solute to form an inclusion complex with CD.

Figure 5.69 shows the MEKC separation of chlorinated benzene congeners without CD. Poor resolution and low efficiency is observed. The low efficiency may be attributable to slow kinetics in the partition equilibrium of the solutes [42]. The addition of 40 mM of γ-CD to the electrophoretic solution permits the separation of all of the chlorinated benzene congeners, as shown in Fig. 5.70. The isomer separation can be ascribed to the differential partition of the isomers to the γ-CD cavity, because these isomers are not resolved in the absence of γ-CD. In CD-MEKC, an isomer included more strongly in the CD cavity is expected to have a shorter migration time than an isomer included less strongly in the cavity.

Enantiomeric separation by CD-MEKC has been investigated by Terabe and co-workers [120–123]. Nishi *et al.* [121] reported the use of SDS and five types of CDs, α-cyclodextrin (α-CD), β-cyclodextrin (β-CD), 2,6-di-O-methyl-β-cyclodextrin (DM-β-CD), 2,3,6-tri-O-methyl-β-cyclodextrin (TM-β-CD) and γ-cyclodextrin (γ-

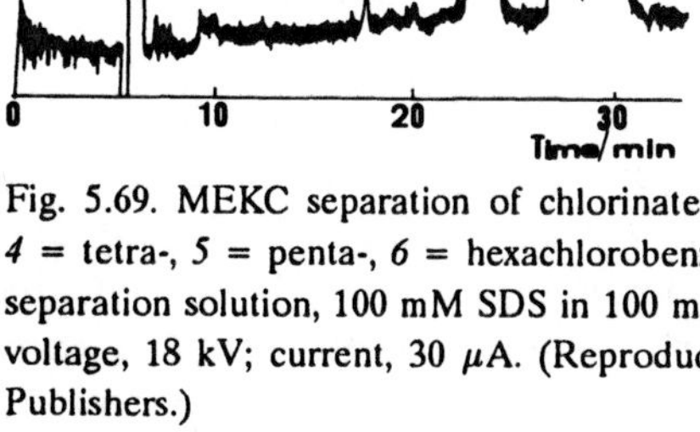

Fig. 5.69. MEKC separation of chlorinated benzene congeners: *1* = mono-, *2* = di-, *3* = tri-, *4* = tetra-, *5* = penta-, *6* = hexachlorobenzene. Capillary, 700 mm (Scientific Class Engineering); separation solution, 100 mM SDS in 100 mM borate buffer (pH 8.0) containing 2 M urea; applied voltage, 18 kV; current, 30 μA. (Reproduced from Ref. 120 with permission of Elsevier Science Publishers.)

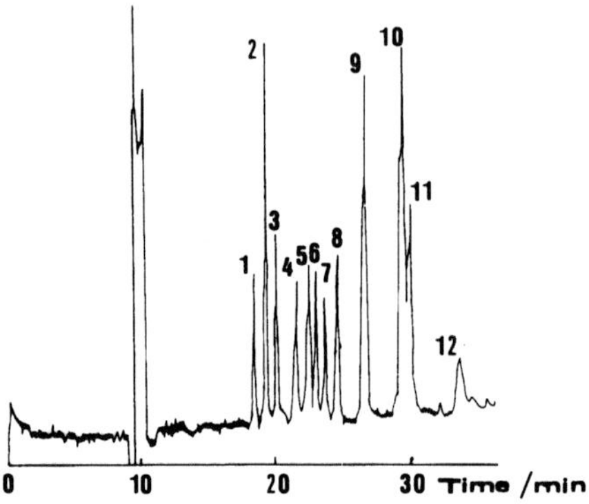

Fig. 5.70. γ-CD-MEKC separation of chlorinated benzene congeners: *1* = 1,2,3,5-tetra-, *2* = 1,2,3-tri-, *3* = 1,3,5-tri-, *4* = 1,2-di-, *5* = 1,2,4,5-tetra-, *6* = mono-, *7* = 1,3-di-, *8* = 1,2,4-tri-, *9* = 1,2,3,4-tetra-, *10* = penta-, *11* = 1,4-di- and *12* = hexachlorobenzene. Capillary, 700 mm (Polymicro Technologies); separation solution, 100 mM SDS in 100 mM borate buffer (pH 8.0), containing 40 mM γ-CD; applied voltage, 15 kV; current, 23 μA. (Reproduced from Ref. 120 with permission of Elsevier Science Publishers.)

CD). The effects of CD concentration on the migration times and chiral recognition are investigated using γ-CD over the concentration range 0–0.06 M. The results are shown in Fig. 5.71, together with the results of CZE and MEKC without addition of CDs. The migration times or capacity factor of some of the solutes, especially of aromatic compounds except two achiral barbiturates, are considerably reduced with an increase in the CD concentration. It is therefore useful to employ CDs in MEKC of lipophilic compounds, which migrate near the migration time of the micelle and cannot be resolved by the SDS solution alone, as reported previously [120,121].

The selectivity in CD-MEKC can be manipulated through the addition of organic modifiers. The effects of methanol addition (10%) on the chiral separation and capacity factors of four enantiomeric solutes using a 0.05 M SDS solution containing 30 mM γ-CD are summarized in Table 5.10. For these lipophilic compounds, the addition of an organic modifier is effective in changing the capacity factors and peak shapes. The capacity factors of the solutes are reduced by the addition of methanol, although their migration times increase because of the reduction of the electroosmotic flow. However, chiral recognition of the enantiomers becomes poorer since methanol can be included in the CD cavity, which is unfavourable for the chiral recognition of the enantiomers.

To improve enantiomeric selectivity, chiral additives such as sodium-d-camphor-10-sulfonate and l-menthoxyacetic acid has been added to the SDS solution containing γ-CD [121]. The effect of l-menthoxyacetic acid on CD-MEKC separation of the four solutes is illustrated in Table 5.11. Optimum resolution is obtained with 60 mM of the chiral additive. Although increasing concentration of l-menthoxyacetic acid is expected to result in enhancements in chiral selectivity, an increase in capacity

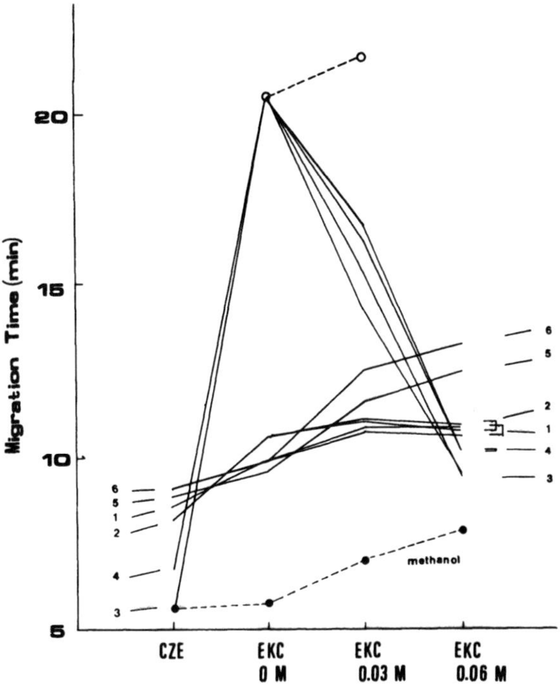

Fig. 5.71. Effects of γ-CD concentrations on the migration times and chiral recognition. *CZE*: 0.02 M phosphate-borate buffer of pH 9.0. *EKC*: 0.05 M SDS added to CZE buffer. Open and closed circles represent the migration times of the micelle and methanol, respectively. *1* = thiopental (sodium); *2* = pentobarbital (calcium); *3* = 2,2,2-trifluoro-1-(9-anthryl)ethanol; *4* = 2,2′-dihydroxy-1,1′-dinaphthyl; *5* = phenobarbital; *6* = barbital (sodium), *7* = 1,1′-binaphthyl-2,2′-diyl hydrogen phosphate. (Reproduced from Ref. 121 with permission of Elsevier Science Publishers.)

TABLE 5.10

EFFECT OF METHANOL ADDITION ON SEPARATION FACTOR AND CAPACITY FACTOR (Adapted from Ref. 121)

Solute	No methanol		10% methanol	
	k'	α	k'	α
Thiopental	1.05	1.06	0.92	1.03
	1.11		0.95	
Pentobarbital	1.15	1.03	0.98	1.00
	1.19		0.98	
Anthrylethanol	3.06	1.34	2.49	1.27
	4.09		3.15	
Dinaphthyl	5.27	1.15	3.89	1.10
	6.06		4.30	

Buffer: 0.05 M SDS buffer solution of pH 9.0 containing 30 mM γ-CD.

TABLE 5.11

EFFECTS OF l-MENTHOXYACETIC ACID ON THE ENANTIOSELECTIVITY (Adapted from Ref. 121)

Solute	l-Menthoxyacetic acid								
	30 mM			60 mM			90 mM		
	k'	α	R_s	k'	α	R_s	k'	α	R_s
Thiopental	1.10	1.07	1.77	1.10	1.07	1.97	1.32	1.06	1.99
	1.18			1.17			1.40		
Pentobarbital	1.39	1.03	0.91	1.46	1.04	1.18	1.84	1.03	0.96
	1.44			1.52			1.90		
Anthrylethanol	9.75	1.49	3.21	11.99	1.27	3.66	17.27	1.24	3.14
	14.56			15.17			21.47		
Dinaphthyl	21.04	1.25	2.31	21.26	1.26	2.40	29.03	1.38	3.05
	26.40			26.70			40.20		

Buffer: 0.05 M SDS containing 30 mM γ-CD and l-menthoxyacetic acid.

factor is also observed, which leads to a decrease in resolution at the highest additive concentration investigated (90 mM).

5.3.2 Crown ethers

Crown ethers are synthetic macrocyclic polyethers which are able to form stable inclusion complexes with various inorganic and organic cations. They contain electron-donor heteroatoms (O, N, S) in the cyclic structure. Cyclic oligomers of epoxyalkanes, derived aromatic and alicyclic crown ethers, thio and aza-crown ethers which contain sulfur and nitrogen atoms in the ring are all examples of crown ethers. The structure of a common crown-ether skeleton, with oxygen atom only, is given in Fig. 5.72, and the cavity diameter of simple crown ethers are given in Table 5.12. The complex formation is based on ion–dipole interactions between the host and guest molecules.

Many crown ethers can form complexes with a wide variety of compounds, including metal ions, organic nitrocompounds, aryldiazonium salts, amines and nitriles in their ionic form. The complex formation depends strongly on the matching of the cavity diameter and the size of the host ion. Maximum stability is achieved when the cation diameter is close to the dimension of the cavity. The number, geometrical position and geometrical symmetry of oxygen-donor atoms in the rings also affect the stability of the complex. The stability of the complex is increased if the number of oxygen atoms which lie in the same plane is high, and if the arrangement of the oxygen atom is symmetrical. By introducing an optically active group into the crown ether, it is possible to provide selectivity for the separation of optically active organic compounds.

TABLE 5.12

CROWN ETHER CAVITY DIAMETERS (Reprinted from Ref. 110)

Name	Internal diameter* (nm)		
	A	B	C
14-Crown-4	0.12	0.15	–
15-Crown-5	0.17	0.22	0.17–0.18
18-Crown-6	0.26	0.32	0.27–0.29
21-Crown-7	0.34	0.43	–
24-Crown-8	0.4	–	–

* A: From Corey Pauling Koltur atomic models; B: from Fisher-Hirschfelder-Taylor atomic models; C: from X-ray crystallographic data.

The utilization of crown ethers in electromigration methods has been restricted to the isotachophoretic separation of alkaline and alkaline-earth metals [124–126]. The rapid progress in CE and in the synthesis of enantioselective crown ethers may open up new possibilities in the separation of enantiomers.

5.4 COMPLEXING ADDITIVES

Complexing additives have been used for the separation of inorganic ions [127–129], neutral compounds [130–136], carbohydrates [137,138], chiral compounds [139,140] as well as bases, nucleosides and oligonucleotides [141,142].

Gebauer *et al.* [127] determined nitrate, chloride and sulfate in drinking water by CZE using complex formation with Cd^{2+} as the counter ion to influence selectively the effective mobilities of the individual ions. The electropherograms obtained are shown in Fig. 5.73.

Separation of 4(2-pyridylazo) resorcinol (par) metal chelates by MEKC was investigated by Saitoh *et al.* [128]. The capacity factor of the chelates varies with the

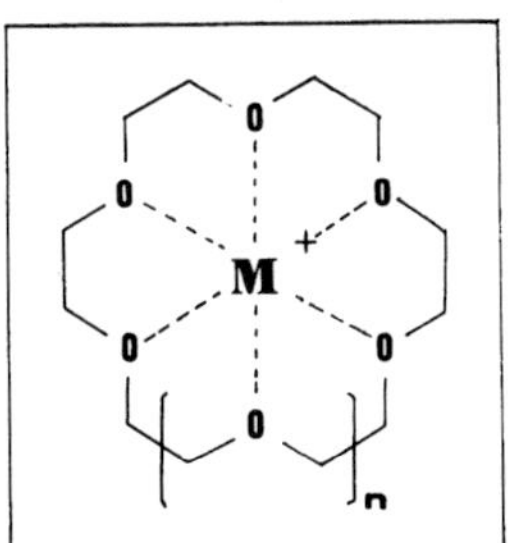

Fig. 5.72. Chemical structure of a crown metal (M^+) complex, e.g., $n = 1$ for 18-crown-6-M^+, $n = 2$ for 21-crown-7-M^+, etc. (Reproduced from Ref. 110 with permission of Elsevier Science Publishers.)

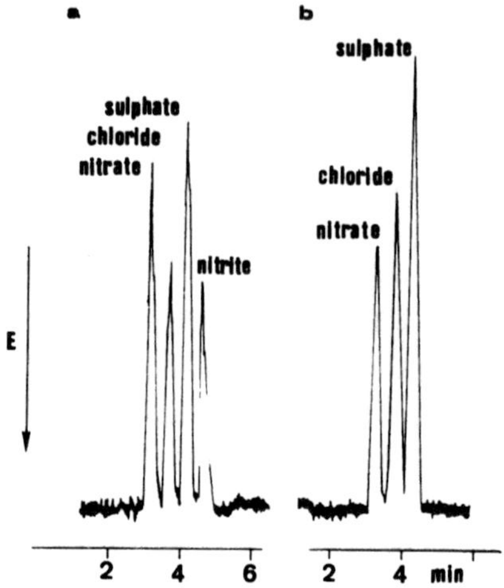

Fig. 5.73. Potential gradient records of zone-electrophoretic analyses of anions. Sample: (a) 0.50 μl of 10^{-3} M nitrate, chloride, sulphate and nitrite; (b) 1.00 μl of drinking water. (Reproduced from Ref. 127 with permission of Elsevier Science Publishers.)

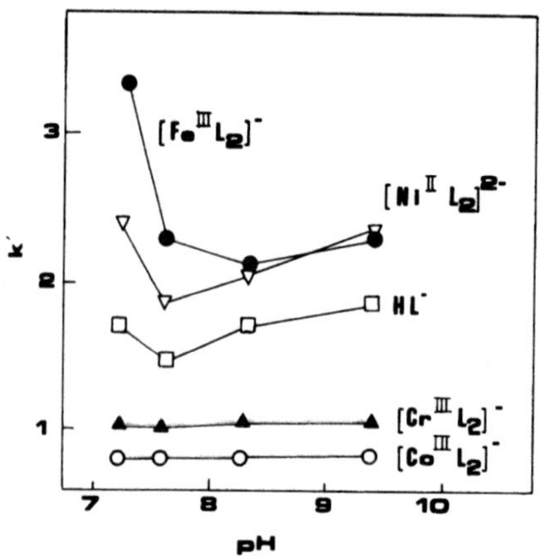

Fig. 5.74. Capacity factor vs. eluent pH plots for par and the chelates. The eluent pH was adjusted by mixing appropriate phosphate and borate buffer solutions to give a constant ionic strength. Temperature: 25±1°C. (Reproduced from Ref. 128 with permission of Elsevier Science Publishers.)

eluent pH, as shown in Fig. 5.74, owing to the acid dissociation equilibria of the par chelates (1-hydroxy groups of par ligands):

$$[M^{n+}(HL)_2]^{(n-2)+} \overset{pK_{a1}}{\rightleftharpoons} [M^{n+}(HL)L]^{(n-3)+} \overset{pK_{a2}}{\rightleftharpoons} [M^{n+}L_2]^{(n-4)+} \tag{5.32}$$

The MEKC separation of metal-par chelates is shown in Fig. 5.75. Very sharp peaks are produced within an acceptable total elution time. The theoretical plate numbers of the chelates are in the range of 105,000–120,000 per 60 cm.

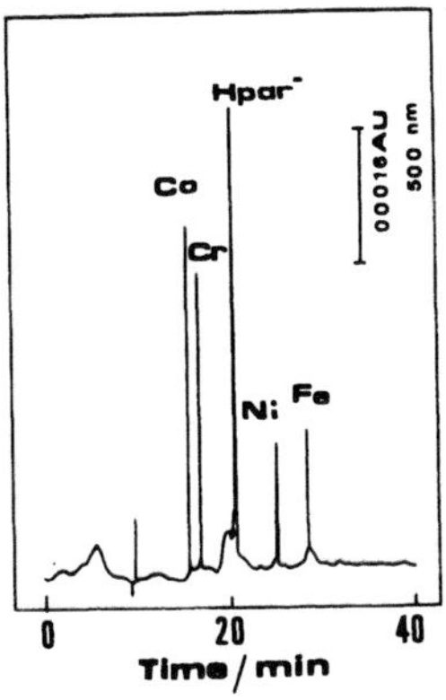

Fig. 5.75. MECC separation of metal-par chelates. Eluent: [SDS] = 0.02 mol dm^{-3}, [NaH_2PO_4] = 0.05 mol dm^{-3}. [$Na_2B_4O_7$] = 0.0125 mol dm^{-3}, [par] = 1.0×10^{-4} mol dm^{-3}. Applied voltage: 16.5 kV (25 μA). Complexation conditions: [par] = 1.0×10^{-3} mol dm^{-3}; [metal ion] (10^{-5} mol dm^{-3}), Co(II) 0.995, Cr(III) 1.23, Fe(III) 1.00, Ni(II) 1.01; [triethanolamine] = 0.025 mol dm^{-3} (pH = 8.8 with HCl), heated at 98°C for 30 min. (Reproduced from Ref. 128 with permission of Elsevier Science Publishers.)

8-Hydroxyquinoline-5-sulfonic acid (HQS^-) was used as an on-column chelating agent for the determination of metal ions (129). The interaction between the metal ion and HQS^- is described by the following expression:

$$K' = \frac{[M(HQS^-)]_n}{[M][HQS^-]_n} \tag{5.33}$$

where K' is the overall conditional formation constant for a particular metal ion and n is the number of ligands. The reaction generally occurs in a stepwise fashion. Several different metal–HQS^- complexes can exist simultaneously. As the HQS^- to metal ratio increases, the charge of the complex becomes effectively more negative and mobility decreases as a result. The net mobility of the ion is determined by the distribution between the various possible forms of the metal complex. The effect of electrophoretic buffer pH on K' is described by:

$$K' = \alpha^n K_f \tag{5.34}$$

where K_f is the overall formation constant for the metal at infinite dilution and α is the degree of protonation of the amine and phenolic functionalities on the HQS^- chelate.

The observed electrophoretic mobility of the metal ion ($\mu_{ep,obs}$) is a combination of the mobility of the free metal and various complexes:

$$(\mu_{ep,obs}) = x_M \mu_M + x_{ML} \mu_{ML} + x_{ML_2} \mu_{ML_2} + \cdots \tag{5.35}$$

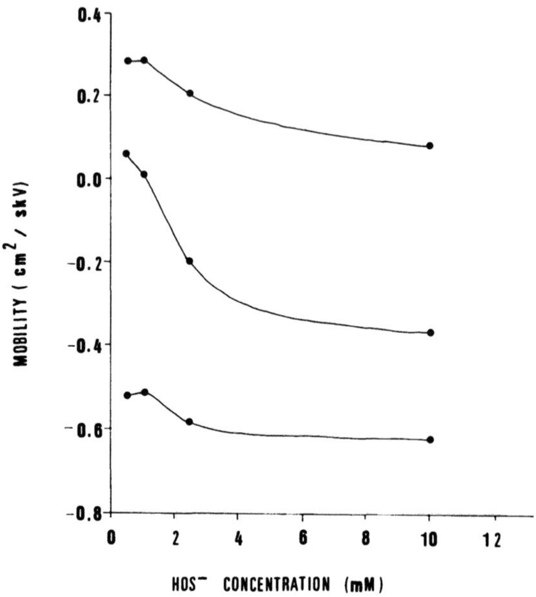

Fig. 5.76. Effects of electrophoretic buffer HQS^- concentration on the mobility of Ca(II) (top curve), Mg(II) (middle curve), and Zn(II) (bottom curve). (Reproduced from Ref. 129 with permission of the American Chemical Society)

where μ_M, μ_{ML} and μ_{ML_2} are the mobilities of the free metal ion, 1:1 HQS^- complex, and 1:2 complex and x_M, x_{ML} and x_{ML_2} are the mole fractions of each species in the capillary. These mole fractions are a function of electrophoretic buffer parameters, which affect the above equilibrium equation. The effects of electrophoretic buffer HQS^- concentration on the mobility of Ca(II), Mg(II), and Zn(II) are shown in Fig. 5.76. The electropherograms obtained are shown in Fig. 5.77.

Another type of application of complexing additives is to enhance the separation of neutral solutes by CE. Neutral solutes can form charged complexes either through solvophobic association [130] or through dynamic complexation [131]. The charged complex can then be separated based on the electrophoretic mobilities of the associated/complex species. Differences in electrophoretic mobilities are caused by the differences in equilibrium constants for the interaction with the solutes.

Separation of neutral organic molecules by solvophobic association with tetraalkylammonium ion by CZE was investigated by Walbroehl and Jorgenson [130]. Electrophoretic solutions consisting of tetrahexylammonium percholate (THAP) dissolved in acetonitrile–water mixtures are used. The general mechanism for solvophobic association is represented as follows:

$$S + L^+ \rightleftharpoons SL^+ \tag{5.36}$$

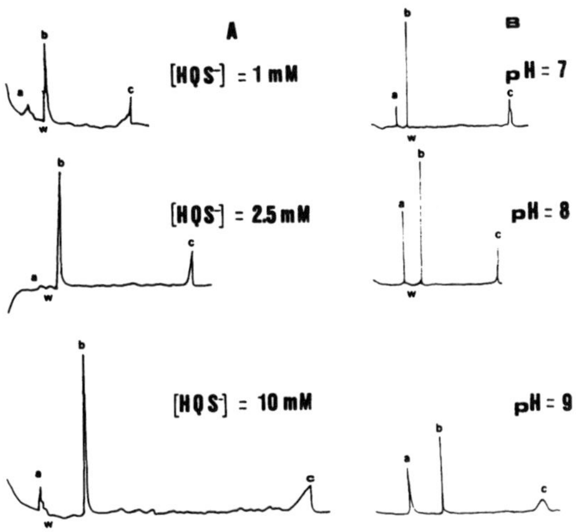

Fig. 5.77. Separation of test metals under various electrophoretic buffer conditions. (A) Electrophoretic buffer pH = 8, applied voltage = 20 kV, capillary length = 50 cm and laser intensity = 1.5 mW. Elution order is Ca(II), Mg(II), Zn(II). All metal ions were 0.3 mM. The arrow denotes a Ca(II) peak with S/N = 2. (B) Electrophoretic buffer containing 2.5 mM HQS^- applied voltage = 20 kV, capillary length = 45 cm and laser intensity = 2.5 mW. Same elution order as in A. Metal concentrations were Ca(II) = 0.63 mM, Mg(II) = 0.08 mM, and Zn(II) = 0.15 mM. (Reproduced from Ref. 129 with permission of the American Chemical Society.)

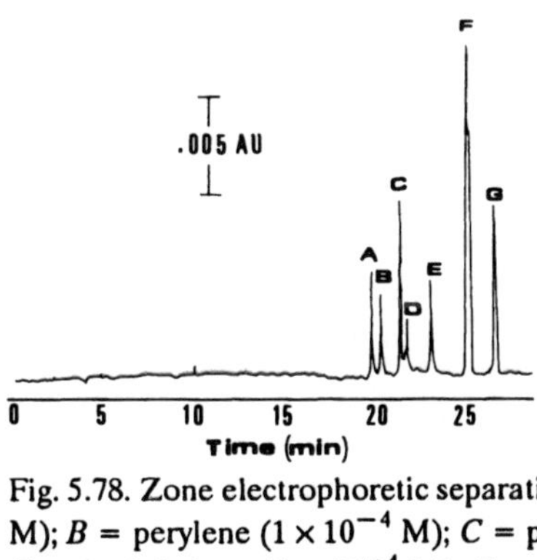

Fig. 5.78. Zone electrophoretic separation of organic compounds: A = benzo[ghi]perylene (1×10^{-4} M); B = perylene (1×10^{-4} M); C = pyrene (1×10^{-4} M); D = 9-methylanthracene (1×10^{-4} M); E = naphthalene (5×10^{-4} M); F = mesityl oxide (1×10^{-4} M); G = formamide. (Reproduced from Ref. 130 with permission of the American Chemical Society.)

$$SL^+ + L^+ \rightleftharpoons SL^+ \tag{5.37}$$

where S is the solute, and L^+ is the THA ion, which may be loosely considered as a ligand. Because of electrostatic repulsion, the formation of complexes with more than one associated THA^+ is unlikely.

Figure 5.78 shows the separation of several organic compounds by CZE with THAP added to the electrophoretic solution. The electrophoretic mobility as a

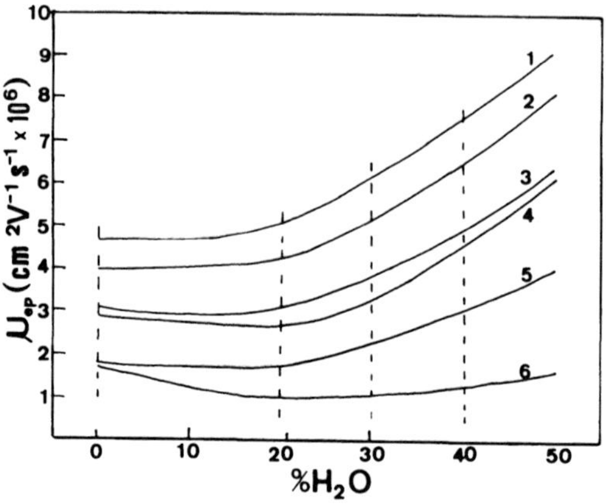

Fig. 5.79. Electrophoretic mobility as a function of percent water in electrophoretic medium for a series of test solutes: *1* = benzo[ghi]perylene; *2* = perylene; *3* = pyrene; *4* = 9-methylanthracene; *5* = naphthalene; *6* = mesityl oxide. (Reproduced from Ref. 130 with permission of the American Chemical Society.)

function of percent water in the electrophoretic medium is shown in Fig. 5.79. The electrophoretic mobilities are found to increase with increasing THAP content to varying degree depending on the molecules. The mobilities of larger, more hydrophobic solutes show a stronger dependence on the THAP concentration, which indicate that stronger interactions exist for these molecules than for a smaller, more polar species.

An alternative approach to separate neutral compounds with CE is to dynamically form charged complexes of these compounds which can be separated electrophoretically [131–136]. A typical example of this approach is the improvement of resolution in CE separation of catechols and catecholamines by complex formation with boric acid. In this system, the catechol–boric acid complexation reaction is represented by the equilibrium:

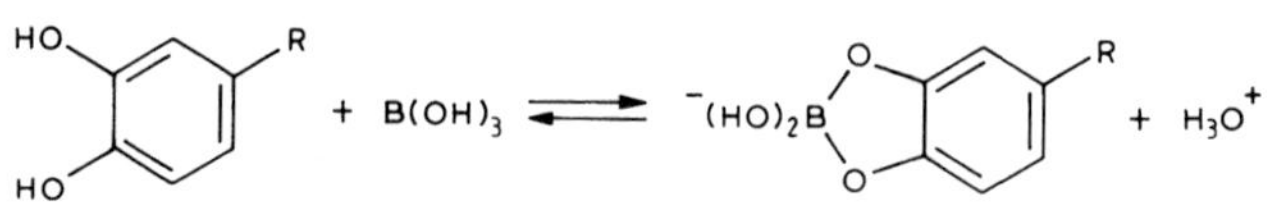 (5.38)

This reaction is specific for compounds possessing an 3,4-dihydroxy functionality, such as catechols and certain carbohydrates. The partition behaviour of neutral catechols is altered due to the formation of the negatively charged borate complexes. The charged complexes can be resolved on the basis of electrophoretic mobility as shown in Fig. 5.80. Selectivity can be altered by changing the boric acid concentration. High borate concentrations result in more negative mobilities and therefore increased migration times.

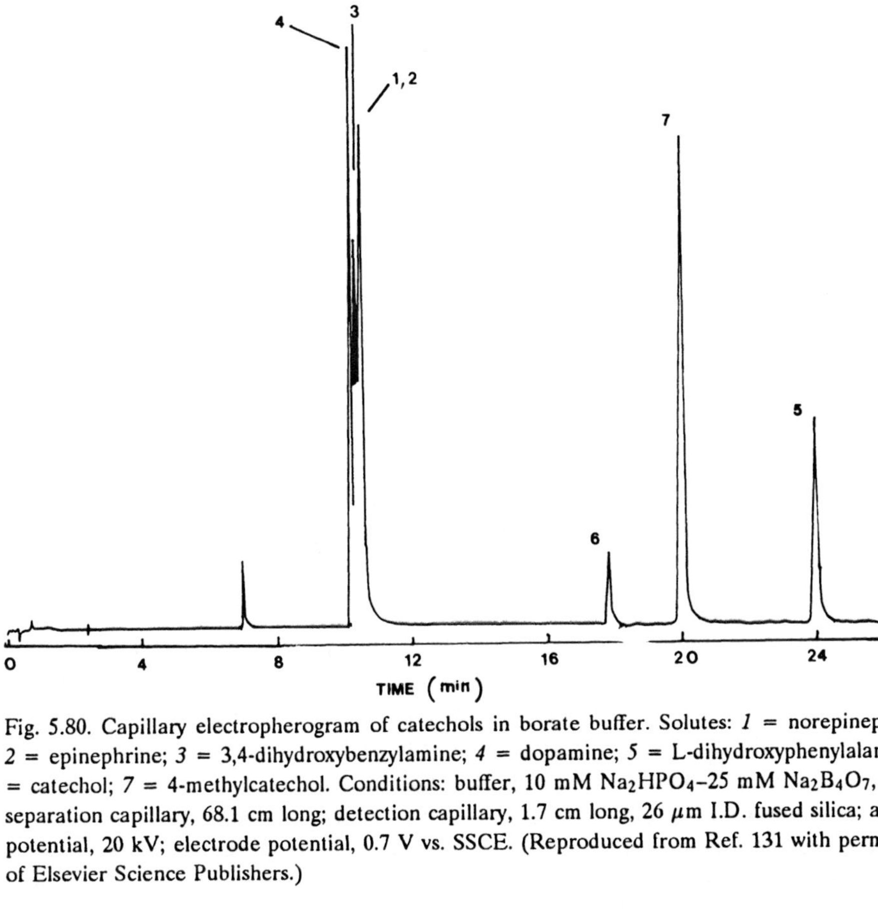

Fig. 5.80. Capillary electropherogram of catechols in borate buffer. Solutes: *1* = norepinephrine; *2* = epinephrine; *3* = 3,4-dihydroxybenzylamine; *4* = dopamine; *5* = L-dihydroxyphenylalanine; *6* = catechol; *7* = 4-methylcatechol. Conditions: buffer, 10 mM Na_2HPO_4–25 mM $Na_2B_4O_7$, pH 7; separation capillary, 68.1 cm long; detection capillary, 1.7 cm long, 26 μm I.D. fused silica; applied potential, 20 kV; electrode potential, 0.7 V vs. SSCE. (Reproduced from Ref. 131 with permission of Elsevier Science Publishers.)

Borate complexation can be combined with MEKC to produce enhanced selectivity for the separation. Figure 5.81 shows the MEKC separation of the catechols at pH 7 in a phosphate–borate buffer containing 10 mM SDS. Excellent selectivity can be obtained with this approach.

Most sugar molecules are neutral and therefore they cannot be separated electrophoretically. However, by adding borate to their aqueous solutions, they are transformed into negatively charged complexes, which can then be separated by CE [138].

The complex formation can be described by the following equations:

$$B^- + L \rightleftharpoons BL^- + H_2O \tag{5.39}$$

$$BL^- + L \rightleftharpoons BL_2^- + H_2O \tag{5.40}$$

where L is the polyol ligand and B^- represents tetrahydroxyborate, $B(OH)_4^-$. In a pH range from 8 to 12, aqueous borate solutions contain not only tetrahydroxyborate

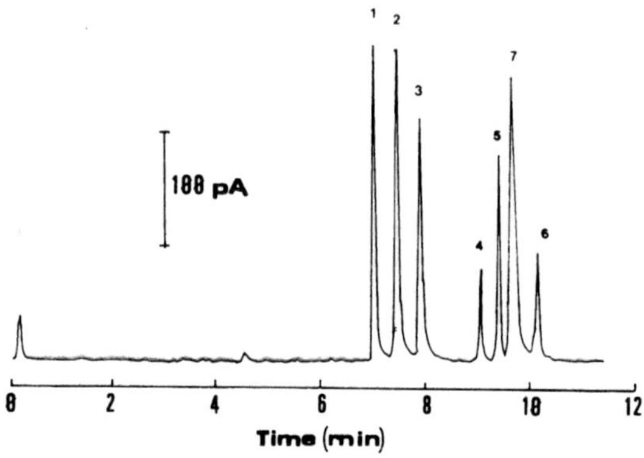

Fig. 5.81. Electrokinetic separation of catechols as borate complexes: 10 mM dibasic sodium phosphate, 6 mM sodium borate at pH 7 with 10 mM SDS; separation capillary, 64.3 cm; detection capillary, 1.7 cm; separation potential, 20 kV (7 μA); injection, 4 s at 20 kV. *1* = norepinephrine (NE); *2* = epinephrine (E); *3* = 3,4-dihydroxybenzylamine (DHBA); *4* = dopamine (DA); *5* = L-3,4-dihydroxyphenylalanine ((L-DOPA); *6* = catechol (CAT); *7* = 4-Methylcatechol (4-MC). (Reproduced from Ref. 131 with permission of Elsevier Science Publishers.)

ions but also more highly condensed polyanions such as triborate, $[B_3O_3(OH)_5]^{2-}$, and tetraborate, $[B_4O_5(OH)_4]^{2-}$. The equilibrium between the different species depends only on the pH and the total borate concentration and is independent on the kind of buffer used [138]. Furthermore, the reaction rate constant is governed by the Arrhenius equation. As a rule of thumb, a temperature increase of 10°C doubles the reaction rate, which means that at higher temperatures the reaction equilibrium is attained faster, thus resulting in narrower peak shapes. Consequently, resolution and efficiency can be improved by performing electrophoresis at elevated temperature. In Fig. 5.82, CE at 60°C of eight of the most common monosaccharides occurring in the carbohydrate moiety is shown.

Honda *et al.* [137] derivatized reducing monosaccharides to N-2-pyridylglycamines and separated them as borate complexes by CZE in a 200 mM borate buffer (pH 10.5). The principle of separation of the borate complexes of N-2-pyridylglycamines by CZE is shown in Fig. 5.83. The electropherogram obtained is shown in Fig. 5.84. The derivatives of 12 saccharides are completely resolved in ca. 25 min with high resolution.

Complexing additives have also been employed for the separation of chiral compounds. Zare and co-workers investigated electrokinetic separation of chiral compounds based on diastereomeric interactions with the Cu(II)-L-histidine complex [139] and the Cu(II)-aspartame complex [140] present in the support electrolyte. Figure 5.85a shows the separation and resolution of a 1:1 mixture of D- and L-dansyl amino acid. Replacement of L-histidine by D-histidine in the support electrolyte reverses the migration order of the DL-amino acids. When a 1:1 mixture of D- and L-histidines is used, no resolution of the D- and L-amino

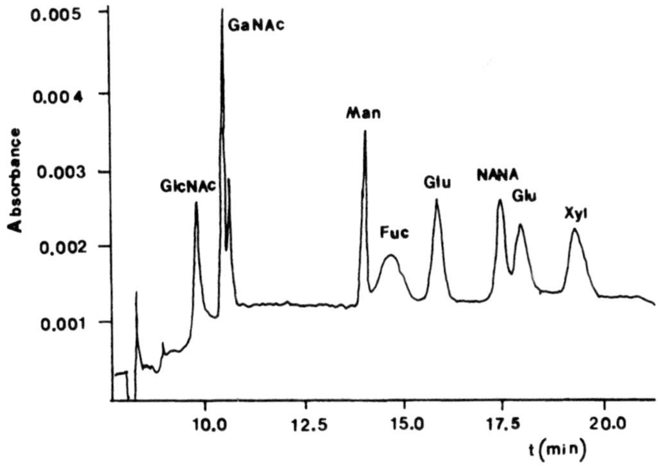

Fig. 5.82. CE of a mixture of underivatized monosaccharides occurring in oligosaccharide moieties of glycoproteins. Sample, N-acetylglucosamine (*GlcNAC*) and N-acetylgalactosamine (*GalNAC*), each 0.75 mM, mannose (*Man*), fucose (*Fuc*), galactose (*Gal*), glucose (*Glu*), and xylose (*Xyl*), each 7.5 mM and sialic acid (*NANA*), 1.5 mM; buffer, 60 mM tetraborate, pH 9.3; temperature, 60°C. (Reproduced from Ref. 138 with permission of the American Chemical Society.)

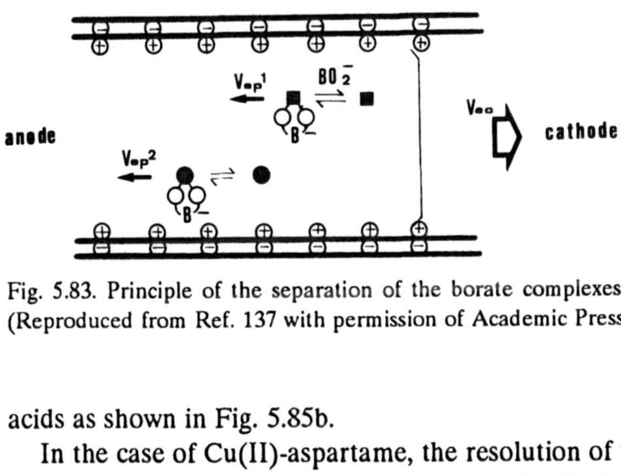

Fig. 5.83. Principle of the separation of the borate complexes of N-2-pyridylglycamines by CZE. (Reproduced from Ref. 137 with permission of Academic Press, Inc.)

acids as shown in Fig. 5.85b.

In the case of Cu(II)-aspartame, the resolution of the amino acids (AA) is based on the formation of a ternary complex of Cu(II) in the electrolyte solution. The configuration of the ternary complex formed with the aspartame and an AA is shown in Fig. 5.86. Stabilizing hydrophobic interaction occurs preferentially with the D-enantiomer, since the interacting groups are closer in this complex than in that formed with the L-enantiomer.

Under neutral pH conditions, the Cu(II) ternary complex is positively charged and will move faster than the neutral free AAs due to the electrophoretic action superimposed on the electroosmotic motion. Thus, the stability of the Cu(II) ternary complex determines the migration times of the two enantiomers and hence the selectivity of the enantiomeric resolution.

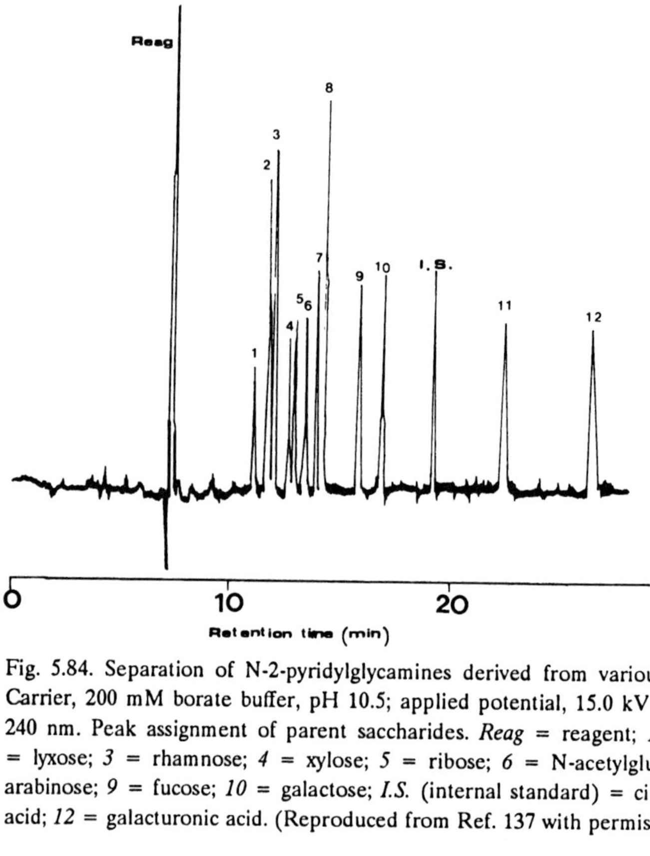

Fig. 5.84. Separation of N-2-pyridylglycamines derived from various reducing monosaccharides. Carrier, 200 mM borate buffer, pH 10.5; applied potential, 15.0 kV; detection, UV absorption at 240 nm. Peak assignment of parent saccharides. *Reag* = reagent; *1* = N-acetylgalactosamine; *2* = lyxose; *3* = rhamnose; *4* = xylose; *5* = ribose; *6* = N-acetylglucosamine; *7* = glucose; *8* = arabinose; *9* = fucose; *10* = galactose; *I.S.* (internal standard) = cinnamic acid; *11* = glucuronic acid; *12* = galacturonic acid. (Reproduced from Ref. 137 with permission of Academic Press, Inc.)

Figure 5.87 shows a typical electropherogram obtained with an electrolyte solution of 2.5 mM $CuSO_4 \cdot 5H_2O$, 5.0 mM aspartame, and 10 mM NH_4OAc at pH 7.4. It can be observed from Fig. 5.87 that sharp and well-resolved peaks are obtained for the enantiomers.

CE with complexing additives has also been employed for the separation of bases, nucleosides and oligonucleotides. Cohen *et al.* [141] investigated the separation of these compounds by MEKC using SDS micelles and metal additives. Metals, such as Cu(II), Zn(II) and Mg(II) form complexes with oligonucleotides, with either the phosphate groups and/or the bases. Manipulation of complexation constant by changing pH or metal or surfactant concentration produced differences in the selectivity of separation. A schematic illustration of the retention mechanism with SDS micelles and metal ions is shown in Fig. 5.88. The effect due to the equilibrium of the solute with free metal present in the solution and the adsorption of oligonucleotide by the fused silica wall may be considered small [138]. The migration mechanism is mainly controlled by oligonucleotide complexation with metal ions attached to the surface of the micelles. The attachment of the nucleotide

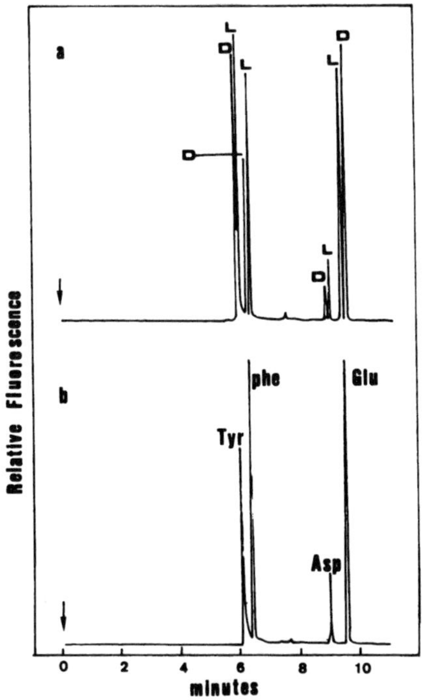

Fig. 5.85. Electropherograms of DL-dansyl amino acids with (a) Cu(II) L-histidine electrolyte at pH 7.0, and (b) 1:1 Cu(II) D- and L-histidine electrolyte at pH 7.0. The concentration of each amino acid is approximately 10^{-4} M. For other conditions refer to the text. The relative standard deviation (R.S.D) for migration times is less than 0.03 and the R.S.D. for relative peak areas is 0.05. (Reproduced from Ref. 139 with permission of the American Association for the Advancement of Science.)

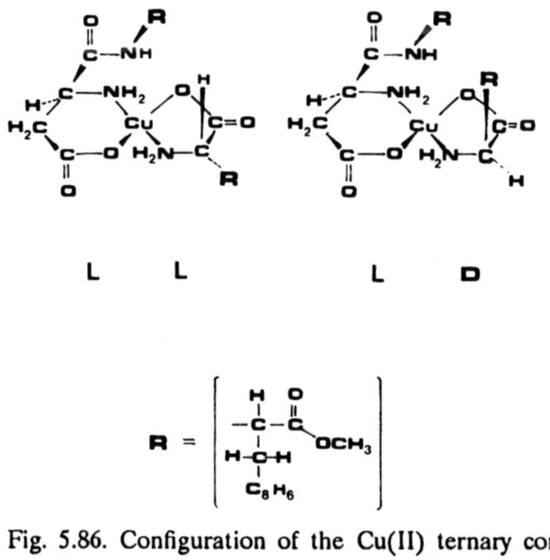

Fig. 5.86. Configuration of the Cu(II) ternary complex formed with the aspartame and an AA. Specific hydrophobic interactions between the aspartame phenylalanine residue (R) and the AA side chains can occur only in the L-D configuration. (Reproduced from Ref. 140 with permission of the American Chemical Society.)

Fig. 5.87. Electropherogram of a mixture of four DNS-DL-AAs. DNS-L-Arg is used as an internal standard. Electrolyte composition is as follows: 2.5 mM $CuSO_4 \cdot 5\,H_2O$, 5.0 mM aspartame and 10 mM NH_4OAc, pH 7.4; capillary, 75μm I.D., 100 cm (75 cm to the detection zone); applied voltage, −30 kV; current, ≅33 μA. (Reproduced from Ref. 140 with permission of the American Chemical Society)

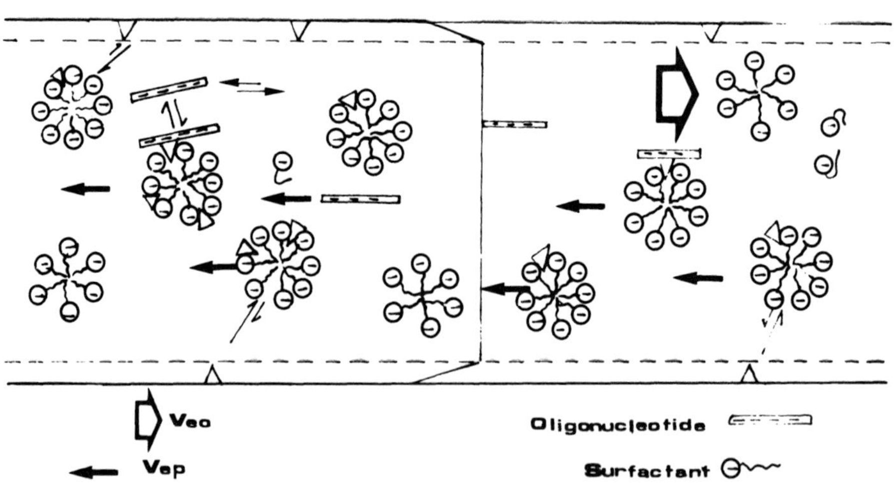

Fig. 5.88. Schematic illustration of retention mechanism with SDS micelles and metal ions. (Reproduced from Ref. 141 with permission of the American Chemical Society.)

to the micelle results in a reduction in the migration velocity of the solute. The time window over which separation is possible is enlarged and hence improved resolution can be obtained. The effects of the addition of 3 mM of different metal ions into the SDS micellar buffer solution on the separation of oligonucleotides, each with eight bases is shown in Fig. 5.89, which shows that baseline separation can be obtained in the case of Zn(II).

Dolnik *et al.* [142] found that the separation of oligonucleotides (polycytidines) was independent of pH or ionic strength but could be influenced with complexation with spermine, with and without SDS.

5.5 OTHER TYPES OF ELECTROPHORETIC MEDIA

Several unusual approaches have been adopted to expand the scope of CE techniques. These include the use of microemulsion [143], supercritical fluid [144] and deuterium oxide [145,146] as the electrolyte system for CE.

5.5.1 Microemulsion capillary electrophoresis (MCE)

Microemulsion capillary electrophoresis (MCE), which employed oil in water microemulsions as the electrophoretic media was demonstrated for the separation of ionic and non-ionic samples by Watarai [143].

Microemulsions are microheterogeneous liquids which have characteristic properties as solvent, such as optical transparency, thermodynamic stability and high solubilization power. The oil in water microemulsion employed by Watarai consisted of water/sodium dodecyl sulfate (SDS)/1-butanol/heptane (89.2%/3.31%/6.61%/0.81% by weight). Negatively charged microdroplets of the organic components are present in the solution. The pH of the solution was controlled by 0.01 M phosphoric acid or 0.01 M carbonate–bicarbonate buffer.

Figure 5.90 shows an example of the electrophoretic separation of some fluorescent aromatic compounds under the acidic (pH = 3.0) conditions. In this case, all of the solutes including neutral and anionic ones migrate to the anodic end. This observation could be explained by the fact that electrophoretic migration of the anionic microemulsion droplets is faster than the electroosmotic flow which is in an opposite direction to that of the droplets. The migration order could be related to hydrophobicity, suggesting that solutes that partition better to the droplets migrate faster. Separation of α- and β-naphthols, which are not resolved in Fig. 5.90, could be achieved by increasing the pH from 3.0 to 6.0, which results in an increase in electroosmotic flow and the resolution is improved, although the migration time is also increased about 2.6 times.

Under the alkaline conditions (pH = 9.5), the electroosmotic mobility of the solution becomes greater than the electrophoretic mobility of the microemulsion. Consequently, solutes migrated towards the cathodic end. In Fig. 5.91, the separation of some ketones and β-diketones in the microemulsion solution is shown. The

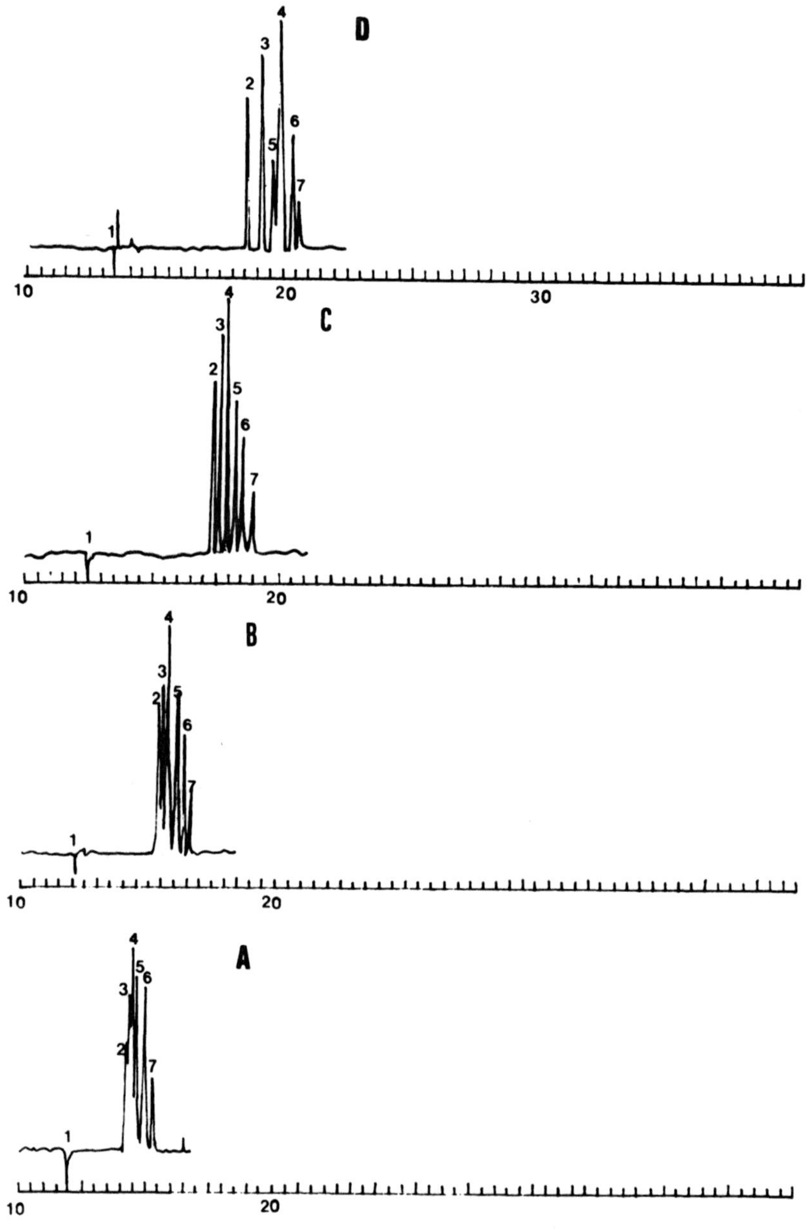

Fig. 5.89. Effect of metal ions on the separation of oligonucleotides, each with eight bases: *1* = water; *2* = CATCGATG; *3* = AACGCGTT; *4* = GGGATCCC; *5* = AAAGCTTT; *6* = CGGGCCCG; *7* = CGCCGGCG. (A) Buffer: 7 M urea, 20 mM Tris, 5 mM Na_2HPO_4, 50 mM SDS. (B) Addition of 3 mM Mg(II) to the buffer. (C) Addition of 3 mM Zn(II) to the buffer. (D) Addition of 3 mM Cu(II) to the buffer. (Reproduced from Ref. 141 with permission of the American Chemical Society.)

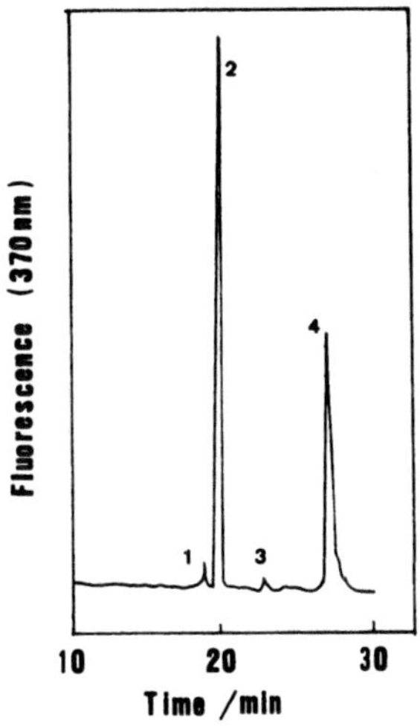

Fig. 5.90. Separation of aromatic compounds detected by the fluorescence of the solute; Excitation at 290 nm, 3.0 kV, 20 μA, 10 s sampling, pH 3.0. *1* = naphthalene: *2* = α- and β-naphthols; *3* = N-acetyl-α-naphthylamine; *4* = 2-naphthol-6-sulfonate. (Reproduced from Ref. 143 with permission of the Chemical Society of Japan.)

migration order is thought to be governed by the hydrophobicity and acidity of the solutes and the electromobility of the anionic dissociation forms.

It was suggested that microemulsion capillary electrophoresis possessed several advantages over micellar electrokinetic chromatography [143]. First, wider range in the migration time between the solvent peak, t_0, and the system peak, t_{me} in MEC and t_{mc} in MEKC could be obtained. The wider elution range could be utilized to achieve better resolution. The higher solubilization power of microemulsions might also be an advantage affording the wider dynamic range in the sample concentration. Despite these claimed advantages, there have been few reports on the use of microemulsion systems in CE to date.

5.5.2 Supercritical capillary electrophoresis (SCE)

Yonker and Smith [144] reported preliminary results on high-pressure and supercritical capillary electrophoresis. The fluid system investigated was methanol (critical temperature (T_c) = 240°C; critical pressure (P_c) = 80.9 bar) containing a small concentration of background electrolyte. The temperature and pressure of the experiments ranged from 21 to 280°C and from 67.1 to 295.7 bar.

There are several advantages and disadvantages in the use of supercritical fluids in CE. One of the advantages is that the use of high pressure permits the use of higher voltages by preventing vapour generation during CE separations. Another advantage is that increased radial mass transfer through the shear layer adjacent

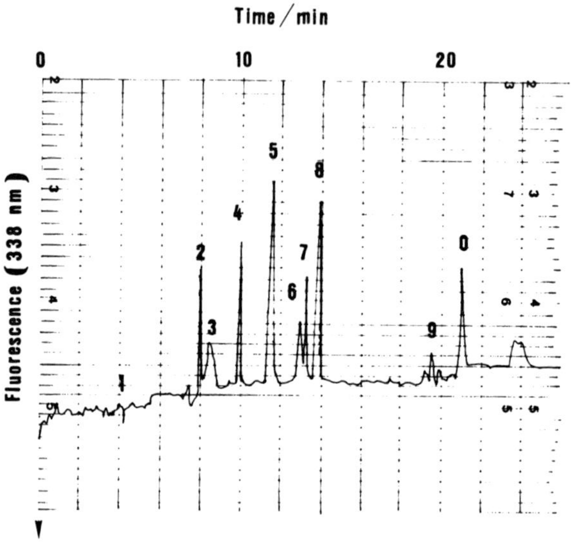

Fig. 5.91. Separation of ketones and β-di-ketones detected by the indirect fluorescence method. Excitation at 279 nm, 10.0 kV, 40 μA, 5 s sampling, pH 9.5. *1* = solvent peak; *2* = 2-acetylthiophene; *3* = acetylacetone; *4* = acetophenone; *5* = benzoylacetone; *6* = pivaloyltrifluoroacetone; *7* = 2-thenoyltrifluoroacetone; *8* = benzoyltrifluoroacetone; *9* = 2-naphthoyltrifluoroacetone; *10* = system peak. (Reproduced from Ref. 143 with permission of the Chemical Society of Japan.)

to the capillary wall could decrease the overall zone broadening in the system [144]. A further advantage is that electrophoretic mobilities increases linearly with molecular diffusion, providing a basis for faster separation. However, in CE where axial diffusion can define the zone broadening limit in some cases, increasing the diffusion coefficient of the solute would not prove beneficial. In the case of mass transfer non-limited situations, zone broadening may be resulted. Another disadvantage is that in order to perform supercritical CE, more sophisticated high-pressure instrumentation would be required.

An example of supercritical CE separation is given in Fig. 5.92, which shows the separation of 1-naphthol-4-sulfonic acid sodium salt and thymol blue.

5.5.3 Deuterium oxide electrolyte systems

Deuterium oxide (D_2O) has properties such as viscosity and ionization which are significantly different to those of water and which can play an important role in CE separations [145,146]. Viscosity of D_2O at 25°C is 1.23 times greater than that of H_2O at the same temperature whereas the ionization constants of H_2O and D_2O are 1.00×10^{-14} and 1.95×10^{-15}, respectively.

Figure 5.93a and b show the CE separation of dansylated amino acids in water and D_2O, respectively. The improvement in resolution is attributable to the lowering

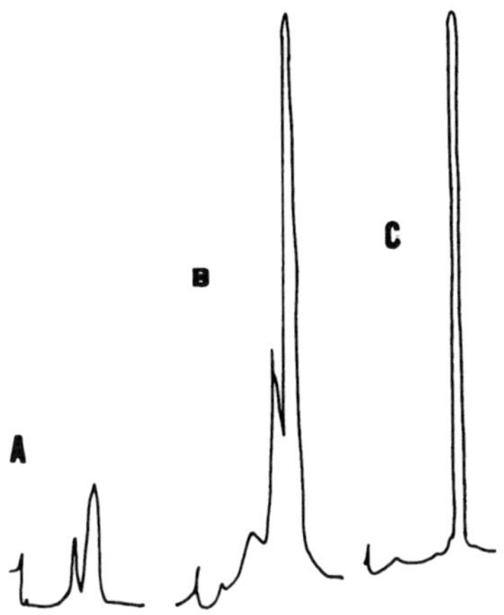

Fig. 5.92. Supercritical fluid electrophoretic separations of 1-naphthol-4-sulfonic acid sodium salt (first) and thymol blue (second peak) at 296 bar and +30 kV. A = 21°C, 42 μA; B = 280°C, 30 μA; C = 280°C, 31 μA. (Reproduced from Ref. 144 with permission of Elsevier Science Publishers.)

of electroosmotic flow, due to a reduction of the zeta potential in agreement with the resolution equation (Eq. 1.12).

Capillary electrophoresis in D_2O-based buffer solution has been shown to give enhanced resolution of a number of nucleosides and dansyl amino acids, compared to electrophoresis carried out in water based buffer of the same pH.

5.6 CONCLUSION

Based on available instrumentation, the most important considerations involved in performing CE separations are the selection of the buffer systems, and the determination of the optimum conditions for their use to achieve high efficiency and selectivity.

The discussions in this chapter provides an insight into the important roles played by the electrophoresis buffer as the basis for different separation chemistries. Through the appropriate choices of the buffer system, the constituent anions and cations, the ionic strength, pH and temperature, the interaction of the buffer with the capillary wall and the analytes can be employed as means to control electroosmotic flow and electrophoretic mobilities, and hence the migration behaviour of the analytes.

Furthermore, by utilizing suitable buffer additives, different separation mechanisms can be introduced to enhance selectivity in CE. The usefulness of this type of techniques is clearly demonstrated by the rapid development and extensive use of micellar electrokinetic chromatography, employing not only anionic but also cationic, non-ionic, zwitterionic and biological surfactants. In addition, many other types of electrokinetic chromatography, with distinctly different separation

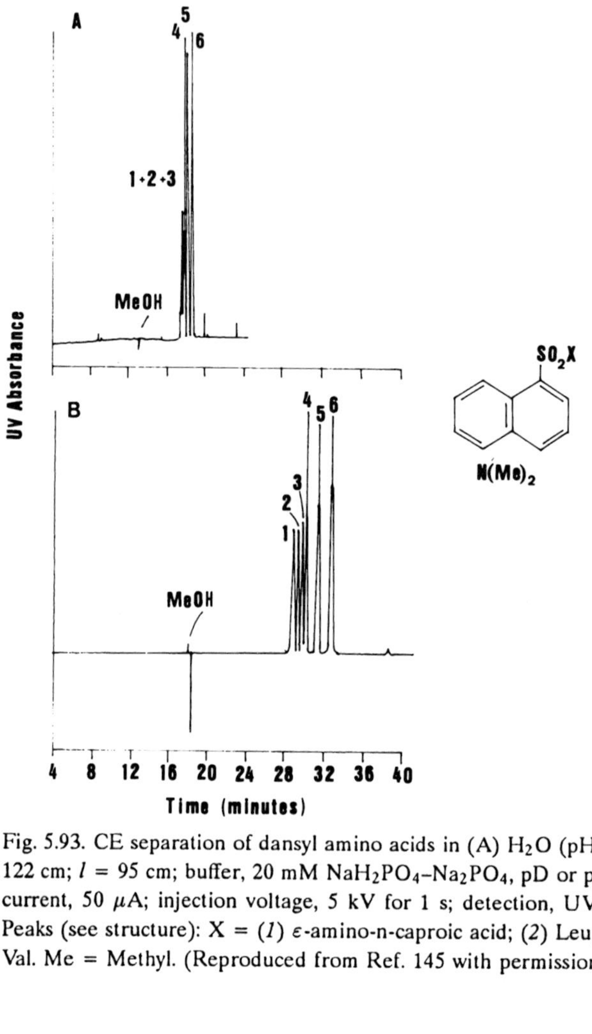

Fig. 5.93. CE separation of dansyl amino acids in (A) H_2O (pH 7.83) and (B) D_2O (pD 7.85), L = 122 cm; l = 95 cm; buffer, 20 mM NaH_2PO_4–Na_2PO_4, pD or pH ≅ 7.8; separation voltage, 40 kV; current, 50 μA; injection voltage, 5 kV for 1 s; detection, UV at 254 nm; temperature, ambient. Peaks (see structure): X = (*1*) ε-amino-n-caproic acid; (*2*) Leu; (*3*) Gly; (*4*) GABA; (*5*) β-Ala; (*6*) Val. Me = Methyl. (Reproduced from Ref. 145 with permission of Elsevier Science Publishers.)

mechanisms from that of micellar electrokinetic chromatography, have also been developed. Examples of these alternative approaches include ion-exchange with polymer ions (Section 5.2.5), inclusion complexation with cyclodextrins and crown ethers (Section 5.3), separation of enantiomers with chiral additives (Sections 5.2.7, 5.2.8, 5.3.2 and 5.4), and complexation with chelating agents for specific interactions (Section 5.4). With appropriate choices of buffer additives, the high efficiency of CE can be extended to the analysis of a wide variety of analytes, including neutral hydrophobic compounds, biopolymers, geometrical isomers and enantiomers.

The relative ease and flexibility in terms of selection, combination and change of the various type of additives in CE provides significant advantages and a wide scope for optimization of separations. Although other separation techniques, such as HPLC, are capable of achieving high selectivity because of the large number of different stationary phases available, in order to employ a different separation mechanism it is usually necessary to change the column and hence prolonged equilibration time in the new mobile phase system may be involved. In the case of CE, different modes of highly efficient separations can be performed by changing the electrophoresis buffer, and hence can be explored rapidly and conveniently, i.e. simply by replacing the buffer solution in the capillary and the reservoirs and equilibrate for a relatively short time.

Although many types of electrolyte systems have been employed as electrophoresis buffers in CE, there should still be immense scope and numerous possibilities for improving the efficiency and selectivity by developing new separation methodologies based on the use of novel additives. In particular, an area which still requires substantial breakthrough is the analysis of compounds that are insoluble in aqueous buffers. Some progress has been made in this respect, by utilizing an oil-in-water microemulsion as the electrophoretic medium (see Section 5.5.1). However, much effort is required to fully exploit the potential of this approach.

As in gradient HPLC, there is also the possibility of improving separation by performing gradient runs in CE, where applied voltage, pH, buffer composition and/or temperature may be varied during the run. In CE, voltage programming can be performed relatively easily (see Section 4.3.6). Attempts on the practical implementations of some of the other techniques are discussed in Section 6.2.

Another aspect concerns the need to optimize the electrophoretic conditions in separations involving multiple parameters, e.g. when several buffer additives are employed. Some examples of the systematic approaches developed for the optimization of CE separations are discussed in Section 6.5.

In view of the progress made so far, and the wide variety of buffers and additives available, it is expected that many innovative separation methodologies based on the manipulation of the electrophoresis buffer will be developed, which will influence enormously the future advance of CE.

5.7 REFERENCES

1 P. Bocek, M. Deml, P. Gebauer and V. Dolnik, Analytical Isotachophoresis, VCH, Weinheim (Germany), (1988) p.90

2 R.J. Wieme, in Chromatography: A Laboratory Handbook of CHromatographic and Electrophoretic Methods, 3rd ed. E. Heftmann, Ed, Van Nostrand Reinhold, New York, (1975) 267

3 E. Huckel, Physik. Z. 25 (1924) 204

4 T. Tsuda, K. Nomura and G. Nakagawa, J. Chromatogr., 248 (1982) 241

5 H. Issaq, I. Atamna, G. Muschik and G. Janini, Chromatographia, 32 (1991) 155

6 E. Kenndler and C. Schwer, Anal. Chem., 63 (1991) 2499

7 F.E. Mikkers, F.M. Everaerts and Th.P.E.M. Verheggen, J. Chromatogr., 169 (1979) 1

8 F.E. Mikkers, F.M. Everaerts and Th.P.E.M. Verheggen, J. Chromatogr., 169 (1979) 11

9 F.E. Mikkers, F.M. Everaerts and J.A.F. Peek, J. Chromatogr., 168 (1979) 293

10 B.B. VanOrman, G.G. Liversidge and G.L. McIntire, J. Microcol. Sep., 2 (1991) 176

11 W.J. Ferguson and N.E. Good, Anal. Biochem., 104 (1980) 300
12 I. Atamna, C.J. Ietral, G.M. Muschik and H.J. Issaq, J. Liq. Chromatogr., 13 (1990) 3201
13 J.S. Green and J.W. Jorgenson, J. Chromatogr., 478 (1989) 63
14 A.D. Tran, S. Park, P.J. Lisi, O.T. Huynh, R.R. Ryall and P.A. Lane, J. Chromatogr., 542 (1991) 459
15 M.W.F. Nielen, J. Chromatogr., 542 (1991) 173
16 K.D. Altria and C.F. Simpson, Chromatographia, 24 (1987) 527
17 K.D. Altria and C.F. Simpson, Anal. Proc., 25 (1988) 85
18 G.J.M. Bruin, J. Chang, R. Kuhlman, K. Zegers, J. Kraak and H. Poppe, J. Chromatogr., 471 (1989) 429
19 W. Nashabeh and Z. El Rassi, J. Chromatogr., 514 (1990) 57
20 M.J. Sepaniak, D.F. Swaile, A.C. Powell and R. Cole, J. High Resolut. Chromatogr., 13 (1990) 679
21 H.T. Rasmussen and H.M. McNair, J. Chromatogr., 516 (1990) 223
22 I. Atamna, H. Issaq, M. Muschik and G. Janini, J. Chromatogr., 588 (1991) 315
23 K.D. Lukacs and J.W. Jorgenson, J. High Resolut. Chromatogr. Chromatogr. Commun., 8 (1985) 407
24 S. Terabe, T. Yashima, N. Tanaka and M. Araki, Anal. Chem., 60 (1988) 1673
25 J. Vindevogel and P. Sandra, J. Chromatogr., 541 (1991) 483
26 A. Vinther and H. Soeberg, J. Chromatogr., 589 (1992) 315
27 C. Schwer and E. Kenndler, Anal. Chem., 63 (1991) 1801
28 J.C. Reijenga, G.V.A. Aben, Th.P.E.M. Verheggen and F.M. Everaerts, J. Chromatogr., 260 (1983) 241
29 F.S. Stover, B.L. Haymore and R.J. McBeath, J. Chromatogr., 470 (1989) 241
30 H.H. Lauer and D. McManigill, Anal. Chem., 58 (1986) 166
31 S. Nathakarnkitkool, P.J. Oefner, A.M. Chin, G. Bonn and G. Bartsch, HPCE '91, San Diego, USA, Feb. 3–7, (1991)
32 H.P. Hopkins and W.D. Wilson, Biopolym., 29 (1990) 449
33 H. Knox, Chromatographia, 26 (1989) 329
34 R.J. Nelson, A. Paulus, A.S. Cohen, A. Guttman and B.L. Karger, J. Chromatogr., 480 (1989) 111
35 S. Hjerten, Chromatogr. Rev., 9 (1967) 122
36 R.S. Rush, A.S. Cohen and B.L. Karger, Anal. Chem., 63 (1991) 1346
37 P. Oroszian, R. Blanko, X.-M. Lu and B.L. Karger, J. Chromatogr., 500 (1990) 481
38 M.J. Kronman. Critical Reviews in Biochemistry & Molecular Biology, CRC Press, Boca Raton, Fl. USA, 24 (1989) 566
39 S. Terabe, K. Otsuka, K. Ichikawa, A. Tsuchiya, T. Ando, Anal. Chem., 56 (1984) 111
40 S. Terabe, K. Otsuka and T. Ando, Anal. Chem., 57 (1985) 834
41 M. Sepaniak and R. Cole, Anal. Chem., 59 (1987) 472
42 S. Terabe, K. Otsuka and T. Ando, Anal. Chem., 61 (1989) 251
43 M.J. Sepaniak, D.E. Burton and M.P. Maskarince, in Ordered Media in Chemical Separations, W.L. Hinze and D.W. Armstrong (Eds), ACS Symp. Ser. 342, Amer. Chem. Soc., Washington, DC, (1987) 142
44 R.A. Wallingford and A.G. Ewing, Anal. Chem., 60 (1988) 258
45 A.S. Cohen, A. Paulus and B.L. Karger, Chromatographia, 24 (1987) 14
46 K. Otsuka, S. Terabe and T. Ando, J. Chromatogr., 332 (1985) 219
47 H. Nishi, T. Fukuyama, M. Matsuo and S. Terabe, J. Chromatogr., 513 (1990) 279

48 S. Terabe, O. Shibata and T. Isemura, J. High Resolut. Chromatogr., 14 (1991) 52
49 H.T. Rasmussen, L.K. Goebel and H.M. McNair, J. High Resolut. Chromatogr., 14 (1991) 25
50 S. Fujiwara and S. Honda, Anal. Chem., 59, (1987) 2773
51 W. Thormann, P. Meher, C. Marcolli and F. Binder, J. Chromatogr., 545 (1991) 445
52 X. Huang, J.A. Luckey, M.J. Gordon and R.N. Zare, Anal. Chem., 61 (1989) 766
53 K. Saitoh, C. Kiyohara and N. Suzuki, J. High Resolut. Chromatogr., 14 (1991) 245
54 S.A. Swedberg, J. Chromatogr., 503 (1990) 449
55 C.P. Ong, C.L. Ng, N.C. Chong, H.K. Lee and S.F.Y. Li, J. Chromatogr., 516 (1990) 263
56 A.T. Balchunas and M.J. Sepaniak, Anal. Chem., 59 (1987) 1466
57 S. Terabe, H. Utsumi, K. Otsuka, T. Ando, T. Inomata, S. Kuze and Y. Hanaoka, J. High Resolut. Chromatogr. Chromatogr. Commun. 9 (1986) 666
58 K. Otsuka and S Terabe, J. Microcol. Sep., 1 (1989) 150
59 M.M. Bushey, J.W. Jorgenson, J. Microcol. Sep., 1 (1989) 125
60 H. Nishi, N. Tsumagari and S. Terabe, Anal. Chem., 61 (1989) 2434
61 S. Terabe, Y. Ishihama, H. Nishi, T. Fukuyama and K. Otsuka, J. Chromatogr., 545 (1991) 359
62 C.P. Ong, H.K. Lee and S.F.Y. Li, J. Chromatogr., 542 (1991) 473
63 C.P. Ong, C.L. Ng, H.K. Lee and S.F.Y. Li, J. Chromatogr., 547 (1991) 419
64 C.P. Ong, S.F. Pang, S.P. Low, H.K. Lee and S.F.Y. Li, J. Chromatogr., 559 (1991) 529
65 T. Tsuda, J. High Resolut. Chromatogr. Chromatogr. Commun., 10 (1987) 622
66 S. Terabe, K. Ishikawa, K. Otsuka, A. Tsuchiya and T. Ando, 26th Int. Liq. Chromatogr. Symp., Kyoto, Japan, Jan 25–26, (1983)
67 S. Terabe and T Isemura, Anal. Chem., 62 (1990) 652
68 S. Terabe and T Isemura, J. Chromatogr., 515 (1990) 667
69 S. Terabe, M. Shibata and Y. Miyashita, J. Chromatogr., 480 (1989) 403
70 H. Nishi, T. Fukuyama, M. Matsuo and S. Terabe, J. Chromatogr., 515 (1990) 233
71 R.O. Cole, M.J. Sepaniak and W.L. Hinze, J. High Resolut. Chromatogr., 13 (1990) 579
72 S. Terabe, H. Nishi, T. Fukuyama and M. Matsuo, J. Microcol. Sep., 1 (1989) 234
73 R.O. Cole, M.J. Sepaniak, W.L. Hinze, J. Gorse and K. Oldiges, J. Chromatogr., 557 (1991) 113
74 T. Balchunas, D.F. Swaile, A.C. Powell and M.J. Sepaniak, Sep. Sci. Technol., 23 (1988) 1891
75 D. Burton, M. Sepaniak and M. Maskarinec, J. Chromatogr. Sci., 25 (1987) 514
76 H. Nishi, N. Tsumagari, T. Kakimoto and S. Terabe, J. Chormatogr., 465 (1989) 331
77 H. Nishi, N. Tsumagari, T. Kakimoto and S. Terabe, J. Chromatogr., 477 (1989) 259
78 W. Griest, M. Maskarinec and K. Row, Sep. Sci. Technol., 23 (1988) 1905
79 S. Fujiwara, S. Iwase and S. Honda, J. Chromatogr., 447 (1988) 133
80 K. Otsuka, S. Terabe and T. Ando, J. Chromatogr., 348 (1985) 39
81 K. Otsuka, S. Terabe and T. Ando, Nippon Kagaku Kaishi, 7 (1986) 950
82 T. Nakagawa, Y. Oda, A. Shibukawa, H. Fukuda and H. Tanaka, Chem. Pharm. Bull., 36 (1988) 1622
83 T. Nakagawa, Y. Oda, A. Shibukawa, H. Fukuda and H. Tanaka, Chem. Pharm. Bull., 37 (1989) 707
84 M.M. Gladdines, J.C. Reijenga, R.G. Trieling, M.J.S. Van Thiel and F.M. Everaerts, J. Chromatogr., 470 (1989) 105
85 K. Otsuka, S. Terabe and T. Ando, J. Chromatogr., 396 (1987) 350
86 S. Terabe, H. Ozaki, K. Otsuka and T. Ando, J. Chromatogr., 332 (1985) 211
87 Y.F. Yik, H.K. Lee and S.F.Y. Li, J. Chromatogr., 585 (1991) 139

88 C.P. Ong, C.L. Ng, H.K. Lee and S.F.Y. Li, J. Chromatogr., 588 (1991) 335
89 J.R. Boyak and J.C. Giddings, Biol. Chem., 235 (1960) 1970
90 L. Emmer, M. Jansson and J. Roeraade, J. Chromatogr., 547 (1991) 544
91 A. Dobashi, T. Ono, S. Hara and J. Yamaguchi, Anal. Chem., 61 (1988) 1984
92 A. Dobashi, T. Ono, S. Hara and J. Yamaguchi, J. Chromatogr., 480 (1989) 413
93 K. Otsuka, J. Kawahara, K. Tatekawa and S. Terabe, J. Chromatogr., 559 (1991) 209
94 K. Otsuka and S. Terabe, J. Chromatogr., 515 (1990) 221
95 K.L. Mittal (Ed.), Micellization, Solubilization and Microemulsions, Vol 1, Plenum, New York, USA, (1977)
96 P Mukerjee and A. Ray, J. Phys. Chem., 67 (1963) 190
97 M.J. Schick, J. Phys. Chem., 68 (1964) 3585
98 H.T. Rasmussen, L.K. Goebel and H.M. McNair, J. Chromatogr., 517 (1990) 549
99 C.P. Ong, C.L. Ng, N.C. Chong, H.K. Lee and S.F.Y. Li, Environ. Monitor. Assess., 19 (1991) 93
100 S.K. Yeo, C.P. Ong and S.F.Y. Li, Anal. Chem., 63 (1991) 2222
101 D. Sybilska and E. Smolkova-Keulemansova, in J.L. Atwood, J.E.D. Davies and D.D. Mac Nicol (Eds), Inclusion Compounds, Vol. 3, Academic Press, London, UK, (1984) p.173
102 E. Smolkova-Keulemansova, in D. Duchene (Ed), Cyclodextrins and Their Industrial Use, de Sante, Paris, (1987)
103 M. Tazaki, M. Takagi and K. Ueno, Chem. Lett., 5 (1982) 639
104 M. Tazaki, T. Hayashita, Y. Fujino and M. Takagi, Bull. Chem. Soc. Jpn., 59 (1986) 3459
105 N. Kuramoto, Chem. Express, 1 (1986) 343
106 N. Kuramato and K. Asao, Osaka-Furitsu Kogyo Gijutsu Kenkyusho Hokoku, 88 (1986) 18
107 I. Jelinek, J. Snopek and E. Smolkova-Keulemansova, J. Chromatogr., 405 (1987) 379
108 I. Jelinek, J. Snopek and E. Smolkova-Keulemansova, J. Chromatogr., 411 (1987) 153
109 I. Jelinek, J. Dohnal, J. Snopek and E. Smolkova-Keulemansova, J. Chromatogr., 435 (1988) 496
110 J. Snopek, I. Jelinek, and E. Smolkova-Keulemansova, J. Chromatogr., 452 (1988) 571
111 D.W. Armstrong, Sep. Purif. Methods, 14 (1985) 213
112 E. Smolkova-Keulemansova, J. Chromatogr., 251 (1982) 17
113 S. Krysl and E. Smolkova-Keulemansova, Chem. Listy, 79 (1985) 919
114 J. Szejtli, Cyclodextrins and their Inclusion Complexes, Akademiai Kiados, Budapest, (1982).
115 S. Fanali, J. Chromatogr., 474 (1989) 441
116 S. Fanali, J. Chromatogr., 545 (1991) 437
117 S. Fanali, J. Chromatogr., 470 (1989) 123
118 J. Snopek, H. Soini, M. Novotny, E. Smolkova-Keulemansova and I. Jelinek, J. Chromatogr., 559 (1991) 215
119 J. Liu, K.A. Cobb and M. Novotny, J. Chromatogr., 519 (1990) 189
120 S. Terabe, Y. Miyashita, O. Shibata, E.R. Barnhart, L.R. Alexander, D. Patterson, B.L. Karger, K. Hosoya and N. Tanaka, J. Chromatogr., 516 (1990) 23
121 H. Nishi, T. Fukuyama and S. Terabe, J. Chromatogr., 553 (1991) 503
122 S. Terabe, Trends. Anal. Chem., 8 (1989) 129
123 H. Nishi, T. Fukuyama and S. Terabe, unpublished data
124 F.S. Stover, J. Chromatogr., 298 (1984) 203
125 F.S. Stover, J. Chromatogr., 368 (1986) 476
126 F.M. Everaerts and J.W. Venema, Isotachoforese als mogelijke analyse methode voor "Glasgemengen", Report, Technical University, Eindhoven, (1987)

127 P. Gebauer, M. Deml, P. Bocek and J. Janak, J. Chromatogr., 267 (1983) 455
128 T. Saitoh, H. Hoshino and T. Yotsuyanagi, J. Chromatogr., 469 (1989) 175
129 D.F. Swaile and M.J. Sepaniak, Anal. Chem., 63 (1991) 179
130 Y. Walbroehl and J.W. Jorgenson, Anal. Chem., 58 (1986) 479
131 R.A. Wallingford and A.G. Ewing, J. Chromatogr., 441 (1988) 299
132 R.A. Wallingford and A.G. Ewing, J. Chromatogr., 61 (1989) 98
133 R.A. Wallingford, P.D. Curry, Jr., and A.G. Ewing, J. Microcol. Sep, 1 (1989) 23
134 S. Tanaka, T. Kaneta and H. Yoshida, Anal. Sci., 6 (1990) 467
135 T. Kaneta, S. Tanaka and H. Yoshida, J. Chromatogr., 538 (1991) 385
136 C.P. Ong, S.F. Pang, S.P. Low, H.K. Lee and S. F Y. Li, J. Chromatogr., 559 (1991) 529
137 S. Honda, S. Iwase, A. Makino and S Fujiwara, Anal. Biochem., 176 (1989) 72
138 S. Hoffstetter-Kuhn, A. Paulus, E. Gassmann and H.M. Widmer, Anal. Chem., 63 (1991) 1541
139 E. Gassmann, J. Kuo and R. Zare, Science, 230 (1985) 813
140 P. Gozel, E. Gassmann, H. Michelsen and R. Zare, Anal. Chem., 59 (1987) 44
141 A. Cohen, S. Terabe, J. Smith and B. Karger, Anal. Chem., 59 (1987) 1021
142 V. Dolnik, J. Liu, J. Banks, M. Novotny and P. Bocek, J. Chromatogr., 480 (1989) 321
143 H. Watarai, Chem. Lett., (1991) 391
144 C.R. Yonker and R.D. Smith, J. Chromatogr., 517 (1990) 573
145 P. Camilleri and G. Okafo, J. Chromatogr., 541 (1991) 489
146 P. Camilleri and G. Okafo, J. Chromatogr., 547 (1991) 551

CHAPTER 6

Special Systems and Methods

6.1 INTRODUCTION

This chapter is mainly devoted to a discussion of some of the features and techniques employed in CE separations which deviate from the basic instrumental set up given in Section 1.2.

First of all, techniques for varying the composition of the buffer during the separation are described [1–5]. Secondly, strategies employed for fraction collection are examined [6–9]. Thirdly, the potential of field effect electroosmosis for achieving direct control of the zeta potential of the capillary/solution interface is discussed [10–21]. On one hand, these techniques are worth considering because they are promising approaches to extend the capability, and to widen the scope of application of CE. In particular, buffer programming may be useful for varying the selectivity during the separation. Voltage programming has been exploited as a means to facilitate fraction collection. The use of an external field to control the zeta potential provides a flexible method to manipulate electroosmotic flow. On the other hand, the implementation of these techniques in CE separations is hampered by their tendency to introduce additional peak broadening, either due to extra connections required, or disturbance to the flow profile. Considering that further improvements in the design of the systems used and the theory of the methods may help to realize the full potential of these approaches, a review of current status of their development is given to provide a basis and some directions for further research efforts.

The next two topics discussed in this chapter are concerned with selected computational and theoretical aspects of CE. A brief discussion on systematic approaches to the optimization of CE separations is included [22–39]. It is anticipated that as improved versions of commercial instruments are being developed, computer assisted optimization systems may become part of the features of a CE instrument. Therefore, recent advances in this area are described.

Another topic covered is the use of CE for the determination of electrophoretic mobilities and diffusion coefficients. Accurate data on relevant physico-chemical properties are useful for the interpretation of migration behaviour and for the checking of the validity of optimization and modelling schemes. CE has been exploited as a method for the determination of some of these properties with remarkable success [40–42].

Isoelectric focusing and isotachophoresis are two widely used techniques in electrophoresis [43–100]. Many excellent texts and reviews have been devoted to these subjects [36,69–72]. Capillary isoelectric focusing (CIEF) and capillary isotachophoresis (CITP) have been classified as two distinct modes of capillary electrophoresis in Section 1.2. The inclusion of these topics in the present chapter as special topics reflects the fact that they have not yet been used in the capillary format extensively. CIEF [43–55] is a relatively new mode of CE. CITP has been performed mainly on instruments similar to those used for conventional isotachophoresis, except that tubings with smaller I.D. are used [69,70,86–100]. A significant development in recent years is that CIEF and CITP have been performed in the presence of electroosmotic flow [50–52,78,83,93–96], which may provide an interesting scope for further exploitation of these techniques.

The final subject discussed in this chapter concerns multidimensional or hyphenated techniques involving CE. The coupling of CE with other separation methods and the coupling of different modes of CE permit different mechanisms to be exploited in a single system [101–120]. Thus potentially higher resolving power can be achieved, although generally more sophisticated instrumentation and interfacing techniques need to be employed to ensure the optimum performance of both systems, and to prevent loss of efficiency.

6.2 BUFFER PROGRAMMING

As in gradient HPLC, programming of the buffer composition in a capillary electrophoresis system can be used as a means to obtain optimum separation conditions. For analyses involving complex mixtures, complete separation of all substances may be difficult to obtain or the analysis time may be too long when the separation is carried out with the buffer composition maintained constant during the run. By programming the buffer pH and/or composition, new operational conditions for the separation can be utilized to affect the migration behaviour of analytes and hence to improve separation.

However, it is important to ensure that the gain in selectivity by performing dynamic variations on the electrophoretic buffer is not obtained at an appreciable loss of efficiency. Due to changes in the electrophoresis buffer, the zeta potential may become a function of distance along the capillary, resulting in local fluctuationsin electroosmotic flow and mobilities. Additional peak broadening can be caused by the disturbance to the plug profile of electroosmotic flow, local variations of migration velocities, as well as the extra connections and manipulations required to produce gradients in the operating conditions. Therefore, the advantageous and successful implementation of buffer programming techniques involve considerations of many factors, such as the optimum design of the experimental setup, the alternative approaches available (e.g. the use of modifiers), and the minimization of loss of efficiency.

The use of buffer programming has been demonstrated for both micellar electrokinetic chromatography, where the composition of the organic modifier used was changed during the run [1], and in capillary zone electrophoresis, where pH gradient [2], step change of counter-ion [3], pulses of counter-ion [4] and a dynamic pH step have been demonstrated [5].

6.2.1 Gradient eluent MEKC

A method for performing gradient elution in micellar electrokinetic capillary chromatography has been described by Balchunas and Sepaniak [1]. A stepwise solvent gradient involving increasing concentrations of 2-propanol and Triton X-100 was used for the separation of a test mixture of derivatized amines.

In this system, the column is first placed in a thermostated water jacket and then filled with the starting mobile phase. The injection end of the column is placed into a 5-ml beaker, which is positioned on a magnetic stirring plate. This inlet reservoir is filled with exactly 2.5 ml of the starting mobile phase. Stepwise gradients are produced by pipetting aliquots of a gradient solvent containing 2-propanol into the 5-ml beaker. A small magnetic stirring bar is used to ensure thorough mixing of the added gradient solvent with the starting mobile phase. Upon injection of sample, the separation voltage is switched on and electrophoresis is initiated. Gradient elution solvent is then manually added in four 0.5 ml increments, performed at 5 min intervals. Prior to solvent addition, it is important to ensure that the power supply is switched off to prevent injury due to accidental contact. The voltage is switched on again after solvent addition.

Figures 6.1 and 6.2 illustrate the chromatograms obtained by using isocratic and gradient elution [1]. Figure 6.1a is the isocratic separation of an 11-component test mixture obtained using an electrophoretic medium containing SDS and the buffer. The applied voltage is 15 kV. The components near the end of the chromatogram in Fig. 6.1a are not resolved satisfactorily. Figure 6.1b is the isocratic separation of the test mixture at the same voltage with 22.5% (v/v) of 2-propanol. The 2-propanol concentration is considered too high as some of the early eluting compounds coeluted. The migration order is found to be different from that in Fig. 6.1a. In Fig. 6.2, results of the gradient elution run is shown. The starting mobile phase consists of exactly 1.5 ml of 0.05 mM SDS, 0.01 M Na_2HPO_4, 0.005 M $Na_2B_4O_7$, and 10% (v/v) 2-propanol. A gradient solution, consisting of identical SDS and buffer salt concentrations, and 50% (v/v) 2-propanol and 2.5% (v/v) Triton X-100 is added in four 0.5-ml increments. Starting voltage for this chromatogram is 15 kV (32 μA) and is adjusted as appropriate to maintain constant current, up to 27 kV. Satisfactory separation could be obtained using this gradient procedure, as a result of an extended migration range and a change in selectivity [1]. However, it

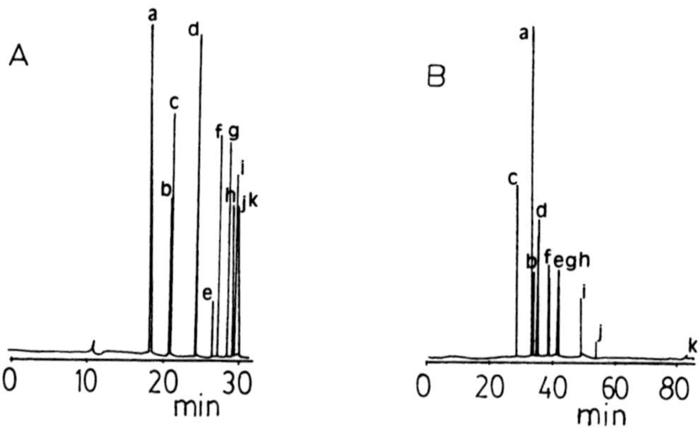

Fig. 6.1. (A) Separation of test mixture by using a 0.05 mm I.D. × 850 mm long column with 0.05 M SDS. 0.01 M Na_2HPO_4–0.005 M $Na_2B_4O_7$ (pH 7) mobile phase. (B) Separation of test mixture with 22.5% 2-propanol in the mobile phase. Peaks: a = NBD methylamine, b = NBD ethylamine, c = NBD dimethylamine, d = NBD n-propylamine, e = NBD diethylaine, f = NBD n-butylamine, g = NBD cyclohexylamine, h = NBD di-n-propylamine, i = NBD n-hexylamine, and j, k = Coumarin 153 and Coumarin 343. (Reproduced from Ref. 1 with permission of the American Chemical Society.)

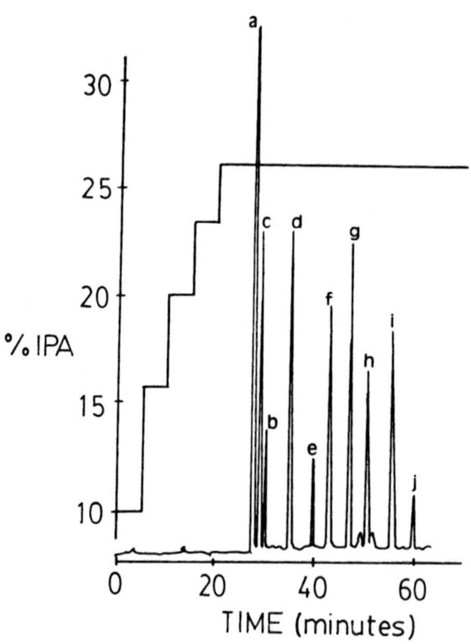

Fig. 6.2. Separation of the test mixture of Fig 6.1 using a stepwise solvent program: Starting mobile phase 1.5 ml of 0.05 M SDS, 0.01 M Na_2HPO_4, 0.005 M $Na_2B_4O_7$, and 102-propanol. A gradient solution identical SDS and buffer salt concentrations and 50% (v/v) 2-propanol and 25% (v/v) Triton-X-100: starting voltage = 15 kV adjusted to maintain constant current up to 27 kV. (Reproduced from Ref. 1 with permission of the American Chemical Society.)

should be noted that the gradient profile shown in the figure represents that of the column inlet reservoir and not the actual profile within the column. Furthermore this gradient dilution procedure involves many manual operations. Automated instrumentation needs to be developed for the technique to be of general use.

6.2.2 pH gradient

Capillary zone electrophoresis in a mobile pH gradient which is dynamically programmed has been performed by Bocek *et al.* [2]. A schematic diagram of the apparatus used is shown in Fig. 6.3. Typically 0.01 M KCl is used as the primary electrolyte. The modifying electrolyte is 0.01 M HCl. The rate of increase of the pH gradient is controlled electrically by setting up the ratio of the two electric currents, I_2/I_1 (see Fig. 6.3) while keeping the total electric current ($I = I_1 + I_2$) constant (200 μA). In Fig. 6.4, the analysis of a model sample containing pyridine and *p*-bromoaniline by isocratic CZE is shown. It can be seen that although the separation of the two substances could be achieved, the analysis time is fairly long and there is considerable broadening of the *p*-bromoaniline peak. Figure 6.5 shows the result of an analysis using a pH gradient [2]. The starting conditions are identical with those in the CZE experiment (background or primary electrolyte: 0.01 M KCl, pH 4.25); the dynamic gradient is then programmed by a stepwise increase in the modifying current (4 μA/min). It can be seen that the analysis time for both substances has decreased and that their peaks have become sharper.

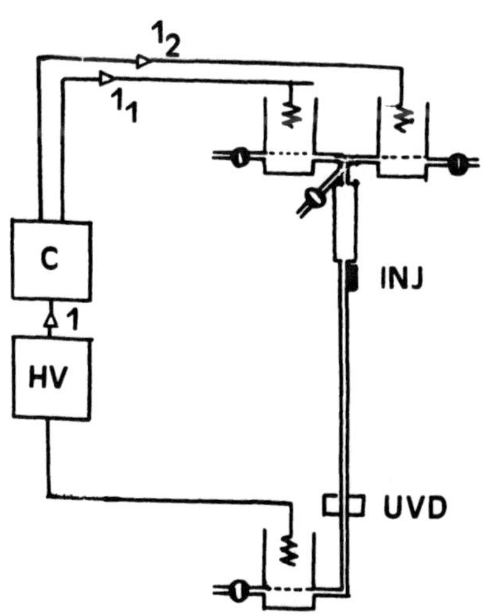

Fig. 6.3. Schematic diagram of the apparatus used for CZE with pH gradient. *C* = electric current ratio controller; *HV* = high-voltage power supply; *INJ* = injection point; *UVD* = UV absorption detector. (Reproduced from Ref. 2 with permission of Elsevier Science Publishers.)

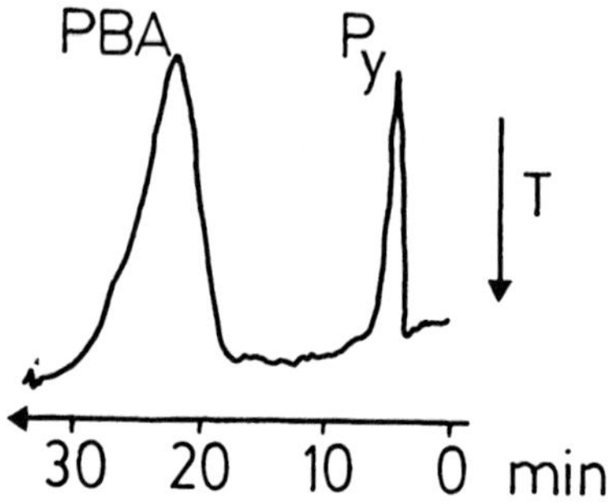

Fig. 6.4. UV detection record of an analysis in the zone electrophoresis mode. Background electrolyte, 0.01 MKCl (pH 4.25); I = 200 μA; *Py* = pyridine; *PBA* = p-bromoaniline; T = transmittance. (Reproduced from Ref. 2 with permission of Elsevier Science Publishers.)

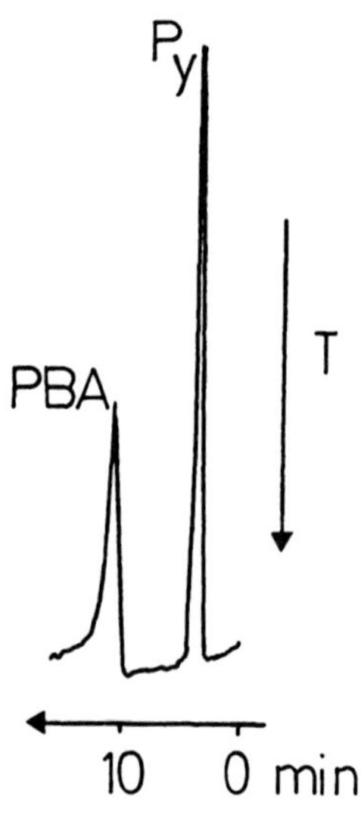

Fig. 6.5. UV detection record of an analysis with a pH gradient. Primary electrolyte, 0.01 M KCl (pH 4.25); modifying electrolyte 0.01 M HCl; I = 200 μA; I_2 was increased and I_3 was decreased at 4 μA/min. Abbreviations as in Fig. 6.4. (Reproduced from Ref. 2 with permission of Elsevier Science Publishers.)

6.2.3 Step change of co-ions

Another method has been developed based on a step change of ionic matrix (IM) composition during the migration of the sample components [3]. The method proposed could be used to expand the scale of p*K*'s of analyzed compounds in a single run and provided additional flexibility for optimizing the separation. A moving, sharp, stepwise change in composition of the ionic matrix is generated by a simple change of the counter-ionic system of the primary electrolyte, made possible by switching the driving current from one chamber (PRIM) to the other (MOD).

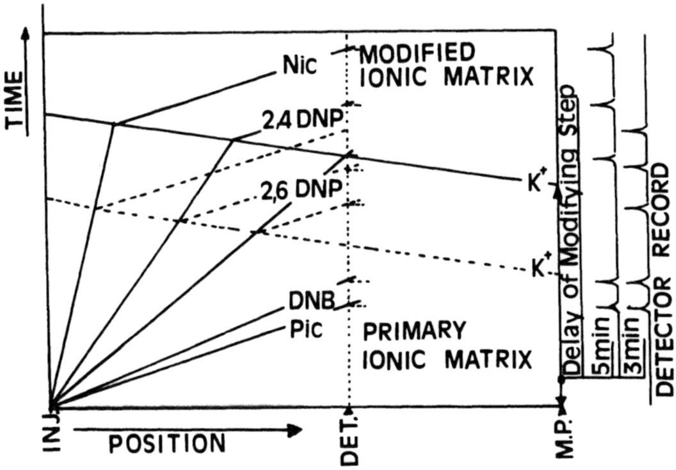

Fig. 6.6. Principle of the method based on step change of the ionic matrix. Trajectories of the sample components in the plane *x-t*. *INJ* = injection; *DET* = detection point; *M.P.* = mixing point. On the right, detection records after a 3- and 5-min delay in starting modification, respectively. Abbreviations: *Nic* = nicotinic acid; *2,4 DNP* = 2,4-dinitrophenol; *2,6 DNP* = 2,6-dinitrophenol; *DNB* = 3,5-dinitrobenzoic acid; *Pic* = picric acid. (Reproduced from Ref. 3 with permission of VCH Publishers, Inc.)

The principle of the method is depicted in Fig. 6.6 where the trajectories of the substances to be separated are shown both in the primary IM, and in the modified IM. The separation proceeds only in the primary IM, whereas in the modified IM all components are fully dissociated and move with velocities proportional to their ionic mobilities. The final separation obtained is not only due to the differences in the effective mobilities of the sample components in the primary IM but also due to the fact that the faster a component is, the sooner it enters the modified IM, being further accelerated [3]. The separations of a model mixture of five acids by isocratic CZE and by CZE with an IM step are shown in Fig. 6.7. In the isocratic CZE cases (Fig. 6.7a–c), satisfactory separation of all components of this mixture in a single operational electrolyte is not observed. Improved resolution is obtained from the stepwise change of the counter-ions, with a 5 min time delay of the step (Fig. 6.8e).

6.2.4 Pulse of counter-ion

An interesting approach is the use of a transient ionic matrix for CZE [4]. Figure 6.8 shows the process where the transient ionic matrix moves through the column against the direction of the sample components. The sample components, proceeding from the sampling point, change their velocities when entering the zone of the transient matrix (notably component B) and are separated. The separated zones continue to migrate with their original velocities towards the detector and

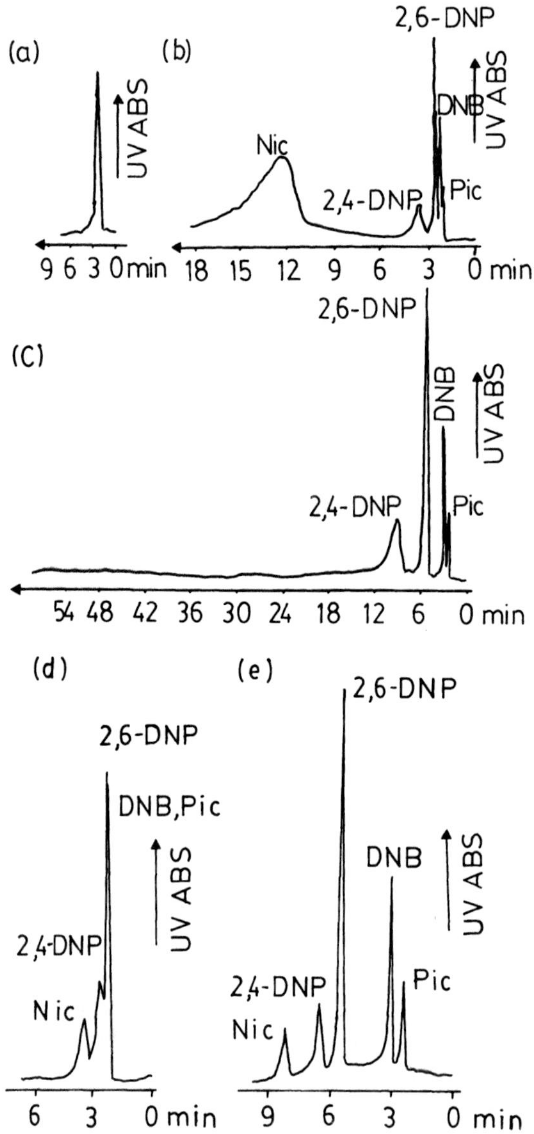

Fig. 6.7. CZE of the model mixture as in Fig. 6.6 at constant pH (a–c) and with pH programming (d and e): (a) 6.5, (b) 3.5, (c) 4.0, and (d) 3-min delay in starting modification; (e) 5-min delay in starting modification. (Reproduced from Ref. 3 with permission of VCH Publishers, Inc.)

are recorded. This scheme is represented, e.g. by the practical case where a pulse of a complexing counter-ion at constant pH is used for the separation of substances. If a pulse of a solvolytic counter-ion, e.g. H^+ or OH^-, is used for the separation of substances, the situation is more complicated and it is necessary to consider not only the effect of pH on the degree of dissociation (and thus mobility) but also the

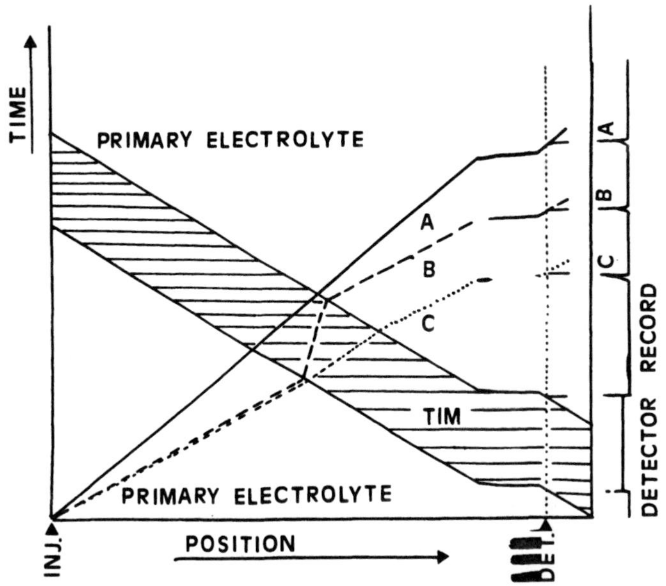

Fig. 6.8. Separation scheme in the transient ionic matrix. Substances A, B, C migrate in the column from injection point *INJ* with constant velocity towards the transient ionic matrix (*TIM*). In the primary electrolyte, substance A with the lowest mobility is separated from B and C. In the transient ionic matrix substances B and C differ in mobilities and are separated: A is not affected. After leaving the transient ionic matrix, the separated substances migrate with their own mobilities in the primary electrolyte and are detected in the detector (*DET*). The detection record is on the right. (Reproduced from Ref. 4 with permission of Elsevier Science Publishers.)

conductivity changes within the pulse (caused by a higher mobility of these ions) and thus the changes in the driving electric field strength [4]. Figure 6.9a shows the dissociation and velocity curves of model substances together with the conductivity of the model transient ionic matrix (system HCl–KCl), as a function of the pH of the transient ionic matrix. In Fig. 6.9b, the trajectories of the sample components which migrate from the sampling point as a function of the pH at the given point in the column at a given time are shown. The trajectories in the transient ionic matrix show increasing distance and thus separation.

The application of the solvolytic transient ionic matrix has been demonstrated by the separation of a model mixture of acid, viz, picric (PIC), 3,5-dinitrobenzoic (DNB), sorbic (SOR) and cinnamic (CIN) acids and 2,4-dinitrophenol (DNP). Figure 6.10 shows the electropherogram of this mixture in the primary electrolyte. The substances are not separated using isocratic CZE. Figure 6.10b illustrates the electrophoretic separation of the same model mixture of substances with the use of a transient ionic matrix. The latter is formed by a 60-s pulse of H^+ into the column just at the beginning of the analysis. Under the influence of the transient ionic matrix, the sample can be separated into the individual components [4].

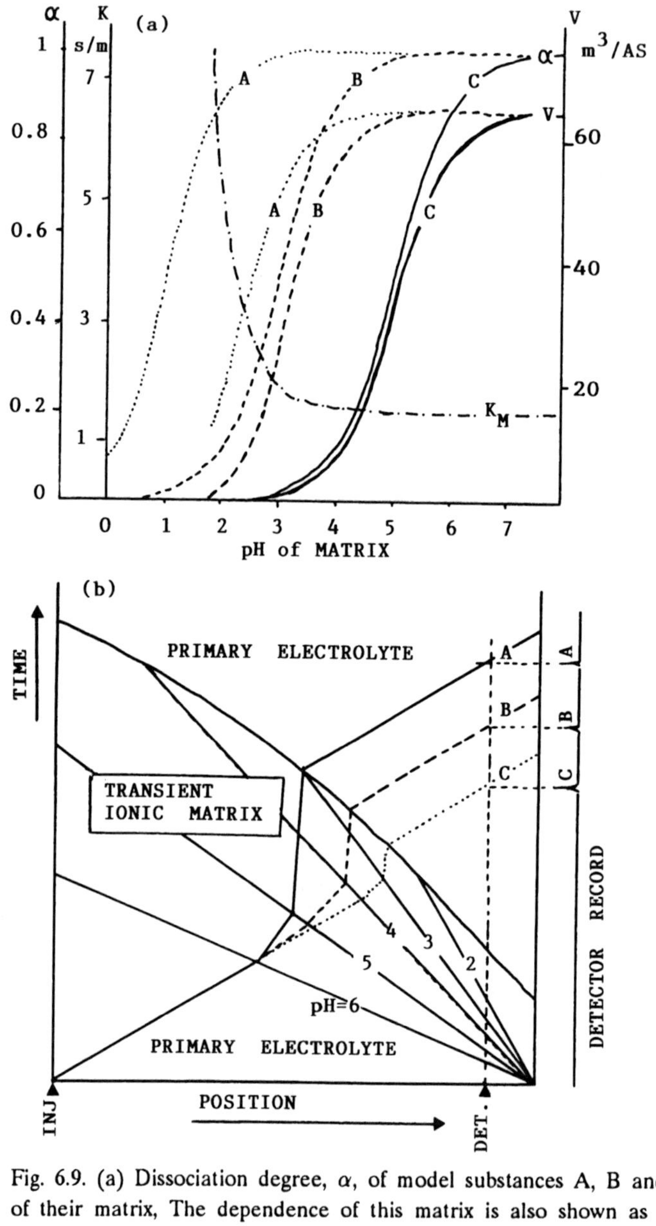

Fig. 6.9. (a) Dissociation degree, α, of model substances A, B and C as a function of the pH of their matrix, The dependence of this matrix is also shown as are the dependencies of the migration velocity, v, of the first three substances (which show the same ionic mobility) on the pH of the matrix adjusted to 0.01 M KCl (primary electrolyte) per unit of electric current density. (b) Simplified separation scheme in the transient ionic matrix. Substances A, B, C migrate in the primary electrolyte from the injection point *INJ* with the same velocities towards the zone of the transient ionic matrix. In this matrix, the substances differ in mobilities and are separated into their own zones. After leaving the transient ionic matrix, the zones migrate again in the primary electrolyte with equal mobilities to the detector (*DET*). The detection record is on the right. (Reproduced from Ref. 4 with permission of Elsevier Science Publishers.)

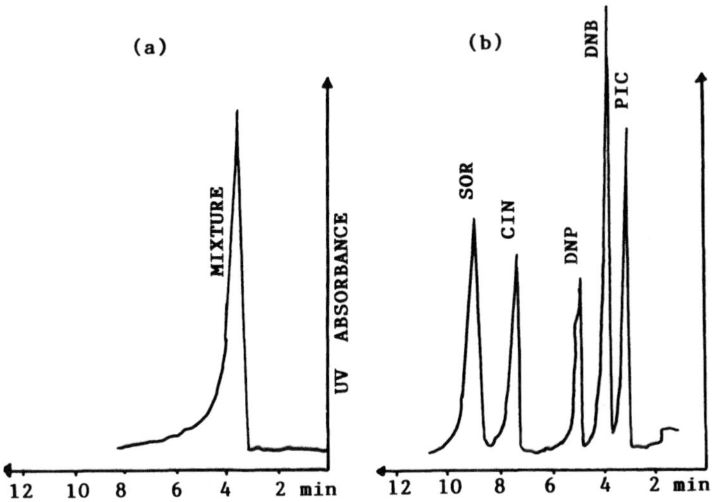

Fig. 6.10. (a) Record of the separation of a mixture of acids (PIC, DNB, DNP, CIN, SOR) in the primary electrolyte (without the use of the transient ionic matrix). The sample was 1 μl of a solution 10^{-4} M in each of the components, The primary electrolyte was 0.01 M KCl (pH 5.6), I = 300 μA. (b) The same separation but with a 1-min pulse of H^+. (Reproduced from Ref. 4 with permission of Elsevier Science Publishers.)

6.2.5 Dynamic pII step

The utilization of dynamic step change of H^+ serving as the co-ion leads to the possibility of using two different pH values during one analytical run. The principle of the performance of stepwise changes of pH of the background electrolyte can be explained in terms of migration trajectories (see Fig. 6.11). It can be seen in Fig. 6.11 that substrates A and B are not separated at pH_1 as their trajectories are identical. The pair of substances A and C cannot be separated at pH_2 as their trajectories show the same slopes (which means that they have the same mobilities under these conditions). A step change in H^+ helps to resolve these species (Fig. 6.11). Once separated, these substances migrate in parallel and do not mix. With the use of the step change of H^+, improvements in separation can be attributed both to the change in selectivity during the separation and focusing of the rear boundary of migrating zones [5].

6.3 FRACTION COLLECTION

Although CE is typically a nanoscale analytical technique, there are several reasons for which fraction collection of the small amounts of separated components may need to be performed [6–12]. First of all, difficulties may be encountered in the purification of valuable products which are available at minute quantities.

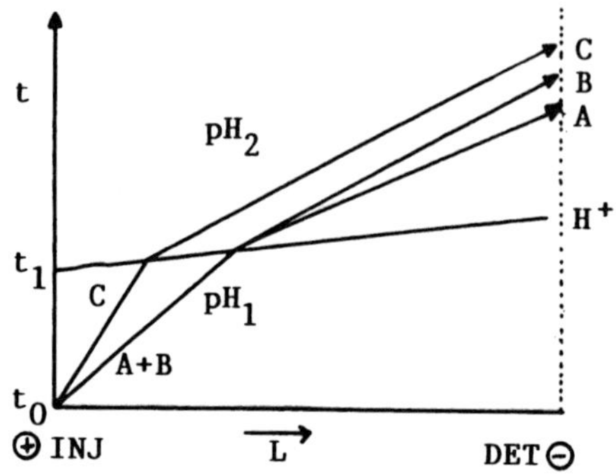

Fig. 6.11. Migration trajectories of cations A, B and C in CZE with a dynamic pH step, t = time of migration; *INJ* = injection; *DET* = detection; L = migration path. (Reproduced from Ref. 5 with permission of Elsevier Science Publishers.)

When conventional separation techniques are employed, the problem would be exacerbated if these techniques are incapable of adequately resolving the components in the sample. In certain applications, the separated peaks from a CE run need to be subjected to additional analysis, such as microsequencing [13] or identification by spectroscopic techniques. In view of these requirements, several approaches developed for fraction collection by CE [6–12] are considered.

The process of fraction collection in CE is fundamentally different from that in liquid chromatography. The end of the capillary must stay in contact with buffer, or kept in contact with an electrode by some other means, during fraction collection in order to maintain the electric field which migrates the zone out of the end of the capillary. Also the electric field applied during electrophoresis may have to be interrupted when moving the capillary from fraction to fraction. Another consideration is that the collected fraction may be diluted by the solution present in the fraction collector, which is necessary for maintaining the electric field. Design of a rugged, automated fraction collection system with increased sample capacity presents interesting challenges and potentially significant benefits in extending the capability of CE.

6.3.1 Stopped-flow techniques

The most common approach for sample collection involves interrupting the electrophoretic run, and transferring the outlet end of the separation capillary from the buffer reservoir to the collection vial. This scheme can be easily performed with the basic instrumental setup for CE. However, there is scope for further improvements [7]. The main problem is that the collected samples tend to be diluted, since the capillary outlet must be immersed in a reservoir containing a suitable buffer medium to complete the circuit. Possible electrochemical reactions at the outlet electrode may also present a problem.

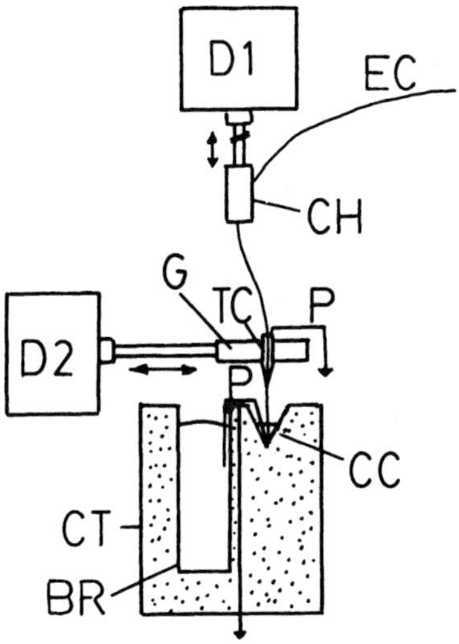

Fig. 6.12. Cross-sectional schematic of collection tray. Key to abbreviations: *D1*, *D2*, *D3* = digital linear actuators; *EC* = electrophoretic capillary; *CH* = capillary holder; *G* = guide; *CT* = collection tray; *CC* collection cone; *BR* = buffer reservoir; *P* = platinum wire electrode; *CW* = copper wire; *TC* = tapered glass capillary. (Reproduced from Ref. 6 with permission of Elsevier Science Publishers.)

An apparatus built in-house for fraction collection in CE has been described by Rose and Jorgenson [6]. Figure 6.12 shows a cross-sectional schematic of the collection tray and collection components. Three digital linear actuators (D1 to D3) are employed to achieve precise movement of the capillary and the collector tray. During normal operation, the electrophoretic capillary (EC) is positioned with its end dipping into the buffer in the reservoir (BR). The sequence of events involved in the collection procedure is as followed: (1) one of the digital linear actuators, D3, moves the collection tray (CT) so that the capillary (EC) is positioned opposite the collection cone (CC) into which the fraction will be collected; (2) D1 lifts the capillary up into the tapered glass capillaries (TC); (3) D2 moves the capillary guide so that the end of the capillary is positioned over the cone, and (4) D1 lowers the end of the capillary into the cone resulting in the configuration shown in Fig. 6.12. The species to be collected then migrates from the end of the electrophoretic capillary into the cone. After collection, steps 2–4 in the sequence above are executed in reverse order. The tapered glass capillary (TC) is filled with buffer which is grounded via a platinum electrode (P). With this arrangement, the current is maintained during the transfer of the capillary from reservoir to cone, thus reducing the amount of time electrophoresis is interrupted [6].

To test the performance of the fraction collector, a sample containing guanosine and adenosine is separated by electrophoresis and collected. In the electropherogram shown in Fig. 6.13, vertical lines representing the zone start and stop times

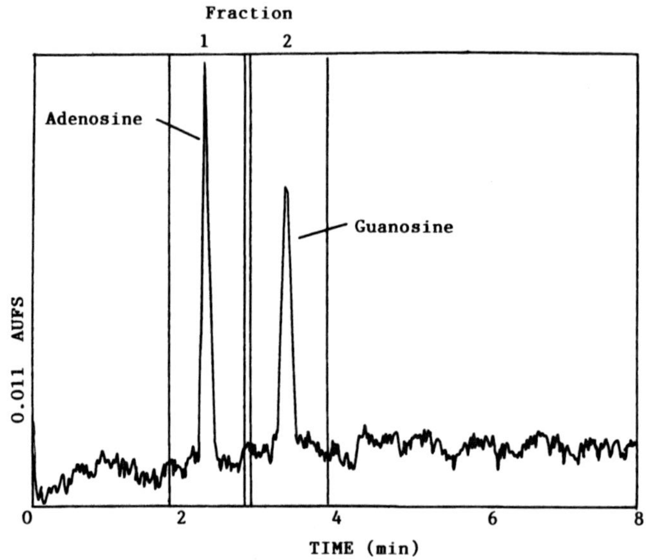

Fig. 6.13. Electropherogram of nucleoside sample. Vertical lines represent unmodified zone start and stop times for fraction collection. Sample introduction: 5 s at 20 kV. (Reproduced from Ref. 6 with permission of Elsevier Science Publishers.)

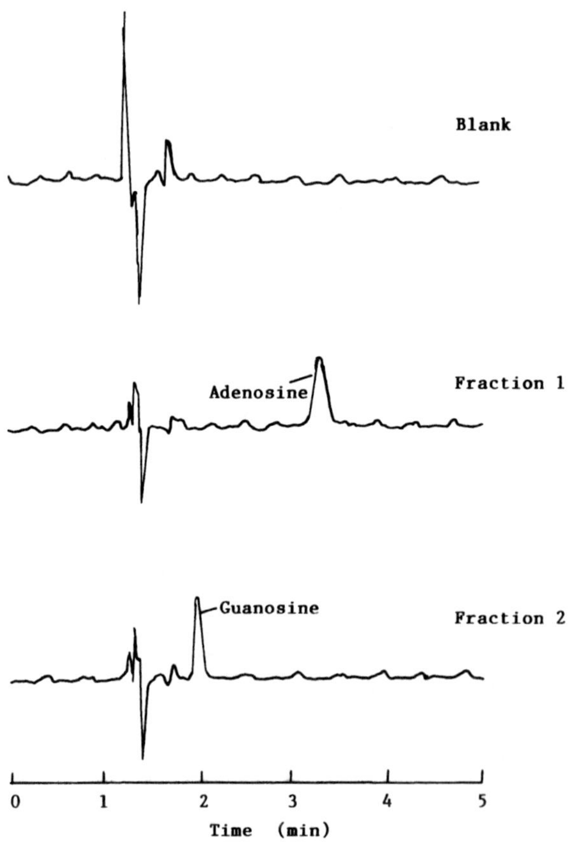

for the fractions are shown. The actual collection start and stop times are 1.5 times greater than these times, if the difference in the total tube length and the effective tube length is taken into account. Figure 6.14 shows the HPLC chromatogram of a fraction collected from electrophoresis (Fig. 6.13). The chromatograms demonstrate discrete collection of a separated two-component mixture.

6.3.2 On-column frit

By using on-column frits similar to those employed in electrochemical detection for decoupling of the electric field (see Section 3.6.2), fraction collection can be performed without interrupting the electrophoretic run [7]. The frit allows electrical connection to be made to the capillary so that the first segment of the capillary (inlet to frit) may be used for electrokinetic separations while the second segment (frit to outlet) is free of applied electric field, facilitating its use for sample collection. The construction steps for fabricating the on-column capillary frit are shown in Fig. 6.15. Although the fabrication of rugged on-column frit with minimum dead volume would require skill, the technique has significant advantages. Most importantly,

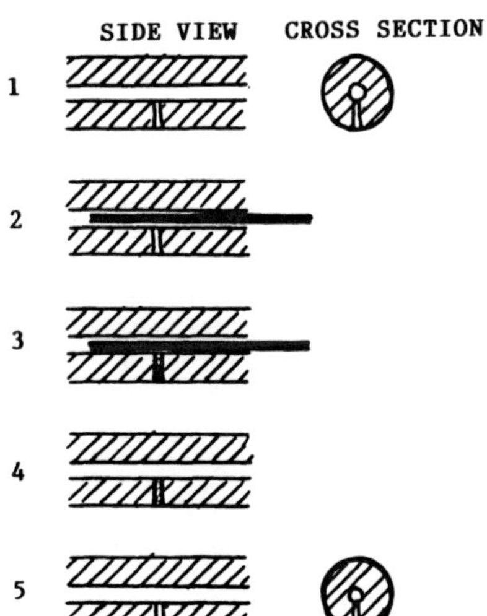

Fig. 6.15. Construction steps for fabricating the on-column capillary frit: (1) prepare hole in wall of capillary; (2) insert tungsten wire to cover hole; (3) add slurry of glass solder and fused silica powder, and heat gently until set; (4) remove tungsten wire; and (5) heat locally frit structure to cause particles to sinter. (Reproduced from Ref. 7 with permission of the American Chemical Society.)

Fig. 6.14. HPLC chromatograms of two fractions collected from electrophoresis run (Fig. 6.13) Top chromatogram is injection of operating buffer only; noise in baseline due to pump pulsations; mobile methanol–0.1 M acetate buffer (15 : 85 v/v), pH 3.2; flow-rate 2 ml/min. (Reproduced from Ref. 6 with permission of Elsevier Science Publishers.)

some of the difficulties encountered in stopped-flow techniques can be alleviated with this system, since the capillary outlet would not need to be immersed into a buffer reservoir to perform fraction collection.

6.3.3 Multiple capillaries

In order to increase the sample quantity isolated in CE, multiple separation capillaries can be used [8,9]. An experimental setup for CE with multiple capillaries is shown in Fig. 6.16. A tee-joint is used to connect a bundle of four capillaries (50 μm I.D.) to a detection/collection capillary (530 μm I.D.). A 10 μm I.D. capillary is glued to a PTFE joint as shown in Fig. 6.16b which is connected to the capillary outlet end to provide electrical grounding while maintaining the separation voltage. Separation of dansylated amino acids (DNS-Val and DNS-Glu) has been demonstrated with this system. As a result of the dead volume caused by the small gap in the tee, low efficiency is observed.

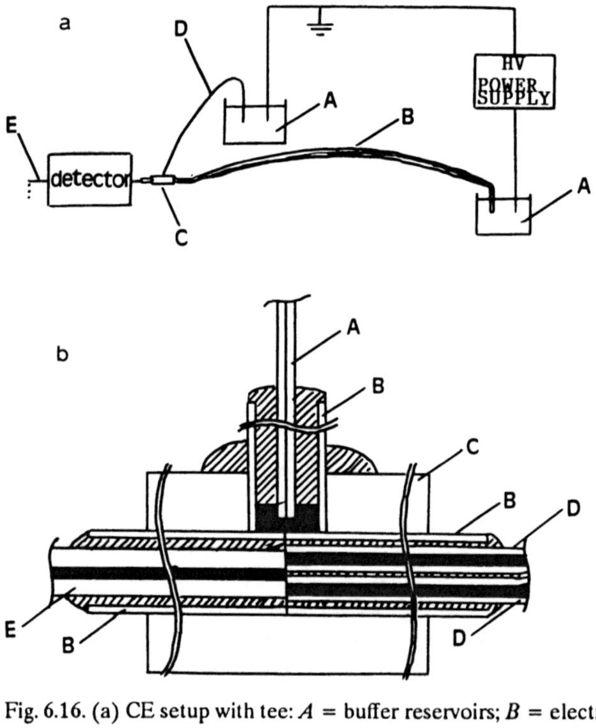

Fig. 6.16. (a) CE setup with tee: A = buffer reservoirs; B = electrophoretic column (a single capillary or a bundle of capillaries); C = tee; D = electrical connection capillary; E = detection/collection capillary. (b) The tee assembly: A = 10 μm I.D., 150 μm O.D. fused silica capillary; B = 530 μm I.D., 660 μm O.D. fused silica capillaries; C = PTFE tubing; D = 100 μm I.D., 200μm O.D. fused silica capillaries; E = 100 μm I.D., 375 μm O.D. fused silica capillary. (Reproduced from Ref. 8 with permission of Dr. Alfred Huethig Publishers.)

Guzman *et al.* [9] described a grounding tee assembly similar in design to that of Fujimoto *et al.* [8] for CE operation with multiple capillaries. However, instead of connecting the bundle of capillaries directly with the detection capillary, a glass connector is used to merge a bundle of five (inlet) capillaries (75 μm I.D., 375 μm O.D.) into a single (outlet) capillary (75 μm I.D., 375 μm O.D.). To merge the capillary bundle, the inlet capillaries are inserted into one side of the glass connector, which has a wider diameter. The outlet capillary is then inserted into the other side of the glass connector which has a smaller diameter. After alignment, the connector tube is heated slowly until the capillaries are fused together to form a single compartment. Subsequently, the outlet capillary and the detector capillary (75 μm I.D., 375 μm O.D.) are both glued inside larger capillaries (530 μm I.D., 660 μm O.D.) for connection in the tee assembly made of PTFE as described earlier for Fig. 6.16b, which is also connected to a 10 μm I.D. capillary which provides grounding. The system has been demonstrated to generate 1 μg of uric acid after eight sample injections [9]. However, the effect of the connections on peak broadening has not been quantified.

6.3.4 Field programming

As discussed in Section 4.3.6, field programming capillary electrophoresis [10–12] has also been employed in micropreparative collection. In this method, separation is initially performed at high separation voltage, and subsequently collection is performed at lower electric field. In principle, the techniques is fully capable of maintaining high separation efficiency, while permitting sample collection to be more readily performed. The practical implementation of this approach, however, involves consideration of the effects of programming on the flow profile and the resulting peak broadening (see Section 6.2). Nevertheless, very high efficiency has been achieved with such a system [10–12]. The technique is especially useful in capillary gel electrophoresis, since in this mode of CE operation, the effect of the introduction of the lower electric field on efficiency is expected to be less significant as the contribution to peak height by diffusion is relatively small. The effect on the flow profile is also reduced, since electroosmosis is practically eliminated in gel-filled columns (see Section 4.3).

6.4 FIELD EFFECT ELECTROOSMOSIS

As discussed in Sections 1.3.2 and 4.2.2, a fundamental challenge in capillary electrophoresis is the control of electroosmosis (see Section 1.3.3). In addition, application of CE to the separation of macromolecules, such as proteins, may be complicated by adsorption of the samples onto the walls of the capillary. Such interactions result in band broadening and tailing, with greatly reduced separation efficiency.

To enhance separation efficiency and to prevent protein adsorption, it would be advantageous to have the ability to directly and dynamically control the polarity and magnitude of the zeta (ζ) potential. A novel concept involving the use of an additional electric field from outside the capillary has been proposed by Lee *et al.* [13–15], and Ghowsi *et al.* [16–18]. This technique vectorially couples the externally applied potential with the potential inside the capillary. Thus, the ζ potential can be controlled with a definite value and made to be positive, zero, or negative. Furthermore, the ζ potential can be changed at any time during the analysis to achieve the desired separation conditions.

The experimental setup employed to demonstrate field effect electroosmosis is shown in Fig. 6.17. In this system, a 20 cm long capillary with 75 μm I.D. (375 μm O.D.) is placed inside a larger capillary (530 μm I.D., 630 μm O.D.) that is 17 cm long. The smaller (inner) capillary is attached between reservoirs 1 and 4

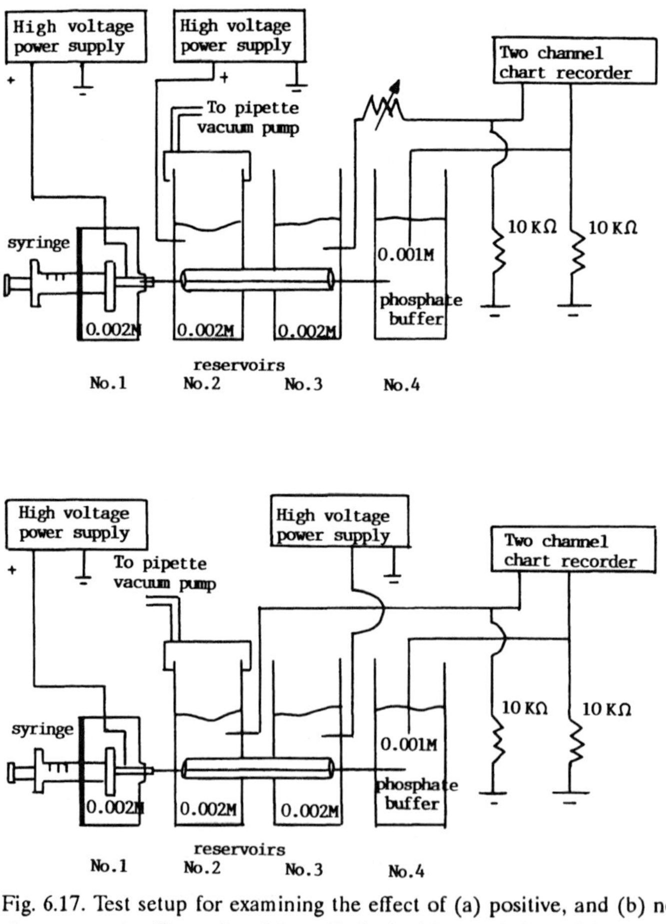

Fig. 6.17. Test setup for examining the effect of (a) positive, and (b) negative external field on the electroosmosis. (Reproduced from Ref. 13 with permission of the American Chemical Society)

while the larger (outer) capillary is attached between reservoir 2 and reservoir 3. The polyimide coating on the exterior surfaces of both capillaries is removed by using concentrated sulfuric acid solutions. A syringe is used as reservoir 1 and as a pumping device for flushing out air bubbles in the inner capillary. Platinum wires immersed in the four buffer reservoirs are used as electrodes [13].

To perform field effect electroosmosis, potassium phosphate buffer at a pH of about 6 is used to fill reservoir 2, reservoir 3, and the annulus between the inner and outer capillaries. One high-voltage power supply is connected to either reservoir 2 or reservoir 3 so that an electric field is applied to the annular space between the two capillaries. Another high-voltage power supply is used to apply an electric field between reservoir 1 and reservoir 4 inside the inner capillary. A small fluid flow is maintained in the annulus between the inner and outer capillaries by a suitable pump. This fluid flow serves to remove efficiently the additional heat generated by the application of external electric field. By means of the variable resistor R_3, the electric field gradients between the inner and outer fields along the annulus can be adjusted. The resulting changes in the direction and speed of electroosmotic flow in the inner capillary due to these electric field gradients is monitored by using the current-monitoring method [19].

Figure 6.18 illustrates the electropherogram showing the change of the direction and flow rate of electroosmosis in the inner capillary with the application of external electric field. The changes in potential drop across the 10 kΩ resistor inserted between the reservoir 4 electrode and ground are recorded. In regions a and b of

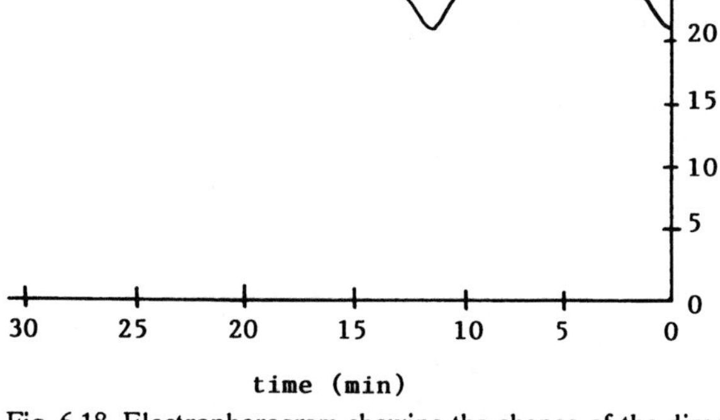

Fig. 6.18. Electropherogram showing the change of the direction and flow rate of electroosmosis in the inner capillary with the application of external electric field. (Reproduced from Ref. 13 with permission of the American Chemical Society.)

Fig. 6.18, the anode end for the inner electric field is in reservoir 1 and reservoir 4, respectively. Regions a and b correspond to the typical electroosmotic flow without the influence of external electric field. From region c to region h, the anode end for the inner electric field is kept in reservoir 1. Between regions c and e, positive potential differences from 0 to 5 kV between the inner and outer fields are applied. A decrease in the flow rate of electroosmosis from reservoir 1 to reservoir 4 is obtained with the application of an external electric field, as observed from the decrease in the slope of electropherogram. The direction of electroosmosis is reversed with a 6 kV positive potential difference between the inner and outer fields in region f. In region g, a 5 kV negative potential difference is applied, which results in an enhancement in the electroosmotic flow, as seen by comparing the slopes of the electropherogram in regions g and h. In regions h and i, no external electrical field is applied [13].

As shown in Section 1.3.3, the electroosmotic velocity (v) is given by Eq. (1.7):

$$v = \left(\frac{\varepsilon}{4\pi\eta}\right) E\zeta$$

where ε is the dielectric constant or relative permittivity of the solution, η is the viscosity, E is the electric field strength, and ζ is the zeta potential [20,21]. The ζ potential is negative if the direction of the electroosmosis is from reservoir 1 to reservoir 4. The ζ potential changes from −29 mV without an external field to −35 mV with a −5 kV potential difference between the outer and the inner capillaries. The absolute value of the ζ potential decreases from −29 mV without an external field to about 0 mV with +5 kV potential difference. The polarity of the ζ potential can be reversed at +6 kV potential difference. These results provide indication that the electroosmosis can be enhanced, decreased, eliminated or reversed by using an external field to control the ζ potential at the aqueous/inner capillary interface.

A capacitor model for the analysis the effect of experimental factors, such as solution condition and capillary dimension on the direct control of electroosmosis has been proposed by Lee *et al.* [14]. In this model, the inverse of the total capacitance across the inner capillary wall is obtained by taking the sum of the reciprocal values of the inner capillary and the electrostatic diffuse layers on both the inside and outside of it. The capacitances of the electrostatic diffuse layer at the inner capillary/inner aqueous interface, the inner capillary tubing, the electrostatic diffuse layer at the inner capillary/outer aqueous interface are represented by ϕ_i, ϕ_c and ϕ_o, respectively. The model implies that the inverse of the total capacitance (ϕ_T^{-1}) is given by $\phi_T^{-1} = \phi_i^{-1} + \phi_o^{-1} + \phi_c^{-1}$. Since the capacitances of the electrostatic layers are much greater than that of the capillary tubing for the accumulation of mobile ions in the electrostatic layers, the total capacitance ϕ_T may be considered as equal to ϕ_c^{-1} (ϕ_i^{-1} and ϕ_o^{-1} are small). Consequently the change in the zeta potential at the inner capillary/inner aqueous interface, Δz, can be represented by [14]:

$$\Delta z = \Delta V / [(\phi_i/\phi_T)] = \Delta V / [\phi_i/\phi_c] \tag{6.1}$$

where ΔV is the applied potential difference across the inner capillary wall. The change in zeta potential can then be used to calculate the change in electroosmotic flow velocity using Eq. (1.7).

The differences between the predictions of the capacitor model and the experimental results ranged between 13 and 59% for different pH, electrolyte conditions and capillary dimensions [14]. Much better agreement ($\simeq$1%) can be obtained with a low buffer pH, i.e. a smaller ζ potential [15]. By using dimethyl sulfoxide as the test compound, 25 μm I.D. and 50 μm I.D. (150 μm O.D.) capillaries, and 10 mM (pH 2.7) phosphate as the electrophoresis buffer, linear relationships are obtained between theoretical plate numbers and electroosmotic flow. Good agreement is obtained between the mobilities and theoretical plate numbers calculated using the capacitor model and the experimental data. Under these conditions, no measurable additional dispersion and peak broadening are observed when an external field is used to control electroosmotic flow. On the other hand, at a higher buffer pH (6.0), poor agreement is observed between the calculated and the experimental values.

A field effect electrokinetic transistor called MIEEKFED, or "Metal-Insulator Electrolyte–Electrokinetic Field Effect Device" has also been postulated [16–18]. In a MIEEKFED, the electrolyte flow may be controlled. A schematic of a MIEEKFED is shown in Fig. 6.19. This device should allow the control of the flow of an electrolyte by two voltages, V, the applied voltage across the capillary and V_G, the gate voltage.

In MIEEKFED, the electroosmotic velocity is given by [16–18]:

$$v_{eo}(z) = \left(\frac{\varepsilon}{4\pi\eta}\right) \frac{V}{L} f^{-1} \left(V_G - \frac{z}{L} V\right) \tag{6.2}$$

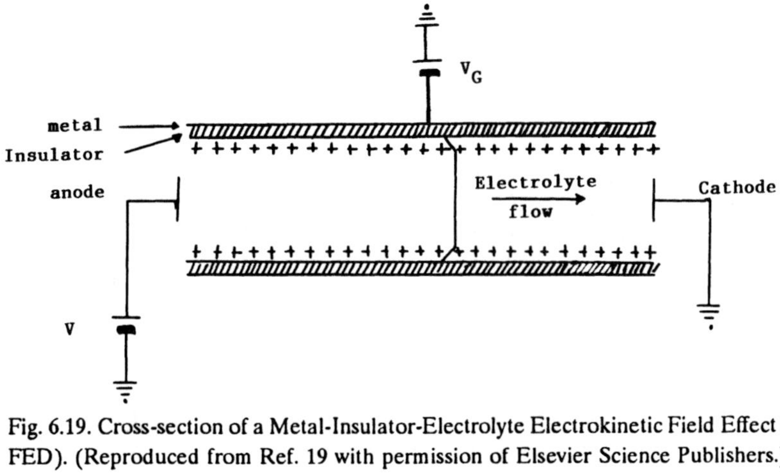

Fig. 6.19. Cross-section of a Metal-Insulator-Electrolyte Electrokinetic Field Effect Device (MIEEK-FED). (Reproduced from Ref. 19 with permission of Elsevier Science Publishers.)

where ε is the dielectric constant, η is the viscosity, L is the length of the capillary and z is the distance along the length of the capillary, and f represents a complex analytical function relating the gate potential to the zeta potential [16].

There are several potential advantages in using MIEEKFED in a capillary electrophoresis apparatus. The zeta potential and therefore the electroosmotic flow, may be flexibly controlled without unwanted restrictions on pH, electrolyte concentration, or the inner layer of the capillary. It could also be used to reduce the degree of tailing commonly encountered in CE of macromolecules. To prevent tailing due to adsorption, one approach based on the concept of MIEEKFED would be to create a low zeta potential and a low charged layer at the solid electrolyte interface.

However, despite the potential advantages of external field programming techniques, the implementation of such techniques presents fundamental challenges. As discussed in Section 6.2, when the zeta potential is not constant along the length of the capillary, local variation in electroosmotic flow and electrophoretic mobilities may occur, which tends to results in additional peak broadening. Furthermore, the instrumental setup required presents another problem. For instance, to permit the control of the zeta potential by an external field, the outer coating of the inner capillary needs to be removed, making it extremely fragile and inconvenient to use. To date, the full potential of external field control and MIEEKFED has not been realized.

6.5 SYSTEMATIC OPTIMIZATION SCHEMES

Usually, optimum separation conditions in CE have been achieved using either trial-and-error approaches, or univariate optimization separation procedures, in which one parameter is varied at a time while the other parameters are kept constant. Even though such procedures have been successfully employed in many previous investigations, because of the trial-and-error nature of such approaches, they are often long and tedious processes. Moreover, such schemes usually result in a local optimum rather than a global optimum. With the increasing degree of complexity of the mixtures to be investigated, there would be an increasing need for multi-modifier systems which can provide additional selectivity for the separation. The optimization of these separations would become extremely complex and time-consuming if trial-and-error type of approaches for optimization of separation are adopted. It would therefore be desirable to develop systematic schemes for the optimization of CE separations.

6.5.1 Plackett-Burman statistical design

Vindevogel and Sandra [22] used the Plackett-Burman statistical design for the optimization of MEKC analysis of testosterone esters. In this method, by combining seven selected parameters (or factors) which occurs at two levels (high or low), eight experimental conditions can be created. The effect E_x of a particular factor is calculated from the difference between the average result at the high level (+) and the average result at the low level (−):

$$E_x = \frac{\Sigma E(+)}{4} - \frac{\Sigma E(-)}{4} \qquad ((6.3)$$

At least one or more of the factors should be a dummy for which the difference between the high level and the low level is zero [23]. The dummy factors can be used to estimate the variability of the system, and the significance of the effects due to true physical parameters:

$$\sigma^2 = \frac{\Sigma E^2_{\text{dummy},i}}{n_{\text{dummy}}} \qquad (6.4)$$

A factor is considered significant if the value of the t-test is above a critical value:

$$t\text{-test:} \quad \frac{|E_x|}{\sigma_x} \geq t_{\text{crit}} \qquad (6.5)$$

The confidence level of the t-test depends on the number of dummies used. When two dummy factors and a t-test value of 1.89 is used, the confidence level is 80% for a two-sided test.

The five factors selected are related to the buffer composition. These factors are pH, concentration of buffer (mM_{buf}), level of organic modifier (% ACN), which is acetonitrile (ACN), surfactant concentration (mM_{SDS}), which is sodium dodecyl sulfate (SDS), and the use of mixed micelles (% SHS), which are SDS and sodium heptyl sulfate (SHS). These factors and the two dummy factors are listed in Table 6.1. For each factor, a high and a low value are selected. These values are chosen arbitrarily [23]. In Table 6.1, the buffer concentration is either 20 or 40 mM. The concentration of ACN is either 40 or 50% (v/v). The pH values are 8 or 9, which correspond to the pH of the pure buffer solutions. The apparent pH values

TABLE 6.1

IMPLEMENTATION OF THE PLACKETT-BURMAN DESIGN (Reprinted from Ref. 22)

Buffer	Dummy 1	pH	% SHS	% ACN	mM_{SDS}	Dummy 2	mM_{buf}
1	*+0*	*9*	*10*	40	*50*	−0	20
2	−0	*9*	*10*	*50*	40	*+0*	20
3	−0	8	*10*	*50*	*50*	−0	*40*
4	*+0*	8	0	*50*	*50*	*+0*	20
5	−0	*9*	0	40	*50*	*+0*	*40*
6	*+0*	8	*10*	40	40	*+0*	*40*
7	*+0*	*9*	0	*50*	40	−0	*40*
8	−0	8	0	40	40	−0	20

TABLE 6.2

PLACKETT-BURMAN t-TEST* (Adapted from Ref. 22)

	t_0	t_L	R_s	N_1
pH	9.75	3.38	12.71	
% SHS	−3.08	−2.06	−9.38	
% ACN	28.77		−25.46	−3.05
mM_{buf}	5.27	3.61	17.69	
mM_{buf}	3.85		5.90	

* For clarity, test values indicating non-significant results (<1.89) are not shown. The sign shows the direction of the effect.

are about 1 unit higher in the presence of 40 or 50% ACN. The SDS and SHS concentrations are either 40 or 50 mM. Experiments are carried out according to the order shown in Table 6.1, which should be as random as possible in order to avoid systematic long-term influence of a particular factor.

Table 6.2 shows the values of the t-test obtained for electroosmotic migration time t_0, the migration time of the last compound t_L, the resolution R_s between a selected pair of peaks, and the efficiency of the first peak N_1. A large value in Table 6.2 indicates that the particular result is significantly influenced by the particular factor corresponding to that test value.

Although the Plackett-Burman design is capable of optimizing several factors simultaneously, there are several drawbacks. First, there are no fixed rules for the selection of the low and high levels [24]. Level selection is based on experience or from preliminary experiments. Secondly, since the t-test only shows which factors are significant, and does not predict the exact values of the experimental conditions for optimum separation, further optimization may be required in order to determine the detailed experimental conditions for achieving optimum separation.

6.5.2 Overlapping resolution mapping scheme

Li and co-workers developed several schemes based on the overlapping resolution mapping (ORM) procedure for the optimization of capillary electrophoresis separations [25–27]. The abbreviation "ORM" refers to the method of overlapping resolution maps to determine the optimum conditions for a separation. In this section, the use of the scheme is demonstrated for three groups of compounds. First, for the separation of sulfonamides, the scheme is employed to optimize two parameters, i.e. the optimum pH and the optimum concentration of sodium dodecyl sulphate (SDS) in MEKC. Secondly, the optimization of three variables is demonstrated for the separation of a group of plant growth hormones, where the scheme is used to optimize the concentrations of α-cyclodextrin, β-cyclodextrin and γ-cyclodextrin, which are added to the electrophoretic buffer to provide additional selectivity based on molecular size and shape in a CE system (see Section 5.3.1).

Thirdly, a more versatile, alternative approach for the simultaneous optimization of three variable is presented. The use of the scheme is demonstrated for the separation of a group of dinitrophenyl derivatized (DNP)-amino acids. In this case, pH, SDS concentration and tetrabutylammonium salt (TBA) concentration are chosen as the parameters to be optimized.

6.5.2.1 Optimization of pH and SDS concentration

As an example of the application of the ORM scheme for the optimization of two parameters, i.e. optimum pH and SDS concentration, in the MEKC separation of sulfonamides is considered.

The first step of the procedure is to define the criteria for the separation. Typical criteria used include the minimum resolution required between all the pairs of peaks, and the maximum run time per analysis. Once the criteria are set, pre-planned experiments which cover the widest possible range of experimental conditions permitted in an analysis are performed. In some cases, narrower range of experimental conditions may be preferable, e.g. if there are practical limits, such as solubility or miscibility limits of a modifier in the electrophoretic medium. The scheme is highly flexible in this respect, since the criteria can be chosen to limit the experimental conditions to prevent such problems. Results of preliminary experiments can also be used to define the maximum range of experimental conditions. For instance, if the migration times for certain species are excessively long under a particular set of experimental conditions, these conditions can be eliminated from the optimization experiments.

For the optimization of two parameters, a set of nine experiments are performed at selected points on a rectangular diagram shown in Fig. 6.20. From the

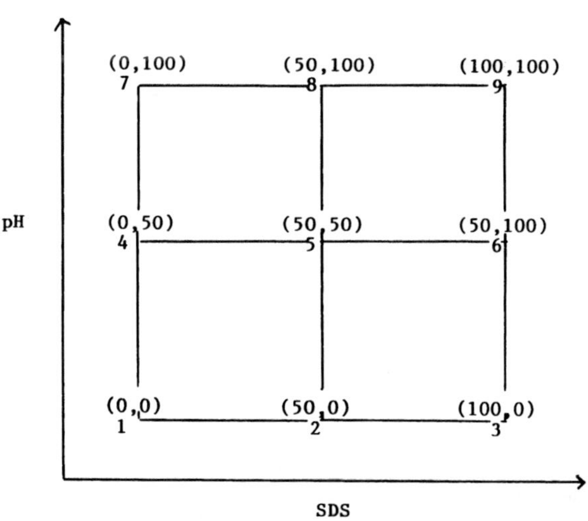

Fig. 6.20. The locations of the nine experiments chosen from the rectangular plot. The compositions at each point is represented as the percentage at the respective apices.

experimental results of migration times, the resolution, R, between every pair of adjacent peaks in the electropherograms are calculated using:

$$R = \frac{2(t_2 - t_1)}{W_1 + W_2}$$

where t_1 and t_2 are the migration times of two adjacent peaks. W_1 and W_2 are the widths of the pair of peaks. The calculated R are then fitted to a polynomial equation [26]:

$$R = a_0 + a_1 x_1 + a_2 x_2 + a_{12} x_1 x_2 + a_{11} x_1^2 + a_{22} x_1^2 + a_{112} x_1^2 x_2 + a_{122} x_1 x_2^2 + a_{1122} x_1^2 x_2^2 \qquad (6.6)$$

where x_i's correspond to the fractions of the parameters in terms of the respective axes and a_i's are coefficients. For the separation of sulfonamides, x_1 and x_2 are used to represent pH and SDS concentration, respectively. A computer program is used for the calculation of the coefficients for each of the peaks. With the known coefficients, the R values for all the experimental conditions within the maximum range selected can be calculated. Using the same program, these values are transformed into a rectangular plot. With the eight sulfonamides, seven rectangular plots are obtained. Subsequently, by overlapping all the rectangular plots and representing each point by the smallest values of resolution among all the individual

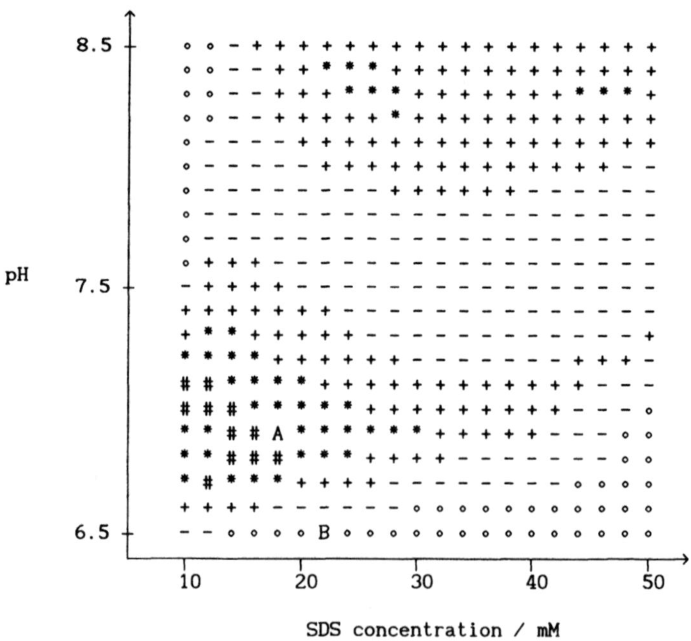

Fig. 6.21. Final overlapped resolution plot for all the seven pairs of peaks. Notation: o = $R < 0.5$; – = $0.5 < R < 1.0$' + = $1.0 < R < 1.5$; * = $1.5 \leq R < 2.0$; # = $R \geq 2.0$.

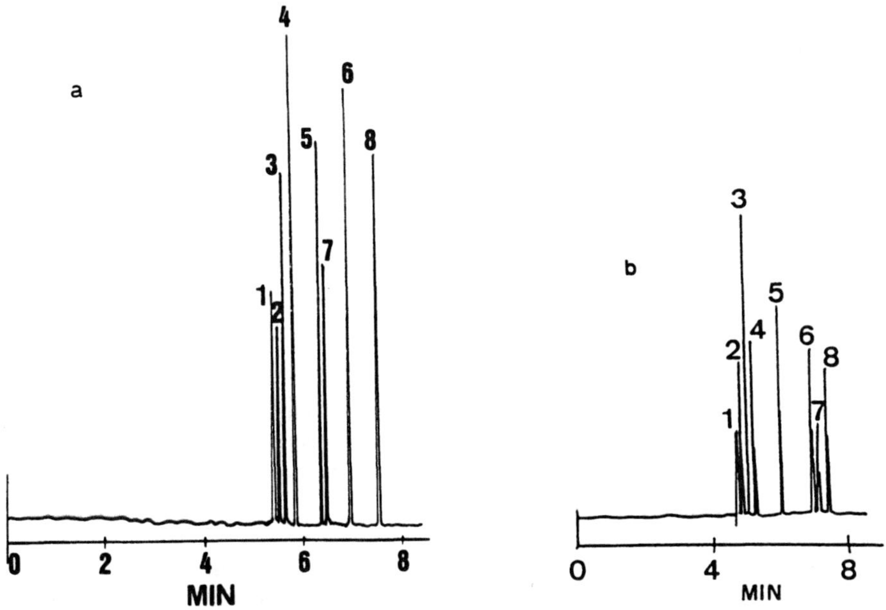

Fig. 6.22. Electropherograms for the eight sulfonamides. (a) Electrophoretic conditions corresponding to point A in Fig. 6.21: 0.05 M phosphate–0.05 M borate at pH 6.4 and 2 mM β-cyclodextrin; separation tube: 50 μm ID × 50 cm; detection wavelength: 210 nm. Peak identification: *1* = MeOH and SG, *2* = SMZ, *3* = SMP, *4* = SM, *5* = SD, *6* = SQX, *7* = SS, *8* = SCP. (b) Electrophoretic conditions corresponding to point B in Fig. 6.21: 0.05 M phosphate–0.05 M borate at pH 6.0 and 3 mM β-cyclodextrin; separation tube: 50 μm I.D. × 50 cm; detection wavelength: 210 nm. Peak identification: *1, 2* = MeOH, SG and SMZ, *3* = SMP, *4* = SM, *5* = SD, *6* = SQX, *7* = SS, *8* = SCP.
Abbreviations: SG = suflaguanidine; SMZ = sulfamethazine; SMP = sulfamethoxypyridazine; SM = sulfamerazine; SD = sulfadiazine; SQX = sulfaquinoxaline; SS = sulfasalazine; SCP = sulfachloropyridazine.

plots, areas which satisfy the minimum desired resolution for all the seven pairs of peaks can be established. Figure 6.21 represents the final overlapped diagram for the eight sulfonamides. From the figure, the common overlapped region marked with # would indicate the region of interest where the optimum conditions can be obtained. Typical electropherograms obtained using the experimental conditions from points A and B are shown in Fig. 6.22. From Fig. 6.22a, it can be seen that all the peaks are baseline separated as predicted. On the contrary, the electropherogram obtained using the experimental conditions corresponding to point B (Fig. 6.22b) shows peaks that are not as well separated as compared to those in Fig. 6.22a.

6.5.2.2 Optimization of concentrations of cyclodextrins

The addition of α, β, or γ-cyclodextrins enhances selectivity in CE separations as a result of the differences in size and shape of the analyte molecules and

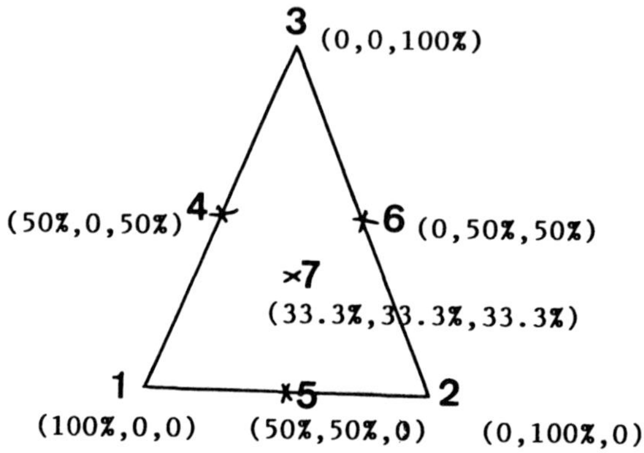

Fig. 6.23. Locations on the composition triangle of the seven experiments to be conducted for the ORM scheme. Compositions are shown as a percentages of compositions at the three vertices. (Reproduced from Ref. 25 with permission of the American Chemical Society.)

their interaction with the cyclodextrin species, which have different cavity sizes (see Section 5.3.1). For the analysis of plant growth regulators, the use of all three cyclodextrins simultaneously serves to provide additional selectivity for the separation [25]. In this case, the parameters to be optimized are the concentrations of the three cyclodextrins. Since three parameters are involved, a triangular diagram is found to be a more convenient representation than the rectangular diagram. Minimum and maximum concentrations of each cyclodextrins used are 0 and 10 mM, respectively. The axes of the triangular diagram are expressed in terms of % of the value at the apexes, which corresponded to the maximum concentrations of the cyclodextrins used. Therefore, 0% represents 0 mM and 100% represents 10 mM of a particular cyclodextrin, respectively. Experiments are then performed using experimental conditions selected from the triangular diagram shown in Fig. 6.23. A set of seven experiments need to be performed. The compositions of the α, β and γ-cyclodextrin modifiers (mM) for the seven experiments are: (1) 10, 0, 0; (2) 0, 10, 0; (3) 0, 0, 10; (4) 5, 0, 5; (5) 5, 5, 0; (6) 0, 5, 5 and (7) 3.3, 3.3, 3.3. From these experiments, the resolutions, R, between every pair of peaks in the electropherograms obtained from the seven experiments are calculated. The calculated R values are then fitted into the following polynomial [25]:

$$R = a_\alpha x_\alpha + a_\beta x_\beta + a_\gamma x_\gamma + a_{\alpha\beta} x_\alpha x_\beta + a_{\beta\gamma} x_\beta x_\gamma + a_{\alpha\gamma} x_\alpha x_\gamma + a_{\alpha\beta\gamma} x_\alpha x_\beta x_\gamma \qquad (6.7)$$

where a_i's are the coefficients and x_i's correspond to the fractions of the modifier compositions. With the aid of a computer program, the coefficients for each pair of peaks are determined. Once the coefficients are known, the R values for all the points in the triangle can be calculated. Individual resolution diagrams are then generated. By overlapping all the individual diagrams, the areas which satisfy the

```
                              0
                            5   o 100
                         10   + o 95
                       15   + + o 90
                     20   + * + o 85
                   25   + * # + o 80
                 30   - + # # + o 75
               35   o + # # # * o 70
             40   o - + # # * + o 65
           45   o - + * # # + + o 60
α-cyclodextrin 50   o - + + * * * + + o 55  γ-cyclodextrin
         55   o - + + + + + + - - o 50
       60   o - - - + + + + - - - o 45
     65   o - - - - - - - - - - - o 40
   70   o - - - - - - - - - - o o - o 35
  75   o - + + - - - - o o o o o - o 30
 80   o - * + + - - - o o o o o o o o 25
 85   o - # * + + - - - o o o o o o o - 20
 90   o + # # # + + - - - o o o N o o - - 15
 95   o + M # * + + - - - o o o o o o o - - 10
100 - * * + + + + - - - - - o o o o o - - - 5
    - - - - - - - - - - - - o o o o - - - + - 0
 0                β-cyclodextrin               100
```

Fig. 6.24. Overlapped resolution diagram for the eight pairs of peaks. Notation: o = $R < 0.1$; − = $0.1 \leq R < 0.4$; + = $0.4 \leq R < 0.8$; $*$ = $0.8 \leq R < 1$; # = $R \geq 1$.

required minimum resolution can be identified. The overlapped resolution diagram is shown in Fig. 6.24.

Figure 6.25a shows an electropherogram obtained using conditions represented by point M, which is selected from the optimum region. It can be seen that complete separation of the peaks can be achieved. In the case of point N, which is selected from a region outside the optimum region, the results show that poor separation is obtained for most of the peaks, as shown in Fig. 6.25b. The results therefore demonstrated that the ORM scheme can be used for the systematic optimization of capillary electrophoretic separations of the plant growth regulators, which involves three parameters.

6.5.2.3 Optimization of pH, SDS concentration and TBA salt concentration

Although the triangular ORM scheme described in the previous section can be used to optimize three parameters, it requires the transformation of the three-dimensional experimental space into a two-dimensional representation (i.e. a triangular diagram). Furthermore, the scheme is subjected to the limitation that the migration behaviour of the solutes depends linearly on the three experimental parameters individually (see Eq. 6.7). The versatility of the triangular scheme for the optimization of three variables simultaneously is limited, since it would not be expected to represent non-linear dependence of migration behaviour on the experimental parameters conveniently and accurately.

Instead, a three-dimensional ORM scheme for the optimization of three

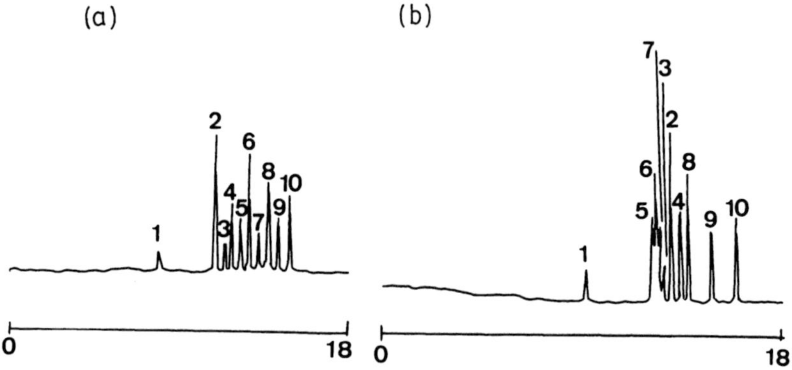

Fig. 6.25. Electropherograms for the nine plant growth regulators. (a) Conditions corresponding point M in Fig. 6.24: 0.05 M phosphate–0.1 M borate buffer; pH 8.09; 8 mM α-cyclodextrin (CD); 1 mM β-CD; 1 mM γ-CD. Peaks: *1* = methanol, *2* = dCPAA, *3* = GA, *4* = pCPAA, *5* = IBA, *6* = tCPAA, *7* = β NAA, *8* = IPA, *9* = αNAA, *10* = IAA. (b) Conditions corresponding to point N in Fig. 6.24: 0.05 M phosphate–0.1 M borate buffer; pH 8.09; 2.0 mM α-CD; 6.5 mM β-CD; 1.5 mM γ-CD. Peaks: *1* = methanol, *5* = IBA, *6* = tCPAA, *4* = pCPAA, *8* = IPA, *9* = αNAA, *10* = IAA. Abbreviations: IAA = indole-3-acetic acid; IBA = indole-3-buteric acid; IPA = indole-3-prprionic acid; αNAA = α-naphthaleneacetic acid; βNAA = β-naphthaleneacetic acid; GA = gibberellic acid; pCPAA = p-chlorophenoxyacetic acid; dCPAA = 2,4-dichlorophenoxyacetic acid. (Reproduced from Ref. 25 with permission of the American Chemical Society.)

interactive experimental parameters can be employed. The use of the scheme has been demonstrated by obtaining the optimum pH, SDS concentration and tetrabutylammonium salt concentration for the separation of sixteen dinitrophenyl derivatized (DNP)-amino acids [27]. In this scheme, eleven pre-planned experiments are carried out at strategic positions on a cubic diagram (see Fig. 6.26). The resolution calculated from the experimental migration times are fitted to a polynomial equation of the form [27]:

$$R = a_0 + a_1x_1 + a_2x_2 + a_3x_3 + a_{11}x_1^2 + a_{22}x_2^2 + a_{33}x_3^2$$
$$a_{12}x_1x_2 + a_{13}x_1x_3 + a_{23}x_2x_3 + a_{123}x_1x_2x_3 \qquad (6.8)$$

where a_i's are coefficients. x_i's correspond to the proportions at the respective axes expressed as percentages and x_1, x_2 and x_3 represent buffer pH, SDS concentration and TBA concentration, respectively in the separation of the DNP-amino acids. Equation (6.8) has been chosen to represent the dependence of resolution on the individual experimental parameters quadratically, and on their interaction terms linearly [27]. As before, the coefficients of the polynomial are determined. Once the coefficients are known, the resolution values for all the points in the cube can be calculated. These values are then represented on the three- dimensional resolution diagrams by the computer program. By overlapping all the individual diagrams, and plotting the lowest resolution in the final overlapped diagram, regions which

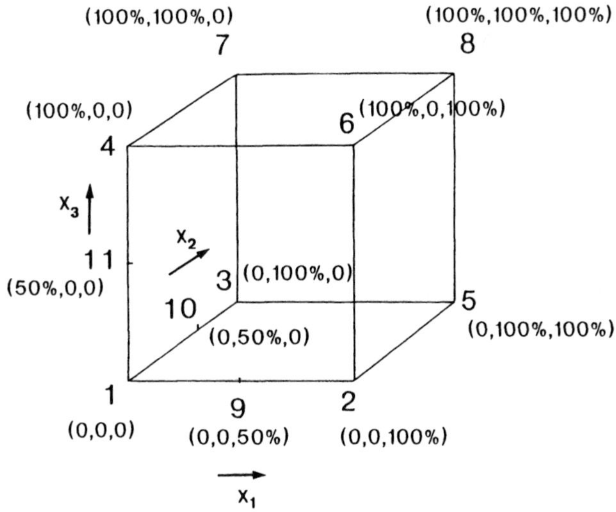

Fig. 6.26. Locations of the eleven experiments to be conducted for the cubic ORM scheme. All experiments performed with 0.05 M phosphate buffer.

Experiment	pH	SDS (mM)	TBA (mM)	Experiment	pH	SDS (mM)	TBA (mM)
1	8	0	20	7	10	0	40
2	8	50	20	8	10	50	40
3	8	0	40	9	8	25	20
4	10	0	20	10	8	0	30
5	8	50	40	11	9	0	20
6	10	50	20				

define the optimum conditions can be identified. Figure 6.27 illustrates locations of the optimum conditions in the overlapped diagram. In Fig. 6.28, electropherograms obtained by performing experiments at points X and Y are shown. Point X is chosen from a region expected to produce high resolution whereas point Y is chosen from a region expected to produce poor separation. As can be seen from Fig. 6.28, the experimental results agree with the prediction of the scheme.

6.5.3 Theoretical approaches for optimization

A theory for the optimization of micellar electrokinetic chromatography has been presented by Foley [28], which is mainly concerned with the optimization of resolution (R_s) and resolution per unit time (R_s/t_R). Equations have been derived that predict the optimum retention factor (k') and the corresponding surfactant concentration to use. Another equation has been derived that relates the analysis time to five parameters: R_s, s_α (selectivity), k', t_0/time, and H/v_{eo} (plate

```
MAXIMUM pH = 100%
TBA  o o o o o o o o o o o o o o o o o o o o o  100%
TBA  o o o o o o o o o o o o o o o o o o o o o  95
TBA  o o o o o o o o o o o o o o o o o o o o o  90
TBA  o o o o o o o o o o o o o o o o o o o o o  85
TBA  o o o o o o o o o o o o o o o o o o o o o  80
TBA  o o o o o o o o o o o o o o o o o o o o o  75
TBA  o o o o o o o o o o o o o o o o o o o o o  70
TBA  o o o o o o o o o o o o o o o o o o o o o  65
TBA  o o o o o o o o o o o o o o o o o o o o o  60
TBA  o o o o o o o o o o o o o o o o o o o o o  55
TBA  o o o o o o o o o o o o o o o o o o o o o  50
TBA  o o o o o o o o o o o o o o o o o o o o o  45
TBA  o o o o o o o o o o o o o o o o o o o o o  40
TBA  o o o o o o o o o o o o o o o o o o o o o  35
TBA  o o o o o o o o o o o o o o o o o o o o o  30
TBA  o o o o o o o o o o o o o o o o o o o o o  25
TBA  o o o o o o o o o o o o o o o o o o o o o  20
TBA  o o o o o o o o o o o o o o o o o o o o o  15
TBA  o o o o o o o o o o o o o o o o o o o o -  10
TBA  o o o o o o o o o o o o o o o o o o o - +  5
TBA  Y o o o o o o o o o o o o o o o o o - X +  0
     0                 SDS                100%
```

Fig. 6.27. Overlapped resolution diagram for the fifteen pair of peaks. Optimum conditions obtained for pH, SDS concentration, TBA concentration as proportion on axes are: 100%, 95% and 0%, respectively, which correspond to pH 10, 47.5 mM SDS and 20 mM TBA, respectively. The plot shows resolution at different SDS and TBA concentrations at a constant pH of 10. Notation: o = $R \leq 0.3$; – = $0.3 < R \leq 0.6$; + = $0.6 < R \leq 0.8$; * = $0.8 < R \leq 1.0$; # = $R > 1.0$.

height/electroosmotic velocity). The relationship of the capacity factor, k' (moles of solute in micelle/moles of solute in bulk water), to surfactant concentration is approximated by [28]:

$$[\text{SURF}] = \frac{k'}{P_{\text{wm}} V} + \text{cmc} \tag{6.9}$$

where P_{wm} is the partition coefficient of the solute between water and the micellar phase, V is the partial molar volume of the surfactant, [SURF] is the molar concentration of surfactant, and cmc is the critical micelle concentration.

For the optimization of resolution (R_s, the optimum k' value is found to be [28]:

$$k'_{\text{opt}(R_s)} = (t_{\text{mc}}/t_0)^{1/2} \tag{6.10}$$

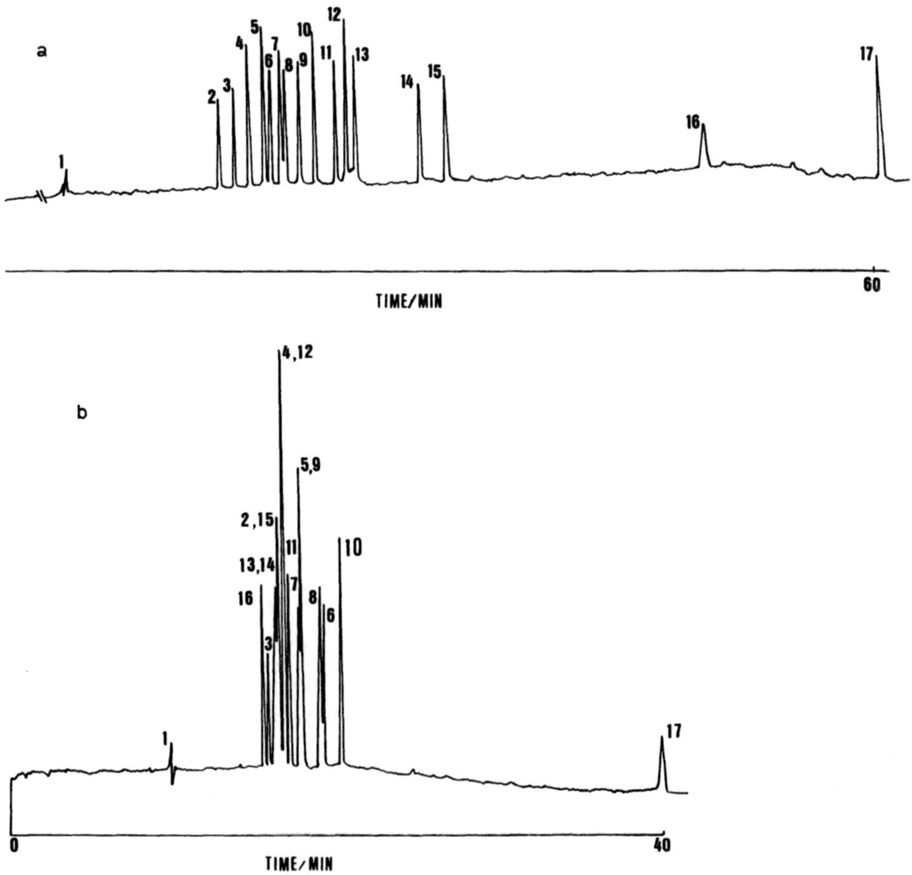

Fig. 6.28. (a) Optimized electropherogram of the separation of the sixteen DNP-amino acids. Electrophoretic condition corresponding to point X in Fig. 6.27: pH 10.0; 47.5 mM SDS; 20.0 mM TBA. (b) Electropherogram of the sixteen DNP-amino acids. Electrophoretic condition corresponding to point Y in Fig. 6.27: pH 10.0; 0 mM SDS; 20.0 mM TBA
Peak identification: *1* = methanol; *2* = N-2,4-DNP-DL-methionine sulfoxide; *3* = N-2,4-DNP-DL-citrulline; *4* = N-2,4-DNP-DL-Methionine sulphone; *5* = N-2,4-DNP-α-aminoisobutyric acid; *6* = N-2,4-DNP-sarcosine; *7* = N-2,4-DNP-γ-amino-*n*-butyric acid; *8* = N-2,4-DNP-β-alanine; *9* = N-2,4-DNP-DL-α-amino-*n*-butyric acid; *10* = N-2,4-DNP-Glycine; *11* = N-DNP-DL-norvaline; *12* = N-2,4-DNP-DL-Methionine; *13* = N-2,4-DNP-ϵ-amino-*n*-caproic acid; *14* = N-2,4-DNP-DL-ethionice; *15* = N-2,4-DNP-DL-norleucine; *16* = N-2,4-DNP-DL-α-amino-*n*-caprylic acid; *17* = N-2,4-DNP-DL-glutamic acid.

and the optimum concentration of surfactant [28]:

$$[\mathrm{SURF}]_{\mathrm{opt}(R_s)} = \frac{(t_{\mathrm{mc}}/t_0)^{1/2} + V_{\mathrm{cmc}}[(t_{\mathrm{mc}}/t_0)^{1/2} + P_{\mathrm{wm}})}{V[(t_{\mathrm{mc}}/t_0)^{1/2} + P_{\mathrm{wm}}]} \tag{6.11}$$

or

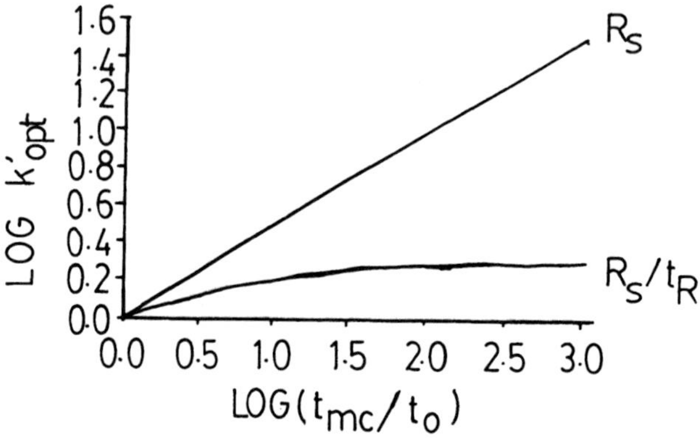

Fig. 6.29. Optimum capacity factor for the best resolution (Eqn. 6.19) and the best resolution per unit time (Eqn. 6.20). (Reproduced from Ref. 28 with permission of the American Chemical Society.)

$$[\mathrm{SURF}]_{\mathrm{opt}(R_s)} = \frac{(t_{\mathrm{mc}}/t_0)^{1/2}}{P_{\mathrm{wm}}V} + \mathrm{cmc} \qquad (6.12)$$

These results show that the concentration of surfactant to use for the best resolution of neutral solutes is essentially predetermined in MEKC and can be calculated from P_{wm}, the average of their partition coefficients, and four parameters: t_0, t_{mc}, V and cmc.

For the optimization of resolution per unit time (R_s/t_R), the following equations has been derived for the capacity factor:

$$k'_{\mathrm{opt}(R_s/t_R)} = \frac{-(1 - t_0/t_{\mathrm{mc}}) + [(1 - t_0/t_{\mathrm{mc}})^2 + 16\,(t_0/t_{\mathrm{mc}})]^{1/2}}{4(t_0/t_{\mathrm{mc}})} \qquad (6.13)$$

and:

$$[\mathrm{SURF}]_{\mathrm{opt}(R_s/t_R)} = \frac{k'_{\mathrm{opt}(R_s/t_R)}}{P_{\mathrm{wm}}V} + \mathrm{cmc} \qquad (6.14)$$

These results show that the concentration of surfactant to use for the best resolution per unit time is essentially predetermined in MEKC and can be calculated from a solute property, the average partition coefficient of two solutes (P_{wm}), and the four parameters: t_0, t_{mc}, V and cmc.

In Fig. 6.29 the optimum capacity factor for the best resolution (Eq. 6.10) and the best resolution per unit time (Eq. 6.12) is plotted against $\log(t_{\mathrm{mc}}/t_0)$.

Equations for resolution, analysis time, and resolution per unit time, which are applicable whenever the capacity factor is optimized, have also been given [28]. For the resolution (R_s) approach, the results are:

$$t_R = (t_{\mathrm{mc}}\,t_0)^{1/2} \qquad (6.15)$$

$$R_s = \frac{N^{1/2}}{4}\left(\frac{s_\alpha - 1}{s_\alpha}\right)\frac{(t_{mc}/t_0)^{1/2} - (t_0/t_{mc})^{1/2}}{2 + (t_{mc}/t_0)^{1/2} + (t_0/t_{mc})^{1/2}} \tag{6.16}$$

$$R_s/t_R = \frac{N^{1/2}}{4}\left(\frac{s_\alpha - 1}{s_\alpha}\right)\frac{1 - t_0/t_{mc}}{t_0\,[2 + (t_{mc}/t_0)^{1/2} + (t_0/t_{mc})^{1/2}]} \tag{6.17}$$

where s_α represents selectivity. For the resolution (R_s/t_R) approach, the following equations have been obtained:

$$t_R = \left(\frac{1 + k'_{opt(R_s/t_R)}}{1 + (t_0/t_{mc})k'_{opt(R_s/t_R)}}\right) t_0 \tag{6.18}$$

$$R = \frac{N^{1/2}}{4}\left(\frac{s_\alpha - 1}{s_\alpha}\right)\left(\frac{k'_{opt(R_s/t_R)}}{1 + k'_{opt(R_s/t_R)}}\right)\frac{1 - t_0/t_{mc}}{1 + (t_0/t_{mc})k'_{opt(R_s/t_R)}} \tag{6.19}$$

$$R_s/t_R = \frac{N^{1/2}}{4}\left(\frac{s_\alpha - 1}{s_\alpha}\right)\left(\frac{k'_{opt(R_s/t_R)}(1 - t_0/t_{mc})}{t_0(1 + k'_{opt(R_s/t_R)})}\right) \tag{6.20}$$

Electrokinetic equations for the resolution and migration time for micellar electrokinetic chromatography have also been derived by Ghowsi *et al.* [29]. These equations permit the determination of optimum resolution for MEKC separation of neutral solutes in three cases, where the migration mobility of the micelle is negative, zero, and positive. For the first case, which is similar to CZE, the condition which makes resolution approach infinity is established. For the two cases of positive and zero migration mobility of micelle, the optimal ranges of the capacity factors for good resolution and resolution per unit time are found to be between 2 and 5.

6.5.4 Computer simulation

A model for predicting migration times in solvent gradient MEKC has been developed by Powell and Sepaniak [30]. The solvent gradient at the inlet of the reservoir is expressed by the exponential gradient equation [31]:

$$\%_{res} = \left[1 - \left(\frac{V_0}{V_0 + \Delta Q t}\right)\frac{Q_{in}}{\Delta Q}\right][\%_{in}] \tag{6.21}$$

where $\%_{res}$ is the percent organic modifier in the reservoir at time t, V_0 is the initial volume of mobile phase in the inlet reservoir, $\%_{in}$ is the percent organic modifier in the solution being pumped into the reservoir, and ΔQ is the difference between the pump rate of solution into the reservoir (Q_{in}) and the pump rate of mixed solution out of the reservoir (Q_{out}).

In Fig. 6.30, separations performed using continuous concave, linear, and convex gradient of acetonitrile concentration are shown. The relationship between the

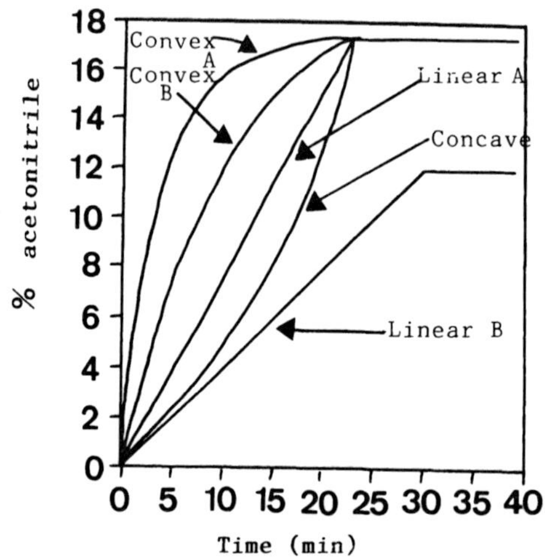

Fig. 6.30. Profiles of acetonitrile concentration gradients (v/v %) generated in the inlet reservoir. (Reproduced from Ref. 30 with permission of Aster Publishing Corp.)

gradient in the inlet reservoir and that in the capillary is determined experimentally. Using the isocratic data, a functional relationship between the capacity factor (k'), and % acetonitrile is determined for each solute. Each solute's function, together with the % acetonitrile matrix (Fig. 6.31) are used to calculate its retention time by summing the transit times for each capillary segment based on:

$$v_b = [1/(1 + k')](k'v_{mc} + v_{eo}) \quad (6.22)$$

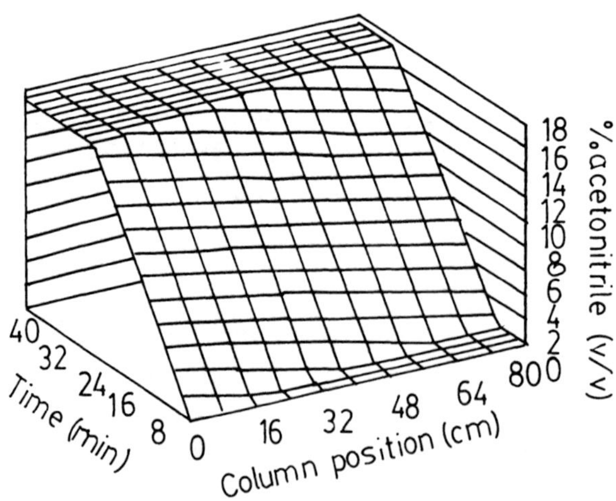

Fig. 6.31. Plot of the temporal/spatial/concentration matrix used to calculate migration times for the linear A gradient. (Reproduced from Ref. 30 with permission of Aster Publishing Corp.)

where v_b is the velocity of a neutral band, v_{mc} is the velocity of the micelle and v_{eo} is the electroosmotic velocity.

A comparison of the results of the calculations and the experimental migration times indicates that the method correctly predicts the elution order of all the components. The % deviation of the calculated values from the experimental values are less than 10% for the concave gradient, 12% for the convex A gradient, 18% for the convex B gradient, 26% for the linear B gradient and 28% for the linear A gradient. The results show that the computational model can be used to optimize a solvent gradient MEKC separation without having to perform trial-and- error experiments to assign the correct gradient.

Dose and Guiochon [32] performed detailed computer simulation of CZE and ITP experiments. The model used is one-dimensional and applies only to strong electrolytes [32–34]. The physical system simulated is reduced to solving the differential mass balance:

$$\frac{\partial C_{z,i}}{\partial t} = \frac{\partial J_{z,i}}{\partial z} + D\,\frac{\partial^2 C_{z,i}}{\partial z^2} \tag{6.23}$$

where t is time, z is position, $C_{z,i}$ is the concentration of ion i at position z, J_i is the migrational flux (whose sign indicates direction) of ion i at z, and D_i is the diffusion coefficient of ion i.

Simulations have been performed for analytical and overloaded injections. The expected profile area, A, for the analytical case is the product of the sample concentration and the displacement during the injection time t_{inj}, given by:

$$A = C_{inj,i}\, t_{inj}\left[v_{eo} - \frac{z_i\mu_i}{|Z_i|}\,E_{inj}\right] \tag{6.24}$$

where Z_i and μ_i are the charge and mobility of ion i, respectively. C_{inj}, t_{inj} and E_{inj} are the concentration of the sample injected, the duration of injection and the applied electric field during injection, respectively. The expected displacement during injection is calculated from:

$$z_i(t) = t\left[v_{eo} - \frac{z_i\mu_i}{|Z_i|}\,E_{inj}\right] \tag{6.25}$$

where E_{inj} is taken to be the average field over the column length. In the case of analytical injection, Eq. (6.23) reduces to:

$$\frac{\partial C_{z,i}}{\partial t} = \frac{\partial J_i^{OH}}{\partial z} + D\,\frac{\partial^2 C_{z,i}}{\partial z^2} \tag{6.26}$$

The flux J_i^{OH} is calculated from:

$$J_{z,i}^{OH} = C_{z,i}\left[v_{eo} - \frac{z_i\mu_i}{|Z_i|}\,E_z^{OH}\right] \tag{6.27}$$

Equation (6.27) assumes that for analytical injections, corrections due to ions' unequal diffusion are negligible and the flux J_i^{OH} is due solely to the Ohm's law contribution of the electric field given by:

$$E_z^{OH} = \frac{\pi r^2 I}{\kappa_z} = \frac{\pi r^2 I}{F \sum_i |Z_i| \mu_i C_{z,i}} \tag{6.28}$$

where r is the radius, I is the current, κ_z is the local solution conductance and F is the Faraday's constant.

The solution of Eq. (6.26) is a Gaussian whose variance σ_i^2 is given by the Einstein relation:

$$\frac{\partial \sigma_i^2}{\partial t} = 2 D_i \tag{6.29}$$

The expected profile variance of ion i at time t is then:

$$\sigma_i^2 = 2 D_i (t - t_{\text{inj}}/2) + \sigma_{\text{inj}}^2 \tag{6.30}$$

where σ_{inj}^2 for a rectangular injection is estimated to be:

$$\sigma_{\text{inj}}^2 = (E_{\text{inj}} \mu_i)^2 \frac{t_{\text{inj}}^2}{12} \tag{6.31}$$

Table 6.3 presents a brief summary of the simulation profile results. For analytical injections, the calculated area and first moment (or migration time) are identical to those of the Gaussian profile, whereas differences of up to 15% are observed for the variance. In the case of the overloaded injections, differences between the calculated values and the Gaussian values amount to 3% for area, 5% for first moment, and 28% for variance, respectively. It is expected that the model can be employed to study migration and separation characteristics in ITP and CZE separations, although a more realistic model may need to be adopted in certain applications.

6.5.5 Miscellaneous optimization techniques

Kuhr and Yeung [35] considered the simplest case where the separation observed in zone electrophoresis could be considered to depend only on the effective mobility of each analyte. They used the Henderson-Hasselbach equation to estimate the mobilities of nucleotides based on literature values for their absolute mobilities [36], pK_a values and the pH of the buffer solution:

$$\text{pH} = \text{p}K_a + \log\left(\frac{1}{\alpha - 1}\right) \tag{6.32}$$

together with the Tiselius equation [36]:

TABLE 6.3

SIMULATION PROFILE RESULTS (Adapted from Ref. 32)

	Analytical injection						Overloaded injection		
	$v_{eo} = 0$			$v_{eo} = 0.05\ cm^{-1}$			$v_{eo} = 0$		
	Gaus	calc	% dif	Gaus	calc	% dif	Gaus	calc	% dif
Mobility = 4×10^{-4}									
Area	0.04	0.04	0	0.09	0.09	0	40.00	41.22	3
$\bar{z}$	0.40	0.40	0	0.90	0.90	0	0.39	0.41	5
σ_z^2	2.04	1.90	7	2.04	1.90	7	2.25	2.85	27
Mobility = 5×10^{-4}									
Area	0.05	0.05	0	0.10	0.10	0	50.00	50.76	2
$\bar{z}$	0.50	0.50	0	1.00	1.00	0	0.49	0.50	2
σ_z^2	2.56	2.30	−10	2.56	2.37	−7	2.89	2.83	−2
Mobility = 7×10^{-4}									
Area	0.07	0.07	0	0.12	0.12	0	70.00	68.96	−1
$\bar{z}$	0.70	0.70	0	1.19	1.19	0	0.69	0.67	−2
σ_z^2	3.58	3.11	−13	3.58	3.34	−6	4.23	4.11	−3
Mobility = 8×10^{-4}									
Area	0.08	0.08	0	0.13	0.13	0	80.00	77.71	−3
$\bar{z}$	0.80	0.80	0	1.29	1.29	0	0.78	0.75	−3
σ_z^2	4.09	3.46	−15	4.09	3.77	−8	4.94	6.36	28

Profiles measured at 10 s after injection start. Analytical injections: electrokinetic, 10^{-5} M × 0.1 s. Overload injection: electrokinetic, 2.5×10^{-3} M × 0.4 s. Simulation parameters: column length 3 cm; voltage 300 V; electrolyte 0.01 M Na^+ (mobility 5.05×10^{-4}), 0.01 M anion of mobility 6×10^{-4}. Mobilities in $cm^2\ V^{-1}\ s^{-1}$; first moments $\bar{z}$ in cm; variance σ_z^2 in cm^2.
% difference calculated as: (calc − Gaus)/Gaus × 100%.

$$\mu_{eff} = \sum_i \alpha_i \mu_i \tag{6.33}$$

where α is the degree of dissociation, μ_{eff} is the effective mobility, and μ_i and α_i are the absolute mobility and the degree of dissociation of the ith ionic form of a molecule. Using tabulated literature values for pK_a values and absolute mobilities [37], a series of simultaneous equations have been constructed for the effective mobility of each nucleotide as a function of pH. These, in turn, can be used to calculate the resolution between adjacent components [38] as shown in Section 1.3:

$$R = 0.177\,(\mu_{eff1} - \mu_{eff2}) \left[\frac{V}{D\,(\bar{\mu} + \mu_{eo})}\right]^{1/2} \tag{6.34}$$

where R is resolution, V is the applied voltage, D is the diffusion coefficient, and $\bar{\mu}$ and μ_{eo} are the average mobility of the two components and the effective mobility of the electroosmotic flow, respectively. An optimum pH for the separation can be found by solving for the best total resolution (sum of resolution between each pair of components) or by specifying a minimum resolution for any pair.

Wren [39] optimized pH for the CE separation of 2-, 3-, and 4- methylpyridines. For the equilibrium:

$$BH^+ \overset{K_a}{\rightleftharpoons} B^+ + H^+ \tag{6.35}$$

$$K_a = \frac{[B][H^+]}{[BH^+]} \tag{6.36}$$

The average charge on a monobasic molecule is given by:

$$Z = \frac{[BH^+]}{[BH^+] + [B]} \tag{6.37}$$

Combining (6.36) and (6.37) gives:

$$Z = \frac{[H^+]}{K_a^+ + [H^+]} = \frac{1}{\dfrac{K_a^+}{[H^+]} + [H^+]} \tag{6.38}$$

In terms of pH and pK_a values, Eq. (6.38) becomes:

$$Z = \frac{1}{10^{(pH - pK_a)} + 1} \tag{6.39}$$

The charge difference between two compounds is given by:

$$\Delta Z = Z_1 - Z_2 = \frac{[H^+]}{K_1 + [H^+]} - \frac{[H^+]}{K_2 + [H^+]} \tag{6.40}$$

Maximum charge difference is obtained when:

$$\frac{d(\Delta Z)}{d([H^+])} = 0, \qquad \text{i.e. } [H^+] = (K_1 K_2)^{0.5} \tag{6.41}$$

or:

$$pH = 0.5\,(pK_1 + pK_2) \tag{6.42}$$

The pK_a values of the 2-, 3- and 4-methylpyridines are 5.97, 5.68 and 6.02 respectively. The charge difference of the methylpyridines as a function of pH is calculated using Eq. (6.39) and plotted in Fig. 6.32. From Fig. 6.32, it can be seen that maximum charge difference occurs at the pH range 5.8 to 6.2. Therefore optimum separation is expected to occur at this pH range. The separation between 3- and 4-methylpyridines at different pH is illustrated in Figs. 6.33 and 6.34.

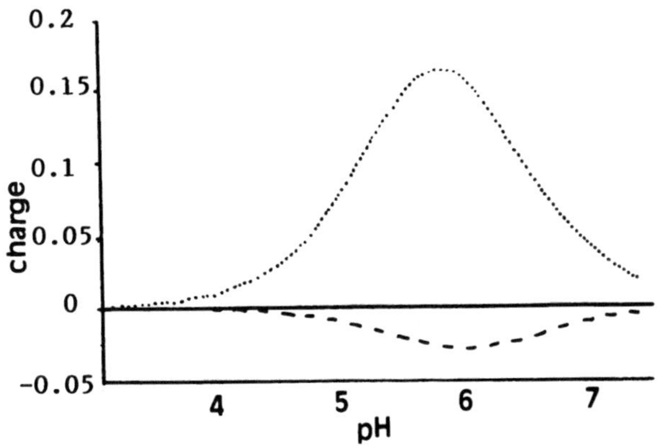

Fig. 6.32. Variation of charge differences between the methylpyridines, (dotted curve) 2-3 isomers and (dashed curve) 2-4 isomers, with pH. (Reproduced from Ref. 39 with permission of Microseparations Inc.)

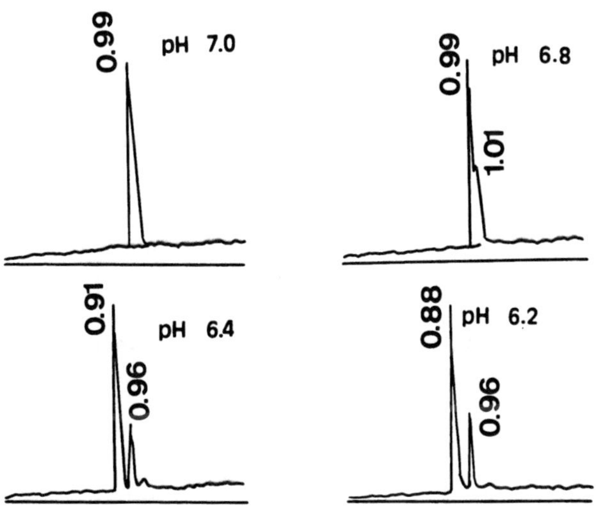

Fig. 6.33. Changes in separation between the 3- and 4-methylpyridines for pH 7.0–6.2. (Reproduced from Ref. 39 with permission of Microseparations Inc.)

6.6 DETERMINATION OF ELECTROPHORETIC MOBILITIES AND DIFFUSION COEFFICIENTS

Experimental data of relevant physico-chemical properties are necessary for the theoretical treatment and optimization of CE separations. CE itself can be exploited as a technique for measuring these physico-chemical constants [40–42], provided that accurate and reproducible electropherograms can be obtained for the system under investigation. The main advantages of methods based on CE in comparison

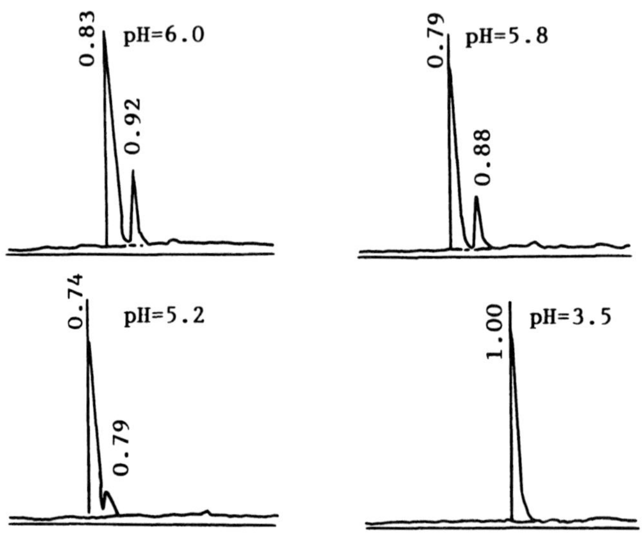

Fig. 6.34. Changes in separation between the 3- and 4-methylpyridines for pH 6.0–3.5. (Reproduced from Ref. 39 with permission of Microseparations Inc.)

with conventional techniques include: (1) A mixture of samples can be measured simultaneously, and (2) the necessary amount for the measurement is usually very small (typically of the order of several nanoliters per run).

CE has been used to determine mobilities and diffusion coefficient for analytes [40–42]. Mobilities can be determined from migration times, provided that correction to electroosmotic flow is made. The correction due to the electroosmotic flow can be determined by measuring the flow magnitude using an uncharged marker substance. The electrophoretic mobility (μ_{ep}) is determined using the following equation:

$$\mu_{ep} = \frac{L_{cap} L}{V t} - \mu_{eo} \tag{6.43}$$

where L_{cap} is the total length of the capillary, L is the length of capillary between the injection and detection point, V is the applied voltage, t the migration time of the solute and μ_{eo} is the coefficient of electroosmotic flow.

Diffusion coefficients can be measured from peak widths of analyte bands, provided that the analyte concentration is not so high that the system is overloaded and the axial homogeneity of the electric field is disturbed. It is also necessary to ensure that the analytes do not adsorb to the surface of the capillary, and the power dissipation is low enough to prevent heating of the capillary.

The calculation of the diffusion coefficient (D) is based on the Einstein equation:

$$D = \frac{\sigma_z^2}{2t} \tag{6.44}$$

where σ_z^2 is the band broadening in the axial direction due to longitudinal diffusion. In practice, two separate experiments are performed to eliminate the external contributions and to evaluate the contribution of molecular diffusion (σ_{dif}^2) to column band broadening. In the first experiment, a sample is injected into the capillary, then a certain high voltage is applied to initiate and complete the analysis. In a second experiment, an equal amount of the same sample is introduced into the capillary; this time the same voltage is applied for a short initial period only. After this short period of separation, the high voltage is shut down, i.e., the electrophoretic migration is interrupted. The solutes remained in the capillary for a defined period of time (Δt) before the high voltage is again switched on to complete the separation.

For this procedure, it may be assumed that the difference in the final band broadening between the two electrophoretic separations is predominantly caused by molecular diffusion during the Δt time period, i.e., band broadening originating from other sources such as extra-column effects, conductivity changes in the solute zone, and Joule heat effects are considered to be the same in both electrophoretic separations.

$$\sigma_{\text{total}}^2 = \sigma_{\text{dif}}^2 + \sigma_{\text{Joule}}^2 + \sigma_{\text{con}}^2 + \sigma_{\text{inj}}^2 + \sigma_{\text{det}}^2 \tag{6.45}$$

$$\sigma_{\text{total}}'^2 = \sigma_{\text{dif}}'^2 + \sigma_{\text{Joule}}^2 + \sigma_{\text{con}}^2 + \sigma_{\text{inj}}^2 + \sigma_{\text{det}}^2 \tag{6.46}$$

Thus, the influence of other contributions than molecular diffusion can be eliminated by substraction of the terms of Eq. (6.45) from those of Eq. (6.46). Hence,

$$\sigma_{\text{total}}'^2 - \sigma_{\text{total}}^2 = \sigma_{\text{dif}}'^2 - \sigma_{\text{dif}}^2 \tag{6.47}$$

Making use of Eq. (6.44), the following equation is obtained:

$$\sigma_{\text{total}}'^2 - \sigma_{\text{total}}^2 = 2D(t' - t) \tag{6.48}$$

and the diffusion coefficients of the analyte molecules can be calculated as follows:

$$D = \frac{(\sigma_{\text{total}}'^2 - \sigma_{\text{total}}^2)}{2\Delta t} \tag{6.49}$$

where σ_{total}^2 is the band broadening (σ = half the peak width at 0.6065 × peak height) in the first electrophoretic run, and $\sigma_{\text{total}}'^2$ is the band broadening in the second run with an interruption of $\Delta t = t' - t$, with D being the diffusion coefficient, and the subscripts "dif", "Joule", "con", "inj" and "det" refers to difference, Joule heating, conductivity changes, injection and detection, respectively.

6.6.1 Diffusion coefficients in free solution

The mobilities and diffusion coefficients of DNS-alanine and DNS-isoleucine was measured by Walbroehl and Jorgenson [40]. The results are given in Table 6.4.

TABLE 6.4

PHYSICAL PROPERTIES OF TEST COMPOUNDS MEASURED BY CZE
(Adapted from Ref. 40)

Compound	$\mu_{ep} \times 10^4$ ($cm^2\ V^{-1}\ s^{-1}$)	$D \times 10^6$ (EL) ($cm^2\ s^{-1}$)	$D \times 10^6$ (SM) ($cm^2\ s^{-1}$)	% diff
DNS-alanine	-1.51 ± 0.02	6.09 ± 0.19	6.01 ± 0.08	1.3
DNS-isolucine	-3751 ± 0.02	5.37 ± 0.33	5.35 ± 0.36	0.4
Myoglobin (horse)	-0.29 ± 0.02	1.15 ± 0.04	1.02 ± 0.65	12.7
Myoglobin (sperm whale)	-0.05 ± 0.02	-	1.13 ± 0.02	-

EL: measured "on-the-flight" by electrophoresis. SM: measured by "stopped migration".
% diff = (EL − SM)/SM × 100%.

Agreement between the values obtained by the CE method and the "stopped migration" method is within 1.3%. The CE method has been extended to the measurement of the mobilities and the diffusion coefficient of proteins. In the case of proteins, precautions need to be taken to prevent adsorption of the proteins by the capillary surface, since adsorption would cause peak distortion and irreproducible results.

6.6.2 Diffusion coefficients in gel-filled column

The diffusion coefficients of oligonucleotides in gel-filled columns have been measured by Yin *et al.* [41]. The oligonucleotide samples consisted of oligomers of the sequence d(GACT)$_n$ and d(GACT)$_n$GT. Therefore, the mixture contained oligomers with 8, 10, 12, 14, 16, 18, 20, 22, 24, 26, 28, 30 and 32 bases in length. Each sequence contained an equimolar ratio of purine and pyrimidines as bases.

The results obtained are shown in Fig. 6.35, where the diffusion coefficients are against the molecular masses of the different oligonucleotide species. The curve obtained indicates that the diffusion coefficients of the oligonucleotides decrease with an increase in the number of base units, i.e., increase in molecular weight.

6.6.3 Back-and-forth capillary electrophoresis

Terabe *et al.* [42] performed band broadening measurements by back-and-forth capillary electrophoresis. A schematic diagram of the experimental setup is shown in Fig. 6.36. A schematic of the back-and-forth electropherogram is shown in Fig. 6.37. The back- and-forth technique can be used to eliminate the band broadening contribution from extra-column effects:

$$\sigma_1^2 = \sigma_{ec}^2 + \sigma_{c1}^2 \tag{6.50}$$

$$\sigma_2^2 = \sigma_{ec}^2 + \sigma_{c2}^2 \tag{6.51}$$

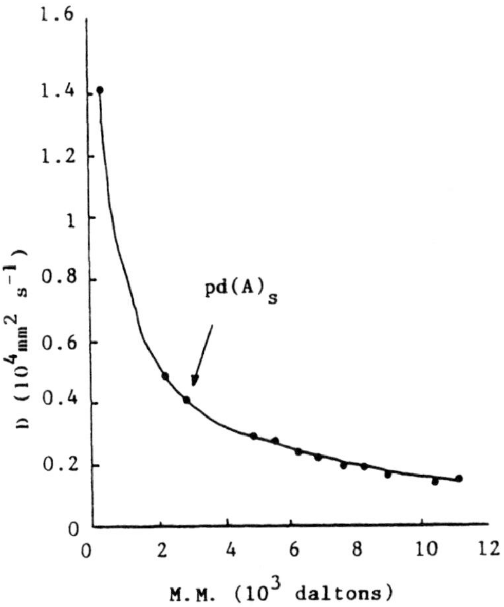

Fig. 6.35. Dependence of diffusion coefficients on oligonucleotide molecular mass. Sample: adenosine-5′-monophosphate, pd(A)$_6$, pd(A)$_8$, and oligonucleotide sizing marker (8-32mer). (Reproduced from Ref. 41 with permission of Microseparations Inc.)

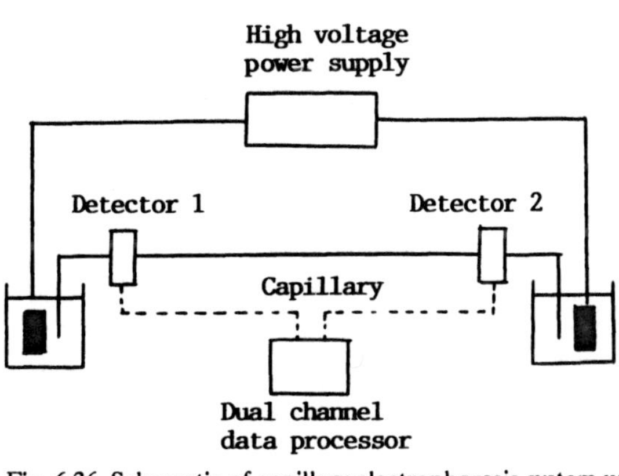

Fig. 6.36. Schematic of capillary electrophoresis system with on-column twin detectors. (Reproduced from Ref. 42 with permission of Dr. Alfred Huethig Publishers.)

where σ_1^2 and σ_2^2 are variances of the peaks observed at t_1 and t_2, respectively, σ_{ec}^2 extra-column variance due to both the injection volume and the detector cell volume, σ_{c1}^2 and σ_{c2}^2 variances generated in the capillary while the zone migrates from the injection end to detector 1 and from the injection end to detector 2, respectively. By subtracting Eq. (6.50) from Eq. (6.51), the extra-column variance is

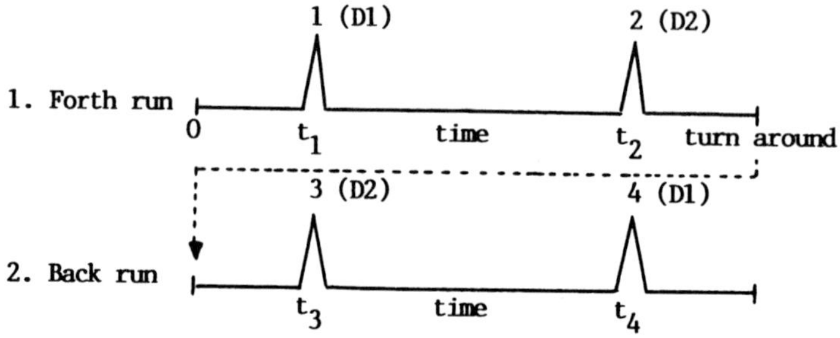

Fig. 6.37. Schematic of back-and-forth capillary electropherogram. (Reproduced from Ref. 42 with permission of Dr. Alfred Huethig Publishers.)

eliminated and the following equation is obtained:

$$\sigma_2^2 - \sigma_1^2 = \sigma_{c2}^2 - \sigma_{c1}^2 = \sigma_{c(2-1)}^2 \tag{6.52}$$

where $\sigma_{c(2-1)}^2$ is the variance generated in the capillary while the zone migrates from the injection end back to detector 1 to detector 2. Similarly, for peaks observed at t_3 and t_4:

$$\sigma_4^2 - \sigma_3^2 = \sigma_{c4}^2 - \sigma_{c3}^2 = \sigma_{c(4-3)}^2 \tag{6.53}$$

When band broadening in the capillary is caused only by irreversible processes, it is noted that $\sigma_{c(4-3)}^2$ is equal to $\sigma_{c(2-1)}^2$. The total variance is divided into reversible and irreversible contributions as follows:

$$\sigma_c^2 = \sigma_{rc}^2 + \sigma_{ic}^2 \tag{6.54}$$

where σ_{rc}^2 and σ_{ic}^2 are variances due to reversible and irreversible band broadening in the capillary, respectively. Consequently, $\sigma_{c(2-1)}^2$ can be divided into a reversible ($\sigma_{rc(2-1)}^2$) and an irreversible ($\sigma_{ic(2-1)}^2$) contribution. Similarly, for $\sigma_{c(4-3)}^2$, the reversible and irreversible contributions would be $\sigma_{rc(4-3)}^2$ and $\sigma_{ic(4-3)}^2$, respectively.

Provided that the migration velocity is constant, the migration times for the forth and back runs are the same. Hence the following relationships are obtained:

$$\sigma_{rc(2-1)}^2 = -\sigma_{rc(4-3)}^2 \tag{6.55}$$

$$\sigma_{ic(2-1)}^2 = -\sigma_{ic(4-3)}^2 \tag{6.56}$$

where the negative sign of $\sigma_{rc(4-3)}^2$ implies that variance decreases with the backward migration. The irreversible and reversible variances can be calculated as:

$$\sigma_{rc(2-1)} = (\sigma_2^2 - \sigma_1^2 + \sigma_4^2 - \sigma_3^2)/2 \tag{6.57}$$

$$\sigma_{ic(2-1)} = (\sigma_2^2 - \sigma_1^2 + \sigma_4^2 - \sigma_3^2)/2 \tag{6.58}$$

TABLE 6.5

VARIANCES (mm^2) AND EFFECTIVE DIFFUSION COEFFICIENTS (10^{-3} mm^{-2} s^{-1}) (Adapted from Ref. 42)

	$\sigma^2_{c(2-1)}$	$\sigma^2_{c(4-3)}$	$\Delta\sigma$	$\sigma^2_{ic(2-1)}$	$\sigma^2_{rc(2-1)}$	$\%\sigma_r$	D_i	D_r	$\%D_r$
Nitrobenzene	1.28	1.19	−7	1.23	0.04	3	1.36	0.09	6
Mesityl oxide	1.08	1.34	24	1.21	−0.13	−12	1.31	−0.06	−5
Micelle (Sudan IV)	0.83	−0.10	112	0.37	0.46	55	0.13	0.15	54
Micelle (timepidium bromide)	0.53	0.24	55	0.39	0.14	26	0.12	0.04	25

Note: Sudan IV and timepidium bromide are used as tracers for the micelles.
$\Delta\sigma = \sigma^2_{c(2-1)} - \sigma^2_{c(4-3)}$; $\%\sigma_r = \sigma^2_{rc(2-1)}/(\sigma^2_{ic(2-1)} + \sigma^2_{rc(2-1)}) \times 100\%$; $\%D_r = D_r/(D_i + D_r) \times 100\%$.

The effective diffusion coefficients are calculated according to the Einstein equation (Eq. 6.44), although it should be noted that these effective diffusion coefficients reflect the total dispersion and are not always identical to the molecular diffusion coefficient [42]. A summary of the data obtained is shown in Table 6.5. It is noted that for nitrobenzene and mesityl oxide, the differences in variances between the forth and back runs, and the reversible contributions to variance and to the diffusion coefficients are relatively small (24% and 12%, respectively). However, for the micelles, large differences are observed between the forth and back runs and the reversible contributions are more significant.

6.7 CAPILLARY ISOELECTRIC FOCUSING (CIEF)

Capillary isoelectric focusing (CIEF) is a relatively new mode of capillary electrophoresis [43–55] introduced by Hjerten and Zhu [44] in 1985. Isoelectric focusing has been commonly used in slab and tube gel electrophoresis [56]. These methods generally requires tedious and time-consuming gel preparation and protein staining procedures. In the capillary format, IEF separations can be run with or without a supporting gel. CIEF in free solution has been demonstrated in glass tubes [43–45], in rectangular cross-section channels [57–61] and in PTFE tubings [60,62,63].

Zwitterionic compounds with different pIs, such as proteins, peptides, amino acids, and various drugs, can be resolved by isoelectric focusing. To achieve complete focusing, the surface of the capillary should be treated to eliminate electroosmotic flow. A pH gradient is generated in the capillary with the cathode at the high-pH side of the gradient and the anode at the low-pH side. The sample is included with the ampholytes that generate the pH gradient. Zwitterionic analytes migrate to the position where they have a net zero charge, their isoelectric point (pI). Any band broadening caused by thermal diffusion is quickly reduced by the existing pH gradient. For instance, if an analyte drifts toward the low-pH side of the capillary, it becomes positively charged and migrates back toward its isoelectric point. One possible problem is that if the analytes are allowed to focus too tightly, they may become so concentrated that they exceed their solubility in solution and precipitate.

The separated analytes can be mobilized after the isoelectric focusing step and removed from the capillary by a number of simple techniques, including pressure differentials (positive pressure or vacuum) and electrophoretic elution, i.e., by replacing the acid at the anode by a base or the base at the cathode by an acid, which caused the pH gradient. The electrophoretic mobilization has the advantage that it is also applicable when the focusing is performed in a gel.

In a more recent approach, small amounts of additives, e.g. hydroxypropylmethylcellulose or methylcellulose, are added to the sample [50,51] or to the buffer [52]. The additives permit isoelectric focusing of proteins to be performed in untreated, open-tubular fused silica capillaries in the presence of an electroosmotic flow.

In the following sections, techniques for mobilization after isoelectric focusing are first described, followed by a brief description of coatings for capillaries used in CIEF. Techniques based on the use of sample or buffer additives to permit CIEF with electroosmotic flow are then discussed. Lastly, detection methods for CIEF are considered.

6.7.1 Hydrodynamic mobilization after isoelectric focusing

In a typical system for the isoelectric focusing and subsequent mobilization of the protein zones by hydrodynamic flow [44], the inside wall is coated with methylcellulose to eliminate zone distortion caused by electroosmosis and possible adsorption of the solutes to the tube wall. After focusing, a pump connected to the cathodic end of the electrophoresis tube via a T-tube is then started to deliver the anolyte into the electrophoresis tube at a slow rate (ca. 0.05 μl/min). The free end of the T-tube, closed with a dialysis membrane to prevent non-controllable liquid flow in the electrophoresis tube, is immersed in the cathodic vessel during both the focusing and the mobilization steps. The focused analyte zones are monitored as they are pumped past the stationary detector while the voltage is maintained during the pumping in order to eliminate distortion of the zones. The electropherogram is shown in Fig. 6.38.

6.7.2 Electrophoretic mobilization after isoelectric focusing

In a system for the isoelectric focusing and subsequent electrophoretic mobilization of the protein zones [44], a methylcellulose-coated electrophoresis tube with an agarose gel plug at the cathodic end of the electrophoresis tube which eliminates the risk of hydrodynamic flow in the tube is used. After focusing, the acid in

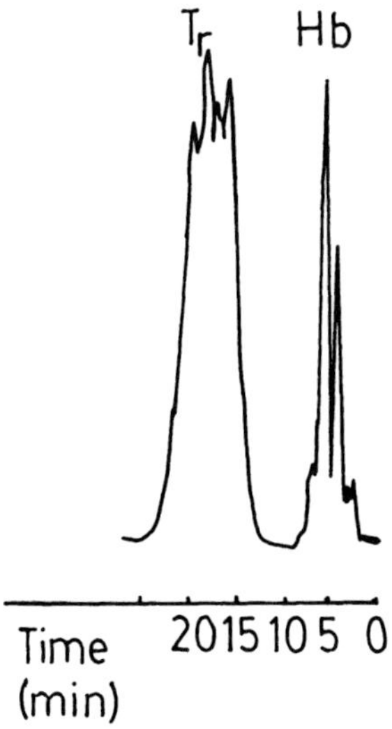

Fig. 6.38. Elution of focused protein zones by hydrodynamic flow and monitoring by on-tube detection. The sample consisted of 10 μg of human hemoglobin (*Hb*) and 15 μg of human transferrin (*Tr*). The protein zones were recorded as they passed a stationary UV detector monitoring at 280 nm. (Reproduced from Ref. 44 with permission of Elsevier Science Publishers.)

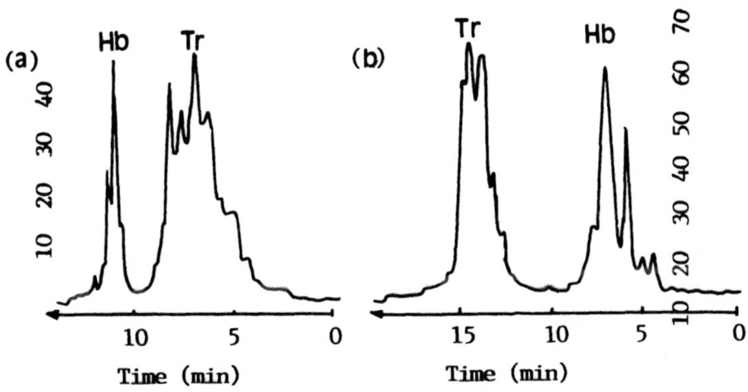

Fig. 6.39. Electrophoretic elution of focused protein zones and their monitoring by on-tube detection. The same sample and equipment were used as in the experiment shown in Fig. 6.38. The elution was achieved by replacing the acid at the anode with a base (a), or the base at the cathode with an acid (b). (Reproduced from Ref. 44 with permission of Elsevier Science Publishers.)

the anode vessel is replaced by a base and the voltage is increased. The current increased gradually during the mobilization. The electropherogram is shown in Fig. 6.39a. Mobilization can also be achieved by replacing the base at the cathode with an acid, as shown in Fig. 6.39b.

Hjerten *et al.* [45] derived expressions which describe the theoretical basis of isoelectric focusing–electrophoretic mobilization. The electroneutrality condition at steady state in the separation tube during focusing is:

$$C_{H^+} + \Sigma C_{NH_3^+} = C_{OH^-} + \Sigma C_{COO^-} \tag{6.59}$$

where C_{H^+}, $\Sigma C_{NH_3^+}$, C_{OH^-}, and ΣC_{COO^-} are the concentrations in equivalents per liter (or Coulomb/cm) of protons, hydroxyl ions and positive and negative groups in the carrier ampholytes, respectively.

The requirement for anodic mobilization of the ampholytes is that $\Sigma C_{NH_3^+} < \Sigma C_{COO}^-$ for any pH, i.e., that they acquire a negative net charge. This can be accomplished only if the pH increases, which is equivalent to a decrease in C_{H^+}; at the same time C_{OH^-} will increase since $C_{H^+} C_{OH^-}$ = constant. Under these conditions, the left side of Eq. (6.59), which applies only for the focusing step, will have a smaller value than the right side. Equality, and thereby mobilization, can be achieved by adding a positive term to the left side of Eq. (6.59), which then takes the form:

$$C_{X^{n+}} + C_{H^+} + \Sigma C_{NH_3^+} = C_{OH^-} + \Sigma C_{COO^-} \tag{6.60}$$

where $C_{X^{n+}}$ (n is the valency) represents a cation. This equation illustrates one way to accomplish anodic mobilization, namely by replacing the anolyte used for focusing with a cation which by electrophoresis can enter the tube.

The analogous expression for cathodic mobilization would be:

$$C_{H^+} + \Sigma C_{NH_3^+} = C_{OH^-} + \Sigma C_{COO^-} + C_{Y^{m-}} \qquad (6.60)$$

where Y^{m-} is an anion.

6.7.3 Coatings for capillary isoelectric focusing

A detailed description of techniques for coating capillaries in CE is given in Section 4.3. An important practical consideration in CIEF is the elimination of electroosmotic flow, which, if present, prevents stable, focused zones from forming. Therefore, coated capillaries are preferred for CIEF. Hjerten employed methylcellulose and polyacrylamide-coated columns for CIEF [43]. The methylcellulose-coated column is prepared by baking at elevated temperatures. The column coated with polyacrylamide is prepared by using 3-methacryloxypropyl-trimethoxysilane as coupling agent, then adding acrylamide monomer as well as polymerization agents potassium sulfate and TEMED, leading to a linear polyacrylamide coating. The polyacrylamide-coated capillary has been used in many applications of CIEF, including the separation of proteins [48] and human transferrin isoforms [47]. A typical electropherogram for the isoelectric focusing of a model protein mixture is shown in Fig. 6.40. However, one potential problem with these type of columns is that the polyacrylamide is attached to the silica via a siloxane bond. Therefore, the hydrolytic stability is not satisfactory at high pH [49]. Other columns described in Section 4.3 may be considered for CIEF applications.

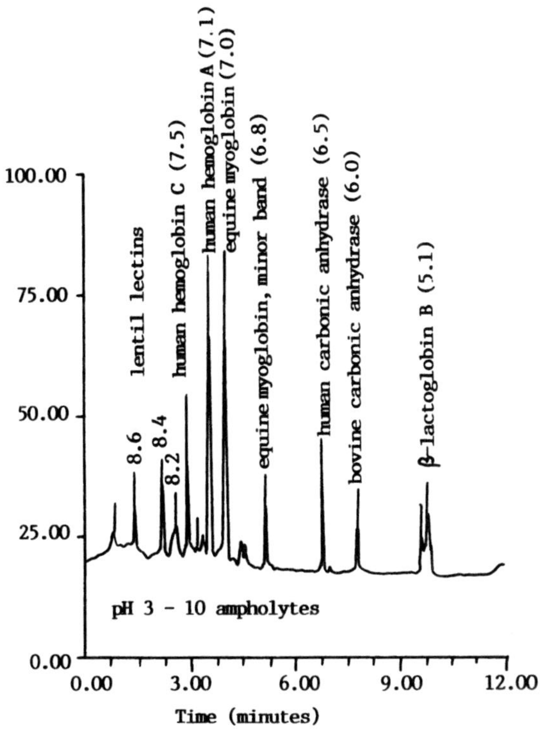

Fig. 6.40. Isoelectric focusing of model protein mixture using a polyacrylamide-coated capillary. (Reproduced from Ref. 51 with permission of Eaton Publishing Co.)

6.7.4 Buffer or sample additives in capillary isoelectric focusing

In CIEF separations, by adding a small amount of a suitable additive into the buffer, e.g. hydroxymethylpropylcellulose (HPMC), a dynamic coating can be formed which serves to reduce the interaction between the analytes and the walls [52]. A schematic representation of the isoelectric focusing process with buffer additives is shown in Fig. 6.41a. In this configuration, migration and focusing occur simultaneously in the presence of a small electrooosmotic flow.

Alternatively, sample additives may also be used to suppress electroosmotic flow and obtain satisfactory separation in uncoated columns or in the case that the column does not totally eliminate electroosmotic flow. Examples of this approach include the use of a PEG-coated capillary with a methylcellulose addition to the sample [50] and the separation of proteins in an uncoated capillary with 0.1% methyl cellulose added to the sample [51].

An example of isoelectric focusing in uncoated capillary with sample additive is shown in Fig. 6.41b. Since the electroosmotic flow is not completely eliminated, the mixture is focused and then eluted by the electroosmotic flow in one step without an extra mobilization step.

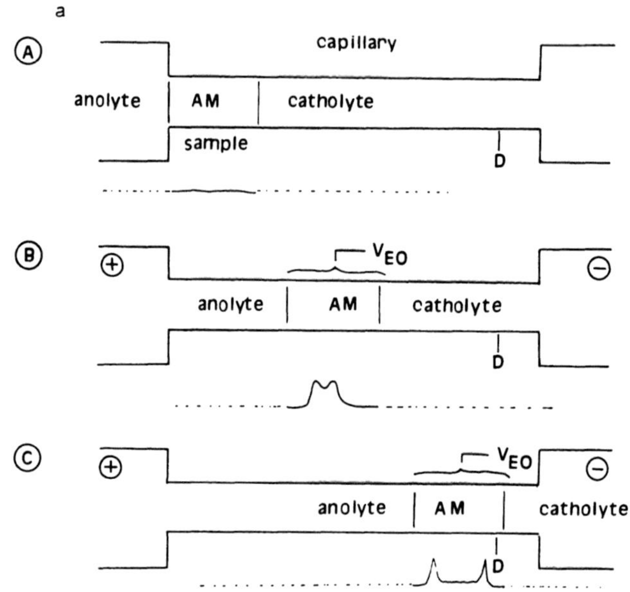

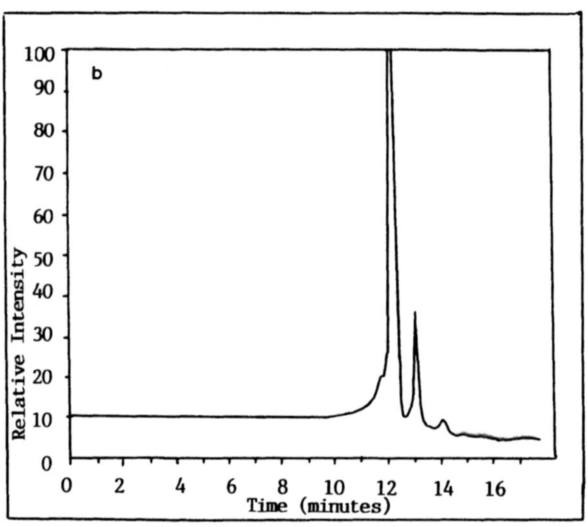

Fig. 6.41. (a) Schematic representation of the initial configuration (A), a transient state (B) and detection for capillary IEF in the presence of electroosmosis (C). *D* represents the point of detection and v_{eo} the electroosmotic displacement. Carrier ampholytes are represented by the region marked by *AM*. The line and peaks below the capillary represent protein distribution. (Adapted from Ref. 52.)

(b) Isoelectric focusing of horse heart myoglobin in uncoated capillary with 0.1% methyl cellulose added to the sample. Field: 300 V/cm; 5% Pharmalyte 3-10 (Sigma chemical), 1 mg/ml; 280 nm UV detection. Anode buffer 10 mM H_3PO_4, cathode buffer 20 mM NaOH. Focusing only, no mobilization. (Reproduced from Ref. 51 with permission of Eaton Publishing Co.)

During CIEF separations with electroosmotic flow, it is important to note that the presence of the electroosmotic flow may result in a decrease in resolving power. In the case of fast eluting peaks, there may be insufficient time for complete focusing before elution by this method [51].

6.7.5 Detection for capillary isoelectric focusing

Several types of detection methods have been employed for CIEF [52,54,55,57–59]. A method for continuous monitoring of CIEF separation [59] is by taking photographs of the focusing processes of blue dye stained proteins inside the capillaries (0.4–0.6 mm I.D.). The main disadvantages of this method include the need to label the protein, and the difficulty to obtain accurate quantitative information from the photographic film.

The focusing process has also been monitored by an electrode array detector [57,58,60,61]. As many as 100-electrode array have been used. However, relatively poor resolution is obtained with this method.

Another method used is based on UV absorbance measurements at the column end. This is usually performed by mobilizing the zones past the detector after focusing [43–45,62,63]. On-column multichannel UV-detectors can also be employed [52]. When CIEF with electroosmotic flow is used, on- line detection can be accomplished without the need for an additional mobilization step.

Wu and Pawliszyn described both a concentration gradient detector [54] for detection of mobilized zones, and a concentration imaging system for detection of zones without mobilization [55]. These detection systems are based on the use of lasers and are similar in principle to the concentration gradient detector described in Section 3.5.2. Detection limits are approximately 10^{-6} M for α-chymotrypsin. The sensitivity is of the same order of magnitude as that of a UV absorbance detector, although the concentration detectors possess the advantages that they are more universal and have smaller detection volumes (see Chapter 3).

6.8 CAPILLARY ISOTACHOPHORESIS

Isotachophoresis is a widely used separation technique which has been employed in many areas of applications [36,56–100]. As described in Section 1.2.6, although capillary isotachophoresis (CITP) separates analytes on the basis of their electrophoretic mobilities, the means of separation and the output derived from the separation are significantly different from other electrophoretic techniques. Typically, conventional ITP is useful only for the separation of ionic materials, and it is not possible to separate anions and cations in the same run. CITP separations usually require capillaries that have been treated so that the electroosmotic flow is eliminated, although CITP with electroosmotic flow has been demonstrated recently [78,83,93–96].

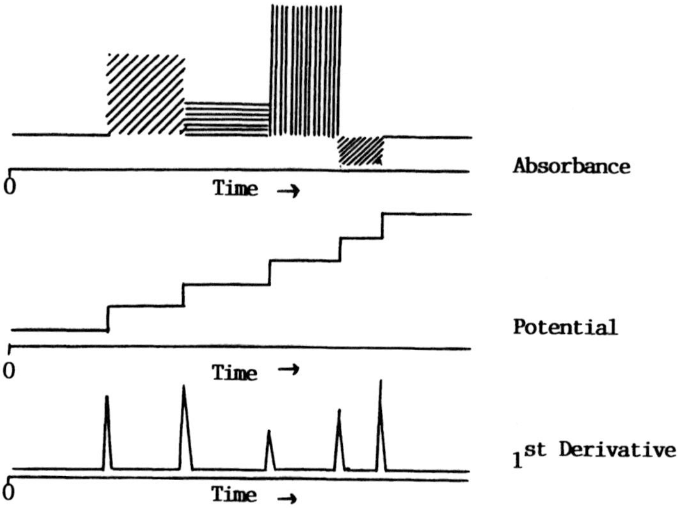

Fig. 6.42. Output of isotachophoresis. (Reproduced from Ref. 100 with permission of International Scientific Communications Inc.)

In ITP, the sample migrates between two solutions of different ionic mobility [36,71,72]. For example, in the separation of anions, a solution containing an anion of high electrophoretic mobility is placed at the anodic side and is used as the leading electrolyte (or buffer). On the cathode side of the sample, a solution containing an anion of low electrophoretic mobility is used as the terminating electrolyte (or buffer). These leading and terminating buffers, act to separate the anions of the sample from each other (see Section 1.2.6). The course of the separation may be monitored by the changes in absorbance, conductivity or temperature (due to Joule heat) as each successive analytes reaches the detector. The first derivative is often taken leading to sharp peaks, which appears similar to chromatograms [100] (see Fig. 6.42). However, the peaks do not represent the analytes themselves but are observed at the interfaces between two adjacent analytes. The peak height is not proportional to the amount of sample and migration time provides no indication of the identity of the analyte. Instead, the time between peaks is proportional to the amount of analyte present.

6.8.1 Theory of capillary isotachophoresis (CITP)

In capillary isotachophoresis, a steady-state configuration is obtained as the result of a separation process that proceeds according to the moving boundary principle [36,56–64]. The migration velocity, v, of a constituent, i, is given by the product of effective mobility, $\overline{\mu}_i$, and the local field strength, E [36,57–59]:

$$v_i = \overline{\mu}_i E \tag{6.62}$$

Since a constituent may consist of several forms of subspecies in rapid equilibrium, the effective mobility represents an ensemble average. The total constituent concentration, $\overline{C}_i$, is given by the summation of all the subspecies concentration, C_n.

$$\overline{C}_i = \sum_n C_{i,n} \tag{6.63}$$

The effective mobility is given by:

$$\mu_i = \sum_n \frac{C_{i,n}\,\mu_{i,n}}{\overline{C}_i} \tag{6.64}$$

where $\overline{\mu}_{i,n}$ is the ionic mobility of the subspecies. In dissociation equilibria, the effective mobility can be evaluated using the degree of dissociation, α:

$$\overline{\mu}_i = \sum_n a_{i,n}\, m_{i,n} \tag{6.65}$$

The degree of dissociation can be calculated once the equilibrium constant, K, for the subspecies and the pH of the solution are known (see Eq. 5.13).

The equation for continuity for electrophoretic process states [36]:

$$\frac{\partial}{\partial t}\overline{C}_i = -\frac{\partial}{\partial t}\left(-\frac{\partial}{\partial t} D_i\overline{C}_i + v_i\,\overline{C}_i\right) \tag{6.66}$$

where t represents time, v_i is the velocity of ion i and D is the diffusion coefficient. Neglecting diffusional dispersion, and applying Eq. (6.66) for each constituent and the overall summation of the constituents gives [36]:

$$\frac{\partial}{\partial t}\sum_i \overline{C}_i = -\frac{\partial}{\partial t} E \sum_i \overline{\mu}_i\overline{C}_i \tag{6.67a}$$

and,

$$\sum_i \overline{C}_i = \text{constant} \tag{6.67b}$$

For monovalent weakly ionic constituents, Eq. (6.66) can be written as [70]:

$$\frac{\partial}{\partial t}\overline{C}_i = -\frac{\partial}{\partial t} E\,\mu_i\,C_i \tag{6.68}$$

where μ_i and C_i are the mobility and the concentration of the charged species, i. Dividing by μ_i and applying the resulting relationship for each constituent yields:

$$\frac{\partial}{\partial t}\sum_i \overline{C}_i/\mu_i = -\frac{\partial}{\partial t} E \sum_i C_i \tag{6.69}$$

For electroneutrality, $\Sigma_i C_i = 0$. Hence

$$\frac{\partial}{\partial t}\sum_i \overline{C}_i/\mu_i = 0 \qquad \text{or} \qquad \sum_i \overline{C}_i/\mu_i = \text{constant} \tag{6.70}$$

Equation (6.70) is known as the Kohlrausch regulating function [63]. In an electrophoretic system, different zones can be present, where a zone is defined as a homogeneous solution separated by moving and/or stationary boundaries. Applying continuity principle (Eq. 6.66) to a boundary leads to the general form of the moving boundary equation [36,70]:

$$\overline{\mu}_i^k\,\overline{C}_i^k\,E^k - \overline{\mu}_i^{k+1}\,\overline{C}_i^{k+1}\,E^{k+1} = v^{k/(k+1)}\,(\overline{C}_i^k - \overline{C}_i^{k+1}) \tag{6.71}$$

where $v^{k/(k+1)}$ represents the drift velocity of the separating boundary between the zones k and $k + 1$. In the case of a stationary boundary, the boundary velocity is zero and Eq. (6.71) reduces to:

$$(\overline{\mu}_i^{k+1}\,\overline{C}_i^{k+1})/(\overline{\mu}_i^k\,\overline{C}_i^k) = E^k/E^{k+1} = \text{constant} \tag{6.72}$$

According to Joule's law, heat generation will occur resulting in different regions that are moving or stationary. In order to reduce the effects of temperature, relative mobilities, μ_{r}, can be introduced. The mobility of the leading ion constituent, μ_{L}, may be used as the reference mobility:

$$\mu_{\mathrm{r}i} = \mu_i/\mu_{\mathrm{L}} \tag{6.73}$$

As discussed in Section 5.1.3, for differential migration methods including isotachophoresis, the criterion for separation in isotachophoresis is that two ionogenic constituents i and j are expected to separate if their migration rates in the mixed state are different. In terms of their effective mobilities, the criterion for separation of ions i and j is given by:

$$\overline{\mu}_i/\overline{\mu}_j \neq 1 \tag{6.74}$$

When the effective mobility of i is higher than that of j the latter constituent will migrate behind the former. Consequently, two monovalent weakly ionic constituents fail to separate, if the pH of the mixed state causes the effective mobilities of these constituents to be equal.

In Fig. 6.43, the resolution lines and the detector outputs for the ITP separation of a relatively simple two-component sample, A and B, is illustrated [74]. Injection of a small amount of sample produces two zones, stacked between the leading constituent, L, and the terminating constituent, T. At a constant load of leading electrolyte, the first boundary is always detected at the same time interval, given by the resolution line between L and A. A sample load of 65 nmol gives a time-based zone length of 124.2 s for constituent A and detection must be started 1112 s after injection. For constituent B, the zone length is 148.1 s. Other sample loads produces proportional zone lengths. Due to the limited load capacity of the separation

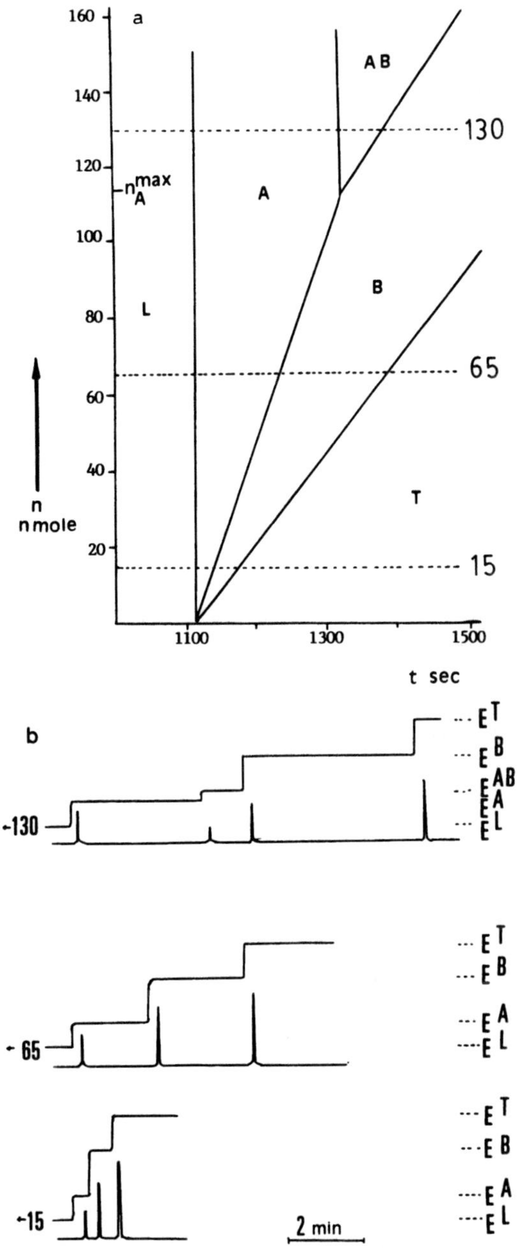

Fig. 6.43. (a) Resolution lines for a two-constituent mixture in isotachophoresis. Operational conditions: pH of leading electrolyte 4.03; leading constituent: 0.01 M Cl^-; counter constituent: 0.005 M C_2H_5COOH; additive to leading electrolyte: 0.05% polyvinyl alcohol. L = chloride; A = formate; B = glycolate; T = propronate; AB = mixed-zone; n = amount sampled. Sample $c_{formate}$ = 0.05 M; $c_{glycolate}$ = 0.05 M; pH_{sample} = 3.00. All constituents are monovalent weak electrolyte. Separation compartment of uniform dimension. Electrical current and temperature remain constant. (b) Isotachopherograms for n = 15, 65 and 130 nmol; $E^{L,A,B,T}$ = electrical field strength. (Reproduced from Ref. 74 with permission of Elsevier Science Publishers.)

compartment, too high a sample load results in a mixed zone, as illustrated by the mixed zone (AB) in Fig. 6.43 for an injected amount of 130 nmol.

6.8.2 Capillary isotachophoresis with electroosmotic flow

Typically in conventional isotachophoretic separations, electroosmotic flow is eliminated or minimized. After the separation process has been completed, all electrophoretic parameters remain constant with time. Assuming a uniform current density, all sample constituents between the leading–terminating electrolyte migrate at identical speeds. Moreover, at constant current density local migration rates will be constant [70].

Recently, capillary isotachophoresis with electroosmotic flow has been demonstrated [78,83,93–96]. Four distinct modes can be observed in CITP with electroosmotic flow: (a) cationic mode, (b) anionic mode, (c) reversed anionic mode, and (d)

TABLE 6.6

CITP MODES WITH ELECTROOSMOTIC FLOW AND CHARACTERISTICS

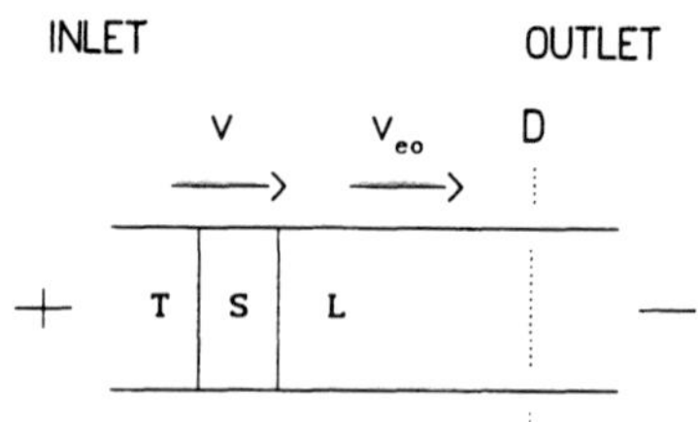

(a) Cationic:
1. T on inlet side
2. Anode on inlet side
3. v and v_{eo} same direction

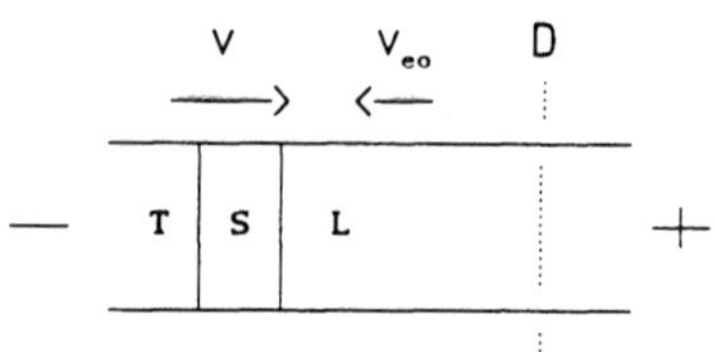

(b) Anionic:
1. T on inlet side
2. Cathode on inlet side
3. v and v_{eo} opposite direction
4. $v > v_{eo}$ for detection

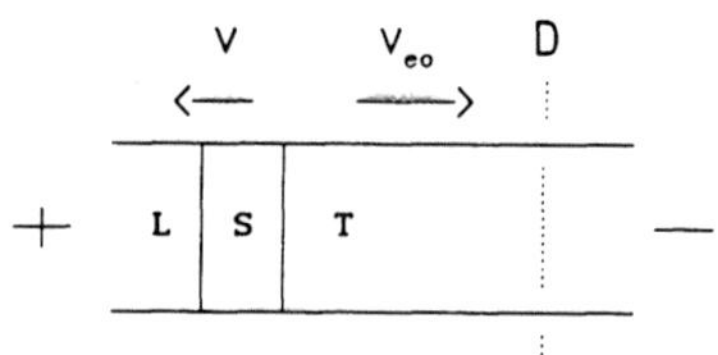

(c) Reversed anionic:
1. L on inlet side
2. Anode on inlet side
3. v and v_{eo} opposite direction
4. $v_{eo} > v$

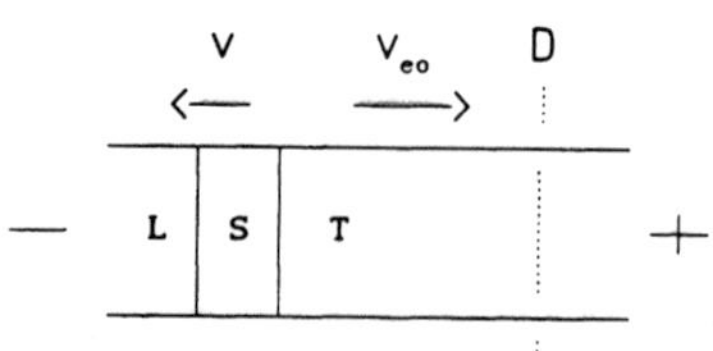

(d) Reversed cationic:
1. L on inlet side
2. Cathode on inlet side
3. v and v_{eo} opposite direction
4. $v_{eo} > v$

reversed cationic mode. In Table 6.6, different modes of CITP with electroosmotic flow and their characteristics are shown.

In the cationic mode, analytes migrate in the same direction as that of the electroosmotic flow, and hence the migration velocity of an analyte is increased. As a result the total separation length may not be used fully, and the separation power may be reduced compared with ITP without electroosmotic flow. Furthermore, in order to prevent strong fluctuations of the net migration velocity, a suitable terminator electrolyte should be chosen [78].

The polarity is reversed in the anionic mode, and the analytes migrate against the electroosmotic flow towards the detector. This mode is useful only if ν_{eo} is low, e.g. when non-aqueous solvents are used as the electrophoresis buffer.

As for the reversed anionic mode, the electroosmotic flow is in an opposite direction to, but faster than the migration velocity of the analyte. This mode of operation can be exploited in separations at high pH (high ν_{eo}).

In the reversed cationic mode, the electroosmotic flow is reversed. This can only be achieved by the use of additives or coated capillaries. A strong ν_{eo} is required. The electroosmotic flow is towards the detector whereas the species migrate in the opposite direction. Therefore, a very strong reversed electroosmotic flow is necessary for this mode of operation to be useful.

Mathematical modelling of CITP with electroosmotic flow has been considered [78]. Results show that the effect of the displacement of the electroosmotic flow is cancelled in all equations. Consequently, the mathematical model for ITP without electroosmotic flow [97,98] is also valid for ITP with electroosmotic flow [78].

Compared with CITP without electroosmotic flow (closed system), the main disadvantage of CITP with electroosmotic flow (open system) is that special precautions need to be taken in the choice of the electrolyte system in order to ensure the optimum operation of the system. Nevertheless, it has the advantage that the presence of the electroosmotic flow provides an easier method for coupling CITP to other systems, e.g. interfacing to mass spectrometry.

6.8.3 Capillary isotachophoresis with additives

In CITP, separation is based on the differential mobilities between analytes [36]. It is therefore important to maximized the differences in the effective mobilities of the sample ions. Additives are commonly used to introduce complex-forming equilibria, which permit the effective mobilities of analyte anions to be controlled, and hence to provide optimum separation.

The use of complexing additives in capillary electrophoresis has been discussed in Section 5.4. Typically, charged ligands or metal ions can be introduced as complexing additives in the electrolyte. Cyclodextrins are also commonly used to form inclusion complexes which provide host–guest interactions to enhance the separation of structurally similar compounds, including optical isomers and structural isomers (see Section 5.3). Figure 6.44a illustrates an example of CITP

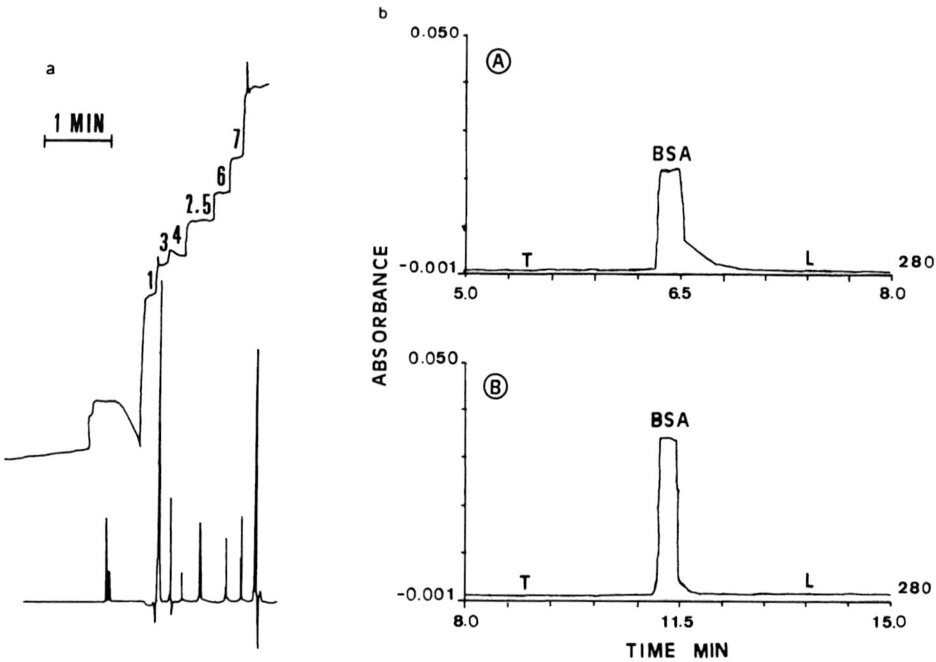

Fig. 6.44. (a) Capillary isotachopherogram of catecholamines. The β-CD concentration is 20 mM. Analytical column: 170 × 0.2 mm I.D. fused silica column. Pre-separation column: 8.0 × 0.7 mm I.D. PTFE. Leading electrolyte: cation, K^+; counter ion, CH_3COO^-; concentration of cation, 5 mM; additive, β-CD; surfactant, 0.05% polyvinyl alcohol or 0.1% Triton-X-100. pH 5.0. Terminating electrolyte: cation: β-aniline; counterion, Cl^-; concentration of cation, 10 mM; pH 1.7. *1* = 3,4-dihydroxybenzylamine; *2* = dopamine; *3* = normetanephrine; *4* = metanephrine; *5* = norepinephrine; *6* = epinephrine; *7* = isoproterenol.
(b) Anionic ITP of bovine serum albumin in (A) a 50 μm I.D. capillary without addition of HPMC, and (B) in a 75 μm I.D. capillary with HPMC. The initial and final currents were 1 and 4, and 1 and 8 μA, respectively. Detection by multiwavelength detector (195–320 nm at 5 nm intervals) Only electropherograms obtained at 280 nm are displayed.
(Reproduced from Refs. 92 and 96 with permission of Elsevier Science Publishers.)

separation employing β-cyclodextrin as an additive to the leading electrolyte for the separation of catecholamines.

Additives can also be used to enhance CITP separations by modifying the capillary surface [96]. Small amounts of hydroxypropylmethylcellulose (HPMC) added to the leading electrolyte forms a dynamic coating which reduces adsorption of proteins by the capillary wall and the electroosmotic flow. The additive has been used successfully in both anionic and cationic systems to improve CITP separations. The concentration of the additive used is typically 0.3%, either in a cationic system consisting of 0.01 M potassium acetate and acetic acid (pH 4.75) as the leader and 0.01 M acetic acid as the terminator, or in an anionic system containing 10 mM formic acid titrated with ammediol to pH 9.1 as the leader and 10 mM

β-alanine- ammediol of pH 9.5 as the terminator. The additive forms a dynamic coating on the surface of fused silica capillary which can be removed easily after each run by rinsing. With this type of coating, although electroosmotic flow is not completely eliminated, high-efficiency separation of proteins can still be achieved. Figure 6.44b shows an isotachopherogram obtained for proteins using CITP with HPMC additive.

Many other additives can be used to enhance CITP separations. As will be discussed in further detail in Chapter 7, capillary isotachophoresis, along with other modes of CE, is gaining popularity for the analysis of a wide variety of compounds.

6.9 HYPHENATED TECHNIQUES

According to Davis and Giddings [101,102], many samples may need to undergo more than one separation mechanism to reduce peak overlaps. For two techniques to be successfully and satisfactorily coupled, several criteria need to be considered. First, the two techniques should be as orthogonal to each other as possible. This means that the two techniques should base their respective separations on as different sample properties as possible. This will reduce the amount of cross-information and hence the two-dimensional (2-D) operation can be fully exploited. Secondly, it is necessary for the second dimension separation to sample the first dimension separation as frequently as possible. Ideally, the second dimension separation should sample the first dimension separation at least several times across each peak's width in order to preserved the first column's resolution. Three-dimensional (3-D) data representations should be employed to take advantage of obtaining peak profiles in two dimensions at the same time.

6.9.1 Coupled HPLC and CE

Many chromatographic techniques can be combined with capillary electrophoretic techniques to provide analytical information not obtainable by any of the techniques on its own. For instance, reversed-phase high-performance liquid chromatography (RP HPLC) and capillary zone electrophoresis (CZE) can be considered to be highly orthogonal separation methods. RP HPLC separates analytes on the basis of hydrophobicity and CZE separates analytes mainly on the basis of charge, and to a lesser degree on size. On this basis, they are good candidates for pairing in a 2-D system.

CZE has been used to analyze collected RP HPLC fractions of enzymatic digests of proteins and to compare the tryptic digest fingerprints of RP HPLC and CZE [103–105]. CZE was found to be able to resolve some peptides that coelute in RP HPLC. On the other hand, it has also been shown that RP HPLC is capable of resolving peptides that comigrate in CZE [106]. However, to fully exploit the high resolving power of the 2-D technique, automated instruments would have to be considered.

Bushey and Jorgenson [107,108] developed an automated instrument for comprehensive 2-D HPLC/CZE separations. In this system, the coupling of a liquid chromatography column with capillary electrophoresis is accomplished through the use of a six-port valve. The two valve configurations, "run" and "inject", are shown in Fig. 6.45. In the "run" position, effluent from the LC column (C1) fills the loop (L) while a piston pump (P2) continuously forces fresh buffer coaxially past the grounded (anode) end of the CZE capillary. A paper wick (PW) carries excess buffer away from the valve. In the "inject" position, effluent from C1 goes directly to waste while P2 forward flushes the contents of the loop past the grounded end of the CZE capillary for electromigration injection. At the end of the injection time, the valve is returned to the "run" position.

Figure 6.46 shows a schematic diagram of the instrumental setup. P1 is a syringe pump fitted with a mixer column (M). V1 is a manually operated six-valve equipped

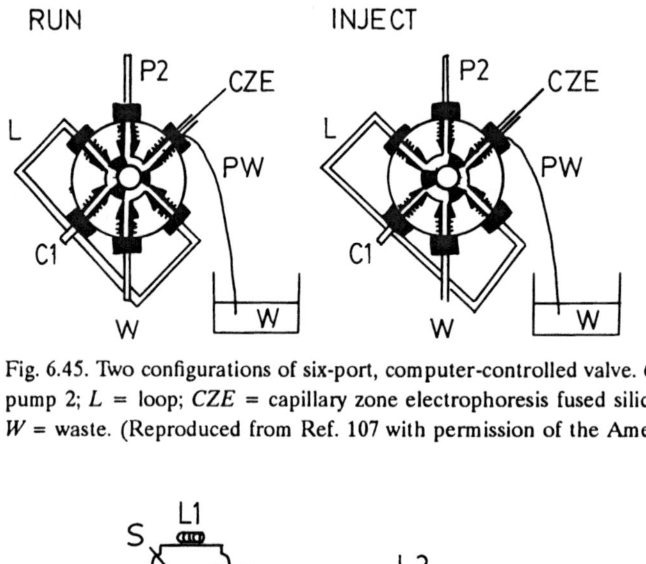

Fig. 6.45. Two configurations of six-port, computer-controlled valve. *C1* = RP HPLC column; *P2* = pump 2; *L* = loop; *CZE* = capillary zone electrophoresis fused silica capillary; *PW* = paper wick; *W* = waste. (Reproduced from Ref. 107 with permission of the American Chemical Society.)

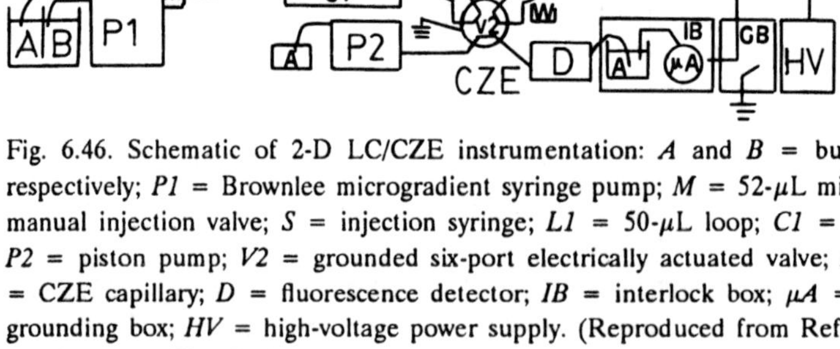

Fig. 6.46. Schematic of 2-D LC/CZE instrumentation: *A* and *B* = buffer A and acetonitrile, respectively; *P1* = Brownlee microgradient syringe pump; *M* = 52-μL mixer; *V1* = valco six-port manual injection valve; *S* = injection syringe; *L1* = 50-μL loop; *C1* = reversed-phase column; *P2* = piston pump; *V2* = grounded six-port electrically actuated valve; *L2* = 10-μL loop; *CZE* = CZE capillary; *D* = fluorescence detector; *IB* = interlock box; *μA* = microammeter; *GB* = grounding box; *HV* = high-voltage power supply. (Reproduced from Ref. 107 with permission of the American Chemical Society.)

with a 50 μl loop. V2 is an electrically actuated six-port valve, which has a 10 μl loop and is grounded. A power supply operating at negative polarity is used to perform CZE experiments. A personal computer is used to control the valve switching of V2 and the power supply, as well as to record data. The recorded data are plotted in 3-D with the aid of a graphics software. Figure 6.47 contains a computer plot of the 3-D chromatoelectropherogram of leucine-enkephalin, Met-Leu-angiotensin I, and methionine enkephalinamide, all of which are labeled with fluorescamine. These plots demonstrate the increased separating power of 2-D operation. Neither method, used alone, could separate all four analytes under these conditions. However, By combining the two methods, all four analytes are separated satisfactorily. The same procedure has been used successfully for the separation of fluorescamine labeled tryptic digest of ovalbumin.

After slight modification as shown in Fig. 6.48, the same system has been used to compare the tryptic digest fingerprints of horse heart cytochrome C and bovine heart cytochrome C. The two- dimensional system proves to be reproducible enough to identify some species differences in the tryptic digest fingerprints.

Gel permeation chromatography (GPC) has also been coupled with CZE. Yamamoto *et al.* [109] developed an apparatus for the analysis of complex protein mixtures in which open column gel permeation chromatography is combined with capillary electrophoresis. Proteins are separated according to their molecular size in the first step, then separated according to their electrophoretic mobility in the second step. The outlet of a microbore column is connected with the sample injection port of a capillary electrophoresis apparatus. A schematic diagram of the combined apparatus is shown in Fig. 6.49. A microbore chromatographic column made from polyethylene tubing packed with fine gel permeation support is set adjacent to the fully automated apparatus for capillary electrophoresis. A microperistaltic pump is employed to introduce the column effluent into the sample injection port of the electrophoresis apparatus. Perfluorinated ethylene-propylene (PFEP) tubing is used to connect the column outlet to the silicone rubber tubing of the microperistaltic pump. Another PFEP tube is used to transfer the column effluent from the silicone-rubber tubing of the pump to a glass capillary tube that has been inserted in the injection port of the electrophoresis apparatus, as shown in Fig. 6.49. Separation of albumin, myoglobin and tyrosine has been demonstrated with the combined apparatus (Fig. 6.50). This system suffers from the disadvantages that the second dimension analysis times are very long at 18 min each, and stopped-flow methods, which interrupt the separation processes, are used on the first column during the second dimension separation.

Subsequently the instrument has been modified for coupling high- performance gel permeation chromatography and capillary electrophoresis [110]. This apparatus has been used successfully for the automatic two-step separation of human serum proteins and water-soluble proteins in bovine brain. Examples of the results obtained with the combined apparatus for a soluble brain fraction is shown in Fig. 6.51.

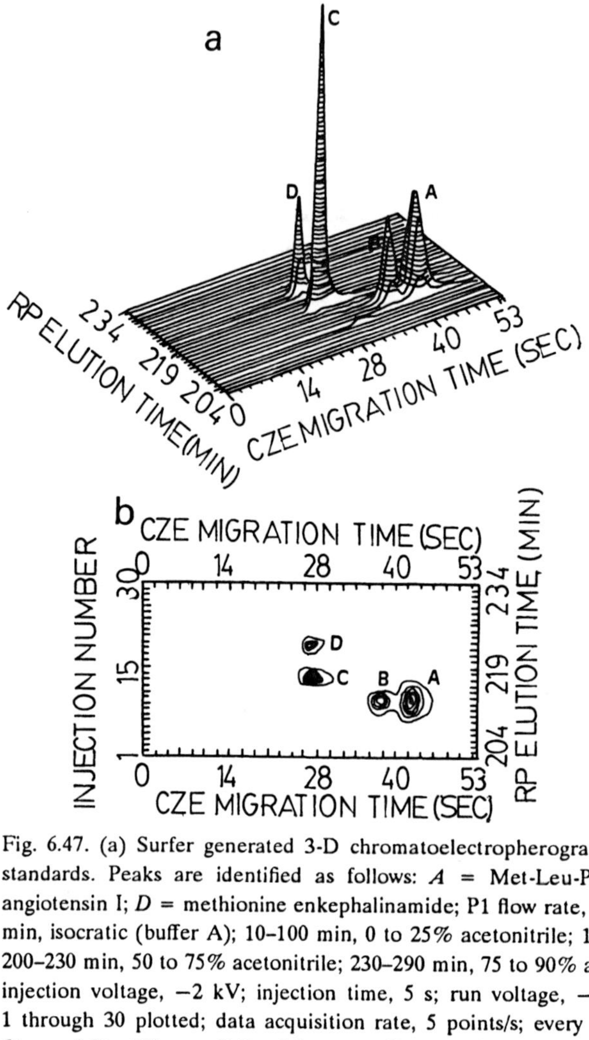

Fig. 6.47. (a) Surfer generated 3-D chromatoelectropherogram of fluorescamine labeled peptide standards. Peaks are identified as follows: *A* = Met-Leu-Phe; *B* = leucine-enkephalin; *C* = angiotensin I; *D* = methionine enkephalinamide; P1 flow rate, 10 μl/min; gradient conditions: 0–10 min, isocratic (buffer A); 10–100 min, 0 to 25% acetonitrile; 100–200 min, 25 to 50% acetonitrile; 200–230 min, 50 to 75% acetonitrile; 230–290 min, 75 to 90% acetonitrile. P2 flow rate, 0.5 ml/min; injection voltage, −2 kV; injection time, 5 s; run voltage, −19 kV; run time, 1 min; injections 1 through 30 plotted; data acquisition rate, 5 points/s; every other point plotted; CZE capillary, 50 μm I.D., 150 μm O.D., 6.5 cm to detector, 38 cm total length. (b) Contour plot of same chromatoelectropherogram. (Reproduced from Ref. 107 with permission of the American Chemical Society.)

6.9.2 Zone electrophoretic sample treatment

Debets *et al.* [111] described a zone electrophoretic sample treatment (ZEST) method based on the different migration velocities of compounds in free solution in an applied electric field. The system consisted of three interconnected pieces of capillary and is coupled on-line to an HPLC instrument. A schematic diagram of the setup for automated zone electrophoretic sample treatment is shown in Fig. 6.52.

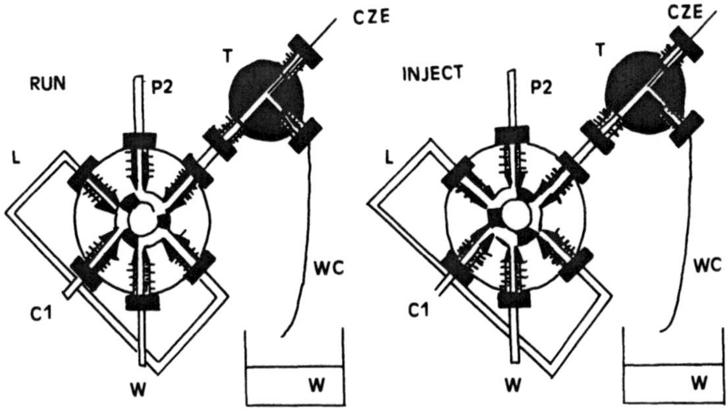

Fig. 6.48. Improved V2 connections. *C1* = RP HPLC column; *P2* = pump 2; *L* = loop; *CZE* = capillary electrophoresis fused silica capillary; *WC* = waste capillary; *T* = Valco low-dead-volume tee; *W* = waste. (Reproduced from Ref. 108 with permission of Aster Publishing Corp.)

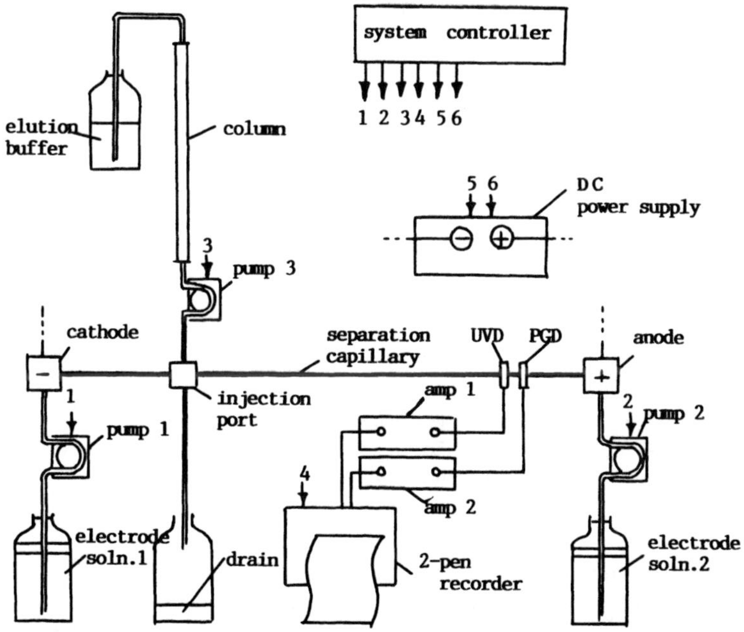

Fig. 6.49. Combined chromatographic and capillary electrophoretic apparatus. The terminating electrolyte (solution 1) and the leading electrolyte (solution 2) are pumped by peristaltic pumps (pumps 1 and 2) to wash the electrodes and separation capillary. The effluent from the column is loaded with the microperistaltic pump (pump 3) through the sample injection port. A d.c. high voltage is applied between the electrodes and the protein zones are detected with the potential gradient and UV detectors during the run. the numbered arrows indicate the output lines connecting the system controller to the equipment. (Reproduced from Ref. 109 with permission of Elsevier Science Publishers.)

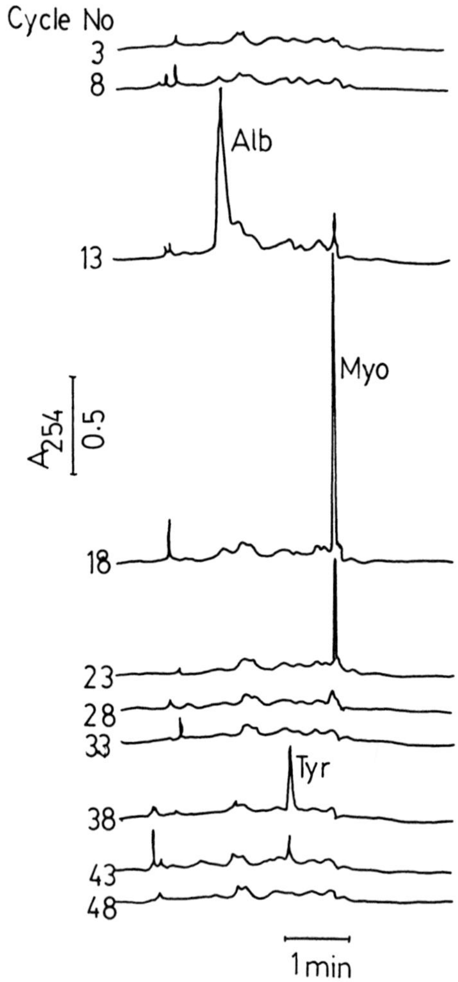

Fig. 6.50. Separation of albumin, myoglobin and tyrosine with the combined apparatus. The UV patterns of the electropherograms were traced every five cycles and the separation of the samples was demonstrated. (Reproduced from Ref. 109 with permission of Elsevier Science Publishers.)

The ZEST procedure consists of four stages: (1) all capillaries are filled with electrophoresis buffer; (2) the first (sample) capillary is filled with sample solution; (3) an electric field is applied over the connected capillaries and migration of ionic compounds with the appropriate charge takes place from the sample capillary, through the second (transfer) piece of capillary towards the injection capillary; and (4) after a given time the electrophoresis is stopped. The compounds that have migrated into the injection capillary are flushed by the mobile phase towards the column and separation by HPLC can be achieved. The recovery and reproducibility of the system has been evaluated using standard solutions of organic acids as model compounds. In Fig. 6.53, the results of recovery studies are shown. The recoveries

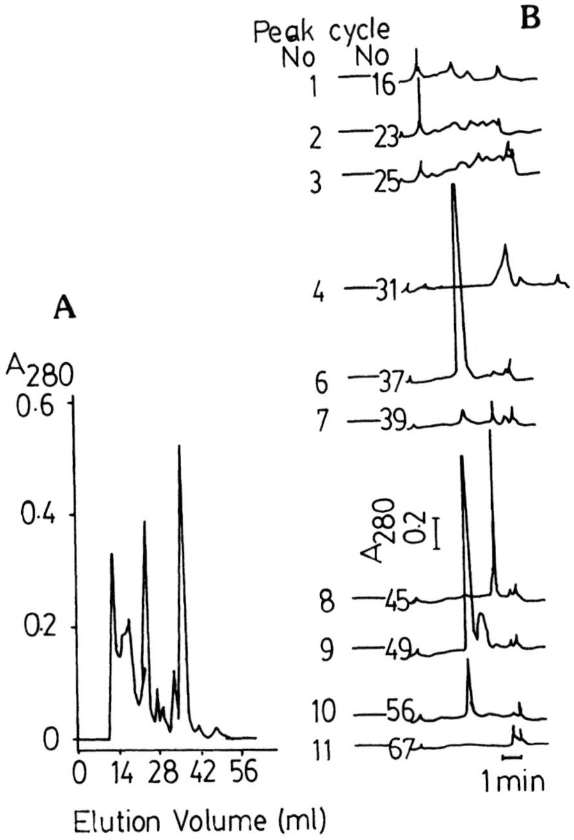

Fig. 6.51. Examples of the electropherograms obtained with the combined apparatus for a soluble brain fraction. Soluble fraction of bovine brain (30 μl) was applied to the combined apparatus. the electropherograms (UV patterns) corresponding to the UV peak positions of gel permeation HPLC (shown in A) are shown in B. (Reproduced from Ref. 109 with permission of Elsevier Science Publishers.)

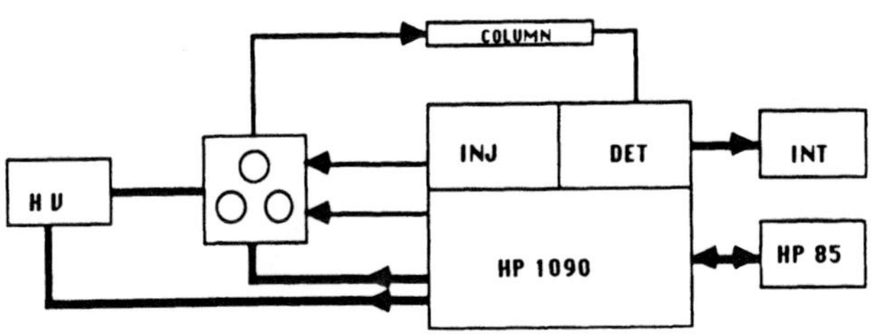

Fig. 6.52. Schematic diagram of set up for automated zone electrophoretic sample treatment. *INJ* = injector; *HV* = high-voltage power supply; *INT* = integrator; *DET* = detector. (Reproduced from Ref. 111 with permission of Elsevier Science Publishers.)

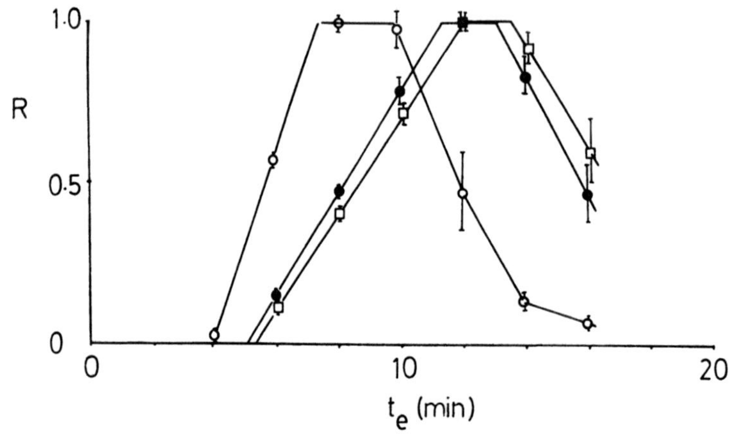

Fig. 6.53. Recovery (R) of (○) phthalic, (□) benzoic and (●) salicylic acid from 0.05 M phosphate buffer (pH 6.5) versus time of electrophoresis (t_e). Each point represents the mean and standard deviation of three experiments. Mobile phase, 0.05 M phosphate buffer (pH 2)–methanol (60 : 40). (Reproduced from Ref. 111 with permission of Elsevier Science Publishers.)

are calculated by comparing peak areas obtained after ZEST with those after direct injection of the standard solution. The plateaus corresponded to complete recovery of the analytes. The reproducibility is considerably better on the plateau than on the slopes of the recovery curve. The relative standard deviation is 2% on the plateau, but can be more than 20% on the slopes.

A new ZEST valve was subsequently described [112], which consists of three discs, held together by stainless-steel back and front discs (Fig. 6.54). Three

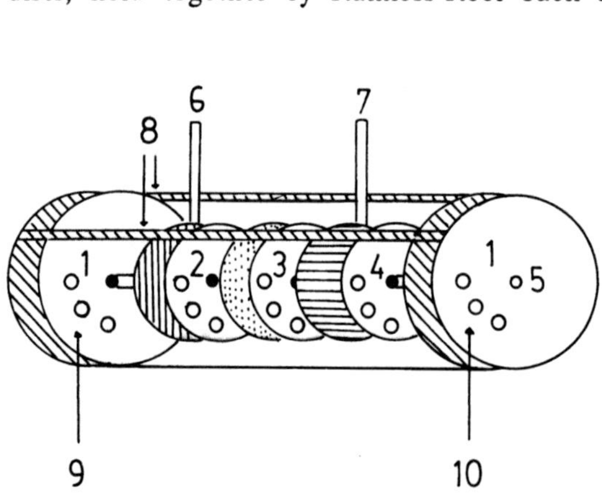

Fig. 6.54. Scheme of the ZEST valve, *1* = Stainless-steel connection disc; *2* = PTFE sample disc; *3* = Kel-F transfer disc; *4* = PTFE injection disc; *5* = central axis; *6* = switching bar of sample disc; *7* = switching bar of injection disc; *8* = switch position of switching bars; *9* = inlet channels; *10* = outlet channels. (Reproduced from Ref. 112 with permission of Friedr. Vieweg & Sohn Verlagsgesellschaft mbH.)

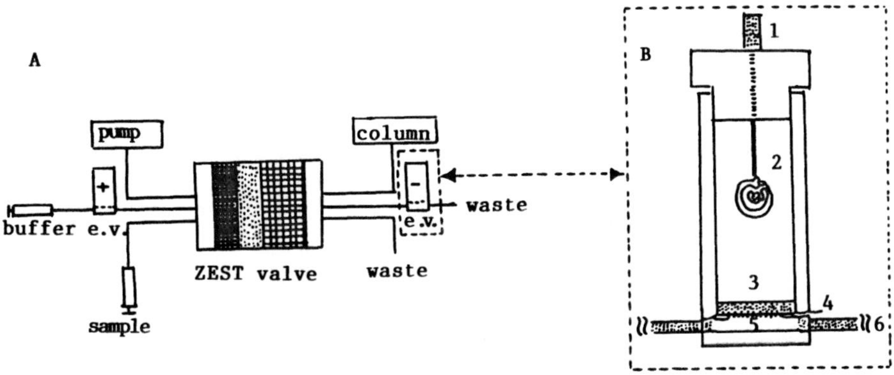

Fig. 6.55. Scheme of the ZEST system (A) (*e.v.* = electrode vessel), and floe-through electrode vessel (B). *1* = electrode connection; *2* = Pt-wire electrode; *3* = PTFE ring; *4* = ion-exchange membrane; *5* = flow-through chamber; *6* = PTFE tubing pressed into the electrode vessel. (Reproduced from Ref. 112 with permission of Friedr. Vieweg & Sohn Verlagsgesellschaft mbH.)

channels are drilled through the discs. Connections in the stainless-steel discs are made with 1/8 inch PTFE tubing (I.D. 1.6 mm) and home-made fingertight fittings. The sample, transfer and injection disc are made of PTFE, Kel-F and PTFE, respectively. The three inner discs are separated from the outer discs by disc-shaped spacers of Kel-F to isolate the metal parts of the valve from the flow channel in which the electrophoresis is performed. The transfer disc has a fixed position, while the sample and the injection disc could be switched to two different positions. This valve can withstand liquid pressures up to 60 bar during operation.

A schematic diagram of the ZEST instrument is shown in Fig. 6.55. The valve is placed between two home-made electrode vessels, each consisting of an electrode and a flow-through compartment. The ZEST valve is positioned in such a way that the channels in the discs can be flushed with buffer solution. The flow-through compartments are isolated from the electrode compartments by ion- exchange membranes (Fig. 6.55).

Figure 6.56 shows an example which illustrates the application of the ZEST procedure. In Fig. 6.56a an HPLC chromatogram of a raw urine sample injected directly onto the HPLC column is shown whereas in Fig. 6.56b shows the chromatogram of the same sample after 12 min ZEST. It can be clearly seen that the ZEST procedure is able to eliminate interfering compounds effectively.

6.9.3 On-line isotachophoretic sample preconcentration

The use of on-line isotachophoretic sample preconcentration for enhancement of zone detectability in capillary electrophoresis was investigated by Foret *et al.* [113], Kaniansky and Marak [114], and Stegehuis *et al.* [115,116]. The methods enhances detectability by several orders of magnitude [113–116].

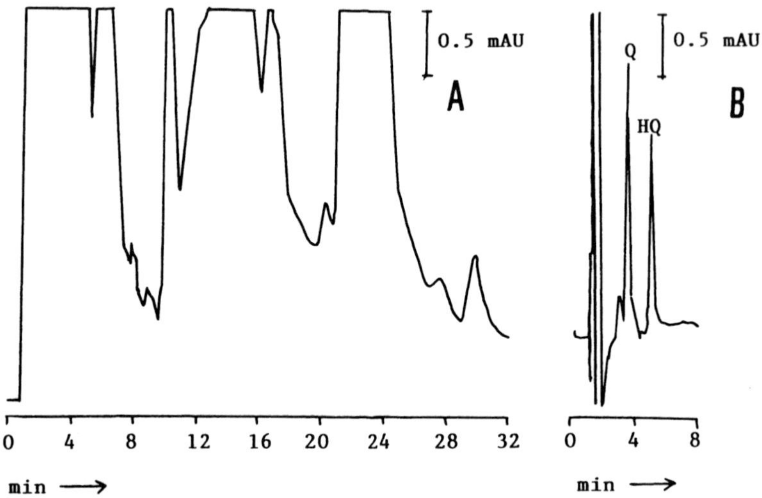

Fig. 6.56. Determination of quinidine (*Q*; 5.70 μg/ml) and hydroquinidine (*HQ*; 5.48 μg/ml) in urine (1 ml urine + 5 ml water) by direction injection of urine sample (A) and after ZEST (B). Wavelength, 245 nm. Time of electrophoresis, 12 min; current, 4.0 mA. (Reproduced from Ref. 112 with permission of Friedr. Vieweg & Sohn Verlagsgesellschaft mbH.)

The main advantage of the combination of capillary ITP (CITP) and CZE is that using CITP a sample having a volume of several microliters can be separated and concentrated into narrow zones of nanoliter volume. The main technical difficulty lies in the design of a column-coupling system with negligible dead volume [117,118].

A schematic diagram of an ITP-CZE apparatus is shown in Fig. 6.57. The preconcentration capillary (e.g. 0.5 mm I.D.) is equipped with an electrode chamber, a septum for the sample injection, and a channel with a valve. The CZE capillary is of smaller I.D. (e.g. 75 μm) and is inserted into the preconcentration column. In the configuration shown in Fig. 6.57a, the sample concentration step is performed and in Fig. 6.57b, the concentrated sample ions are already in the CZE capillary [113].

The first step of the sample introduction procedure is to fill both capillaries with the leading electrolyte, which contains an ion that had an electrophoretic mobility higher than any sample component of interest. Then the sample mixture is injected through the septum into the volume between the injection port and the electrode chamber. The electrode chamber is then filled with the terminating electrolyte, which contains an ion having an electrophoretic mobility lower than any sample component of interest. When the separation current is switched on, the sample species separate and concentrate into narrow isotachophoretic zones. The concentrations of these zones would not vary over time and would be given by the Kohlrausch regulating function [54]. For a monovalent ionic species A, the

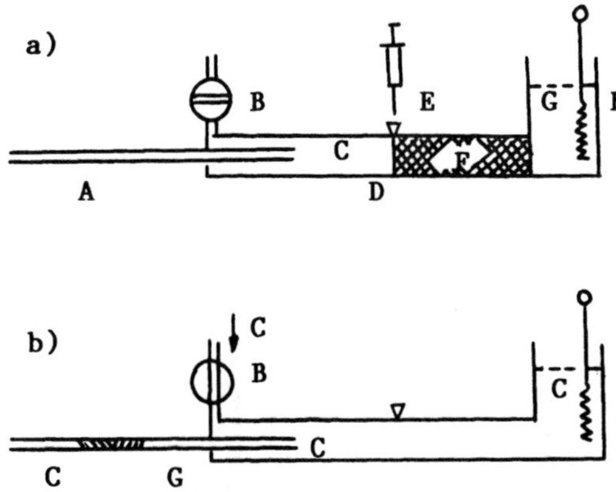

Fig. 6.57. Schematic of the ITP-CZE apparatus (a) during the injection of a low-concentration sample, and (b) after ITP preconcentration. A = fused silica capillary; B = valve; C = leading electrolyte; D = preconcentration column; E = injection port; F = sample mixture; G = terminating electrolyte; and H = electrode chamber. (Reproduced from Ref. 113 with permission of Aster Publishing Corp.)

concentration can be expressed as:

$$C_A = C_L(\mu_A/\mu_L)(\mu_L + \mu_R)/(\mu_A + \mu_R) \tag{6.75}$$

where C_A and C_L are the concentrations and μ_A and μ_L are the ionic mobilities of the separated and the leading ion, respectively, and μ_R is the mobility of the counter-ion. If the mobilities of the counter-ion and the sample species are comparable, Eq. (6.75) can be approximated as follows:

$$C_A = a_A C_L \tag{6.76}$$

where $a_A \cong 0.5$–0.9.

Thus, the ions present in the original sample at low concentrations are concentrated to nearly the level of the concentration of the leading ions. After the concentration step, the sample ions would have entered the CZE separation capillary and are followed by a short zone of terminating electrolyte. The ITP separation current is then switched off, and the terminating electrolyte is washed from the pre-separation capillary with leading electrolyte. The CZE driving current is then switched on, and the separation continued in the CZE mode and the leading electrolyte became the background electrolyte.

To illustrate the advantages of ITP-CZE the separation of nucleotides using CZE and ITP-CZE is described [113]. Figure 6.58a is an electropherogram obtained from the CZE separation of 45 nl of a mixture of nucleotides and an azo dye, amaranth. The concentration of each nucleotide is 300 μM, and the concentration of amaranth is 60 μM. Figure 6.58b shows the separation of 10.6 μl of the same model mixture. The concentration of each nucleotide is only 1 μM, and the concentration of

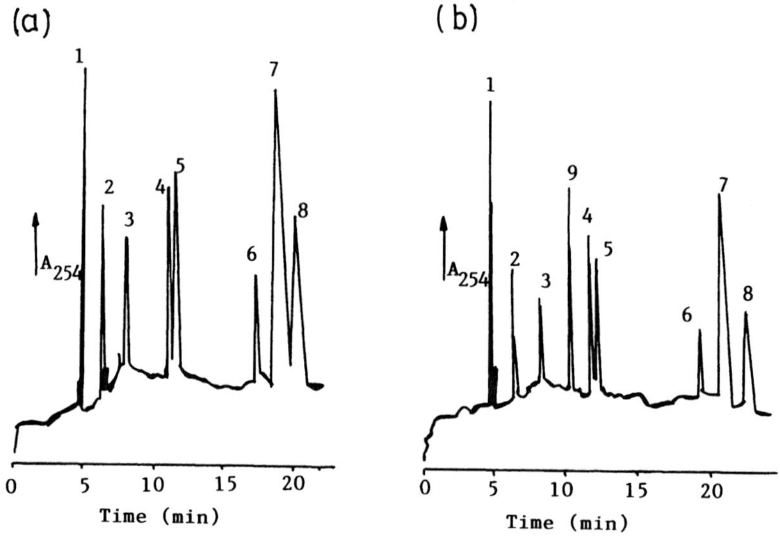

Fig. 6.58. Comparison of CZE and ITP-CZE, including (a) the CZE separation of a model mixture introduced by siphoning, and (b) the CZE separation of the mixture using ITP preconcentration. Background electrolyte in (a): 20 mM formic acid (0.1% octoxynol) adjusted to pH 5.85 with histidine. Separation current in (a): 36 μA at 13 kV. The sample volume in (b) was 10.6 μl, and its concentration was 300-fold lower than in (a). The leading electrolyte in (b) was the same as that used in (a). Terminating electrolyte in (b): 5 mM MES adjusted to pH 6.0 with histidine. Separation current in (b): 36 μA at 13 kV. Detection for both systems: UV, 254 nm. Peaks: *1* = amaranth; *2* = adenosine triphosphate; *3* = adenosine diphosphate; *4* = reduced nicotinamide adenine dinucleotide; *5* = nicotinamide adenine nucleotide diphosphate; *6* = cytidine monophosphate; *7* = adenosine monophosphate; *8* = guanosine monophosphate; *9* = impurity. (Reproduced from Ref. 113 with permission of Aster Publishing Corp.)

amaranth is 200 nM. In this example, ITP-CZE achieves approximately 200-fold better detectability than CZE at the same signal-to-noise ratio.

Kaniansky and Marak [114] coupled ITP and CE using fluorinated ethylene propylene (FEP) capillaries of the same diameter (0.3 mm I.D., 0.65 mm O.D.). A schematic diagram of the ITP-CE system is shown in Fig. 6.59. The ITP stage reduces the volume of the injected sample so that CZE separation with high efficiencies (2–3 μm plate height) could be achieved. Detection limits of 10^{-7} M for a 50-μl injection volume have been obtained by using a UV photometric detector.

Stegehuis *et al.* [115] employed an alternative design of the coupled ITP-CE system (Fig. 6.60). In this design, the CE capillary is inserted into the PTFE capillary of the ITP system as close as possible to the UV detector cell. As a result, the ITP system is hardly affected by the coupling to CE. The CE part of the coupled system performs as a separate CE system, except for the fact that the CE anode is now the electrode of the leading electrolyte buffer vial of the ITP system, which is connected with earth. For the ITP-CE operation, a reversible power supply is used.

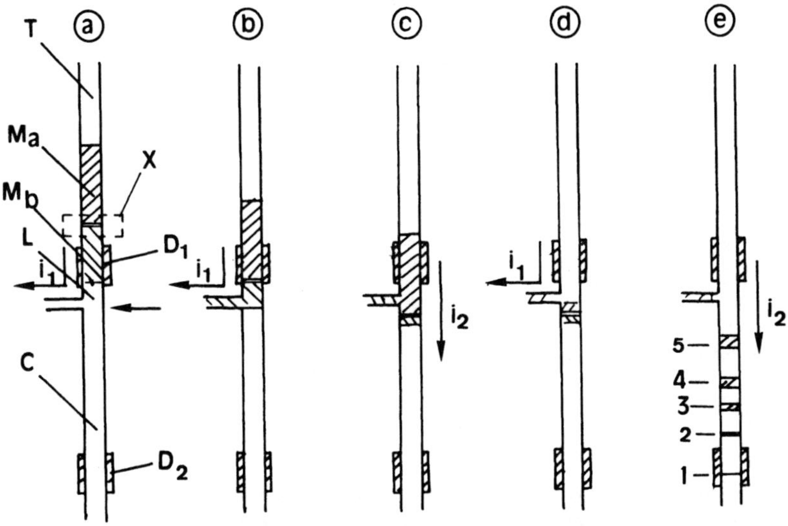

Fig. 6.59. Schematic illustration of the separation phases in combining ITP with ZE in the column-coupling configuration of the separation unit. (a) ITP separation in the first column (ITP stage); (b) removal of matrix constituent(s), M_b, from the separation compartment; (c) transfer of the sample fraction containing the analyte(s), X, into the second column (ZE stage); (d) removal of matrix constituent(s), M_b, from the separation compartment; (e) ZE separation in the second column (ZE stage). L, T = leading and terminating zones, respectively; C = carrier electrolyte; D_1, D_2 = detectors for ITP and ZE stages, respectively; i_1, i_2 = directions of the driving currents; *1–5* = symbols for the separated constituents in the ZE stage. The arrow on the right in (a) indicates the bifurcation point. (Reproduced from Ref. 114 with permission of Elsevier Science Publishers.)

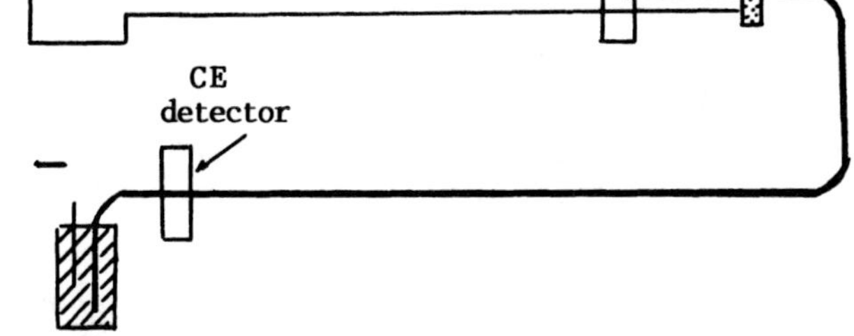

Fig. 6.60. Scheme of coupled ITP-CE system. (Reproduced from Ref. 115 with permission of Elsevier Science Publishers.)

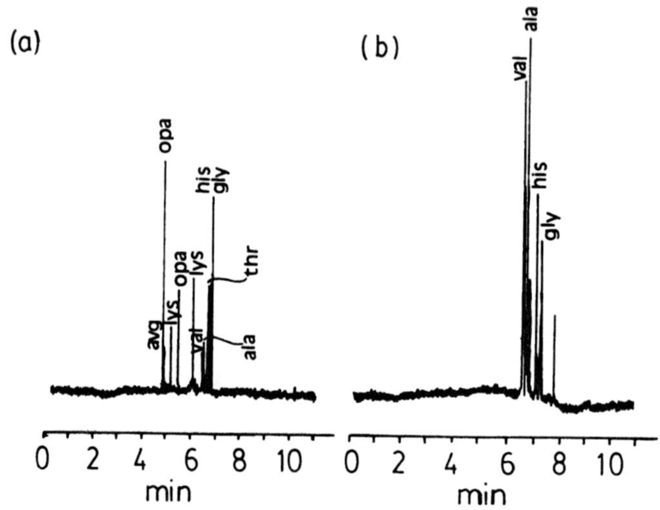

Fig. 6.61. (a) Single capillary electropherogram of some OPA-derivatized amino acids. Electrokinetic injection (5 kV, 5 s) of a test mixture with each amino acid at a concentration of about 100 ng/ml. Applied voltage: 25 kV; detection: UV absorbance at 234 nm. (b) Coupled ITP-CE of OPA-derivatized amino acids. ITP injection; 25 ml of a test mixture with each amino acid at the concentration of about 1 ng/ml. Injection in CE electrokinetic (5 kV, 5 s). Applied voltage in ITP 8 kV and in CE 25 kV. (Reproduced from Ref. 115 with permission of Elsevier Science Publishers.)

Figure 6.61a shows the electropherogram of a mixture of a number of amino acids derivatized with o-phthaldialdehyde (OPA) obtained with CE alone, and Fig. 6.61b is the result of the coupled ITP-CZE system. Enhancement in detectability of over two orders of magnitude is observed.

Subsequently, an improved design of the ITP-CZE system has been developed [116]. In this design, the exact position of the capillary is determined by inserting the CE capillary just before blocking the light beam of the UV source. Minimum detectable concentrations are three orders of magnitude lower than that obtained for conventional CE without coupling to the ITP system.

6.9.4 Combined open-tubular and packed capillary columns

The use of packed columns in CE has been described in Section 4.4. A possible problem with filling the entire capillary for CE with packing materials is that excessively high pressure drop may be obtained, which tends to result in operational difficulties such as the need to use high-pressure pumps for the filling and replacement of electrolyte in the capillary. An alternative approach is to combine an open-tubular column with a short section of capillary packed with stationary materials [9,119,120]. This approach has the advantages that the pressure drop can be greatly reduced, a wide variety of stationary phase materials can be used to enhance selectivity, and on-line sample concentration can be achieved.

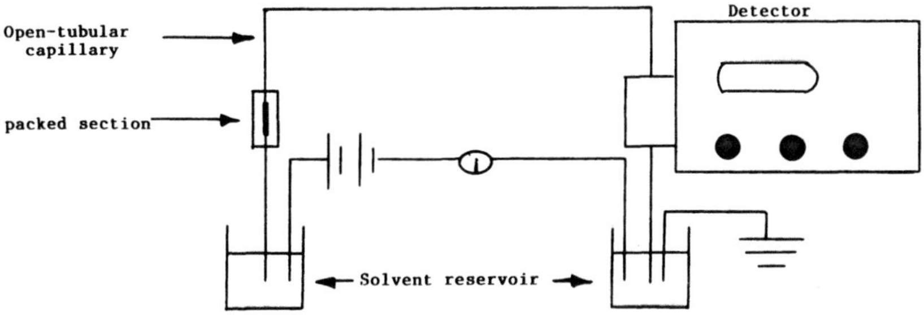

Fig. 6.62. Schematic of the combined packed and open tubular capillary electrophoresis system.

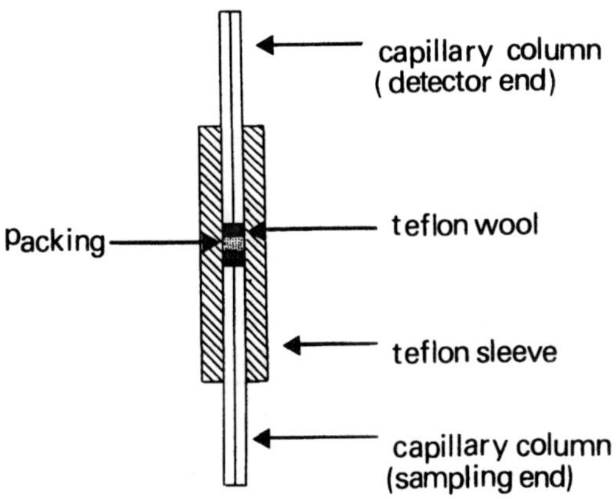

Fig. 6.63. Schematic diagram of the mini packed column used in this investigation.

A schematic diagram of a CE system employing a combined open-tubular and packed capillary column is shown in Fig. 6.62. A schematic diagram of the packed section is shown in Fig. 6.63. Preliminary results have been obtained for the separations of a group of nitrosamines (see Fig. 6.64). Improved resolution compared with separation by micellar electrokinetic chromatography is observed [119].

Merion *et al.* [120] constructed a fused silica capillary that contained a 1–2 mm packed bed of C_{18} chromatographic material at the injection end. A schematic diagram of the preconcentration capillary is shown in Fig. 6.65. To utilize this system, large volumes of dilute peptide samples are injected using conditions where the sample binds to the C_{18}. After sample loading, a small volume of organic solvent is injected and electrophoresis is performed in the normal way. Using this technique, peptides have been detected at original sample concentration down to approximately 10^{-8} M. In Fig. 6.66, improvement in detection sensitivity of trace impurities using this pre-concentration capillary is demonstrated.

Another approach for coupling packed and open tubular columns is based on the use of an analyte concentrator containing an antibody covalently bound

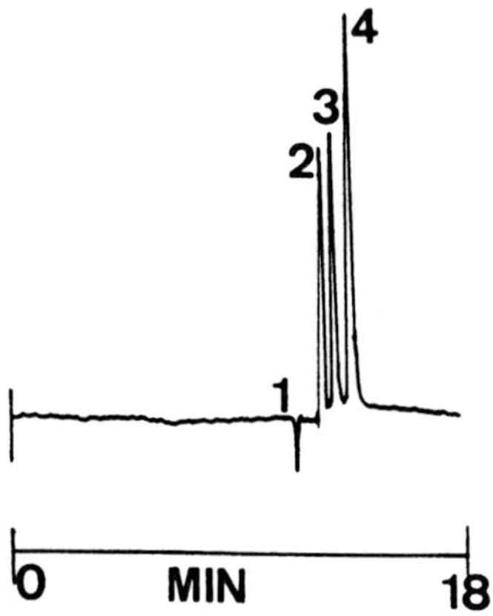

Fig. 6.64. Electropherogram of nitrosamines obtained using a combined packed and open tubular column with 80% C_{18} and 20% silica packing. Packing thickness: 1 mm. Electrophoretic conditions: 0.05 M borate–0.05 M phosphate buffer at pH = 6.6 with 40 mM SDS; 75 μm × 50 cm fused silica tubing; detection wavelength: 254 nm; voltage : 12.5 kV. Peak identification: *1* = MeOH; *2* = N-nitrosodiethanolamine; *3* = N-nitrodimethylamine; *4* = N-nitrosomorpholine.

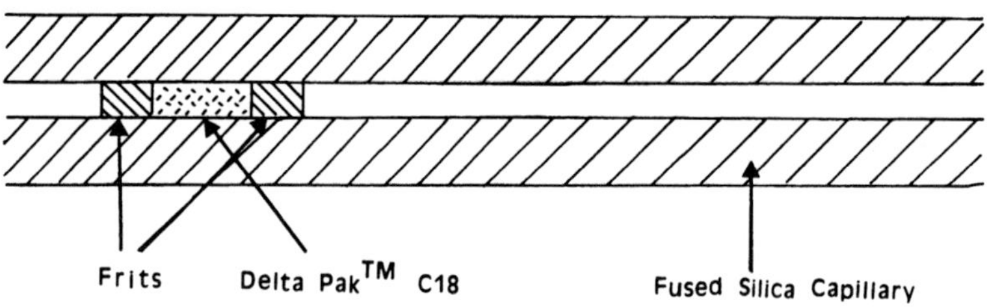

Fig. 6.65. Pre-concentration capillary (Reproduced from Ref. 120 with permission of Waters Division of Millipore.)

to a solid material [9]. The system has been employed for the concentration of urinary components. Monoclonal antibodies are conjugated to 1,4-phenylene diisothiocynate (DITC)-glass beads. The conjugated glass beads are held inside the column between two porous glass frits made by sintering of borosilicate beads. Using one analyte concentrator, it is possible to collect approximately 200 ng of methamphetamine spiked into urine from a single injection [9]. Larger amounts can be collected when more than one analyte concentrations are used simultaneously.

6.10 CONCLUSION

The interests in the techniques discussed in this chapter stem from their potential to further enhance the power, and to widen the scope of application of CE.

Buffer programming and external field control techniques are promising approaches to enhance separation by permitting more flexible, or direct control of

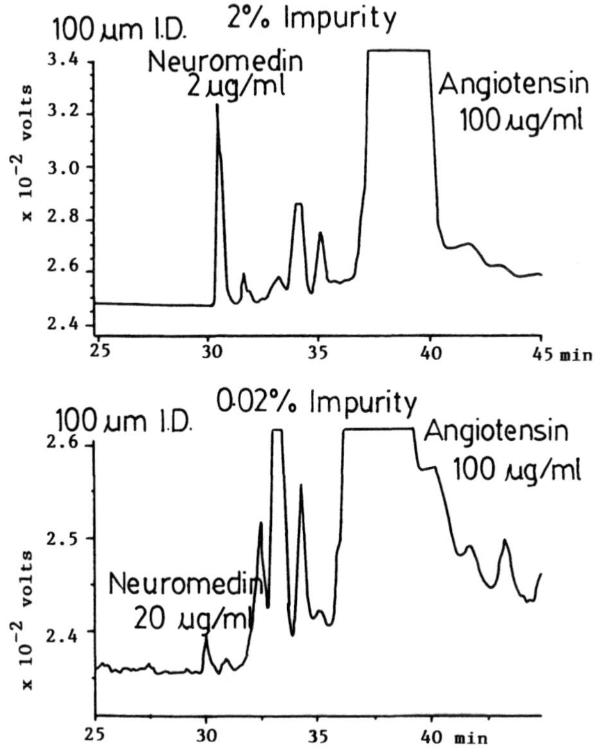

Fig. 6.66. Detection of trace impurities using the pre-concentration capillary. Conditions: (a) capillary = 100μm $\times$ 60 cm; buffer = 25 mM Na citrate, pH 4.0; eluent = 25 mM Na citrate, pH 4.0/acetonitrile, 25:75; samples, as indicated in electropherogram in the buffer; injection = 500 s at 8 kV; elution = 20 s at 7 kV; run = 8 kV. (b) Capillary = 100 μm $\times$ 60 cm, 1.5, 1.5 mm Delta Pak^{TM}; other conditions as in (a). (Reproduced from Ref. 120 with permission of Waters Division of Millipore.)

mobilities and electroosmotic flow. These techniques can be exploited as alternative techniques to improve selectivity. However, the implementation of these techniques generally requires more elaborate experimental setups than the standard CE instruments. Buffer and field programming techniques also tend to reduce efficiency as a result of their disturbance to the flow profile. From a practical point of view, these techniques must attain a level of sophistication, in terms of selectivity, efficiency and ease of operation, equivalent to that of alternative approaches, such as the use of buffer additives and column coatings, in order for them to become widely accepted.

Interest in improvements in the design of fraction collection systems for CE is likely to grow rapidly. Although the sample capacity of CE is relatively small due to the use of small I.D. capillaries, the high efficiency of the technique may be used advantageously in some applications to permit micropreparative scale separation. Several interesting approaches have already been developed. It is feasible, and certainly worthwhile to exploit and to improve the capability of CE for the separation, detection and collection of components from very complex mixtures.

Based on advances already made, CE analysis of increasingly more complex samples will be attempted. In order to determine the optimum conditions for these separations, more experimental parameters may need to be considered. To reduce the method development time and to ensure that global optimum conditions are found, systematic optimization schemes and migration prediction models are preferable to trial-and-error approaches. Several versatile optimization schemes and computer models have been developed for CE. In view of the fact that these approaches have been frequently adopted in other separation techniques, such as HPLC, it is believed that they are likely to become important tools for optimizing CE separations in the future.

For the development of optimization schemes and for the understanding of the fundamental phenomena governing migration behaviour, data for relevant physico-chemical properties are required. With minimum modifications to the instrument, CE itself has been successfully exploited as rapid and reliable methods for the measurement of some of these properties. Investigation on these aspects represents a unique area of application of CE.

Advances in CE have also been made in areas which promise to extend the capabilities of conventional electrophoretic techniques, such as isoelectric focusing and isotachophoresis. These techniques are themselves widely used separation methods. Their use in the capillary format, although only demonstrated recently, has already shown excellent potential. Originally, these separation methods are performed only in systems where electroosmotic flow is eliminated or minimized. Recently developed techniques to exploit the electroosmotic flow demonstrate the potential to facilitate the operation of these modes of separation, e.g. by eliminating the mobilization step after focusing in IEF, and to permit easy interfacing to other systems. Together with other modes of CE, such as capillary zone electrophoresis and micellar electrokinetic chromatography, CIEF and CITP are expected to develop rapidly as separation techniques capable of achieving remarkable resolving power.

Ultimately, the aim in an analysis is to obtain as much relevant information regarding the sample as possible, and as quickly as possible. The coupling of CE with other separation techniques, and the coupling of different modes of CE provide interesting possibilities to further enhance the resolving power, to increase detection sensitivity, and to provide additional information from a single analysis. Generally, the combination of multiple separation mechanisms requires special systems and methods. However, the potential benefits are significant, and therefore further advances in these areas are likely to be made.

In short, although the special systems and methods described in this chapter are not expected to be used in routine analyses, their unique advantages in special applications and their potential to further enhance the capability of CE justify increasing efforts on their development and utilization.

6.11 REFERENCES

1 A.T. Balchunas and M.J. Sepaniak, Anal. Chem., 60 (1988) 617
2 P. Bocek, M. Deml, J. Pospichal and J. Sudor, J. Chromatogr., 470 (1989) 309
3 J. Sudor, Z. Stransky, J. Pospichal, M. Deml and P. Bocek, Electrophoresis, 10 (1989) 802
4 P. Bocek, M. Deml and J. Pospichal, J. Chromatogr., 500 (1990) 673
5 F. Foret, S. Fanali and P. Bocek, J. Chromatogr., 500 (1990) 673
6 D.J. Rose and J.W. Jorgenson, J. Chromatogr., 438 (1988) 23
7 X. Huang and R.N. Zare, Anal. Chem., 62 (1990) 443
8 C. Fujimoto, Y. Muramatsu, M. Suzuki and K. Jinno, J. High Resolut. Chromatogr., 14 (1991) 178
9 N.A. Guzman, M.A. Trebilcock and J. Advis, J. Chromatography, 14 (1991) 997
10 A. Guttman, A. Cohen, D.N. Heiger and B.L. Karger, Anal. Chem., 62 (1990) 137
11 B.L. Karger, A.S. Cohen, A. Guttman, J. Chromatogr., 492 (1989) 585
12 A.S. Cohen, D.R. Najarian, A. Paulus, A. Guttman, J.A. Smith and B.L. Karger, Proc. Natl. Acad. Sci. USA, 85 (1988) 9660
13 C.S. Lee, W.C. Blanchard and C.-T. Wu, Anal. Chem., 62 (1990) 1550
14 C.S. Lee, D. McManigill, C.-T. Wu and B. Patel, Anal. Chem., 63 (1991) 1519
15 C.S. Lee, C.-T. Wu, T. Lopes and B. Patel, J. Chromatogr., 559 (1991) 95
16 K. Ghowsi and R.J. Gale, in Biosensor Technology, (R.P. Buck, W.E. Hatfield, M. Umana and E.T. Bowden, eds.) Marcel Dekker, New York, (1990) 55
17 K. Ghowsi and R.J. Gale, J. Chromatogr., 559 (1991) 95
18 K. Ghowsi, PhD Dissertation, Louisiana State University, Dec 1990
19 X. Huang, M. Gordon and R.N. Zare, Anal. Chem., 60 (1988) 1837
20 V. Pretorius, B.J. Hopkins and J.D. Schieke, J. Chromatogr., 26 (1974) 385
21 A. Tsuda, K. Normura and G. Nakagawa, J. Chromatogr., 264 (1983) 385
22 J. Vindevogel and P. Sandra, Anal. Chem., 63 (1991) 1530
23 R.A. Stowe and R.P. Mayer, Ind. Eng. Chem., 58 (1966) 36
24 D.L. Massart, B.G. Vandeginste, S.N. Deming, Y. Michotte and L. Kaufman, Chemometrics: A Text Book; Elsevier, Amsterdam, (1988) Ch. 6 and 7
25 S.K. Yeo, C.P. Ong and S.F.Y. Li, Anal. Chem., 63 (1991) 2222
26 C.L. Ng, C.P. Ong, H.K. Lee and S.F.Y. Li, J. Chromatogr., 598 (1992) 133
27 Y.F. Yik and S.F.Y. Li, J. Chromatogr. (submitted)
28 J.P. Foley, Anal. Chem., 62 (1990) 1302
29 K. Ghowsi, J.P. Foley and R.J. Gale, Anal. Chem., 62 (1990) 2714
30 A.P. Powell and M.J. Sepaniak, J. Microcol. Sep., 2 (1990) 278
31 R. Stock and C.B. F. Rice, Chromatographic Methods, Wiley & Sons, New York, (1974) 25
32 E.V. Dose and G.A. Guiochon, Anal. Chem., 63 (1991) 1063
33 D.A. Saville and D.A. Palusinski, AIChE J., 32 (1986) 207
34 O.A. Palusinski, A. Graham, R.A. Musher, M. Bier and D.A. Saville, Anal. Chem., 61 (1989) 362
35 W.G. Kuhr and E.S. Yeung, Anal. Chem. 60 (1988) 2642
36 F.M. Everaerts, J. L. Beckers and T.P. Verheggen, Isotachophoresis: Theory, Instrumentation and Applications, Journal of Chromatography Library, Vol. 6, Elsevier, Amsterdam, 1976
37 T. Hirokawa, S. Kobayashi and Y. Kiso, J. Chromatogr., 318 (1985) 195
38 J. W. Jorgenson and K. D. Lukacs, Science (Washington D.C.), 222 (1983) 266
39 S. Wren, J. Mircrocol. Sep., 3 (1991) 147
40 Y. Walbroehl and J. W. Jorgenson, J. Microcol. Sep., 1 (1989) 41
41 H.F. Yin, M.H. Kleenib, J. A. Lux and G. Schomburg, J. Microcol. Sep., 3 (1991) 331

42 S. Terabe, O. Shibata and T. Isemura, J. High Resolut., 14 (1990) 52
43 S. Hjerten, K. Elenbring, F. Kilar, J.L. Liao, J.C. Chen, C.J. Siebert and M.D. Zhu, 403 (1987) 47
44 S. Hjerten and M.D. Zhu, J. Chromatogr., 346 (1985) 265
45 S. Hjerten, J.L. Liao and K. Yao, J. Chromatogr., 387 (1987) 127
46 S. Hjerten, J. Chromatogr., 347 (1985) 191
47 F. Kilar and S. Hjerten, Electrophoresis, 10 (1989) 23
48 F. Kilar and S. Hjerten, J. Chromatogr., 480 (1989) 351
49 K.A. Cobb, V. Dolnik and M. Novotny, Anal. Chem., 62 (1990) 2478
50 T.J. Kasper and M. Melera, Poster presented at 27th Eastern Analytical Symp., New York, 1988
51 J.R. Mazzeo and I.S. Krull, BioTechniques, 10 (1991) 638
52 W. Thormann, J. Caslavska, S. Molteni and J. Chmelik, J. Chromatogr., 589 (1992) 321
53 C.A. Bolger, M. Zhu, R. Rodrigues and T. Wehr, J. Liq. Chromatogr., 14 (1991) 895
54 J. Wu and J. Pawliszyn, Anal. Chem., 63 (1992) 219
55 J. Wu and J. Pawliszyn, Anal. Chem., 63 (1992) 224
56 P.G. Righetti, Isoelectric Focusing: Theory, Methodology and Applications, Elsevier, Amsterdam, 1983
57 W. Thormann, R.A. Mosher and M. Bier, J. Chromatogr., 351 (1986) 17
58 W. Thormann, N.B. Egen, R.A. Mosher and M. Bier, J. Biochem. Biophys. Methods, 11 (1985) 287
59 W. Thormann, A. Tsai, J.P. Michaud, R.A. Mosher and M. Bier, J. Chromatogr., 389 (1987) 75
60 R.A. Mosher, W. Thormann and M. Bier, J. Chromatogr., 436 (1988) 191
61 R.A. Mosher, W. Thormann, R. Kuhn and H. Wagner, J. Chromatogr., 478 (1989) 39
62 M.A. Firestone and W. Thormann, J. Chromatogr., 436 (1988) 309
63 W. Thormann and M.A. Firestone, in J.C. Janson and L. Riden (Editors), Protein Purification, VCH Weinheim, (1989) 469
64 W. Thormann and R.A. Mosher, Adv. Electrophoresis, 2 (1988) 45
65 M. Zhu, D.L. Hansen, S. Burd and F. Gannon, J. Chromatogr., 480 (1989) 311
66 J.R. Mazzeo and I.S. Krull, J. Microcol. Sep., 4 (1992) 29.
67 T. Wehr, M. Zhu, R. Rodrigues, D. Burke and D. Duncan, Amer. Biotechnol. Lab., 8 (1990) 22
68 K.C. Chan, L.B. Koutny and E.S. Yeung, Anal. Chem., 63 (1991) 746
69 P. Bocek, M. Deml, P. Gebauer and V. Dolnik, Anal. Isotachophoresis, VCH Verlagsgesellschaft, Weinheim, (1988).
70 F.M. Everaerts and P.E.M. Verheggen, in New Directions in Electrophoretic Methods, J.W. Jorgenson and M. Phillips, Amer. Chem. Soc., Symp. Vol. 335, Washington D.C., (1987) Ch. 14
71 Z. Deyl, F.M. Everaerts, Z. Prusik and P.J. Svenden (Eds), Electrophoresis (Part A: Theory), J. Chromatogr. Lib., Vol. 18A, Elsevier, Amsterdam, 1979
72 A. Deyl, A. Chrombach, F.M. Everaerts and Z. Prusik, (Eds), Electrophoresis (Part B: A survey of techniques and applications), J. Chromatogr. Lib., Vol. 18B, Elsevier, Amsterdam, 1983
73 F.E.P. Mikkers, F.M. Everaerts and P.A.F. Peek, J. Chromatogr., 168 (1979) 293
74 F.E.P. Mikkers, F.M. Everaerts and P.A.F. Peek, J. Chromatogr., 168 (1979) 317
75 F. Kohlrausch, Ann. Phys. (Leipzig) 62 (1897) 209
76 Z. Prusik (Guest Editor), J. Chromatogr., Vol. 320, No. 1 (1985) 1–268
77 J.C. Reijenga, Th. P.E.M. Verheggen, F.M. Everaerts, J. Chromatogr., 267 (1983) 75
78 J.L. Beckers, F.M. Everaerts and M.T. Ackermans, J. Chromatogr., 537 (1991) 429

79 S. Tanaka, T. Kaneta, K. Nishimura and H. Yoshida, J. Chromatogr., 540 (1991) 475
80 D. Tsikas, A. Hofrichter and G. Brunner, Chromatogr., 30 (1990) 657
81 L. Krivankova and P. Bocek, J. Microcol. Sep., 2 (1990) 80
82 E. Kenndler, Chromatogr., 30 (1990) 713
83 P. Gebauer and W. Thormann, J. Chromatogr., 545 (1991) 299
84 D. Tsikas, J. Fauler, G. Brunner and J. C. Frolick, J. Chromatogr., 545 (1991) 373
85 P. Kopacek, J. Fauler, G. Brunner and J.C. Frolick, J. Chromatogr., 545 (1991) 461
86 T. Hirokawa, Y. Kokotu and Y. Kiso, J. Chromatogr., 538 (1991) 403
87 D. Kaniansky, P. Rajec, A. Svec, J. Marak, M. Koval, M. Lucka, S. Franko and G. Sabanos, J. Radio. Nucl. Chem., 129 (1989) 305
88 A. Petru, P. Rajec, R. Ceck and J. Kuruc, J. Radio. Nucl. Chem., 129 (1989) 229
89 A. Widhalm and E. Kenndler, Anal. Chem., 63 (1991) 645
90 L. Krivankova and P. Bocek, J. Microcolumn Sep., 2 (1990) 80
91 S. Matsushita, M. Sugita, I. Motooka and Y. Kanaji, J. Chromatogr., 586 (1991) 363
92 S. Tanaka, T. Kaneta, M. Taga, H. Yoshida and H. Ohtaka, J. Chromatogr., 587 (1991) 364
93 W. Thormann, M.A. Firestone, J.E. Sloan, T.D. Long and R.A. Mosher, Electrophoresis, 11 (1990) 298
94 W. Thormann, J. Chromatogr., 516 (1990) 211
95 Y. Tanaka and W. Thormann, Electrophoresis, 11 (1990) 760
96 P. Gebauer and W. Thormann, J. Chromatogr., 558 (1991) 423
97 F.S. Stover, J. Chromatogr., 470 (1989) 201
98 J.L. Beckers and F.M. Everaerts, J. Chromatogr., 480 (1989) 69
99 J.L. Beckers and F.M. Everaerts, J. Chromatogr., 508 (1990) 3
100 J.D. Olecho, J.M.Y. Tso, J. Thayer and A. Wainright, Amer. Lab., Dec. (1990) 30
101 J.M. Davis and J.C. Giddings, Anal. Chem., 57 (1985) 2168
102 J.M. Davis and J.C. Giddings, Anal. Chem., 57 (1985) 2178
103 P. Puma, P. Young and M. Fuchs, Poster M-P-116, presented at HPCE '89, Boston, MA, April 10–12, 1989
104 R.G. Nielsen, R.M. Riggin and E. Rickard, J. Chromatogr., 4809 (1989) 393
105 P.D. Grossman, J.C. Colburn, H.K. Lauer, R.G. Nielsen, R.M. Riggen, G.S. Sittampalam and E.C. Richard, Anal. Chem., 61 (1989) 1186
106 P.S.L. Janssen, J.W. van Nispen, M.J.M. van Zeeland, P.A.T. Melgers, J. Chromatogr., 470 (1989) 171
107 M.M. Bushey and J.W. Jorgenson. Anal. Chem., 62 (1990) 978
108 M.M. Bushey and J.W. Jorgenson, J. Microcol. Sep., 2 (1990) 293
109 H. Yamamoto, T. Manabe and T. Okuyama, J. Chromatogr., 480 (1989) 277
110 H. Yamamoto, T. Manabe and T. Okugamo, J. Chromatogr., 515 (1990) 659
111 A.J.J. Debets, R.W. Frei, K.-P. Hupe and W.Th. Kok, J. Chromatogr., 465 (1989) 315
112 A.J.J. Debets, K.-P. Hupe, U.A.Th. Brinkman and W.Th. Kok, Chromatogr., 29 (1990) 217
113 F. Foret, V. Sustacek and P. Bocek, J. Microcol. Sep., 2 (1990) 229
114 D. Kaniansky and J. Marak, J. Chromatogr., 498 (1990) 191
115 D.S. Stegehuis, H. Irth, U.R. Tjaden and J. van der Greef, J. Chromatogr., 538 (1991) 393
116 D.S. Stegehuis, U.R. Tjaden and J. van der Greef, J. Chromatogr., 591 (1992) 341
117 Th. P.E.M. Verheggen, F.E.P. Mikkers, D.M.J. Kroonenberg and F.M. Everaerts, Anal. Chem. Symp. Ser., 5 (1990) 41
118 F.M. Everaerts, Th.P.E.M. Verheggen and F.E.P. Mikkers, J. Chromatogr., 169 (1979) 21
119 C.L. Ng, C.P. Ong, H.K. Lee, K.Y. Sim and S.F.Y. Li (unpublished data)
120 M. Merion, H. Ruedi, M.F. Aebersold, Poster presented at HPCE '91, San Diego, Calif., Feb. 3–7, 1991

CHAPTER 7

Applications

Although practical analytical applications of capillary electrophoresis (CE) have appeared only in the past few years, there is already a significant amount of information available in various literature sources. Furthermore, since there is currently a tremendous interest in CE among researchers, it is expected that many new applications based on this technique will be developed in the near future. The increasing need for powerful analytical techniques and the availability of commercial CE instruments recently have provided further stimulus to advance the technique, and to explore new areas of practical applications.

In this chapter, selected examples of CE applications have been grouped into separate categories to illustrate the methodologies, and approaches developed based on this rapidly growing analytical technique. The main aims are to furnish information on the applicability of CE in practice; to give a rich compendium of the technical data on specific applications, with adequate references to lead the reader to additional information; and to provide an insight into areas which have potential for further development.

It is inevitable that in some of the applications, the principles and instrumental features described in earlier chapters would be involved. For the sake of brevity, details concerning these aspects are omitted in this chapter. Instead, basic data for initial experiments are included. After gaining an understanding of the principles and separation methodologies described in earlier chapters, the readers should find that these data would provide sufficient information for developing analytical procedures to solve the problems encountered in specific areas of application.

In addition to the discussions on the separations of individual types of compounds, a survey of selected commercial instruments is included to serve as a quick guide to instrumental features available in the current generation of CE instruments. Detailed specifications and price information should be readily available from the respective manufacturers or their representatives.

It would be impossible to cover in detail all applications of CE in this chapter. Even the classification of these applications into distinct categories may present a problem in some cases, especially in the case of complicated mixtures which contain components from several different groups of compounds. In such cases, cross references are included where necessary to indicate the general applicability of the technique in different areas. Moreover, in order to make the task of method development easier, a summary of various applications developed based on CE is included. It is hoped that the reader will be equipped with sufficient technical information to choose the most appropriate methodologies and instrumentation for his/her analytical applications.

7.1 AMINO ACIDS

In their pioneering works on capillary zone electrophoresis, Jorgenson and Lukacs demonstrated high resolution of fluorescent derivatives of amino acids (AAs) [1–3]. Since then the analysis of amino acids by CE has been widely studied. These studies include the separation of dansylated (DNS) derivatives [1–17], phenylthiohydantoin (PTH) derivatives [18–21], naphthalene dicarboxaldehyde (NDA) derivatives [22–24], 4,4-dimethylamino azobenzene-4′-sulfonyl chloride (DABSYL) derivatives [25–27], fluorescein isothiocyanate (FITC) derivatives [24,28–30], ortho-phthaldiadehyde (OPA) derivatives [24,29–31], fluorescamine [29,32,33], and 9-fluorenylmethylchloroformate (FMOC) derivatives [29, 34]. Comparison of different derivatizing agents were made in some of these studies [24,29]. In addition, detection without derivatization has also been performed. The strategies employed included concentration gradient detection [35], indirect fluorescence detection [36], indirect amperometric detection [37]. More detailed descriptions of the detection systems are given in Chapter 3. In this section, the main emphasis of the discussion is on the separation conditions for the amino acids and their derivatives.

7.1.1 Dansylated (DNS)-amino acids

Separation of dansylated amino acids (DNS-AAs) by capillary electrophoresis can be accomplished with high efficiency using an electrophoretic buffer at around neutral pH. This is because most of the amino acids (AA) have a net negative charge under these conditions and hence can be separated based on differences in charge [6].

The role of buffer anion on the separation of DNS-AA has been investigated [12]. It was found that the anions used affect not only electroosmotic flow and mobility times, but also the resolution and selectivity of DNS-AAs. The use of deuterium oxide (D_2O) instead of water in the electrophoretic solution was found to improve the resolution of DNS-AAs [14].

Mixed chiral surfactants were used as buffer additives to resolve enantiomers of DNS-AA [4]. DNS-DL-AAs were separated by MEKC with and without cyclodextrins [17]. The separation of DNS-DL-AAs by using gel-filled columns containing cyclodextrins has also been investigated [11]. High efficiency separation of DNS-DL-AAs was obtained by adding 75 mM β-CD to the buffer and the gel. The gel composition was 5% T, 3.3% C and 7 M urea. The buffer consisted of 0.1 M Tris–0.25 M boric acid at pH 8.3. An alternative approach employed for the separation of enantiomers of DNS-AA was based on the diastereomeric interaction between DL-amino acids and chiral Cu(II)-L-histidine [7] and Cu(II)-aspartame [9] complexes, which were added as modifiers to the electrophoretic media.

A system employing multiple capillaries for capillary electrophoretic separation to facilitate fraction collection was demonstrated with the separation and collection of DNS-AAs [15].Rectangular capillaries have also been used to enhance detection sensitivity for CE separation of DNS-AAs [13]. Another method to improve detection sensitivity involved the use of an immersed flow cell for fluorescence detection [16]. The use of capillary zone electrophoresis as an experimental method for the determination of electrophoretic mobilities and diffusion coefficients has also been demonstrated for selected dansylated amino acids [5].

7.1.2 Phenylthiohydantoin (PTH)-amino acids

Phenylthiohydantoin (PTH)-amino acids, which are derivatives of amino acids resulting from the Edman degradation of peptides and proteins, are important for determining amino acid sequences.

Micellar electrokinetic chromatography (MEKC) was used for the separation of PTH-AAs. Both sodium dodecyl sulfate (SDS) and dodecyltrimethylammonium bromide (DTAB) have been used [19]. The addition of organic modifiers to the micellar solution was found to improve resolution of the AAs [18].

Determination of PTH-phosphotyrosine by CE was investigated by Meyer *et al.* [20]. Solid phase sequencing of phosphotyrosine-containing peptides was followed by CE of PTH-phosphotyrosine. The procedure permitted unambiguous identification of this phosphoamino acid without any need for radioactive labelling.

The separation of enantiomers of PTH-AAs by MEKC with chiral surfactants has been studied [21]. The non-ionic chiral surfactant, digitonin and the anionic chiral surfactant, sodium-N-dodecanoyl-L-valinate (SDVal), were used with anionic SDS to form mixed micelles in the MEKC separations.

7.1.3 Naphthalene dicarboxaldehyde (NDA)-amino acids

Naphthalene dicarboxaldehyde (NDA) has been used as a pre-column fluorescent label for CE with laser-induced fluorescence detection [23]. Mixtures of NDA derivatives were separated. The electrophoretic buffer was 75/25 by volume of borate buffer/methanol. Separation of eight NDA-labeled AAs could be accomplished in 71 s with limit of detection of 0.42 amol ($S/N = 3$). The analysis of NDA derivatives of AA of single cells by CE has been demonstrated on neurons from the land snail *Helix aspersa* [22].

7.1.4 4-(dimethylamino)azobenzene-4′-sulfonyl chloride (DABSYL)-amino acids

Amino acids derivatized with 4-(dimethylamino) azobenzene-4′-sulfonyl chloride (DABSYL) were separated by CE and detected by thermooptical absorbance detection [25–27]. The buffer was an equal volume mixture of acetonitrile with 20 mM phosphate buffer (pH 7) containing 5 mM SDS. The thermooptical system consisted of an argon ion pump laser (458 nm, 130 mW), and a helium-neon probe laser (633 nm, 1 mW). The probe beam intensity was monitored with a silicon photocell. Detection limits in the attomole range were obtained for the DABSYL-amino acids.

7.1.5 Fluorescein isothiocyanate (FITC)-amino acids

Fluorescein isothiocyanate (FITC) derivatives of amino acids were separated by CE and detected by laser-induced fluorescence with a sheath flow cuvette [28]. Separation was performed in 5 mM carbonate buffer. A 1-W argon ion laser operated at 488 nm was used as the excitation source. The sensitivity achieved using this system was in the subattomole range for the FITC amino acids.

The use of FITC as a pre-column derivatization reagent for tagging of amine groups on amino acids was investigated together with other reagents, such as OPA and NDA [24]. For the separation of FITC derivatives, the electrophoretic buffer used was 0.01 M $NaHCO_3$ (pH 9.0). Laser induced fluorescence with a helium-cadmium laser (442 nm) was used for detection. Minimum detection amount of 1.1 amol (0.22 nanomolar) was reported for FITC-phenylalanine.

CE with conventional fluorescence detection of FITC, FMOC, OPA, fluorescamine derivatives of amino acids was performed [29]. Despite the fact that conventional fluorescence detectors are less sensitive than laser-induced fluorescence detection systems, they are superior in terms of the availability of spectral lines to match analyte absorption bands. For the separation of the FITC amino acids, a borate buffer was used in the electrophoretic buffer. Pre-column derivatization was performed. It was noted that FITC provided good sensitivity for primary and secondary amines, but the derivatization reaction was slow. Furthermore, the free FITC gave a large signal, which may interfere with the analyte peaks.

Isotachophoresis (ITP) was used as an on-line concentration pretreatment technique for the CE analysis of FITC derivatized amino acids [30]. Two orders of magnitude improvement in detection sensitivity could be achieved as a result of the concentrating and separating power of ITP as an injection/pretreatment method for CE. However, there are also limitations with the technique, such as the need to consider the optimum buffer systems for the combined ITP/CE system and the separate ITP and CE systems, and the discrimination of samples during injection.

7.1.6 o-Phthaldialdehyde (OPA)-amino acids

o-Phthaldialdehyde (OPA) is a commonly used pre-column derivatizing agent for amino acids. In the presence of a reducing agent, such as ethanethiol, OPA reacts rapidly with primary amino acids to form highly fluorescent thio-substituted isoindoles. The separation of OPA-AAs has been investigated [18,29–31].

MEKC was found to provide high selectivity of the OPA amino acids [18]. Figure 7.1 shows an electrokinetic chromatogram of OPA-amino acids for the total hydrolysis of cytochrome C. OPA-AAs were separated by CZE in 0.2 M borate buffer (pH 7.8) [31]. Laser-induced fluorescence was used for detection. By using an on-line connector for microcolumns, both pre-column and on-column derivatization could be performed. It was found that although the on-column method was less sensitive for analytes that formed stable adducts, it provided higher sensitivity for analytes that formed unstable adducts.

The use of OPA as a derivatizing agent for CE with conventional fluorescence detection has also been investigated [29]. Although OPA would react rapidly with primary amines, and excess reagent would not fluoresce, the derivatives tend to be unstable and its use as a post-column derivatizing reagent was preferred. A post-column liquid junction reactor [29] was developed which produced a detection limits of around 60 ng/ml for OPA-amino acids.

On-line isotachophoretic pretreatment for CZE separation of OPA-amino acids has also been investigated [30]. Serum samples spiked with OPA-amino acid were analyzed. It was found that the trace amount of amino acids were concentrated, whereas the major components were diluted, resulting in enhanced selectivity of the system.

7.1.7 Fluorescamine-amino acids

Fluorescamine derivatives of amino acids were separated by CE [29]. A sodium borate buffer (pH 9.5, 20 mM) containing 100 mM SDS was used. Fluorescamine was found to react quickly with primary amines, and excess reagent hydrolyzed to form non-fluorescent products. However, broad peaks were obtained and satisfactory resolution was not obtained for the six derivatives investigated.

Fluorescamine-derivatized amino acids have been used to characterize a microinjector for CZE [32]. The performance of a fluorescence detector based on a fluorescence microscope was tested with fluorescamine-AAs [33]. A 0.05 M sodium tetraborate buffer (pH 6.0) was used for the separation.

7.1.8 9-Fluorenylmethylchloroformate (FMOC)-amino acids

9-Fluorenylmethylchloroformate (FMOC) is another derivatizing agent which has been used to derivatize amino acids for CE analysis [29]. FMOC reacts rapidly with primary and secondary amines. Although the free reagent is fluorescent, it can be

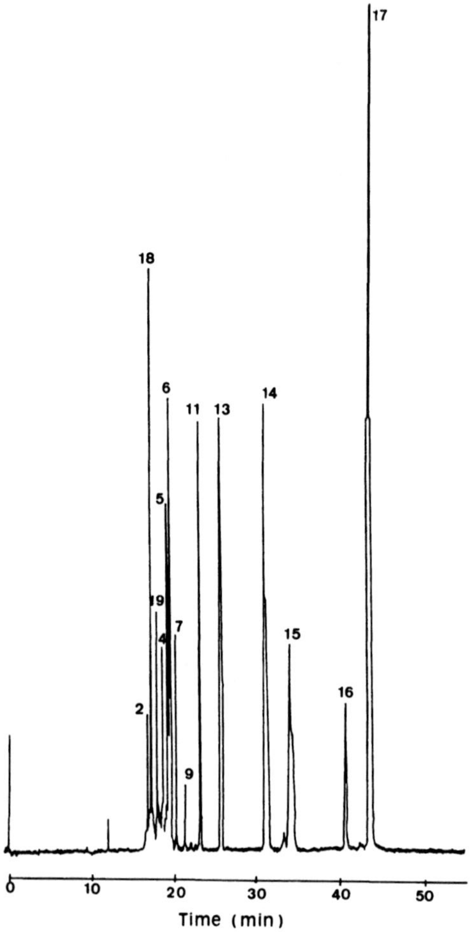

Fig. 7.1. Electropherogram of OPA-amino acids obtained from the total hydrolysis of cytochrome c. Buffer, 0.05 M borate buffer (pH 9.50)–15% methanol–2% THF–0.05M SDS; capillary, 86 cm × 50 μm I.D. (53 cm to the detector); injection time, 40 s; applied voltage, 23 kV. *1* = Glutamine; *2* = Threonine; *3* = Serine; *4* = Histidine; *5* = Alanine; *6* = Glycine; *7* = Valine; *8* = γ-aminobutyric acid; *9* = Methionine; *10* = Tourine; *11* = Isoleucine; *12* = Tryptophan; *13* = Leucine; *14* = Lysine; *15* = Glutamic acid; *16* = Arginine; *17* = 2-Aminoethanol (internal standard); *18* = Aspartic acid, *19* = Tyrosine. (Reproduced with permission from Ref. 18. Copyright Elsevier Science Publishers.)

extracted with pentane readily. This reagent was used for precapillary derivatization of amino acids for CE separation with a xenon lamp based fluorescence detection system. The electrophoretic buffer used was 20 mM sodium tetraborate at pH 9.5, with 25 mM SDS. The electropherogram obtained is shown in Fig. 7.2. CE analysis of amino acids after pre-column derivatization with FMOC has also been reported [34].

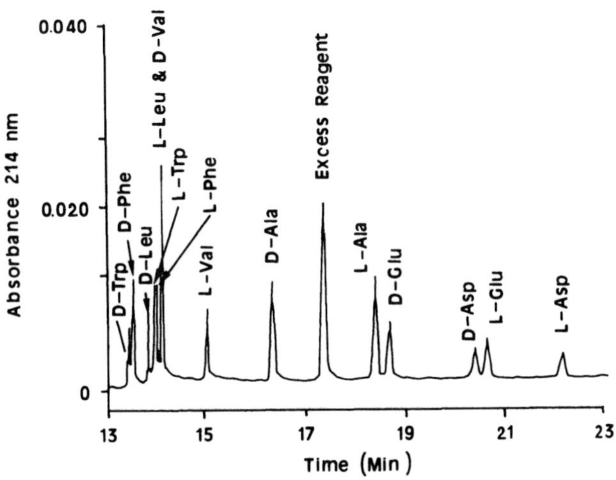

Fig. 7.2. Electropherogram of racemic mixtures of Ala, Asp, Glu, Val, Leu, Phe and Trp derivatized with L-Marfey's reagent. The electropherogram was obtained by micellar electrokinetic capillary chromatography using a 75 μm I.D. capillary: 50 mM ammonium phosphate buffer, pH 3.3; 10 kV, 60 mA. 25 $\pm$ 0.1°C; detection at 214 nm. (Reproduced with permission from Ref. 29. Copyright Elsevier Science Publishers.)

7.1.9 Underivatized amino acids

The analysis of CE of underivatized amino acids by CE could be accomplished with various detection techniques, such as concentration gradient detection [35], indirect fluorescence detection [36], and indirect electrochemical detection [37]. Details of these techniques are given in Chapter 3. The concentration gradient detection system was based on differential refractive index gradient, which is proportional to the concentration of the analyte [35]. Indirect amperometric detection was performed by adding an electrophore (e.g. 3,4-dihydroxybenzylamine) into the electrophoretic medium [37]. The use of indirect fluorescence detection for the analysis of amino acids was demonstrated using coumarin 343 as the fluorophore [36].

7.2 PEPTIDES

The analysis of peptides by CE is an area of great interest. Several excellent reviews discuss the applications of CE to the separation of peptides and proteins [38–46].

In this section, selected examples of CE separations of peptides are first described [1–3,6,20,47–52]. The migration behaviour and methods to enhance selectivity in CE of peptides [53–67] are then discussed. This is followed by a description of recent advances in the use of capillary electrophoresis–mass spectrometry for the analysis of peptides [68–81]. Finally, other instrumental developments for CE separation of peptides [82–90] are described.

7.2.1 Examples of CE separation of peptides

Jorgenson and Lukacs demonstrated in their early work on CZE the high efficiency capability of this technique. The separation of fluorescamine derivatives of dipeptides [1], and peptides from tryptic digests of ovalbumin were investigated [1–3]. CZE separation of fluorescamine labeled peptides from a tryptic digest of chicken albumin was also investigated by Green and Jorgenson [6,47].

CZE has been used for high sensitivity peptide mapping by Cobb and Novotny [48]. By using trypsin immobilized on agarose gel for protein digestion, a decrease in sample size approximately 3 orders of magnitude was achieved compared with conventional tryptic digestion methods. The relative standard deviation of migration times was less than 1% for CZE. Satisfactory separation of tryptic digests of β-casein was obtained by CZE using a 0.04 M Tris/0.04 M Tricine buffer (pH 8.1).

Ludi *et al.* employed CZE for the analysis of recombinant insulin-like growth factor (rIGF) and recombinant hirudin (r-hirudin) [50]. The separation of r-IGF product and r-hirudin with different numbers of amino acids was demonstrated.

Selected examples of applications of CE for the separation of peptides and proteins were investigated by Josic and Zeilinger [51], which included the separation of human and animal neurotensine. A comparison of CE and HPLC separation of sulfate and non-sulfated forms of peptides was made [52]. CE was performed using phosphate buffer at pH 6.5, and was found to provide more rapid and efficient separations of the peptides corresponding to segments of cholecystokinin, enkephalin and hirudin.

7.2.2 Migration behaviour of peptides in CE

Grossman *et al.* [53] investigated the effect of buffer pH and peptide composition on the selectivity of CZE separation of peptides. A series of 10 synthetic peptides which contain different degree of charge and hydrophobicity were studied. Buffer pH was found to be the primary parameter affecting the selectivity of the separation of the peptide mixture. The results also showed that differences in the neutral amino acid composition and sequence of a peptide could be detected by CZE.

The effects of peptide size, charge and hydrophobicity on its electrophoretic mobility (μ) have also been investigated [54]. A semi-empirical model was developed based on the measurement of the electrophoretic mobility of 40 peptides varying in size from 3 to 39 amino acids and varying in charge from 0.33 to 14.0. Figure 7.3 illustrates the dependence of electrophoretic mobility on the group: $D \ln(Z + 1)/n^{0.43}$, where D is a constant, Z is the charge and n is the number of amino acids in the polypeptide chain.

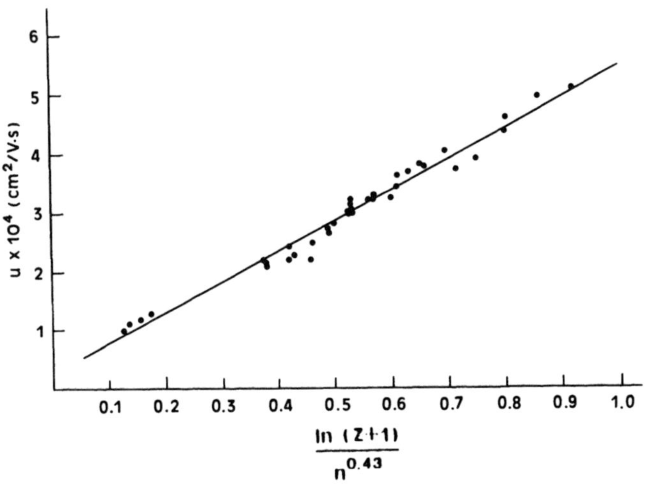

Fig. 7.3. Electrophoretic mobility vs. $D \ln(Z + 1)/n^{0.43}$ for 40 different peptides ranging in size from 3 to 39 amino acids and ranging in charge from 0.33 to 14. Y intercept = 2.47×10^{-5}, slope = 5.23×10^{-4}, r = 0.989. (Reproduced with permission from Ref. 54. Copyright Academic Press.)

The separation of a modified adrenocorticotropic hormone [4–9] fragment (Org 2766) and several of its fragments was performed by van de Goor *et al.* [55]. The mobility of the peptides could be predicted reasonably well using the pK_a values of the separate amino acids. Factors affecting the choice of buffer systems were discussed.

Migration times of peptides and proteins in CE may be correlated with physicochemical properties [56]. The electrophoretic mobility was found to be proportional to $Z/MM^{2/3}$, where Z is the charge and MM is the molecular mass. Figure 7.4 shows the linear fit obtained for human growth hormone (hGH), biosynthetic insulin-like growth factor II (IGF-II), biosynthetic human insulin (BHI), human proinsulin (hPI) and bovine somatostatin (bST).

In a separation of tryptic digests of bovine, chicken, horse, and rabbit cytochrome C [57], differences in selectivity were observed with subtle changes in pH of the electrolyte. By adjusting the pH, methods could be developed to detect single amino acid differences with a panel of homologous peptide. In Fig. 7.5, the separation of tryptic peptides of horse heart cytochrome C at pH 4.0 is shown.

The effects of different buffers and pH on the migration times and resolution of a gut hormone, motilin, were studied [58]. Buffers investigated include 20 mM sodium citrate (pH 2.5), 50 mM sodium phosphate (pH 7.0) and 50 mM sodium borate (pH 9.4). Voltage used was 30 kV and a capillary of 50 μm I.D. and 72 cm length (48 cm to detector) was employed. Detection was by UV absorbance at 200 nm and the temperature was maintained at 30°C. The citrate buffer at pH 2.5 was found to provide the best resolution for the motilin peptide. Figure 7.6 shows an electropherogram of fourteen motilin peptides.

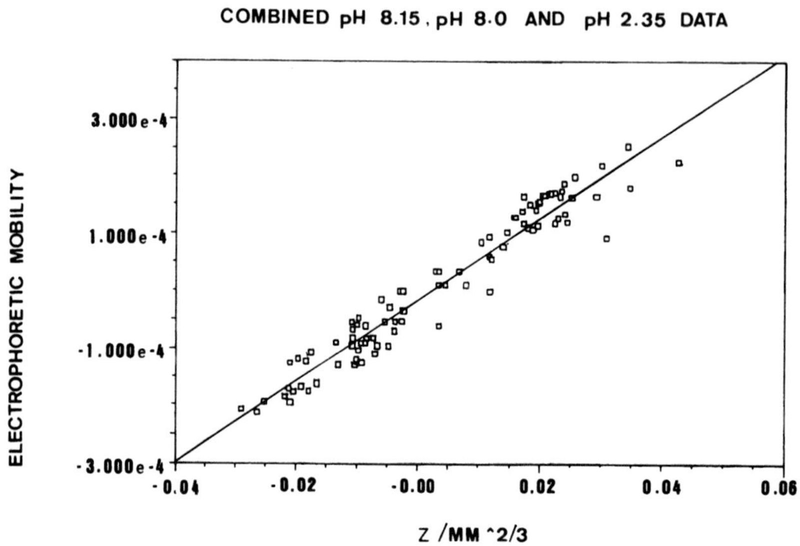

Fig. 7.4. Fit of electrophoretic mobility for hGH digest (pH 2.35, pH 8.0 and pH 8.15) and IGF-II digest, hGH, BHI, IGF-II, hPI and bST (pH 2.35 and pH 8.15). $y = 7.08 \times 10^{-3} - 1.65 \times 10^{-5}$, $r = 0.948$. (Reproduced with permission from Ref. 56. Copyright Eli Lily & Company.)

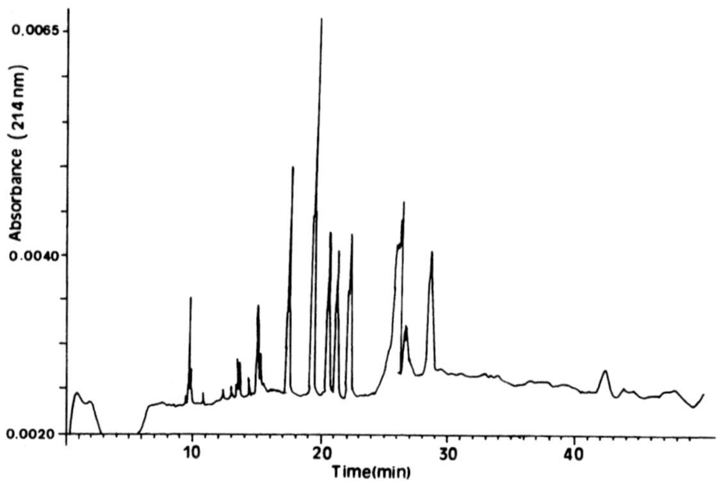

Fig. 7.5. CE Separation of tryptic peptides of horse heart cytochrome c, pH 4.0 electrophoresis was performed in 0.025M sodium citrate, pH 4.0, in a 75 μm × 60 cm capillary. following a 10 s hydrostatic injection, a 20 kV potential was applied. The separation was monitored at 214 nm. (Reproduced with permission from Ref. 57. Copyright Marcell Dekker) Publishers.)

Multiple antigen peptides (MAPs) were separated by CZE [59]. The MAPs were denoted with a superscript which refers to the number of branches and a sequence in parenthesis which represents the sequence of amino acid of the lysine residues.

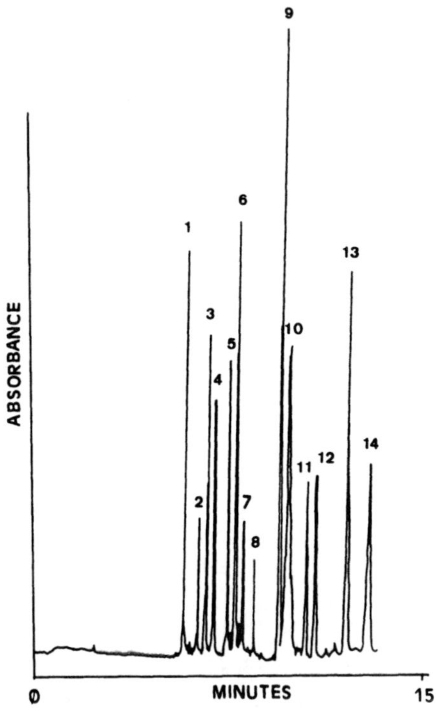

Fig. 7.6. Electropherogram of fourteen motilin peptides. *1* = Mot_{15-22}; *2* = Mot_{10-22}; *3* = Mot_{1-14}; *4* = Mot_{3-22}; *5* = Mot_{1-22}, *6* = Mot_{1-19}; *7* = Mot_{1-16}; *8* = Mot_{1-12}; *9* = Mot_{3-14}; *10* = Mot_{2-14}; *11* = Mot_{1-15}; *12* = Mot_{1-5}; *13* = Mot_{1-7}; *14* = Mot_{1-9}. Conditions: pH 2.5 citrate buffer at 20 kV. Absorbance range 0–0.01 a.u.f.s. Each motilin peptide was run individually to correlate migration times in the mixture. (Reproduced with permission from Ref. 58. Copyright Elsevier Science Publishers.)

The separation of peptides differing only in the Ala-Ser substitution is shown in Fig. 7.7, where the peaks are: 2-$MAP^8(PGTHLPALP)_2$; 3-$MAP^8(PGTHLPSLP)_2$; 4-$(PGTHLPALP)_2$-NH_2; 5-$(PGTHLPSLP)_2$-NH_2.

The action of endoproteinase Arg C on the cleavage of adrenocorticotrophic hormone (ACTH) peptide bonds were investigated by CE [60]. Uncoated capillaries (50 μm I.D., 50 cm long) were employed for the separation. Several buffers systems were investigated. The buffer systems which gave high separation efficiencies included: 40 mM phosphate–2 M betaine–0.1 M K_2SO_4, pH 7.6 (plate number = 201,000) and 40 mM 2-(N-morpholino)ethanesulfonic acid (MES)–2 M betaine–40 mM K_2SO_4, pH 5.5 (plate number = 221,700). Addition of salts and zwitterions to the electrolyte minimized capillary–peptide interaction and resulted in high efficiency separation.

CZE has been used for the separation of histine-containing compounds, including angiotensins, cationic heptapeptides and model histine derivatives [61].

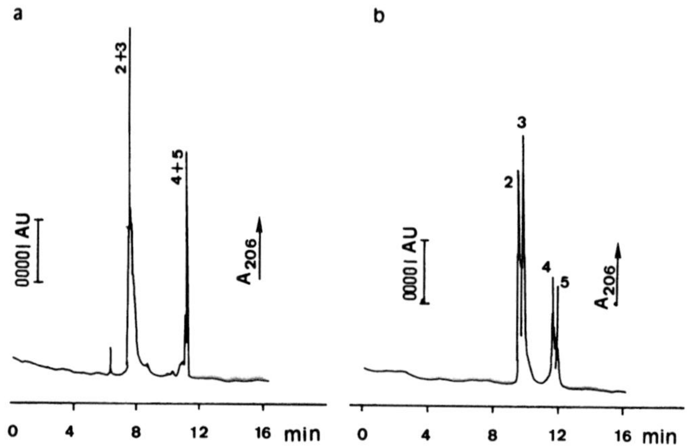

Fig. 7.7. Electrophoretic separation of peptides differing only in the Ala-Ser substitution. Sample: mixture compounds 2, 3, 4 and 5, 0.1 mg/ml each. (a) BGE, 50 mM phosphate buffer (pH 2.5); sampling, 4 kV, 6 s; electrophoresis at 7 kV (constant), 11 μA; (b) BGE, 40% ACN in 50 mM phosphate buffer (pH 2.5); electrophoresis at 7 kV (constant), 7 μA. (Reproduced with permission from Ref. 59. Copyright Elsevier Science Publishers.)

For compounds with net charge greater than +2, the addition of putrescine to pH 6 buffer was found to be effective in suppressing adsorption by the fused silica wall. Addition of Zn^{2+} permitted separation based on metal binding when operating at pH values at which histidine groups are neutral.

Metal ion-containing buffers were used to enhance resolution of peptide in CZE separations [63]. Cu(II) and Zn(II) salts were added to the electrophoretic buffer. The separation of dipeptides containing histidine was investigated. The metal ion differentially retarded the migration of peptides, and hence caused their separation.

A commercially coated capillary with a hydrophilic polymer coating (20 cm length × 25 μm I.D.) has been used for the analysis of peptides by CZE [64]. The electrolyte was 0.1 M phosphate buffer (pH 2.5). Applied voltage was between 5 and 8 kV and detection was by UV absorbance at 200 nm. Figure 7.8 shows the separation of peptides by CZE.

The effect of buffer ions in CE was studied by using peptides as test substances [62]. It was found that if the electroosmotic flow exceeded the mobility of the fastest catholytic cations, it would be unnecessary to match anolytes and catholytes, i.e. discontinuous buffer systems could be used. However, in the case of slow electroosmotic flow, the use of continuous buffer systems was recommended.

The sample matrix has been shown to affect CE separations of peptides [65]. The electrolytes used were 100 mM phosphate buffer at pH 2.5 for dynorphin, and 100 mM borate buffer at pH 9.2 for the bioactive peptides, including bradykinin, neurotensin and angiotensin I. Separation was performed at 12 kV and 17 kV for dynorphin and the bioactive standard peptide mixture, respectively. Detection was by UV at 200 nm and the oven temperature was maintained at 25°C. The peak

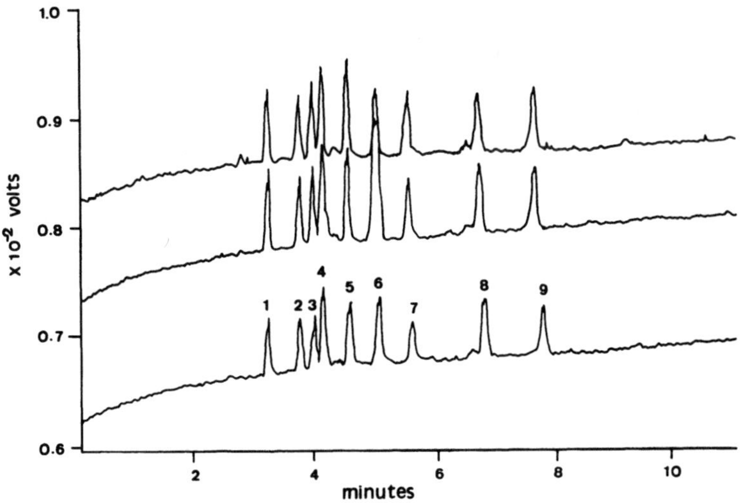

Fig. 7.8. CZE of peptides, with coated capillary, coefficients of variation for migration time and peak area. (Reproduced with permission from Ref. 64. Copyright Friedr. Vieweg & Sohn Verlagsgesellschaft mbH.)

height and resolution as a function of injection time and salt concentration are shown in Fig. 7.9.

The separation of chiral dipeptides by CZE and MEKC was achieved by derivatization with L- and D-Marfey's reagent [66] or simply by adding the Marfey's reagent into the electrophoretic buffer [67]. Subsequently, the chirality of the amino acid constituents of peptides was used to determine the quality of peptide analogues.

7.2.3 Peptide analysis by capillary electrophoresis–mass spectrometry

Capillary electrophoresis–mass spectrometry (CE-MS) with electrospray ionization (ESI) at atmospheric pressure has been used for the analysis of peptides [68–81]. Both CE-MS and CE-MS-MS have been employed. Details of the electrospray interface are given in Chapter 3. Some examples of the types of compounds investigated are shown in Figs. 7.10 and 7.11. Figure 7.10 shows the multiple-ion detection of two ions characteristic of B-chain insulin. Figure 7.11 shows the separation of a tryptic digest of tuna cytochrome C.

Capillary electrophoresis–atmospheric pressure ionization mass spectrometry (CE-API-MS) was used for the characterization of a synthetic peptide mixture and a tryptic digest of human hemoglobin [75]. An uncoated fused silica capillary was used for the CE separation. Volatile buffers (ammonium acetate or ammonium formate) of low concentration (15–20 mM) and containing high organic modifier content (5–30% acetonitrile or methanol) were found to be more suitable for

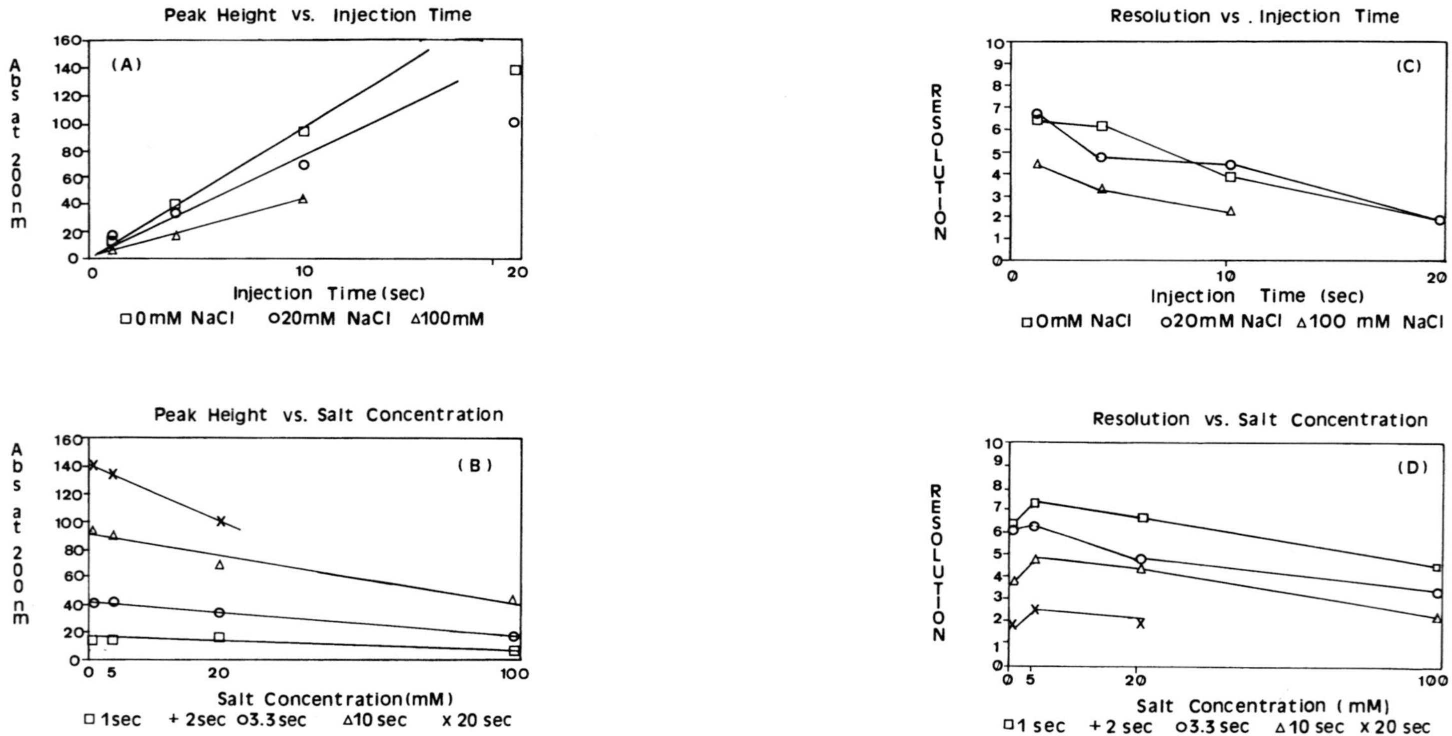

Fig. 7.9. Peak height and resolution as a function of injection time or salt concentration: plots of the peak height (×1000) of Neurotensin (A, C) and the resolution of the Bradykinin and Neurotensin pair (B, D) against injection time or salt concentration. (Reproduced with permission from Ref. 65. Copyright Dr. Alfred Huethig Publishers.)

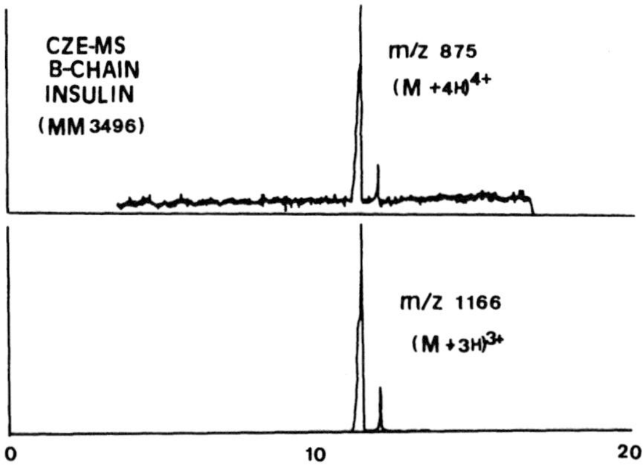

Fig. 7.10. Multiple-ion detection of the two ions characteristic of B-chain insulin in ESI. The electrophoresis was done in 10 mM Tris (pH 8.3) at 150 V/m. The response of both values of m/Z suggests that the second peak is likely a dimer, or alternatively is a digopeptide variant of similar molecular weight to B-chain insulin. (Reproduced with permission from Ref. 72. Copyright Elsevier Science Publishers.)

electrospray ionization. Figure 7.12 shows the selected ion monitoring (SIM) CE-MS total selected ion electropherogram for eight components in a human hemoglobin tryptic digest. In addition, amino acid sequence information for peptides were obtained by on-line tandem MS. Structural information of peptides were obtained by collision-induced dissociation of protonated peptide molecules.

CE-MS with continuous-flow fast atom bombardment (CF-FAB) is an alternative interfacing technique frequently employed for the analysis of peptides [76–81].

An on-line coaxial CF-FAB interface was used by Moseley *et al.* [76,77,80] for the coupling of CZE and MS. The interface was successfully employed for the acquisition of both MS and MS-MS spectra of femtomole level of non-volatile analytes, including dipeptides, tripeptides and decapeptides. Bioactive peptides were analyzed by the on-line coaxial CF-FAB interface [80]. High separation efficiency (plate number = 410,000) and high sensitivity (femtomoles) were obtained. Capillary columns modified with (3-aminopropyl)trimethoxysilane (APS) were used to minimize zone broadening due to adsorption effects in the analysis of basic peptides. Volatile buffers were employed to achieve stable operation of the FAB-MS system. The buffers used included 0.005 M ammonium acetate and 0.01 M acetic acid, with pH adjustment made with ammonium hydroxide. Chemotactic peptides and neuropeptides were analyzed by the CE-MS system. Both positive-ion detection and negative-ion detection were performed. CZE-MS-MS data were acquired in real time. Figure 7.13 shows the CZE-MS-MS analysis of three neuropeptides.

Caprioli *et al.* [78] performed CZE with continuous-flow fast atom bombardment mass spectrometry. Synthetic peptides and protease digests of recombinant human growth hormone and horse heart cytochrome C were separated with high efficiency.

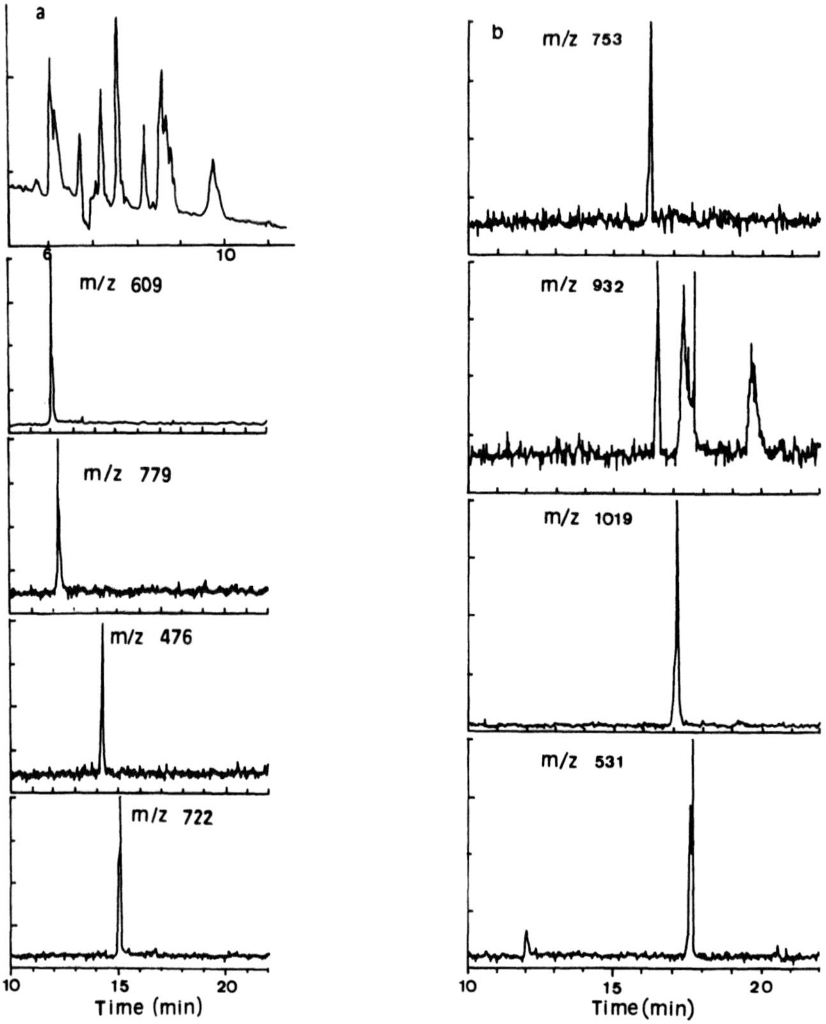

Fig. 7.11. CE-MS separation with on-line UV detection of a tryptic digest of tuna cytochrome c. The UV trace and several selected ion electropherograms are given in (a), superimposed with adjusted time scales. (b) Four later eluting ions are shown. Most of the peaks in the UV signals in mass spectrometry. Buffer: 10 mM Tris (pH 8.3) voltage: 120 V/cm. Approximately 0.6 pmol per component was injected. (Reproduced with permission from Ref. 72. Copyright Elsevier Science Publishers.)

Very small amounts of samples (0.1–20 pmol) were needed. Reinhoud *et al.* [79] performed CZE/MS using CF-FAB and a liquid junction coupling. Static and scanning array detection based on position and time-resolved counting was employed to enhance sensitivity. Detection limits of 1–5 fmol were obtained for peptides. Samples investigated included β-endorphin fragments, galanin and magainin.

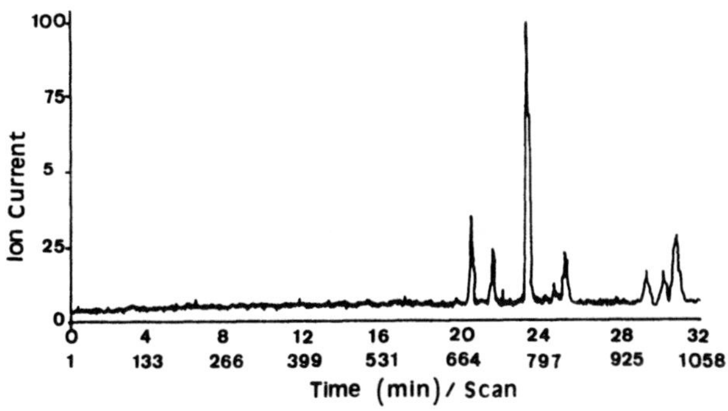

Fig. 7.12. SIM CE-MS total selected ion electropherogram for eight components in a human hemoglobin tryptic digest. Injection: 20 s at 10 kV of a 154 pmol/μl solution (8 pmol). Capillary: uncoated fused silica 100 cm × 75 μm I.D.; buffer: 15 mM ammonium acetate and TFA to pH 2.5 with 15% methanol; voltage: 26 kV. (Reproduced with permission from Ref. 75. Copyright Elsevier Science Publishers.)

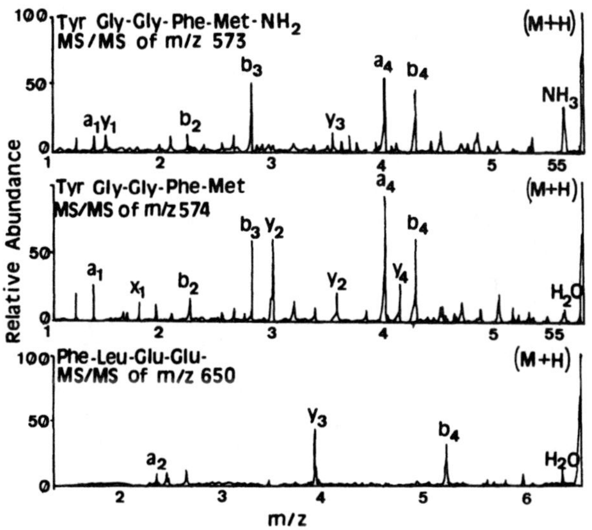

Fig. 7.13. CZE-MS-MS analysis of three neuropeptides. The peptides were separated as negative ions (CZE buffer pH of 8.5) and desorbed, fragmented, and detected as positive ions (FAB matrix pH of 3.5). (Reproduced with permission from Ref. 80. Copyright American Chemical Society.)

Both continuous-flow fast atom bombardment and electrospray ionization (ESI) interfaces were used by Deterding *et al.* [81] for CE-MS of peptides and proteins, such as neuropeptides, angiotensin-related peptides, myoglobin, pigeon heart cytochrome C and horse heart cytochrome C. For CZE-CF-FAB-MS, a column of 13 μm I.D., untreated or coated with aminopropylsilane were used. A sheath column of 160 μm I.D. and 365 μm O.D. with a flow of glycerol–0.5 mM aqueous

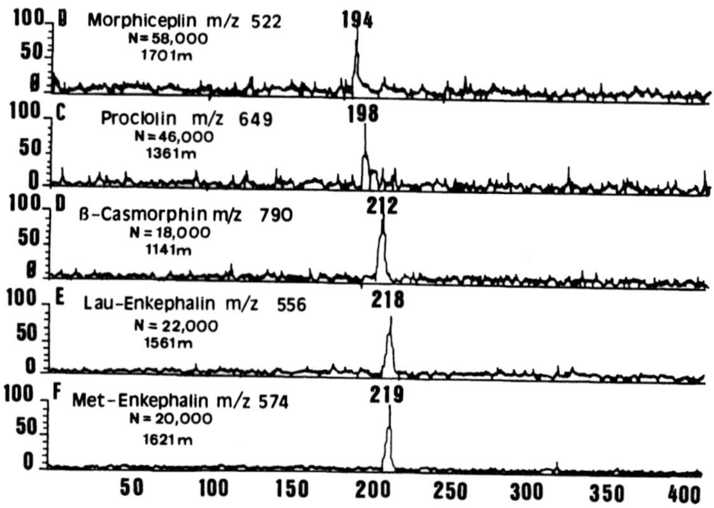

Fig. 7.14. CZE-CF-FAB-MS separation and analysis of neuropeptides. Selected ion chromatograms of $(M + H)^+$ from full scan data. The CZE capillary was 1 m long, 13 μm I.D. The buffer was 5 mM ammonium acetate (pH 8.5) containing 1% 2-propanol. The voltage drop across the column was 38 kV. *N* = plate number. (Reproduced with permission from Ref. 81. Copyright Elsevier Science Publishers.)

heptafluoroyutyric acid (25:75) was used. The buffer composition was 5 mM ammonium acetate adjusted to pH 8.5 with ammonium hydroxide or 0.01 M acetic acid adjusted to pH 3.4–3.5 with ammonium hydroxide. For CZE-ESI, 75 μm I.D. × 150 μm O.D. columns which are either untreated or coated with aminopropylsilane were used. Sheath column was a 600 μm I.D. stainless steel tube, containing a sheath flow of 5–10 μl/min of methanol–3% aqueous acetic acid (50:50). Buffer used was ammonium acetate adjusted to pH 8.5 with ammonium hydroxide or 0.01 M acetic acid adjusted to pH 3.8 with ammonium hydroxide. In both cases, a separation voltage of 30 kV was used. Figure 7.14 shows the CZE-CF-FAB-MS analysis of neuropeptides.

7.2.4 Other instrumental developments for CE of peptides

A simple apparatus was constructed and used for CZE analysis of proteins [82]. The separation of a standard peptide mixture was demonstrated. Indirect fluorescence detection of tryptic digests separated by CZE was performed [83,84]. Detection was based upon charge displacement of a fluorophore from the detection region. The detection system has the advantages of universality and high sensitivity. CE with indirect amperometric detection was demonstrated for the analysis of mixtures of amino acids and peptides, and catecholamines and [37]. Simultaneous direct and indirect amperometric detection could also be performed. Indirect electrochemical detection was accomplished by addition of a cationic electrophore,

3,4-dihydroxybenzylamine (DHBA) to the electrophoretic medium. The approach extended the scope of electrochemical detection by permitting the use of these methods for the detection of non-electroactive species.

CZE of fourteen test peptides was studied by Strickland and Strickland [85]. 100 mM phosphate buffer (pH 2.7) was used as the electrolyte. UV detection at 200 nm was employed. Figure 7.15 shows a comparison of CE and reversed phase chromatography (RP-HPLC) of the peptide mixture. From the figure, it can be seen that CE can be used as an efficient and orthogonal separation technique for reversed-phase chromatography.

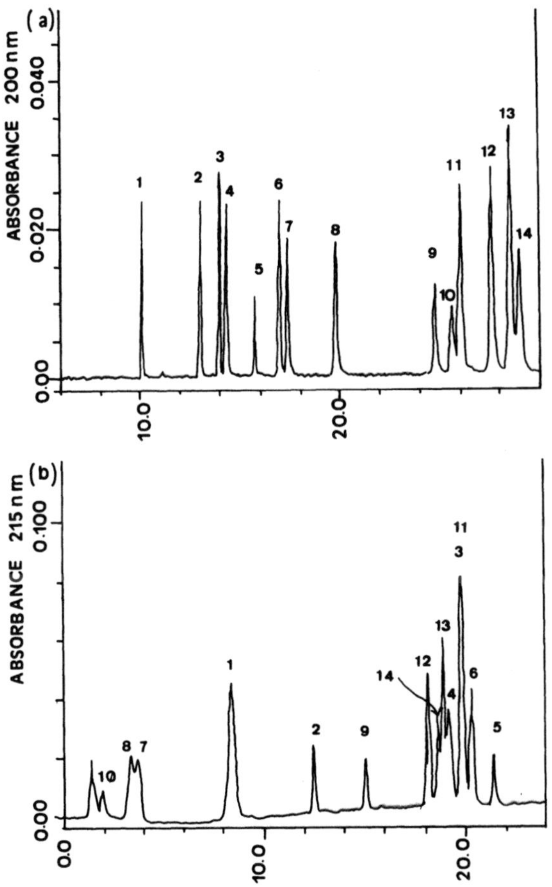

Fig. 7.15. Comparision of capillary electrophoresis and reversed phase chromatography of the peptide mixture. (a) Electropherogram performed under the following conditions: capillary: 57 cm × 75 μm; buffer: 100 mM sodium phosphate (pH 2.7), 20 nl injection of the 14 peptide mixture. (b) Chromatogram performed on a 2.1 × 100 mm C_8 column. Gradient: 0–50% B in 25 min. Flow rate: 200 μl/min. Eluant A: 0.1% trifluoroacetic acid (TFA); eluant B: 0.086% TFA in 80% acetonitrite. Detection UV: 215 nm. Sample load was 1 μl of the same mixture used before. (Reproduced with permission from Ref. 85. Copyright International Scientific Communications.)

A two-dimensional separation system based on the combination of high performance liquid chromatography with capillary zone electrophoresis was developed [88]. The system was used for the analysis of peptide standards and fluorescently labeled peptide products from a tryptic digest of ovalbumin. The two-dimensional system demonstrated greater resolving power and peak capacity than the two separate systems used individually.

Collection of separated peptides and amino acids from a CE system was performed by Bergman *et al.* [89]. The materials collected after just a few runs

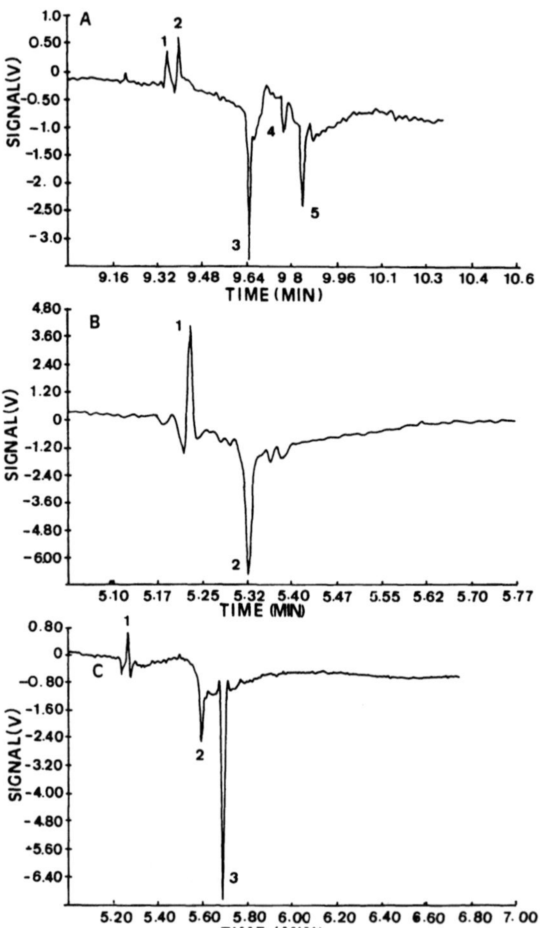

Fig. 7.16. CITP of crude synthetic peptide. Leading ion is chloride ion 2 mM, counter ion is ammediol pH 8.9, tailing ion is β-alanine 1.6 mM adjusted to pH 10.3 with $Ba(OH)_2$. (A) Injection of 0.5 μl of 0.6 mM concentration of a crude preparation of the linear peptide, 12 kV. (B) Injection of 0.5 μl of 0.3 mM concentration of the purified linear peptide, 12.5 kV. (C) Injection of 0.5 μl of 0.6 mM concentration of two other components of the crude preparation from (A), 13.5 kV. (Reproduced with permission from Ref. 87. Copyright Elsevier Science Publishers.)

provided sufficient material for direct sequence analysis. Electrostacking was used to concentrate dilute peptide samples for analysis by CE [90]. More than 5 times improvement in concentration sensitivity was obtained by this method.

Capillary isotachophoresis has been used for CE separation of peptides [86,87]. Firestone used CZE and CITP for purity control of synthetic peptides [86]. PTFE capillaries (0.5 mm I.D. × 22.2 or 44.4 cm length), on-line UV (206 nm or 254 nm) and conductivity detection were employed. Capillary isotachophoresis (CITP) with concentration gradient detection has also been used to determine the purity of several synthetic peptides [87]. The concentration gradient detection system consisted of a helium-neon laser, a focusing lens and a position sensor. The laser beam was focused directly into the separation capillary and was then intercepted by a photodiode position sensor. Figure 7.16 shows the CITP of a crude synthetic peptide.

7.3 PROTEINS

The separation of proteins by CE represents an area of tremendous interest in recent years. Many approaches have been developed to achieve reproducible and highly efficient separation of proteins. Several reviews have appeared which described the application of CE in protein separations [41,45,46]. In this section, the approaches to minimize adsorption of protein in the inner wall of the separation capillaries are described and some examples of protein separation by CE are given. The use of CE-MS for the analysis of protein is discussed next. The applications of CE in process control and monitoring are then considered. Subsequently, techniques for the modelling of CE separations are presented. The use of capillary isoelectric focusing and capillary isotachophoresis for the analysis of proteins was also considered.

7.3.1 Minimization of protein adsorption

Green and Jorgenson [91] demonstrated that protein adsorption by the capillary wall could be reduced by adding metal salts to the electrophoretic buffer. In their study, the buffer used was 0.1 M 2-(N-cyclohexylamino)ethanesulfonic acid (CHES) at pH 9.0, containing varying amounts of KCl, NaCl, LiCl and CsCl. Addition of sodium and potassium salts was found to be effective in minimizing adsorption of proteins [91]. However, the increase in conductivity required the use of lower voltages and capillaries of smaller diameter to provide adequate heat dissipation. Nevertheless, high efficiency separations of proteins, including hen-egg lysozyme, bovine pancreatic trypsinogen, horse heart myoglobin, bovine milk β-lactoglobulin B and bovine milk β-lactoglobulin A were obtained.

Zwitterionic buffers were used to enhance separation efficiency in CZE of proteins in untreated fused silica capillaries [92]. The main advantage of the use of zwitterionic surfactant noted was that the method could be used to reduce protein adsorption as with the use of high salt concentration but it would be less likely to cause excessive Joule heating during separation.

Bullock and Yuan [93] separated basic proteins by CZE in uncoated capillaries with 1,3-diaminopropane modifier at 30–60 mM, and moderate levels of akali metal salts (20–40 mM) to suppress protein–capillary wall interactions. The proteins investigated included lysozyme, cytochrome C ribonuclease A, trypsinogen, α-chymotrypsinogen A and recombinant human interleukin-4 (rhuIL-4). Optimum buffer systems at different pH for the analysis of basic proteins were developed. The effect of pH on the migration order of protein was investigated.

A cationic fluorosurfactant buffer additive was used to reduce the adsorption of basic proteins in CZE [94]. Charge reversal occurred at the surface of the capillary upon addition of the fluorosurfactant. Proteins at a pH below their pI were repelled from the wall. High efficiency separation of a number of model proteins could be obtained at low ionic strength.

Hydrophobic interaction electrophoresis was performed in the presence of amphiphilic polymers, e.g. steaoryl dextran, together with ethylene glycol [95]. The additives can be used to change the degree of hydrophobic interaction of proteins and thereby the resolution. The analysis of human serum albumin and transferrin was investigated.

Metal-binding proteins were analyzed by CZE and MEKC in electrophoretic media containing modifiers [96]. Calcium-binding proteins (calmodulin, paralubumin, thermolysin and proteolytic peptides of calmodulin), zinc-binding proteins (carbonic anhydrase and thermolysin), and internal standard proteins (carbonic anhydrase and lactoglobulin) were separated. For CZE, buffers of 0.1 M tris (hydroxymethyl) aminomethane (Tris)–0.1 M tricine (pH 8.3) or 50 mM Tris–384 mM glycine (pH 8.3) with several additives (2 mM calcium chloride, zinc chloride or EDTA) were used. For MEKC, 0.1% sodium dodecyl sulfate was added to 0.1 M Tris–0.1 M tricine (pH 8.3). Interaction between the hydrophobic region of calmodulin and hydrophobic probes, trifluoperazine (Trif) and N-(6-aminohexyl)-5-chloro-1-naphthalene sulphonamide (W-7) was observed in CZE. The binding shift of the metal-binding proteins was found to depend on cations in the electrophoretic buffer.

McCormick [97] utilized two different approaches for protein separations. For the CZE separation of protein mixtures in low pH buffer, a buffer of 150 mM H_3PO_4 (pH 1.50) was employed. For the CZE separation of protein mixture with coated capillaries, poly(vinylpyrrolidinone)-modified (PVP) capillaries were employed. Fifteen proteins with molecular mass ranging from 12 kDa to 77 kDa and pI values of 4.5–11 were separated in less than 25 min. However, insulin, ovalbumin and chymotrypsinogen could not be separated by this method because of protein adsorption. The proteins investigated include: β-lactoglobulin, lysozyme (ovine egg), albumin (human serum), albumin (bovine serum), cytochrome C (horse), trypsinogen (bovine), myoglobin (whale), transferrin, conalbumin, (horse), carbonic

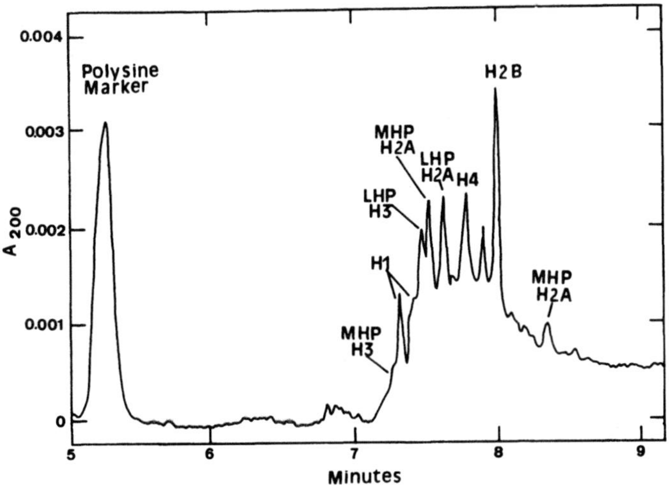

Fig. 7.17. Assignment of histone fractions obtained by CE of a whole histone sample. HPCE was performed in a 35 cm × 50 μm I.D. coated capillary. The sample was dissolved in 0.2% TFA and electroinjected for 10 s at 10 kV. Electrophoresis was performed at 10 kV in 0.1 M phosphate buffer (pH 2.5). Resolution was monitored at 200 nm with a detector response time of 1 s and a chart speed of 6 cm/min. Polylysine was added to the sample as an internal mobility standard. (Reproduced with permission from Ref. 98. Copyright Elsevier Science Publishers.)

anhydrase B (bovine), carbonic anhydrase A (bovine), hemoglobin (human), and paralubumin (rabbit).

CE of histones employing a 0.1 M phosphate buffer (pH 2.5) in a coated capillary was performed [98]. The histone preparation was separated into its components: (MHP)H3, H1 (major variant), H1 (minor variant), (LHP)H3, (MHP)H2A (major variant), (LHP)H2A, H4, H2B and (MHP)H2A (minor variant), where MHP is the more hydrophobic component, and LHP is the less hydrophobic component. Figure 7.17 shows the electropherogram for a whole histone sample. The order of separation was found to be different from those in acid-urea polyacrylamide gel electrophoresis and in reversed-phase HPLC. Therefore, CE provided a new dimension for the analysis of histone samples.

Low pH buffer (pH 1.5) containing 150 mM phosphoric acid, 10 mM Tris, 10 mM phosphate was used for the separation of a trypsin/cyanogen bromide digest of cytochrome C [99]. Buffers in the pH range of pH 7–8 were used for the separation of β-casein dephosphorylated, with ammonium acetate at pH 7.2 and Tris/tricine at pH 8.2, respectively [99].

Proteins from the fluid lining of the lungs, bronchoalveolar lavage fluid (BALF), of rats exposed to perfluoroisobutylene (PFIB) were analyzed by CE [100]. The buffer used was 0.1 M sodium phosphate (pH 2.5). Figure 7.18 shows the electropherogram of BALF proteins from control and PFIB-treated rats. Profound changes in the CE profile were observed. In addition, the use of CE as a second-dimensional analytical technique following a first-dimensional fractionation of BALF protein was explored. Unexpected contaminants in HPLC fractions were revealed.

CE was used to analyze the degradation products of a preparation of bovine aprotinin and bovine pancreatic trypsin inhibitor [101]. A 20 mM citrate buffer (pH 2.5) was used as the running buffer. An uncoated 50 μm I.D. fused silica capillary was used. The potential applied was 20 kV and the temperature was maintained at 27°C. Detection was by UV absorbance at 200 nm. The resolution obtained with CE was better than that with HPLC for aprotinin purity control.

By using borate buffer of high pH (9.5) and washing the capillary between runs with 1.0 M sodium hydroxide, selected model proteins and proteins in human serum could be determined by CZE [102]. The proteins studied included: trypsinogen, myoglobin, carbonic anhydrase I, carbonic anhydrase II, trypsin inhibitor, β-lactoglobulin and amyloglucosidase. The coefficients of variation in migration times and area were less than 0.5% and 4%, respectively, except for carbonic anhydrase

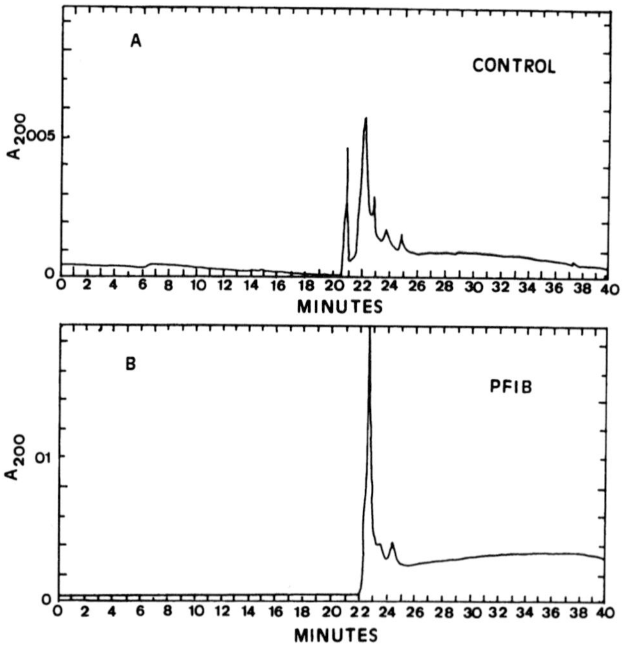

Fig. 7.18. CE of BALF proteins from control and PFIB-treated rats. Protein were precipitated with acetone from the BALF of control (air-exposed) and PFIB-exposed rats. The amounts of BALF used to prepare the protein sample from the PFIB-exposed rat was one tenth of that used for the preparation of the control sample. (A) CE of BALF proteins from control rats; (B) CE of BAIF proteins from PFIB-treated rats. (Reproduced with permission from Ref. 100. Copyright Elsevier Science Publishers.)

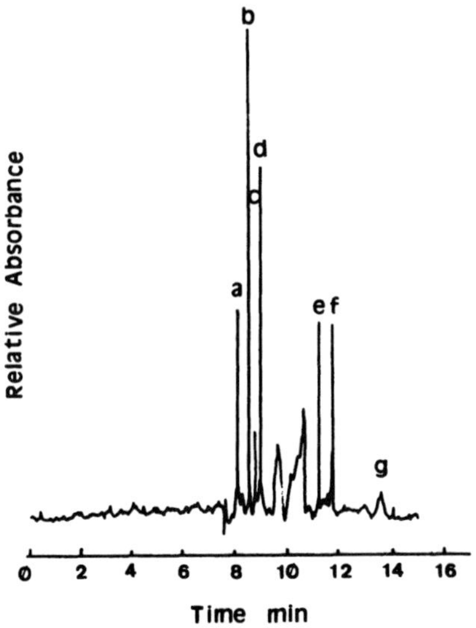

Fig. 7.19. Zone electrophoretic separation of seven model proteins on a 100 cm × 50 μm I.D. bare fused silica column. Peaks: *a* = Trypsinogen, pI = 9.3 mol. mass = 24.5 × 10^3; *b* = myoglobin, pI = 6.8, 7.2 mol. mass = 17.5 × 10^3; *c* = carbonic anhydraseI. pI = 6.6 mol. mass = 29.7 × 10^3; *d* = carbonic anhydrase II, pI = 5.9 mol. mass = 29.7 × 10^3; *e* = trypsin inhibitor, pI = 4.6. mol. mass = 7.5 × 10^3; *f* = β-lactoglobulin A, pI = 5.1 mol. mass = 17.5 × 10^3; *g* = amyloglucsidase, pI = 3.6, mol. mass = 170 × 10^3. Conditions: buffer, 50 mM sodium borate buffer (pH 9.5); applied voltage, 22kV; temperature, ambient; detection wavelength, 200 nm. (Reproduced with permission from Ref. 102. Copyright Elsevier Science Publishers.)

and amyloglucosidase. Figure 7.19 illustrates the separation of model serum proteins using this method.

Rapid analysis of proteins by CZE was investigated [103] using short capillaries with small I.D. (25 or 37 cm length × 25 μm I.D.). Electropherogram of a serum protein is shown in Fig. 7.20. The compounds investigated include: γ-globulin, transferrin, β-lipoprotein, heptoglobulin, α_2-macroglobulin, α_1-antitrypsin, α_1-lipoprotein, albumin, prealbumin and complements.

Fused silica capillaries modified first with γ-glycidoxypropyltrimethoxysilane and then polyethylene glycol 600 (PEG 600) were used for the separation of proteins [104]. Satisfactory performance was obtained at low pH (3–5) with the PEG-600 coated column. However, at higher pH (5–8), peak distortion and decrease in resolving power were observed. The proteins investigated included cytochrome C, lysozyme, myoglobin, trypsin, ribonuclease, trypsinogen and chymotrypsinogen. Capillaries modified with γ-glycidoxypropyltri- methoxysilane and coated with diol by acid hydrolysis were found to be effective in reducing protein adsorption at pH < 5. However, at higher pH, the coating was not effective. A covalent maltose coating has been used to reduce protein adsorption in the pH range of 5–7.

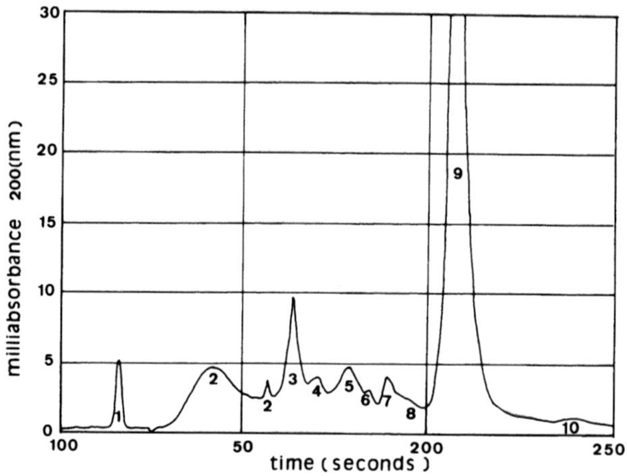

Fig. 7.20. Electropherogram of a normal serum protein: untreated fused silica capillary 25 cm × 25 μm I.D. Applied potential; 10 kV; buffer 150 mM borate; buffer strength, pH 10.0. Peaks: *1* = DMF; *2* = γ-globulin; *3* = transferrin; *4* = β-lipoproteins; *5* = haptoglobin; *6* = α_2-macroglobulin; *7* = α_1-antitrypsin; 8 = α_1-lipoproteins; *9* = albumin; *10* = prealbumin; *2'* = complements. (Reproduced with permission from Ref. 103. Copyright Elsevier Science Publishers.)

Capillaries coated with polyethyleneimine has been used for the separation of proteins [105]. Electroosmotic flow was reversed due to the positive charges of the coating. The coating was stable in the pH range 2–12. Separation of proteins can be optimized by variation of pH from 3 to 11 using this type of coated capillaries.

Capillaries coated with non-ionic surfactants (TWEEN series, BRIJ series) were found to be stable over the pH range 4–11 [106]. The addition of 0.01% surfactant in the running buffer improved separation performance. The proteins investigated included: lysozyme, cytochrome C, ribonuclease, α-chymotrypsinogen, myoglobin, conalbumin, carbonic anhydrase, β-lactoglobulin B, β-lactoglobulin A, ovalbumin and pepsin.

Capillaries deactivated with a terminal aryl pentafluoro (APF) group have been used for the separation of proteins [107,108]. High efficiencies over a wide pI range at neutral pH and moderate ionic strength were obtained. In one study, the separation of proteins, such as lysozyme, ribonuclease, trypsinogen, whale myoglobin, horse myoglobin, human carbonic anhydrase B and bovine carbonic anhydrase B [107] was investigated. In another study, the separation of lysozyme, cytochrome C, ribonuclease, α-chymotrypsinogen and myoglobin was performed [108].

Three types of stationary phase coatings were used in CE separations of proteins [109]. Two of the coatings were bonded hydrophobic phases (C_8 and C_{18}) and one was a bonded hydrophilic phase. These coatings were capable of reducing electroosmotic flow and inhibiting the interaction of proteins with the fused silica surface, thus permitting protein separation to be performed at neutral pH's [109].

Cobb *et al.* [110] used vinyl-bonded polyacrylamide coated capillaries to reduce protein adsorption. Electroosmotic flow could be eliminated and the coating was stable over a wide range of pH conditions (2–10.5). The separation of insulin chain A (porcine), serum albumin (bovine), ovalbumin (chicken egg), insulin (porcine), α-lactalbumin (bovine milk), β-casein (bovine milk) and insulin chain B (porcine) was investigated.

Polymethylglutamate (PMG)-coated capillaries were used for the separation of proteins, such as myoglobin, carbonic anhydrase B, lactalbumin, bovine serum albumin and cellulase [111]. The electroosmotic flow rate was reduced compared with uncoated columns. However, the flow rate at pH 7 was still high enough to elute anion sample as well as cationic samples. Significant improvements in separation efficiencies were obtained (2–10 times) for proteins with neutral or weakly acidic pI values. However, for more acidic and hydrophobic proteins such as bovine serum albumin, the coating was found to be inefficient. In Fig. 7.21, the separation of cellulase is shown.

Hydrophilic coatings with hydroxylated polyether functions were used for the separations of proteins [112]. Two types of coatings were investigated. The first type was referred to as fuzzy coatings which consisted of two layers: a glyceropropylpolysiloxane sublayer covalently attached to the inner surface and a polyether layer formed on top of the sublayer. The second type was referred to as interlocked coating which consisted of polysiloxane polyether chains, whose monomeric units at both ends were covalently attached to the capillary inner surface with possible interconnection. The polyether chains were found to be effective in shielding the unreacted surface silanols and hence minimized adsorption of the proteins. High separation efficiencies were obtained in the pH range 4.0–7.5. After

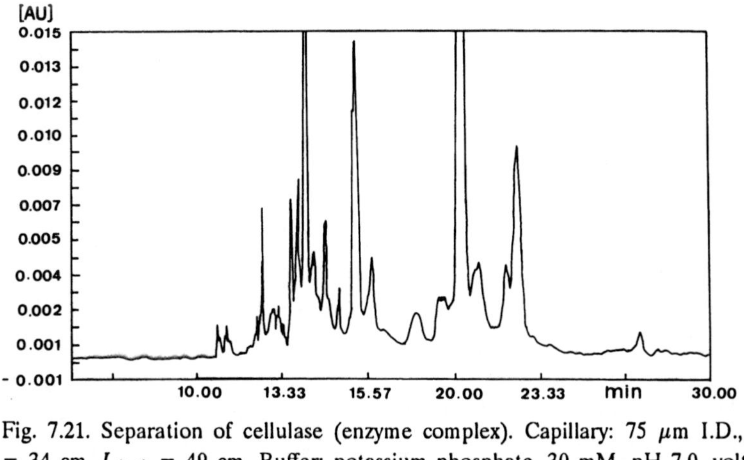

Fig. 7.21. Separation of cellulase (enzyme complex). Capillary: 75 μm I.D., PMG-coated, $L_{(\mathrm{eff})}$ = 34 cm, $L_{(\mathrm{tot})}$ = 49 cm. Buffer: potassium phosphate, 30 mM, pH 7.0, voltage: 8 kV (42 μA). (Reproduced with permission from Ref. 111. Copyright Friedr. Vieweg & Sohn Verlagsgesellschaft mbH.)

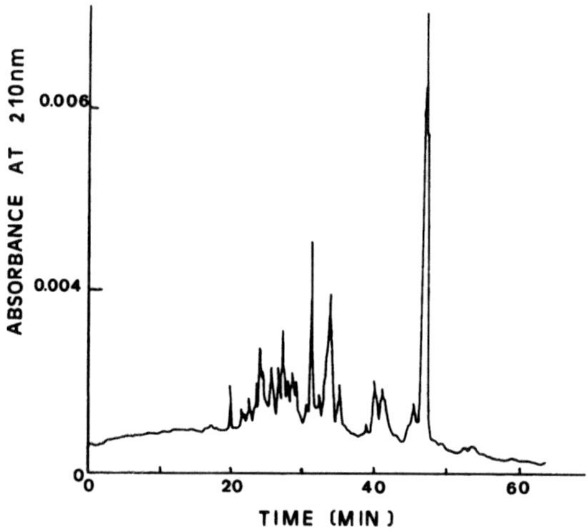

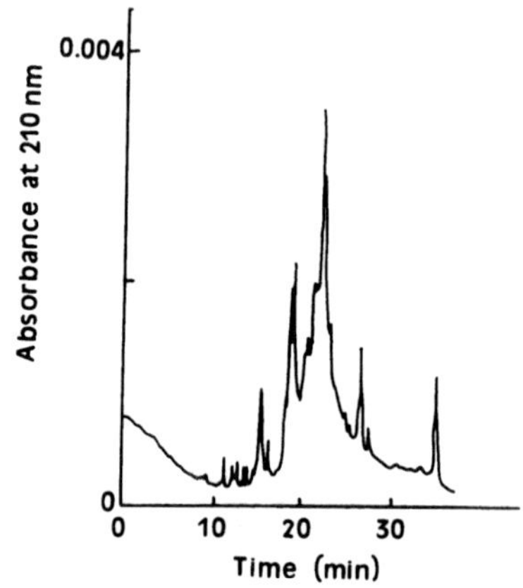

Fig. 7.22 (left). Electropherogram of crude soybean trypsin inhibitor. Capillary, I-200, 80 cm × 50 μm I.D. (50 cm to the detection point); electrolyte, 0.1 M phosphate solution, pH 6.5; running voltage, 17 kV. (Reproduced with permission from Ref. 112. Copyright Elsevier Science Publishers.)

Fig. 7.23 (right). Electropherogram of commercial soybean liposidase. Capillary, I-200, 0.1 M phosphate solution, pH 7.0; running voltage 17 kV. All other conditions as Fig. 7.22. (Reproduced with permission from Ref. 112. Copyright Elsevier Science Publishers.)

prolonged use, the coating which had been damaged could be removed by aqueous solutions of sodium hydroxide, and the columns could then be recoated. Proteins investigated included lysozyme, cytochrome C, ribonuclease A, α-chymotrypsinogen A, myoglobin, carbonic anhydrase, transferrin, α-lactalbumin, and trypsin inhibitor lipoxidase. Electropherograms obtained using capillary with interlocked PEG 200 polyether chains (I-200) for crude soybean trypsin inhibitor and commercial soybean lipoxidase are shown in Figs. 7.22 and 7.23.

Flat perfluorinated poly(ethylene-propylene) tubing (0.2 × 0.8 mm internal cross-section), with hydroxypropylmethylcellulose (HPMC) coating was used for the CITP separation of a commercial immunoglobulin (IgG) preparations for intravenous administration [113]. The leading electrolyte was 5 mM hydrochloric acid–9.3 mM 2-amino-2-methyl-1-propanol (pH 9.9). The terminating electrolyte was 50 mM tranexamic acid trans-4-(aminomethyl) cyclohexane carboxylic acid–15 mM potassium hydroxide (pH 10.8).

A charge-reversal technique [114], in which the negative charges of the capillary surface were changed to positive charges by introducing a dynamic coating onto standard fused silica capillary, was employed to characterize polyethylene glycol (PEG) modified proteins. The electropherograms for myoglobin conjugated with p-nitrophenyl (PNP) esters are shown in Fig. 7.24. Separation was found to be governed by the size of the PEG and the net positive charge on the conjugate at the separation pH.

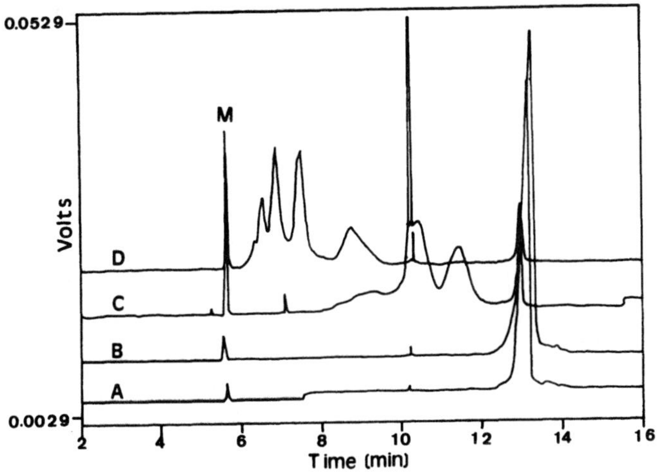

Fig. 7.24. Capillary electrophoresis of myoglobin conjuated with PNP esters. The PEG ester-to-protein molar ratio in these reactions was 5 : 1. (A) Unmodified myoglobin, (B) 0.15 kDa PEG, (C) 2 kDa PEG, (D) 10 kDa PEG. The electropherograms were obtained using a 72 cm × 50 μm I.D. fused silica capillary tube; 100 mM phosphate, 10% ethylene glycol, pH 2.0; 30 kV, 45 μA; UV detection at 215 nm. Hydrolyzed ρ-nitrophenol (*) was used as a neutral marker. The peak at 10.1 min is not identified, and may be a system-related artifact. (Reproduced with permission from Ref. 114. Copyright Elsevier Science Publishers.)

The use of coated capillaries to minimize electroosmosis and adsorption of solutes was demonstrated with several examples [115], including the separation of tryptic digests of bovine serum albumin, polymerase chain reaction product analogues and the measurement of human growth hormone.

A comparison of uncoated and coated capillaries for CZE and CIEF separation of proteins was made [116]. A mixture of six standard proteins with pI between 4.2 to 7.4 and molecular mass between 18,000 and 64,000 Da were used for comparison. The proteins included α-lactoalbumin, phycocyanin, β-lactoglobulin A, β-lactoglobulin B, hemoglobin A, hemoglobin F, hemoglobin S and hemoglobin C. Under acidic conditions using 0.1 M sodium phosphate (pH 2.5) as the electrophoretic buffer, although adsorption was reduced effectively, the resolution was not satisfactory with both the uncoated and coated capillaries. Under alkaline conditions using a 0.3 M sodium borate buffer (pH 8.5) with zwitterionic additives, the coated capillaries provided satisfactory separation reproducibly.

Three strategies for the analysis of proteins under alkaline conditions were investigated [117], including the use of capillaries coated with a linear hydrophilic polymer, the washing of the capillary with acidic solution between runs, and the addition of buffer modifiers, e.g. zwitterionic species, to minimize adsorption. Improved performance could be obtained by combining the use of the three strategies.

A number of patents have been granted which described methods for the analysis of proteins by CE [118–120]. A patent was obtained for a flow-rate controlled surface charge coating for CE based on the charge-reversal concept [118]. A compound, e.g. polybrene, which is capable of altering the surface charge of the capillary in a stable manner was electrophoretically drawn into the tube until a desired flow rate was obtained. Separation of three isoforms of lactate dehydrogenase could be achieved after coating a capillary with 0.01% polybrene for 10 min. A patent on a method for coating capillaries with aryl pentafluoro (APF) has also been granted [119]. Another patent was on an apparatus and a method for protein and peptide sequencing [120].

The use of gel-filled columns for the separation of proteins by CE has been investigated [121–126]. Hjerten and co-workers [121,122] reported the use of gel-filled capillaries for separation based on molecular sizes with sodium dodecyl sulfate-polyacrylamide gels (SDS-PAGE). A butanol extract of a culture filtrate of actinomycetes SP. 913 was separated in a vertical water cooled capillary (0.2 mm I.D., 0.4 mm O.D.) containing a polyacrylamide gel of composition 3% T, 4% C [122].

Cohen *et al.* [123] performed separation of human growth hormone (hGH) by capillary gel electrophoresis (CGE). Separation of a genetically engineered hGH, with methionine added to the N-terminal amino acid of growth hormone (Met-hGH), from a minor product called two chain was performed under non-denaturing conditions (no SDS and no urea added to the gel). Columns with composition of 10% T and 3.3% C were used. The buffer was 0.1 M Tris–borate at pH 7.8.

Non-crosslinked polyacrylamide gel was used as a sieving medium for proteins in CE [124]. Four proteins in the molecular mass range of 17,800 and 77,000 Da were separated as SDS complexes. The electrophoretic medium consisted of a phosphate buffer (pH 5.5) with 0.5% SDS and 10% liquid linear polyacrylamide gel. Migration times of the proteins correlated with the logarithm of their molecular masses.

CE separation of recombinant biotechnology-derived proteins was performed with capillaries filled with sodium dodecyl sulfate-polyacrylamide gel containing ethylene glycol [125]. The gel compositions were either 5.1% T and 2.6% C or 3.1% T and 2.6% C. The presence of 1.8–2.7 M ethylene glycol significantly improved the stability of the gel. Figure 7.25 shows the electropherogram obtained for a recombinant biotechnology-derived protein.

Capillary tubes (130 μm I.D.) filled with ultramicrogradient gels were used for the determination of proteins in kidney tubule fluid, and urine from normal rats and those with glomerulonephritis [126]. The acrylamide gel concentration varies from 4 to 27%. The gels were extruded after electrophoresis from the capillaries stained with fast green, destained and scanned in a microdensitometer. The method was capable of quantitating low molecular mass proteins in biological fluids.

Additives in protein samples to be analyzed by CE have been found to influence electrophoretic behaviour of the proteins [127,128]. A method for preventing protein adsorption developed is based on the addition of ethylene glycol to the

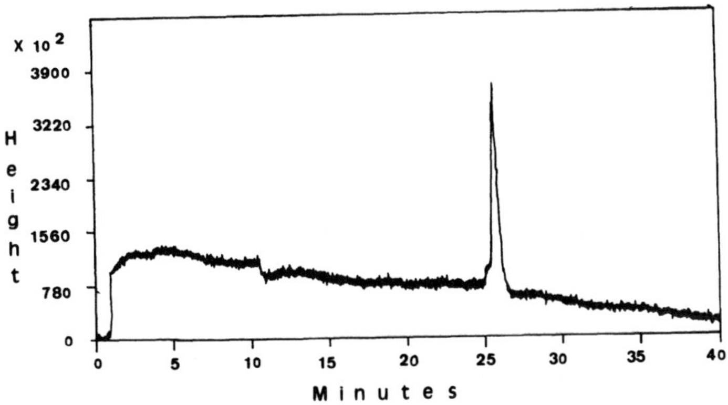

Fig. 7.25. Electropherogram of a recombinant biotechnology-derived protein (soluble CD_4-PE) as monitored at 214 nm. Conditions: −83 V/cm; 12 μA; column temperature, 25°C; migration distance, 7 cm; fused silica capillary, 75 μm I.D.; running buffer, 375 mM Tris (pH 8.8)–0.1% SDS-ethylene glycol. (Reproduced with permission from Ref. 125. Copyright Elsevier Science Publishers.)

protein sample [127]. In this method, the pH and molarities of the running buffer and the protein sample were kept at different values. A mixture of six proteins was dissolved in 20 mM boric acid (pH 4.0), containing 20% ethylene glycol. The running buffer was 50 mM sodium borate decahydrate (pH 10.0). The proteins investigated included carbonic anhydrase, urease, α-lactalbumin, ovalbumin contaminant, ovalbumin and bovine serum albumin. The method was applied to the analysis of proteins in human serum from a normal individual and from a patient suffering form multiple myeloma.

The addition of sodium dodecyl sulfate to protein samples before separation by CZE was found to significantly affect electrophoretic behaviour [128]. Even at SDS concentration of 0.1% (w/w), a complete loss in resolution of protein samples (conalbumin and ovalbumin) was observed, although an increase in efficiency was obtained. It has been shown that these effects were caused mainly by conductivity differences generated in the system.

7.3.2 Examples of protein separation by CE

The use of CE for the separation of the charge variants of recombinant DNA-derived (rDNA) proteins, such as human growth hormone (rhGH), a soluble form of T4 receptor protein (rCD4) and tissue plasminogen activator (rt-PA), as well as a glycoprotein has been investigated [129]. A coated silica capillary was used. It was found that the pH of the electrophoretic buffer was an important parameter for obtaining optimum separation.

Biosynthetic human insulin (BHI),human growth hormone (hGH), their derivatives and related proteins were analyzed by CE [130]. The separation was performed in a buffer containing 0.01 M tricine, 0.058 M morpholin and 0.02 M sodium chloride, adjusted to pH 8 with 1 M NaOH. UV detection was performed at 200 nm. An electric field strength of approximately 300 V/cm was used.

Tryptic digest fragments of human growth hormone (hGH) by CZE has been investigated [131]. The effects of pH, concentration of buffer (tricine), ionic strength adjuster (sodium chloride), and mobile phase additives (morpholine) on separation efficiency were studies. The pH and the concentration of the buffer were found to be the most important parameters affecting the separation of the hGH digest. The buffer modifier also helped to enhance resolution by influencing the electroosmotic flow.

The separation of antibody-antigen complexes formed by an immunochemical reaction by CZE was performed [132]. The antigen was a recombinant human growth hormone (hGH) which reacted with a monoclonal antibody (MoAb, immunoglobulin G (IgG) class). A 0.1 M tricine buffer (pH 8), an applied voltage of 30 kV and UV detection at 200 nm were used. The temperature of the column was maintained at 30°C. Figure 7.26 shows the separation of the IgG, hGH and IgG-$(hGH)_n$ complexes.

The effects of pH, buffer type, organic additives and capillary equilibration time on the separation of the glycoform of recombinant human erythropoietin (r-HuEPO) were investigated by Tran *et al.* [133]. Optimum conditions for the resolution of the glycoforms of r-HuEPO employed a mixed buffer (100 mM acetate–phosphate) at pH 4.0, with 10 h capillary pre-equilibration time in the running buffer.

CZE and CIEF were used to fractionate the human recombinant tissue plasminogen activator (rtPA) glycoforms [134]. The sialic acids were removed from the carbohydrate chains of the rtPA by treating with neuraminidase. The result of desialylated and untreated rtPA were compared. For IEF, 20 mM sodium hydroxide was used as catholyte and 10 mM phosphoric acid was used as anolyte. The mobilizer was 10 mM sodium hydroxide and 80 mM sodium chloride. For CZE, the running buffer consisted of 0.1 M ammonium phosphate (pH 4.6) with 0.01% Triton X-100 (reduced UV absorbance) and 0.2 M ϵ-aminocaproic acid (EACA).

Both analytical scale and micropreparative scale separation of human recombinant interleukin-3 (rhIL-3) and related proteins were performed [135,136]. Untreated fused silica capillaries were used. For the analytical separation, the electrophoretic buffer employed was 20 mM cyclohexylethane sulfonic acid (CHES) at pH 9.0 containing 10 mM KCl. Separation of rhIL-3 and human serum albumin in a commercial sample is shown in Fig. 7.27. For the micropreparative separation, 10 mM ammonium bicarbonate (pH 7.2) was used as the electrophoretic buffer. Nanogram quantities of rhIL-3 were collected and then identified by amino acid sequencing and SDS-gel electrophoresis.

Protein fractions of carbonic anhydrase from a micropreparative electrophoresis system was analyzed by CE for purity [137]. Very small amounts of sample and short separation times (less than 10 min.) were required for the CE analysis.

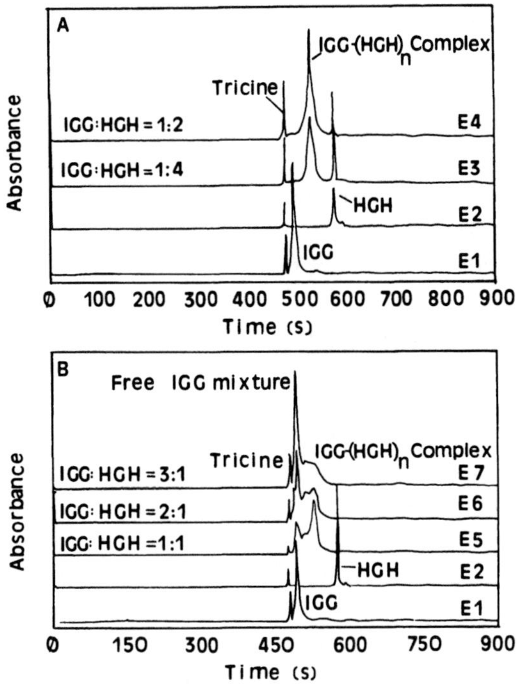

Fig. 7.26. Separation of IgG, hGH and IgG-(hGH)$_n$ complexes by CZE. (A) Electropherograms of IgG, hGH and mixtures containing an excess of hGH. (B) Electropherograms of IgG, hGH and mixtures containing an excess of IgG. Capillary: 100 cm, 50 μm I.D. (80 cm to detector) Buffer: 0.1 M tricine (pH 8.0); applied voltage: 30 kV, current: 19 μA; injection volume: 9 nl (3-s injection at 0.169 bar vacuum); detection at 200 nm; temperature: 30 $\pm$ 0.1°C. (Reproduced with permission from Ref. 132. Copyright Elsevier Science Publishers.)

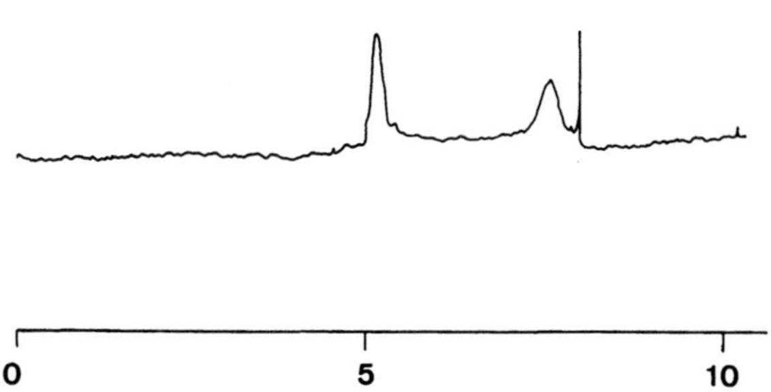

Fig. 7.27. Separation of rh IL-3 (5.3 min) and human serum albumin (8 min) in a commercial sample containing 0.2 mg/ml rh IL-3. Injection at 15 kV for 5 s. Capillary: 70 cm $\times$ 75 μm I.D. (40 cm to detector); buffer: 20 mm cyclohexylethane sulfonic acid (CHES) plus 10 mM KCl, adjusted to pH 9.0 with 1 M KOH. Detection at 220 nm. (Reproduced with permission from Ref. 135. Copyright Marcel Dekker Inc.)

A CE system with analyte velocity modulation and refractive index detection was used to analyze carbonic anhydrase [138]. An ac voltage at 11 kV (390 Hz) was superimposed on the dc separation voltage during electrophoresis. The base-line noise of the detection system was estimated to be 5×10^{-8} refractive index units, hence permitting highly sensitive, universal detection.

The separation of adrenocorticotropic hormone (ACTH) and its fragments was investigated by CZE [55]. Several buffer systems were employed. It was found that the migration order could be predicted by plotting charge divided by the two-third power of the molecular mass against pH in the different systems. Figure 7.28 shows the electropherograms of Org 2766, a modified ACTH-(4–9) fragment (I) and its related fragments. Electrophoresis was performed by buffers containing 100 mM borate/KOH (pH 8.3) and varying concentration of SDS.

CZE of streptavidin (60 kDa) and a polymeric protein (5×10^6 Da) was performed using a short, small I.D. (20 cm × 25 μm) coated capillary [64]. A phosphate buffer (0.1 M, pH 2.5) was used. Applied voltage was between 5 and 8 kV and detection was by UV at 200 nm.

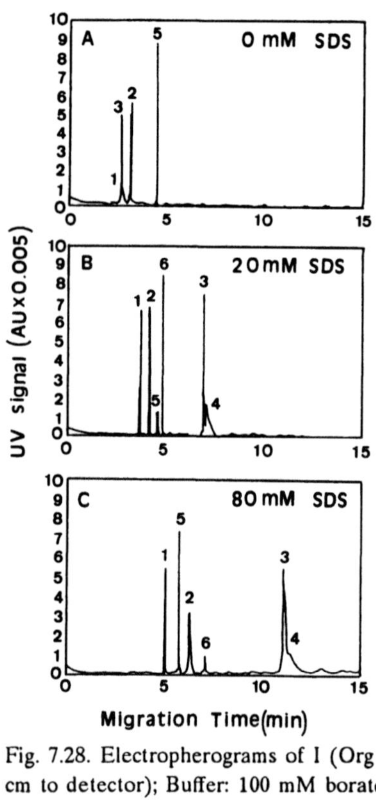

Fig. 7.28. Electropherograms of I (Org 2766) and fragments. Capillary: 57 cm × 75 μm I.D. (50 cm to detector); Buffer: 100 mM borate adjusted to pH 8.3 with KOH. UV detection at 214 nm. (Reproduced with permission from Ref. 55. Copyright Elsevier Science Publishers.)

Three laser-induced fluorimetric detection schemes were developed for the analysis of proteins by CZE [139]. The methods included detection of native fluorescence of the protein, pre-column labelling with fluorescein isothiocyanate (FITC), and on-column labelling with acrylaminonaphthalene-sulfonates (ANS) and 2-p-toluidinonaphthalene-6-sulfonate (TNS). Detection of native proteins was performed using 257 nm (9 mW) radiation produced by frequency doubling of the 514 nm line of an argon ion laser (7 W). The main advantage of this method is that no labelling is required. However, the technique required expensive instrumentation and long stabilization times. Detection limits of 14–25 nM were obtained using this method. Detection of FITC labeled proteins was performed with an argon ion laser (488 nm, 10 mW). Detection limits of 0.1 nM was achieved. The major advantage of this method is its high sensitivity. On the other hand, the method involved time-consuming sample preparation, and the integrity of the CZE bands was found to be affected due to the formation of mixtures of derivatives. In the case of the on-column method, excitation of ANS and TNS derivatives was performed using a helium-cadmium laser (325 nm, 15 mW). Although this method preserved the high efficiency of separation, it suffered from high background noise and hence resulted in lower sensitivity. Detection limits of 360 nM for TNS labelling and 615 nM for ANS labelling were obtained.

The effects of temperature on the electrophoretic behaviour of myoglobin and α-lactalbumin type III (calcium depleted) in CE were investigated [140]. Shorter migration times were obtained at higher temperature due to a decrease in viscosity. Specific temperature effects were also observed. In the case of myoglobin, different rate of interconversion of the Fe^{3+} and the Fe^{2+} in the heme group at different temperatures resulted in either a single peak or two peaks in the electropherogram. For α-lactalbumin type III, a conformational transition resulted in asymmetrical peaks at 35 and 40°C. Between 20–35°C and 40–50°C, a single sharp peak was obtained. The column temperature also affected the amount injected. As much as 70% increase in the volume injected could be observed when the temperature increased from 20 to 45°C.

The separation of proteins by CZE was found to be improved by applying a dynamic change of the buffer pH [141]. This was achieved by means of time-controlled pH changes in the anodic electrode chamber. The method was demonstrated for the cationic separation of proteins.

7.3.3 Protein separation by capillary electrophoresis–mass spectrometry

Capillary zone electrophoresis–mass spectrometry (CZE-MS) with electrospray ionization (ESI) has been used for the analysis of proteins [69–73,81,142]. The ESI interface produced multiply charged molecular ions and high molecular mass proteins could be analyzed. Loo *et al.* [69] investigated the separation of proteins, such as sperm whale myoglobin, porcine insulin, bovine insulin, leucine enkephalin, horse cytochrome C and chicken cytochrome C. The separation of bovine insulin,

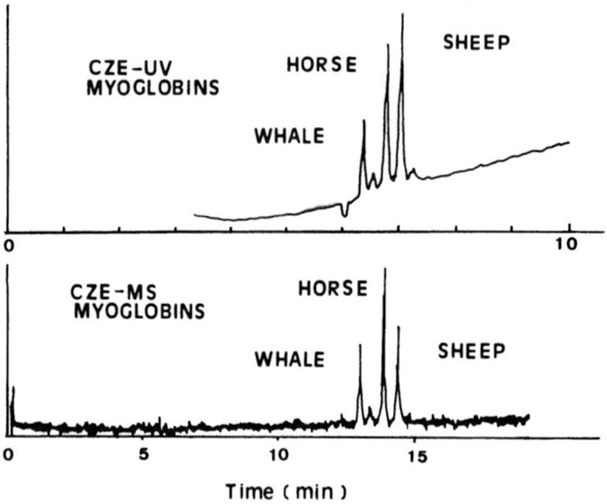

Fig. 7.29. CE-MS separation of a mixture of whale (M_r = 17,199.1), horse (M_r = 16,950.7) and sheep (M_r = 16,923.3) myoglobins. The on-line UV response and late MS reconstructed ion signal have been superimposed with adjsted time scales. The separation as done in 10 mM Tris(pH 8.3) at 120 V/m. Approximately 100 fmol per component were injected. (Reproduced with permission from Ref. 72. Copyright Elsevier Science Publishers.)

porcine insulin, egg lysozyme, horse heart myoglobin, bovine apotransferrin and bovine hemoglobin was demonstrated by Edmonds *et al.* [73]. Factors affecting the sensitivity of electrospray ionization (ESI) interface for the analysis of large molecules were discussed by Smith *et al.* [142] based on ESI-MS measurements for cytochrome C. A commercial CE instrument has been modified for interfacing with mass spectrometry [72]. A laboratory designed liquid sheath electrospray source was used in the electrospray ionization mass spectrometric interface. High separation efficiencies were obtained using the modified instrumentation. Figure 7.29 shows the separation of a mixture of whale, horse and sheep myoglobin. Deterding *et al.* [81] employed both electrospray ionization and continuous-flow fast atom bombardment interfaces for CE-MS of peptides and proteins. The proteins investigated included myoglobin, pigeon heart cytochrome C and horse heart cytochrome C.

7.3.4 Process control by CE analysis of proteins

CE was employed as a fast in-process control method for enzyme-labeled monoclonal antibody conjugates [143]. The running buffer consisted of 0.5% methylcellose in 0.1 M borax with 0.5 mM SDS (pH 10). Figure 7.30 shows the separation of unpurified conjugate of alkaline phosphatase (AP) and immunoglobulin (IgG) and the unreacted AP and IgG. Analysis times of less than 5 min were required.

Analysis of crude fermentation broth from the fermentation of Aspergillus oryzae by CZE was performed [144]. For the analytical separation of the broth, an alkaline running buffer, 25 mM phosphate buffer at pH 9.5 was used. For fraction collection,

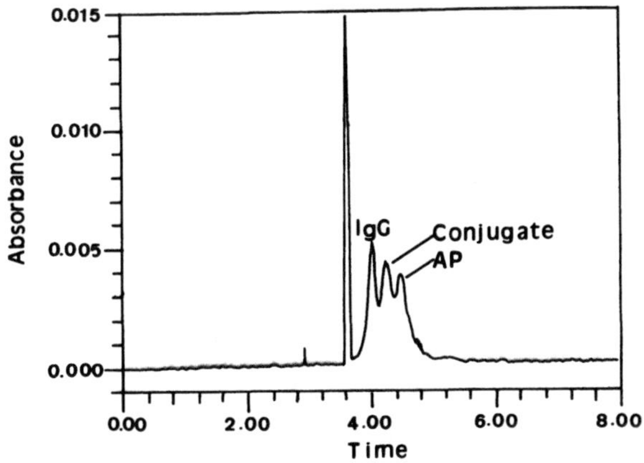

Fig. 7.30. Separation of unpurified AP-IgG conjugate containing the conjugate and unreacted AP and IgG. Capillary: 27 cm × 75 μm I.D.; sample buffer: 0.1 M borax with 0.5 mM SDS at pH 10; running buffer: 0.5% methylcellulose on 0.1 M borax with 0.5 mM SDS (pH 10); voltage: 5 kV; temperature: 15°C; detection: 280 nm. (Reproduced with permission from Ref. 143. Copyright Elsevier Science Publishers.)

the strength of the running buffer was increased to 35 mM. The increase in buffer strength resulted in longer sample zones which facilitated sample collection. On the other hand, increases in conductivity and Joule heating occurred, and hence separation was less efficient. Figures 7.31 and 7.32 illustrate the electropherogram obtained under the separation and collection conditions respectively. The collected fraction 2 was further investigated for protease activity. Proteolytic conversion of Suc-Ala-Ala-Pro-Phe-p-nitroanilide to p-nitroanilide and Suc-Ala-Ala-Pro-Phe could be studied quantitatively.

In-process assay of eminase by CZE using a low pH (2.5) buffer and hydroxypropylmethylcellulose (HPMC) additive was performed [145]. Rapid and highly efficient separations were obtained. A multi-compartment electrolyser with immobiline membranes was used to monitor r-DNA purification by isoelectric focusing [146]. CZE was used to monitor the content of each chamber. Isoforms of human monoclonal antibodies against the gp-41 of AIDS virus and of human recombinant superoxide dismutase (rhSOD) were studied.

7.3.5 Modelling of CE separation of proteins

For a large number of proteins, an almost linear relationship was obtained between their migration times and pI values in a broad pH range (6.86 to 10.5). On the other hand, migration times of proteins of very high molecular mass (100,000 to 600,000 Da) show a linear relationship to relative molecular mass. Figures 7.33 and 7.34 illustrate the relationship between the migration time with pI and with relative molecular mass respectively [147].

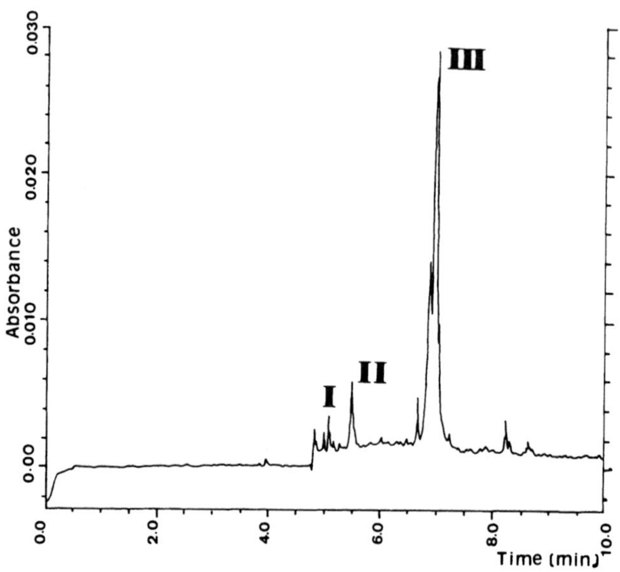

Fig. 7.31. CE pattern of fermentation broth from Aspergillus oryzae diluted three times with deionized water. The running buffer was 25 mM phosphate (pH 9.5), and a constant field of 15 kV (263 V/cm) was applied during separation. A 6 nl aliquot was injected, and detection was at 200 nm. *I* = an alkaline protease identified by "spiking" with a pure component; *II* = peak investigated for protease activity; *III* = main product of the fermentation, a neutral α-amylase consisting of three or more non-baseline-separated peaks, caused by isoenzymes. (Reproduced with permission from Ref. 144. Copyright Elsevier Science Publishers.)

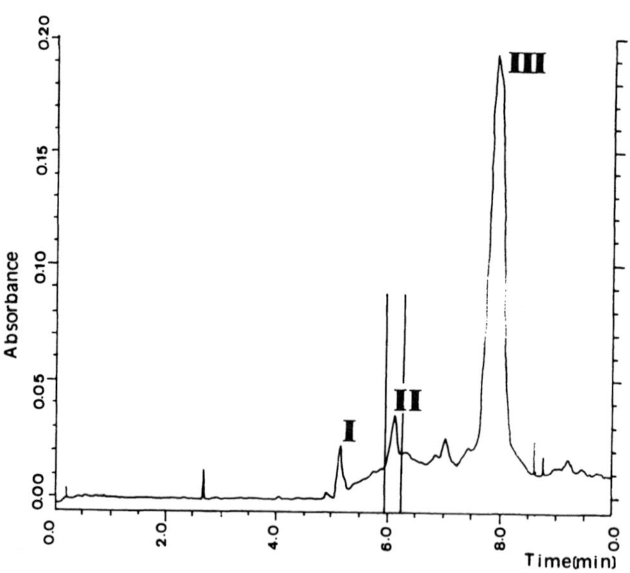

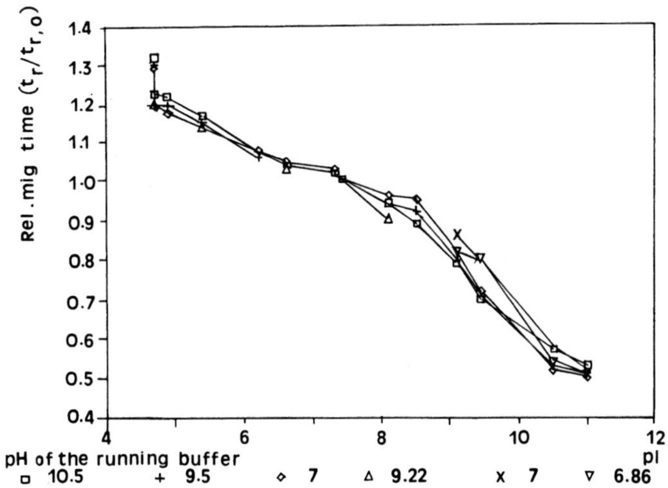

Fig. 7.33. Relation between pI and migration time (relative to horse skeletal muscle myoglobin) at different pH's. (Reproduced with permission from Ref. 147. Copyright Marcel Dekker Inc.)

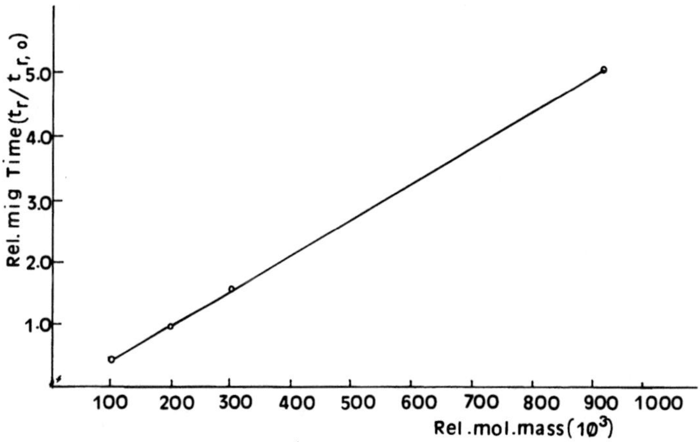

Fig. 7.34. Relation between relative migration times of collagen chain polymers and their molecular mass. (Reproduced with permission from Ref. 147. Copyright Marcel Dekker Inc.)

Fig. 7.32. CE pattern of fermentation broth from Aspergillus oryzae under fraction collection conditions. The running buffer was 33 mM phosphate (pH 9.5) and a constant field of 15 kV (263 V/cm) was applied during separation. A 30-nL aliquot was injected, and detection was at 200 nm. *I* = an alkaline protease; *II* = peak investigated for protease activity; *III* = main product of the fermentation, a neutral α-amylase. The time window for collection of the peak with proteolytic activity is shown. (Reproduced with permission from Ref. 144. Copyright Elsevier Science Publishers.)

The electrophoretic mobilities of a number of proteins in CE was found to be proportional to $Z/MM^{2/3}$, where Z is the charge and MM is the molecular mass [56]. The compounds investigated included human growth hormone, biosynthetic insulin-like growth factor II, biosynthetic human insulin, human proinsulin and bovine somatostatin.

Compton performed modelling of electrophoretic mobilities of proteins in CZE [149,150] based on physicochemical properties of proteins. CZE has also been used to determine the electrophoretic mobilities and diffusion coefficients of proteins [5]. Protein standards, including sperm whale myoglobin, horse heart myoglobin, carbonic anhydrase A and β-lactoglobulin were investigated. Diffusion coefficients were determined from peak width of analyte bands. The advantages of the method include small sample requirements and the capability to perform measurements on several compounds simultaneously.

7.3.6 Capillary isoelectric focusing of proteins

Non-crosslinked polyacrylamide methylcellulose coatings were used to perform isoelectric focusing of human hemoglobin and human transferrin [151]. A CE system was adapted to perform isoelectric focusing with hydrodynamic mobilization and electrophoretic mobilization. The analysis of proteins, such as human transferrin, was investigated [152]. Zhu *et al.* [153] investigated methods to improve separation and detection of proteins by capillary isoelectric focusing. By adding a basic compound into the ampholytic mixture, the segment between the monitor point and the capillary end could be blocked during focusing, permitting detection of basic proteins during mobilization. Low-pI and neutral pH zwitterionic mobilization agents resulted in improved detection of acidic proteins, and improved resolution of neutral and basic proteins, respectively. Non-ionic detergents (e.g. Triton X-100) in the sample and the ampholyte mixture reduced precipitation of γ-globulin [153].

The use of coated capillaries and additives for the separation of protein by CZE and CIEF was reviewed by Mazzeo and Krull [46]. Capillary isoelectric focusing of proteins in uncoated fused silica capillaries with polymeric additives in the sample/ampholyte mixture was performed [154]. Methylcellulose and tetramethylethylene diamine (TEMED) were used as additives. The additives provided a dynamic coating of the silica wall, hence preventing adsorption of proteins. Electroosmotic flow could be controlled by varying the concentration of the additive. Since the electroosmotic flow was not totally eliminated, there was no need to perform salt mobilization after focusing.

7.3.7 Capillary isotachophoresis of proteins

Separation of hydrophobic proteins, such as serum lipoproteins and membrane proteins by capillary isotachophoresis has been investigated by Josic *et al.* [155,156]. In one study, PTFE capillaries were used. The leading electrolyte consisted of 5 mM

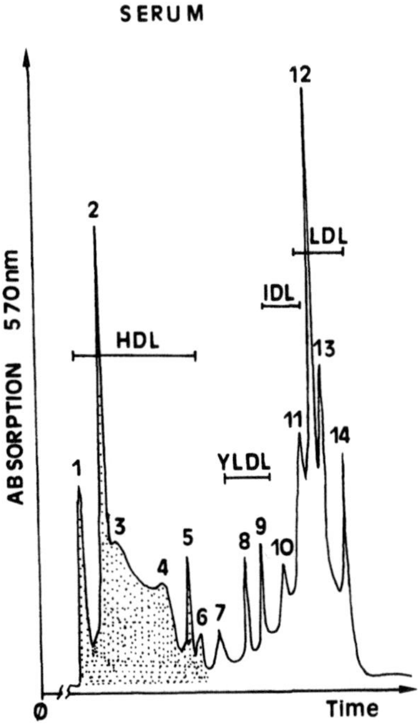

Fig. 7.35. Capillary isotachophoresis of normal serum lipoproteins. Serum was incubated with Sudan Black B for 30 minutes at 4°C and mixed with spacers (2:1, v/v). 2 μl of the mixture were used for the analysis. Separation was performed using a 15 cm × 400 μm I.D. PTFE capillary. The analysis lasted 15 minutes; temperature, 10°C. (Reproduced with permission from Ref. 155. Copyright Friedr. Vieweg & Sohn Verlagsgesellschaft mbH.)

H_3PO_4, 0.25% (w/v) hydroxypropylmethylcellulose (HPMC), 20 mM ammediol, pH 9.2. Figure 7.35 shows the pattern obtained for serum lipoprotein of a healthy person. The main groups of the liproteins are high-density lipoprotein (HDL), very low-density lipoproteins (VLDL), intermediate-density lipoproteins (IDL) and low-density lipoproteins (LDL). In another study, both CZE and CITP were employed [156] for the separation of hydrophobic membrane proteins. For CZE, both coated and uncoated capillaries quartz were used. The addition of detergents and/or denaturing reagents, e.g. urea, was found to be necessary in order to obtain reproducible results. For CITP, PTFE capillaries were used. Separation of calcium-binding proteins from liver and Morris hepatoma 7777 plasma membrane by CZE with the addition of urea is shown in Fig. 7.36a and b, respectively.

Capillary isotachophoresis with conductometric detection was employed for the analysis of blood serum proteins [157]. Separation of albumin, immunoglobin and ceruloplasmin was performed. The effect of buffer strength on the separation time was investigated.

Proteins were separated by isotachophoresis in open-tubular fused silica capillaries [159]. Small amounts of hydroxypropylmethylcellulose (HPMC) were added

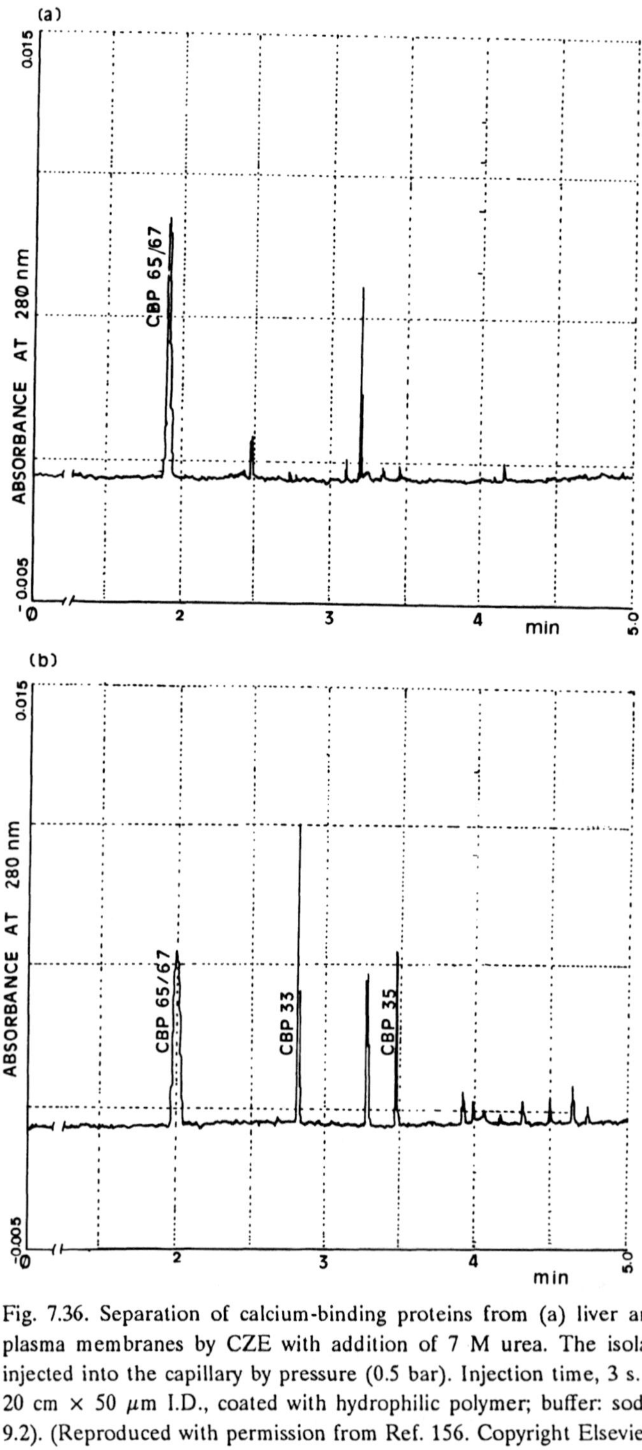

Fig. 7.36. Separation of calcium-binding proteins from (a) liver and (b) Morris hepatoma 7777 plasma membranes by CZE with addition of 7 M urea. The isolated membrane proteins were injected into the capillary by pressure (0.5 bar). Injection time, 3 s. Other conditions — capillary: 20 cm × 50 μm I.D., coated with hydrophilic polymer; buffer: sodium borate buffer (0.2 M, pH 9.2). (Reproduced with permission from Ref. 156. Copyright Elsevier Science Publishers.)

to the leading electrolyte to form a dynamic coating which drastically reduces both the electroosmotic flow and the interaction between proteins and the wall. The dynamic coating was renewed for each run. High resolution isotachophoretic determination of both cationic and anionic proteins were performed. An on-column multi-wavelength absorbance detector was used. In the cationic system, 0.01 M potassium acetate and acetic acid, with 0.3% HPMC was used as the leader and 0.01 M acetic acid was used as the terminator. The leader of the anionic system consisted of 10 mM formic acid titrated with ammediol to pH 9.1, with 0.3% HPMC, and 10 mM β-alanine-ammediol of pH 9.5 was used as the terminator.

7.4 NUCLEIC ACIDS

Nucleic acids are the fundamental genetic material of all living organisms. They are generally long chained polymers consisting of several basic components, including phosphoric acid, a sugar component (deoxyribose or ribose) and purine and pyrimidine bases (adenine, guanine and cytosine, thymine or uracil). While total hydrolysis of the polymer leads to the basic components, partial hydrolysis of the polynucleotide chain leads to fragments in the form of nucleosides, nucleotides and oligonucleotides. Nucleic acids also have widely varied arrangements (native and single-stranded DNA, circular and supercoiled DNA, viral and phage RNAs, rRNA, mRNA, tRNA, DNA-RNA hybrids, natural and synthetic mono- and polythematic polymers, etc.). The procedures for separating nucleic acids are generally complex processes that differ according to the type of the material to be separated. The use of CE for the separation of nucleic acids have been considered in several reviews [4,159,160]. Many recent investigations have been devoted to the development of CE methods for the isolation, purification and sequencing of nucleic acids and their constituents.

7.4.1 Nucleotides and nucleosides

MEKC was used by Row *et al.* [161] for the separation of modified nucleic acid constituents. They investigated the effect of SDS concentration on migration of nucleotides and derivatives, including normal and modified deoxyribonuclease, deoxyribomononucleotides, a ribonucleoside and a pyrimidine. Optimum resolution was obtained with 0.075 M SDS in 0.01 M phosphate–0.006 M borate buffer (pH 8.9), at a separation voltage of 10 kV.

MEKC separation of phosphorylated nucleosides was investigated by Liu *et al.* [162]. The effects of surfactant type and concentration were investigated. The surfactants used included an anionic surfactant (SDS), and two cationic surfactants, dodecyltrimethyl ammonium bromide (DTAB), and hexadecyltrimethylammonium bromide (HTAB). CE separation of bases, nucleotides and oligonucleotides was investigated by Cohen *et al.* [163]. For bases and nucleotides, MEKC with SDS

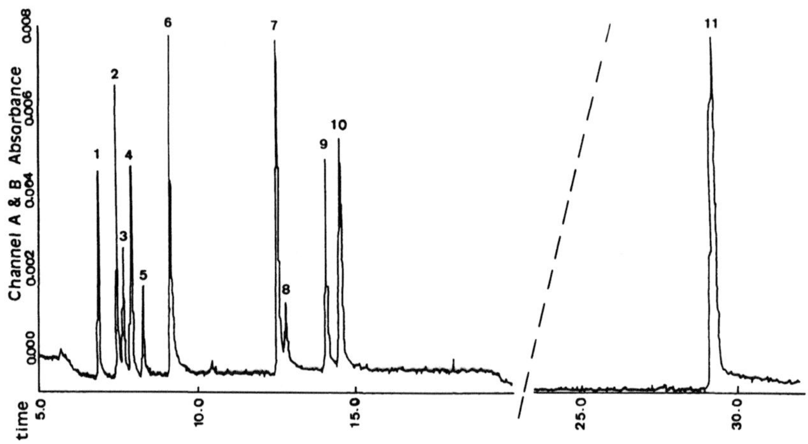

Fig. 7.37. Separation of normal and modified bases: *1* = Ura; *2* = 6metUra; *3* = Cyt; *4* = Thy; *5* = Gua; *6* = Ade + 7metGua; *7* = Nmetade; *8* = 1N6ethenoAde; *9* = 9ethGua; *10* = 9ethAde; *11* = 1metAde. Buffer, 20 mM sodium dihydrogen phosphate containing 100mM SDS, pH 7; separation voltage, 20 kV; temperature, 20°C; detection, UV at 254 nm. (Reproduced with permission from Ref. 165. Copyright Dr. Alfred Huethig Publishers.)

micelles (pH 7) was also employed. Kaspar *et al.* [164] investigated the MEKC separation of nucleic acid bases with SDS added to the phosphate–borate buffer (pH 7.0).

MEKC with SDS was used for the analysis of normal and chemically modified nucleobases, nucleosides and nucleotides [165]. Figure 7.37 shows the separation of normal and modified bases. The separation of normal and modified nucleoside and nucleotide 3′-monophosphate is shown in Fig. 7.38. MEKC with SDS solution was also used to separate mixtures of nucleic acid derivatives and analogues. Sixteen nucleic acids and their analogues were separated in 25 min. The superior separation power of MEKC over ion-pairing liquid chromatography (IPLC) for the separation of 14 nucleosides and nucleotides was demonstrated [166]. MEKC was able to resolve all the compounds within 40 min while optimized IPLC failed to resolve satisfactorily half of the compounds investigated.

Analysis of cyclic-AMP, AMP, ADP, ATP and guanosine nucleotides was performed in a neutral buffer (pH 7) containing 0.5% ethyl glycol [167]. UV absorbance was used for detection and sensitivity in the nanogram range was obtained. The utilization of a dynamic pH gradient in CZE extended its separation power for the analysis of compounds with large differences in pK_a values [168].

The pH gradient was generated by modifying the composition of an electrolyte in the electrode chamber. The technique was demonstrated for the separation of eleven purines and pyrimidine bases.

Two versions of on-line radioactivity detectors for CZE were developed [169,170]. One consisted of a planar semi-conductor (Cd-Te) detector and another consisted

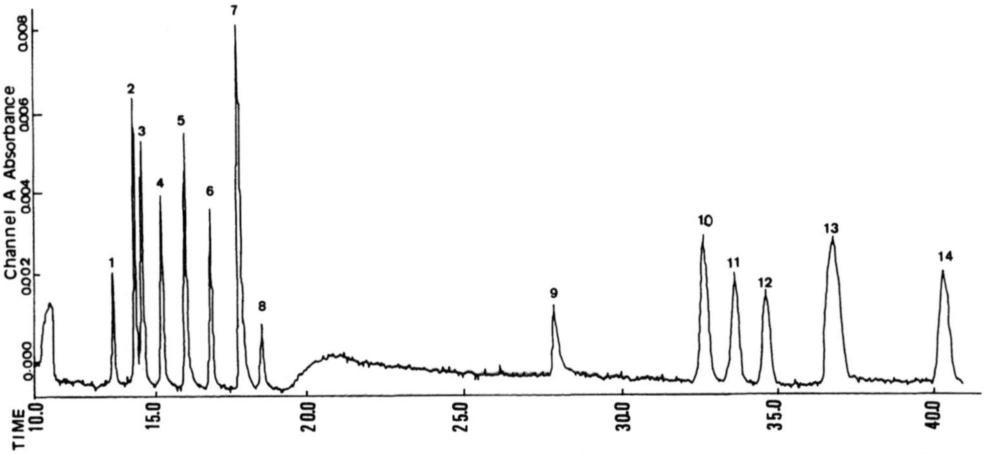

Fig. 7.38. Separation of normal and modified nucleosides and nucleotide nucleotide 3′-monophosphates: *1* = 2′dCyd; *2* = 2′dThd + 5met2′dCyd; *3* = 2′dAdo; *4* = 2′dUrd; *5* = 2′dGuo; *6* = 2′dLno; *7* = O6met2′dGuo; *8* = 1N4etheno2′dAdo; *9* = 3met2′dCyd; *10* = 2′dAdo3′mP; *11* = 2′dCyd3′mP; *12* = 2′dThd3'mP; *13* = 2′dGuo3′mP; *14* = 2′dUrd3'mP. Buffer: 20 mM sodium dihydrogen phosphate + 20 mM sodium borate containing 100 mM SDS and 10% acetonitrile, pH 9.2. (Reproduced with permission from Ref. 165. Copyright Dr. Alfred Huethig Publishers.)

of a cylindrical plastic scintillator. By using voltage programming to increase the residence time of labeled sample components in the detection volume, detection limits of subnanocurie level were obtained for ^{32}P labeled analytes.

CZE with laser-induced fluorescence has been used for the separation of nucleotides, mono, di, and triphosphates [172]. The separation was performed from the cathode to anode. Detection response was found to depend on the sample matrix composition. The largest response was obtained with injection from water. A linear relationship between detector response and resistance of the sample solution was obtained. CZE with indirect fluorescence detection has been demonstrated for the analysis of nucleotides and nucleosides [171]. Salicylate was used as the background fluorophore in the electrophoretic media. Detection limits of 50–100 amol of sample injected was obtained with this system.

Hjerten [173] investigated the CE separation of nucleic acids and their degradation products. The difference between theoretical and actual plate heights are discussed. Optimization of pH in the CZE separation of 2-, 3-, 4- methylpyridine was performed by Wren [174]. A cationic surfactant suppress electroosmotic flow and improved the resolution. The electrophoretic mobilities of the 3- and 4- isomers were found to depend upon their calculated charges.

The performance of a CE system with a split-flow sample introduction mechanism was tested with the separation of nucleotides [99]. Run-to-run variation for a mixture of 5 nucleotides were on the order of 0.3 to 0.5% and 1–3% for migration time and peak height, respectively. CE has been used to quantitate the purine nucleotides involved in the decomposition of adenosine-5′-triphosphate (ATP) of fish tissues [175]. The major compounds included inosine monophosphate (IMP), inosine (HxR), and hypoxanthine (Hx). The electrophoretic solution was 100 mM CAPS buffer at pH 11.

Separation of ribonucleoside triphosphate (NTP) and deoxyribonucleoside triphosphate (dNTP) by CZE has been investigated [176]. Capillaries with a polyacrylamide coating were used to reduce electroosmotic flow. Reversed polarity was employed (i.e. the injection end was held at high voltage). Electropherogram of NTP and dNTP mixtures is shown in Fig. 7.39.

The performance of a modified UV detector was tested by detection of nucleotides separated by CZE [177]. The detector was modified by replacing the original flow cell with a holder which housed optical fibres for bringing light from the UV source to the capillary and from the capillary to the photomultiplier.

Cyclic nucleotides have been separated by CE [178]. The cyclic nucleotides investigated included 3′,5′-cyclic adenosine monophosphate (c-AMP),3′,5′-cyclic guanosine monophosphate (c-GMP), and 3′,5′-cyclic inosine monophosphate (c-IMP). A 0.05 M sodium tetraborate buffer, adjusted with 1 M HCl to pH 8.3, was used as the electrophoretic buffer. The analytes were detected by UV absorbance at 210 nm.

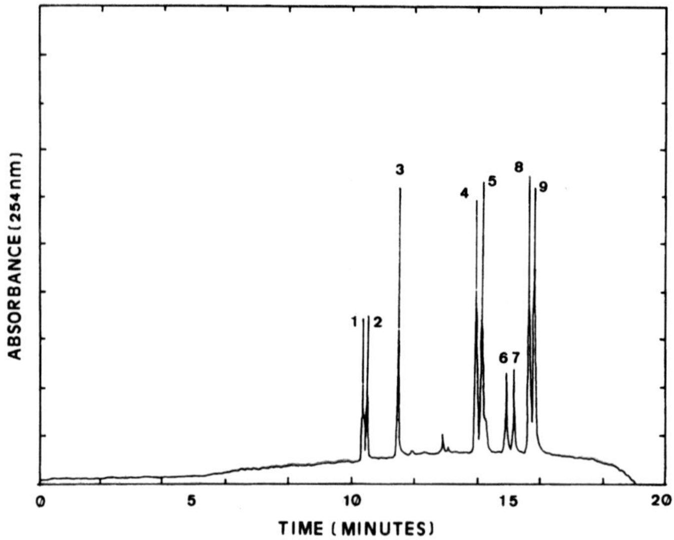

Fig. 7.39. Electropherogram of NTP and dNTP mixture. Concentration of original mixture was 0.031 mg/ml per nucleotide, ITP was 0.063 mg/ml. Capillary: 69.5 cm × 75 μm I.D., silylated with γ-methacryloxypropyltrimethoxy-silane, length to detector 62.8 cm; pressure injection for 5 s, 20 kV applied voltage. Buffer: 0.05 M phosphate (pH 2.7) with 0.002 M EDTA. Peaks: *1* = UTP; *2* = dTTP; *3* = ITP; *4* = GTP; *5* = dGTP; *7* = CTP; *8* = dATP; *9* = ATP. (Reproduced with permission from Ref. 176. Copyright Elsevier Science Publishers.)

The use of heavy water (D_2O) based buffer solutions for CE separation resulted in improved resolution of nucleosides compared with that obtained with normal water based buffer solutions [14]. Suppression of electroosmotic flow was demonstrated to improve the resolution of nucleotides [179]. Fixed membranes were used to separate the electrode reservoirs for the separation capillaries in order to reduce electroosmotic flow.

7.4.2 Oligonucleotides and DNA fragments

CZE with UV detection was used for the analysis of DNAs [180]. Ampholine was used in the separation buffer for the separation of oligonucleotides and enzymatic digests of calf thymus DNA. Separation of oligonucleotides has also been performed by MEKC [129]. The buffer used consisted of 50 mM SDS, 2 M urea, 5 mM Tris–Na_2HPO_4 and 3 mM Zn(II). Separation was based on different complexation with the metal-micelle surface.

Capillary gel electrophoresis (CGE) has been used extensively for the separation of oligonucleotides [181–183]. Oligodeoxynucleotides (polydeoxyadenylic acids, $(dA)_{40-60}$) were resolved in less than 8 min using a gel of composition 7.5% T and 3.3% C. For the separation of 70-mer and 99-mer, a 5% T and 3.3% C gel column was used. In addition, micropreparative scale separation of 800 ng of a 20-mer primer was demonstrated. By plotting base number vs migration time, a straight line was obtained. The slope of the line indicated the number of peaks which would pass through the detector in one minute and could be used to calculate the average peak width for each oligonucleotide. Separation of oligodeoxythymidylic acids $pd(T)_{20-160}$ has been performed using a 2.5% T, 3.3% C gel column [183]. The buffer pH was 8.6 and contained 89 mM Tris, 69 mM boric acid, 7 M urea and 2 mM EDTA.

Gel-filled columns were used for the separation of a mixture of 24- to 36-mer oligonucleotides [184]. Excellent resolution and peak shape were obtained. The effects of column length and applied voltage on efficiency and resolution of two oligonucleotides in CGE were investigated [185]. The oligonucleotides were: CTT.GTG.GTG.GG (XI) and CTT.GTG.GTG.GGc (XII). Best resolution was obtained by using a long column in combination with optimum electric field. Gel composition was 6% T and 5% C. Buffer used was 0.1 M Tris–0.2 M boric acid.

Gel-filled capillaries and counter migration CE were used for the separation of nucleic acids [186]. Separation of polydeoxyadenylic acid mixtures dA_{12-30} and dA_{40-60} was performed in gel-filled columns. Separation of DNA fragments (Phi × 174/Hae III fragments) was performed by counter-migration CE. These techniques were found to be more than 4 times faster than HPLC for the separation of oligonucleotides up to the 60-mer. Resolution was also better with CE. The capabilities to perform single-base resolution of polynucleotides by HPLC and CE were compared [187]. Reversed-phase, ion-exchange and mixed-mode columns were

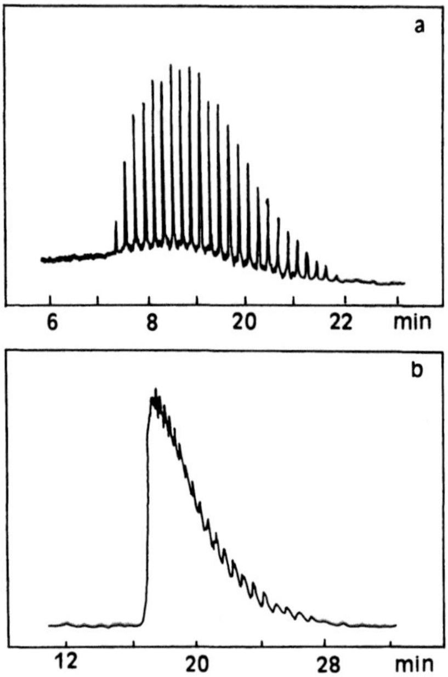

Fig. 7.40. (a) CGE and (b) anion exchange HPLC separation of poly(dA)$_{40\text{-}60}$. Gel electrophoretic conditions: gel matrix (7% T and 5% C). Chromatographic conditions: 0.1 M sodium chloride in 20 mM Tris–HCl buffer (pH 9.0); gradient programme. 0–4 min from 0 to 20% B, 4–12 min from 20 to 40% B, 12–55 min from 40 to 55% B at 40°C. (Reproduced with permission from Ref. 187. Copyright Elsevier Science Publishers.)

used for HPLC. CGE separation of poly (A) enzymatic partial digest hydrolysate and poly(dA)$_{12–18}$ was performed with a gel matrix of 5% T and 5% C. For the separation of poly(dA)$_{40–60}$, a gel matrix of 7% T and 5% C was used. Figure 7.40 shows the separation of poly(dA)$_{40–60}$ by CGE and HPLC. The superior resolving power of CGE was clearly demonstrated. CGE was able to resolve chain length up to 250 within 60 min, whereas reversed-phase, ion-exchange and mixed mode HPLC could resolved only 30, 40 and 40 nucleotides, respectively within the same time interval. The efficiency of CGE was found to be about 10 times larger than HPLC.

Wang *et al.* [188] used polyacrylamide gel-filled capillaries prepared by photopolymerization with riboflavin as initiator for the separation of pd(A)$_{15–80}$. A gel of composition 7.5% T, 3.3% C was used. No significant loss in performance was observed after 50 injections, although above 200 V/cm of applied voltage, bubble formation was found to occur.

The effects of instrumental and sample matrix factors on the resolution, sensitivity and detectability of oligodeoxynucleotides were investigated [189]. The results showed that numerous operating parameters affected the separation of oligonucleotides in CGE in a mutually dependent manner. For instance, by

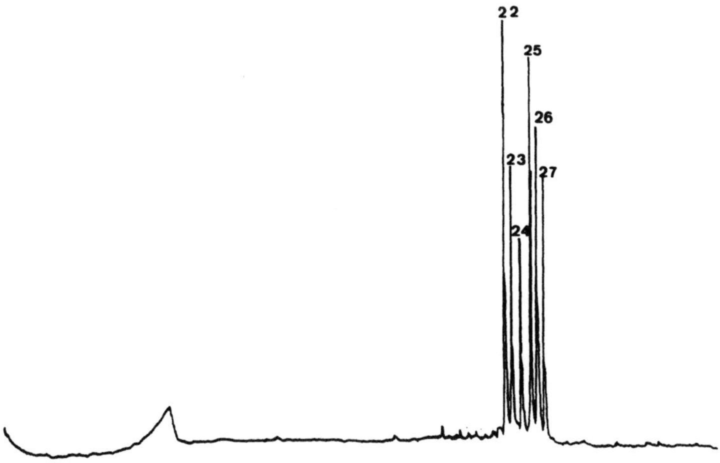

Fig. 7.41. CGE of normal mix (22–27) phosphodiesters, 6ng/μl. Micro-Gel$_{100}$ capillary gel column (100 μm I.D., 50 cm), Micro-Gel$_{100}$ buffer; Injections: electrokinetic, 5 s (−ve polarity); voltage: 10 kV; detection: 260 nm; temperature: 55°C. (Reproduced with permission from Ref. 190. Copyright Applied Biosystems.)

increasing the field strength, decreasing the separation length or increasing the column temperature, analysis time could be decreased. However, the decrease in migration time could result in poorer resolution and possible decrease in detectability due to increased peak velocity and decreased signal-to-noise ratio. By decreasing the conductivity of the sample, or increasing the sample concentration, the detectability could be improved. However, resolution could be affected when the column was overloaded.

CGE was used for the analysis of phosphodiester oligonucleotides and their phosphorothioate analogues [190]. A commercially available gel-filled capillary was used. Electropherograms for the phosphodiester (Normal-mix) and phosphorothioate (Thioate-mix) mixes are shown in Fig. 7.41 and 7.42, respectively. Baseline resolution was not obtained for the thioate-mix. It was suggested that phosphorothioates would need harsher denaturing condition than that for the phosphodiesters. Commercially prepacked polyacrylamide gel column containing 7 M urea has also been used successfully for the separation of polydeoxyadenylic acid mixture [191].

Separation of nucleotides by MEKC with UV detection in silylated capillaries was performed by Dolnik *et al.* [192]. The migration of oligonucleotide (polycytidines) was found to be independent of pH or ionic strength but could be influenced by complexation with spermine, with and without SDS.

The effect of pore size of the gel on the number of base/minute and resolution in CGE has been investigated [193]. The pore size was related to the monomer concentration. Gels with monomer concentration in the range of 2.5%, 4%, 5%, 6% and 8% were used. Figure 7.43 shows the gel composition vs base/minute and

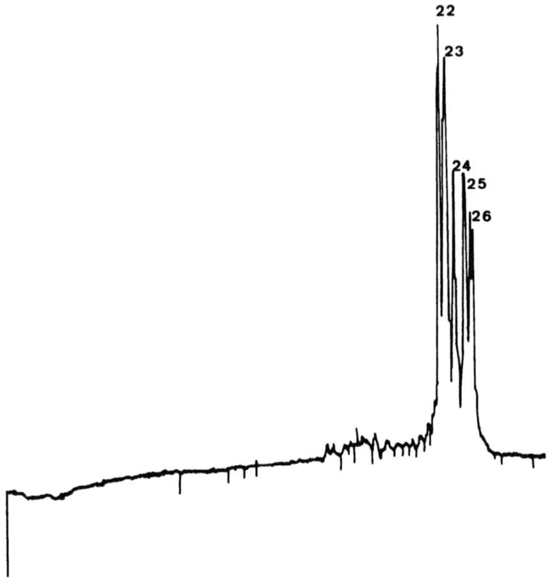

Fig. 7.42. CGE of thioate-mix (22–26) phosphorothioates 3 ng/μl. Conditions as in 7.41. (Reproduced with permission from Ref. 190. Copyright Applied Biosystems.)

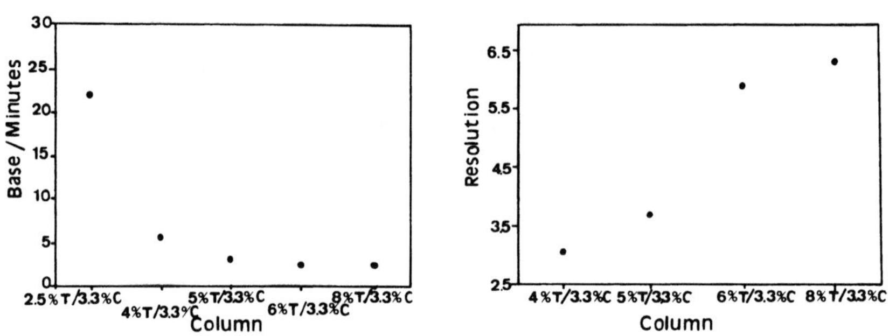

Fig. 7.43. Gel composition versus resolution and bases/minute. Capillary: 40 cm, 100 μm I.D.; buffer: 0.1 M boric acid, 0.1 M Tris, 2 mM EDTA 7 M urea. Samples: polydeoxythymidylic acid pd(T)$_{12\text{-}30}$ and pd(T)$_{40\text{-}60}$. Injection: 7 kV, 6 s. (Reproduced with permission from Ref. 193. Copyright Ciba-Geigy Ltd.)

resolution. The resolution of oligonucleotides was found to increase with increasing monomer concentration of the gel. Pore size did not show a linear dependence on monomer concentration. However, pore size could be used to optimize separation of a given sample size range.

The transportation, storage, conditioning and analytical performance of polyacrylamide gel-filled capillaries were studied by Weinberger and Schlabech [62]. The study showed that these gel-filled capillaries could be manufactured, shipped long distances and stored at room temperature with minimal special effort. It was recommended that the gel-columns should be prepared with extra length which could be trimmed off to remove bubbles formed at the ends. A linear voltage gradient was suggested for the conditioning of the column before use to elute residue TEMED. Gel columns stored for three months could be used for oligonucleotide analysis for as many as 100 runs.

CGE separation of poly(A) and poly(dA) digested by nuclease P1 was performed by Baba *et al.* [194], using gels containing 5% T and 5% C, and 3% T and 3% C, respectively. A simple, well-designed injection device was employed to prevent formation of bubbles during gel preparation.

The use of an on-column frit for sample collection in CE has been demonstrated [195]. The system was utilized for the separation and collection of fractions of pyriodoxamin and adenosine. Another instrument capable of collecting fractions from a CZE separation has been developed [196], which was used for collection of separated zones of nucleosides and α-chymotrypsin.

CZE was used for the separation of restriction fragments of DNA with a Tris–borate buffer containing 7 M urea and 0.1% SDS [197]. Sample pretreatment and injection of heated sample (60° C for 20 min) produced better results. A DNA restriction fragment mixture for 72 to 23130 base pairs, with molecular mass range from 4.6×10^4 to 1.5×10^7 Da was separated in 10 min on a column of 15 cm effective length. Tris–borate buffer provided better resolution than tris–phosphate buffer did, since borate complexation with the sugar moieties affected mobility in a manner to alter electrophoretic mobility differences.

The use of low and zero crosslinked polyacrylamide in CGE for the separation of DNA restriction fragments has been investigated [198]. The size selectivity of linear polyacrylamide capillary was characterized by means of a Ferguson plot. Linear relationships were obtained in the plot of log mobility vs. %T at constant %C. The effect of applied electric field on resolution was investigated. Pulsed field CGE with a symmetrical square wave of frequency between 0.1 and 1000 Hz was performed. A 20% increase in peak separation was observed in pulsed-field operation (100-Hz) relative to continuous field operation. Field programming CGE was also used by Guttman *et al.* [199] for the micropreparative scale separation of polydeoxyoligonucleotides.

An affinity ligand (ethidium bromide) was incorporated into the gel of a gel-filled capillary to permit selectivity manipulation in capillary gel affinity electrophoresis [200]. The complexing ligand intercalated between the strands of double-stranded DNA, and decreased the electrophoretic mobility of the DNA-ligand complex by reducing the ionic charge. The effects of various parameters, such as ligand concentration and applied electric field, on the retention and selectivity of DNA molecules of different sizes were investigated. Figure 7.44 illustrates the effect of ethidium bromide on the CGE separation of a DNA restriction fragment mixture. Two types of capillary systems were employed. The first type were capillaries filled with linear non-crosslinked polyacrylamide, which was bond to the capillary wall by a bifunctional agent, (methacryloxypropyl)trimethylsilane. The second column type

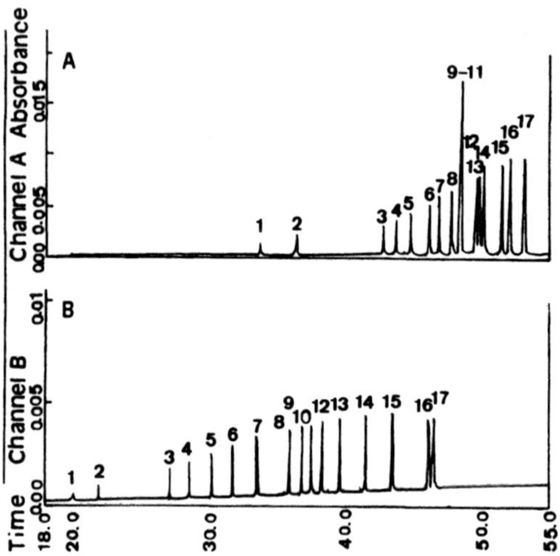

Fig. 7.44. Effect of the ethidium bromide on the capillary gel electrophoretic separation of pBR322 DNA-Msp I digest restriction fragment mixture: (A) no ethidium bromide; (B) 1 μg/ml ethidium bromide in the gel-buffer system. Peaks (base pairs): *1* = 26; *2* = 34; *3* = 67; *4* = 76; *5* = 90; *6* = 110; *7* = 123; *8* = 147; *9* = 147; *10* = 160; *11* = 160; *12* = 180; *13* = 190; *14* = 201; *15* = 217; *16* = 238; *17* = 242. Conditions: isoelectrostatic 100 V/cm (A), 200 V/cm (B); high-viscosity polyacrylamide gel column, effective length 20 cm, total length 27 cm; buffer, 100 mM TBE; injection 3 s, 30 mW. (Reproduced with permission from Ref. 200. Copyright American Chemical Society.)

contained low-viscosity linear polyacrylamide without binding to the capillary wall, which could be rinsed away easily.

The effect of temperature on the separation of DNA restriction fragments in CGE has been studied [201]. In the isoelectrostatic (constant electric field) mode, the migration time and resolution decreased as a function of column temperature between 20 and 50°C. The separation of ϕX-174 DNA fragments by CGE at different temperatures in the isoelectrostatic mode is shown in Fig. 7.45. In the isorheic (constant current) mode, the maximum migration temperature and resolution decreased as temperature increased. Figure 7.46 shows the isorheic separation of ϕx-174 DNA restriction fragment mixture.

CGE was used for the separation of fluorescently labeled DNA fragments generated in enzymatic sequencing reactions [202]. Composition of gel was 4% T and 5% C with 8.3 M urea. Laser induced fluorescence detection was performed. The system was demonstrated for the separation of fluorescently labeled T and G reaction products of M13 mp19 DNA.

Agarose gel was used for the separation of DNA restriction fragments [203]. The resolution of the agarose gel (AG)-filled columns were not as good as those obtained with a linear polyacrylmide gel (PAG), although the separation in the former was faster. The agarose gel columns were found to be stable only for a few

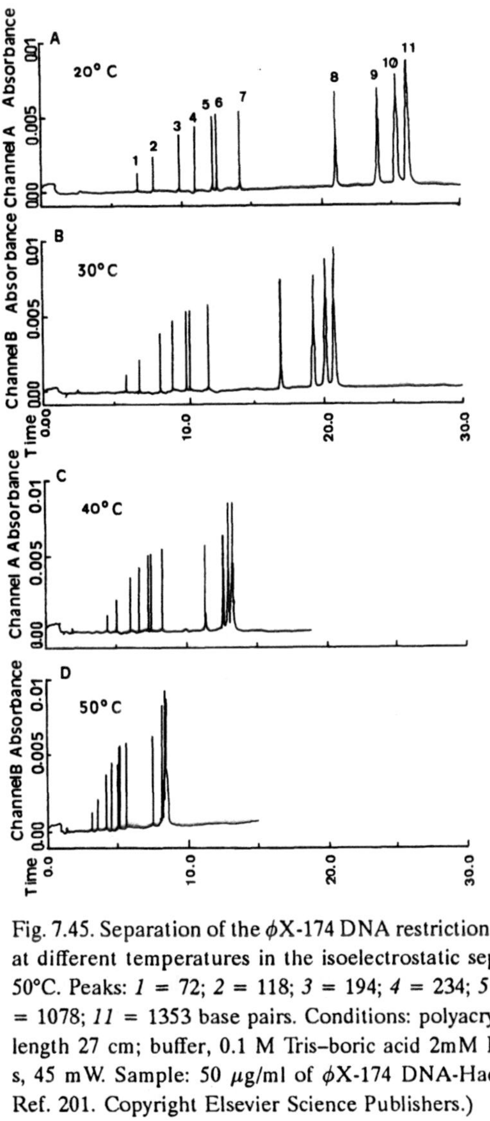

Fig. 7.45. Separation of the ϕX-174 DNA restriction fragment mixture by capillary gel electrophoresis at different temperatures in the isoelectrostatic separation mode: (A) 20, (B) 30, (C) 40, and (D) 50°C. Peaks: *1* = 72; *2* = 118; *3* = 194; *4* = 234; *5* = 271; *6* = 281; *7* = 310; *8* = 603; *9* = 872; *10* = 1078; *11* = 1353 base pairs. Conditions: polyacrylamide gel column, effective length 20 cm, total length 27 cm; buffer, 0.1 M Tris–boric acid 2mM EDTA (pH 8.5); voltage, 400 V/cm; injection, 2 s, 45 mW. Sample: 50 μg/ml of ϕX-174 DNA-Hae III digest. (Reproduced with permission from Ref. 201. Copyright Elsevier Science Publishers.)

days when used continuously.

By using an OV-17 coated capillary and buffer systems containing a polymeric additive hydroxypropylmethylcellulose), the analysis of DNA restriction fragments and polymerase chain reaction products were performed [204]. Resolution and improved peak shape were obtained by adding ethidium bromide to the buffer. Ultrafiltration was employed in the sample preparation step to remove low-molecular mass material.The system was applied to the detection of AIDS (HIV-1) virus in blood.

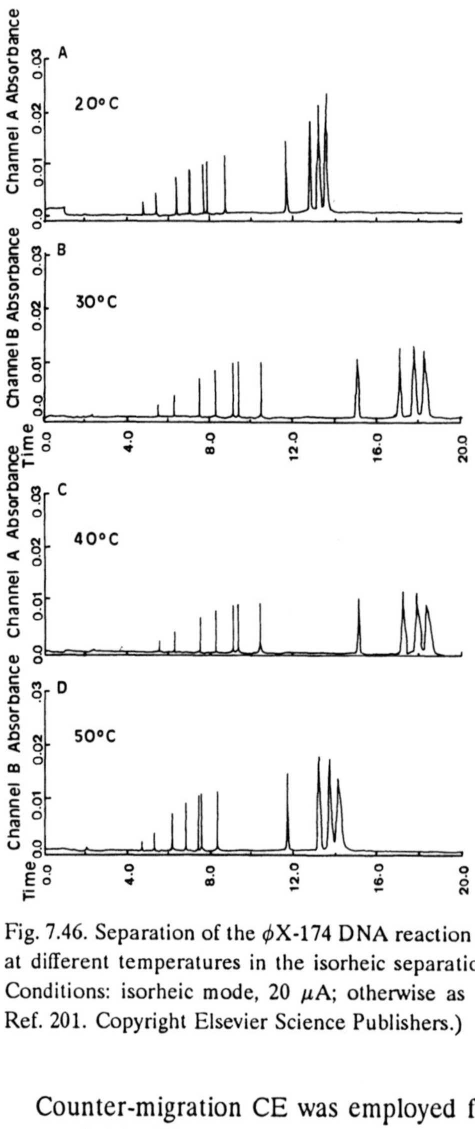

Fig. 7.46. Separation of the ϕX-174 DNA reaction fragment mixture by capillary gel electrophoresis at different temperatures in the isorheic separation mode: (A) 20, (B) 30, (C) 40, and (D) 50°C. Conditions: isorheic mode, 20 μA; otherwise as in Fig 7.45. (Reproduced with permission from Ref. 201. Copyright Elsevier Science Publishers.)

Counter-migration CE was employed for the analysis of DNA restriction fragments [186,205]. Separation and quantitation of double-stranded DNA fragments in the size range of 100–7000 kb was demonstrated [205].

Solutions of hydroxyethylcellulose (HEC) has been used as electrophoretic medium for the CE separation of DNA restriction fragments [206]. It was found that low-viscosity polymer solutions could provide a promising alternative matrix to gel-filled capillaries for size-based separation in CE, although at high electric fields, a saturated, size-independent mobility which limited resolution was obtained.

Molecular sieving in the absence of electroosmotic flow was performed by using a capillary coated with polyacrylamide to eliminate electroosmotic flow and a buffer containing methylcellulose as a sieving media [207,208].

Analyte velocity modulation was performed by superimposing a sinusoidal ac field over the dc voltage during the electrophoretic separation [209]. With the superimposed field, the RMS voltage in the capillary was:

$$V_{\mathrm{RMS}} = V_{\mathrm{dc}}\,[1 + (V_{\mathrm{dc}}/V_{\mathrm{ac}})^2\,]^2 \qquad (7.1)$$

where V_{dc}, and V_{ac} are the dc and peak ac voltages, respectively. Separation of nucleic acids in gel-filled capillaries was investigated. The polyacrylamide gel composition was 3.5% T and 3.3% C. It was found that for CGE separation of nucleic acid fragments, sinusoidal variation of the driving electric field could improve resolution or decrease separation times of both large and small fragments. The field variation could have caused fragment reorientation in the gel. Other wave forms may also be useful in influencing the reorientation and hence migration behaviour of nucleic acids.

The use of CGE to perform Southern blotting with on-column UV or laser-induced fluorescence detection was studied [210]. In particular, parameters that affected the hybridization of DNA molecules in solution were investigated. Figure 7.47 illustrates the identification of DNA molecule by hybridization with a fluorescence-tagged oligonucleotide probe using CGE. The dissociation of the hybridization species in gels containing urea by heat was investigated. The effects of probe concentration on hybridization and the incubation temperature for hybridization were also studied. The use of the method was demonstrated for small single-stranded DNA molecules.

CZE with UV detection was used for the separation of the DNA adducts, benzo(a)pyrene deoxy guanosyl-5-monophosphate adduct, commonly found in environmentally exposed organisms [211]. The electropherogram obtained for 2-deoxy-5′-nucleotide is shown in Fig. 7.48. A 0.04 M, pH 9.6 sodium carbonate buffer was used. The separation voltage was 15 kV.

The quantitation of ribonucleotides from base-hydrolyzed RNA was investigated by Huang *et al.* [212,213]. Two different sets of conditions were used. The first employed 50 mM formate, at pH between 3.7 and 4.0, as the running buffer. The alternative system employed a running buffer which contained 12.5 mM formate at pH 3.8 with 0.1 M cetyltrimethylammonium bromide (CTAB). The cationic surfactant, CTAB, changed the charge associated with the inner wall of the fused silica capillary from net negative to net positive. Hence the high voltage was applied at the detector end (anode) to ensure that direction of electroosmotic flow was from cathode to anode. It was found that there were several advantages in the use of CZE for the analysis of ribonucleotides, including the small amount of sample required, the high resolution, and the short time of analysis. Figure 7.49 shows the electropherograms of several examples of the CZE separation of RNA samples.

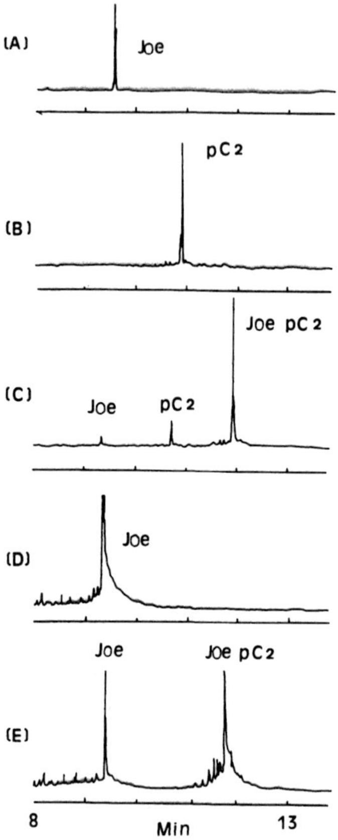

Fig. 7.47. Identification of DNA molecule by hybridization with a fluorescence-tagged oligonucleotide probe using capillary gel electrophoresis. (A) Joe-labelled 17-mer alone (5 μg/ml in 10 mM TBE). (B) pC2 alone (5 μg/ml in 10 mM TBE). (C) Equal amounts of Joe-labelled primer (10 μg/ml in 10 mM TBE) and pC2 (10μg/ml in 10 mM TBE) were annealed prior to electrophoresis and were analyzed by UV absorption. (D) The same sample as in (A) was analyzed by the laser system. (E) The same sample as in (C) was analyzed by the laser system. The hybridized species was clearly identified either by UV (C) or by the laser system (E). The capillary was filled with 9% T, 0% C polyacrylamide gel with no urea. Total length of the capillary was 45 cm with an effective length of 25 cm. Injection was done electrokinetically at 13.5 kV for 5 s, and the electric field for electrophoresis was 300 V/cm. (Reproduced with permission from Ref. 210. Copyright Elsevier Science Publishers.)

7.4.3 DNA sequencing

CGE was used for the separation and detection of DNA sequencing samples [214]. Gel composition was 6% T, 5% C. A buffer containing 8 M urea, 89 mM Tris, 89 mM boric acid, 2 mM EDTA was used. Laser-induced fluorescence was used for the detection of fluorescently labeled oligonucleotide primers. Limit of detection was 6000 molecules per peak. Baseline resolution of peaks which differed by one

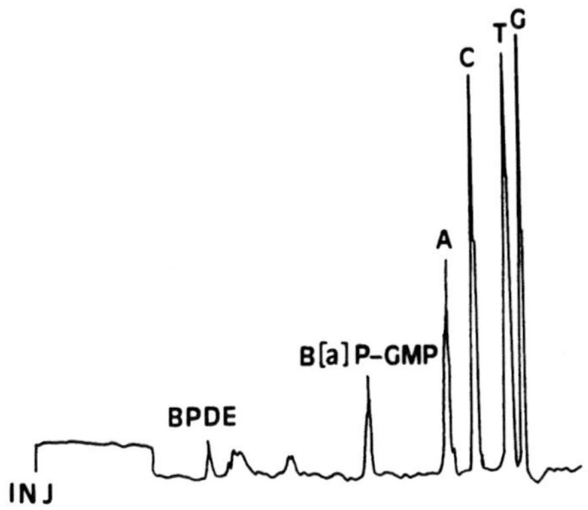

Fig. 7.48. Separation of 2-deoxy-5′-nucleotides. Electrokinetic injection was 5 s at 5 kV and monitored at 279 nm. Separation was at 15 kV in 0.04 M, pH 9.6 sodium carbonate buffer. A 75 μm I.D. silica column was used. *A*, *C*, *T*, *G* represent the adenosine, cytosine, thymidine, and quanosine 2-deoxy-5′-nucleotides respectively. B[a]P-GMP is the suspected benzo[a]pyrene guanosine mono phosphate peak and BPDE peak is either the benz[a]pyrene diol epoxide or the B[a]P tetrols. (Reproduced with permission from Ref. 211. Copyright Dr Alfred Huethig Publishers)

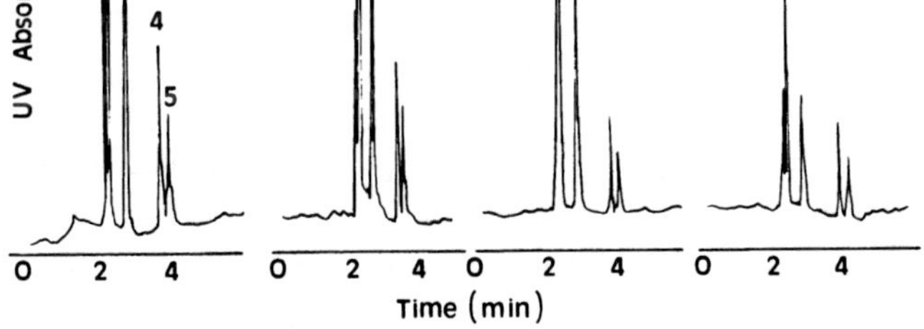

Fig. 7.49. Electropherograms of (a) equimolar mixture of 2′ and 3′ isomers of AMP, GMP, UMP, and CMP; (b) base-hydrolyzed rabbit intestinal mucosa RNA; (c) base-hydrolyzed calf liver RNA; (d) base-hydrolyzed yeast RNA. For each electropherogram, peaks are as follows: *1* = 2′- and 3′-UMP; *2* = 2′ and 3′-GMP; *3* = 2′ and 3″-AMP; *4* = 3′-CMP; *5* = 2′-CMP. Buffer was 12 mM sodium formate, 0.1 mM CTAB, pH 3.8. Sample was injected for 5 s at the cathode, which was at −26 kV with respect to the anode. The capillary had a 50 μm I.D. and was 20 cm from cathode to detector, 42 cm from cathode to anode. Concentration of each nucleoside monophosphate in part a was 0.5 mM. (Reproduced with permission from Ref. 213. Copyright American Chemical Society.)

residue was obtained from C_{56} to C_{157} and doublet peaks could be identified up to nearly 300 bases. The sensitivity of laser induced fluorescence detection was enhanced by using a sheath flow cuvette [215]. A mass detection limit of 10^{-20} mol of fluorescein-labeled DNA fragments was obtained.

Swerdlow *et al.* [216] investigated three methods for DNA sequencing based on CGE with laser-induced fluorescence. The methods included four-spectral-channel sequencing, two-spectral-channel sequencing, and one-spectral-channel sequencing, which employed four, two and a single fluorescent dye, respectively. In the four channel method, 6-carboxy-fluorescein (FAM), 2′,7′-dimethoxy-4′,5′-dichloro-6-carboxy-fluorescein (JOE), N,N,N′,N′-tetramethyl-6-carboxyl-rhodamine (TAMRA), and 6-carboxy-x-rhodamine (ROX) were used to label primers to be used with each dideoxynucleotide reaction. In the two channel method, succinnylfluorescein dyes were used to label the four dideoxynucleotides and the ratio of the fluorescence intensity in the two spectral channel was used to identify the terminating dideoxynucleotides. In the single-channel method, a TAMRA-labeled primer was used, and variation of the amount of dideoxynucleotide in the reaction mixture resulted in differences in peak height which could be used to identify the dideoxynucleotides. Figures 7.50 to 7.52 illustrate the results obtained for M13 mp 18. Sequencing rates up to 1000 bases/hour were obtained at electric field strength of 465 V/cm. Detection limits of 200, 20 and 2 zmol were obtained with the four-, two-and one-spectral-channel techniques, respectively.

The use of CGE with laser-induced fluorescence for DNA sequencing was investigated by Cohen *et al.* [217]. An argon ion laser (0.03–0.05 W, 488 nm) was used. Gel composition used was 3% T, 5% C. Synthetic oligonucleotides and single stranded DNA were used as templates. The fluorescent dye JOE, was used to label the 5′ end of primer molecules. Baseline resolution of a dA extended primer from 18 to 81 bases long was obtained. In addition, the sequence reaction products containing all four bases was analyzed in four separate runs. Baseline resolution was obtained for fragments different in length by a single nucleotide up to 330 base total length in the separation of sequence reaction products from single-stranded M13 mp 18 phase DNA template.

DNA sequencing by capillary gel electrophoresis with laser induced fluorescence was also investigated by Chen *et al.* [218]. Tetramethylrhodamine isothiocyanate-labeled DNA fragments were excited by a low cost helium-neon laser (534.5 nm, 0.75 mW). A detection limit of 300 analyte molecules was reported. By employing 6% T, 5% C acrylamide, 7 M urea gel columns, the sequencing rate achieved was 300 bases/hour at 200 V/cm, although compression of some peaks was observed. The addition of 20% formamide to the sequencing gel eliminated peak compression. However, a lower sequencing rate of 70 base/hour was obtained. Figure 7.53 shows a typical electropherogram for M13 mp 18 reaction fragments. The combination of CGE with laser-induced fluorescence was also investigated by Zheng *et al.* [219]. Considerations of the CGE/LIF system for DNA sequencing were discussed.

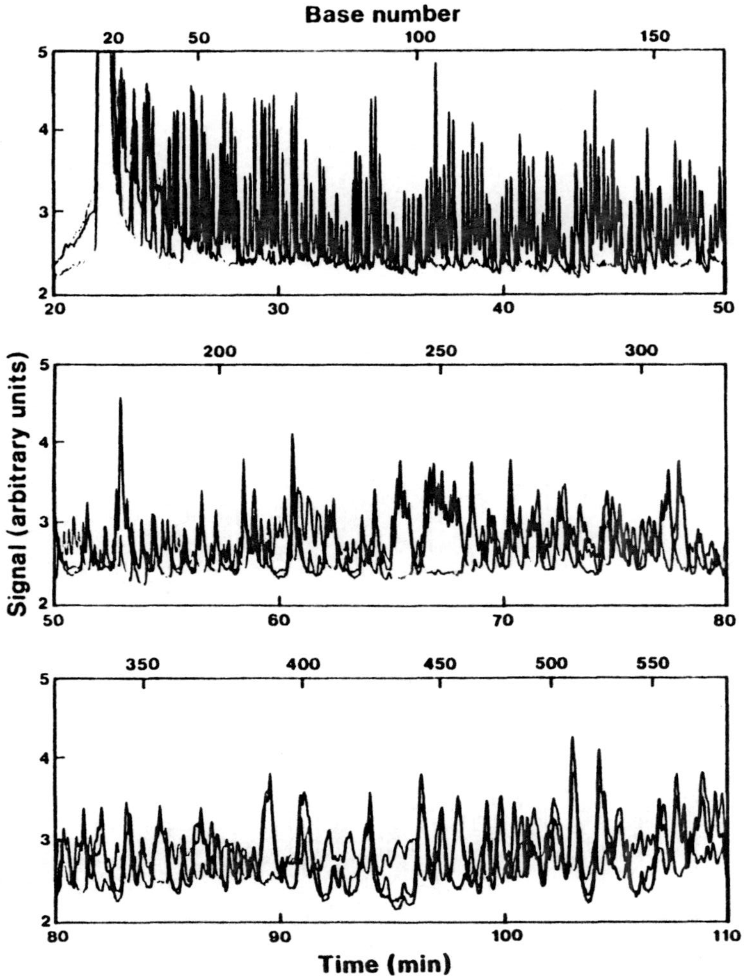

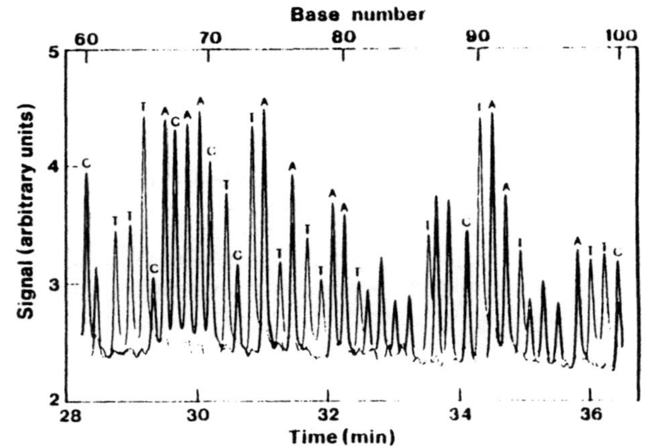

Fig. 7.50. Four-spectral channel sequencing of M13mp 18 with a histidine tRNA insert: (a, left) extended run; (b, right) expanded region corresponding to nucleotides 60–100; the known sequence is printed above. Sample preparation: four labelling reactions were performed separately and products were pooled in the ratio C:A:T:G 1:2:1:1. The 41 cm long (27 cm to the detector), 50-μm I.D., 375-μm O.D. capillary was filled with 6% T, 5% C gel and prerun for 60 min at 150 V/cm at room temperature. The sample was introduced electrophoretically for 30 s at 150 V/cm; after injection, the sample was replaced with a fresh vial of 1 × TBE. The gel was run for 1 min, trimmed at the injection end by 2 mm, and the run was continued at 150 V/cm. Time is measured from the injection; the numbers at the top of the figure correspond to the nucleotide length including the universal-21 18-mer primer. The four traces correspond to emission centered at 540, 560, 580, and 610 nm. (Reproduced with permission from Ref. 216. Copyright American Chemical Society.)

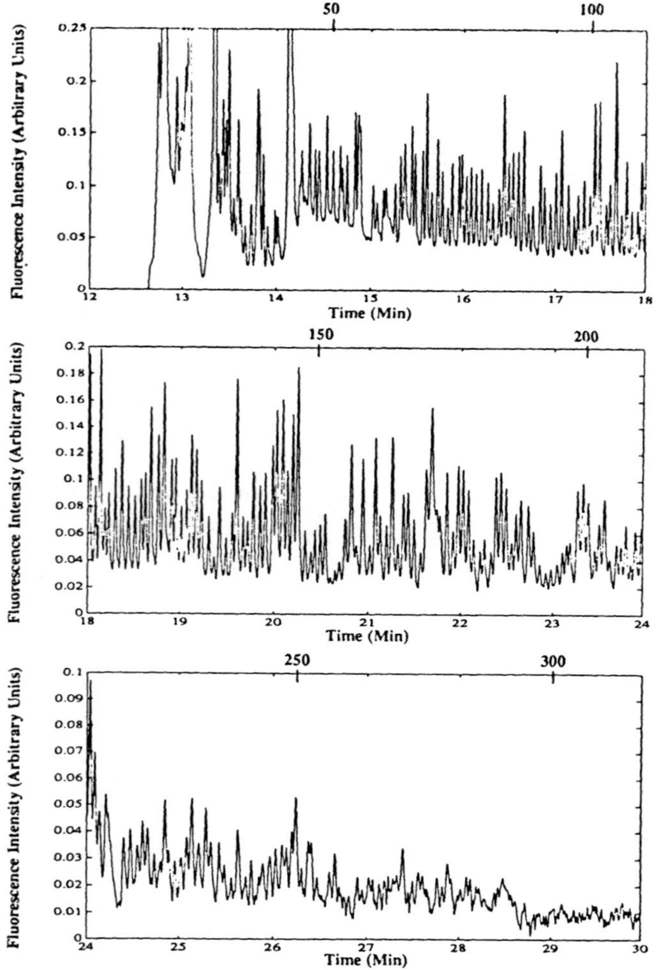

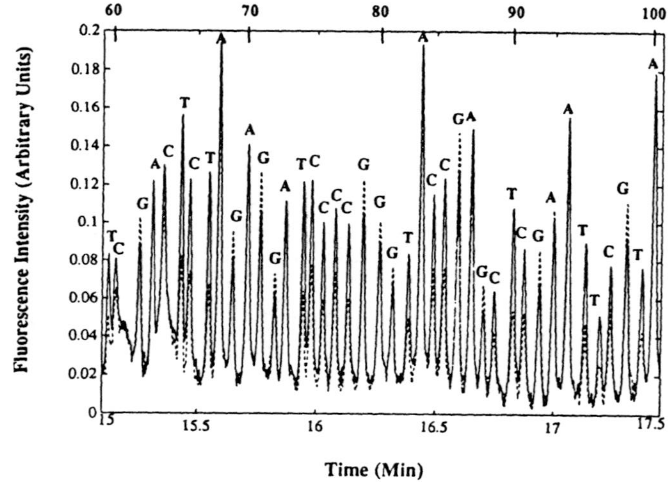

Fig. 7.51. Two-spectral-channel sequencingof M13mp 18: (a, left) extended run showing only the long-wavelength data. (b, right) expanded region corresponding to nucleotides 60–100 with the known sequence printed above. The sample was prepared from a Du Pont genesis 2000 protocol using 3 μg of M13mp 18 single stranded DNA, 15 ng of -40 17-mer M13 primer, and 1 μl of Sequenase (U.S. Biochemicals) in a standard reaction mix. They were ethanol precipitated and washed and resuspended in 3 μl of a 49:1 mixture of formamide: 0.5 M EDTA at pH 8.0. The 34 cm long, 50-μm-I.D., 190-μm-O.D. capillary was filled with 4% T, 5% C acrylamide gel that was covalently bound to the capillary wall through use of γ-[(methacryloxy)propyl] trimethoxysilane. The sample was injected at 100 V/cm for 30 s; after injection, the sample was replaced with a fresh vial of 1 × TBE. The electrophoresis continued for 1 min at 100 V/cm. The capillary was then trimmed at the injection end by 1 mm, and the voltage was increased in 1-kV increments over a 14-min period to a total electric field strength of 465 V/cm. The sheath stream was 1 × TBE at a flow rate of 0.16 ml/h. Time is measured from the injection; the numbers at the top of the figure correspond to the nucleotide length including the 17-mer primer. The solid trace corresponds to emission at wavelengths longer than 525 nm and the dashed trace coresponds to emission at wavelengths shorter than 525 nm. (Reproduced with permission from Ref. 216. Copyright American

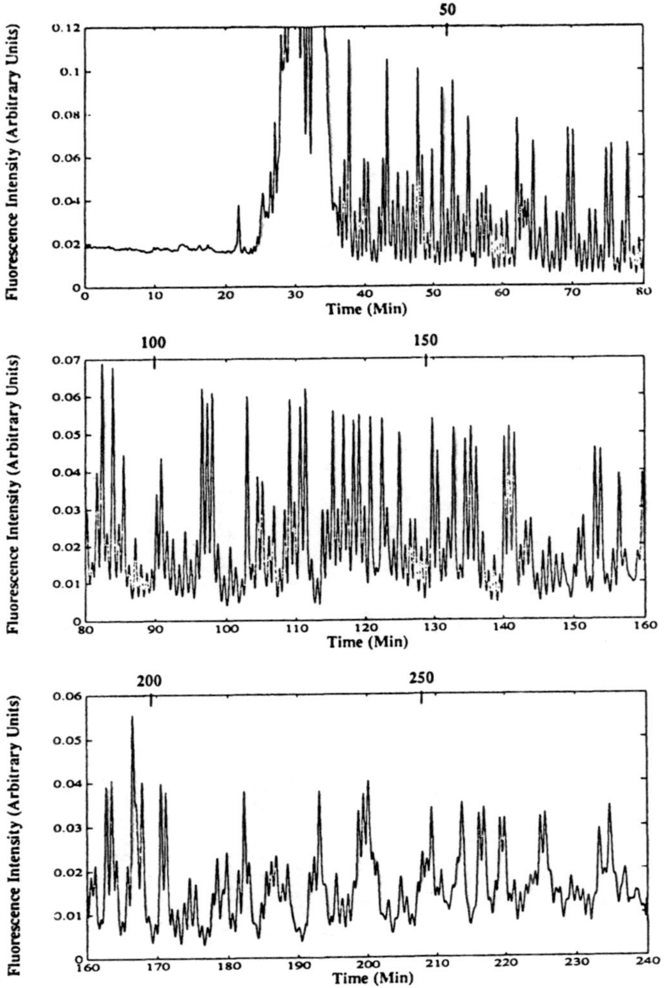

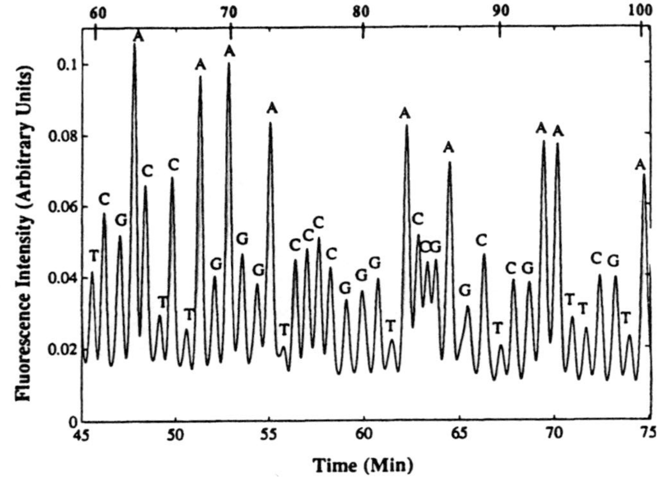

Fig. 7.52. One-spectral-channel sequencing of M13mp 18: (a, left) extended run; (b, right) expanded region corresponding to nucleotides 60–100. The sequencing reaction was carried out in 40 mM MOPS buffer, pH 7.5, 50 mM NaCl, 10 mM $MgCl_2$, and 15 mM sodium isocitrate. Dye-labeled primer (Applied Biosystems 21M13 TAMRA, 1.6 pM) was annealed to 1 μg of M13mp18 single-stranded DNA at 65°C for 2 min followed by slow cooling. A mixture of deoxy- and dideoxynucleoside triphosphates was added to give an average nucleotide ratio (dNTP/ddNTP) of 1200:1 with 7-deaza-2′-deoxyguanosine 5′-triphosphate used in place of dGTP. The ratios of nucleotides were adjusted to yield a normal peak height ratio of 8:4:2:1 for A:C:G:T. The mixture was warmed to 37°C for 30 min, after which the DNA was precipitated with ethanol. The 37 cm long, 50-μm-I.D., 190-μm-O.D. capillary was filled with 6% T, 5% C, 30% formamide, 7 M urea gel that was covalently bound to the capillary wall through use of γ-[(methacryloxy)propyl] trimethoxysilane. The sample was injected at 200 V/cm for 30 s; after injection, the sample was replaced with a fresh vial of 1 × TBE. The capillary tip was not trimmed after injection, and separation proceeded at 200 V/cm. The sheath stream was 1 × TBE at a flow rate of 0.16 ml/h. The numbers at the top of the figure corrspond to the nucleotide length including the −21 18-mer primer. (Reproduced with permission from Ref. 216. Copyright American Chemical Society.)

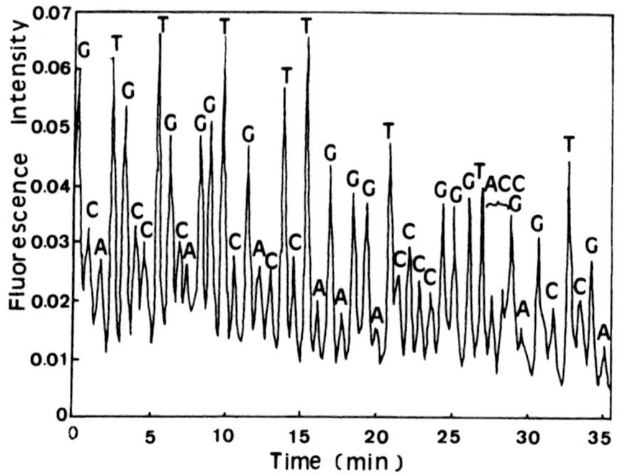

Fig. 7.53. Capillary gel electropherogram of M13mp18 reaction fragments: 6% T, 5% C-7M urea-20% formamide acrylamide gel; fragments 28–73 nucleotides in length are shown. Time is arbitrarily set to zero for the fragment 27 nucleotides in length. (Reproduced with permission from Ref. 218. Copyright Elsevier Science Publishers.)

7.5 PHARMACEUTICALS AND DRUGS

Although many techniques have been developed for the analysis of pharmaceuticals and drugs, such as radioimmunoassay (RIA), high-performance liquid chromatography (HPLC), polyacrylamide gel electrophoresis (PAGE), bioassay and enzyme-linked immunosorbent assay (ELISA), most of these techniques still suffer from certain limitations. Some of the limitations include the complexity and time-consuming nature, the high degree of variability, the low separation efficiency or the lack of sensitivity of some of the methods. Consequently, there has recently been a growing interest in the use of CE for the analysis of pharmaceuticals and drugs [220–262]. In Table 7.1, a list of selected types of pharmaceutical compounds which has been analyzed by CE is given.

Fujiwara and Honda separated cinnamic acid and its analogues [220,221] by capillary electrophoresis. Typically a 0.025 M phosphate buffer at pH 9.2 was used as the electrophoretic medium. It was found that higher pH resulted in longer migration times. Higher salt concentration of carrier resulted in longer retardation and higher resolution of peaks.

The principal ingredients of antipyretic analgesic preparation was determined by micellar electrokinetic chromatography [222]. Sodium dodecyl sulphate (0.05 M) was used as the electrophoretic medium. Application of the method to a commercial antipyretic analgesic tablet demonstrated the usefulness of the method.

The effect of buffer pH, ionic strength, applied voltage, and SDS concentration were systematically evaluated for their effects on migration and selectivity in the separation of some common analgesic standards [223]. The linearity, sensitivity and

TABLE 7.1

SELECTED TYPES OF PHARMACEUTICAL COMPOUNDS ANALYSED BY CE

Type of compounds	Reference
1. Cinnamic acid and analogues	220,221
2. Analgesic preparations	222,223
3. Crude drug	224
4. Antibiotics	225–233
5. Chiral drugs	234–237
6. Corticosteroids	238,239
7. Benzodiazepine and benzothiazepine	232,238
8. Barbiturates	227,240,241
9. Cefpiramide	242,243
10. Anticancer drug	8
11. Basic drugs	95
12. Antidepressant	244,245
13. Water-soluble vitamins	223,225,246,247
14. Echistain	248
15. Anti-inflammatory	227,249
16. Recombinant cytokine	250
17. Flavonoids	251,252
18. Vaccine production process monitoring	253
19. Selected drugs insoluble in water	254
20. Urinary porphyrins	255,256
21. Peptide drugs	227
22. Polyamines in urine	258
23. Cimetidine	259,260
24. Radiopharmaceuticals	261
25. Robitussin	223
26. Aminophenol and diaminophenol	257
27. Cough/cold formulation	223
28. Antihistamines	262

reproducibility for these types of compounds were also investigated. The optimum separation of five analgesics by CE is shown in Fig. 7.54.

Honda *et al.* [224] analyzed the components of *Paeonia radix*, which has been traditionally used as a sedative, lenitive or antispasmolytic agent. A 100 mM borate buffer (pH 10.5) was used. The applied voltage was 20 kV.

Nishi and co-workers have developed several different methods for the separation of drugs [225,226,228,234,235,238] using capillary electrophoretic techniques. It was found that the addition of tetraalkylammonium salts (TAA) to the SDS solution in micellar electrokinetic chromatography of ionic substances remarkably improved the resolution of ionic drugs, such as penicillin and cephalosporin antibiotics [225]. The separation of nine cephalosporin antibiotics was achieved with 0.04 M tetramethylammonium bromide (TMAB) added to an electrophoretic medium

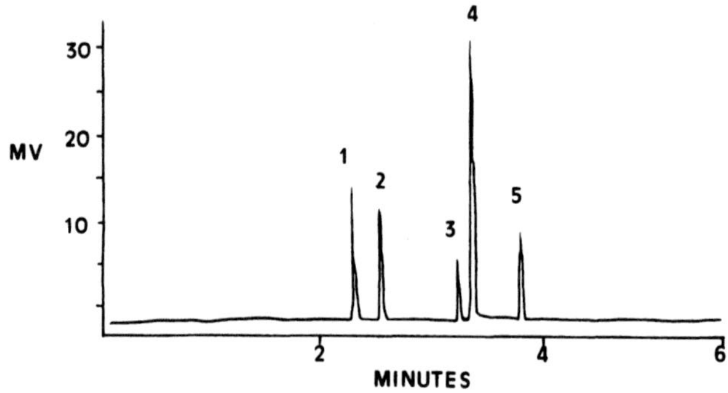

Fig. 7.54. Separation of analgesics by CE: optimized conditions. Conditions — mode: MEKC; buffer: 15mM $NaHPO_4$, pH 11.0; modifier: 25 mM SDS; capillary: 50 μm by 60 cm; voltage: 30 kV; detector: UV, 214 nm; injection: hydrostatic, 10 s. Sample: 0.1 mg/ml each in water. Peak identifications: *1* = caffeine; *2* = acetaminophen; *3* = acetylsalicylic acid; *4* = salicylamide; *5* = salicylic acid. (Reproduced with permission from Ref. 223. Copyright Millipore Waters Chromatography Division.)

containing 0.02 M phosphate–borate buffer (pH 9.0) and 0.05 M SDS. The addition of the TAA salts to the micellar system decreased the migration times of cationic solutes and, on the contrary, increased that of anionic solutes by an ion-pairing mechanism. Although TAA salt addition made the separation mechanism more complex, it provided a wider scope for the optimization of separation.

The migration behaviour of β-lactam antibiotics in micellar electrokinetic chromatography has been investigated [226]. Sodium dodecyl sulphate (SDS) and sodium N-lauroyl-N-methyltaurate (LMT) were used as anionic surfactants at concentrations of 0.05–0.3 M. A difference in the migration behaviour of cationic substances were observed between the two anionic surfactants, which have different groups neighbouring the charge-bearing groups. Seven penicillin antibiotics and nine cephalosporin antibiotics were successfully separated within 15 min by MEKC with SDS or LMT solutions at concentrations of 0.1–0.3 M.

Capillary isotachophoresis was used for the analysis of β-lactam antibiotics and their precursors, including penicillins and cephalosporins [229]. The leading electrolyte which provided satisfactory separation consisted of 5 mM HCl adjusted to pH 7 by the addition of Tris, and 0.25%. hydroxypropylmethylcellulose (HPMC). The terminating electrolyte was 10 mM phenol, adjusted to pH 10 by adding freshly prepared and filtered barium hydroxide.

The separation and determination of aspoxicillin in human plasma by MEKC have been investigated [228]. A direct sample injection method was employed. One analysis of a plasma sample required less than 20 min without pretreatment. Good linearity (r = 0.999) and recovery (94–100%) were obtained in the range of plasma level typical for clinical analysis. A limit of detection of 1.3 μg/ml (S/N = 3) was obtained for aspoxicillin. The separation of seven penicillins by CZE and MEKC

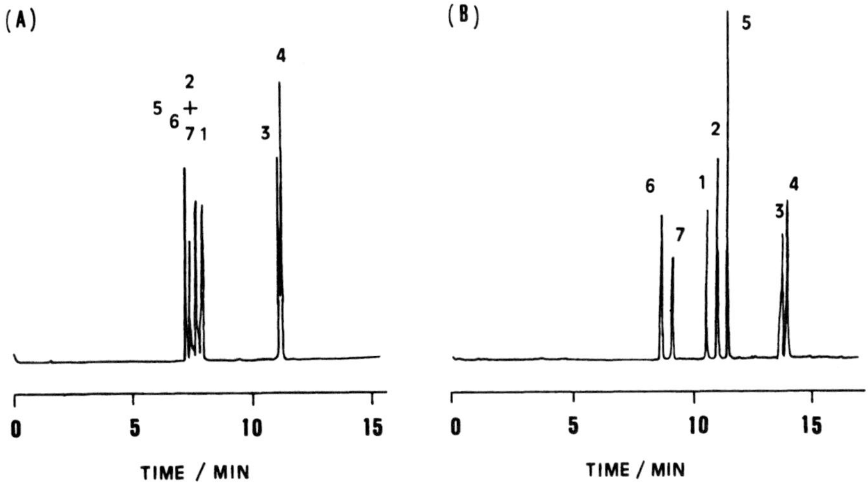

Fig. 7.55. Separation of seven penicillins by (A) CZE and (B) MEKC. (A) 0.02 M phosphate-borate buffer of pH 8.5; (B) 0.1 M SDS added to A. Applied voltage, 20 kV; temperature, ambient; detection, 210 nm; attenuation, 0.04 a.u.f.s. Peaks: *1* = benzylpenicillin; *2* = ampicillin; *3* = carbenicillin; *4* = sulbenicillin; *5* = piperacillin; *6* = aspoxicillin; *7* = amoxicillin. (Reproduced with permission from Ref. 228. Copyright Elsevier Science Publishers.)

is shown in Fig. 7.55, which shows that MEKC provides higher selectivity for these drugs. The MEKC analysis of aspoxicillin with acetaminophen as internal standard is shown in Fig. 7.56.

Separation conditions for some commonly used antibiotics, including sulfonamides, cephalosporins and penicillins were investigated by Wainwright [227]. For the separation of nine cephalosporins, a pH of 7.0 and ionic strength corresponding to 30 mM phosphate in 9 mM tetraborate were used. An electropherogram showing the separation of the cephalosporin by CZE is shown in Fig. 7.57.

Highly ionizable sulfonamides were separated by MEKC in a polymethylsiloxane coated column [231]. Phosphate and β-alanine buffers were used and both gave satisfactory separation of the sulfonamides. The use of photodiode-array detection in the CE analysis of antibiotics has also been investigated [233]. Simultaneous detection at different wavelengths and characterization of separated components by spectral analysis were performed.

Capillary electrophoresis–mass spectrometry provides a powerful tool for the analysis of drugs [232]. Determination of small drug molecules by CE–atmospheric pressure ionization mass spectrometry has been demonstrated. Figure 7.58 shows the electropherogram obtained with (A) UV and (B) mass spectrometric detection for the separation of mixture of sulfonamides using an electrophoretic buffer consisting of 20 mM ammonium acetate (pH 6.8) and 20% (v/v) of methanol. The relative response of the sulfonamides by ion-spray–MS operated under positive and negative ion modes of detection are shown. Sulfamethazine gave the highest response among

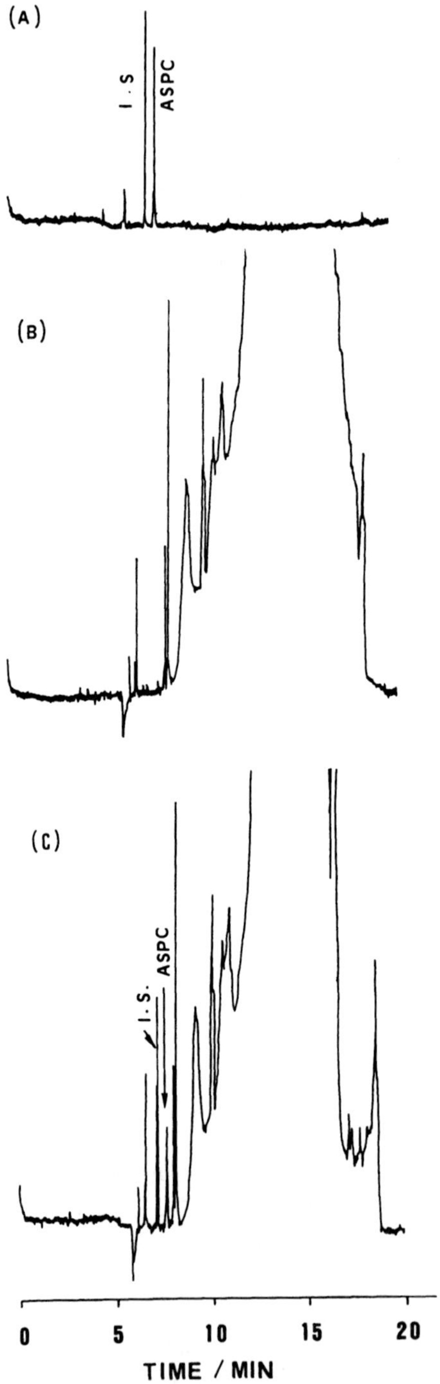

Fig. 7.56. MEKC saparation of aspoxicillin (ASPC) and acetaminophen (I.S.). (A) Standard solution; (B) blank plasma; (C) plasma spiked with ASPC and I.S. Buffer: 0.02 M phosphate-borate buffer (pH 8.5) containing 0.05 M SDS. Other conditions as in Fig 7.55. (Reproduced with permission from Ref. 228. Copyright Elsevier Science Publishers.)

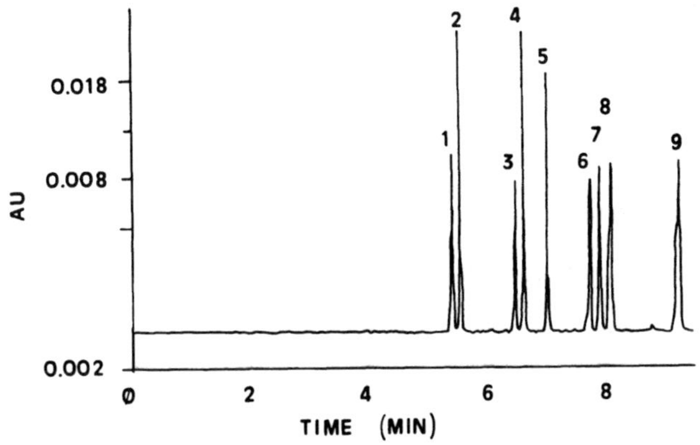

Fig. 7.57. Separation of cephalosporins by CZE at pH 7.0. Samples: *1* = cephaloridine; *2* = D-(−)-hydroxyphenylglycine; *3* = cephradine; *4* = cephadroxil; *5* = cephalexin; *6* = cephalosporin C; *7* = cefazolin; *8* = cephalothin, *9* = 7-aminocephalosporic acid. Conditions — effective capillary length: 60 cm; applied voltage: 275 V/cm (I = 70 μA); detection wavelength: 215 nm; injection: by gravity for 10 s at 50 mm; and buffer: 30 mM NaH_2PO_4-$Na_2B_4O_7$, pH 7.0. (Reproduced with permission from Ref. 227. Copyright Aster Publishing Corp.)

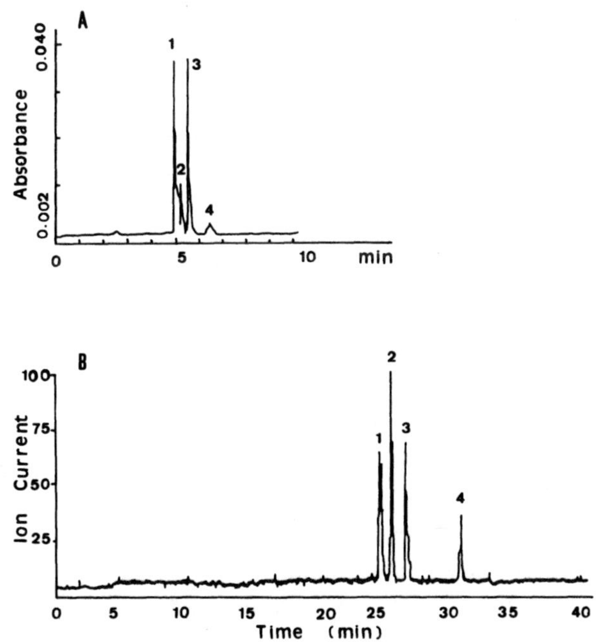

Fig. 7.58. (A) CE-UV and (B) CE-SIM-MS total selected ion current electropherograms from the separation of four benzodiazepines. Capillary: uncoated fused silica, 100 cm × 75 μm I.D.; voltage 26 kV; buffer, 15 mM ammonium acetate adjusted to pH 2.5 with TFA and containing 15% (v/v) of methanol; injection, 28 nl of a 20 μg/ml solution of each drug (1.4–2.0 pmol). Peaks: *1* = chlordiazepoxide, m/Z 300 $[M + H]^+$; *2* = flurazepam, m/Z 388 $[M + H]^+$; *3* = diazepam, m/Z 286 $[M + H]^+$; *4* = prazepam, m/Z 325 $[M + H]^+$. (A) CE-UV electropherogram at 254 nm, detection after 20-cm capillary; (B) CE-SIM-MS total selected ion electropherogram, detection after 100-cm capillary. (Reproduced with permission from Ref. 232. Copyright Elsevier Science Publishers.)

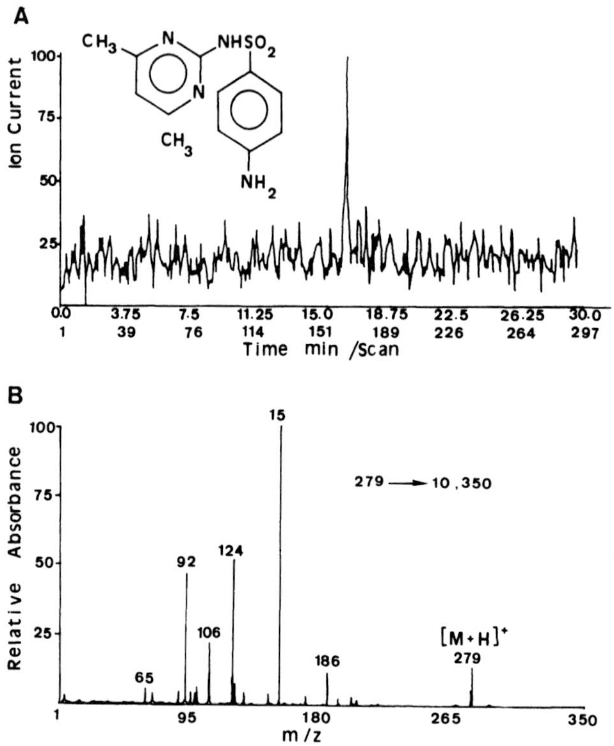

Fig. 7.59. (A) CE-MS-MS total ion electropherogram and (B) daughter-ion mass spectrum for sulfamethazine from a synthetic mixture of sulfonamides containing 600 pg each (2 pmol). Capillary: uncoated fused-silica, 100 cm × 75 μm I.D.; voltage: 26 kV; buffer: 20 mM ammonium acetate (pH 6.8) containing 20% (v/v) of methanol; injection: 5.5 nl of a 10^{-4} M solution (2 pmol). (Reproduced with permission from Ref. 232. Copyright Elsevier Science Publishers.)

the sulfonamides studied and was normalized to 100%. By using tandem mass spectrometry (MS-MS), additional structural information could be obtained for the analytes of interest. This is illustrated in Fig. 7.59 for sulfamethazine. The daughter-ion mass spectrum for 2 pmol (600 pg) of sulfamethazine was acquired to produce the data shown in Fig. 7.59. The CE-MS-MS trace shown in Fig. 7.59B displays structurally important fragment ions at m/Z 186, 156, 124, 108, 92 and 65, which are characteristics of many sulfa drugs. Simultaneous detection by UV and mass spectrometry was also demonstrated to be a versatile combination for CE analysis.

Optical isomer resolution is an important and attractive field of research, especially in the separation of chiral drugs. This is the case since the effectiveness and possible toxicity of a drug may be affected by the presence of the racemates. Several different approaches have been adopted for the chiral separation of optical isomer drugs [234,237]. One of the approaches involved the use of bile

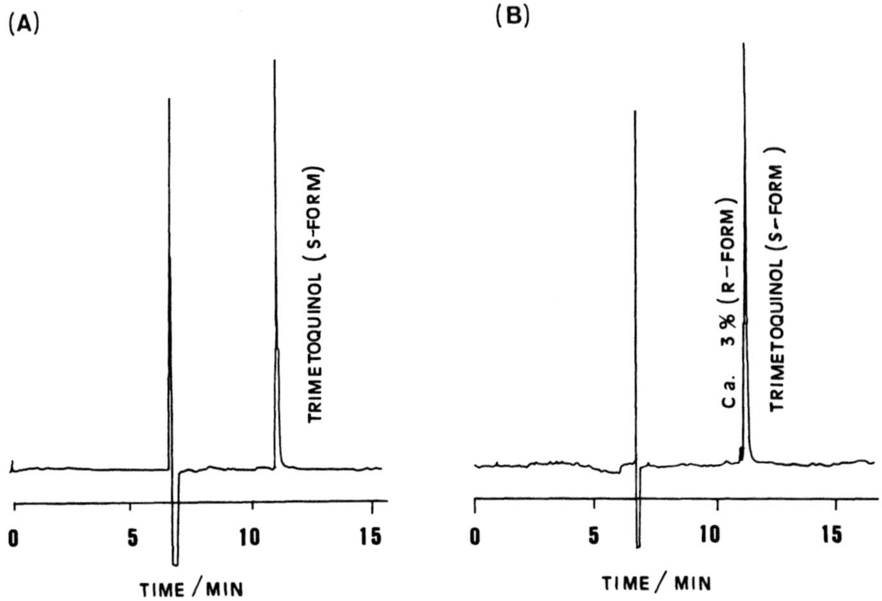

Fig. 7.60. Optical purity testing of trimetoquinol hydrochloride. (A) Authentic trimetoquinol hydrochloride [(S)-form] and (B) ca. 3% of (R)-form added to A. Conditions: buffer, 0.05 M sodium taurodeoxycholate (SDTC) in 0.02 M phosphate–borate buffer (pH 7.0). (Reproduced with permission from Ref. 235. Copyright Elsevier Science Publishers.)

salts as surfactants in MEKC separations [234,235]. Optical isomers of diltiazem hydrochloride, trimetoquinol hydrochloride, and other drugs were separated using chiral bile salts under neutral or alkaline conditions. A buffer containing 0.05 M taurodeoxycholate in a 0.02 M phosphate–borate buffer solution at pH 7 was used. The technique was applied to the purity testing of trimetoquionol hydrochloride. In Fig. 7.60, electropherogram for an authentic trimetoquionol hydrochloride sample ([s]-form) and a sample with 3% of [R]-form added are shown.

Host-guest complexation with cyclodextrins (CDs) provided another method for the separation of optical isomers [236,237]. The enantioselectivity of CDs arises from the chiral atoms present in the glucose units and depends on the stability of the complexes formed with the compounds studied. Complete separation of five racemic sympathomimetic amines could be obtained within 4 min of analysis time. Capillary isotachophoresis and CZE with cyclodextrin modifiers were used to separate chloramphenicol drug enantiomers, thioridazine enantiomers, ketotifein enantiomers, and ketotifein and its synthetic intermediate [237]. It was noted that complexation of different solutes occurred by slightly different solutes occurred by slightly different mechanisms. Figure 7.61 shows both the ITP and the CZE separation of thioridazine enantiomers.

Cyclodextrin modified MEKC has also been employed for the separation of

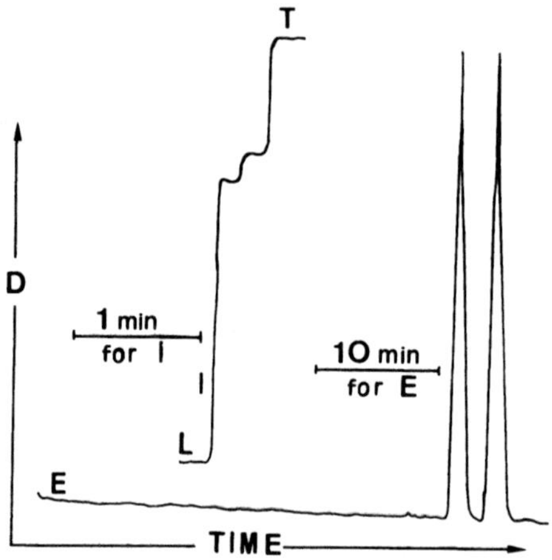

Fig. 7.61. ITP (*I*) and CZE (*E*) separations of thioridazine enantiomers. ITP conditions: capillary, 22 cm; LE, 10 mM NaAc + HAc to pH 5.47 + 0.08% HEC + 5 mM γ-CD; TE, 10 mM β-Ala; current, 100 μA (13 min), 50 μA for detection. L = leading zone; T = terminating zone. CZE conditions: capillary, 65 cm (45 cm to the detector); BE, 20 mM Tris + H_3PO_4 to pH 2.50 + 5 mM γ-CD; voltage, 18 kV; 14 μA.; D = response of the conductivity detector (for I) and UV detector (254 nm) (for E). (Reproduced with permission from Ref. 237. Copyright Elsevier Science Publishers.)

corticosteroids. Eight corticosteroids were successfully separated by the addition of γ-CD [239]. In certain applications, enantioselectivity and separation efficiency were found to be influenced by the addition of soluble alkylhydroxyalkylcellulose derivatives to the CD-modified electrolytes [237]. The effect of methylhydroxyethylcellulose (MHEC) on the CD-based chiral separation of the isomers of chloramphenical is shown in Fig. 7.62. Baseline separation was observed for the electrolyte containing MHEC.

Separation and determination of lipophilic corticosteroids and benzothiozepin analogues by micellar electrokinetic chromatography using bile salts was investigated by Nishi *et al.* [238]. Sodium cholate was found to be the most suitable for these applications. Satisfactory separations of eight corticosteroids and twelve benzothiozpin analogues were obtained using 0.1 M sodium cholate (SC) and 0.05 M sodium taurocholate (STC) in a 0.02 M phosphate–borate buffer (pH 9.0), respectively. The method was applied to the determination of the drug substances in tablets and cream using the internal standard method and to purity testing of drug substances and tablets.

The separation of benzothiazepin and its related compounds were found to be improved by the use of bile salts in MEKC [238]. The separation of twelve benzothiazepin analogues was achieved with 0.05 M sodium taurochloate in a 0.02 M phosphate–borate (pH 9.0) buffer. Capillary electrophoresis–atmospheric pressure ionization mass spectrometry has been used for the analysis of benzodiazepines

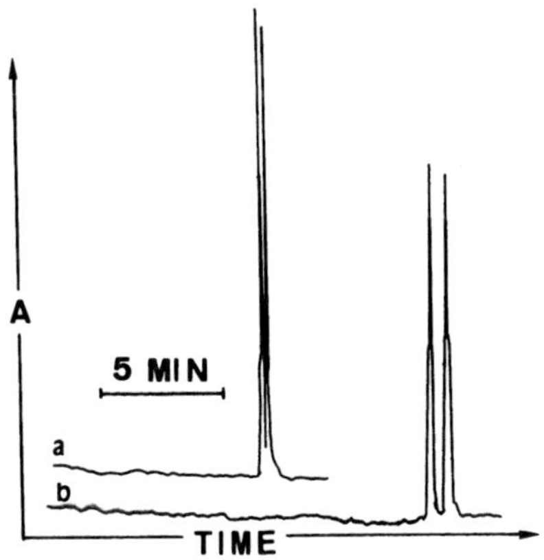

Fig. 7.62. Influence of MHEC on the chiral separation of chloramphenicol drug enantiomers with cyclodextrins. Record (a) without MHEC; (b) with 0.1% MHEC. BE: 20 mM Tris + CAc to pH 3.50 + 10 mM DM-β-CD; capillary, 65 cm (45 cm to the detector); voltage, 18 kV; curent, 6 μA. A = response of UV detector (254 nm). (Reproduced with permission from Ref. 237. Copyright Elsevier Science Publishers.)

[232]. Four benzodiazepines were separated using a 15 mM ammonium acetate buffer adjusted with TFA to pH 2.5 and containing 15% (v/v) of methanol. Simultaneous UV detection and CE-MS were performed. The mass spectrometer was operated under selected ion monitoring (SIM) conditions. In the case of CE-MS, the electropherogram shows slightly improved resolution than the UV case as a result of the longer effective separation length for the analytes. Figure 7.63 shows the CE-UV and CE-SIM-MS total selected ion current electropherograms from the separation of four benzodiazepines. The method was used to determine the metabolite of flurazepam, N-1-hydroxyethylflurazepam in human urine.

The use of MEKC for the separation of barbiturates has been investigated [227,240]. Separation could be achieved based on differences in hydrophobicity. On-column multi-wavelength detection was employed for the analysis of barbiturates in human and urine by MEKC [240]. Absorbance detection between 195 and 320 nm was performed. A phosphate–borate buffer of pH 7.8 containing 50 mM SDS was used.

MEKC with sodium dodecyl sulfate (SDS) was applied to the separation and determination of cefpiramide (CPM) in human plasma with the use of antipyrine (AP) as an internal standard [242,243]. A phosphate buffer solution (pH 8) containing 10 mM SDS was employed. The calibration plot for CPM in plasma sample showed good linearity in the range over 10 to 300 μg/ml. The limit of detection was 5 μg/ml at S/N = 3 with UV detection at 280 nm.

Separation and quantitation of the anti-cancer drug, methotrexate (MTX) and its major metabolite, 7-hydroxymethotrexate (7-OHMTX) by CE was performed with laser-induced fluorescence detection [8]. The buffer used contained 5 mM

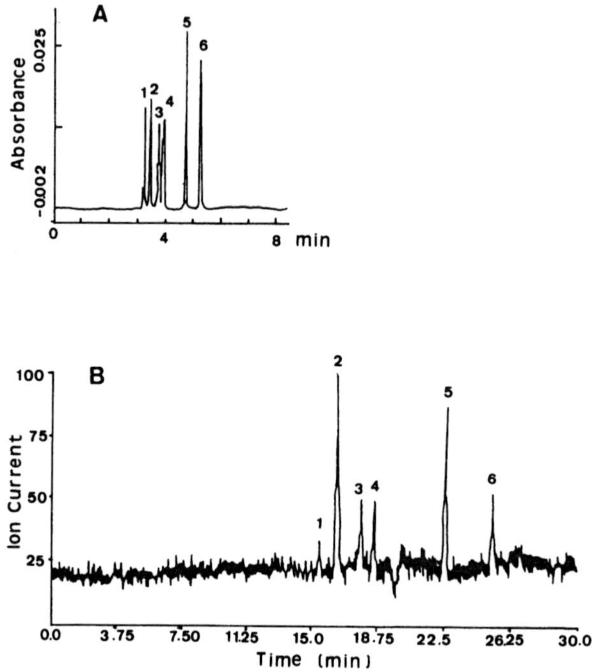

Fig. 7.63. (A) CE-UV and (B) CE-SIM-MS total selected ion current electropherograms for the separation of six sulfonamides. Capillary, uncoated fused silica, 100 cm × 75 μm I.D.; voltage, 26 kV; buffer, 20 mM ammonium acetate (pH 6.8) containing 20% (v/v) of methanol; injection, 5.5 nl of a 4×10^{-4} M solution (2 pmol) of each drug. Peaks: *1* = sulfanilamide, m/Z 190 $[M + NH_4]^+$ (pK_a = 10.4); *2* = sulfamethazine, m/Z 279 $[M + H]^+$ (pK_a = 7.4); *3* = sulfathiazole, m/Z 256 $[M + H]^+$ (pK_a = 7.27); *4* = sulfamerazine, m/Z 265 $[M + H]^+$ (pK_a = 6.90); *5* = sulfadimethoxine, m/Z 32 $[M + H]^+$ (pK_a = 5.9); *6* = sulfamethoxazole, m/Z 254 $[M + H]^+$ (pK_a = 5.6). (A) CE-UV electropherogram at 254 nm; detection after 20-cm capillary. (B) CE-SIM-MS total selcted ion current electropherogram; detection after 100-cm capillary. (Reproduced with permission from Ref. 232. Copyright Elsevier Science Publishers.)

2-(N-morpholino) ethanesulfonic acid (MES), 16 mM sodium sulfate, 30% (v/v) methanol, at pH 6.7. Detection limits for methotrexate of 5×10^{-10} M (S/N = 5) was obtained. The analysis of a clinical sample from a patient undergoing high-dose MTX therapy was performed.

Several basic drugs were separated by CE with 2% octyl glycoside (a non-ionic surfactant) added as a modifier [95]. The analysis of the tricyclic antidepressant, desipramine, was also performed with the addition of octyl glycoside into the electrophoretic buffer [244]. The electropherograms obtained for different concentrations of the surfactant are shown in Fig. 7.64.

Seven tricyclic antidepressants were separated using capillary electrophoresis [245]. Full resolution was achieved by the addition of methanol to the buffer which decreased both the electroosmotic flow and the electrophoretic mobilities

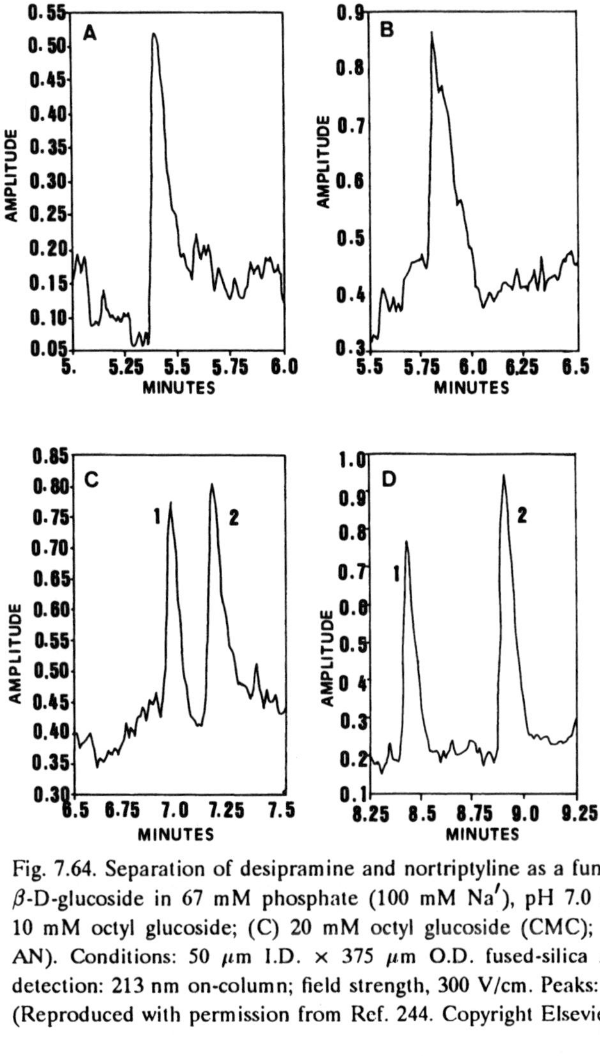

Fig. 7.64. Separation of desipramine and nortriptyline as a function of increasing amounts of octyl β-D-glucoside in 67 mM phosphate (100 mM Na′), pH 7.0 buffer. (A) No octyl glucoside; (B) 10 mM octyl glucoside; (C) 20 mM octyl glucoside (CMC); (D) 30 mM octyl glucoside (above AN). Conditions: 50 μm I.D. × 375 μm O.D. fused-silica APF capillary, 60 cm to detection; detection: 213 nm on-column; field strength, 300 V/cm. Peaks: *1* = desipramine; *2* = nortriptyline. (Reproduced with permission from Ref. 244. Copyright Elsevier Science Publishers.)

of the samples. The electropherograms obtained are shown in Fig. 7.65. The buffer used was 3-(cyclohexylamino)-2-hydroxy-1-propanesulfonic acid (CAPSO). A 50 mM CAPSO buffer with variable concentration of NaOH for each methanol concentration, i.e. 13.8, 16.2 and 18.0 mM of NaOH for 4%, 7.9% and 15.8% methanol, respectively, was used. The order of elution was: (1) protriptyline, (2) desipramine, (3) nortriptyline, (4) nordoxepin, (5) imipramine, (6) amitriptyline, and (7) doxepin.

The effect of tetraakylammonium salts (TAA) salts on the separation of cationic vitamin B1 and zwitterionic vitamin B12 has been investigated [225,247]. The addition of TAA salts caused a decrease in the migration times and this effect was more remarkable when TAA salts with longer alkyl chain were used.

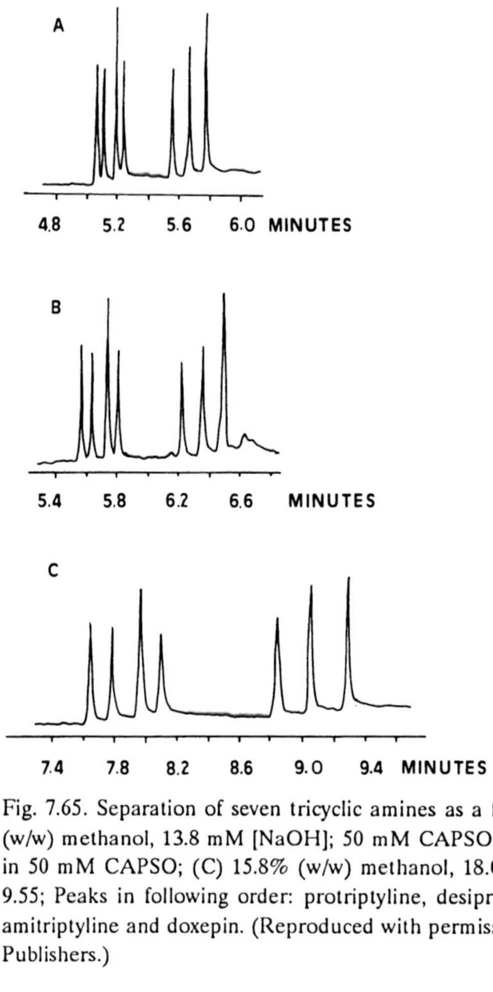

Fig. 7.65. Separation of seven tricyclic amines as a function of methanol concentration. (A) 4.0% (w/w) methanol, 13.8 mM [NaOH]; 50 mM CAPSO; (B) 7.9% (w/w) methanol, 16.2 mM [NaOH] in 50 mM CAPSO; (C) 15.8% (w/w) methanol, 18.0 mM [NaOH] in 50 mM CAPSO. Buffer pH 9.55; Peaks in following order: protriptyline, desipramine, nortriptyline, nordoxepin, imipramine, amitriptyline and doxepin. (Reproduced with permission from Ref. 245. Copyright Elsevier Science Publishers.)

A mixture of seven water and two fat-soluble vitamins could be separated simultaneously by MEKC with SDS solution containing γ-cyclodextrin as modifier [246]. The electrokinetic chromatogram obtained is shown in Fig. 7.66. This investigation demonstrated that both fat- and water-soluble compounds could be separated in a single analysis by CE.

The separation of anti-inflammatory drugs by CZE at several pH was investigated by Wainwright [227]. An uncoated capillary and a capillary coated with linear polyacrylamide were employed. The advantage of the coated capillary is that regardless of electrolyte pH, all negatively or positively charged analytes can be eluted. On the other hand, the uncoated capillary could be used to separate both charged and uncharged compounds, provided that the electroosmotic flow is high enough to elute the negative species. Micellar electrokinetic capillary has been used to separate a group of non-steroidal anti-inflammatory drugs [249] with a buffer

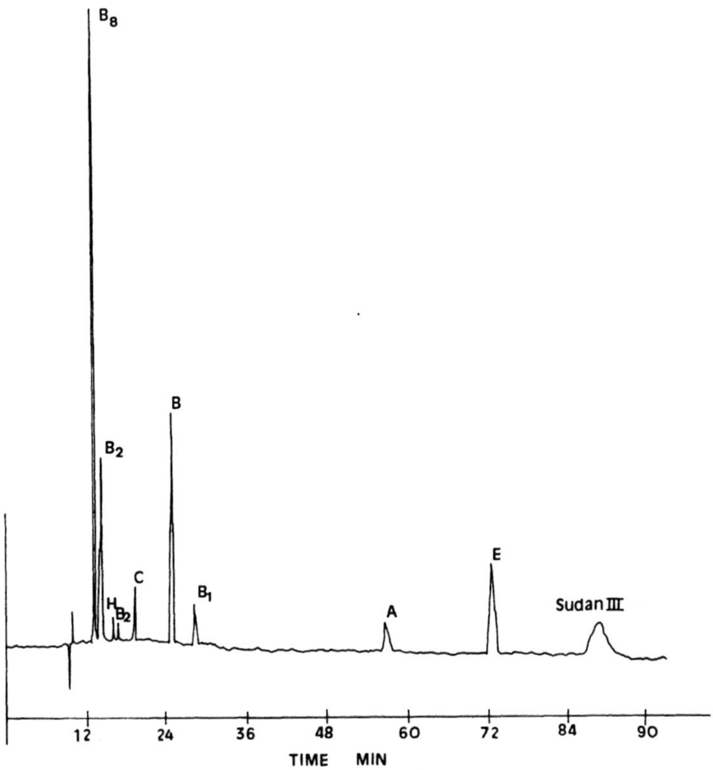

Fig. 7.66. Electrokinetic chromatogram of the vitamins with γ-cyclodextrin. Electrophoretic solution: 3 mM γ-cyclodextrin and 30 mM SDS in 0.1 M borate–0.05 M phosphate; pH 7.6; separation tube: 50 cm × 50 μm I.D. fused silica capillary; voltage: 20 kV; amount injected: 0.75 nl. (Reproduced with permission from Ref. 246. Copyright Elsevier Science Publishers.)

containing 20 mM phosphate and 25 mM SDS.

The use of CE for the analysis and quantification of a recombinant cytokine in an injectable dosage form was demonstrated [250]. Separations were performed using a 0.05 M sodium tetraborate buffer (pH 8.3) containing 0.025 M lithium chloride. The applied voltage was 10 kV. Detection was by UV absorbance at 200 nm. Figure 7.67 shows the electropherogram of a pharmaceutical dosage form containing recombinant interleukin-1α.

The separation of flavonoids by MEKC has been investigated [251,252]. For the analysis of flavonol-3-O-glycosides, sodium dodecyl sulfate–20 mM sodium borate buffer (pH 8.3) were used for the electrophoretic separation. Baseline separation was obtained for a mixture consisting of quercetin-3-O-glycosides (rutin, isoquercitrin, hyperosid, quercitrin and auicularin), kaempferol-3-O-glycosides (kaempferol-3-O-rutinoside and astragalin) and isohamnetin-3-O-glycosides (isohamnetin-3-O-rutinoside and isohamnetin-3-O-glucoside).

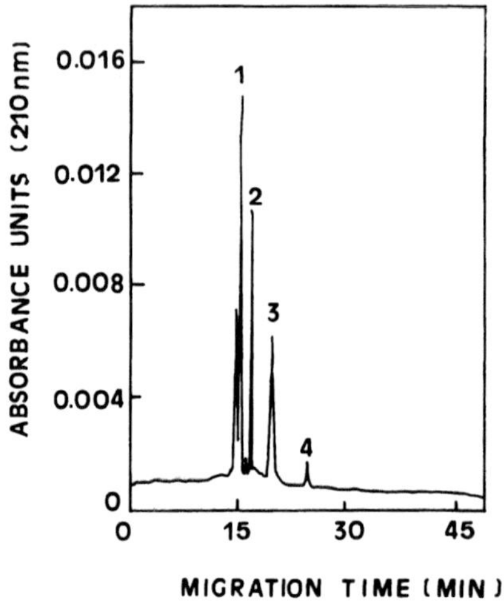

Fig. 7.67. Capillary electrophoresis profile of an actual pharmaceutical dosage form containing recombinant interleukin-1α (peak *2*), human serum albumin (peak *3*), N-acetyltryptophan (peak *4*) and spiked with recombinant leukocyte A interferon (peak *1*). Applied voltage: 10 kV; detection wavelength: 200 nm; capillary: 57 cm (50 cm to detector) $\times$ 75 μm I.D.; buffer: 0.05 M sodium tetraborate buffer (pH 8.3) containing 0.025 M lithium chloride. (Reproduced with permission from Ref. 250. Copyright Elsevier Science Publishers.)

The use of capillary zone electrophoresis for the analysis of samples from the individual steps in a vaccine purification process for the production of recombivax-1 (B) hepatitus B vaccine has been investigated [253]. The samples were analyzed at 30°C using 25 mM sodium phosphate running buffer at pH 7.25. An uncoated open capillary of 50 μm I.D. $\times$ 72 cm length (50 cm effective length) was used. Sample detection was by UV absorption at 200 nm. By using this technique, the process parameters could be monitored in real-time to ensure that they were within process-control limits.

MEKC was used for the analysis of drugs which were uncharged and almost insoluble in water [254]. Instead of using the capacity factor as a screening parameter, the pseudo-effective mobility ($\mu^{ps}_{eff,s}$) was used, which could be calculated even when the migration time of the micelle was not known:

$$\mu^{ps}_{eff,s} = \mu^{ps}_{app,s} - \mu_{eo} = \frac{L L_{cap}}{V t_s} - \frac{L L_{cap}}{V t_o} \tag{7.2}$$

where L_{cap} and L are the total length of the capillary and the length of the capillary from the injection point to the detection cell window, respectively, and V is the applied voltage, $\mu^{ps}_{eff,s}$ and $\mu^{ps}_{app,s}$ are the pseudo-effective mobilities and the pseudoapparent mobilities respectively, μ_{eo} is the mobility of the electroosmotic flow, t_s and t_o are the migration times of the solute and the marker (insolubilized compound) for the determination of the electroosmotic flow. In these investigations,

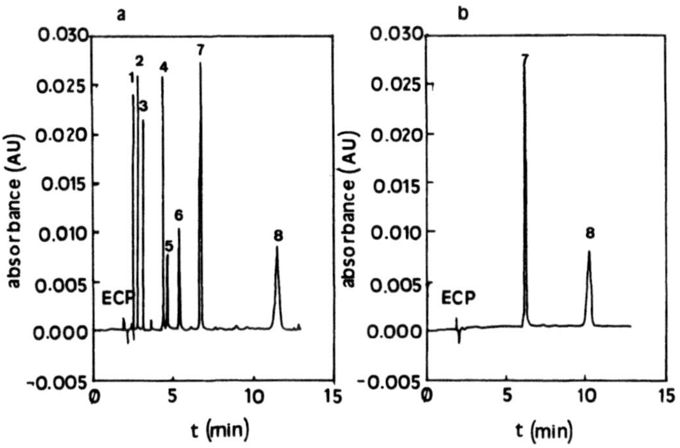

Fig. 7.68. Electropherogram of (a) a standard sample of (*1*) nicarbazin, (*2*) dimetridazole, (*3*) sulphadimidine, (*4*) sulphadiazine, (*5*) carbadox, (*6*) furaltadone, (*7*) dapsone and (*8*) fenbendazole, and (b) a sample of a dapsone with fenbendazole added as t_{MC} marker. The injection volume was about 11 nl. Capillary: 50 μm I.D. 27.65 cm (20.85 cm to detector), capillary treated with 10 M hydrochloric acid for 5 h at 160°C to produce high μ_{eo}. Detection: 214 nm; buffer: 0.02 M tris (hydroxymethyl) aminomethane (tris) with 100 mM SDS (pH 8.5) adjusted by adding boric acid. (Reproduced with permission from Ref. 254. Copyright Elsevier Science Publishers.)

the capillary used (50 μm I.D. $\times$ 27.65 cm total length, 20.85 cm effective length) was treated with 10 M hydrochloric acid for 5 h at 160°C to produce a high μ_{eo}. UV absorbance at 214 nm was used for detection. The experiments were carried out using a constant voltage of 10 kV. 0.02 M tris (hydroxymethyl) aminomethane (Tris) with 100 mM sodium dodecyl sulfate at pH 5.5 adjusted by adding boric acid was used. Figure 7.68 shows the electropherogram of a standard sample and a dapsone tablet obtained using these conditions.

Analysis of urine provides information for the diagnosis of diseases, especially renal or urinary tract diseases [255,256]. Human urine samples were analyzed using CE [255] with an untreated fused silica capillary column (75 μm $\times$ 150 cm). Separation was performed at 15 kV with 0.05 M sodium tetraborate buffer, pH 8.3. The temperature was maintained at 25°C and separated components were monitored at 210 nm. Typical electropherograms obtained for a freshly collected urine sample and the same urine sample after decantation and thawing are shown in Fig. 7.69. One of the peaks (peak 8) shows a decrease in peak height. This peak was thought to be urinary orthophosphate. The separation of groups of free and acetylpolyamines in human urine has been investigated [258]. The amount of polyamines in human body fluids is known to be related to cancer.

CE has been employed for the analysis of a wide variety of commercial pharmaceutical products. Robitussin preparations were separated by MEKC [223]. Figure 7.70 shows a typical electropherogram obtained. The components of a

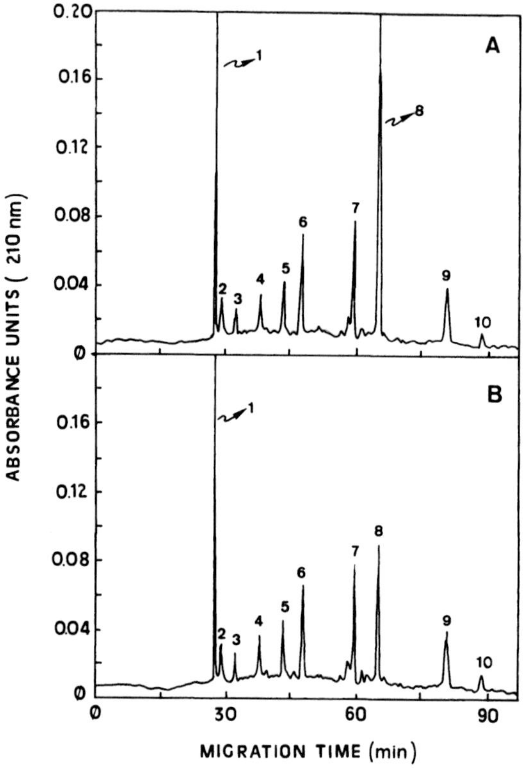

Fig. 7.69. Typical electropherograms of human urine using ultraviolet detection methods. Approximately 10 major components were observed in the capillary electrophoresis profiles of normal urine specimens. Electropherogram of a freshly-collected clean-catch urine specimen obtained from a normal person (A). Electropherogram of the supernatant of the same urine specimen after decantation and following sample thawing from a −70°C storage (B). Capillary: 150 cm × 75 μm I.D., voltage: 15 kV; buffer: 0.05 M sodium tetraborate, pH 8.3; temperature: 25°C. (Reproduced with permission from Ref. 255. Copyright Marcel Dekker Inc.)

cough/cold formulation have been separated by MEKC. Capillary isotachophoresis has been used for the determination of aminophenol and diaminobenzene isomers in permanent colour [257]. Antihistamines in pharmaceuticals were determined by CE [262]. A mixed carrier system containing sodium dodecyl sulfate, β-cyclodextrin and tetrabutylammonium hydrosulfate (TBA) as modifier was used. Figure 7.71 shows the electropherogram obtained for the antihistamines. The analysis of dinenhydrinate and promethazine drug samples were performed. Cimetidine in pharmaceutical preparations was determined by CZE [260]. Separation was performed in 20 mM phosphate buffer at pH 7. More than sixty commercially available formulation were analyzed. Relative standard deviation ranging between 1.9 and 6.4% were obtained for different types of samples, such as injectable, liquid and tablet preparations.

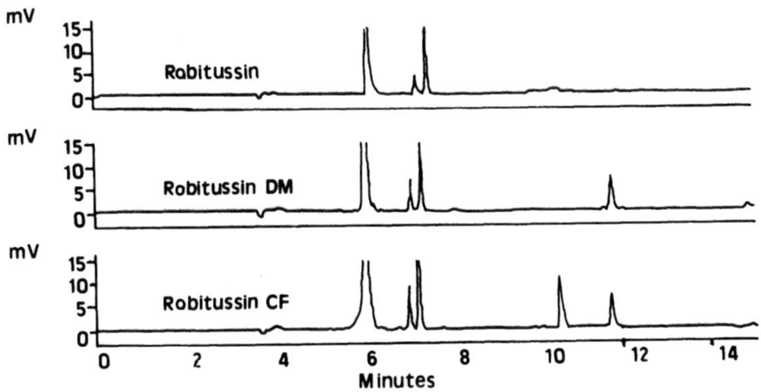

Fig. 7.70. Comparison of CE separation of three different Robitussin preparations. Conditions — mode: MEKC; buffer: 20 mM phosphate–borate 9.0; modifier: 50 mM SDS; capillary: 75 μm by 60 cm; voltage: 20 kV; detector: UV, 214 nm; injection: hydrostatic, 10 s. Sample: 1/100 dilution each. (Reproduced with permission from Ref. 223. Copyright Millipore Waters Chromatography Division.)

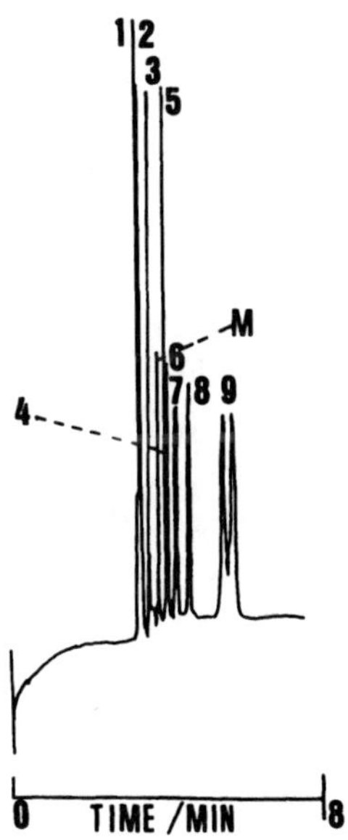

Fig. 7.71. Electropherogram of antihistamines. Electrophoretic solution, 10 mM SDS in 0.05 M borate–0.05 M phosphate buffer with 10 mM TBA and 10 mM β-cyclodextrin (pH 7.5); separation tube, 47 cm × 50 μm I.D. capillary tube kept at 25°C; voltage, 21.5 kV; detection wavelength, 214 nm; Peak numbers: *1* = pheniramine; *2* = doxylamine; *3* = methapyrilene, *4* = thonylamine; *5* = triprolidine HCl; *6* = dimenhydrirate; *7* = cyclizine; *8* = promethazine; *9* = (±)-chloropheniramine. *M* = methanol. (Reproduced with permission from Ref. 262. Copyright Elsevier Science Publishers.)

An electrochromatographic solid-phase extraction method was developed for the preconcentration of cimetidine in serum [259]. MEKC was used to analyze the preconcentrated samples. The preconcentration was performed with a C_{18} cartridge, followed by elution at 150 V in the eluting solvent (THF-buffer (50:50)). 10–15 fold increase in sample concentration for cimetidine was obtained. For MEKC separation, negative voltage, −18 to −23 kV were used. An uncoated capillary was employed (60 cm × 50 μm I.D.). Detection was by UV absorbance at 228 nm. The buffer system contained a cationic detergent, hexadecyl trimethylammonium bromide (HTAB), aminomethane (TRIZMA base, 3.3 mM) and sodium hydrogenphosphate (9.4 mM) at pH 6.4.

A gamma-ray detector for CZE was employed for the detection of radiopharmaceuticals [261]. The system was demonstrated on the analysis of ${}_{43}T_c^{99m}$ ion in the form of sodium pertechnetate and the chelates derived from it.

7.6 CELLS, VIRUS AND BACTERIA

The electrophoresis of Tobacco mosaic virus (TMV) was performed by Hjerten *et al.* [121] in fused silica tubes (115 mm × 0.1 mm I.D.) coated with methylcellulose or linear polyacrylamide. Buffer used was 0.02 M Tris HCl (pH 7.5). The applied voltage was 800 V and detection was by UV absorbance at 260 nm. The analysis of a bacteria, Lactobacillus casei NCTC 10302, has also been studied. The buffer used was 0.1 M Tris HOAc (pH 8.6). The applied voitage was 4000 V and detection was performed at 220 nm.

Human, chicken, porcine and rabbit red blood cell (RBC) were analyzed by CE [263]. Fluorinated ethylene-propylene copolymer (FEP) tubings of 0.45 mm I.D. were used. The capillary were positioned vertically, with semi-permeable membrane sealing the outlet to prevent the solution from flowing by gravity. Both cell suspensions and fixed cells were analyzed. Addition of hydroxypropylmethylcellulose (HPMC) helped to prevent linear agglomeration of RBC. Inner surface of the FEP capillary was found to degrade during electrophoresis, resulting in irreproducibility in migration times. Human RBC lysate was analyzed using 0.1 M phosphate buffer (pH 9.0), which contained a zwitterionic additive [263].

Direct sampling of plasma from a single nerve cell of a giant land snail using microinjectors for CE has been demonstrated [32,264,265]. Electrochemical detection was employed, which provided extremely high sensitivity.

7.7 ANALYSIS OF BODY FLUIDS

Body fluids contain many different groups of chemical substances, including amino acids, peptides, proteins, nucleic acids, drugs, virus, bacteria, carboxylic acids, organic ions, and etc. The analysis of many of these groups of compounds are

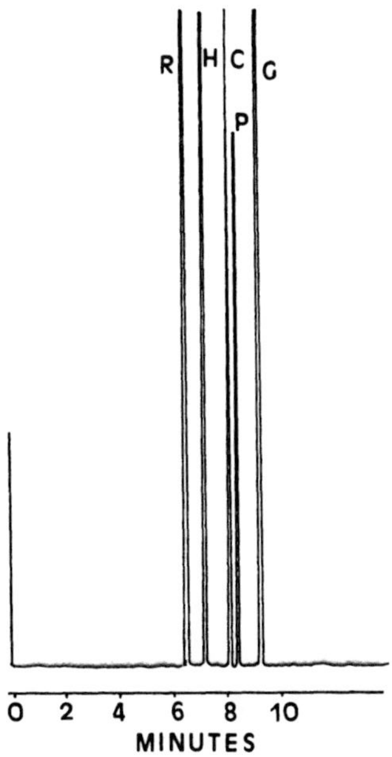

Fig. 7.72. Separation of thiol-containing compounds by CE. the standard compounds were derivatized with monobromobimane (MB). Fluorescence detection: 375 nm (excitation), 480 nm (emission) and the capillary was 40 cm × 0.075 mm I.D. A voltage of 25 kV gave a current of about 150 μA (0.05 mol/l phosphate buffer, pH 7.5). Peaks: R = reagent peak; H = homocysteine; C = cysteine; P = penicillamine; G = glutathione. (Reproduced with permission from Ref. 266. Copyright Elsevier Science Publishers.)

discussed in other sections of this chapter. In this section, only those compounds which are not covered in any of the other sections are considered. Furthermore, attention is mostly concentrated on the analysis of blood serum (or plasma) and urine. The utilization of CE for the analysis of other types of body fluids, such as tissue homogenates, cerebrospinal fluid, saliva, semen, bile, aqueous humor, and urinary calculi, have rarely been studied to date.

Analytical methods based on CE for diagnosis and studies of several types of human diseases have been developed [266]. Thiol-containing amino compounds was analyzed by CE for the diagnosis of diseases such as homocystinuriaand cystinura. The separation of thiol-containing compounds derivatized with monobromobimane for fluorescence detection is shown in Fig. 7.72. Glutathione synthetase deficiency was diagnosed by the absence of the tripeptide GSH (τ-Glu-Cys-Gly) in cells. Figure 7.73 shows the electropherogram for the analysis of GSH in red blood cells. Adenylosuccinase deficiency was determined by the analysis of succinylaminoimidazoyl carboxyamid ribotide and adenylsuccinate, as shown in Fig. 7.74. Determination of taurine in biopsies of heart muscle was also performed. 3-amino-1-propanesulfonic acid was used as internal standard. Derivatization with

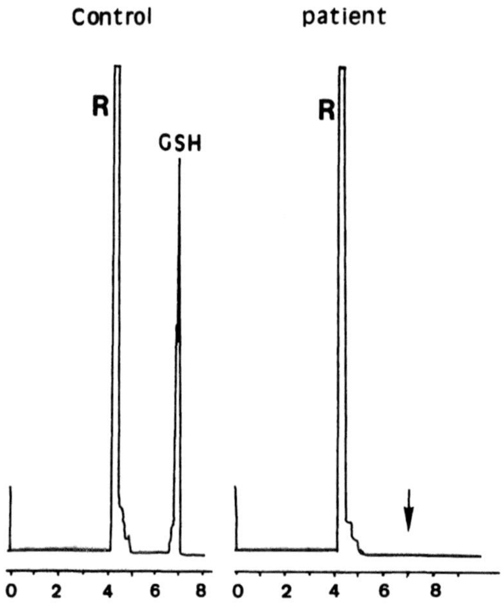

Fig. 7.73. CE analysis of GSH in red blood cells from a control (left) and a patient with GSH synthetase deficiency (right). The capillary was 40 cm × 0.10 mm I.D.. Conditions — buffer: 0.05 mol/l sodium phosphate buffer, pH 7.5; voltage: 25 kV; detection: fluorescence excitation 375 nm and detection setting emission 480 nm. (Reproduced with permission from Ref. 266. Copyright Elsevier Science Publishers.)

9-fluorenylmethylchloroformate (FMOC) permitted detection by fluorescence at high sensitivity. Figure 7.75 shows the separation of taurine and the internal standard.

By using a polysiloxane-coated capillary, and polymeric buffer additives (e.g. hydroxypropylmethylcellose), the possibility of applying CE to the analysis of specific co-amplified DNA sequences (HIV-1 and HLA-DQ-alpha) was explored [204]. The buffer used was 89 mM tris borate, containing 2 mM EDTA (pH 8.5) and 10 μM ethidium bromide.

MEKC has been employed for the analysis of thiopental in human serum and plasma [240]. Separation was performed using a running buffer consisting of 50 mM SDS, 9 mM sodium borate and 15 mM sodium dihydrogenophosphate at 40°C. Detection was by UV absorbance at 290 nm. Figure 7.76 shows the eletropherogram of two calibration samples and two patient samples.

Separation of urinary porphyrins by MEKC has been performed [256]. An uncoated fused silica capillary was used. The electrophoretic medium consisted of 100 mM sodium dodecyl sulfate and 20 mM 3-(cyclohexylamino)-1-propanesulfonic acid at pH 11. Absorbance detection at 400 nm, or fluorescence detection with excitation at 400 nm and emission at 550 nm and above were employed. Injection of samples prepared in buffer without surfactant resulted in trace enrichment.

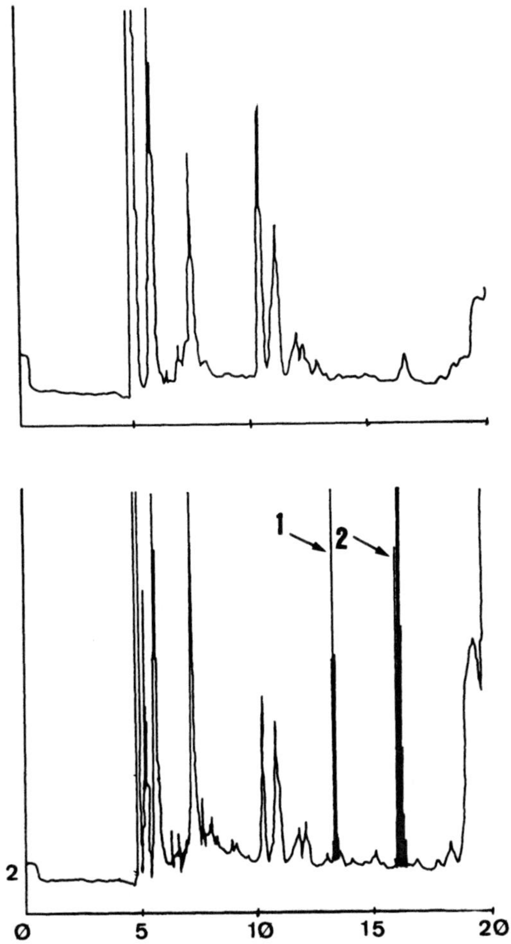

Fig. 7.74. CE profiles of urine from a control (upper panel) and from a patient with adenylosuccinase deficiency (lower panel). Absorbance detector set at 214 nm. The sucinyl purines are marked with arrows. Peaks: *1* = succinylaminoimidazole carboxyamid ribotide, *2* = adenylsuccinate. Voltage: 20 kV; buffer: 0.05 M sodium phosphate buffer; pH = 2.5. (Reproduced with permission from Ref. 266. Copyright Elsevier Science Publishers.)

Limits of detection of 100 pmol/ml were obtained. The method was applied to the determination of porphyrins in clinical urine samples. Figure 7.77 shows the electropherogram of a urine sample from a patient with porphyria cutanea tarda.

Amino acids in human urine derivatized with 4-(di-methylamino) azobenzene-4′-sulfonyl (DABSL) were separated by CE and detected by thermooptical absorbance detection [25]. CZE of urine samples was performed in a coated capillary with 0.1 M phosphate buffer (pH 2.5) by Dulffer *et al.* [64]. The separation of theophylline and uric acid was also demonstrated with SDS surfactants in MEKC.

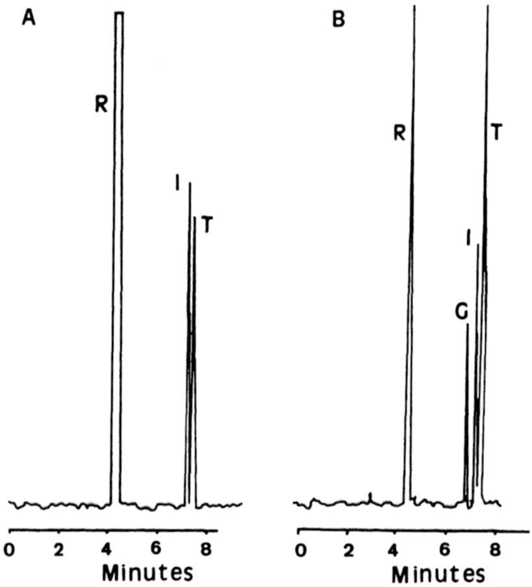

Fig. 7.75. (A) CE separation of taurine and the internal standard. (B) CE analysis of a sub-milligram biopsy of human heart muscle, to which the internal standard was added before derivatization with FMOC and electrophoresis. Peaks: R = reagent peak; I = internal standard; T = taurine; G = tentatively identified as GSH. The taurine concentration was calculated to be about 5 mmol/kg wet weight of tissue. Capillary: 40 cm × 0.075 mM I.D.; buffer: 0.05 mol/l sodium phosphate buffer at pH 7.5; detection: fluorescence, excitation at 265 nm, emission at 305 nm. (Reproduced with permission from Ref. 266. Copyright Elsevier Science Publishers.)

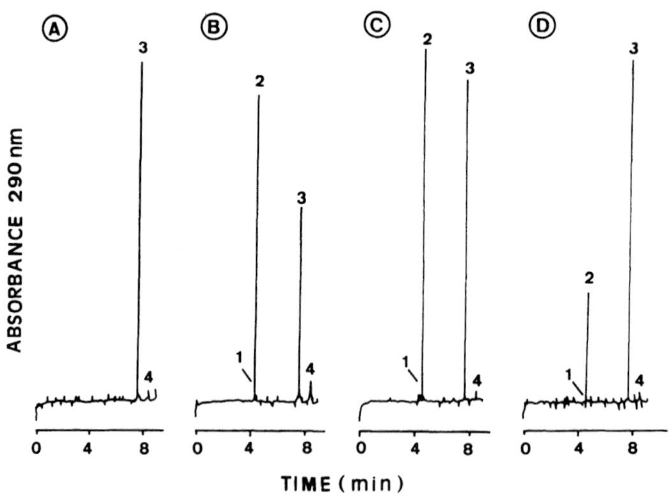

Fig. 7.76. Electropherograms of two calibration samples. (A) bovine plasma spiked with carbamazepine (0.4 mg/ml): (B) bovine plasma spiked with internal standard and 40 μg/ml thiopental] and two patient samples [(C) 23.6 μg/ml: (D) 6.4 μg/ml thiopental]. All experiments were performed with a 2-s injection and a constant 30-kV voltage (55–60 μA). Peaks: *1* = isomer of thiopental: *2* = thiopental; *3* = carbamazepine; *4* = unknown compound. (Reproduced with permission from Ref. 240. Copyright Elsevier Science Publishers.)

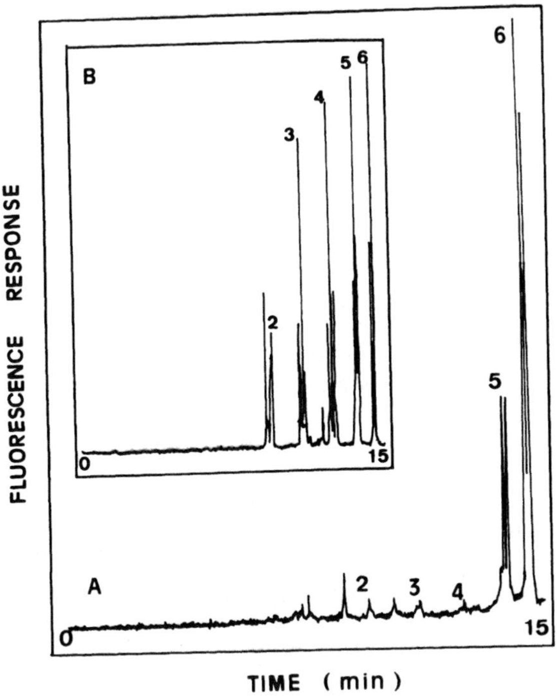

Fig. 7.77. (A) Electropherogram of a urine sample from a patient with porphyria cutanea tarda. Sample preparation: centrifugation for 1 min. Injection: 2 s vaccum. Other conditions: run buffer: 20 mM CAPS, pH 11, detection: fluorescence, excitation 400 nm, emission >550 nm; temperature: 45°C. Electropherogram of a partially decomposed porphyrin standard. Initial sample concentration, 5 nmol/ml. Peaks: *1* = mesoporphyrin (dicarboxyl); *2* = coproporphyrin (tetracarboxyl); *3* = pentacarboxyl porphyrin; *4* = hexacarboxyl porphyrin positional isomers (two peaks); *5* = heptacarboxyl porphyrin; *6* = uroporphyrin (octacarboxyl). (Reproduced with permission from Ref. 256. Copyright Elsevier Science Publishers.)

Creatinine and uric acid in human plasma and urine were separated by MEKC [267]. Phosphate buffer (ionic strength 0.02) containing 50 mM SDS and 5% isopropanol was used as the electrophoretic medium. The pH of the buffer was 9 for the analysis of the serum samples and 6 for the analysis of urine samples, respectively.

Guzman *et al.* [268] developed several methods to concentrate urinary constituents. One method employed bundles of capillaries which were coupled to a single capillary through a glass connector. When a five-capillary bundle was used, approximately 140 ng of uric acid was collected in 95 min. Another method used an analyte concentrator which contained an antibody covalently bonded to a solid support material. With one analyte preconcentrator, approximately 200 ng of material could be collected. Sufficient amount of samples were collected to permit further analysis by mass spectrometry.

7.8 CAPILLARY ION ANALYSIS

Long before the development of modern CE techniques, Hjerten [269] reported the separation of cations by electrophoresis in tubes in 1967. In 1979, Mikkers *et al.* [270] described the separation of anions by electrophoresis in small tubes. Since then there have been many reports on the use of CE for the analysis of anions [121,145,171,172,179,270–284] and cations [29,271,275,279,280,285–294]. The analysis of anions of inorganic acids [279,280,295–298] and salts [138,299–302] have also been described.

In CE separations performed in fused silica capillaries, the electroosmotic flow is towards the negative terminal when a positive potential is applied at the injection end across the fused silica capillary to the detection end. Consequently, cations move towards the negative terminal with the apparent velocity, ν_{app}:

$$\nu_{app} = \nu_{eo} + \nu_{ep} \tag{7.3}$$

where ν_{eo} and ν_{ep} are the electroosmotic flow velocity and the electrophoretic velocity, respectively. In the case of anions, due to the strong attraction by the positive electrode, they flow towards the positive terminal against the electroosmotic flow. The apparent velocity then become:

$$\nu_{app} = \nu_{eo} - \nu_{ep} \tag{7.4}$$

Under these conditions, the separation of anions usually requires longer times compared to the separation of cations due to the smaller apparent velocity. To elute a given sample, it may be necessary to control the electroosmotic flow velocity. Furthermore, it can be noted from Eqn. (7.4) that to achieve optimum separation, an electrolyte anion with a mobility closely matched to the ions of interest should be used. Methods to control the velocity of electroosmosis include the change of the concentration of the electrolyte, the material of the capillary tubing and the use of additives. Highly efficient separation of ions could be achieved by CE, provided that several conditions are fulfilled. These conditions are: (1) The analytes ions should migrate in the same direction as the electroosmotic flow from the injection end to the detection end to permit detection; (2) the mobility of the electrolyte closely matches the mobility of the analyte ion to ensure symmetrical peak shapes [183]; (3) if hydrostatic injection is employed, the sample zones must have much lower ionic strength than the electrophoretic buffer to effect electrostacking (see Chapter 2) and hence sample concentration and improved sensitivity; and (4) if electrokinetic injection is employed, an additive with a lower mobility than any analytic ion of the same polarity should be added into a dilute sample so that it functions as an isotachophoretic terminating electrolyte. Consequently, sample preconcentration by isotachophoresis could be achieved.

7.8.1 Anions

Several groups of investigators have employed conductivity detection in the CE analysis of anions [272,273,275,276]. Tsuda *et al.* [271] described the separation of sulfonic acids by C.E. Yeung and co-workers developed indirect laser-induced fluorescence detection systems which could be used for the analysis of anions in CE separations [171,172]. In this type of methods, a fluorophore is introduced into the electrophoretic buffer. Detection of non-fluorescent species is achieved by charge displacement of the fluorophore by analyte molecules at the detection zone.

Indirect UV detection has been used successfully in many applications of CE in the analysis of anions [272,276,282–284]. In this type of methods, an electrolyte containing a UV absorbing anion plus an electroosmotic flow modifier is used. The displacement of the chromophore, e.g. chromate ion by the analyte ions permit indirect photometric detection. The addition of a suitable additive (e.g. a cationic surfactant) in the electrophoretic medium caused the electroosmotic flow to move toward the anode, forcing the anions to migrate in the same direction as the electroosmotic flow from the injection end toward the detection end. With this approach, separation of alkylsulfonate, sulfates, carboxylates, and inorganic anions

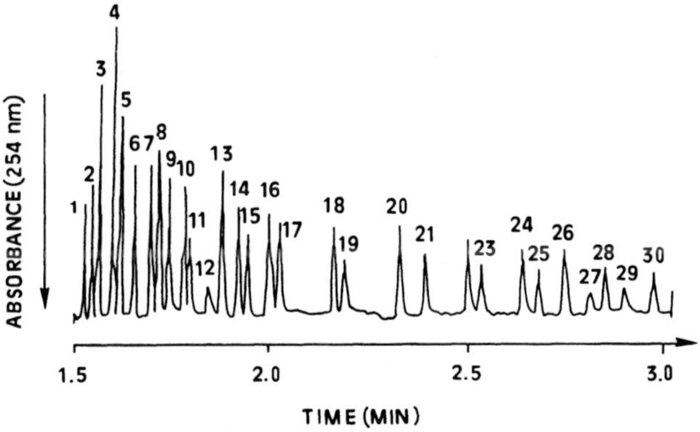

Fig. 7.78. Electropherogram showing the separation of 30 anions. Capillary I.D.: 50 μm (fused silica); carrier electrolyte: 5 mM chromate, 0.5 mM OFM Anion-BT (Waters); voltage: −30 kV; injection: electromigration, 15 s at 1 kV; detection: indirect UV, 254 nm. Peaks: *1* = thiosulfate (4 ppm); *2* = bromide (4 ppm); *3* = chloride (2 ppm); *4* = sulfate (4 ppm); *5* = nitrite (4 ppm); *6* = nitrate (4 ppm); *7* = molybdate (10 ppm); *8* = azide (4 ppm); *9* = tungstate (10 ppm); *10* = monofluorophosphate (4 ppm); *11* = chlorate (4 ppm); *12* = citrate (2 ppm); *13* = fluoride (1 ppm); *14* = formate (2 ppm); *15* = phosphate (4 ppm); *16* = phosphite (4 ppm); *17* = chlorite (4 ppm); *18* = galactarate (5 ppm); *19* = carbanate (4 ppm); *20* = acetate (4 ppm); *21* = ethanesulfonate (4 ppm); *22* = propionate (5 ppm), *23* = propanesulfonate (4 ppm); *24* = butyrate (5 ppm); *25* = butanesulfonate (4 ppm); *26* = valerate (5 ppm); *27* = benzoate (4 ppm); *28* = l-glutamate (5 ppm); *29* = pentanesulfanate (4 ppm); *30* = d-gluconate (5 ppm). (Reproduced with permission from Ref. 284. Copyright Elsevier Science Publishers.)

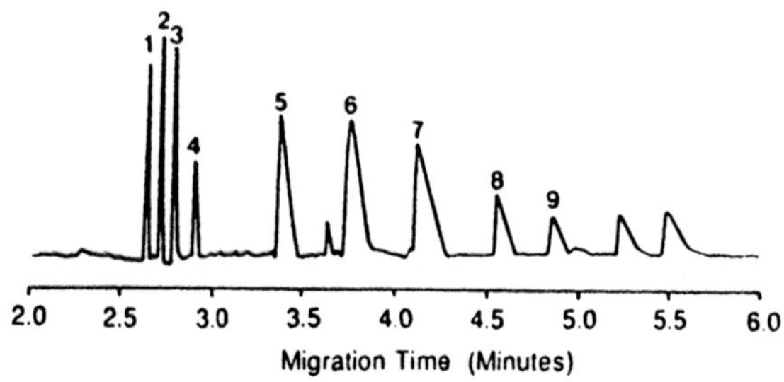

Fig. 7.79. Electropherogram of Kraft black liquor. 60 cm × 75 μm I.D. fused silica; power supply: negative; electrolyte: 5 mM chromate with Nice-Pak OFM Anion BT (Waters) at pH 10.0; injection: hydrostatic for 20 s; detection: indirect UV at 254 nm. Sample preparation: 1000 × diluted in deionized water. Solutes: *1* = thiosulfate; *2* = chloride; *3* = sulfate; *4* = oxalate; *5* = formate; *6* = carbonate; *7* = acetate; *8* = proprionate; *9* = butyrate. (Reproduced with permission from Ref. 278. Copyright Elsevier Science Publishers.)

could be achieved rapidly, typically within 2–4 min. By selecting an electrolyte anion with a mobility closely matched to the ions of interest, optimum separation of the anions could be more readily achieved. An electropherogram showing the separation of 30 anions with a buffer containing chromate is shown in Fig. 7.78. Figure 7.79 shows the electropherogram of a black liquor sample (diluted 1:1000 followed by C_{18} Sep-Pak clean up). For the separation of anions and organic acids with lower mobilities, phthalate was found to provide better selectivity [278]. The analysis of dental plaque extract and human saliva were demonstrated.

Further improvements in peak shape and efficiency could be obtained by using an electrolyte with even lower mobility, such as benzoate. The separation of the same anionic and organic acids with a benzoate electrolyte is shown in Fig. 7.80. The separation of butyric extract of an air sample was demonstrated. The separation of C_1–C_7 alkylsulfonate standards with benzoate electrolyte was also investigated and the analysis of an isopropyl alcohol process extract from a petroleum refinery is shown in Fig. 7.81.

For the elution and separation of alkylsulfonates with very long chains, which have extremely low mobilities, conventional CZE employing an electrolyte with very low mobility, e.g. naphthalenesulfonate, was found to provide satisfactory separation. In Fig. 7.82, the electropherograms of C_4–C_{12} alkylsulfonate standards and an alkylamido glycinate shampoo base is shown. For the analysis of the shampoo base, no sample pretreatment except for dilution was required. The analysis of reactive anions, such as hypochlorite, persulfate and permanganate by CE has also been reported [283]. A high pH chromate electrolyte permitted the analysis of these reactive anions.

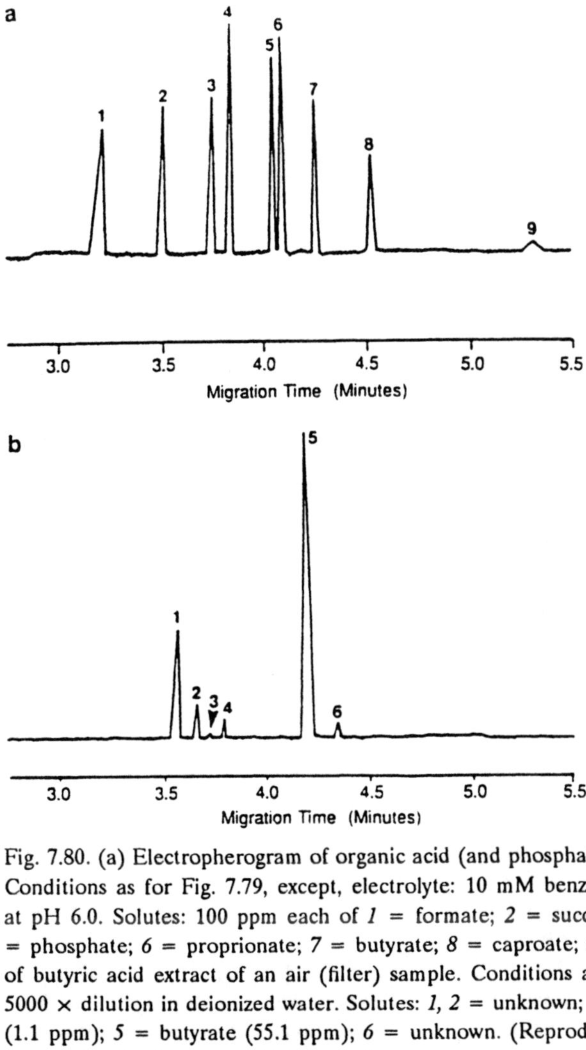

Fig. 7.80. (a) Electropherogram of organic acid (and phosphate) standard with benzoate electrlyte. Conditions as for Fig. 7.79, except, electrolyte: 10 mM benzoate with Nice-Pak OFM Anion BT at pH 6.0. Solutes: 100 ppm each of *1* = formate; *2* = succinate; *3* = glycolate; *4* = acetate; *5* = phosphate; *6* = proprionate; *7* = butyrate; *8* = caproate; *9* = caprylate. (b) Electropherogram of butyric acid extract of an air (filter) sample. Conditions as in (a) except, sample preparation: 5000 × dilution in deionized water. Solutes: *1, 2* = unknown; *3* = glycolate (0.4 ppm); *4* = acetate (1.1 ppm); *5* = butyrate (55.1 ppm); *6* = unknown. (Reproduced with permission from Ref. 278. Copyright Elsevier Science Publishers.)

7.8.2 Cations

The use of CE for the analysis of cations was investigated by Tsuda *et al.* [271]. The separation of cupric and ferric ions was performed. Zare and co-workers employed conductivity detection for the analysis of cations in several studies [285–287]. Monovalent cations were separated by Beckers *et al.* [275]. The separation of the lanthanide series cations was investigated by Bocek *et al.* [295]. CZE with laser-induced fluorescence was used by Gross and Yeung to analyze alkali and alkaline earth metal ions [288]. An electropherogram showing the simultaneous

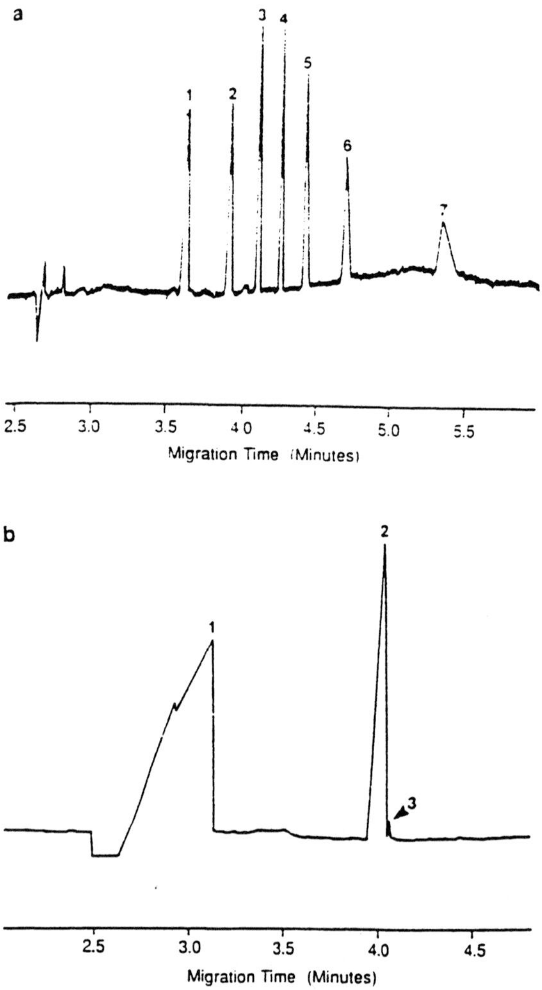

Fig. 7.81. Electropherogram of C_1–C_7 alkylsulfonate standards with benzoate electrolyte. Conditions as in Fig. 7.80. Solutes: *1* = methanesulfonate; *2* = ethanesulfonate; *3* = propanesulfonate; *4* = butanesulfonate; *5* = pentanesulfonate; *6* = hexanesulfonate; *7* = heptanesulfonate. (b) Electropherogram of an isopropyl alcohol process extract from a petroleum refinery. Conditions as in (a) except, sample preparation: 1000 × dilution in deionized water. Solutes *1* = sulfate; *2* = propanesulfonate (137.7 ppm); *3* = butanesulfonate (2.4 ppm). (Reproduced with permission from Ref. 278. Copyright Elsevier Science Publishers.)

separation of alkali, alkali earths, and lanthanides is illustrated in Fig. 7.83. Practical examples of CE of cations include the analysis of a fermentation broth sample (Fig. 7.84) and an industrial waste-water sample (Fig. 7.85).

CITP was used for the determination of calcium ion in the presence of phosphate anion and collagen [294]. The leading electrolyte contained 1×10^{-2} M potassium acetate–acetic acid (pH 5.4) and the terminating electrolyte contained 1×10^{-2} M

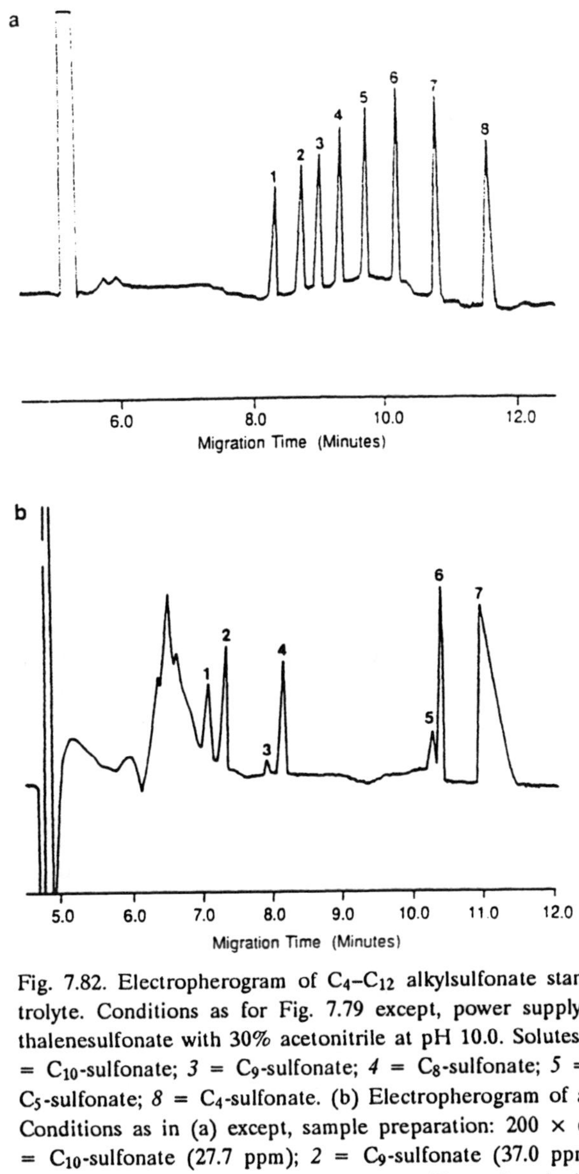

Fig. 7.82. Electropherogram of C_4–C_{12} alkylsulfonate standard with napththalenesulfonate electrolyte. Conditions as for Fig. 7.79 except, power supply: positive; electyrolyte: 10 mM naphthalenesulfonate with 30% acetonitrile at pH 10.0. Solutes: 25 ppm each of *1* = C_{12}-sulfonate; *2* = C_{10}-sulfonate; *3* = C_9-sulfonate; *4* = C_8-sulfonate; *5* = C_7-sulfonate; *6* = C_6-sulfonate; *7* = C_5-sulfonate; *8* = C_4-sulfonate. (b) Electropherogram of an alkylamido glycinate shampoo base. Conditions as in (a) except, sample preparation: 200 × dilution in deionized water. Solutes: *1* = C_{10}-sulfonate (27.7 ppm); *2* = C_9-sulfonate (37.0 ppm); *3* = C_7-sulfonate (3.4 ppm); *4* = C_6-sulfonate (25.6 ppm); *5–7* = unknown. (Reproduced with permission from Ref. 278. Copyright Elsevier Science Publishers.)

n-hexanioc acid in the presence of collagen.

The separation by CZE and detection by electrospray ionization mass spectrometry of quaternary ammonium salts has been investigated [299–302]. The quaternary ammonium salts gave good electrospray signals with the dominant peaks in the mass spectrum being the quaternary ammonium cations.

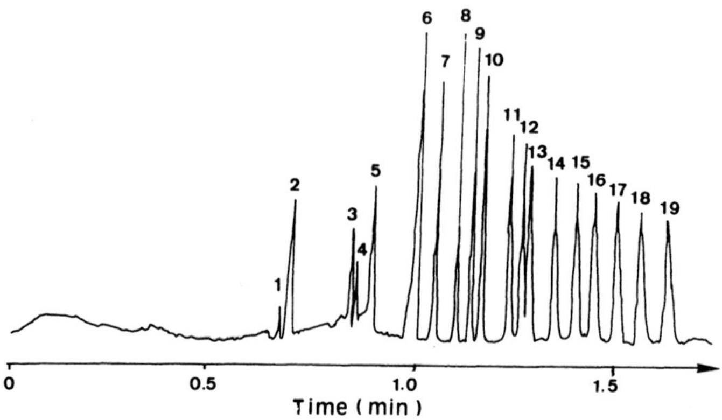

Fig. 7.83. Simultaneous separation of alkali, alkaline earths, and lanthanides. Capillary: 36.5 cm × 75 μm fused silica; carrier electrolyte: 10 mM UVCat-1 (Waters Chromatography Division of Millipore, Milford, Massachusetts), 4.0 mM α-hydroxyisobutyric acid (adjusted to pH 4.4 with acetic acid); voltage: +30 kV; injection: hydrostatic, 20 s at 10 cm; detection: indirect UV, 214 nm. Peaks: *1* = rubidium (2 ppm); *2* = potassium (5 ppm); *3* = calcium (2 ppm); *4* =sodium (1 ppm); *5* = magnesium (1 ppm); *6* = lithium (1 ppm); *7* = lanthanum (5 ppm); *8* = cerium (5 ppm); *9* = praseodymium (5 ppm); *10* = neodymium (5 ppm), *11* = samarium (5 ppm); *12* = europium (5 ppm); *13* = gadolinium (5 ppm); *14* = terbium (5 ppm); *15* = dysprosium (5 ppm); *16* = holmium (5 ppm); 17 = erbium (5 ppm); *18* = thulium (5 ppm), *19* = ytterbium (5 ppm). (Reproduced with permission from Ref. 284. Copyright Aster Publishing Corp.)

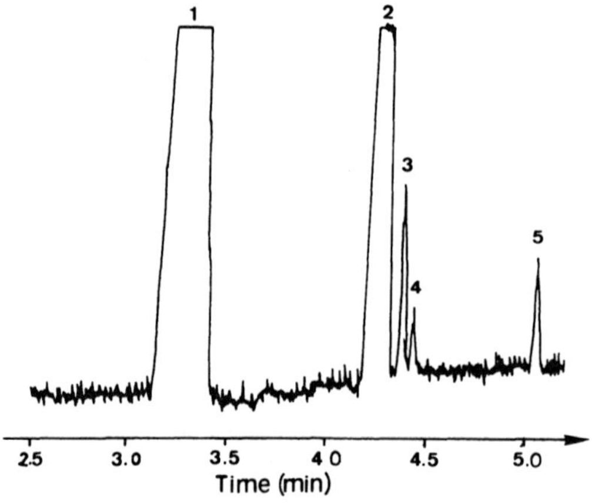

Fig. 7.84. Capillary ion analysis of a fermentation broth sample. Capillary: 60 cm × 75 μm fused silica; carrier electrolyte: 5 mM UVCat-1, 6.5 mM α-hydroxyisobutyric acid (pH 4.0); voltage: +20 kV; injection: hydrostatic, 30 s at 10 cm; detection: indirect UV, 214 nm. The sample was diluted 1:100 before analysis to obtain concentrations of potassium and sodium of 100-1000 ppm and concentrations of magnesium, manganese, and zinc of 10–100 ppb. Peaks: *1* = potassium, *2* = sodium, *3* = magnesiun, *4* = manganese, *5* = zinc. (Reproduced with permission from Ref. 284. Copyright Aster Publishing Corp.)

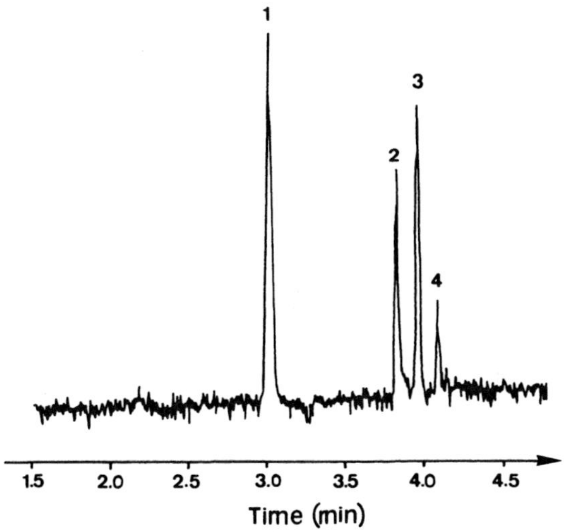

Fig. 7.85. Capillary ion analysis of an industrial waste-water sample. Conditions were the same as for Fig. 7.84. The sample was diluted 1;100 with water. Peaks: *1* = potassium (260 ppb); *2* = calcium (58 ppb); *3* = sodium (50 ppb); *4* = magnesium (8 ppb). (Reproduced with permission from Ref. 284. Copyright Aster Publishing Corp.)

7.9 METAL CHELATES

There have been several investigations on the use of MEKC for the separation of metal chelates [303–305]. The use of chelating agent in CE for the analysis of metal ions has also been studied [289]. Capillary isotachophoresis presents an alternative approach for the separation of metal complexes [291].

The first application of MEKC to the separation of 4-(2-pyridylazo) resorcinolato (par) metal chelates by MEKC was reported by Saitoh *et al.* [303]. Excellent resolution of the peaks was obtained. The electrophoretic solution contained a 0.02 M sodium dodecylsulfate. Subsequently MEKC was used for the separation of porphinato chelates [305]. α, β, γ, δ-tetrakis (4-carboxylphenyl) porphine (TCPP) was found to be the most promising reagent to use as pre-column labeling agent. The electrophoretic medium contained 0.02 M sodium dodecyl sulfate, 0.05 M imidazole, 0.05 M NaH_2PO_4 and 0.0125 M $Na_2B_4O_7$ at pH 7.0. Figure 7.86 shows the MEKC separation of metal TCPP chelates.

MEKC with SDS was used successfully to separate electrically neutral acetylacetonato complexes of chromium (III), cobalt (III), rhodium (III), and platinum (II) [304]. The capacity factor is related to the distribution coefficient by the equation:

$$k' = KV_{SDS}([SDS] - cmc) \tag{7.5}$$

where k' is the capacity factor, defined as the mole ratio of the solute included by the micelle phase to that in the aqueous bulk phase, and K is the distribution

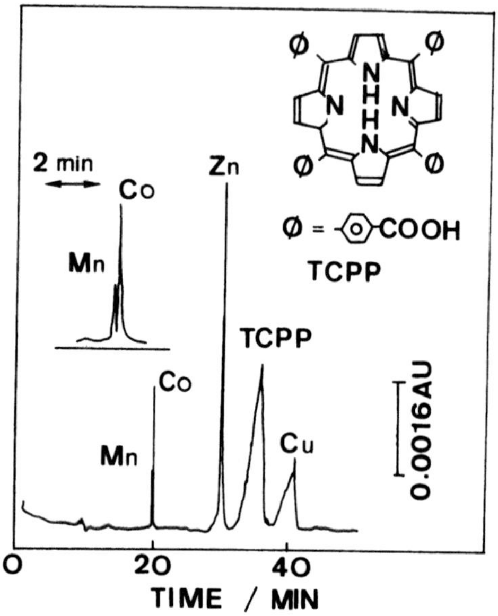

Fig. 7.86. MEKC separation of metal TCPP chelates. Filling solution: 0.02 M SDS, 0.05 M imidazole, 0.05 M NaH_2PO_4 and 0.0125 M $Na_2B_4O_7$ at pH 7.0. Applied voltage and current: 16.5 kV and 25 μA; injection volume: 6 nl; detection wavelength: 422 nm. Concentration of each metal ion: 1×10^{-5} M. (Reproduced with permission from Ref. 305. Copyright The Japan Society for Analytical Chemistry.)

coefficient, given by the ratio of the concentration of the solute in the micelle to that in the aqueous phase. V_{SDS} is the specific partial molar volume of SDS in the micelle. cmc is the critical micelle concentration. [SDS] is the total concentration of SDS. Figure 7.87 shows the MEKC separation of the metal chelates. In addition, a linear log–log relationship between the KV_{SDS} and the partition coefficient in the dodecane/water system ($P_{DOD/W}$) was obtained. Figure 7.88 shows the log–log plot of KV_{SDS} vs. $P_{DOD/W}$. The linear relationship was used to predict the distribution coefficients and migration times of other metal complexes, such as palladium (II) acetylacetonato and chromium (III) 3-methylacetlyacetonato.

Chelation of metal ions with 8-hydroxyquinoline-5-sulfonic acid (HQS^-) was used for the determination of the metal ions by CZE [290]. Detection was by laser-induced fluorescence using a helium-cadmium laser (λ = 325 nm). Limits of detection for Ca(II), Mg(II) and Zn(II) were 613 ppb (105 fmol), 46 ppb (14 fmol), and 205 ppb (21 fmol) respectively. The method was applied to the detection of Ca(II) and Mg(II) in blood serum.

Fig. 7.88. Log–log plot of KV_{SDS} against $P_{DOD/W}$. V_{SDS} = specific partial molar volume of SDS in the micelle. $P_{DOD/W}$ = partition coefficient of solute between dodecane and water. (Reproduced with permission from Ref. 304. Copyright Dr. Alfred Huethig Publishers.)

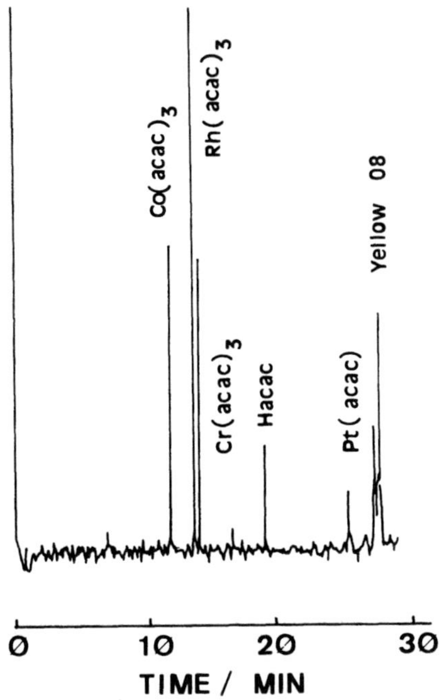

Fig. 7.87. MEKC separation of $Cr(acac)_3$, $Co(acac)_3$, $Rh(acac)_3$, and $Pt(acac)_2$. Carrier electrolytes: 20 mM $Na_2B_4O_7$, 100 mM SDS; potential difference: 230 V/cm, 32 μA. Metal β-diketonato complexes: tris-(acetylacetonato) chromium(III), -cobalt(III) and -rhodium(III): $Cr(acac)_3$, $Co(acac)_3$, $Rh(acac)_3$; and bis-(acetylacetonato) palladium(II), and -platinum(II): $Pd(acac)_2$, $Pt(acac)_2$; and tris(3-methylacetylacetonato) chromium(III): $Cr((maa)_3)$. (Reproduced with permission from Ref. 304. Copyright Dr. Alfred Huethig Publishers.)

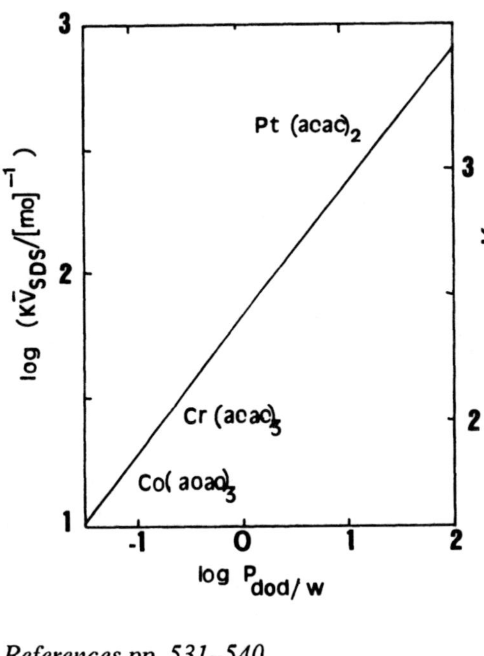

Capillary isotachophoresis has been used for the separation of complexes of rare earth elements [291]. Ethylene diaminetetraacetic acid (EDTA) and triethylenetetraamine hexaacetic acid (TTHA) were used as the terminating electrolytes and 10 mM HCl containing 45% (v/v) acetone as the leading electrolyte. The EDTA complexes of the anions of Gd, Eu, Sm, Pr, Ce and La were separated in accordance with the order of their stability constants.

7.10 ORGANIC COMPOUNDS

This section is concerned mainly with the analysis of organic compounds which do not belong to any specific groups discussed in other sections. The types of compounds include hydrocarbons, phenols, nitroaromatics, organic acids, ketones and alcohols. Various methods, such as MEKC, the use of coated capillary, the use of bile salts, and the use of packed columns have been employed for the analysis of organic compounds.

7.10.1 Hydrocarbons

Since hydrocarbons are generally highly hydrophobic, micellar electrokinetic chromatography has been commonly employed for their separation. By introducing cyclodextrin modifiers, enhancement in selectivity could be obtained, especially for the separation of isomers.

EKC with 2-O-carboxymethyl-β-cyclodextrin (β-CMCD) was used to separate substituted benzenes. Migration times, capacity factors and distribution coefficients were obtained for acetophenone, anisole, methyl benzoate, ethyl benzoate, propyl benzoate, butyl benzenes and the positional isomers of cresol, nitroaniline, chloroaniline, dinitroaniline, nitrophenol, xylidine and xylenol [306].

Cyclodextrin modified MEKC was also used for the separation of highly hydrophobic compounds, including chlorinated benzene congeners, polychlorinated biphenyl congeners, tetrachlorodibenzo-p-dioxin isomers and polycyclic aromatic hydrocarbons [307]. Figure 7.89 illustrates the separation of chlorinated benzene cogeners with a 100 mM borate buffer (pH 8.0) containing 100 mM SDS, 2 M urea and 40 mM γ-CD. Figure 7.90 shows the separation of trichlorobiphenyl isomers by γ-CD-MEKC using a phosphate buffer (pH 8.0) containing 100 mM SDS and 2 M urea. Figure 9.91 shows the separation of aromatic hydrocarbons by γ-CD-MEKC employing a borate buffer (pH 9.0), containing 30 mM γ-CD, 5 M urea and 100 mM SDS.

Photodiode array detection was utilized in CE separations of several groups of compounds, including selected aromatic hydrocarbons [308]. Capillaries of 100 μm and 75 μm I.D. were used. On-line measurement of the UV spectra could be obtained. The use of the photodiode array detector facilitated peak identification.

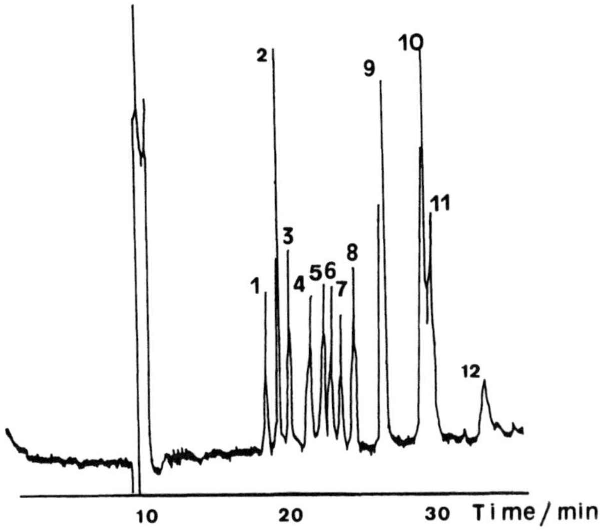

Fig. 7.89. γ-CD-MEKC separation of chlorinated benzene congeners: *1* = 1,2,3,5-tetra-, *2* = 1,2,3-tri-, *3* = 1,3,5-tri-, *4* = 1,2,-di-, *5* = 1,2,4,5-tetra-, *6* = mono-, *7* = 1,3-di-, *8* = 1,2,4-tri-, *9* = 1,2,3,4-tetra-, *10* = penta-, *11* = 1,4-di- and *12* = hexachlorobenzene. Capillary: 700 mm (Polymicro Technologies); separation solution: 100 mM SDS in 100 mM borate buffer (pH 8.0), 2 M urea, 40 mM γ-CD; applied voltage: 15 kV; current: 23 μA. (Reproduced with permission from Ref. 307. Copyright Elsevier Science Publishers.)

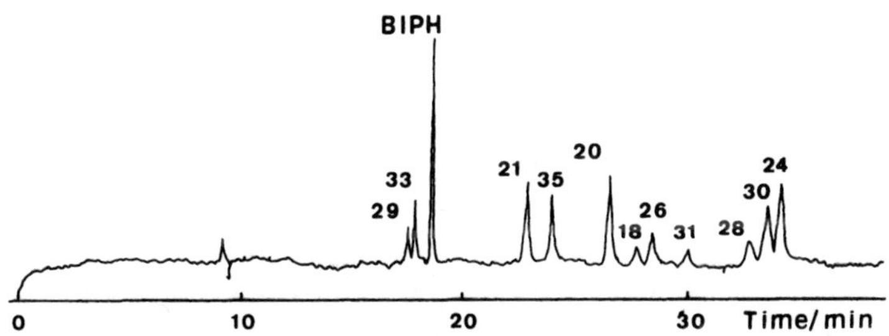

Fig. 7.90. Separation of eleven trichlorobiphenyl isomers by γ-CD-MEKC. Peaks are identified with the IUPAC number: *18* = 2,2′,5-, *20* = 2,3,3′-, *21* = 2,3,4-, *24* = 2,3,6-, *26* = 2,3′,5-, *28* = 2,4,4′-, *29* = 2,4,5-, *30* = 2,4,6-, *31* = 2,4′,5-, *33* = 2′,3,4- and *35* = 3,3′,4-trichlorobiphenyl; *BIPH* = biphenyl. Capillary: 650 mm (Scientific Glass Engineering); separation solution,:60 mM γ-CD, 100 mM SDS and 2 M urea in 100 mM borate–50 mM phosphate buffer 9pH 8.0); applied voltage: 15.4 kV; current: 50 μA. (Reproduced with permission from Ref. 307. Copyright Elsevier Science Publishers.)

MEKC with bile salt surfactants was used for the separation of hydrophobic molecules [309]. The separation of polyaromatic hydrocarbons (PAH) was performed using this method. Sodium cholate was used as the surfactant. Addition of methanol increased resolution without significant loss of resolution. The separation

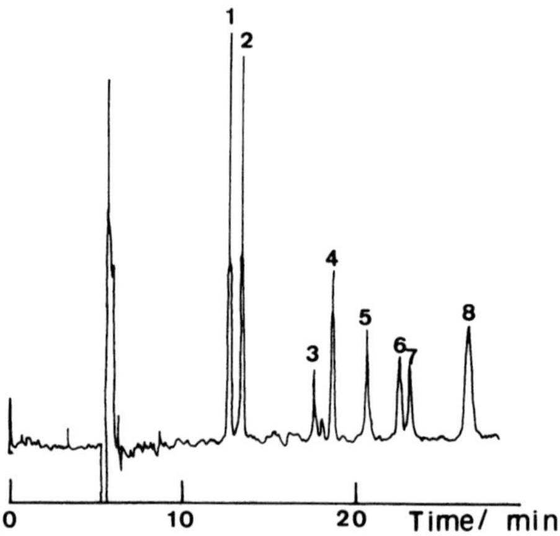

Fig. 7.91. γ-CD-MEKC separation of a mixture of naphthalene and four tricyclic and three tetracyclic aromatic hydrocarbons: *1* = naphthalene; *2* = acenaphthene; *3* = anthracene; *4* = fluorene; *5* = phenanthrene; *6* = chrysene; *7* = pyrene; *8* = fluoranthene. Capillary: 700 mm (Polymicro Technologies); separation solution: 30 mM γ-CD, 100 mM SDS and 5 M urea in 100 mM borate buffer (pH 9.0); applied voltage: 20 kV: current: 41 μA. (Reproduced with permission from Ref. 307. Copyright Elsevier Science Publishers.)

of hydroxyl derivatives of PAHs by MEKC with bile salt surfactants has also been investigated.

Instead of employing MEKC with anionic surfactants, neutral organic molecules were separated by solvophobic association with tetraalkylammonium ions [310]. The separation of mesityl oxide formaldehyde and five aromatic hydrocarbons: benzo [ghi] perylene; perylene, pyrene, p-methylanthracene and naphthalene was performed. The electrophoretic medium consisted of 50% acetonitrile and 50% water (v/v) with 0.025 M tetrahexylammonium phosphate (THAP).

An alternative separation strategy involved the use of electrically driven open tubular liquid chromatography (ED-OTLC) with porous silica-layered capillaries coated with octadecylsilane (ODS) [311]. Acetonitrile–phosphate buffer (0.05 or 0.1 M, pH 7.0) (2:3, v/v), and methanol–phosphate buffer (0.05 M, pH 7.0) (1:1, v/v) were used as mobile phases. The separation of eight polycyclic aromatics is shown in Fig. 7.92. ED-OTLC was found to provide higher efficiency than pressure-driven OTLC by a factor of more than 2. Tsuda also used glass capillaries (30 μm I.D.) coated with octadecylsilane (ODS) for the separation of aromatics [312]. The eluent used consisted of acetonitrile and water.

Electrochromatography in packed capillary column has also been employed for the separation of aromatic compounds [313]. Separation was performed in a column packed with 5 μm silica particles, derivatized to ODS-Hypersil. The electrolyte used was 70:30 acetonitrile:water containing 2 mM sodium dihydrogen phosphate. The chromatogram obtained is shown in Fig. 7.93. Efficiencies of 500,000 plates/m was obtained.

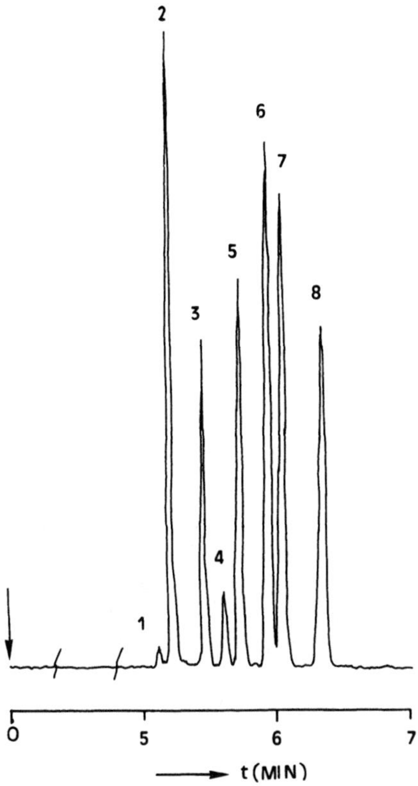

Fig. 7.92. Chromatogram of eight polycyclic aromatics obtained with a 10-μm I.D. PSL-ODS capillary. Applied voltage: 20 kV, L = 49.0 cm, $l_{\text{inj-det}}$ = 26.5 cm; current: 0.7 μA; T = 25°C; V_{inj} = 9 pl. Mobile phase: 0.05 M phosphate buffer (pH 7.0)–methanol (1 : 1). Peaks: *1* = naphthoquinone (k' = 0); *2* = 9-anthracenemethanol (k' = 0.03); *3* = 9-anthracenecarbonitrile (k' = 0.06); *4* = anthracene (k' = 0.09); *5* = 7,8-benzoflavone (k' = 0.12); *6* = fluoranthene (k' = 0.16); *7* = pyrene (k' = 0.19); *8* = 9-vinylanthracene (k' = 0.24). (Reproduced with permission from Ref. 311. Copyright Elsevier Science Publishers.)

An interesting approach was the use of microemulsion in CE, which employed oil in water microemulsions as electrophoretic media for the separations of aromatic compounds, and ketones and diketones [314]. The microemulsion used was water (89.28%)/SDS (3.51%)/1-butanol (6.61%)/heptane (0.81%) by weight. Fluorescence detection was used to detect the aromatic compounds, which included naphthalene, α- and β-naphthols, N-acetyl-α-naphthylamine and 2-naphthol-6-sulfonate. Indirect fluorescence detection was employed for the detection of the ketone and β-diketones, which included 2-acetylthiophene, acetophenone, benzoylacetone, pivalonyl trifluoro acetone, 2-thenolyltrifluoroacetone, benzoyl trifluoroacetone and 2-naphthozyltrifluoroacetone.

The separation of phenols by MEKC has been reported [315,316]. MEKC employing SDS and polyoxyethylene (23) dodecanol (BRIJ-35) was used to

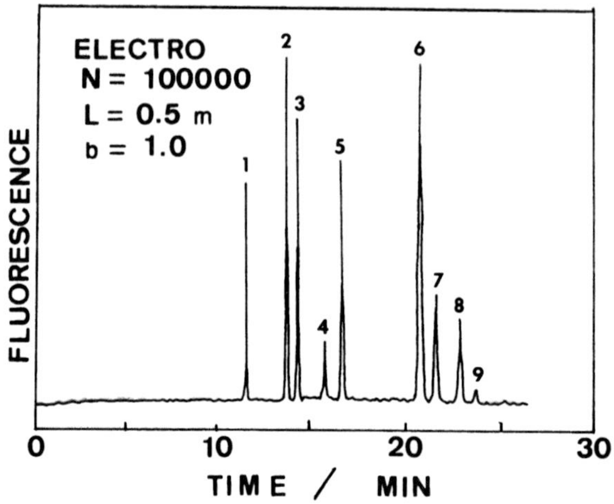

Fig. 7.93. Separation by electrochromatography of aromatic test mixture on a particularly efficient capillary packed with 5 μm Hypersil subsequently derivatized by ODS. Column length: 500 mm, column bore: 40 μm. Solutes in order of elution: (1) naphthalene; (2) = 2-methylnaphthalene; (3) fluorene, (4) phenanthrene, (5) anthracene, (6) pyrene, (7) 9-methylanthracene, (8) and (9) unknown. Electrolyte 70:30 acetonitrile: water containing 2 mM sodium dihydrogen phosphate. (Reproduced with permission from Ref. 313. Copyright Friedr. Vieweg & Sohn Verlagsgesellschaft mbH.)

separate ASTM test mixture LC-79-2, which contains benzyl alcohol, benzene and benzaldehyde, acetophenone, methylbenzoate and dimethyl terephthalate [317]. The addition of BRIJ-35 to SDS caused the formation of mixed micelles, and hence improved selectivity to permit separation of benzene and benzaldehyde, which were not separated when SDS alone was used. Since BRIJ-35 is a non-ionic surfactant, its addition does not cause an increase in Joule heating.

Mixed micelle solutions with different ratios of sodium dodecyl sulfate (SDS) to sodium dodecyl benzene sulfonate (SDBS) were used in MEKC separation of aliphatic alcohols and carboxylate anions [318]. Indirect UV detection was employed with the UV-active surfactant (SDBS) providing the background absorbance. Figure 7.94 illustrates the MEKC separation of aliphatic alcohols.

7.10.2 Organic acids

The use of CE for the analysis of lower carboxylic acids has been described in several reports [95,121,151,265,319–324]. Huang *et al.* [319] investigated the separation of low molecular mass carboxylic acids by CZE with on-column conductivity detection [319]. 0.2–0.5 mM TTAB (tetradecyltrimethylammonium bromide) was added to reverse the electroosmotic flow, so that all the anions passed through the detector. Electrolytic solutions containing 2-morpholinoethane (MES) or tris (hydroxymethyl) aminomethane (TRIS) were used. In a subsequent

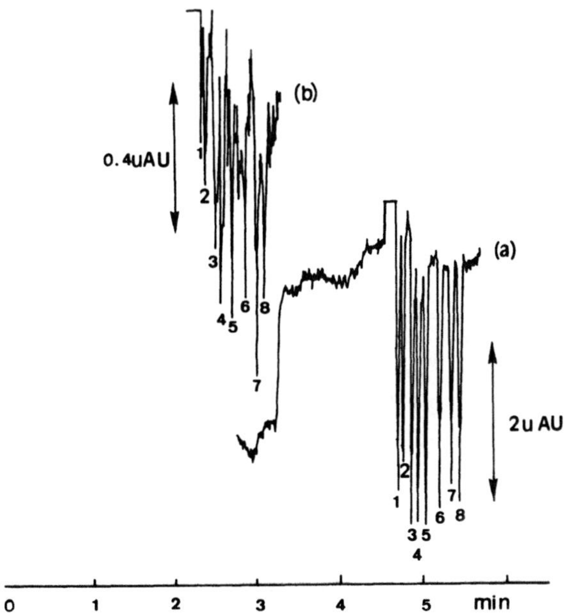

Fig. 7.94. MEKC separation of aliphatic alcohols: (a) surfactant, 50 mM SDBS; sample, 1% (v/v) of each alcohol; (b) surfactant, 4 mM SDBS + 46 mM SDS; sample, 0.05% (v/v) of each alcohol. *1* = methanol; *2* = ethanol; *3* = 2-propanol; *4* = 1-propanol; *5* = t-butanol; *6* = 2-butanol; *7* = 2-methyl-1-propanol; *8* = 1-butanol. Capillary: 57 cm (50 cm effective length) × 50 μm I.D. deactivated fused silica; injection: 2 s by pressure; applied voltage: 15 kV; detection: 214 nm, buffer: 12.5 mM borax buffer containing surfactants. (Reproduced with permission from Ref. 318. Copyright Dr. Alfred Huethig Publishers)

investigation, an end-column conductivity detector for CZE was developed [321]. The system was tested with the separation of six carboxylic acids.

Hjerten and co-workers performed several investigations on the separation of carboxylic acids [95,121,151,322]. Rotating quartz tubes of 3 mm I.D. coated with polyacrylamide or methylcellulose were used for the separation of aromatic carboxylic acids [151]. The coating eliminated electroosmotic flow and prevented adsorption of the analyte molecules. Glass tubes of 0.05–0.3 mm I.D. which contained polyacrylamide gel of the composition T = 10%, C = 3% have been used for CE separations of carboxylic acids [322]. Electrophoretic solution was a 0.1 M Tris–acetic acid buffer (pH 8.6). The effects of buffer additives, such as sodium dodecyl sulfate, heptaoxyethylene lauryl ether (G3707, an non-ionic surfactant), and dextrin (molecular weight 40,000), in CE separation of carboxylic acids have been investigated [95]. The buffer additives were found to increase resolution and alter the order of migration of the analytes. Better resolution was obtained with the addition of dextrin modifier.

CE with amperometric detection was used for the separation of several biogenic

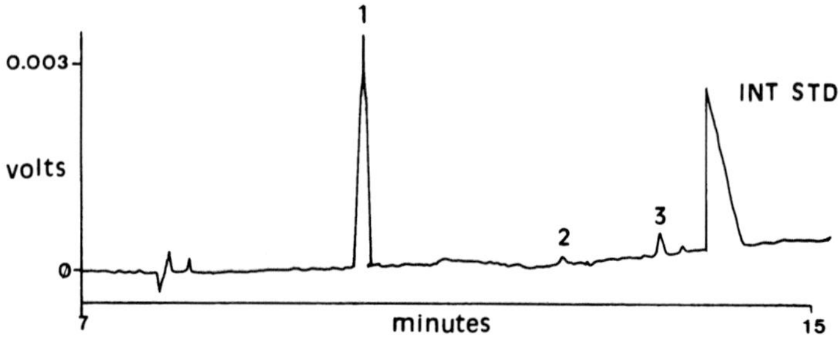

Fig. 7.95. Electropherogram of apple juice. Sample preparation: dilute and inject. Peaks: *1* = malic acid, 4352 μg/ml; *2* = acetic acid, 48 μg/ml; *3* = lactic acid, 254 μg/ml; *INT STD* = internal standard. Capillary: 100 cm × 75 μm I.D., electrolyte: 0.5 mM OFM™ Anion-BT (osmotic flow modifier Anion-BT)–5 mM potassium phthalate (pH 7.0), voltage: −20 kV; injection time: 45 s hydrostatic injection (10 cm). (Reproduced with permission from Ref. 323. Copyright Elsevier Science Publishers.)

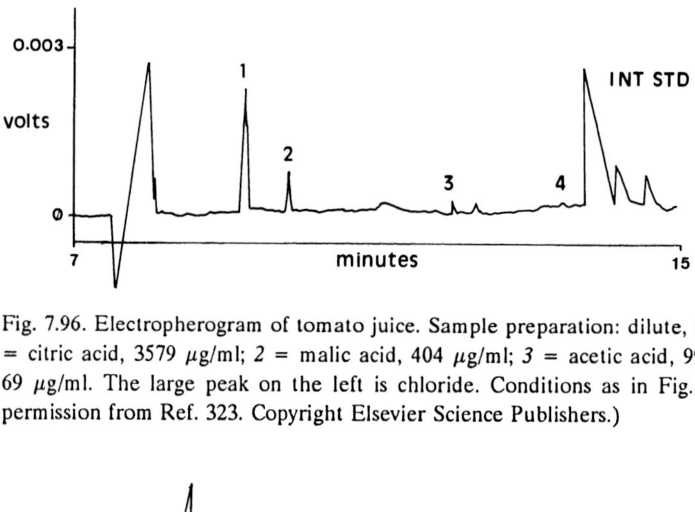

Fig. 7.96. Electropherogram of tomato juice. Sample preparation: dilute, filter and inject. Peaks: *1* = citric acid, 3579 μg/ml; *2* = malic acid, 404 μg/ml; *3* = acetic acid, 99 μg/ml; *4* = lactic acid, 69 μg/ml. The large peak on the left is chloride. Conditions as in Fig. 7.95. (Reproduced with permission from Ref. 323. Copyright Elsevier Science Publishers.)

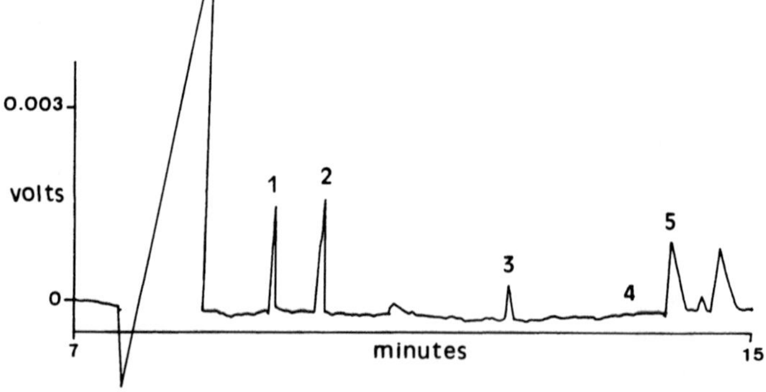

Fig. 7.97. Electropherogram of soy sauce. sample preparation: dilute and inject. Peaks: *1* = citric acid; *2* = tartaric acid; *3* = acetic acid; *4* = lactic acid; *5* = butyric acid (not added as an internal standard). The large peak on the left is chloride. Conditions as in Fig 7.95. (Reproduced with permission from Ref. 323. Copyright Elsevier Science Publishers.)

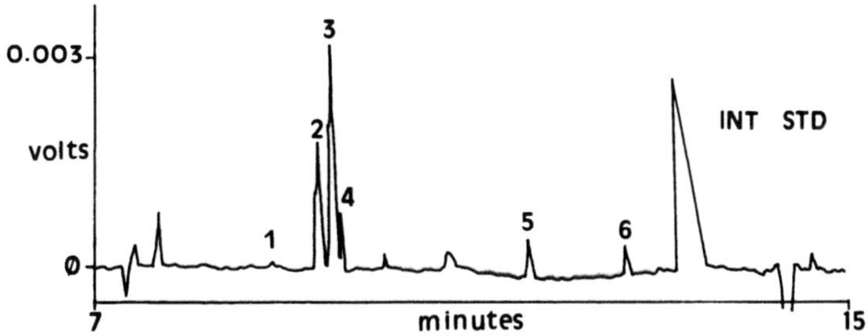

Fig. 7.98. Electropherogram of Chablis wine. Sample preparation: dilute and inject. Peaks: *1* = citric acid, 127 μg/ml; *2* = tartaric acid, 2645 μg/ml; *3* = malic acid, 3291 μg/ml; *4* = succinic acid, 300 μg/ml; *5* = acetic acid, 260 μg/ml; *6* = lactic acid, 296 μg/ml. Conditions as in Fig. 7.95. (Reproduced with permission from Ref. 323. Copyright Elsevier Science Publishers.)

amines and carboxylic acid metabolites [265]. Small I.D. (12.7 μm) capillaries were used. The system was adapted to perform direct sampling from a single nerve cell.

Organic acids in a variety of food matrices were determined by CE [323]. Indirect UV detection was employed. The electropherograms for apple juice, tomato juice, soy sauce and Chablis wine are shown in Figs. 7.95 to 7.98, respectively. The CE system was found to be a low cost, quantitative and selective method for the analysis of organic acids in food samples.

CZE has also been used to analyze organic anions in blood serum of patients with chronic renal failure [324]. PTFE capillaries drawn to 0.2 mm I.D. were used as separation tube. Two electrolyte systems were used. One system contained acetate (0.01 M) and β-alanine at pH 3.8 with 0.5 g/l of methyl hydroxyethylcellulose (MHEC) which acted as a surface-active agent. Another system contained 0.01 M 2-(N-morpholino) ethane sulfonic acid (MES) and histidine at pH 6.1 with 0.5 g/l MHEC.

7.10.3 Amines

Fluorescamine labeled *n*-alkyl amines were separated by CZE. The separations of *n*-hexyl, *n*-butyl and *n*-propylamines (1) and *n*-octyl, heptyl, hexyl, pentyl, butyl and propyl amines (3) were demonstrated. CZE with 5×10^{-2} M phosphate buffer (pH 7) was employed. The use of gradient elution MEKC involving increasing concentration of 2-propanol and Triton X-100 was used for the separation of NBD-derivatized amines [325]. MEKC with columns treated with TMCS (trimethylchlorosilane) using a mobile phase containing 10% by volume of 2-propanol for the separation of NBD-amines [326] has also been demonstrated. The effects of voltage, column dimension, concentration of buffer, and surfactant on column efficiency for the MEKC of NBD-amines were studied [327]. NBD-amines have been separated by MEKC with linear and concave acetonitrile gradients [328]. Fluorescently (NBD) labeled alkylamines were separated by solvent gradient MEKC [329]. Models for predicting migration times for linear, convex and concave gradients have been proposed. NBD-derivatized alkylamines were separated in bile salt surfactants [309]. The electrophoretic medium contained 0.05 M sodium cholate, 0.01 M Na_2HPO_4, and 0.006 M $Na_2B_4O_7$. Addition of 20% methanol improved resolution of the amines.

Indirect fluorometric detection using quinine sulfate as the background fluorescent buffer was demonstrated by Gross and Yeung [288]. The system was applied to the separation and detection of substituted amines. The composition of the buffer used was 0.38 mM quinine sulfate and 0.58 mM H_2SO_4 (pH 3.7). Polyamines in human urine and organ in rats were separated by CE with post-column derivatization for fluorescence detection [258].

7.11 CARBOHYDRATES

Carbohydrates are usually neutral and do not possess chromophores. They therefore pose problems for CE separation and optical detection methods. One common strategy to enhance separation and detectability is to perform derivatization to introduce charged groups and chromophores [336–342]. Indirect detection methods have also been employed to improve the detectability of these compounds [36].

In one of the CE procedures reported for the separation of sugars, reducing monosaccharides were derivatized to N-2-pyridylglycamine, which formed charged borate complexes with the borate in the buffer solution [336]. A 200 mM borate buffer was used as the electrolyte. Detection limits at the 10-pmol level could be achieved by UV detection. Another method for the analysis of carbohydrate is based on the determination of iodate and periodate by CE [337]. The C–C bonds in 1,2-diol and 1,2,3-triol compounds can be readily cleaved with periodate in aqueous media. The periodate consumed is equimolar to the oxidized C–C bond and also to the iodate formed. By determining the iodate and periodate simultaneously, the analysis procedure could be applied to periodate oxidation analysis of various carbohydrates.

Pyridylamino (PA) derivatives of maltooligosaccharide were separated by CE in the pH range of 3.0–4.5, with 0.1 M phosphate buffer as the running electrolyte [340]. The electropherogram obtained is shown in Fig. 7.99. The migration of the PA-maltooligosaccharides followed a linear function of the number of glucose residues in a homologous series.

Human α_1-acid glycoprotein (AGP) contains oligosaccharides chains which amount to as much as 45% of the total weight of the proteins [341]. Human AGP oligosaccharides cleaved from the glycopeptide fragments of the tryptic digest by PNGase F (peptide-N-glycosidase F) was derivatized with 2-aminopyridine (2-AP). Figure 7.100 shows an electropherogram of the pyridylamino derivatives of human and bovine AGP oligosaccharides by CZE with 0.1 M phosphate solution (pH 5.0)

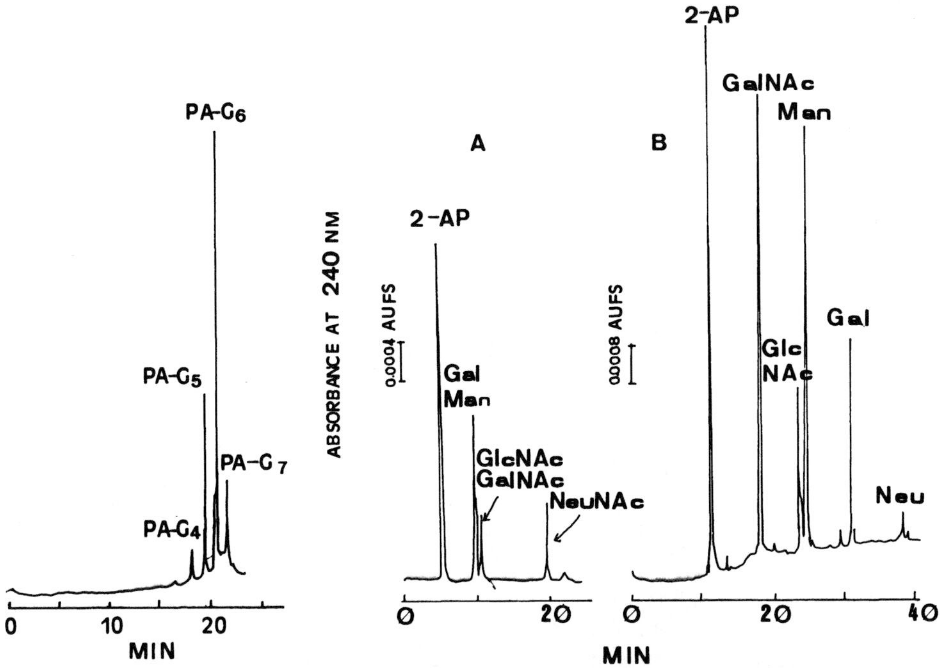

Fig. 7.99 (left). Separation of PA-maltooligosaccharides. Electrolyte 0.1 M phosphate, pH 4.0. UV detection at 240 nm. Capillary: 80 cm (50 cm to detector) × 50 μm I.D.; voltage: 20 kV (current 60 μA); injection: electromigration for 15 s at 18 kV; temperature: 25°C; *PA* = pyridylamino; $PA-G_4$ = PA-maltotetraose; $PA-G_5$ = PA-maltopentaose; $PA-G_6$ = PA-maltohexaose; $PA-G_7$ = PA- maltoheptaose. (Reproduced with permission from Ref. 340. Copyright Elsevier Science Publishers.)

Fig. 7.100 (right). CZE of Pyridylamino derivatives of standard monosaccharides. (a) Capillary: fused silica tube with hydrophilic coating on the inner wall, 35 cm (to the detection point), 70 cm total length × 50 μm I.D.; electrolyte: 0.1 M sodium phosphate solution, pH 5.0 containing 50 mM tetrabutylammonium bromide; running voltage: 15 kV; current was ca. 90 μA; injection by electromigration for 8 s at 15 kV. (b) Capillary: uncoated fused silica tube, 50 cm (to the detection point), 80 cm total length × 50 μm I.D.; electrolyte: 0.2 M sodium borate, pH 10.5; running voltage: 18 kV; current was 75 μA; injection by electromigration for 1 s at 10 kV. (Reproduced with permission from Ref. 341. Copyright Elsevier Science Publishers.)

containing 50 mM tetrabutylammonium bromide as the electrophoretic buffer.

Aldose oligosaccharides derivatized with fluorogenic reagents, such as 3-(4-carboxybenzoyl)-2-quinolinecarboxaldehyde or 3-benzoyl-2-naphthaldehyde were separated by capillary electrophoresis [338,339]. On-column laser-induced fluorescence detection at 457 nm was employed. Both open-tubular columns and gel-filled columns have been investigated. The separations of partially hydrolyzed polysaccharides (dextrin 15), enzymatically digested chondroitin sulfate A and enzymatically digested hyaluronic acid were demonstrated. Figures 7.101 and 7.102 illustrate the

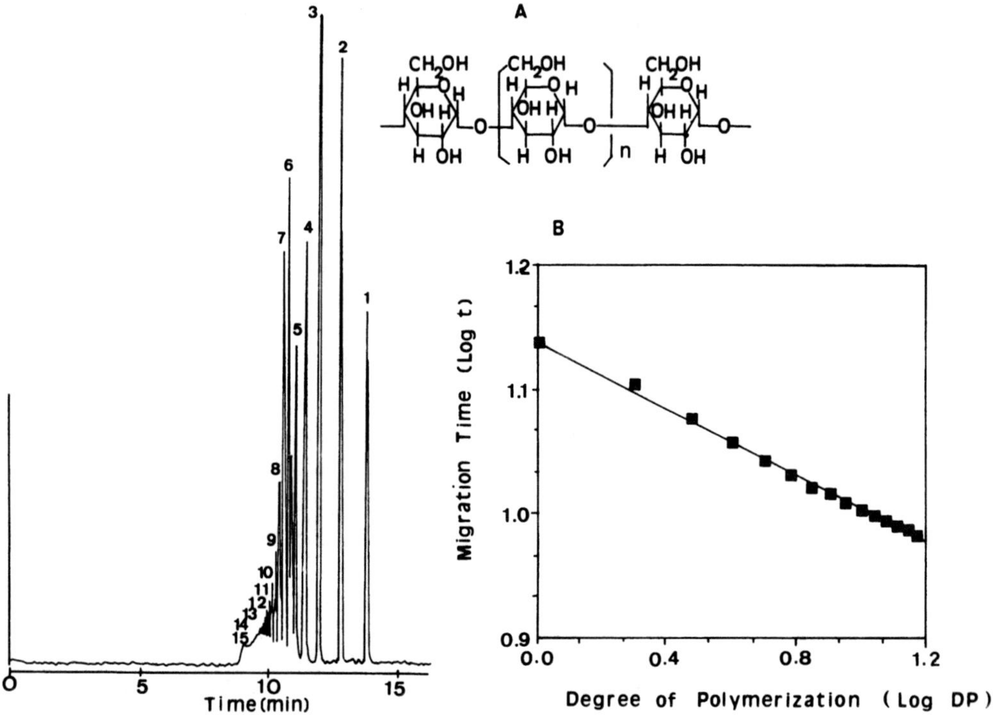

Fig. 7.101. (A) Electrophoretic separation of a partially hydrolyzed polysaccharide (Dextrin 15) in the open tubular system. (B) Correlation (log–log) between molecular mass and migration time. Capillary 88 cm (58 cm effective length) × 50 μm I.D.; buffer: 10 mM Na_2HPO_4–10 mM $Na_2B_4O_7 \cdot 10H_2O$ (pH 9.40); applied voltage: 20 kV (12 μA). (Reproduced with permission from Ref. 339. Copyright Elsevier Science Publishers.)

separation of a partially hydrolyzed dextrin 15 in an open-tubular column and in a polyacrylamide gel-filled column, respectively.

The effect of different borate buffers on the migration behaviour of carbohydrates in CE was investigated [343]. Borate forms negatively charged complexes with sugar molecules, which permitted their separation by CE. Improved resolution and efficiency were obtained at elevated temperature up to 60°C. The addition of borate to aqueous solutions of mono- and oligosaccharides resulted in an increase of absorbance at 195 nm. Consequently, the sugars could be detected without derivatization. In Figs. 7.103 and 7.104, the electropherograms of underivatized monosaccharides and disaccharides, respectively, at several temperatures are shown.

CZE was used for the separation of chondroitin sulfate disaccharides and hyaluronan oligosaccharides [344]. The method developed could be applied to the assay of these molecules for digests of connective tissues.

CE was used for the separation of Maillard reaction products, i.e. products arising from the reaction of free amino acids with aldehydic sugars [345]. The products

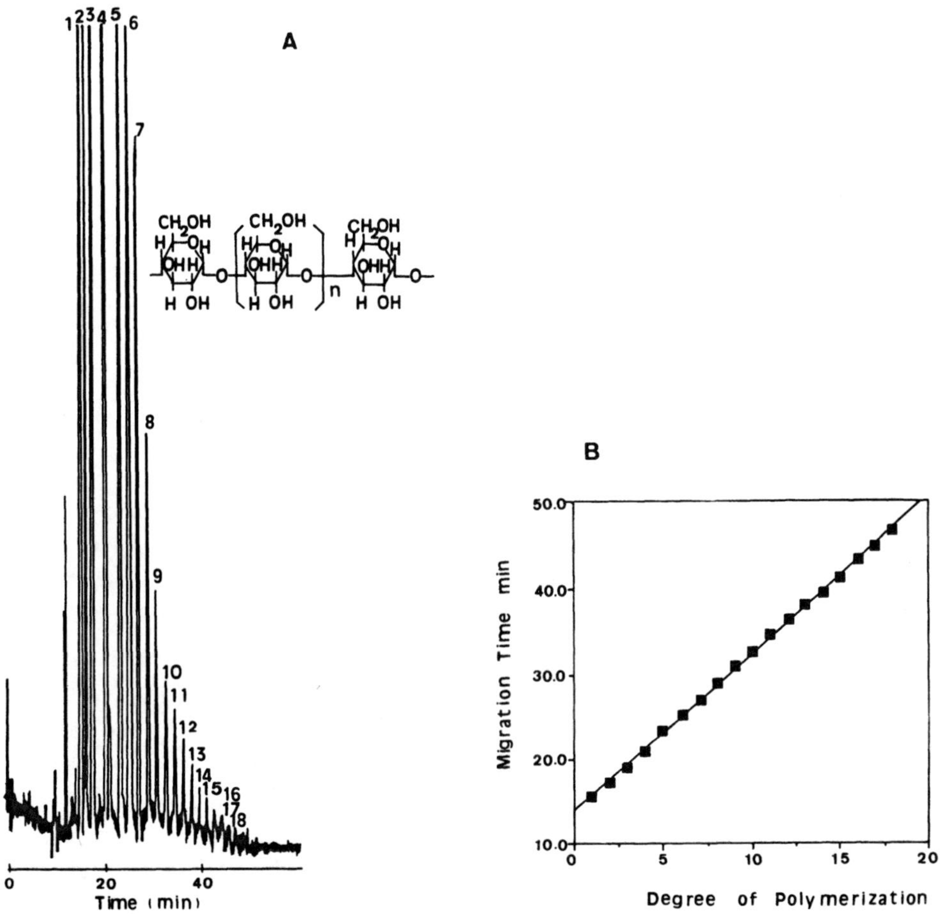

Fig. 7.102. (A) Electrophoretic separation of a partially hydrolyzed Dextrin 15 in a polyacrylamide gel-filled column. (B) Correlation between molecular masses of the oligomers from maltodextrin and their corresponding migration times. Gel concentration: 10% T, 3% C; capillary: 26 cm (19 cm effective length) × 50 μm I.D.; buffer: 0.1 M Tris–0.25 M borate–7 M urea (pH 8.33); applied field, 269 V/cm (20 μA). (Reproduced with permission from Ref. 339. Copyright Elsevier Science Publishers.)

from the reaction of glucose or ribose with glycine, alanine and isoleucine were studied. The separation of underivatized products was performed in an untreated capillary (50 cm × 75 μm I.D.) with a 0.02 mol/l phosphate buffer (pH 7.5) and a separation voltage of 18 kV. Detection was by UV absorbance at 220 nm. For the phenylthiocarbonyl (PTH) derivatives, 0.005 mol/l borate buffer (pH 9.6) was used. For the 3,4-dinitrophenyl hydrazones, MEKC in a solution containing 0.01 mol/l Na_2HPO_4–0.006 mol/l tetraborate, 0.05 mol/l SDS was employed.

The use of indirect fluorescence detection for the CE analysis of sugars has been investigated [36]. The method was based on charge displacement of a chromophore

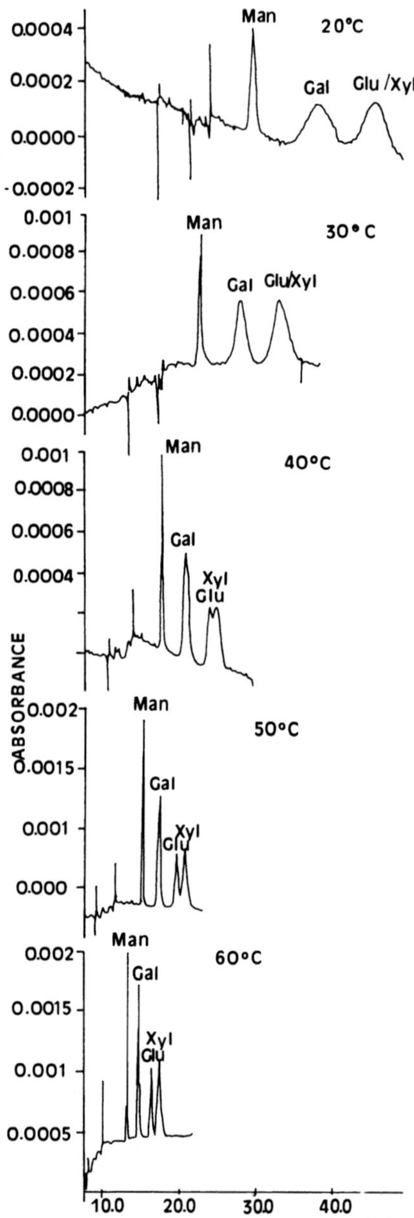

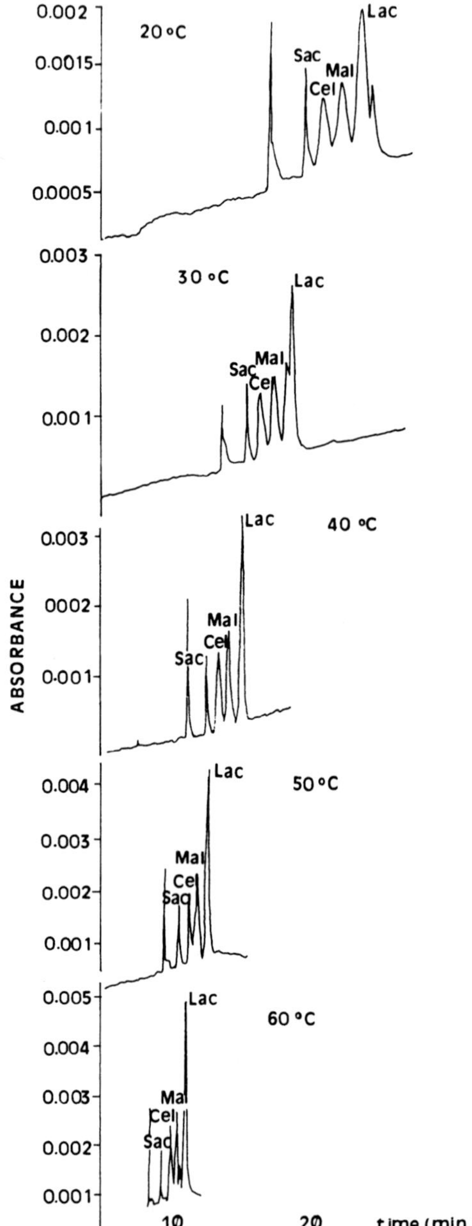

Fig. 7.103 (left). Effect of temperature of CE of underivatized monosaccharidesin 50 mM tetraborate, pH 9.3. Sample: mannose (*Man*), galactose (*Gal*), glucose (*Glu*), and xylose (*Xyl*), each 10 mM, dissolved in water; temperature: 20–60°C; capillary length: 94 cm, 87 cm to detector; voltage: 20 kV. (Reproduced with permission from Ref. 343. Copyright American Chemical Society.)

Fig. 7.104 (right). Effect of temperature on CE of underivatized disaccharides. Buffer: 50 mM tetraborate, pH 9.3. Sample: saccharose (*Sac*), celloblose (*Cel*), maltose (*Mal*), and lactose (*Lac*), each 25 mM, dissolved in water. Other conditions are the same as in Fig 7.103. (Reproduced with permission from Ref. 343. Copyright American Chemical Society.)

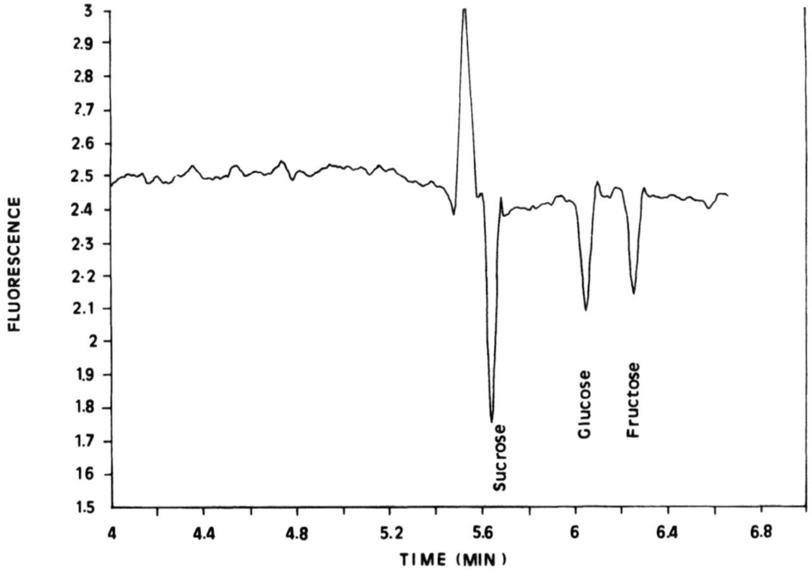

Fig. 7.105. Separation of a sugar mixture. Indirect detection of 640 fmol of each sugar in 1 mM coumarin at pH 11.5. The column is 90 cm × 18 μm I.D. × 150 μm O.D. (Reproduced with permission from Ref. 36. Copyright Elsevier Science Publishers.)

present in the background electrolyte. Laser beam of a helium-cadmium laser (442 nm, 1 mW) or an argon ion laser (488 nm, 1 mW), which are stabilized by a laser power stabilizer to within 0.04% were used as excitation sources. Fluorescein and Coumarin 343 were used as the fluorophores. Figure 7.105 shows the separation of a sugar mixture. The buffer used was 10 mM sodium bicarbonate of pH 11.5. A detection limit of 2 fmol ($S/N = 3$) was reported for fructose.

Capillaries filled with agarose gels and linear polyacrylamide were used for CE separation of unsaturated sulfonated disaccharides [203]. Figure 7.106 illustrate the electropherograms obtained.

Separation of reducing carbohydrates as 1-phenyl-3-methyl-5-pyrazolone (PMP) derivatives was performed by CZE [342] with on-column UV detection at 245 nm. Electrophoretic buffers containing alkaline earth metal salts were used. A separation voltage of 10 kV at reversed polarity was used. Figure 7.107 illustrates the separation of pentose-PMPs in aqueous solutions of calcium acetate, barium acetate and strontium acetate.

7.12 FOOD ANALYSIS

Some aspects of food analysis have already been covered in other sections. In particular, the analysis of amino acids, proteins, vitamins, organic acids and carbohydrates have been described in Sections 7.1, 7.3, 7.5, 7.10 and 7.11, respectively.

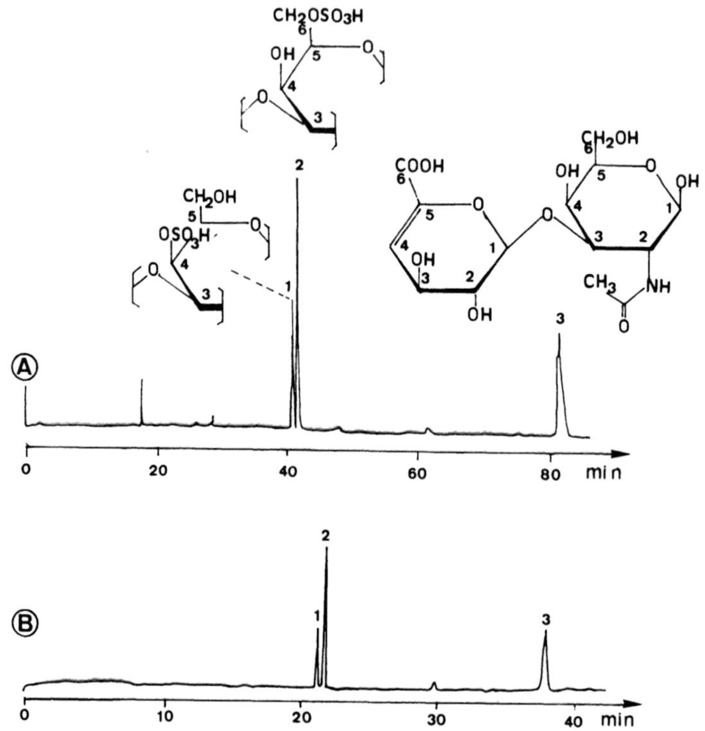

Fig. 7.106. Separation of unsaturated disaccharides on (A) agarose gel (2.5% in 0.01 M disodium hydrogen phosphate + 0.01 M sodium dihydrogen phosphate buffer; 75 cm effective length, 75 μm I.D. MOT-AG-19; ambient temperature; injection at 20000 V for 10 s; 100 V/cm separation voltage; detection by UV at 232 nm), and (B) linear polyacrylamide gel (8% T in 0.1 M Tris + 0.25 M boric acid buffer; 41 cm effective length, 100 μm I.D.; 25°C; injection at 9000 V for 5 s; 170 V/cm separation voltage; detection by UV at 250 nm): *1* = ΔDi 4S; *2* = ΔDi 6S; *3* = ΔDi 0S. (Reproduced with permission from Ref. 203. Copyright Friedr. Vieweg & Sohn Verlagsgesellschaft mbH.)

Additional examples of the use of CE in food analysis include the analysis of caffeine, food additives, amine-containing compounds in red wine, bitter acids in hop extract, milk proteins, organic acids in food samples, and caramel color.

The separation of water-soluble vitamin and caffein by MEKC with SDS and photodiode detection has been demonstrated by Kobayashi *et al.* [308]. A mixture of food additives were separated by CE with on-column UV detection at 190 nm [330]. A phosphate–borate buffer containing 10 mM tetrabutylammonium hydrogen sulfate (TBA) was used as the electrophoretic medium. The method was applied to the determination of food additives in soy sauces.

A post-capillary fluorescence detection scheme using o-phthaldialdehyde as the fluorescent label has been described [331]. The method was employed for the separation and detection of amine-containing compounds in red wine. No sample

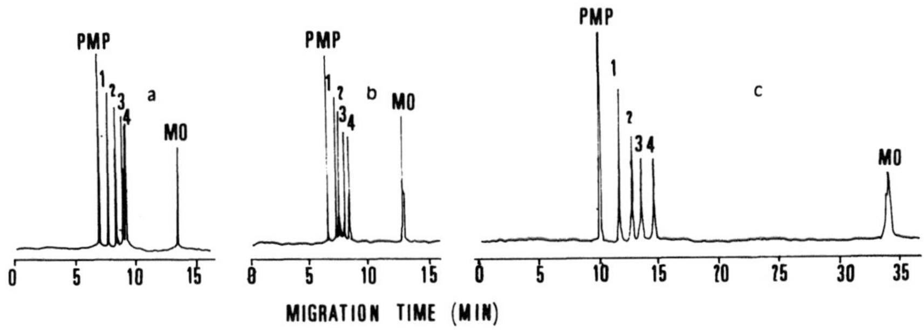

Fig. 7.107. Separation of pentose-PMPs in aqueous 100 mM solutions of (a) calcium acetate, (b) barium acetate and (c) stronium acetate. Capillary fused silica (53 cm × 50 μm I.D.). An aqueous sample solution was introduced from the cathodic end of the tube. Peaks: *1* = ribose-PMP; *2* = lyxose-PMP; *3* = arabinose-PMP; *4* = xylose-PMP. (Reproduced with permission from Ref. 342. Copyright Elsevier Science Publishers.)

preparation was required. The sample was simply diluted four times before injection into the CE capillary. Because of the high selectivity and sensitivity of the detection scheme, the quantification of the minor components, histamine, in a complex sample could be performed.

The analysis of hop bitter acids in hop extract is important in the quality control of the taste of beer. Due to the use of fused silica capillaries, CE possesses the unique advantage that the problem of complexation of the bitter acids with metal ions which affects the accuracy of quantitation could be eliminated. Separation of hop bitter acids by MEKC with UV-diode array detection was performed [332]. Figure 7.108 shows the electrokinetic chromatogram of hop extract and the on-line UV-spectra of the hop acids. The analysis of beer iso-α-acids by MEKC with multi-wavelength UV detection was also performed [333]. Figure 7.109 shows the chromatogram obtained for a chloroform extract of hopped beer.

The separation and analysis of milk proteins is important because protein composition affects the nutritional quality and physical properties of milk-based foods. The major components of whey (α-lactalbumin, β-lactoglobulins) were resolved using pH 8.5 sodium borate buffer [333a].

Capillary isotachophoresis [334] was used for the determination of organic acids in food samples (e.g. fermented cabbage and red wines). Figure 7.110 illustrates the organic acids contents in fermented cabbage, which include lactic, acetic, phosphoric and ascorbic acids. The contents of organic acids in a decanted wine during maturation were also determined.

The determination of 4-methylimidazole in caramel color by capillary isotachophoresis was studied [335]. The method required no pretreatment of the sample and detection limit of 5 ppm was obtained. A PTFE separation capillary (180 mm × 0.3 mm I.D.) was coupled with a PTFE pre-separation capillary (180 mm × 0.8 mm I.D.). Detection was by conductivity measurement. The leading electrolyte was 5

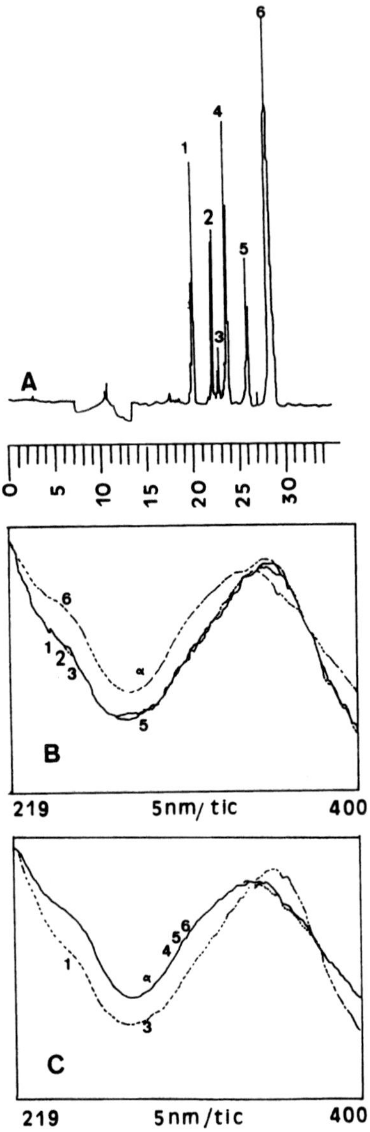

Fig. 7.108. (A) MEKC-analysis of hop extract (25 mM TRIS/HAc + 25 mM SDS, pH 9.0, 10 kV), UV 220 nm. (B) On-line UV-spectra of the β-acids (*1–3*), compared with an α-acid (*6*). (C) On-line UV-spectra of the α-acids (*4–6*), compared with a β-acid (*1*). (Reproduced with permission from Ref. 332. Copyright Dr. Alfred Huethig Publishers.)

mM KOH, 20 mM N-2-hydroxyethylpiperazine-N′-2-ethane sulfonic acid (HEPES), 0.1% polyvinylpyrrolidone (PVP) in the pre-separation capillary, and 1 mM KOH, 4 mM HEPES, 0.1% PVP in the separation capillary. The terminating electrolyte was a solution of 10 mM Tris and 5 mM acetic acid.

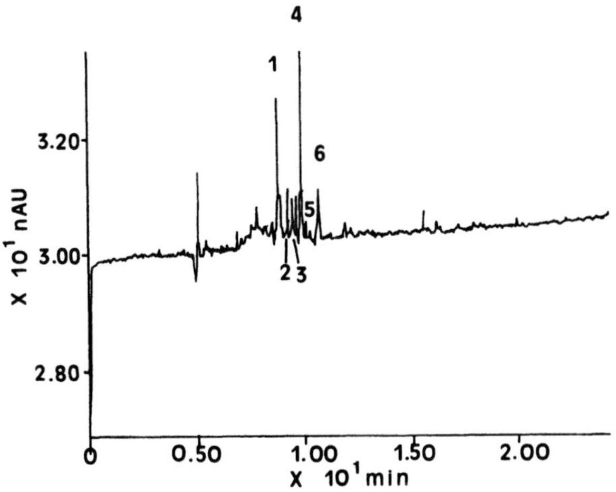

Fig. 7.109. MEKC analysis of a chloroform extract of hopped beer. Capillary: 60 cm × 75 μm I.D.; detection: 254 nm; buffer: 30 mM phosphate (pH 7.6) containing 40 mM sodium dodecyl sulfate. (Reproduced with permission from Ref. 333. Copyright Dr. Alfred Huethig Publishers)

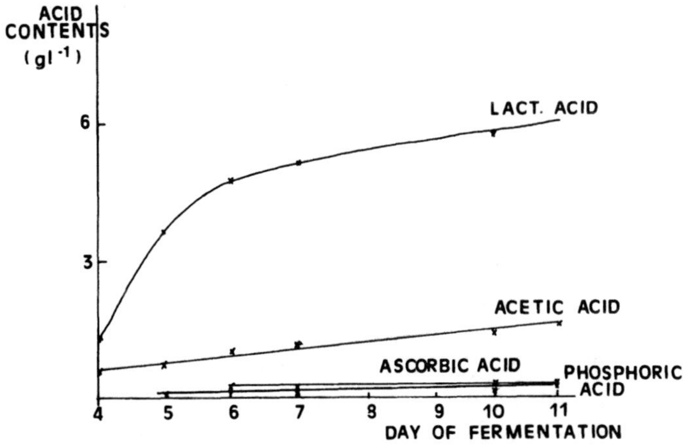

Fig. 7.110. Organic acids in fermented cabbage. *LACT* = lactic. Leading electrolyte: 0.02 M hydrochloric acid, counter ion: β-alanine, pH 3.8, 0.1% polyvinylpyrrolidone; terminating electrolyte: 5 mM caproic acid. (Reproduced with permission from Ref. 334. Copyright Elsevier Science Publishers.)

7.13 ENVIRONMENTAL ANALYSIS

CE has been used in several studies for the analysis of environmental pollutants [70,316,346–352]. The types of compounds investigated include phenols and polycyclic aromatic hydrocarbons listed by the United States Environmental

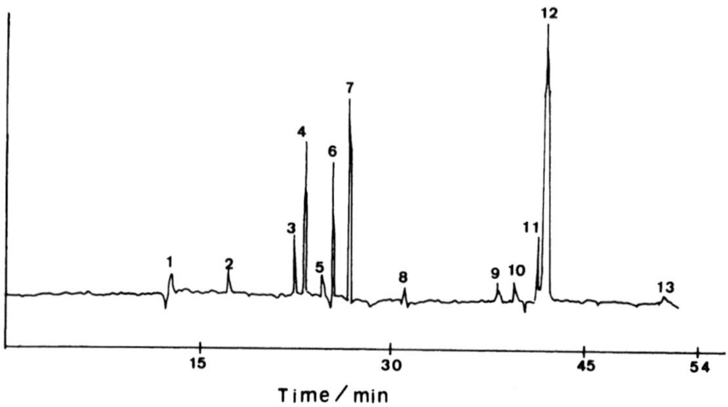

Fig. 7.111. Electrokinetic chromatogram of the eleven phenols: *1* = methanol; *2* = phenol; *3* = 4-NP; *4* = 2-NP; *5* = 2-CP; *6* = DNOC; *7* = 2,4-DNP; *8* = 2,4-DMP; *9* = 4C, 3-MP; *10* = 2,4-DCP; *11* = 2,4,6-TCP; *12* = PCP; *13* = Sudan III. Electrophoretic solution: 0.05 M SDS in 0.1 M borate–0.05 M phosphate buffer, pH 6.5; separation tube: 60 cm × 50 μm I.D. fused silica capillary; voltage: −15 kV; current: 20 μA; detection wavelength: 254 nm. Abbreviations: 4-NP = 4-nitrophenol, 2-NP = 2-nitrophenol, 2-CP = 2-chlorophenol, DNOL = 2-methyl-4,6-dinitrophenol, 2,4-DNP = 2,4-dinitrophenol, 2,4-DMP = 2,4-dimethylphenol, 4C,3-MP = 4-chloro-3-methylphenol, 2,4-DCP = 2,4-dichlorophenol, 2,4,6-TCP = 2,4,6-trichlorophenol, PCP = pentachlorophenol. (Reproduced with permission from Ref. 346. Copyright Kluwer Academic Publishers.)

Protection Agency (USEPA) as priority pollutants, nitroaromatics, chlorophenols and pesticides.

The separation of eleven substituted phenols listed by USEPA as priority environmental pollutants by MEKC was performed [316,346]. The electrophoretic solution used was 0.05 M SDS in 0.1 M borate/0.05 M phosphate buffer (pH 6.5). Figure 7.111 shows the electropherogram of the eleven phenols.

Cyclodextrin modified MEKC has been used for the separation of polycyclic aromatic hydrocarbons [307,350]. For the analysis of PAH of environmental concern, MEKC with γ-cyclodextrin provided excellent selectivity.

CZE was used for the separation of chlorophenols in industrial waste water [347]. On-column electrochemical detection was performed. A small-diameter capillary (25 μm) was used to enhance sensitivity. 10μm diameter carbon fibres were used as electrodes. The analysis of an industrial waster water sample is illustrated in Fig. 7.112.

The separation of nitroaromatic compounds by CE has been demonstrated using micellar electrokinetic chromatography [349]. A solution containing a phosphate–borate buffer (pH 7.04) was used as the electrophoretic medium. The migration behaviour of the nitroaromatics at different pH and SDS concentration was studied.

Lee *et al.* [70] performed CZE with mass spectrometric detection in the selected ion monitoring (SIM) mode. A liquid junction coupling was used, which served to

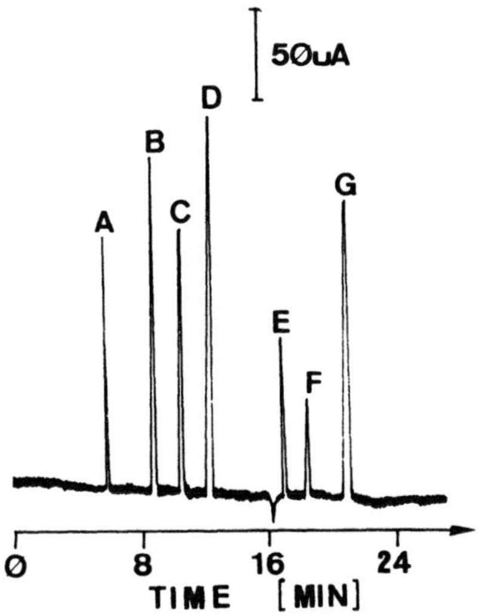

Fig. 7.112. Electropherogram of an industrial waste water sample with a 2-chlorophenol concentration of 100 ng/ml. Conditions — capillary: 65 cm (35 cm to detector), 25 μm I.D., buffer: phosphate (0.045 M)/borate (0.015 M) mixed buffer at pH 8.0; applied voltage: 20 kV; carbon fibre: 1.4 V Vs SCE. (Reproduced with permission from Ref. 347. Copyright Elsevier Science Publishers.)

compensate for the different eluent flow required by the ion-spray interface and the CZE column. The analysis of acid pesticides was demonstrated.

Capillary isotachophoresis was used to determine the pyrethroid insecticides, alphamethrine and cypermethrine, in water and soil samples [352]. The leading electrolyte contained 0.01 mol/l HCl, creatinine and 0.05% polyvinyl alcohol at pH 4.80. The terminating electrolyte was 0.003 mol/l morpholino ethane sulphoric acid (MES). Pesticides from water was extracted by double extraction with diethyl ether. For the extraction of soil samples, acetonitrile : water (4 : 1) was used. Figure 7.113 shows the determination of alphamethrine in water and soil samples.

7.14 POLYMER AND PARTICLE ANALYSIS

CE has been used for the separation of polystyrene particles [353–355], poly(oxalkylene)diamine polymers [356], surfactants [357], silica sols [358], accelerators [359] and plasticizers [360] used in the manufacture of rubber and plastic products.

The separation of polystyrene nanospheres by CE has been investigated [353,354]. Rapid, sensitive resolution of polystyrene particles ranging from 39 to 683 nm in diameter could be obtained. The fused silica capillary was pretreated with a 0.5 mM solution of cetyltrimethylammonium bromide (CTAB) before use. A 1 mM ACES buffer (N-[2-acetamido]-2-aminoethane sulfonic acid) at pH 5.8 was used as the electrolyte. Detection was by UV absorbance at either 254 nm or 225 nm.

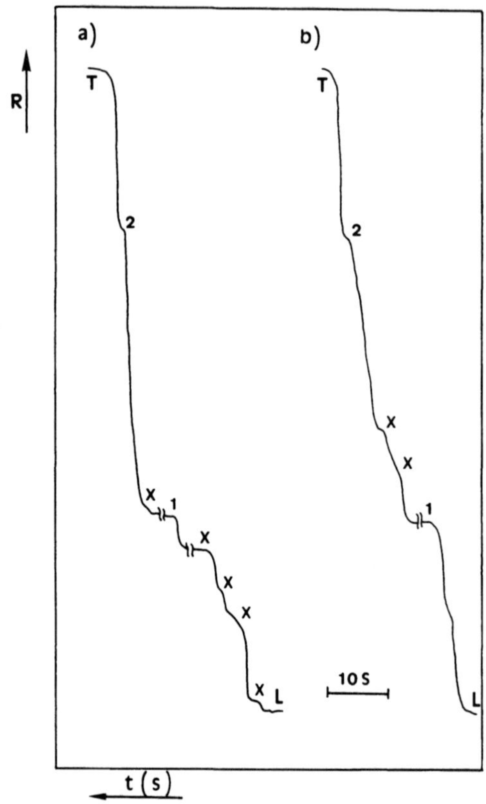

Fig. 7.113. Determination of alphamethrine (a) in water (0.025 mg/l), and (b) in soil (4 mg/kg). One-capillary array (150 × 0.3 mm I.D.). I = 30 μA; volume injected: 30 μl; chart speed: 1 mm/s. *1* = PO_4; *2* = cis-dichlorochrysanthemic acid; × = unidentified zones. (Reproduced with permission from Ref. 352. Copyright Elsevier Science Publishers.)

Separation was found to be based upon particle size, with the smallest particles migrating the fastest through the capillary.

Polymer latex particles with different numbers of attached carboxylate or sulfate functional groups were separated by CE [355]. For the CE separation, a 75 μm I.D., 40 cm effective length (55 cm total length) capillary column was used. The applied voltage was 30 kV. Electropherogram of the seven-component latex mixture at pH 10.71 is shown in Fig. 7.114. The phosphate buffer used for this separation consisted of 5 mM Na_2HPO_4 and 1 mM NaOH.

The characterization by CE of poly(oxyalkylene)diamine (Jeffamine) polymers, ranging in molecular mass from 600 to 2000 Da has been investigated [356]. A fluorescent label, 2,3-naphthalene dialdehyde (NDA), was used to perform pre-column derivation for fluorescence detection of the oligomers prior to separation. Excitation wavelength was 442 nm and emission wavelength was 475 nm. Electropherogram

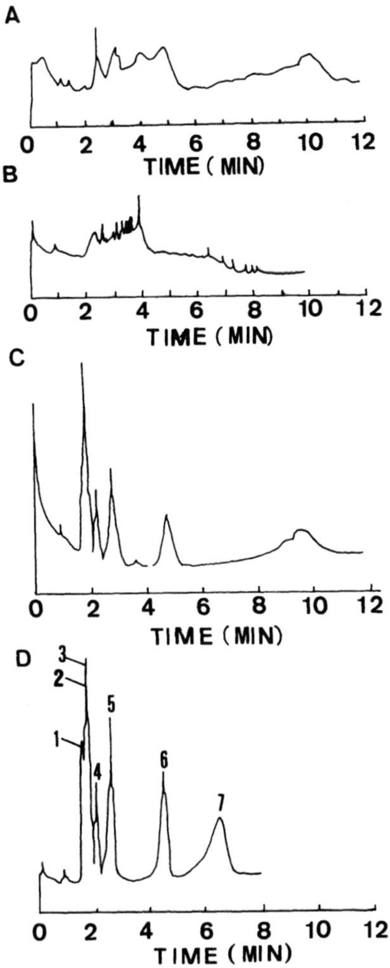

Fig. 7.114. Electropherograms of seven-component latex mixture at four pH values: (A) 6.64, (B) 7.21, (C) 8.49, (D) 10.71. Peaks *1, 2, 3, 4, 5, 6* and *7* correspond to particles of 0.030-, 0.079-, 0.070-, 0.100-, 0.200-, 0.500-, and 1.16-μm diameter, respectively. Phosphate buffers: pH 6.64 (2.5 mM NaOH, 7.5 mM KH_2PO_4), pH 7.21 (4.5 mM NaOH, 7.5 mM KH_2PO_4), pH 8.49 (3.0 mM NaOH, 3.75 mM KH_2PO_4), pH 10.71 (1.0 mM NaOH, 5.0 Na_2HPO_4). Capillary: 55 cm (40 cm to detection), 75 μm I.D.; voltage: 30 kV; detector: 254 nm (190 nm for neutral marker, acetone). (Reproduced with permission from Ref. 355. Copyright American Chemical Society.)

showing oligomer separation of Jeffamine ED 600 and ED 2001 are shown in Figs. 7.115 and 7.116, respectively.

The separation of a surfactant containing both anion and non-ionic components (Triton 770) by CE and MEKC has been investigated [357]. Increasing ionic strength was found to increase resolution.

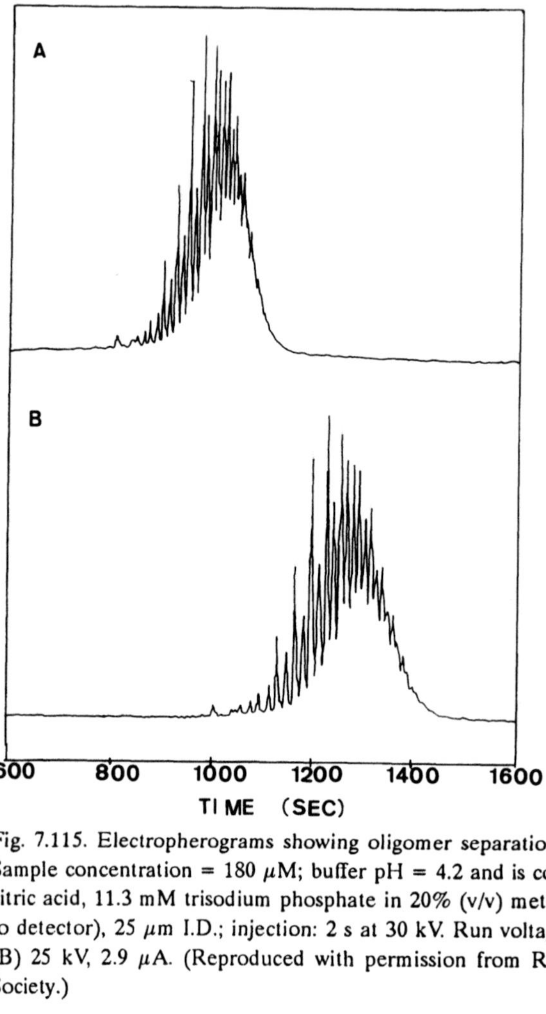

Fig. 7.115. Electropherograms showing oligomer separation of Jeffamine ED 600 by CZE/LIFD. Sample concentration = 180 μM; buffer pH = 4.2 and is composed of 77 mM boric acid, 19.4 mM citric acid, 11.3 mM trisodium phosphate in 20% (v/v) methanol solution. Capillary: 75 cm (60 cm to detector), 25 μm I.D.; injection: 2 s at 30 kV. Run voltages were as follows: (A) 30 kV, 3.1 μA; (B) 25 kV, 2.9 μA. (Reproduced with permission from Ref. 356. Copyright American Chemical Society.)

Capillary zone electrophoresis has been used to separate silica sols ranging from 5 to 500 nm [358]. Fused silica capillary of 50 μm I.D. were used. Detection was by on-column turbidity measurements at 190 nm. Figure 7.117 shows the separation of the silica sols at different buffer concentrations. The ionic strength of the buffer was

---

Fig. 7.117. Effect of buffer concentration on elution time of silica sols in CZE. (A) 0.5 × buffer (2.5 mM NH_4OH, 4.65 mM NH_4Cl, pH 9.0). (B) 2 × buffer (10 mM NH_4OH, 18.6 mM NH_4Cl, pH 9.0). Separations were done in a 101 cm length (61 cm anode section before the detector) of 50 μm I.D. fused silica capillary. Sample was injected hydrostatically for 5–10 s at 15 cm head pressurre. Detection was by light-scattering at 190 nm. Separation voltage was 0 to 30 kV linear ramp in 30 s; then held at 30 kV. (Reproduced with permission from Ref. 358. Copyright Marcel Dekker Inc.)

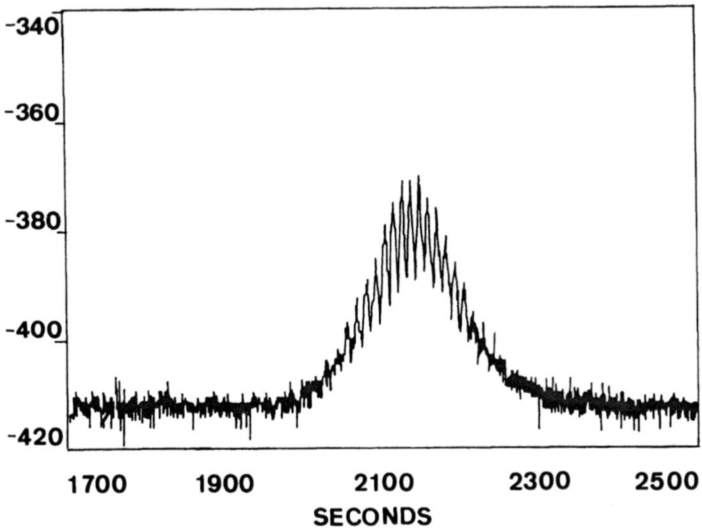

Fig. 7.116. Electropherogram showing oligomer separation of EDJeffamine ED 2001 by CZE/LIFD. CZE parameters: sample concentration = 180 μm; buffer pH = 4.2 and is composed of 77 mM boric acid, 19.4 mM citric acid, 11.3 mM trisodium phosphate in 20% (v/v) methanl solution. Capillary: 100 cm (85 cm to detector), 25 μm I.D.; injection: 3 s at 30 kV; run, 25 kV; current: 1.8 μA. (Reproduced with permission from Ref. 356. Copyright American Chemical Society.)

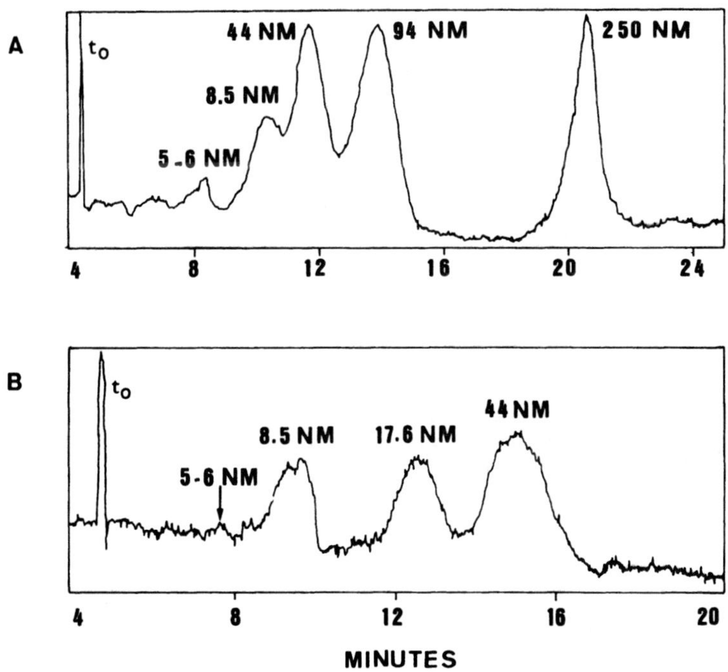

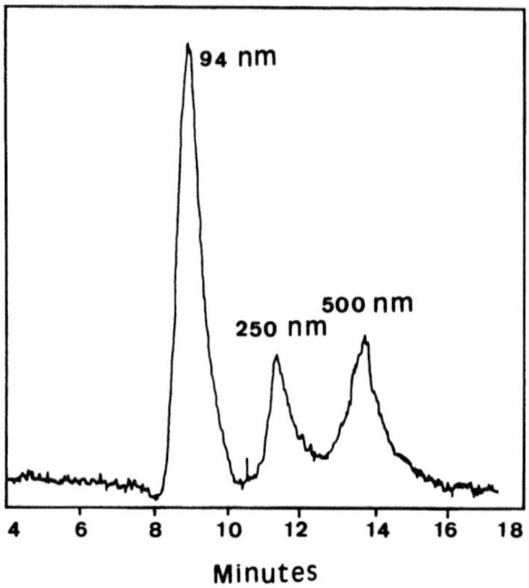

Fig. 7.118. CZE separation of colloidal silica sols of larger particle diameters. 0.1 × buffer — 0.5 mM NH_4OH, 0.93 mM NH_4Cl, pH 9.0. Separations were done in a 100 cm length of 50 μm I.D. fused silica with 60 cm anode section located before the detector. Injection was hydrostatic at 15 cm head pressure for 5 s. Other conditions same as Fig. 7.117. (Reproduced with permission from Ref. 358. Copyright Marcel Dekker Inc.)

found to have a significant effect on the resolution and electrophoretic mobilities of the silica sols. High ionic strength resulted in better resolution of smaller silica sols, but greatly increased the elution times of the larger silica sols. For the separation of larger particles, a buffer of lower ionic strength was used, as shown in Fig. 7.118.

Benzothiazole sulfenamides are commonly used as accelerators in the manufacturing of rubber products [359]. These compounds tend to have poor stability in polar solvents. MEKC with methanol and urea as buffer modifiers was employed for the separation of the benzothiazole sulfenamides. The separation of N-t-butyl-2-benzothiazole sulfenamide (TBBS) and N-cyclohexyl-2-benzothiazole sulfenamide (CBS) was demonstrated.

The analysis of polymer additives by CE has also been investigated. An example is the analysis of phthalates by MEKC [360]. Phthalates are widely used as plasticizers in the formulation of polymers. The separation of the phthalates is shown in Fig. 7.119. Sodium dodecyl sulfate in borate–phosphate buffer (pH 6.0) was used as the electrophoretic medium.

7.15 NATURAL PRODUCTS

Some aspects concerning the isolation of specific compounds from natural products have been discussed in other sections, e.g. drugs and pharmaceuticals. In

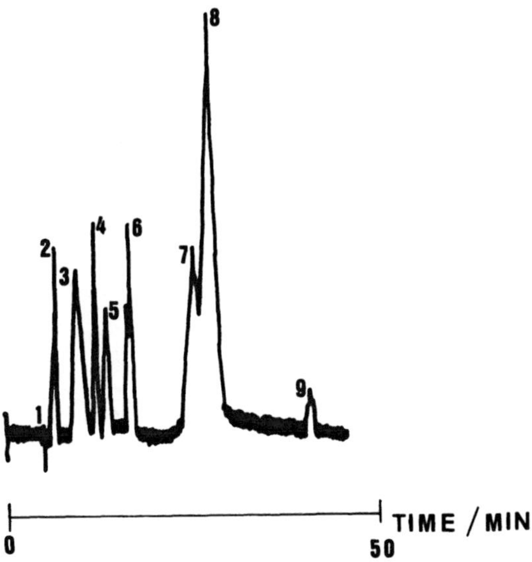

Fig. 7.119. Electrokinetic chromatogram of the six phthalate esters: *1* = methanol; *2* = dimethylphthalate (DMP); *3* = diethylphthalate (DE); *4* = diallylphthalate (DAP); *5* = benzyl-n-butylphthalate (BBP); *6* = dibutylphthalate (DBP); *7* = bis(2-ethylhexyl)phthalate (BEHP); *8* = Sudan III. Electrophoretic solution: 10 mM SDS in 0.1 borate–0.05 M phosphate buffer, pH, 6.0; separation tube: 50 cm × 50 μm I.D. fused silica capillary; voltage: 25 kV; detection wavelength: 210 nm; volume of sample injected: 1.5 nl. (Reproduced with permission from Ref. 360. Copyright Elsevier Science Publishers.)

this section, the emphasis is on the methodologies employed for the analysis of natural products.

CZE was used to determine arbutin (hydroquinone β-D-monogluco pyranoside) in *Uvae ursi folium* (bearberry) leaves [361]. The aqueous extract of the crude drug acts as a urinary disinfectant. Typical electropherogram of an aqueous extract of the leaves is shown in Fig. 7.120. A sodium borate buffer (0.1 mol/l, pH 9.5) was used as the electrophoretic medium.

The methanol extract of *Paeonia radix*, which is commonly used as a sedative, lenitive or antispasmolytic agent, was analyzed by capillary zone electrophoresis in a 100 mM borate buffer (pH 10.5) [224]. The analysis of the methanol extract of *Paeonia radix* is shown in Fig. 7.121. Micellar electrokinetic capillary chromatography was used for the determination of flavonol-3-O-glycosides [251]. A 20 mM sodium borate buffer (pH 8.3) with 50 mM SDS was used for the electrophoretic separation. Figure 7.122 shows the MEKC separation of a standardized extract of *Ginkgo biloba* leaves.

The use of capillary isotachophoresis for the analysis of flavonoids and phenol carboxylic acids has been investigated [362]. These compounds are of phytopharmaceutical importance. The isotachopherogram of a methanolic extract of Sambuci

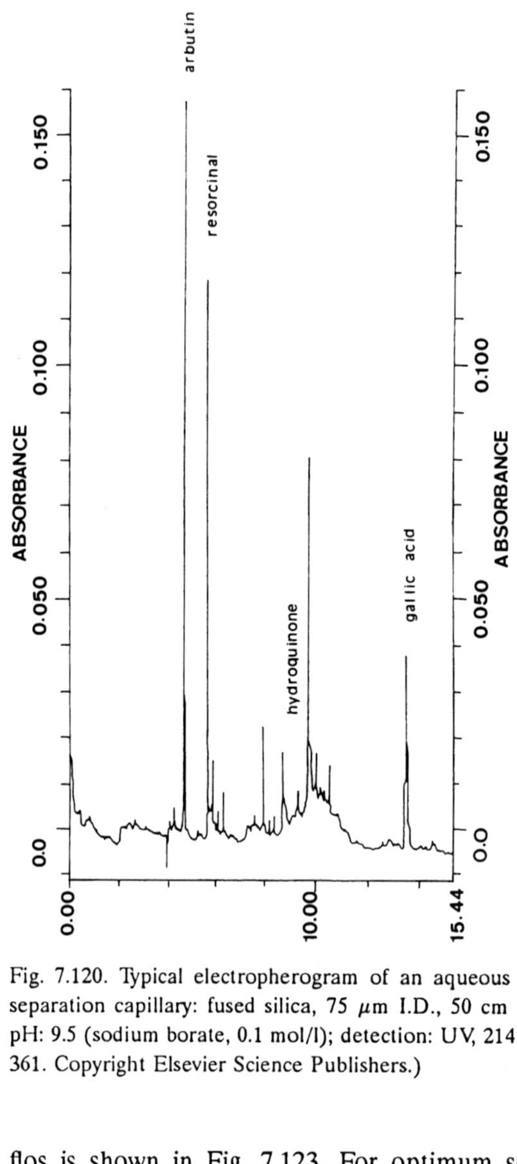

Fig. 7.120. Typical electropherogram of an aqueous extract of Uvae-ursi folium. Conditions — separation capillary: fused silica, 75 μm I.D., 50 cm length to detecor; 350 V/cm, 70 μA; buffer pH: 9.5 (sodium borate, 0.1 mol/l); detection: UV, 214 nm. (Reproduced with permission from Ref. 361. Copyright Elsevier Science Publishers.)

flos is shown in Fig. 7.123. For optimum separation, a leading electrolyte (pH 9.5) containing 15 mM hydrochloric acid, 30% methanol and 0.2% hydroxypropyl methylcellulose was used. Separation was performed using 250 mm long PTFE capillaries of 0.5 mm I.D.

7.16 CHIRAL SEPARATION

A review on the application of CE to the separation of chiral compounds has been given [363]. The approaches used for the separation of chiral compounds by

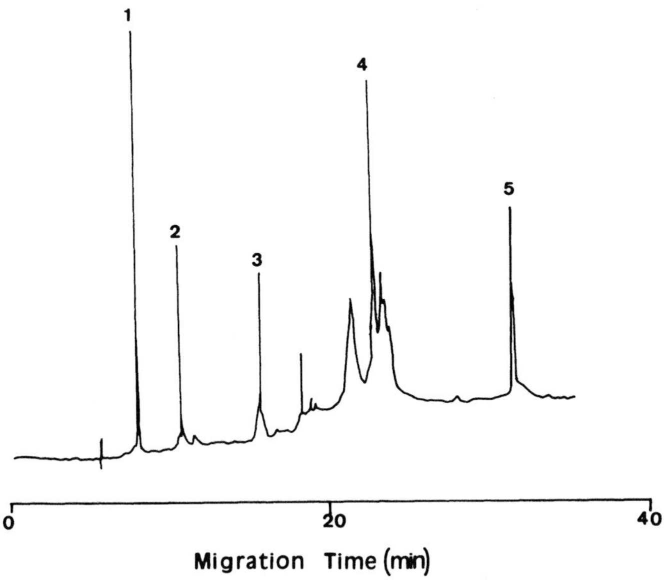

Fig. 7.121. Typical electropherogram of the methanol extract of *Paeonia radix*. Capillary: fused silica (80 cm × 50 μm I.D.); carrier: 100 mM borate buffer (pH 10.5); applied voltage: 20 kV; detection: UV, 254 nm. Peaks: *1* = paeoniflorin; *2* = oxypaeoniflorin; *3* = methyl gallate; *4* = tannic acid; *5* = gallic acid. (Reproduced with permission from Ref. 224. Copyright Elsevier Science Publishers.)

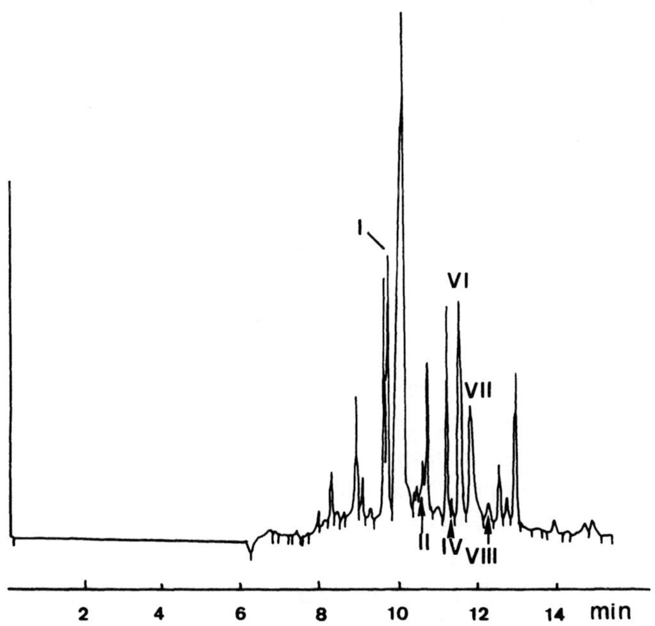

Fig. 7.122. MECC separation of a standardized extract of *Ginkgo biloba* leaves. Conditions — capillary: 72 cm × 50 μm I.D.; voltage: 277 V/cm; buffer: 50 mM SDS–20 mM borate (pH 8.3); injection: 1 s aspiration; detection: 260 nm. Peaks: *I* = Royin; *II* = isoquercitrin; *III* = hyperosid; *IV* = quercitrin; *V* = avicularin; *VI* = kaempferol-3-rutinoside; *VII* = isorhamnetin-3-O-rutinoside; *VIII* = astragalin; *IX* = isorhamnetin-3-O-glucoside. (Reproduced with permission from Ref. 251. Copyright Elsevier Science Publishers.)

Fig. 7.123 Isotachophoretic analysis of a methanolic extract of Sambuci flos. Leading electrolyte: 15 mM hydrochloric acid-ammediol–0.2% HPMC in methanol–water (30 : 70, v/v), pH 9.5; terminating electrolyte: 10 mM glycine–barium hydroxide–0.2% hydroxypropylmethylcellulose (HPMC) in methanol–water (30 : 70, v/v), pH 10.6. Detection: conductivity (*COND*) and differential conductivity (*DIFF*). Current: 60 μA; chart speed: 10 mm/min; injection volume: 3 μl. Zone identification: R = run. (Reproduced with permission from Ref. 362. Copyright Elsevier Science Publishers.)

CE include the use of chiral surfactants, bile salts, or cyclodextrin modifiers in MEKC, the addition of chiral complexing agents or L- and D-Marfey's reagent into the electrophoretic buffer, and the incorporation of a complexing agent into a gel-filled column.

Enantiomeric separation of phenylthiohydantoin (PTH)-DL-amino acids has been achieved by MEKC with sodium dodecanoyl-L-valinate (SDVal) surfactant [367]. Addition of methanol improved resolution although peak tailing was observed. The addition of urea improved peak shapes. Figure 7.124 shows the MEKC separation of PTH derivatives of several DL-amino acids.

The use of MEKC with mixed micellar (SDS and SDVal) solutions containing methanol and urea for the separation of PTH amino acids was performed [365]. The chiral separation of six PTH-DL-amino acids by this method is shown in Fig. 7.125. Separation of other chiral compounds was also investigated with this electrophoretic system. The resolution of enantiomers of 2-hydroxy-1, 2-diphenylethanone (benzoin)

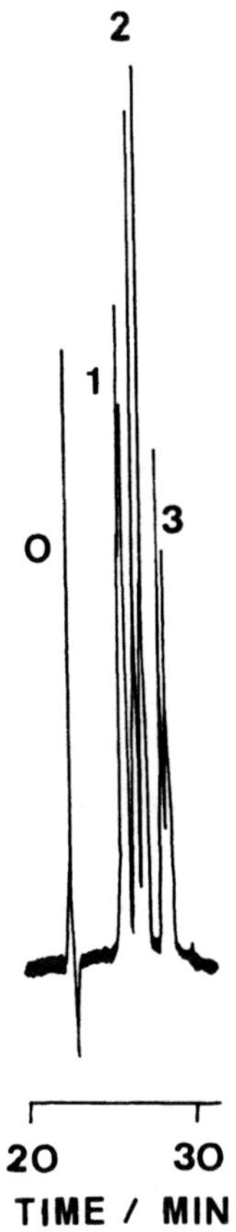

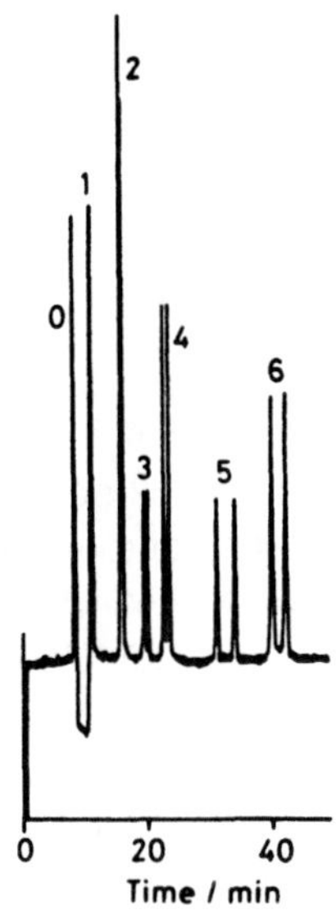

Fig. 7.124 (left). Micellar electrokinetic chromatogram of PTH derivatives of three DL-amino acids: Peaks (corresponding amino acids). *O* = acetonitrite, *1* = Nva, *2* = Trp, *3* = Nle. Micellar solution: 25 mM SDVal, pH 7.0, containing 10% methanol and 5 M urea; current, 7.3 μA. Capillary: 65 cm × 50 μm I.D.; voltage: 15 kV (10 μA); detection: 260 nm. (Reproduced with permission from Ref. 364. Copyright Dr. Alfred Huethig Publishers.)

Fig. 7.125 (right). Chiral separation of six PTH-DL-AA by MEKC. Corresponding amino acids: *1* = Ser, *2* = Aba; *3* = Nva; *4* = Val; *5* = Trp; *6* = Nle. *O* = acetonitrile. Micellar solution: 50 mM SDVal–30 mM SDS–0.5 M urea (pH 9.0) containing 10% (v/v) methanol; separation column: 650 mm × 0.05 mm I.D.; length of the tube used for separation: 500 mm; total applied voltage: 20 kV; current: 17 μA; detection wavelength: 260 nm; temperature, ambient. (Reproduced with permission from Ref. 365. Copyright Elsevier Science Publishers.)

and 4-hydroxy-3-(3-oxo-1-phenylbutyl)-2H-1 benzopyran-1-one (warfarin) is shown in Fig. 7.126.

MEKC with bile salt surfactants was used to separate optical isomeric drugs [234,364,366]. Enantiomers of diltiazem hydrochloride, trimetoquinol hydrochloride, carboline derivative A and B, 2, 2′-dihydroxy-1, 1′-dinaphthyl and 2, 2, 2-trifluoro-1-(9-anthryl)ethanol were separated [234]. The bile salts investigated included sodium

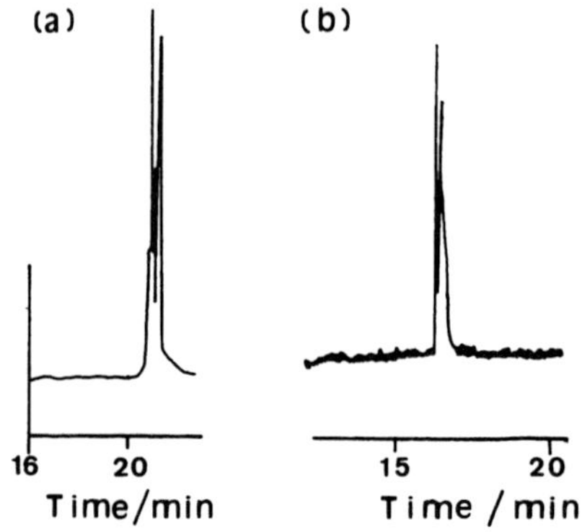

Fig. 7.126. Optical resolution of (a) benzoin and (b) warfarin by MEKC. Current: (a) 20 μA and (b) 19 μA; other conditions as in Fig. 7.125. (Reproduced with permission from Ref. 365. Copyright Elsevier Science Publishers.)

cholate, sodium taurocholate, sodium deoxycholate and sodium taurodeoxycholate. For the separation of a bronchodilater (Inolin) as its hydrochloride and three related compounds [366], a 0.05 M sodium taurodeoxycholate solution of pH 7.0 was found to provide optimum separation. MEKC with bile salts and organic modifiers, e.g. methanol, was used for chiral separation [364] of binaphthyl enantiomers, including 1,1′bi-2-naphthol,1,1′ binaphthyl dicarboxylic acid, biphenanthrene dihydroxide and 1,1′ binaphthyldiyl hydrogen phosphate.

Cyclodextrins have been used as modifiers in MEKC for the separation of chiral compounds [369]. Among the CDs, γ-CD was found to be the most effective for this purpose. The addition of organic solvent or a chiral compound such as sodium d-camphor-10-sulfonate or l-menthoxyacetic acid to the SDS micelle solution containing CD was found to enhance the enantioselectivity. The separation of chiral drugs, such as thiopental, pentabarbital, dinaphthyl and anthrylethanol was performed. Separation of enantiomers of ephedrine, norephedrine, epinephrine, norepinephrine and isoprotererenol has also been investigated [236]. These compounds are amines which acts as sympathomimetic drugs. CD were added to the electrolyte as chiral agents. Satisfactory separation was obtained by using low pH and 18 mM heptakis (2,6-di-O-methyl-β- cyclodextrin).

MEKC with SDS surfactant was used to separate isomers of 2,3,4,6-tetra-O-acetyl-β-D-glucopyranosyl isothiocyanate (GITC)-derivatized DL-amino acids [368]. The MEKC separation of 14 GITC derivatized DL-amino acids is shown in Fig. 7.127. The resolution of enantiomers of tryptophan was achieved by using α-cyclodextrin as an additive to the electrophoretic buffer [370]. For the separation of racemic epinephrine, heptakis (2,6-di-O-methyl-β-cyclodextrin) was used as the modifier.

The use of cyclodextrin for the separation of terbutaline and propanolol enantiomers has been investigated by Fanali [371]. Satisfactory separation of the

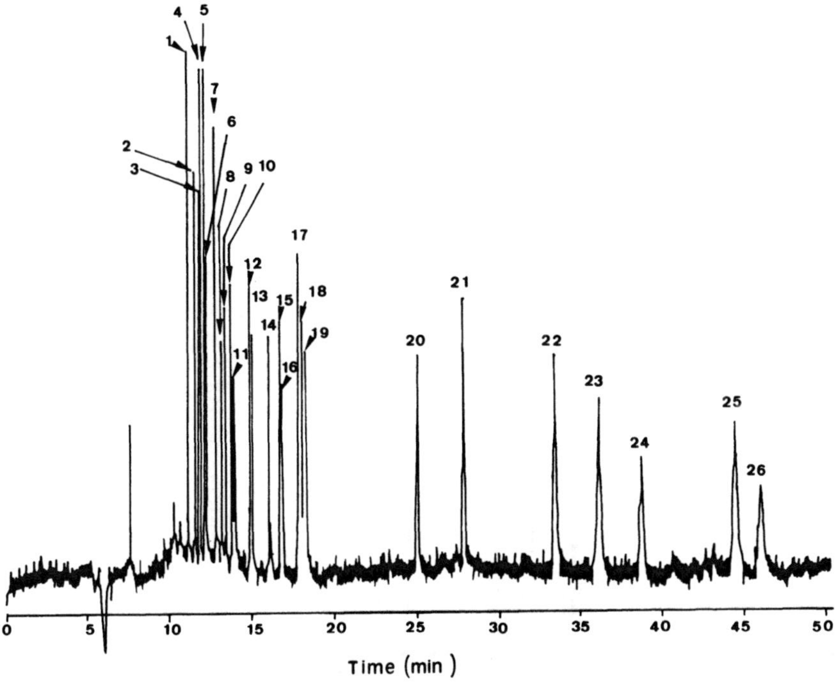

Fig. 7.127. Micellar EKC separation of 14 GITC-derivatized DL-amino acids. Buffer: 0.02 M phosphate buffer solution of pH 9.0 containing 0.25 M SDS; applied voltage: 20 kV. Peaks: *1* = L-Pro + L-Ala; *2* = L-Ala; *3* = D-Pro; *4* = L-His; *5* = D-Ala + L-Val; *6* = D-Thr; *7* = D-His; *8* = L-Tyr; *9* = L-Met; *10* = D-Val; *11* = L-Ile; *12* = L-PheG; *13* = L-Leu; *14* = D-Met; *15* = D-Tyr; *16* = D-Ile; *17* = D-PheG; *18* = L-Phe; *19* = D-Leu; *20* = D-Phe; *21* = L-Trp; *22* = D-Trp; *23* = L-Lys; *24* = L-Arg; *25* = D-Lys; and *26* = D-Arg. (Reproduced with permission from Ref. 368. Copyright Aster Publishing Corp.)

racemic terbutaline was achieve using a phosphate buffer at pH 2.5, containing either 5 mM heptakis (2,6-di-O-methyl-β-cyclodextrin) or 15 mM β-cyclodextrin. For the separation of propanolol enantiomers, a 50 mM phosphate buffer (pH 2.5) which contained 4 M urea, 40 mM β-cyclodextrin and 30% (v/v) methanol provided the best resolution.

Separation of naphthalene-2,3-dicarboxaldehyde (NDA)-labeled amino acid enantiomers was performed by MEKC with cyclodextrin modifiers [372]. NDA reacts with primary amines in the presence of cyanide to form 1-cyano-2-substituted-benz[f]isoindole (CBI) derivatives. Laser-induced fluorescence was performed with a He-Cd laser (442 nm). γ-CD was found to provide better separation than β-CD did for the CBI-DL-amino acids. Figure 7.128 shows an electropherogram of a mixture of five CBI-DL-amino acids obtained with an electrolyte containing 10 mM γ-CD and 50 mM SDS in a 100 mM borate buffer at pH 9.0.

The separation of amino acids enantiomers and peptide isomers has been performed by derivatization with L- and D-Marfey's reagent [66,67]. Both CZE and

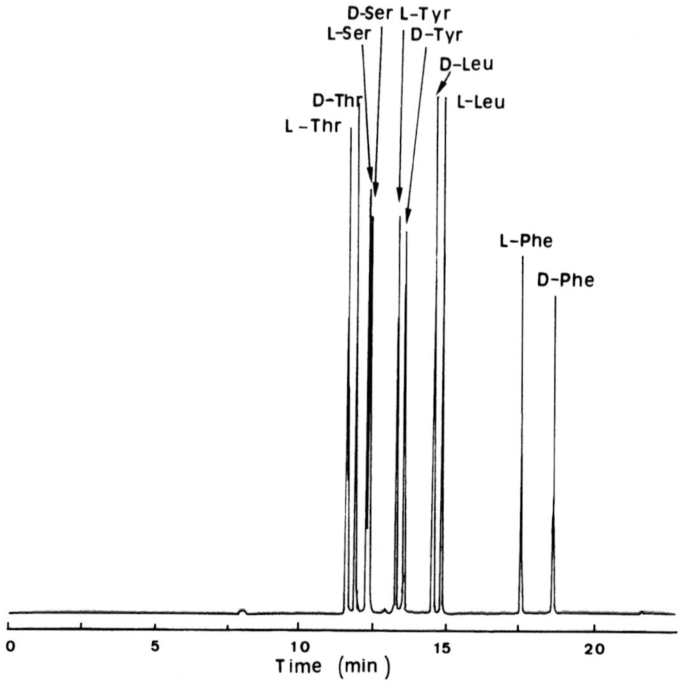

Fig. 7.128. Electropherogram of a mixture of five CBI-DL-amino acids. Electrolyte composition is as follows: 10 mM SDS, and 100 mM borate buffer, pH 9.0. The capillary is 50 μm I.D. (290 μm O.D.), 70 cm in length (50 cm to the detection); the concentration for each CBI-DL-amino acid is 200 nM. The applied voltage is 15 kV and current is 35 μA. (Reproduced with permission from Ref. 372. Copyright American Chemical Society.)

MEKC were employed. DL-dansyl-amino acids were resolved by the diastereomeric interaction between the amino acids and copper(II) complex of L-histidine present in the supporting electrolyte [7]. Resolution of amino acid enantiomers was also achieved by using an electrolyte containing a chiral Cu(II)-aspartame complex [9]. By incorporating a complexing agent (β-cyclodextrin) within a polyacrylamide gel column, chiral separation of dansylated amino acids by CE could be achieved [11].

7.17 SEPARATION OF GEOMETRICAL AND POSITIONAL ISOMERS

Several investigations on the use of CE for the separation of geometrical and positional isomers have been reported [179,373–375].

The effects of experimental parameters on the separation of positional isomers of aminobenzoic acid by CZE have been investigated [373]. Experiments were performed to study the influence of pH, electrolyte, ionic strength, addition of alcohols, the choice of the counter-ion and the temperature on separation efficiency and electroosmotic flow. For the CZE separation of the aminobenzoic acid

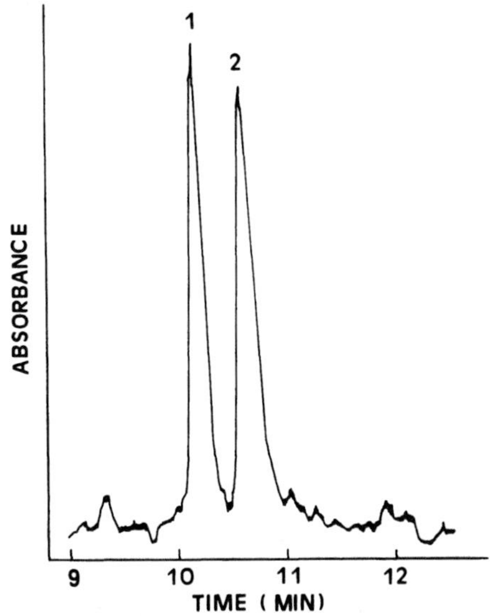

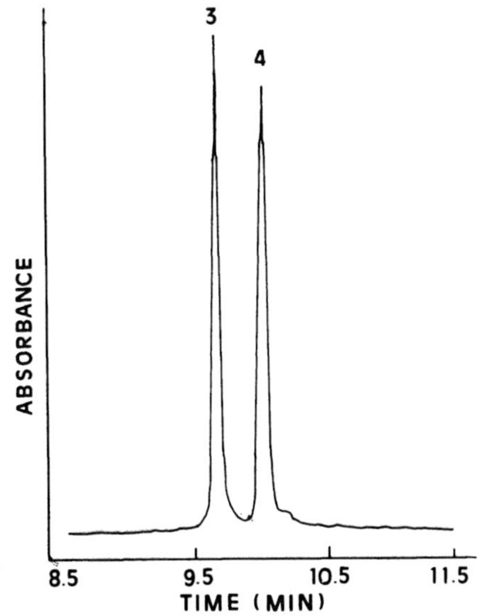

Fig. 7.129 (left). Expanded electropherogram of the mixture of 0.045 mM fumaric acid and 0.076 mM maleic acid. Electrophoretic conditions: 10 mM $Na_2B_4O_7 \cdot 10H_2O$, 50 mM H_3BO_3, pH = 8.5; capillary: 75 μm I.D., 375 μm O.D., 54.8 cm to detector, 60.0 cm total length; +30 kV at 64 μA; 200 nm at 0.002 a.u.f.s. (Reproduced with permission from Ref. 374. Copyright American Chemical Society.)

Fig. 7.130 (right). Expanded electropherogram of the mixture of 0.1 mM all-trans-retinoic acid and 1.2 mM 13-cis-retinoic acid. Electrophoretic conditions: 10 mM $Na_2B_4O_7 \cdot 10H_2O$, 50 mM H_3BO_3 in 50/50 (v/v) CH_3CN/H_2O; capillary: 75 μm I.D., 375 μm O.D., 59.6 cm to detector, 64.8 cm total length; +30 kV at 38 μA; 345 nm at 0.02 a.u.f.s. (Reproduced with permission from Ref. 374. Copyright American Chemical Society)

isomers, pH was found to have the strongest impact on resolution and selectivity. Temperature regulation, use of an appropriate counter ion (potassium in acetate buffer), the addition of alcohols improved resolution, although the addition of alcohol also resulted in an increase in analysis time.

Separation of cis and trans double-bond isomers by CZE was investigated by Chadwick and Hsieh [374]. Double bond isomers, fumaric acid (trans-butenedioic acid) and maleic acid (cis-butenedioic acid), and prototypical retinoid isomers all-trans-retinoic acid and 13-cis-tetinoic acid were separated in free solution CZE. Figures 7.129 and 7.130 illustrate the separations obtained. Since the double-bond isomers which have the larger hydrodynamic radii were expected to migrate with a higher velocity due to reduced attraction by the anodic end, it was deduced that the trans isomers had larger hydrodynamic radii than the cis isomers in both cases.

MEKC using methanol-modified surfactant solutions has been employed for the separation of isotopically substituted compounds [375]. The separation of dansylated methylamine and dansylated methyl-d_3-amine was demonstrated. The electrolyte contained 25 mM SDS and 4.9 M methanol. Rather than using a single marker, the migration time of micelles was determined based upon the capacity factor of a homologous series of amine standards. CZE has also been used for the separation of the isomers of chondroitin sulfate disaccharides and hyaluronan oligosaccharides [371].

7.18 COAL AND FUELS

In addition to the methods described in Section 7.10 for the analysis of organic compounds, there have been reports devoted specifically to the use of CE for the analysis of coal and fuels [376,377].

The use of CZE for the separation of the basic fraction of solvent refined coal has been demonstrated [376]. A mixed water-organic medium containing 0.05 M tetramethylammonium perchlorate, 0.01 M HCl in acetonitrile:water (75:25) was used. Figure 7.131 shows the electropherogram obtained. The potential of capillary zone electrophoresis for the separation of polar and high molecular weight fuel-related materials was investigated by Wright *et al.* [377]. Figure 7.132 shows an electropherogram of the tetrahydrofuran (THF) extract of a diesel fuel sediment. It was noted that the neutral species coeluted whereas negatively charged compounds (such as phenol and acid-substituted moieties) could be resolved by CZE. A heavy petroleum residue sample was also investigated. Positively charged

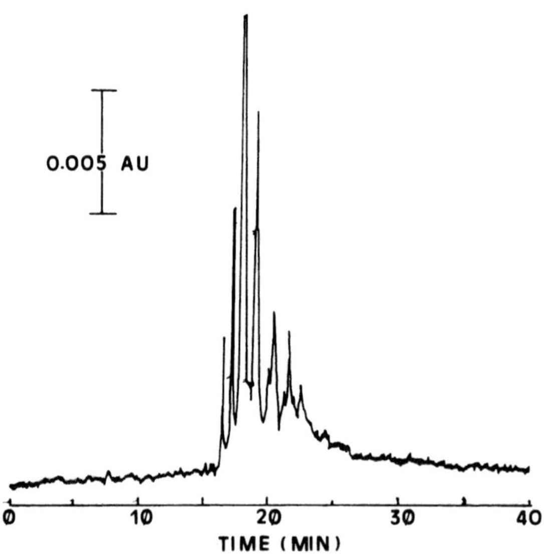

Fig. 7.131. Electropherogram of basic fraction of solvent refined coal in mixed solvent medium. Untreated capillary, 100 cm × 75 μm I.D., filled with 0.05 M tetraethylammonium perchlorate, 0.01 M HCl in acetonitrile water (75:25); applied voltage: 20 kV; detection wavelength: 229 nm. (Reproduced with permission from Ref. 376. Copyright Elsevier Science Publishers.)

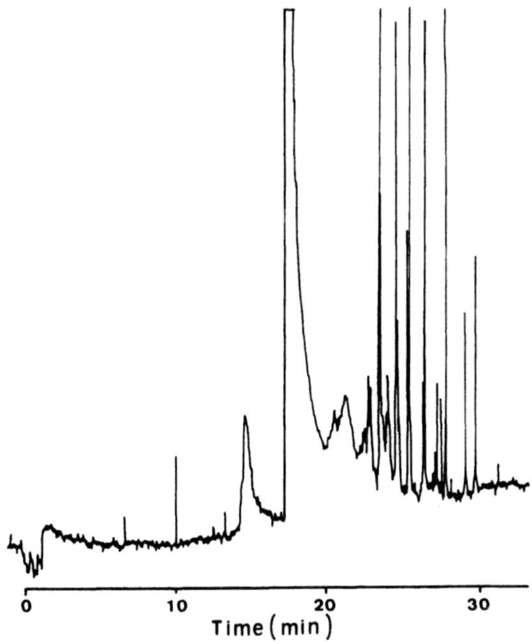

Fig. 7.132. Capillary electropherogram of the THF extract of a diesel fuel sediment. Conditions: +50-kV potential; 10-μA current; buffer solution composed of 5% THF and 95% pH 10.7 phosphate buffer (3×10^{-3} M). (Reproduced with permission from Ref. 377. Copyright American Chemical Society.)

species, coeluting neutral species and negatively charged species were observed. In both cases, the addition of THF enhanced resolution of the compounds. In the case of a Soxhlet extract of a bituminous coal, the use of THF did not improve the separation, although some of the negatively charged species were still resolved [337].

7.19 TEXTILE AND DYES

Fibre manufacturing involves the use of a wide range of materials and compounds, such as fabrics, dyes, flame retardants and lubricants. CE is potentially a useful alternative analytical tool for the monitoring of the chemicals involved in various steps of the production process.

CZE with tandem mass spectrometry was used for the analysis of sulfonated azo dyes [70,378]. Figure 7.133 illustrates the full scan CZE-MS-MS total daughter ion count electropherogram and the collision-induced (CID) mass spectrum of Acid Yellow 49 from municipal waste water.

Capillary isotachophoresis was used for the determination of morpholine and N-methyl morpholine degradation products [379] in cellulose fibre-production.

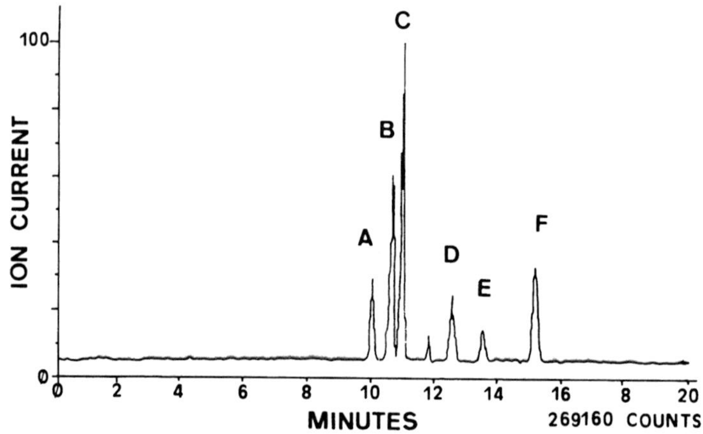

Fig. 7.133. SIM CZE/MS total selected ion current electropherogram for 3 pmol per component of A = Acid Red 151 (m/Z 431); B = Acid Red 88 (m/Z 377); C = Acid Orange 7 (m/Z 327); D = Acid Blue 113 (m/Z 317); E = Acid Black 1 (m/Z 285); and F = Acid Red 14 (m/Z 228). Peak C exhibits 50,000 theoratical plates measured at half height. (Reproduced with permission from Ref. 378. Copyright John Wiley & Sons.)

The leading electrolyte was 0.01 mol/l K^+, with morpholine propanesulfonate as counter-ion. The pH of the leading electrolyte was 7. Conductivity detection was performed. A detection limit of 30 ppm was obtained.

7.20 EXPLOSIVES

Micellar electrokinetic chromatography was used for the analysis of organic gunshot and explosive constituents [380]. Separation of twenty six of these constituents could be achieved within 10 min. A 100 μm I.D. capillary was used. The electrophoretic medium contained 2.5 mM borate, 25 mM SDS. Electrokinetic injection was performed at 5 kV for 2 s. The applied voltage was 20 kV and separation was performed at ambient temperature. Figure 7.134 shows the separation of the gunshot and high explosive constituents by MEKC.

7.21 SURVEY OF COMMERCIAL CE INSTRUMENTS

There is currently a tremendous interest in the development of separation methodologies based on CE for various types of applications, mainly as a result of the numerous advantages of the technique, such as high efficiency, small sample requirement, rapid analysis and low running cost. Another important factor contributing to the rapid progress of CE could be attributed to the availability of commercial instruments for this analytical technique recently. Since the introduction of the first commercial CE systems in 1989, more than ten models have become available from various manufacturers. A basic instrument generally would consist

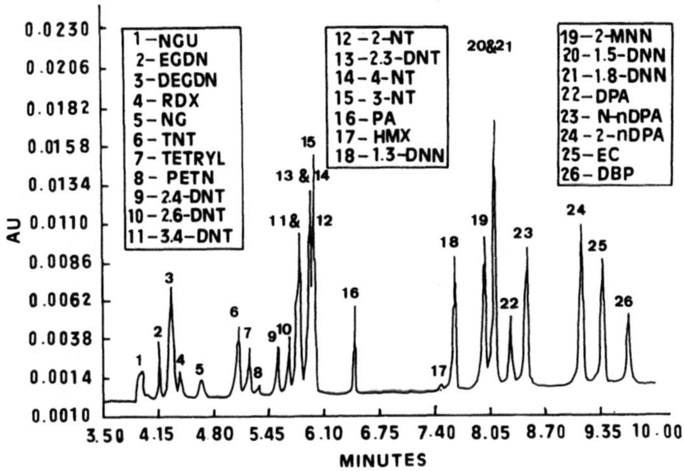

Fig. 7.134. Mixture of gunshot and high-explosive constituents standards at 10^{-4} mol/l. Conditions — capillary: 100 μm I.D. × 67 cm (62 cm to detector); buffer: 2.5 mM borate, 25 mM SDS; injection at 5 kV for 2 s; voltage: 20 kV. Compounds: *1* = nitroguanidine (NGU); *2* = ethylent glycol dinitrate (EGDN); *3* = diethylene glycol dinitrate (DEGDN); *4* = 1,3,5-trinitro-1,3,5-triazacyclohexane (RDX); *5* = 1,2,3-propanetriol trinitrate (nitroglycerine) (NG); *6* = 2,4,6-trinitrotoluene (TNT); *7* = 2,4,6,N-tetronitro-N-methylaniline (Tetryl); *8* = pentaerythritol tetraitrate (PETN); *9* = 2,4-dinitrotoluene (2,4-DNT); *10* = 2,6-dinitrotoluene (2,6-DNT); *11* = 3,4-dinitrotoluene (3,4-DNT); *12* = 2-nitrotoluene (2-NT); *13* = 2,3-dinitrotoluene (2,3-DNT); *14* = 4-nitrotoluene (4-NT); *15* = 3-nitrotoluene (3-NT); *16* = picric acid (PA); *17* = 1,3,5,7-tetranitro-1,3,5,7-tetrazacyclooctane (HMX); *18* = 1,3-dinitronaphthalene (1,3-DNN); *19* = 2-nitronaphthalene (2-MNN); *20* = 1,5-dinitronaphthalene (1,5-DNN); *21* = 1,8-dinitronaphthalene (1,8-DNN); *22* = diphenylamine (DPA); *23* = N-nitrosodiphenylamine (N-nDPA); *24* = 2-nitrodiphenylamine (2-nDPA); *25* = N,N′-diethyl-N,N′-diphenylurea (ethylcentralite) (EC); *26* = dibutyl phthalate (DBP). (Reproduced with permission from Ref. 380. Copyright American Chemical Society.)

of a high voltage power supply, an injection system, a detection system, and the separation capillary. Typically, the power supply would be capable of delivering +30 kV. Some models permit the use of negative polarity (−30 kV) and voltage programming. The simplest systems do not have any elaborate injection devices, since both hydrodynamic and electrokinetic injection can be performed on-column using the separation capillary. Nevertheless, fully and semi-automated injection systems, based on various combinations of pressure, vacuum, gravity and voltage driven injection modes, are available. Most instruments are supplied with a UV absorbance detector, with either single wavelength or multi-wavelength detection capability. Fluorescence detection is available in some of the instruments, although this mode of detection is not a standard feature in most of the models. Laser-induced fluorescence detectors have recently been introduced. An interface for CE-MS has also become available. The separation capillary can be made simply from fused silica columns available from suppliers of capillary tubings. Some instruments utilize capillary cartridges, which also contain the same type of fused silica columns. In the more sophisticated instruments, temperature control of the capillary can be accomplished either with liquid circulation through a capillary cartridge, or with convection cooling of the exposed capillary. In Table 7.2, a list of selected commercial CE instruments is given.

Commercial instrumentation has no doubt accelerated the growth of CE in recent years. Responding to demand for enhanced features, manufacturers have been developing new products with improvements in design, capability and versatility. In fact, new models are introduced so quickly that the features listed in Table 7.2 may soon be outdated. Nevertheless, a few recent developments are considered especially significant, including the introduction of modular equipment, the availability of an interface to mass spectrometry, and the commercial exploitation of laser-induced fluorescence.

The significance of the modular concept can be related to the growth of HPLC. In their early stages of development in the 1970s, HPLC instruments were based mainly on integrated designs. Subsequently, modular systems have almost dominated the market completely as HPLC reaches maturity as an analytical technique. Although their separation mechanisms can be entirely different, CE is remarkably similar to HPLC from the instrumentation point of view, except that high voltage instead of high pressure is used to drive the separation process, and that capillary columns instead of packed columns are more commonly employed. Modular CE instruments are expected to gain popularity as the technique continues to develop.

The availability of a commercial CE-MS interface is significant not only because it presents the power of mass spectrometry to CE for on-line compound identification and structural determination, but also because it demonstrates the successful hyphenation of CE to other analytical techniques. Interfacing of CE to well developed instrumental methods, such as Fourier transform infrared (FT-IR) spectrometry, inductively coupled plasma (ICP) spectrometry and thermogravimetry (TG) may be interesting areas to be explored in the near future.

Although UV absorbance detectors have been the most popular detection systems in the current generation of commercial CE instruments, many other universal or highly sensitive detectors, including electrochemical detection, refractive index detection and fluorescence detection, have been successfully employed in CE separations. The commercialization of the laser-induced fluorescence detector is not only an important development for CE, but may also herald the imminent arrival of some of the alternative detection techniques. Since lasers produce intense light which can be easily focused into very small beams, they should be very suitable for CE detection, where small I.D. capillaries need to be used by necessity to permit adequate heat dissipation and high separation efficiency. Nevertheless, despite the advantages of lasers as light sources, there are very few laser-based analytical instruments, except for laser Raman spectrometers. The situation is partly attributable to the fact that the use of expensive, bulky and difficult to maintain lasers in on-line, routine analyses is likely to results in technical problems at times.

Another reason is that there are only very limited number of wavelengths available for direct detection by lasers. However, with rapid developments in laser technology in recent years, the situation is beginning to change. There is now growing interest

TABLE 7.2

SELECTED COMMERCIAL CE INSTRUMENTS

	Inj.[a]	Det.[b]	Volt.[c] (kV)	Therm.[d]	Cap.[e]	Cost[f]
1. Applied Biosystems						
270A-HT	H(A)	UV	±30	A	T	***
2. Beckman Instruments						
PACE/2100	H(A)	UV/LIF	±30	L	C	***
3. Bio-Rad						
HPE-100	S(M)	UV	+12	N	C	**
Biofocus-3000	H(A)	UV	+30	L	C	***
4. Dionex	H(A)	UV/FL	±30(P)	A	T	****
5. Europhore	H(A)	LIF	±30	A	T	****
6. Isco						
3850 (manual)	S(M)	UV	+30	A	T	**
3140 (automated)	H(A)	UV	±30	A	T	***
7. Jasco	H(M)	UV	+30	N	T	**
8. Lauerlabs						
Prince-Modular	H(A)	–	–	–	–	*
9. Otsuka Electric Ltd.						
CAPI-3000	H(A)	UV	±30	A	T	***
10. Spectra-Physics	H(A)	UV	±30	A	T	***
11. Spectrovision	H(M)	UV	+30	N	T	*
12. Waters						
Quanta 4000	H(A)	UV	±30	A	T	**

[a] Injection system: All systems can perform electrokinetic injection. Hydrodynamic injection included pressure, vacuum and gravity injection methods. Automated systems are denoted H(A) and manual systems are denoted H(M). Syringe injection systems are denoted S(M).
[b] Detection system: All systems include UV detection as standard. Fluorescence (FL) detection, including laser-induced fluorescence (LIF), is available in some of the models. Commercial interface for mass spectrometry is available (Finnigan Electrospray MS).
[c] Voltage supply: All systems provide positive voltage supply, mostly up to 30 kV. Reversed polarity is available on some of the models. Voltage programming (P) is an additional feature.
[d] Thermostating: Thermostating of capillary is either by air (A) or liquid (L) circulation. Some models have no such facility (N).
[e] Capillary format: Capillaries are installed either in the tube (T) or cartridge (C) format.
[f] Approximate price category (in US$): * = 10,000–20,000; ** = 20,000--30,000; *** = 30,000–40,000; **** = 40,000–50,000.

in exploiting low-cost, compact and durable lasers, such as air-cooled gas lasers, helium-neon lasers or the ubiquitous diode lasers extensively used in electronic and office equipment for detection in CE. Furthermore, with the advent of indirect detection techniques, some of the advantages of laser-induced fluorescence may be extended to the detection of a wider range of analytes, including non-fluorescent compounds.

Since several important new features have been incorporated into commercial CE instruments recently, it would be worthwhile to consider some other features and their potential commercial viability in future generations of CE instruments. The possible new developments will probably not be confined only to those discussed here, but the discussion should serve to indicate the minimum expectations of which improvements can be made.

A technique which is currently generating a tremendous amount of excitement is the use of an external field to control the electroosmotic flow. Most of the commercial instruments are not equipped to perform such a task. Although it seems premature at the present stage of development to design an integrated system to permit the use of external field effects, there would be advantages in incorporating features to permit their exploitation as a separate module in the future.

Some other voltage programming techniques, including pulsed electric field and analyte velocity modulation, can be implemented relatively easily. The simplest solution is to provide a power supply with programmable capabilities as an option.

Temperature programming may be another feature worth considering. The effect of temperature can be utilized in several mechanisms to enhance selectivity, such as pH dependent mobilities, partitioning with micelles and complexation equilibria (see Chapter 5). Due to the use of small capillaries in CE, rapid response to temperature change would be expected and hence temperature programming can be an additional useful tool for optimizing CE separations. To implement rapid and precise temperature control, solid-state devices may be utilized.

Several other features are also worth considering. First, a system with increased sample capacity, e.g. based on multiple capillaries, may be supplied with a micropreparative scale option. On-line sample preconcentration or derivatization can be incorporated in the injection system to enhance detection. Finally, computer-assisted optimization software and migration prediction systems would be useful additional features as CE develops into a more sophisticated level.

7.22 SUMMARY OF APPLICATIONS OF CE

For both new and experienced users, the most time-consuming task in method development is often the determination of the optimum conditions for a separation. To make this task easier, a list of methods developed based on CE is given in Table 7.3. Analytes are grouped according to the classification in earlier parts of this chapter. Additional information relevant to the separation of each groups of compounds can be found in the earlier sections.

TABLE 7.3

SUMMARY OF CE APPLICATIONS

Compounds	Separation conditions	Ref.
AMINO ACIDS (AAs):		
DNS-AAs	phosphate buffer (0.0125 M), pH 6.86	6
DNS-AAs	phosphate buffer in D_2O (0.02 M), pH 7.8	14
DNS-AAs (enantiomers)	sodium acetate (20 mM), SDS (20 mM) N,N-didecyl L-alanine (5 mM) Cu(II) (25 mM), 10% (v/v) glycerol	4
DNS-AAs (enantiomers)	CGE: Gel: 5% T, 3.3% C, 7% urea, 100 mM β-CD Buffer: Tris (0.1 M), boric acid (0.25 M) urea (7 M), pH 8.3	11
DNS-AAs (enantiomers)	L-histidine (5 mM), $CuSO_4 \cdot 15H_2O$ (2.5 mM), ammonium acetate (10mM), adjusted to pH 7 by NH_4 OH	7
DNS-AAs (enantiomers)	L-aspartame (5 mM), $CuSO_4 \cdot 15H_2O$ (2.5 mM), ammonium acetate (10mM), adjusted to pH 7 by NH_4OH	9
PTH-AAs	(1) M phosphate/0.0125 borate (pH 7), 0.05 M SDS (2) 0.1 M Tris, 0.1 M HCl (pH 7), 0.05 M DTAB	19
PTH-AAs	sodium acetate (pH 9.5) 15% methanol, 1% THF	18
PTH-AAs	(1) sodium phosphate (20 mM), pH 5 (2) sodium citrate (20 mM), pH 5	20 20
PTH-AAs (enantiomers)	phosphate (50 mM), pH 3, 50 mM SDS, 20 mM digitonin or 20 mM SDVal,	21
NDA-AAs	75/25 by volume buffer/methanol, buffer: 0.01 M boric acid, 0.02 M KCl pH 9.5	23
NDA-AAs	0.01 M borate, 0.04 M KCl, pH 9.5	22
DABSYL-AAs	20 mM phosphate (pH 7), 5 mM SDS	25–27
FITC-AAs	5 mM carbonate, pH 10	28
FITC-AAs	0.01 M $NaHCO_3$, pH 9	24
FITC-AAs	20 mM $Na_2B_4O_7$, pH 9.5	29
FITC-AAs	ITP: Leading: 5 mM borate (pH 9.5) Terminating: 5 mM ACES (pH 10.0)	30
OPA-AAs	0.2 M borate (pH 7.8)	31

TABLE 7.3 (continued)

Compounds	Separation conditions	Ref.
OPA-AAs	0.05 M sodium acetate (pH 9.5) 15% methanol, 1% THF	
OPA-AAs	20 mM $Na_2B_4O_7$ (pH 9.5), 100 mM SDS	29
OPA-AAs	ITP: Leading: 5 mM borate (pH 9.5) Terminating: 5 mM ACES (pH 10.0) CZE: same buffer as leading electrolyte	30
Fluorescamine-AAs	20 mM sodium borate (pH 9.5), 100 mM SDS	29
Fluorescamine-AAs	50 mM sodium tetraborate (pH 6)	33
FMOC-AAs	20 mM tetraborate, pH 9.5, 25 mM SDS	29
Underivatized AAs	10 mM sodium carbonate, pH 9.5, 1 mM coumarin 343	36
Underivatized AAs	25 mM MES, pH 5.65, 0.01 mM DHBA	37
PEPTIDES:		
Tryptic digests of chicken albumin (fluorescamine derivatives)	0.0125 M phosphate (pH 7)	47
Tryptic digests of β-casein rIGF	0.04 M Tris, 0.04 M tricine, pH 8.1 20 mM tricine, 10 mM $Na_2B_4O_7$, 1 mM EDTA, pH 8.2	48 50
r-hirudin	16.7 mM PIPES, 12 mM $Na_2B_4O_7$, 1 mM EDTA, pH 6.7	50
Neurotensin, sulfate and non-sulfate forms of peptide	phosphate (pH 6.5)	52
Modified adrenocorticotropic hormone fragment (Org 2766)	25 mM phosphate + KOH (pH 2.2) 20 mM formate + alanine (pH 3.8) 20 mM ϵ-aminocaproate, acetic acid (pH 4.4) 20 mM histidine, MES (pH 6.2) 40 mM imidazole, MOPS (pH 7.5) 100 mM borate, KOH (pH 8.3)	55
Tryptic digests of horse heart cyctochrome C	0.025 M sodium citrate, pH 4.0	57
Motilin peptide	20 mM sodium citrate, pH 2.5	58
Multiple antigen peptides	50 mM phosphate, pH 2.5	59
Adrenocorticotrophic hormone peptide cleaved by endo-proteinase Arg C	(1) 40 mM phosphate, 2 M betaine, 0.1 M K_2SO_4, pH 7.6 (2) 40 mM MES, 2 M betaine, 0.1 M K_2SO_4, pH 5.5	60

TABLE 7.3 (continued)

Compounds	Separation conditions	Ref.
Histine-containing compounds, e.g. angiotensins, cationic heptapeptides	20 mM MES, 10 mM KCl, 1 mM $Zn(ClO_4)_2$	61
Peptides	0.1 M phosphate buffer (pH 2.5) coated capillary (Bio-Rad)	64
Dynorphin	100 mM phosphate, pH 2.5	65
Bioactive peptides, e.g. bradykinin, neurotensin and angiotensin I	100 mM borate, pH 9.2	65
Chiral dipeptides derivatized with L- and D-Marfey's reagent	(1) 50 mM ammonium phosphate, 200 mM SDS pH 3.3, reversed polarity (2) 100 mM sodium borate, 200 mM SDS, 5% acetonitrile, pH 8.5	66
Synthetic peptide mixture and tryptic digest of human hemoglobin	ammonium acetate or ammonium formate (15–20 mM), adjusted to pH 5 with acetic acid, or pH 2.5 with trifluoroacetic acid, 5–30% acetonitrile or methanol	75
Bioactive peptides	0.005 M ammonium acetate, 0.01 M acetic acid, ammonium hydroxide, pH 8.5 column modified with APS	80
Synthetic peptides, protease digests of rhGH and horse heart cytochrome C	40 mM citric acid, ammonium hydroxide, pH 3.6	78
Peptides and proteins, e.g. neuropeptides, angiotensin-related peptides, myoglobin, pigeon heart cytochrome C and horse heart cytochrome C	column coated with aminopropylsilane. For CZE-CF-FAB-MS: (1) 5 mM ammonium acetate, ammomium hydorxide, pH 8.5 (2) 0.01 M acetic acid, ammonium hydroxide, pH 3.4–3.5 For CZE-ESI-MS: (1) ammonium acetate, ammonium hydroxide, pH 8.5, (2) 0.01 M acetic acid, ammonium hydroxide, pH 3.8 Sheath flow for ESI: 5–10 μl/min of methanol–3% aqueous acetic acid (50:50)	81
Fourteen test peptides	100 mM phosphate (pH 2.7)	85
Synthetic peptides	CITP: Leading: 2 mM chloride; counter ion, ammediol, pH 8.9. Terminating: β-alanine adjusted to pH 10.3 with $Ba(OH)_2$	87

TABLE 7.3 (continued)

Compounds	Separation conditions	Ref.
PROTEINS:		
Proteins, e.g. lysozyme, trypsinogen, myoglobin, β-lactoglobulin B, β-lactoglobulin A	0.1 M CHES, 0.25 M K_2SO_4, 1 M EDTA, pH 9.0, varying amounts of sodium and potassium salts were added to prevent adsorption	91
Proteins	(1) 0.04 M phosphate, 2 M sarcosine, pH 6.7 (2) 0.04 M phosphate, 2 M betaine, pH 7.5 (3) 0.04 M phosphate, 0.1 M K_2SO_4, pH 7.0 (4) 0.04 M phosphate, 0.25 M K_2SO_4, pH 7.0 (5) 0.04 M CHES, 0.25 M K_2SO_4, pH 9.0 (6) 0.04 M phosphate, 2 M betaine, 0.1 M K_2SO_4, pH 7.6	92
Proteins: lysozyme, cytochrome C, ribonuclease A, trypsinogen, α-chymotrypsinogen A, rhuIL-4	50 mM phosphate, 40 mM Na_2SO_4, 50 mM 1,3-diaminopropane, pH 7	93
Basic proteins	0.05 M phosphate, 100 μg/ml FC-134, pH 7	94
Human serum albumin and transferrin	(1) 0.01 M or 0.1 M Tris-HAc, pH 8.6 (2) 0.01 M or 0.03 M sodium phosphate, 0–20% steaoryl dextrin, (pH 6.8)	95
Metal-binding proteins: calmodulin, parvalubumin, thermolysin, proteolytic peptides of calmodulin, carbonic anhydrase, lactoglobulin	For CZE: 2 mM additives (calcium chloride, zinc chloride, or EDTA) in: (1) 0.1 M Tris, 0.1 M tricine (pH 8.3) (2) 50 mM Tris, 384 mM glycine, (pH 8.3) For MEKC: 0.1% SDS in 0.1 M Tris–0.1 M tricine, pH 8.3	96
Fifteen proteins	(1) 150 mM H_3PO_4, pH 1.5, uncoated capillary (2) Polyvinylpyrrolidinone-modified capillary, 38.5 mM H_3PO_4, 20 mM NaH_2PO_4, pH 2.0,	97
Histones	0.1 M phosphate (pH 2.5)	98
Trypsin/cyanogen bromide digest of cytochrome C	150 mM phosphoric acid, 10 mM Tris, 10 mM phosphate, pH 1.5	99
β-casein dephosphorylated	(1) ammonium acetate (pH 7.2) (2) Tris-tricine (pH 8.2)	99
Proteins-bronchoalveolar lavage fluid (BALF) exposed to perfluoroisobutylene (PFIB)	0.1 M sodium phosphate, pH 2.5	100
Aprotinin and trypsin inhibitor (bovine): degradation products	20 mM citrate, pH 2.5	101
Proteins: model mixtures and in serum	50 mM borate, pH 9.5 (wash with 1 M NaOH between runs)	102

TABLE 7.3 (continued)

Compounds	Separation conditions	Ref.
Serum proteins	150 mM borate (pH 10.5)	103
Proteins	Coated capillaries (polyethylene glycol 600 coating, diol coating for pH i 5) 0.03 M KH_2PO_4, pH 3.8, (maltose coating for pH 5–7)	104
Proteins	Coated capillaries (polyethyleneimine, for pH 2–12) 0.02 M hydroxylamine–HCl, pH 7.0	105
Proteins	Coated capillaries (nonionic surfactants: TWEEN series, BRIJ series, for pH 4–11), 0.01 M phosphate, pH 7, 0.1% surfactant	106
Proteins	Coated capillaries (aryl pentafluoro, APF), 250 mM $(NH_4)HPO_4$, pH 7	107 108
Proteins	Coated capillaries, C_8, C_{18} coatings: 10 mM $Na_2H_2PO_4$, 6 mM $Na_2B_4O_7$, pH 6.2; polar coating: pH 9.0	109
Proteins	Coated capillaries, (vinyl-bonded polyacrylamide), (1) 0.05 M glutamine, triethylamine, pH 9,5; (2) 0.03 M citric acid, adjusted with 1 M NaOH, pH 2.7	110
Proteins	Coated capillaries (polymethylglutamate coating), 10 mM sodium phosphate, pH 7.0	111
Proteins	Capillaries with fuzzy coating, (pH 4–7.5), 0.1 M phosphate, pH 6.0	112
Proteins	Capillaries with interlocked coating, (pH 4–7.5), 0.1 M phosphate, pH 6.0	112
Immunoglobulin preparations for intravenous administration	CITP: Leading: 5 mM hydrochloric acid, 9.3 mM 2-amino-2-methyl-1-propanol, pH 9.9 Terminating: 50 mM tranexamin acid trans-4-(aminomethyl)cyclohexane carboxiylic acid, 15 mM potassium hydroxide (pH 10.8)	113
Polyethylene glycol modified proteins	Charge-reversal by dynamic coating, 100 mM phosphate, 10% ethylene glycol pH 2.0	114
Proteins	0.3 M sodium borate (pH 8.5), with coated column and proprietary zwitterionic additives	116
Actinomycetes SP 913: butanol extract of culture filtrate	Gel-filled capillary: 3% T, 4% C. 0.02 M sodium phosphate (pH 8.2). Water cooled capillary (0.2 mm i.d.).	122

TABLE 7.3 (continued)

Compounds	Separation conditions	Ref.
Human growth hormone	Gel-filled capillary: 10% T, 3.3% C. 0.1 M Tris–borate, pH 7.8	123
Proteins (17,800–77,000 kDa)	10% liquid linear polyacrylamide gel. phosphate (pH 5.5), 0.5% SDS	124
Recombinant biotechnology derived-proteins	Gel-filled capillary: 5.1% T, 2.6% C or 3.1% T, 2.6% C, with 1.8–2.7 M ethylene glycol to improve stability of gel Buffer: 375 mM Tris (pH 8.8), 3.2 mM SDS and ethylene glycol	125
Six proteins	50 mM sodium borate decahydrate (pH 10). Samples dissolved in 20 mM boric acid (pH 4.0), with 20% ethylene glycol.	127
Proteins: conalbumin, ovalbumin	0.1 M borate (pH 10)	128
Recombinant DNA-derived human growth hormone (rhGH) T4 receptor protein (rCD4) tissue plasminogen activator (rt-PA)	Coated capillary, phosphate buffer (pH 6.5) for rhGH, phosphate buffer (pH 5.5) for rCD4, and phosphate buffer (pH 4.5) for rt-PA	129
Biosynthetic human insulin, human growth	0.01 M tricine, 0.058 M morpholin, 0.02 M sodium chloride, adjusted with 1 M NaOH to pH 8hormone	130
Antibody-antigen complexes: antigen is recombinant human growth hormone, antibody is monoclonal	0.1 M tricine buffer (pH 8.0)	132
Glycoform of recombinant human erythropoietin	100 mM acetate–phosphate, pH 4.0	133
Human recombinant tissue plasminogen activator	CIEF: 20 mM sodium hydroxide (catholyte), 10 mM phosphoric acid (anolyte), 10 mM sodium hydorxide and 80 mM sodium chloride (mobilizer). CZE: 0.1 M ammonium phosphate (pH 4.6), with 0.01% Triton X-100 (reduced UV absorbance) and 0.2 M ϵ-aminocaproic acid	134
Human recombinant interleukin-3	20 mM cyclohexylethane sulfonic acid, pH 9.0, with 10 mM KCl. Micropreparative separation: 10 mM ammonium bicarbonate, pH 7.2.	135, 136
Adrenocorticotropic hormone	100 mM borate/KOH (pH 8.3), with varying concentrations of SDS	55
Streptavidin and polymeric protein	Short, small i.d. (20 cm × 25 μm) coated capillary (Bio-Rad), 0.1 M phosphate (pH 2.5)	64

TABLE 7.3 (continued)

Compounds	Separation conditions	Ref.
Proteins	0.1 M sodium phosphate, sodium hydroxide pH 11	73
Proteins	water–methanol (50:50), 5% acetic acid	72
Mixture of whale, horse and sheep myoglobin	0.01 M acetic acid, pH 3.4	81
Enzyme-labelled monoclonal antibody conjugates (in-process control)	0.5% methylcellulose in 0.1 M borax with 0.5 mM SDS (pH 10)	143
Crude fermentation broth	25 mM phosphate, pH 9.5 Fraction collection: 35 mM phosphate, pH 9.5	144
Eminase (in-process assay)	40 mM sodium phosphate, pH 2.5, 0.01% hydroxymethylcellulose	145
Isoforms of human monoclonal antibodies against gp-41 AIDS virus	100 mM phosphate, pH 5.8	146
Human recombinant superoxide dismutase	20 mM sodium citrate, pH 4.0	146
Proteins	CIEF: Anolyte: 10 mM phosphoric acid. Catholyte: 20 mM sodium hydroxide. Mobilizer: pI 3.22 zwitterionic ion in basic solution. Ampholyte (pH 3–10), with 0.5% TEMED	153
Proteins: hemoglobin	CIEF: 40 mM glutamic acid (anolyte), 40 mM arginine (catholyte).	46
Proteins:	CIEF: 10 mM H_3PO_4 (anolyte), 20 mM NaOH. Samples contained 5% pharmalyte 3–10 and 0.05% methylcellulose	46
Serum lipoproteins and membrane proteins	CITP: Leading: 5 mM H_3PO_4, 0.25% (w/v) HPMC, 20 mM ammediol, pH 9.2 Terminating: 100 mM valine, adjusted to pH 9.4 with ammediol	155 156
Hydrophobic membrane proteins	0.2 M sodium borate, pH 9.2, 7 M urea	156
Proteins	CITP: Cationic system: Leading: 0.01 M potassium acetate, and acetic acid, with 0.3% HPMC; Terminating: 0.01 M acetic acid Anionic system: Leading: 10 mM formic acid titrated with ammediol to pH 9.1, with 0.3% HPMC; Terminating: 10 mM β-alanine-ammediol, pH 9.5.	159

TABLE 7.3 (continued)

Compounds	Separation conditions	Ref.
NUCLEIC ACIDS:		
Modified nucleic acid constituents	0.075 M SDS in 0.01 M phosphate–0.006 M borate, pH 8.9	161
Phosphorylated nucleosides	0.1 M DTAB in 10 mM Tris, 10 mM Na_2HPO_4 pH 7.05	162
Bases and nucleotides	7 M ruea, 5 mM Tris, 5 mM Na_2HPO_4, pH 7 50 mM SDS, 0.3 mM Cu(II)	163
Nucleic acid bases	0.1 M SDS in 50 mM phosphate–25 mM borate (pH 7)	164
Normal and chemically modified nucleobases, nucleosides and nucleotides	20 mM sodium dihydrogen phosphate, 100 mM SDS	165
Sixteen nucleic acids and their analogues	20 mM sodium phosphate, pH 7.1, 200 mM SDS	166
Nucleotides, mono, di, and tri-phosphates	0.5 mM salicylate, pH 3.5	172
Methylpyridines	20 mM potassium hydroxide (pH 6.2), with cationic surfactant	174
Purine nucleotides from decomposition of fish tissue	100 mM CAPS, pH 10–11.	175
Ribonucleoside triphosphate	polyacrylamide-coated capillary, 0.05 M phosphate (pH 2.7), with 0.002 M EDTA	176
Cyclic nucleotides	0.05 M sodium tetraborate, adjusted with 1 M HCl to pH 8.3	178
Nucleosides	40 mM tricine in D_2O, pD 9.3	14
Nucleotides	0.01 M β-alanine acetate, pH 3.8, 0.05% MHEC, PTFE capillary (0.2 mm i.d.)	179
Oligonucleotides and enzymatic digests of calf thymus DNA	50 mM SDS, 2 M urea, 5 mM Tris–Na_2HPO_4, 3 mM Zn(II)	180
Oligodeoxy-nucleotides	CGE: gel: 4% T, 3.3% C; buffer: 100 mM Tris, 100 mM boric acid, 7 M urea, 2 mM EDTA, pH 8.6	181 182 183
Oligonucleotides:	CGE: gel: 6% T, 5% C; buffer: 0.1 M Tris, 0.2 M boric acid, 7 M urea, pH 7.5	184
Nucleic acids	CGE: gel: proprietary, buffer: Tris–borate (Kodak buffer EZE Formula 3)	186

TABLE 7.3 (continued)

Compounds	Separation conditions	Ref.
Poly(A) enzymatic partial digest hydrolysate and poly$(dA)_{12-18}$	CGE: gel: 5% T and 5% C, buffer: 0.1 M Tris, 0.1 M boric acid, 7 M urea, pH 8.8	187
Poly$(dA)_{12-18}$	CGE: gel: 7% T and 5% C, buffer: 0.1 M Tris, 0.1 M boric acid, 7 M urea, pH 8.8	187
Pd$(A)_{15-80}$	CGE: gel: 100 mM boric acid, 2 mM EDTA, 7 M urea, pH 8.7	188
Phosphodiester oligonucleotides and phosphorothioate analogues	CGE: gel: Micro-Gel_{100} capillaries; buffer: Micro-Gel_{100}	190
Nucleotides: (polycytidines)	60 mM Tris, 30 mM glutamic acid, 50 mM SDS, 3 mM spermine, pH 8.1	192
Oligonucleotides	CGE: gel: 2.5% to 8% T, 3.3% C; buffer: 0.1 M boric acid, 0.1 M Tris, 2 mM EDTA 7 M urea	193
Poly(A) and poly(dA) digested by nuclease P1	CGE: gel: 5% T and 5% C for poly(A), 3% T and 3% C for poly(dA); buffer: 0.1 M Tris–borate, 7 M urea, pH 8.6	194
Restriction fragments of DNA	Tris–borate buffer, 7 M urea, 0.1% SDS	197
DNA restriction fragments	CGE: gel: 3% T, 0.5% C, 1 μg/ml ethidium bromide; buffer: 0.1 M Tris–boric acid, 2 mM EDTA, pH 8.5	200
DNA restriction fragments	CGE, 0.1 M Tris–boric acid, 2 mM EDTA (pH 8.5)	201
Fluorescently labelled DNA Fragments from seqencing reactions	CGE: gel: 4% T, 5% C, 8.3 M urea Buffer: 90 mM Tris, borate, pH 8.3, 0.2 mM EDTA	202
DNA restriction fragments	CGE: agarose gel, 2% in buffer; buffer: 0.01 M disodium hydrogen phosphate, 0.01 M sodium dihydrogen phosphare	203
DNA restriction fragments and polymerase chain reaction products	89 mM Tris–borate, 2 mM EDTA, pH 8.5, 0.5% HPMC-4000, 10 μ ethidium bromide, OV-17 coated capillary	204
DNA fragments: (doubled strand 100–7000 kb)	SuperGene™ 500 (counter-migration CE)	186
DNA restriction fragments	0.25% HEC in 89 mM Tris, 89 mM boric acid, 5 mM EDTA	206
DNA restriction fragments	0.089 M TBE, 7 M urea, 0.5% HMC, pH 8.0	207

TABLE 7.3 (continued)

Compounds	Separation conditions	Ref.
DNA restriction fragments	0.05 M Tris–borate, 25 mM EDTA, 7 M urea, 0.5% methycellulose, pH 8.0	208
Nucleic acid fragments	GCE: Gel: 3.5% T, 3.3% C. Buffer: 1 × TBA (90 mM Tris TBE buffer), 0.5 μg/ml ethidium bromide. Voltages: 180 V/cm DC, 52–107 Hz, 50–120% modulation depth	209
Southern blotting	CGE: gel: 9% T, 0% C, polyacrylamide with no urea, buffer: 25 mM Tris–borate, 0.25 mM EDTA, pH 8.0	210
DNA adducts	0.04 M, pH 9.6 sodium carbonate buffer	211
Ribonucleotides	(1) 50 mM formate, pH 3.7 to 4 (2) 12.5 mM formate, pH 3.8, 0.1 M CTAB	212
DNA SEQUENCING:		
DNA sequencing	CGE: gel: 6% T, 5% C; buffer: 8 mM urea, 89 mM Tris, 89 mM boric acid 2 mM EDTA, pH 8.3	214 215
Four-spectral channel sequencing	CGE: gel: 6% T, 5% C; buffer: 1 × TBE	216
Two-spectral channel sequencing	CGE: gel: 4% T, 5% C; buffer: 1 × TBE	216
One-spectral channel sequencing	CGE: gel: 6% T, 5% C; buffer: 1 × TBE 30% formamide, 7 M urea	216
DNA sequencing	CGE: gel: 3% T, 5% C; buffer 0.1 M Tris–borate, 2.5 mM EDTA, 7 M urea, pH 7.6 or 8	217
Tetramethylrhod-amine isothio-cynate-labelled DNA fragments	CGE: gel: 6% T, 5% C, 7 M urea, with or without 20% formamide; buffer: 1.08% Tris, 0.55% boric acid, 0.07% EDTA in 100 ml water	218
PHARMACEUTICALS AND DRUGS:		
Cinnamic acid and analogues	0.025 M phosphate, pH 9.2	220, 221
Antipyretic analgesic preparation	0.05 M sodium dodecyl sulfate in 0.02 M phosphate, pH 11	222
Analgesics	15 mM $NaHOP_4$ (pH 11), 25 mM SDS	223
Paeonia radix (sedative, lenitive or antispasmolytic agent)	100 mM borate, pH 10.5	224
Drugs: penicillin and cephalsporin	0.04 M TMAB in 0.02 M phosphate–borate pH 9.0 and 0.05 M SDS	225

TABLE 7.3 (continued)

Compounds	Separation conditions	Ref.
β-lactum antibiotics	0.02 M phosphate–borate (pH 9.0), 0.15 M SDS or 0.15 M LMT	226
β-lactum antibiotics	CITP: Leading: 5 mM HCl adjusted to pH 7 by Tris, with 0.25% HPMC. Terminating: 10 mM phenol, adjusted to pH 10 with freshly prepared and filtered barium hydroxide	229
Aspoxicillin	0.02 M phosphate–borate, pH 8.5, 0.1 M SDS	228
Cephalosporins	30 mM phosphate in 9 mM tetraborate, pH 7	227
Sulfonamides	0.05 M phosphate, 0.05 M borate, pH 6.5	230
Sulfonamides	polymethylsiloxane coated column, 0.02 M phosphate, 0.05 M SDS, pH 7, or 0.04 M β-alanine , 0.1 M SDS, pH 4.2	231
Antibiotics	0.05 M phosphate, 0.1 M borate, pH 7.06	233
Small drug molecules	20 mM ammonium acetate (pH 6.8) and 20% (v/v) of methanol	232
Optical isomers of diltiazem hydrochloride, trimetoquinol, hydrochloride, etc. .	0.05 M taurodeoxycholate in 0.02 M phosphate–borate, pH 7	234 235
Racemic sympathomimetic amines	10 mM Tris–H_3PO_4, pH 2.4, 18 mM di-OMe-β-CD	236
Chloramphenicol drug enantiomers	0.1% MHEC in 20 mM Tris + citric acid (CAc) to pH 3.5 + 10 mM DM-β-CD	237
Thioridazine enantioners	20 mM Tris + H_3PO_4 to pH 2.5 + 5 mM γ-CD	237
Ketotifein enantiomers	20 mM Tris + CAc to pH 3.5 + 10 mM β-CD + 0.05% MHEC	237
Ketotifein and its synthetic intermediates	20 mM Tris + CAc to pH 3.5 + 20 mM γ-CD	237
Corticosteroids	0.02 M phosphate/borate (pH 9), 4 M urea 0.03 M β-CD	239
Corticosteroids	20 mM Tris + CAc (pH 3.5) + 10 mM DM-β-CD + 0.1% MHEC	237
Corticosteroids	0.1 M sodium cholate and 0.05 M sodium taurocholate in 0.02 M phosphate–borate (pH 9.0)	238
Benzothiazepin	0.05 M sodium taurocholate in 0.02 M phosphate–borate (pH 9.0)	238

TABLE 7.3 (continued)

Compounds	Separation conditions	Ref.
Barbiturates	phosphate–borate (pH 7.8), with 50 mM SDS	227 240
Cefpiramide	phosphate (pH 8), 10 mM SDS	242, 243
Anti-cancer drug (methotrexate)	5 mM MES, 16 mM sodium sulfate, 30% (v/v) methanol, pH 6.7	8
Basic drugs	0.03 M sodium phosphate, pH 6.8, 2% octyl glycoside	95
Tricyclic antidepressant (desipramine)	67 mM phosphate, 100 mM Na^+, pH 7.0, 30 mM octyl glycoside	244
Tricyclic antidepressants	50 mM CAPSO, with different concentrations of methanol (0 to 23.7%) and NaOH (12.5 to 19.8 mM), pH 9.55	245
Vitamins (fat- and water-soluble)	3 mM γ-CD and 30 mM SDS in 0.1 M borate–0.05 M phosphate (pH 7.6)	246
Anti-inflammatories	(1) 80 mM MES + 30 mM Tris, pH 6.1 (2) 30 mM phosphate + 9 mM borate, pH 7 (3) 20 mM borate + 100 mM boric acid, pH 8.4	227
Non-steroidal anti-inflammatories	20 mM phosphate, 25 mM SDS	249
Recombinant cytokine	0.05 M sodium tetraborate (pH 8.3) with 0.025 M lithium chloride	250
Flavonoids	20 mM sodium borate (pH 8.3) with SDS	251
Flavonoids	50 mM sodium phosphate, 50 mM sodium borate, pH 7.5, 42 mM SDS	252
Hepatitus B vaccine process samples production	25 mM sodium phosphate, pH 7.25	253
Drugs sparingly soluble in water	0.02 M Tris with 100 mM SDS at pH 5.5	254
Urine	0.05 M sodium tetraborate, pH 8.3	255
Robitussin	20 mM phosphate–borate (pH 9), 50 mM SDS	223
Antihistamines	10 mM SDS in 0.05 M borate–0.05 M phosphate with 10 mM TBA and 10 mM β-CD (pH 7.5)	262
Cimetidine	20 mM phosphate at pH 7	260
Cimetidine	sodium hydrogenphosphate (9.4 mM), HTAB, aminomethane (TRIZMA base, 3.3 mM), pH 6.4	259
Radiopharmaceuticals	0.002 M CTAB, pH 7	261

TABLE 7.3 (continued)

Compounds	Separation conditions	Ref.
CELLS, VIRUSES AND BACTERIA:		
Tobacco mosaic virus	0.02 M Tris-HCl (pH 7.5), column coated with methyl cellulose or linear polyacrylamide	121
Bacteria: Lacto bacillus casei NCTC 10302	0.1 M Tris–HOAc (pH 8.6)	121
Red blood cell	FEP tubings, 0.1 M phosphate buffer (pH 9), with zwitterionic additive	263
BODY FLUIDS:		
Thiol-containing amino compounds	0.05 M phosphate, pH 7.5	266
Co-amplified DNA sequences (HIV-1 and HLA-DQ-alpha)	89 mM Tris borate, 2 mM EDTA (pH 8.5), 10 μM ethidium bromide, 0.5% HPMC-4000, OV-17 coated capillary	204
Thiopental in humna serum and plasma	50 mM SDS, 9 mM sodium borate, 15 mM sodium dihydrogenphosphate, pH 7.8	240
Urinary porphyrins	10 mM SDS, 20 mM CAPS, pH 11	256
Amino acids in human urine	Derivatized to DABSYL-AAs; buffer: 0.1 M phosphate (pH 2.5)	64
Creatinine and uric acid	phosphate buffer (ionic strength 0.02), 50 mM SDS and 5% isopropanol, pH 9 for serum and pH 6 for urine	267
CAPILLARY ION ANALYSIS:		
30 anions	5 mM chromate, 0.5 mM OFM Anion-BT (Waters) (pH 8.0)	280
Kraft black liquor	5 mM chromate, 0.5 mM OFM Anion-BT (Waters) (pH 10.0)	278
Dental plaque extract and human saliva	5 mM phthalate, with Nice-Pak OFM Anion BT (pH 5.6)	278
Organic acids and phosphate	10 mM benzoate with Nice-Pak OFM Anion BT (pH 6)	278
Alkylsulphonates (C_1–C_7)	10 mM benzoate with Nice-Pak OFM Anion BT (pH 6)	278
Alkylsulphonates (C_4–C_{12})	10 mM naphthalenesulphonate, 30% acetonitrile, pH 10	278
Hypochlorite, persulfate and permanganate	5 mM chromate, pH 11	283
Pyridinium salts, sulfonic acids	0.05 M Na_2HPO_4, pH 7	271

TABLE 7.3 (continued)

Compounds	Separation conditions	Ref.
Cupric and ferric ions	0.05 M acetic acid	271
Cations	20 mM MES/His, pH 6	286, 287
Monovalent cations	0.01 M MES and 0.02 M His, pH 6.4	275
Alkali and alkaline earth metal ions	0.38 mM quinine sulfate and 0.58 mM H_2SO_4 pH 3.7	288
Alkali, alkaline earths, and lanthanides	10 mM UVCat-1 (Waters), 4 mM α-hydroxyisobutyric acid (adjusted to pH 4.4 with acetic acid)	284
Cations in fermentation broth and industrial wastewater	5 mM UVCat-1, 6.5 mM α-hydroxyisobutyric acid (pH 4.2)	284
Calcium ion	CITP: Leading: 0.01 M potassium acetate–acetic acid (pH 5.4). Terminating: 0.01 M n-hexanioc acid	294
METAL CHELATES:		
Par-metal chelates	0.02 M SDS, 0.05 M phosphate, 0.125 M borate, 1×10^{-4} M par, pH 9.2	303
Porphinato chelates as	0.02 M SDS, 0.05 M imidazole, 0.05 M phosphate, 0.125 M borate, pH 7	305
TCPP derivatives		
Acetylacetonato complexes of chromium, cobalt, rhodium and platinum	20 mM borate, 100 mM SDS	304
Metal ions as chelates with HQS	10 mM phosphate, 6 mM borate, 2.5 mM HQS^-, 0.63 mM Ca(II), or 0.08 mM Mg(II), or 0.15 mM Zm(II)	290
Complexes of rare earth elements	CITP: Leading: 10 mM HCl, 45% (v/v) acetone. Terminating: 5 mM EDTA	291
ORGANIC COMPOUNDS:		
Hydrocarbons:		
Substituted benzenes	0.025 M β-CMCD in 0.1 M phosphate, pH 7	306
Chlorinated bezene congeners, trichloro-biphenyl congeners	100 mM borate, pH 8, 100 mM SDS, 2 M urea 40 mM γ-CD	307
	60 mM γ-CD, 100 mM SDS, 2 M urea, 100 mM borate–50 mM phosphate, pH 8	307
Tetrachlorodibenzo-p-dioxin isomers	60 mM γ-CD, 100 mM SDS, 2 M urea, 100 mM borate–50 mM phosphate, pH 8	307
Polycyclic aromatic hyrocarbons	30 mM γ-CD, 100 mM SDS, 5 M urea, borate, pH 9	307

TABLE 7.3 (continued)

Compounds	Separation conditions	Ref.
Aromatic hydrocarbons	50 mM SDS, 10 mM Tris–phosphate, pH 6.9	308
Polyaromatic hydrocarbons and their hydroxyl derivatives	0.05 M NaCH, 0.01 M phosphate, 0.006 M 100 mM borate, 10% methanol	309
Aromatic hydrocarbons	50% acetonitrile, 50% water (v/v) with 0.025 M THAP	310
Polycyclic aromatics	Acetonitrile–phosphate buffer (0.05 or 0.1 M, pH 7), and methanol–phosphate buffer (0.05 M, pH 7) (1 : 1, v/v), capillary coated with octadecylsilane	311
Aromatic compounds	Column packed with 5 μm silica particles, acetonitrile:water (70 : 30 v/v), 2 mM sodium dihydrogen phosphate	313
Chlorophenols	70 mM SDS in 0.025 M borate-0.05 M phosphate	315
Phenols	50 mM SDS in 0.025 M borate–0.05 M phosphate	316
Phenols	0.025 M SDS, 0.025 M BRIJ-35, in 0.01 M Na_2HPO_4 (pH 7)	317
Aliphatic alcohols	12.5 mM borax buffer, 4 mM SDBS, 46 mM SDS	318
Organic acids:		
Lower carboxylic acids	0.5 mM TTAB, 10 mM MES/His, pH 5.9	319
Carboxylic acids	20 mM MES, 20 mM His, pH 6, 1 mM TTAB	321
Carboxylic acids	Gel: 10% T, 3% C. Buffer: 0.1 M Tris–acetic acid, pH 8.6	322
Carboxylic acid metabolites	0.025 M MES, pH 5.55, 5% (v/v) 2-propanol	265
Organic acids	0.5 mM OFM™ Anion-BT–5 mM potassium phthalate (pH 7)	323
Organic anions in blood serum from patients with renal failure	(1) 0.01 M acetate and alanine at pH 3.8 (2) 0.01 M MES and histidine at pH 6.1 with 0.5 g/l MHEC	324
Amines:		
Fluorescamine *n*-alkyl labelled amines	0.05 M phosphate buffer (pH 7)	3
NBD-derivatized amines	0.025 M SDS, 5 mM phosphate, 10% (v/v) 2-propanol	326
NBD-derivatized amines	0.05 M sodium cholate, 0.01 M phosphate 0.006 M borate, 20% methanol	309

TABLE 7.3 (continued)

Compounds	Separation conditions	Ref.
Substituted amines	0.38 mM quinine sulfate, 0.58 mM H_2SO_4, pH 3.7	288
Carbohydrates:		
N-2-pyridyl-glycamine derivatives of reducing monosaccharides	200 mM borate, pH 10.5	336
Carbohydrate by determining iodate and periodate	100 mM acetate, pH 4.5	337
Pyridylamino derivatives of maltooligosaccharide	0.1 M phosphate, pH 3–4.5	340
Human α_1-acid glycoprotein	0.1 M phosphate, pH 5, 50 mM TBAB	341
Aldose oligosaccharides derivatized with fluorogenic reagents	CZE: 20 mM phosphate, 20 mM borate, pH 9.12 50 mM SDS	339
Aldose oligosaccharides derivatized with fluorogenic reagents	CZE: 10 mM phosphate, 10 mM borate, pH 9.4 CGE: Gel: 10% T, 3% C. Buffer: 0.1 M Tris–0.25 M borate, 7 M urea (pH 8.33)	339
Mono- and oligosaccharides	5 mM tetraborate, pH 9.3	343
Maillard reaction products	0.02 M phosphate, pH 7.5	345
PTH-derivatives of Maillard reaction products	0.005 M borate buffer, pH 9.6	345
3,4-dinitrophenyl hydrazones	0.01 M phosphate, 0.006 M tetraborate, 0.05 M SDS	345
Sugars	10 mM sodium bicarbonate, pH 11.5, 1 mM coumarin	36
Unsaturated sulfonated dissacharides	CGE: Gel: 2.5% agarose in buffer. Buffer: 0.01 M disodium hydrogen phosphate, 0.01 M sodium dihydrogen phosphate	203
Reducing carbohydrates as PMP-derivatives	100 mM calucium acetate	342
FOOD ANALYSIS:		
Vitamins and caffein	50 mM SDS in 50 mM Tris-borate, pH 8.4	308
Food additives	0.05 M phosphate–0.05 M borate buffer containing 10 mM TBA	330
OPA-derivatives of amine-containing compounds in red wine	50 mM borate, 50 mM KCl, pH 9.5	331
Hop extract	25 mM Tris/HAc, 25 mM SDS, pH 9	332

TABLE 7.3 (continued)

Compounds	Separation conditions	Ref.
Beer iso-α-acids	30 mM phosphate, pH 7.6, 40 mM SDS	333
Milk proteins	sodium borate buffer, pH 8.5	333a
Organic acids in food samples	CITP: Leading: 0.01 M HCl, counter ion, β-alanine, pH 3.8, 0.1% PVP. Terminating: 0.005 M caproic acid	334
4-methylimidazole in caramel color	CITP: Leading: 5 mM KOH, 20 mM HEPES, 0.1% PVP in preseparation capillary, 1 mM KOH, 4 mM HEPES, 0.1% PVP in separation capillary. Termiating: 10 mM Tris and 5 mM acetic acid	335
ENVIRONMENTAL ANALYSIS:		
Eleven priority phenols	0.05 M SDS in 0.1 M borate/0.05 M phosphate	348
Polycyclic aromatic hydrocarbons	10 mM SDS and 2 mM γ-CD in 0.05 M phosphate/0.1 M borate (pH 7)	350
Polycyclic aromatic hydrocarbons	100 mM borate, 30 mM γ-CD, 100 mM SDS, 5 mM urea (pH 10)	307
Chlorophenols	0.045 M sodium dihydrogen phosphate, 0.015 M sodium borate, pH 8	347
Nitroaromatic compounds	0.05 M phosphate, 0.1 M borate, pH 7.04, 30 mM SDS	349
Acid pesticides	90:10 acetonitrile–0.1 M NH_4 OAc	70
Pyrethroid insecticides	CITP: Leading: 0.01 M HCl, creatinine 0.05% polyvinyl alcohol, pH 4.8. Terminating: 0.003 M MES	352
POLYMER AND PARTICLE ANALYSIS:		
Polystyrene nanospheres	1 mM ACES buffer at pH 5.8, capillary treated with 0.5 mM CTAB	353, 354
Polymer latex particles	5 mM Na_2HPO_4 and 1 mM NaOH, pH 10.71	355
Poly(oxyalkylene) diamine (Jeffamine), (NDA derivatives)	77 mM boric acid, 19.4 mM citric acid, 11.3 mM trisodium phosphate in 20% (v/v) methanol, pH 4.2	356
Silica sols	(1) 2.5 mM NH_4OH, 4.65 mM NH_4Cl, pH 9 (2) 10 mM NH_4OH, 18.6 mM NH_4Cl, pH 9 (3) 0.5 mM NH_4OH, 0.93 mM NH_4Cl, pH 9, for colloidal silica sols of large particles	358
Benzothiazole sulfenamides (accelators)	10 mM phosphate, 50 mM SDS, 20% methanol (pH 7), or 6 M urea (pH 8.7)	359

TABLE 7.3 (continued)

Compounds	Separation conditions	Ref.
Phthalates (plastizicers)	10 mM SDS, 0.1 M borate–0.05 M phosphate (pH 6)	360
NATURAL PRODUCTS:		
Uvae ursi folium	0.1 M sodium borate buffer, pH 9.5	361
Paeonia radix	0.1 M borate buffer, pH 10.5	224
Ginkgo biloba leaves	20 mM sodium borate buffer, pH 8.3	251
Phytopharmaceut icals: flavonids and phenol carboxylic acid	CITP: Leading: 15 mM HCl, 30% methanol, 0.2% HPMC, adjusted to pH 9.5 with ammediol. Terminating: 10 mM glycine in methanol–water (30 : 70, v/v), adjusted to pH 10.6 by adding barium hydroxide	362
CHIRAL SEPARATION (see also amino acids and drugs):		
PTH-amino acids	25 mM SDVal, pH 7, 10% methanol, 5 M urea	367
PTH-amino acids, benzoin, warfarin	50 mM SDVal, 30 mM SDS, 0.5 M urea, pH 9, 10% (v/v) methanol	365
Optical isomeric drugs: diltiazem hydrochloride, trimetoquinol, hydro-chloride, etc.	bile salts: sodium cholate, sodium taurodeoxycholate, e.g. 0.05 M sodium taurodeoxycholate, in 0.05 M phosphate–borate buffer (pH 7)	234
Inolin	0.05 M sodium taurodeoxycholate, pH 7	366
Thiopental, pentabarbital, dinaphthyl, anthrylethanol	(1) 0.02 M phosphate–borate, pH 9, 0.05 M SDS, 30 mM γ-CD for lipophilic compounds, add 10% methanol (2) 20 mM sodium-d-camphor-10-sulfonate in (1) (3) 0.06 M l-methoxyacetic acid in (1) (4) 15 mM TM-β-CD, 4 M urea, 40 mM sodium-d-camphor-10-sulfonate in (1)	369
Amines which acts as sympathomimetic drugs	18 mM di-OMe-β-CD, 10 mM Tris–H_3PO_4, pH 2.4	236
GITC-amino acids	0.02 M phosphate–borate buffer of pH 9, 0.25 M SDS	368
Propanolol enantiomers	50 mM phosphate, pH 2.5, 4 M urea, 40 mM β-CD, 30% (v/v) methanol	368
CBI-amino acids	10 mM γ-CD, 50 mM SDS, 100 mM borate, pH 9	372

TABLE 7.3 (continued)

Compounds	Separation conditions	Ref.
GEOMETRICAL AND POSITIONAL ISOMERS:		
Positional isomers of amino-benzoic acids	40 mM ammonium acetate, pH 6.8	373
Cis and trans double-bond isomers	10 mM borate, 50 mM phosphate, in 50% (v/v) methanol/water	374
Isotopically substituted compounds	25 mM SDS, 25 mM phosphate, 0.625 M borate, (pH 8) with 2.5, 4.9 or 7.4 M methanol	375
COAL AND FUELS:		
Basic fraction of solvent refined coal	0.05 M tetramethylammonia perchorate, 0.01 M HCl in acetonitrile/water (75:25)	376
Diesel fuel sediment	5% THF and 95% 0.003 M phosphate buffer (pH 10.7)	377
TEXTILE AND DYES:		
Sulfonated azo dyes	60:40 acetonitrile/20 mM NH_4OAc, pH 6.8	70, 378
Morpholine and degradation products in cellulose fibre-production	CITP: Leading: 0.01 M K^+ with morpholine propansulfonate as counter ion, pH 7. Terminating: 0.01 M TBA with ClO_4^-, pH 7.	379
EXPLOSIVES:		
Gunshot and explosives constituents	2.5 or 5 mM borate, 25 mM SDS, pH 7.8 to 8.9	380

7.23 REFERENCES

1 J.W. Jorgenson and K.D. Lukacs, Anal. Chem., 53 (1981) 1298
2 J.W. Jorgenson and K.D. Lukacs, Clin. Chem., 27 (1981) 1551
3 J.W. Jorgenson and K.D. Lukacs, J. Chromatogr., 218 (1981) 209
4 A.S. Cohen, A. Paulus and B.L. Karger, Chromatographia, 24 (1987) 15
5 Y. Walbroehl and J.W. Jorgenson, J. Microcol. Sep., 1 (1989) 41
6 J. Green and J. Jorgenson, J. High Resolut. Chromatogr., 7 (1984) 529
7 E. Gassmann, J. Kuo and R. Zare, Science, 230 (1985) 813
8 M. Roach, P. Gozel and R. Zare, J. Chromatogr., 426 (1988) 129
9 P. Gozel, E. Gassman, H. Michelsen and R.N. Zare, Anal. Chem. , 59 (1987) 44
10 J.W. Jorgenson and K.D. Lukacs, J. High Resolut. Chromatogr., 4 (1981) 230
11 A. Guttman, A. Paulus, A. Cohen, N. Grinberg and B. Karger, J. Chromatogr., 448 (1988) 41
12 I.Z. Atmana, C.J. Metral, G.M. Muschik and H.J. Isaaq, J. Liq. Chromatogr., 13 (1990) 3201
13 T. Tsuda, J.V. Sweedler and R.N. Zare, Anal. Chem., 62 (1990) 2149
14 P. Camilleri and G. Okafo, J. Chromatogr., 541 (1991) 489
15 C. Fujimoto, Y. Muramatsu, M. Suzuki and K. Jinno, J. High Resolut. Chromatogr., 14 (1991) 178

16 Y. Kurosu, K. Hibi, T. Sasaki and M. Saito, J. High Resolut. Chromatogr., 14 (1991) 186
17 Y. Miyashita and S. Terabe, Chromatogram, 11 (1990) 6
18 J. Liu, K. Cobb and M. Novotny, J. Chromatogr., 468 (1988) 55
19 K. Otsuka, S. Terabe and T. Ando, J. Chromatogr., 332 (1985) 219
20 H.E. Meyer, E. Hoffmann-Posorske, H. Korte, A. Donella-Deana, A.M. Brunati, L.A. Pinna, J. Coull., J. Perich, R.M. Valerio and R.B. Johns, Chromatographia, 30 (1990) 691
21 K. Otsuka and S. Terabe, J. Chromatogr., 515 (1990) 221
22 R.T. Kennedy, M.D. Oates, B.R. Cooper, B. Nickerson and J.W. Jorgenson, Science, 246 (1989) 57
23 B. Nickerson and J. Jorgenson, J. High Resolut. Chromatogr., 11 (1988) 533
24 B. Nickerson and J. Jorgenson, J. High Resolut. Chromatogr., 11 (1988) 878
25 M. Yu and N. Dovichi, Anal. Chem., 61 (1989) 37
26 M. Yu and N. Dovichi, Mickrochim. Acta, 111 (1988) 27
27 M. Yu and N. Dovichi, J. Appl. Spectrosc., 43 (1989) 196
28 N.J. Dovichi and Y.F. Cheng, Am. Biotechnol. Lab., 7 (1989) 10
29 M. Albin, R. Weinberger, E. Sapp and S. Morig, Anal. Chem., 63 (1991) 417
30 D.S. Stegehuis, H. Irth, U.R. Tjaden and J. van der Greef, J. Chromatogr., 538 (1991) 393
31 S. Pentoney, X. Huang, D. Burgi and R. Zare, Anal. Chem., 60 (1988) 2625
32 R. Wallingford and A.G. Ewing, Anal. Chem., 59 (1987) 678
33 L. Hernandez, R. Marquina, J. Escalona and N.A. Guzman, J. Chromatogr., 502 (1990) 247
34 J.T.K. Pang, D.J. Kramer and T.R. Tullsen, Curr. Res. Protein Chem.: Tech., Struct., Funct., [Pap. Annu. Symp. Protein Soc.], 3rd, Meeting Date 1989, 11–22. Edited by: J. J. Villafranca, Academic: San Diego, Calif., USA.
35 J. Pawliszyn, Anal. Chem., 60 (1988) 2796
36 T.W. Garner and E.S. Yeung, J. Chromatogr., 515 (1990) 639
37 T.M. Olefirowizc and A.G. Ewing, J. Chromatogr., 499 (1990) 713
38 J.W. Jorgenson and K.D. Lukacs, Science, 222 (1983) 266
39 B.L. Karger, A.S. Cohen and A. Guttman, J. Chromatogr., 492 (1989) 585
40 B.L. Karger, Res. Natl. Bur. Stand., 93 (1988) 406
41 B. Karger, Nature, 339 (1989) 641
42 S. Compton and R. Brownlee, Biotechniques, 6 (1988) 432
43 P.D. Grossman, J.C. Colburn, H.H. Lauer, R.G. Nielsen, R. Riggin, G. Sittampalam and E. Rickard, Anal. Chem., 61 (1989) 1186
44 D. McManigill, S.A. Swedberg, Tech. Protein Chem., T.E. Hugli, (Ed.), Academic: San Diego, CA, (1989) 468
45 W. Steuer, I. Grant and F. Erni, J. Chromatogr., 507 (1990) 125
46 J.R. Mazzeo and I.S. Krull, Biotechniques, 10 (1991) 638
47 J. Green and J. Jorgenson, J. Chromatogr., 352 (1986) 337
48 K.A. Cobb and M. Novotny, Anal. Chem., 61 (1989) 2226
49 K.A. Cobb and M. Novotny, Anal. Chem., 61 (1989) 2226
50 H. Ludi, E. Gassman, H. Grossenbacher and W. Marki, Anal. Chim. Acta, 213 (1988) 215
51 D. Josic and K. Zeilinger, LaborPraxis, 13 (1989) 1111
52 G.L. Hortin, T. Griest and B.M. Benutto, Biochromatogr., 5 (1990) 118
53 P. Grossman, K. Wilson, G. Petrie and H. Lauer, Anal. Biochem., 173 (1988) 265
54 P. Grossman, J. Colburn and H. Lauer, Anal. Biochem., 179 (1989) 28
55 T.A.A.M. van de Goor, P.S.L. Janssen, J.W. van Nispen, M.J.M. van Zeeland and F.M. Everaerts, J. Chromatogr., 545 (1991) 379
56 E. Rickard, M. Strohl, P. Farb and R. Nielsen, Poster presented at HPCE '91, San Diego, Calif., USA, Feb. 3–6, 1991

57 T.E. Wheat, P.M. Young and N.E. Astephen, J. Liq. Chromatogr. , 14 (1991) 987
58 J.R. Florance, Z.D. Konteatis, M.J. Macielag, R.A. Lessor and A. Galdes, J. Chromatogr., 559 (1991) 391
59 A. Pessi, E. Bianchi, L. Chiappinelli, A. Nardi and S. Fanal, J. Chromatogr., 557 (1991) 307
60 R.J. Krueger, T.R. Hobbs, K.A. Mihal, J. Tehrani and M.G. Zeece, J. Chromatogr., 543 (1991) 451
61 F. Stover, B. Haymore and R. McBeath, J. Chromatogr., 470 (1989) 241
62 S.R. Weinberger and T. Schlabech, Poster presented at HPCE '91, San Diego, Calif., USA, Feb. 3–6, 1991
63 R.A. Mosher, Electrophoresis, 11 (1990) 765
64 T. Dulffer, R. Herb, H. Herrmann and V. Kobold, Chromatographia, 30 (1990) 675
65 T. Satow, A. Mechida, K. Funakushi and R. Palmieri, J. High Resolut. Chromatogr., 14 (1991) 276
66 A.D. Tran and T. Blanc and E.J. Leopold, J. Chromatogr., 516 (1990) 241
67 E. Leopold and L. Gouesclou, Spectra 2000, 156 (1991) 27
68 R.D. Smith, J.A. Olivares, N.T. Nguyen and H.R. Udseth, Anal. Chem., 60 (1988) 436
69 J. Loo, H. Udseth and R. Smith, Anal. Biochem., 179 (1989) 404
70 E.D. Lee, W. Mueck, J. Henion and T. Covey, Biomed. Environ. Mass Spectrom., 18 (1989) 844
71 R. Smith, J. Loo, C. Barinaga, C. Edmonds and H. Udseth, J. Chromatogr., 480 (1989) 211
72 R.D. Smith, H.R. Udseth, C.J. Barinaga and C.G. Edmonds, J. Chromatogr., 559 (1991) 197
73 C.G. Edmonds, J.A. Loo, C.J. Barinaga, H.R. Udseth and R.D. Smith, J. Chromatogr., 474 (1989) 21
74 R.D. Smith, J.A. Loo, R.R.O. Loo, M. Busman and H.R. Udseth, Mass Spectrom Rev., 10 (1991) 359
75 I.M. Johansson, E.C. Huang, J.D. Henion and J. Zweigenbaum, J. Chromatogr., 554 (1991) 311
76 M. Moseley, L. Deterding, K. Tomer and J. Jorgenson, J. Chromatogr., 480 (1989) 197
77 M. Moseley, L. Deterding, K. Tomer and J. Jorgenson, Rapid Comm. Mass Spectrom., 3 (1989) 87
78 R. Caprioli, W. Moore, M. Martin, B. Dague, K. Wilson and S. Moring, J. Chromatogr., 480 (1989) 247
79 N.J. Reinhoud, E. Schroder, U.R. Tjaden, W.M.A. Niessen, M.C. ten Noever de Brauw and J. van der Greef, J. Chromatogr., 516 (1990) 147
80 M.A. Moseley, L.J. Deterding, K.B. Tomer and J.W. Jorgenson, Anal. Chem., 63 (1991) 109
81 L.J. Deterding, C.E. Parker, J.P. Perkins, M.A. Moseley, J.W. Jorgenson, J. Chromatogr., 554 (1991) 329
82 V. Rohlicek and Z. Deyl, J. Chromatogr., 494 (1989) 87
83 B.L. Hogan and E.A. Yeung, J. Chromatogr. Sci., 28 (1990) 15
84 E. Yeung, Acc. Chem. Res., 22 (1989) 125
85 M. Strickland and N. Strickland, Amer. Lab., Nov. (1990) 60
86 M. Firestone, P. Michaud, R. Carter and W. Thormann, J. Chromatogr., 407 (1987) 363
87 T. McDonnell and J. Pawliszyn, J. Chromatogr., 559 (1991) 489
88 M.M. Bushey and J.W. Jorgenson, Anal. Chem., 62 (1990) 978
89 T. Bergman, B. Agerberth and H. Joernvall, FEBS Lett., 283 (1991) 100
90 R. Aebersold and H.D. Morrison, J. Chromatogr., 516 (1990) 79
91 J. Green and J. Jorgenson, J. Chromatogr., 478 (1989) 63
92 M. Bushey and J. Jorgenson, J. Chromatogr., 480 (1989) 301
93 J.A. Bullock and L.C. Yuan, J. Microcol. Sep., 3 (1991) 241

94 L. Emmer, M. Jansson and J. Roeraade, J. Chromatogr., 547 (1991) 544
95 S. Hjerten, L. Valtcheva, K. Elenbring and D. Eaker, J. Liq. Chromatogr., 12 (1989) 2471
96 H. Kajiwara, J. Chromatogr., 559 (1991) 345
97 R. McCormick, Anal. Chem., 60 (1988) 2322
98 J.R. Gurley, J.E. London and J.G. Valdez, J. Chromatogr., 559 (1991) 431
99 J. Tehrani, R. Macomber and L. Day, J. High Resolut. Chromatogr., 14 (1991) 10
100 J.R. Gurley, J.S. Buchanan, J.E. London, D.M. Stevert and B.E. Lehnert, J. Chromatogr., 559 (1991) 411
101 A. Vinther, S.E. Bjorn, H.H. Sorensen and H. Soeberg, J. Chromatogr., 516 (1990) 175
102 K.J. Lee and G.S. Heo, J. Chromatogr., 559 (1991) 317
103 F.-T. A. Chen, J. Chromatogr., 559 (1991) 445
104 G. Bruin, J. Chang, R. Kuhiman, K. Zegers, J. Kraak and H. Poppe, J. Chromatogr., 471 (1989) 429
105 J.K. Towns and F.E. Regnier, J. Chromatogr., 516 (1990) 69
106 J.K. Towns and F.E. Regnier, Anal. Chem., 63 (1991) 1126
107 S.A. Swedberg, Anal. Biochem., 185 (1990) 51
108 Y.-F. Maa, K.J. Hyver and S.A. Swedberg, J. High Resolut. Chromatogr., 14 (1991) 65
109 A.M. Dougherty, C.L. Wolley, D.L. William, D.F. Swaile, R.O. Cole and M.J. Sepaniak, J. Liq. Chromatogr., 14 (1991) 907
110 K.A. Cobb, V. Dolnik and M. Novotny, Anal. Chem., 62 (1990) 2478
111 D. Bentrop, J. Kohr and H. Engelhard, Chromatographia, 32 (1991) 177
112 W. Nashabeh and Z. El Rassi, J. Chromatogr., 559 (1991) 367
113 T. Izumi, T. Nagahori and T. Okuyama, J. High Resolut. Chromatogr., 14 (1991) 351
114 R.L. Cunico, V. Gruhn, L. Kresin, D.E. Nitecki and J.E. Wiktorowicz, J. Chromatogr., 559 (1991) 467
115 L. Christiansen, T. Wehr, D. Hansen, J.M. Brosse and J. Le Gall, Spectra 2000, 149 (1990) 35
116 C.A. Bolger, M. Zhu, R. Rodriguez and T. Wehr, J. Liq. Chromatogr., 14 (1991) 895
117 M. Zhu, R. Rodriguez, D. Hansen and T. Wehr, J. Chromatogr., 516 (1990) 123
118 J.E. Wiktorowicz, US 5015350 A 14 Mar., (1991) Applied Biosystems
119 S.A. Swedberg, WO 8912225 A1, 14 Dec., (1989) Hewlett Packard
120 M. Rusnak, J.E. Shively, J.R. Calaycay, WO9009595 A1, 23 Aug., Beckman Res. Ind. (1990)
121 S. Hjerten, K. Elenbring, F. Kilar, J. Liao, A. Chen, C. Siebert and M. Zhu, J. Chromatogr., 403 (1987) 47
122 S. Hjerten and M.D. Zhu, Protides Biol., 33 (1985) 537
123 A. Cohen, A. Paulus and B. Karger, Chromatographia, 24 (1987) 15
124 A. Widhalm, C. Schwer, D. Blaas and E. Kenndler, J. Chromatogr., 549 (1991) 446
125 K. Tsuji, J. Chromatogr., 550 (1991) 823
126 N. Manz, R.G. Galaske, J. Cho and J.B. van Liew, Electrophor., 7 (1986) 183
127 M.J. Gordon, K.J. Lee, A.A. Arias and R.N. Zare, Anal. Chem., 63 (1991) 69
128 E. Kenndler and K. Schmidt-Beiwl, J. Chromatogr., 545 (1991) 397
129 S.-L. Wu, G. Teshima, J. Cacia and W.S. Hancock, J. Chromatogr., 516 (1990) 115
130 R. G. Nielsen,. Sittampalam and E. Rickard, Anal. Biochem., 177 (1989) 20
131 R.G. Nielsen and E.C. Rickard, J. Chromatogr., 516 (1990) 99
132 R.G. Nielsen, E.C. Rickard, P.F. Santa, D.A. Sharknas and G. Sittampalam, J. Chromatogr., 539 (1991) 177
133 A.D. Tran, S. Park, P.J. Lisi, O.T. Huynh, R.R. Ryall and P.A. Lane, J. Chromatogr., 542 (1991) 459
134 K.W. Yim, J. Chromatogr., 559 (1991) 401
135 R.I. Hecht, J. Morris, G.D. Searle, F. Stover, L. Fossey and C. Demarest, Prep. Biochem., 19

(1989) 201
136 R.I. Hecht, J.F. Coleman, J.C. Morris, F. Stover and C. Demarest, Prep. Biochem., 19 (1989) 363
137 B.B. Rosenblum, J. Liq. Chromatogr., 14 (1991) 1017
138 C.Y. Chen, T. Demana, S.D. Huang and M.D. Morris, Anal. Chem., 61 (1989) 1590
139 D.F. Swaile and M.J. Sepaniak, J. Liq. Chromatogr., 14 (1991) 869
140 R.S. Rush, A.S. Cohen and B.L. Karger, Anal. Chem., 63 (1991) 1346
141 F. Foret, S. Fanali and P. Bocek, J. Chromatogr., 516 (1990) 219
142 R.D. Smith, J.A. Loo, C.G. Edmonds, C.J. Barinaga and H.R. Udseth, J. Chromatogr., 516 (1990) 157
143 S.J. Harrington, R. Varro and T.M. Li, J. Chromatogr., 559 (1991) 385
144 N. Banke, K. Hansen and I. Diers, J. Chromatogr., 559 (1991) 325
145 H.C. Birrell and M.D. Brightwell, Poster presented at Int. Conf. Electrophor., York, UK, Aug. 22–24,1990
146 E. Wenisch, C. Tauer, A. Jungbauer, H. Katinger, M. Faupel and P.G. Righetti, J. Chromatogr., 516 (1990) 133
147 Z. Deyl, V. Rohlicek and R. Struzinsky, J. Liq. Chromatogr., 12 (1989) 2515
148 F. Nyberg, M.D. Zhu, J. Liao and S. Hjerten, in Electrophoresis'88, C. Shaefer-Nielsen (Editor), VCH Publishers, New York, (1988) p. 141
149 B.J. Compton and E.A. O'Grady, Anal. Chem., 63 (1991) 2597
150 B.J. Compton, J. Chromatogr., 559 (1991) 357
151 S. Hjerten, J. Chromatogr., 347 (1985) 191
152 S. Hjerten and M. Zhu, J. Chromatogr., 346 (1985) 265
153 M. Zhu, R. Rodriguez and T. Wehr, J. Chromatogr., 559 (1991) 479
154 J.R. Mazzeo and I.S. Krull, Anal. Chem., 63 (1991) 2852
155 D. Josic, A. Bottcher and G. Schmitz, Chromatographia, 30 (1990) 703
156 D. Josic, K. Zeilinger, W. Reutter, A. Bottcher and G. Schmitz, J. Chromatogr., 516 (1990) 89
157 O.N. Kovalenko, E.I. Zorin, I.V. Kucher and E.G. Larskii, Lab. Delo, 10 (1989) 47 (in Russian)
158 J.D. Olechno, J.M.Y. Tso, J.D. Thayer and A. Wainwright, Amer. Lab., Dec. (1990) 30
159 P. Gebauer and W. Thormann, J. Chromatogr., 558 (1991) 423
160 W. Kuhr, Anal. Chem., 62 (1990) 403R
161 K.H. Row, W.H. Griest, M.P. Maskarinec, J. Chromatogr., 409 (1987) 193
162 J. Liu, J. Banks and M. Novotny, J. Microcol. Sep., 1 (1989) 136
163 A. Cohen, S. Terabe, J. Smith and B. Karger, Anal. Chem., 59 (1987) 102
164 T. Kasper, M. Melera, P. Gozel and R. Brownlee, J. Chromatogr., 458 (1988) 303
165 A.-F. Lecoq, C. Leuratfi, E. Marafante and S. DiBiase, J. High Resolut. Chromatogr., 14 (1991) 667
166 A. Lahey and R.L. St. Clare, III, Amer. Lab., Nov. (1990) 68
167 T. Tsuda, G. Nakagawa, M. Sato and K. Yagi, Appl. Biochem., 5 (1983) 330
168 V. Sustacek, F. Foret and P. Bocek, J. Chromatogr., 480 (1989) 271
169 S. Pentoney, R. Zare and J. Quint, Anal. Chem., 61 (1989) 1642
170 S. Pentoney, R. Zare and J. Quint, J. Chromatogr., 480 (1989) 259
171 W.G. Kuhr and E.S. Yeung, Anal. Chem., 60 (1988) 2642
172 L. Gross and E. Yeung, J. Chromatogr., 480 (1989) 169
173 S. Hjerten, Nucleosid. Nucleot., 9 (1990) 319
174 S. Wren, J. Microcol. Sep., 3 (1991) 147
175 A.L. Nguyen, J.H.T. Wong and C. Masson, Anal. Chem., 62 (1990) 2490

176 R. Takigiku and R.E. Schneider, J. Chromatogr., 559 (1991) 247
177 F. Foret and P. Bocek, Chem. Listy, 83 (1989) 191
178 L. Hernandez, B.G. Hoebel and N.A. Guzman, ACS Symp. Ser. 434, (1990) 50
179 Th. P.E. M. Verheggen, A.C. Schoots and F.M. Everaerts, J. Chromatogr., 503 (1990) 245
180 H. Yamamoto, T. Manabe and T. Okuyama, J. Chromatogr., 480 (1989) 331
181 A. Cohen, D. Nagarian, A. Paulus, A. Guttman, J. Smith and B. Karger, Proc. Natl. Acad. (USA), 85 (1988) 9660
182 A. Paulus and J. Ohms, J. Chromatogr., 507 (1990) 113
183 A. Paulus, E. Gassmann and M.J. Field, Electrophor., 11 (1990) 702
184 J.A. Lux, H.-F. Yin and G. Schomburg, J. High Resolut. Chromatogr., 13 (1990) 436
185 J. Macek, U.R. Tjaden and J. van der Greef, J. Chromatogr., 545 (1991) 177
186 Y. Baba, M. Tsuhako, S. Enomoto, A.M. Chin and R.S. Dubrone, J. High Resolut. Chromatogr., 14 (1991) 204
187 Y. Baba, T. Matsuura, K. Wakamoto and M. Tsuhako, J. Chromatogr., 558 (1991) 273
188 T. Wang, G. Bruin and J.C. Kraak, Anal. Chem., 63 (1991) 2207
189 D. Demorest and R. Dubrow, J. Chromatogr., 559 (1991) 43
190 L.A. DeDionsio, HPCE '91, San Diego, Calif., USA, Feb. 3–6, 1991
191 A. Guttman and N. Cooke, Int. Biotech. Lab., May (1991) 24
192 V. Dolnik, J. Liu, J. Banks, M. Novotny and P. Bocek, J. Chromatogr., 480 (1989) 321
193 G. Tschopp, A. Paulus and E. Gassman, Int. Conf. Cap. Electrophor., York, UK, Aug. 22–24, 1990
194 Y. Baba, T. Matsuura, K. Wakamoto, Y. Morita, Y. Nishitsu, and Tsuhako, Anal. Chem. (in press)
195 X. Huang and R.N. Zare, Anal. Chem., 62 (1990) 443
196 D. Rose and J. Jorgenson, J. Chromatogr., 438 (1988) 23
197 A. Cohen, D. Nagarian, J. Smith and B. Karger, J. Chromatogr., 458 (1988) 323
198 D.N. Heiger, A.S. Cohen and B.L. Karger, J. Chromatogr., 516 (1990) 33
199 A. Guttman, A. Cohen, D. Heiger and B. Karger, Anal. Chem., 62 (1990) 137
200 A. Guttman and N. Cooke, Anal. Chem., 63 (1991) 2038
201 A. Guttman and N. Cooke, J. Chromatogr., 559 (1991) 285
202 H. Drossman, J.A. Luckey, A.J. Kostichka, J.D'Cunha and L.M. Smith, Anal. Chem., 62 (1990) 900
203 S.R. Motsch, M.-H. Kleemib and G. Schomburg, J. High Resolut. Chromatogr., 14 (1991) 630
204 H.E. Schwartz, K. Ulfelder, F.J. Sunzeri, P. Busch and R.O. Brownlee, J. Chromatogr., 559 (1991) 257
205 A.M. Chin and J.C. Colburn, Amer. Biotechnol. Lab., 7 (1989) 16
206 P.D. Grossman and D.S. Saone, J. Chromatogr., 559 (1991) 257
207 M. Zhu, D. Hansen, S. Burd and F. Gannon, J. Chromatogr., 480 (1989) 311
208 M. Strege and A. Lagu, Anal. Chem., 63 (1991) 1233
209 T. Demana, M. Lanan and M.D. Morris, Anal. Chem., 63 (1991) 2795
210 J.W. Chen, A. Cohen and B.L. Karger, J. Chromatogr., 559 (1991) 295
211 E. Jackim and C. Norwood, J. High Resolut. Chromatogr., 13 (1990) 195
212 X. Huang, J. B. Shear and R.N. Zare, Poster presented at HPCE '91, San Diego, Calif., USA, Feb. 3–6, 1991
213 X. Huang, J.B. Shear and R.N. Zare, Anal. Chem., 62 (1990) 2049
214 H. Swerdlow and R. Gesteland, Nucl. Acid Res., 18 (1990) 1415
215 H. Swerdlow and S. Wu, H. Harke and N.J. Dovichi, J. Chromatogr., 516 (1990) 61
216 H. Swerdlow, J.Z. Zhang, D.Y. Chen, H.R. Harke, R. Grey, S. Wu and N.J. Dovichi, Anal.

Chem., 63 (1991) 2835
217 A.S. Cohen, D.R. Najarian and B.L. Karger, J. Chromatogr., 516 (1990) 49
218 D.Y. Chen, H.P. Swerdlow, H.R. Harke, J.Z. Zhang and N.J. Dovichi, J. Chromatogr., 559 (1991) 237
219 T.Z. Zheng, D.Y. Chen, S. Wu, H. Harke and N.J. Dovichi, Clin. Chem., 37 (1991) 1492
220 S. Fujiwara and S. Honda, Anal. Chem., 58 (1986) 1811
221 S. Honda, S. Iwase and S. Fujiwara, J. Chromatogr., 404 (1987) 313
222 S. Fujiwara and S. Honda, Anal. Chem., 59 (1987) 2773
223 M.E. Swartz, HPCE '91, San Diego, Calif., USA, Feb. 3–6, 1991
224 S. Honda, K. Suzuki, M. Kataoka, A. Makino and K. Kakehi, J. Chromatogr., 515 (1990) 653
225 H. Nishi, N. Tsumagari and S. Terabe, Anal. Chem., 61 (1989) 2434
226 H. Nishi, N. Tsumagari, T. Kakimoto and S. Terabe, J. Chromatogr., 477 (1989) 259
227 A. Wainwright, J. Microcol. Sep., 2 (1990) 169
228 H. Nishi, T. Sukuyama, M. Matsao and S. Terabe, J. Chromatogr., 515 (1990) 245
229 D. Tsikas, A. Hofrichter and G. Brunner, Chromatographia, 30 (1990) 657
230 C.L. Ng, H.K. Lee and S.F.Y. Li, J. Chromatogr., 598 (1992) 133
231 J.A. Lux, H.-F. Yin and G. Schomburg, Chromatographia, 30 (1990) 7
232 I.M. Johansson, R. Pavelka and J.D. Henion, J. Chromatogr., 559 (1991) 515
233 S.K. Yeo, H.K. Lee and S.F.Y. Li, J. Chromatogr., 585 (1991) 133
234 H. Nishi, T. Fukuyama, M. Matsuo and S. Terabe, J. Microcol. Sep., 1 (1989) 234
235 H. Nishi, T. Fukuyama, M. Matsuo and S. Terabe, J. Chromatogr., 515 (1990) 233
236 S. Fanali, J. Chromatogr., 474 (1989) 441
237 I. Snopek, H. Soini, M. Novotny, E. Smolkova-Keulemansova and I. Jelinek, J. Chromatogr., 559 (1991) 215
238 H. Nishi, T. Fukugama, M. Matsuo and S. Terabe, J. Chromatogr., 513 (1990) 279
239 H. Nishi and M. Matsuo, J. Liq. Chromatogr., 14 (1991) 973
240 P. Meier and W. Thormann, J. Chromatogr., 559 (1991) 505
241 W. Thormann, P. Meier, C. Marcolli and F. Binder, J. Chromatogr., 545 (1991) 445
242 T. Nakagawa, Y. Oda, A. Shibukawa, H. Fukuda and H. Tanaka, Chem. Pharm. Bull., 36 (1988) 1622
243 T. Nakagawa, Y. Oda, A. Shibukawa, H. Fukuda and H. Tanaka, Chem. Pharm. Bull., 37 (1989) 707
244 S.A. Swedberg, J. Chromatogr., 503 (1990) 449
245 K. Salomon, D.S. Burgi and J.C. Helmer, J. Chromatogr., 549 (1991) 375
246 C.P. Ong, C.L. Ng, H.K. Lee and S.F.Y. Li, J. Chromatogr., 547 (1991) 419
247 S. Fujiwara, S. Iwase and S. Honda, J. Chromatogr., 447 (1988) 133
248 I.M. Mutton, Poster presented at HPCE '91, San Diego, Calif., USA, Feb. 3–6, 1991
249 R. Weinberger and M. Albin, J. Liq. Chromatogr., 14 (1991) 953
250 N.A. Guzman, H. Ali, J. Moschera, K. Igbal and A.W. Malick, J. Chromatogr., 559 (1991) 307
251 P.G. Pietta, P.L. Mauri, A. Rava and G. Sabbatini, J. Chromatogr., 549 (1991) 367
252 C.L. Ng, C.P. Ong, H.K. Lee and S.F.Y. Li, Chromatographia, in press
253 W.M. Hurni and W.J. Miller, J. Chromatogr., 559 (1991) 337
254 M.T. Ackermans, F.M. Everaerts and J.L. Beckers, J. Chromatogr., 585 (1991) 123
255 N.A. Guzman, L. Hernandez and J.P. Advis, J. Liq. Chromatogr. , 12 (1989) 2536
256 R. Weinberger and E. Sapp and S. Moring, J. Chromatogr., 516 (1990) 271
257 S. Fanali, J. Chromatogr., 470 (1989) 123
258 T. Tsuda, Y. Kobayashi, A. Hori, T. Matsumoto and O. Suzuki, J. Chromatogr., 456 (1988) 375

259 H. Soini, T. Tsuda and M.V. Novotny, J. Chromatogr., 559 (1991) 547
260 S. Arrowood and A.M. Hoyt, J. Chromatogr., 586 (1991) 177
261 K.D. Atria, C.F. Simpson, A.K. Bharij and A.E. Theobald, Electrophor., 11 (1990) 732
262 C.P. Ong, C.L. Ng, H.K. Lee and S.F.Y. Li, J. Chromatogr., 588 (1991) 335
263 A. Zhu and Y. Chen, J. Chromatogr., 470 (1989) 251
264 A.G. Ewing, R.A. Wallingford and T.M. Olefirowicz, Anal. Chem., 61 (1989) 292A
265 R. Wallingford and A.G. Ewing, Anal. Chem., 60 (1988) 23
266 E. Jellum, A.K. Thorsrud and E. Time, J. Chromatogr., 559 (1991) 455
267 M. Miyake, A. Shibukawa and T. Nakagawa, J. High Resolut. Chromatogr., 14 (1991) 181
268 N.A. Guzman, M.A. Trebilcock and J.P. Advis, J. Liq. Chromatogr., 14 (1991) 997
269 S. Hjerten, Chromatogr. Rev., 9 (1967) 122
270 F. Mikkers, F. Everaerts and T. Verheggen, J. Chromatogr., 169 (1979) 11
271 T. Tsuda, K. Nomura and G. Nakagawa, J. Chromatogr., 264 (1983) 385
272 F. Foret, S. Fanali, L. Ossicini and P. Bocek, J. Chromatogr., 470 (1989) 299
273 M. Deml, F. Foret and P. Bocek, J. Chromatogr., 320 (1985) 159
274 T. Verheggen, J. Beckers and F. Everaerts, J. Chromatogr., 452 (1988) 615
275 J. Beckers, T. Verheggen and F. Everaerts, J. Chromatogr., 452 (1988) 591
276 P. Gebauer, M. Deml, P. Bocek and J. Janak, J. Chromatogr., 267 (1983) 455
277 S. Terabe and T. Isemura, Anal. Chem., 62 (1990) 650
278 T. Romano, P. Jadik, W.R. Jones and P.E. Jackson, J. Chromatogr., 546 (1991) 411
279 J. Sudor, J. Pospichal, M. Deml and P. Bocek, J. Chromatogr., 545 (1991) 331
280 W.R. Jones and P. Jandik, J. Chromatogr., 546 (1991). 445
281 W.R. Jones and P.J. Jandik, Poster presented at HPCE '91, San Diego, Calif., USA, Feb. 3–6, 1991
282 M.T. Ackermans, F.M. Everaerts and J.L. Beckers, J. Chromatogr., 549 (1991) 345
283 W.R. Jones and P.J. Jandik, Int. Ion Chromatogr. Symp. 91, Oct. 6–9, (1991) Denver, CO, USA
284 P. Jandik, W.R. Jones, A. Weston and P.R. Brown, LC-GC, 9 (1991) 634
285 X. Huang, M.J. Gordon and R.N. Zare, Anal. Chem., 60 (1988) 357
286 X. Huang, T. Pang, M. Gordon and R. Zare, Anal. Chem., 59 (1987) 2747
287 X. Huang, M. Gordon and R. Zare, J. Chromatogr., 425 (1988) 385
288 L. Gross and E.S. Yeung, Anal. Chem., 62 (1990) 427
289 T. Kaneta, S. Taneka, M. Taga and H. Yoshida, Anal. Chem., 64 (1992) 798
290 D.F. Swaile and M.J. Sepaniak, Anal. Chem., 63 (1991) 179
291 S. Tanaka, T. Kaneta, K. Nishimura and H. Yoshida, J. Chromatogr., 540 (1991) 475
292 C. Haber, I. Silvestri, S. Roosli, and W. Simon, Chimia, 45 (1991) 117
293 A. Weston, P.R. Brown, P. Jandik, W.R. Jones and A.L. Heckenberg, Poster presented at HPCE '91, San Diego, Calif., USA, Feb. 3–6, 1991
294 S. Maysushita, M. Sushita, I. Motooka and Y Kanagi, J. Chromatogr., 586 (1991) 363
295 P. Bocek, M. Deml and J. Pospichal, J. Chromatogr., 500 (1990) 673
296 P. Oefner, R. Hafele, G. Bartsch and G. Bonn, J. Chromatogr., 516 (1990) 251
297 S. Terabe and T. Isemura, J. Chromatogr., 515 (1990) 667
298 S.J. William, D.M. Goodall and K.P. Evans, Poster presented at Int. Conf. Cap. Electrophor., York, UK, Aug. 22–24, 1990
299 J. Olivares, N. Nguyen, C. Yonker and R. Smith, Anal. Chem., 59 (1987) 1230
300 R. Smith, H. Udseth, J. Loo, B. Wright and G. Ross, Talanta, 36 (1989) 161
301 R. Smith, C. Barinaga and H. Udseth, Anal. Chem., 60 (1988) 1948
302 R. Hallen, C. Shumate, W. Siems, T. Tsuda and H. Hill, J. Chromatogr., 480 (1989) 233
303 T. Saitoh, H. Hoshino and T. Yotsuyanagi, J. Chromatogr., 469 (1989) 175
304 K. Saitoh, C. Kiyohara and N. Suzuki, J. High Resolut. Chromatogr., 14 (1991) 245

305 T. Saitoh, H. Hoshino and T Yotsuyanagi, Anal. Sci., 7 (1991) 495
306 S. Terabe, H. Ozaki, K. Otsuka and T. Ando, J. Chromatogr., 332 (1985) 211
307 S. Terabe, Y. Miyashita, O. Shibata, E.R. Barnhart, L.R. Alexandra, D.G. Paterson, B. L. Karger, K. Hosaya and N. Tanaka, J. Chromatogr., 516 (1990) 23
308 S. Kobayashi, T. Ueda and M. Kikumoto, J. Chromatogr., 480 (1989) 179
309 R.O. Cole, M.J. Sepaniak, W.L. Hinze, J. Gorse and K. Oldiges, J. Chromatogr., 557 (1991) 113
310 Y. Walbroehl and J. Jorgenson, Anal. Chem., 58 (1986) 479
311 G.J.M. Bruin, P.P.H. Tock, J.C. Kraak and H. Poppe, J. Chromatogr., 517 (1990) 557
312 T. Tsuda, K. Nomura and G. Nakagawa, J. Chromatogr., 248 (1982) 241
313 J.H. Knox and I.H. Grant, Chromatographia, 32 (1991) 317
314 H. Watari, Chem. Lett., (1991) 391
315 S. Terabe, K. Otsuka, K. Ichikawa, A. Tsuchiya and T. Ando, Anal. Chem., 56 (1984) 111
316 C.P. Ong, C.L. Ng, H.K. Lee and S.F.Y. Li, J. Chromatogr., 542 (1991) 473
317 H.T. Rasmussen, L.K. Goebel and H.M. McNair, J. Chromatogr., 517 (1990) 549
318 R. Szucs, J.Vindevogel and P. Sandra, J. High Resolut. Chromatogr., 14 (1991) 692
319 X. Huang, J.A. Luckey, M.J. Gordon and R.N. Zare, Anal. Chem. , 61 (1989) 766
320 M.J. Gordon, X. Huang,. L. Pentoney and R.N. Zare, Science, 242 (1988) 224
321 X. Huang, R.N. Zare, S. Sloss and A.G. Ewing, Anal. Chem., 63 (1991) 189
322 S. Hjerten, J. Chromatogr., 270 (1983) 1
323 B.F. Kenney, J. Chromatogr., 546 (1991) 423
324 A.C. Schoots, T.P.E. Verheggen, P.M. De Vries and F.M. Everaerts, Clin. Chem., 36 (1990) 435
325 A.T. Balchunas and M.J. Sepaniak, Anal. Chem., 60 (1988) 617
326 A.T. Balchunas and M.J. Sepaniak, Anal. Chem., 59 (1987) 1446
327 M. Sepaniak and R. Cole, Anal. Chem., 59 (1987) 472
328 M.J. Sepaniak, D.F. Swaile, A.C. Powell and R.O. Cole, J. High Resolut. Chromatogr., 13 (1990) 679
329 A.C. Powell and M.J. Sepaniak, J. Microcol. Sep., 2 (1990) 278
330 C.L. Ng, H.K. Lee and S.F.Y. Li, J. Chromatogr. Sci. (in press)
331 D.R. Rose and J.W. Jorgenson, J. Chromatogr., 447 (1988) 117
332 J. Vindevogel, P. Sandra and L.C. Verhagen, J. High Resolut. Chromatogr., 13 (1990) 295;
333 J. Vindevogel, R. Szucs and P. Sandra, J. High Resolut. Chromatogr., 14 (1991) 584; J. Vindevogel and P. Sandra, High Resolut. Chromatogr., 14 (1991) 795
333a R. Rodrigues, M. Zhu and T. Wehr, HPCE '91, San Diego, Calif., USA, Feb. 3–6, 1991
334 J. Karovicova, M. Drdak and J. Polonsky, J. Chromatogr., 509 (1990) 283
335 C.W. Earle and N.J. Dovichi, J. Liq. Chromatogr., 12 (1989) 2575
336 S. Honda, S. Iwase, A. Makino and S. Fujiwara, Anal. Biochem., 176 (1989) 72
337 S. Honda, K. Suzuki and K. Kakehi, Anal. Biochem., 177 (1989) 62
338 J. Liu, O. Shirota and M. Novotny, Anal. Chem., 63 (1991) 413
339 J. Liu, O. Shirota and M. Novotny, J. Chromatogr., 559 (1991) 223
340 W. Nashabeh and Z. El Rassi, J. Chromatogr., 514 (1990) 57
341 W. Nahsabeh and Z. El Rassi, J. Chromatogr., 536 (1991) 31
342 S. Honda, K. Yamamoto, S. Suzuki, M. Ueda and K. Kakihi, J. Chromatogr., 588 (1991) 327
343 L. N. Amankwa and W.G. Kuhr, Anal. Chem., 63 (1991) 1733
344 S.L. Carney and D.J. Osborne, Anal. Biochem., 195 (1991) 132
345 Z. Deyl, I. Miksik and R. Struzinsky, J. Chromatogr., 516 (1990) 287
346 C.P. Ong, C.L. Ng, N.C. Chong, H.K. Lee and S.F.Y. Li, Environ. Monitor. Asses., 19 (1991) 93

347 C.D. Gaitonde and P. Pathak, J. Chromatogr., 514 (1990) 389
348 C.P. Ong, C.L. Ng, N.C. Chong, H.K. Lee and S.F.Y. Li, J. Chromatogr., 516 (1990) 263
349 Y.F. Yik, H.K. Lee, S.F.Y. Li, J. High Resolut. Chromatogr., 15 (1992) 198
350 Y.F. Yik, C.P. Ong, S.B. Khoo, H.K. Lee and S.F.Y. Li, Environ. Monit. Asses., 19 (1991) 73
351 F. Foret, V. Sustacek and P. Bocek,Electrophoresis, 11 (1990) 95
352 V. Dombek and Z. Stransky, Anal. Chim. Acta, 256 (1992) 69
353 B. VanOrmann and G. McIntire, J. Microcol. Sep., 1 (1989) 289
354 B.B. VanOrman and G.L. McIntire, Amer. Lab., Nov. (1990) 66
355 H.K. Jones and N.E. Ballou, Anal. Chem., 62 (1990) 2484
356 L.N. Amankwa, J. Scholl and W.G. Kuhr, Anal. Chem., 62 (1990) 2189
357 J. Zweigenbaum, Chromatogram, 11 (1990) 9
358 R.M. McCormick, J. Liq. Chromatogr., 14 (1991) 939
359 M.W. F. Nielen and M.J. A. Mensink, J. High Resolut. Chromatogr., 14 (1991) 417
360 C.P. Ong, H.K. Lee and S.F.Y. Li, J. Chromatogr., 542 (1991) 473
361 E. Kenndler, Ch. Schwer, B. Fritsche and M. Pohm, J. Chromatogr., 514 (1990) 383
362 V. Seitz, G. Bonn, P. Oefner and M. Popp, J. Chromatogr., 559 (1991) 499
363 K. Otsuka and S. Terabe, Jasco Rep., 33 (1991) 1
364 R.O. Cole, M.J. Sepaniak and S.L. Hinze, J. High Resolut. Chromatogr., 13 (1990) 579
365 K. Otsuka, J. Kawahara, K. Takekawa and S. Terabe, J. Chromatogr., 559 (1991) 209
366 H. Nishi, T. Fukuyama, M. Matsuo and S. Terabe, Anal. Chim. Acta, 236 (1990) 281
367 K.O. Otsuka and S. Terabe, Electrophor., 11 (1990) 982
368 H. Nishi, T. Fukuyama and M. Matsuo, J. Microcol. Sep., 2 (1990) 234
369 H. Nishi and T. Fukuyama, J. Chromatogr., 553 (1991) 503
370 S. Fanali, P. Bocek, Electrophoresis, 11 (1990) 757
371 S. Fanali, J. Chromatogr., J. Chromatogr., 545 (1991) 437
372 T. Ueda, F. Kitamura, R. Mitchell, T. Metcalf, T. Kuwana and A. Nakamoto, Anal. Chem., 63 (1991) 2981
373 M.W.F. Nielen, J. Chromatogr., 542 (1991) 173
374 R.R. Chadwick and J.C. Hsieh, Anal. Chem., 63 (1991) 2377
375 M. Bushey and J. Jorgenson, J. Microcol. Sep., 1 (1989) 125
376 Y. Walbroehl and J.W. Jorgenson, J. Chromatogr., 315 (1984) 135
377 B. Wright, G. Ross and R. Smith, Energy and Fuels, 3 (1989) 428
378 E.D. Lee, W. Mueck, J. Henion and T. Covey, Biomed. Environ. Mass Spectrom, 18 (1989) 253
379 A. Widhalm and E. Kenndler, Anal. Chem., 63 (1991) 645
380 D.M. Northrop, D.E. Martire and W.A. MacCrehan, Anal. Chem., 63 (1991) 1038

CHAPTER 8

Recent Advances and Prospect for Growth

There is no doubt that CE is at a stage of fast growth. Exciting advances are being made rapidly at the present moment. In this section, the latest (January–April, 1992) developments are described. The main emphasis is to cover aspects not discussed in earlier chapters in order to provide a most up-to-date and comprehensive treatment of techniques based on CE.

In addition to technical advantages, a variety of factors may contribute to the growth of an analytical technique. In the last section of this chapter, the criteria which are considered to be the most important for future growth of CE will be examined.

8.1 RECENT REVIEWS ON CE

Two reviews on CE have been published recently. The theory, mechanism, instrumentation and separation modes were discussed by Carchon and Eggermont [1]. Historical developments that have occurred in Europe were the focus of the paper by Campos and Simpson [2]. Topics covered included sample introduction, separation capillary and detection.

8.2 ADVANCES IN INJECTION TECHNIQUES

On-column sample concentration using field amplified CZE (see Section 2.2.2) have been discussed by Chen and Burgi [3]. Sample stacking in continuous buffer systems, in large volume injections, in electroinjections, and in polarity-switch injections are described. With these techniques, 100-fold improvements in detectability of analytes have been obtained compared with injections with no stacking. The concentration limits of detection achievable are comparable to the best obtainable in HPLC.

Computer simulations of electrokinetic injection have been performed [4]. The effects of diffusion due to concentration gradient at the capillary inlet, and inadvertent hydrodynamic flow caused by a difference in levels of liquids in the two buffer reservoirs on the effective injection volumes in CZE were determined.

The results of the simulation show that the effects of diffusion on injection are less significant for larger molecules, and for fast injection sequences. The use of gel-filled capillaries reduces the effect of diffusion. In addition, calculations show that inadvertent hydrodynamic flow effects can be reduced by decreasing the injection time and the delays between the injection steps. The use of gel-filled columns would also serve to eliminate problems caused by inadvertent hydrodynamic flow.

8.3 NOVEL DETECTION TECIINIQUES

Significant new developments in CE detection include the first demonstrations of chemiluminescence detection [5], semiconductor laser fluorimetry [6] and a laser-excited confocal fluorescence detection system [7]. A review of fluorescence detection in CE has been given [8]. A concentration gradient detector [9] and a concentration imaging system [10] have been developed for detection in capillary isoelectric focusing (see Section 6.7.5). Electrochemical detection with an on-column joint made of a conductive polymer (Nafion) has been studied [11]. CE with electrochemical detection has been reviewed [12]. Analyses of the quantitative aspects of indirect absorption detection [13] and the use of binary buffers for this mode of detection [14] have been investigated. The role of CE in electrospray ionization-mass spectrometry of large polypeptides and proteins was discussed in a recent review [15]. In the following sections, some of the interesting developments on detection techniques for CE are described.

8.3.1 Chemiluminescence detection

In chemiluminescence detection, analytes react with a labelling agent to emit light, without the need for an excitation source as in fluorescence detection. As with fluorescence, chemiluminescence has the potential to be exploited as a highly sensitive and specific detection method.

A system with a post-column reactor was used to demonstrate the feasibility of chemiluminescence detection in CE [5]. The basis of the detection scheme is the reaction between luminol and its derivatives with hydrogen peroxide in the presence of a catalyst in an alkaline solution which results in light emission in the wavelength range 425–435 nm. The electrophoretic buffer consisted of 10 mM phosphate buffer, adjusted to pH 9.9 with 1 M potassium hydroxide and 0.2 M hydrogen peroxide. Detection limits of 100 amol (3×10^{-9} M) for luminol and 400 amol (7×10^{-9} M) for N-(4-aminobutyl)-N-ethylisoluminol (ABEI) were obtained.

8.3.2 Semiconductor laser fluorimetry

Laser-induced fluorescence has demonstrated excellent sensitivity in CE detection (see Chapter 3). Commercial CE instruments with this type of detectors have recently become available. Parallel developments in semiconductor laser technology have been taking place. Currently, semiconductor lasers are widely used in printers, compact disk players, and bar-code readers. These lasers are orders of magnitude cheaper than conventional laser sources. They are extremely compact, durable and easy to replace or to maintain. With suitable fluorophores and frequency-doubling to shorter wavelengths when necessary, laser-fluorescence detection with semiconductor laser sources is potentially a highly sensitive and inexpensive detection method for CE.

An investigation has been performed to study the use of semiconductor lasers in different detection arrangements [6]. Direct determination of chlorophyll could be achieved with a diode laser emitting at 670 nm (2 mW). Indirect fluorometry with fluorescent additives, e.g. methylene blue, rhodamine 800 and oxzine 750, in the buffer produced a detection limit of approximately 1 pmol (10^{-5} M) for amino acids. A labelling reagent was synthesized which contained a thiazine chromophore for fluorescence detection and a succinimidyl ester for combination with an amino acid. With the 670 nm diode laser, detection limits of 10 pmol for amino acids labelled with this reagent was obtained. Another semiconductor laser frequency doubled to 415 nm (5 μW) was used for the detection of the same amino acids. Detection limit achieved was about 100 amol (10^{-9} M).

8.3.3 Laser-excited confocal fluorescence detection

Laser-excited confocal fluorescence detection in CE has been demonstrated recently [7]. In this approach, an argon-ion laser is scanned across an array of capillaries. Detection is facilitated by the use of epi-illumination, i.e. the same microscope objective is employed to focus the laser onto the capillary, and to collect the emitted fluorescence. This method is more suitable for detection of multiple capillaries compared with conventional laser-induced fluorescence detection techniques. Confocal excitation and detection ensures that the depth of view is confined to within the interior of the capillary. A limit of detection of 2×10^{-12} M ($S/N = 3$) has been obtained for fluorescein in the scanning mode, compared with 1×10^{-12} M for the stationary mode obtained by focusing the laser on a single capillary.

8.3.4 Indirect UV detection

Indirect UV detection is another detection method currently receiving a lot of interest [13]. In indirect detection, the dynamic range is limited, i.e. high sensitivity is obtained only for analytes which have similar electrophoretic mobility to that of the background chromophore ion. Hence the response factors are different for different analytes. Another problem is that at low concentration of the background, poor stability may be observed, due to the effect of impurities and pH shifts on the electroosmotic flow [16]. For the analysis of C_9–C_{13} primary sodium alkyl sulfates, it was found that by selecting UV-absorbing ions with the appropriate buffering capability under the pH conditions used (e.g. veronal buffer for the sulfates), uniform response factor and linear dynamic range of two orders of magnitude could be obtained. Detection limits of about 10^{-5} M were observed.

When monobuffer systems are used for indirect detection, the pH range is limited by the pK of the acid and the base. Peak shape may become poor when the mobility of the analyte differs greatly from that of the buffer. By using binary buffer containing two visualization agents, peak sharpness can be improved. The theoretical background of this approach was examined. Practical demonstration was performed by using 0.25 mM 1,3,5-benzenetricarboxylic acids (BTA) and 0.25 mM 1-naphthylacetic acid (NAA) as the binary buffer for the analysis of a mixture containing fast and slow migrating species. Fast migrating species have similar mobilities as BTA, whereas slow migrating species have similar mobilities as NAA. Consequently, improved peak shapes are observed.

8.3.5 Electrochemical detection

An electrochemical detection system was developed which employed an on-column Nafion joint for decoupling the high voltage from the detection system [11]. The design is similar to previous on-column fracture electrochemical detection systems (see Section 3.6), except that it appears to be easier to construct and more durable. With this system, a detection limit of 6×10^{-9} M (34.8 amol) was obtained for hydroquinone. Analyses of phenolic acids in apple juice and naphthalene-2,3-dicarboxaldehyde derivatized amino acids in a brain homogenate were performed using this system.

8.4 ADVANCES IN COLUMN TECHNOLOGY

Interesting developments have been taking place in column technology [17–23]. A recent demonstration of capillary electrophoresis on a chip has been described in Section 4.5. Several new procedures for coating capillaries have been developed [17–19]. Capillary gel electrophoresis has also been investigated [20–23].

8.4.1 New types of column coatings

Enantiomeric separation was achieved by using capillaries coated with Chiral-Dex [17]. The separation of enantiomers of (±)-1,1′- binaphthyl-2,2′-diylhydrogenphosphate and (±)-1-phenylethanol was achieved with high efficiency.

A new procedure [18] for coating capillaries involved first bonding a cross-linked polymer containing reactive functional groups (epoxy) onto the capillary surface. Subsequently, a hydrophilic coating, either polyethylene glycol (PEG) or polyethyleneimine (PEI), was bonded to this reactive layer. These coatings were stable over the pH range 3–9 [18]. Improvements in separation of a substituted benzoic acid mixture, and basic peptides were obtained compared with the use of uncoated columns. Satisfactory separation of proteins with pI $\geq$ 7 was obtained using a PEI coated column. For proteins with pI $\leq$ 7, PEG coated columns were found to be more suitable.

In an alternative approach, enzymes were immobilized on the inner walls of short fused-silica capillaries [19]. The short sections of coated capillary was coupled to an uncoated fused silica capillary for separation. After the analytes were introduced into the capillary, they first reacted with the immobilized enzymes in the absence of an electric field. Subsequently, the voltage was turned on and separation commenced. For the analysis of ATP, adenosine, and RNA, the enzyme used included hexokinase, adenosine deamidase and ribonuclease T1.

8.4.2 Progress in capillary gel electrophoresis

The use of gel-filled capillaries for nucleic acid separation was discussed in a recent review [20]. Peak dispersion and separation efficiency in capillary gel electrophoresis was described on the basis of the Einstein-Nernst equation between diffusion coefficient and ionic mobility [21]. It was found that the effective charge number would be the only analyte-specific parameter which influenced the plate height or the plate number [21].

A method for the production of polyacrylamide gel-filled columns was described by Baba *et al.* [22]. A new injection device for introducing the gel was described. Plate numbers in the range $3–7 \times 10^6$ per meter was obtained for the separation of polyadenylic acids.

Migration behaviour of oligonucleotides in capillary gel electrophoresis has been studied by Guttman *et al.* [23]. An equation was proposed which can be used to predict relative migration times under denaturing and non-denaturing conditions. Observed and predicted relative migration times of various homo- and hetero-oligodeoxyribonucleotides were in good agreement.

8.5 PROGRESS ON ELECTROLYTE SYSTEMS

A review on micellar electrokinetic chromatography has been presented by Terabe [24]. Separation principles and resolution in MEKC were first described. The effects of surfactant type, mixed micelles, pH, temperature, and additives on selectivity were then discussed. Some typical applications were also presented.

Little and Foley [25] employed SDS and a mixed micellar system consisting of SDS and polyoxyethylene [23] dodecanol (Brij 35) for the separation of PTH-amino acids. Higher efficiency and selectivity, and lower operating currents were obtained with the mixed micellar system.

8.6 NEW SYSTEMS AND METHODS

As CE develops into a more sophisticated level, more advanced systems and methods are being explored. There is now a growing interest in exploring various programming techniques. Temperature programming [26], pH gradient [27], external field control [28,29], analyte velocity modulation [30] and pulsed-field CE [31] have been investigated recently. The use of CE and related techniques for the measurements of physico-chemical properties has also been reported [32,33].

8.6.1 Programming techniques

Temperature programming influences migration behavior by affecting both pH and viscosity of the buffer. For certain analyte/buffer system, an appropriate temperature program can also be employed to speed up the analysis. Temperature variations with respect to time or to position along the capillary can be introduced. The application of temperature programming techniques to the separation of a group of organic acids was demonstrated [26]. This method has several advantages over other programming techniques. Firstly, problems caused by inhomogeneity in mixing are eliminated. There is no limitation imposed by the magnitude or direction of electroosmotic flow. Another advantage is that temperature programming can be implemented relatively easily. On the other hand, precautions must be taken to ensure that the system is able to respond rapidly to temperature changes, and that Joule heat can still be dissipated efficiently at higher temperatures.

Tsuda developed a solvent program delivery system for capillary electrophoresis [27]. The design was based on a HPLC gradient pump and column couplings. Typical examples of pH gradient were demonstrated. The separation of a mixture containing quinin, 5-bromouracil, dansylated L-serine and dansylated L-cysteric acid was performed for different gradients. Migration order could be changed or shortened with suitable choices of pH gradients.

By coating the external surface of the capillary with a conductive polymer sheath, a radial voltage can be applied to control electroosmotic flow in CZE [28]. An inverse hyperbolic sine function was derived to represent the trends in electroosmotic flow as a function of the radial voltage. By varying the applied radial voltage, electroosmotic flow velocity can be changed, e.g. from 8 cm/min at −14 kV to about 4 cm/min at 6 kV.

Lee and co-workers [29] investigated the effect of external field control on peptide and protein separations by CE. An external field can be applied to influence directly the zeta potential at the buffer/capillary interface (see Section 6.4). An application of the technique is to reduce adsorption of peptides and proteins by the capillary wall. Different external applied voltages were used and the effect on the migration behavior was studied. The results demonstrate that separation efficiency can be enhanced by controlling the zeta potential and the electroosmotic flow with the use of an external field.

Control of electroosmotic flow can also be achieved by modulating the separation voltage, i.e. by superimposing an AC voltage on the DC separation voltage. This technique is referred to as analyte velocity modulation, which has been employed to enhance sensitivity in concentration gradient detection (see Section 3.5). During analyte velocity modulation, a radial concentration gradient is generated, which may be employed as a means to affect electroosmosis [30].

Alternatively, a pulsed electric field can be used to improve resolution and to facilitate fraction collection (see Section 4.3.6). In a recent study, the wave form in CGE employing pulsed field was found to be unsymmetrical [31]. Methods to improve the waveform have been proposed.

8.6.2 Capillary array electrophoresis

By using a scanning laser-excited confocal fluorescence detection system (see Section 8.3), simultaneous detection of separation in multiple capillaries can be performed readily [7]. The technique is termed capillary array electrophoresis (CAE). Images have been obtained by scanning a four-capillary array containing a sample of fluorescent primer-labelled M13mp 18 DNA "G" fragments, and a 24-capillary array containing a sample of the "T" fragments of the same DNA. With multiple sample capacities, and high sensitivity detection, capillary array electrophoresis is likely to emerge as an important technique for DNA sequencing. Furthermore, it is a promising approach to extend the capability of CE for micropreparative separations.

8.6.3 Determination of physico-chemical properties

Capillary zone electrophoresis and isotachophoresis have been employed to determine physico-chemical properties. The ionization constants of weak bases, such as aniline and p-anisidine, and ampholytes, e.g. p-aminobenzoic acid, were determined by CZE. Good agreement was obtained between the experimental data and literature values [32]. Electrophoretic mobilities and dissociation constants of twenty tripeptides and several oligoglycines have been measured by isotachophoresis in the pH range 8.1–9.5. The results were in good agreement with theoretical values [33].

8.7 ADDITIONAL APPLICATIONS BASED ON CE

New applications developed based on CE are listed in Table 8.1. Analyses of amino acids [34–38], peptides [39,40], proteins [41–43], body fluids [44,45], pharmaceuticals and drugs [46–51], ions [52–54], carbohydrates [55,56], food [57,58], natural products [59,60], environmental pollutants [61,62] and toxins [51] have been investigated.

TABLE 8.1

ADDITIONAL CE APPLICATIONS

Analytes	Separation conditions	Ref.
AMINO ACIDS (AAs):		
DNP-AAs (dinitrophenyl-amino acids)	0.05 M phosphate, 47.5 mM SDS, 20 mM TBA (pH 10)	34
NDA-AAs (naphthalene-2,3-dicar-boxy-aldehyde-amino acids)	100 mM borate, 50 mM SDS, 10 mM γ-CD (pH 9)	35
CBQCA-AAs (3-(4-carboxybenzoyl)-2-quinolinecarboxaldehyde-amino acids)	0.05 M TES buffer, 50 mM SDS (pH 7.02)	36
CBQCA-AAs	20 mM sodium phosphate/borate, 0.1 mM diamopentane (pH 9.5)	37
FTH-AAs (fluorescein thiohydan-hydantoin-amino acids)	10 mM phosphate (pH 7)	38
PEPTIDES:		
Tryptic peptides	25 mM sodium citrate (pH 4)	39
Tryptic and chymotryptic digests of trypsinogen	0.05 M CAPS, pH 9.5	40
Cyanogen bromide digests of trypsinogen	Capillary coating: linear polyacrylamide, buffer: 0.04 M citric acid, pH 3.0	40
Tryptic digests of reduced and alkylated human serum albumin	(1) Capillary coating: linear poly-acrylamide. 0.05 M CHES, pH 9.7 (2) Uncoated capillary. 0.05 M CAPS, pH 9.1, 0.06 M SDS, 10% acetonitrile	40
Cyanogen bromide digests of human serum albumin with intact disulfide bridges or reduced and alkylated disulfide bridges	Capillary coating: linear polyacrylamide, buffer: 0.05 M glutamine, triethylamine (pH 9.7)	40
Tryptic digests of oxidized lysozyme	0.05 M sodium phosphate (pH 2.3)	40
PROTEINS:		
Native tryptophane or tyrosine-containing proteins	5 mM phosphate (pH 10.2)	41
Recombinant chimeric glycoprotein	Column coated with Micro-Coat (Applied Biosystems), 50 mM sodium citrate-acetate (pH 5.2)	42
Isoforms of monoclonal antibody	CIEF: Anolyte: 10 mM phosphoric acid Catholyte: 20 mM sodium hydroxide Mobilizer: 20 mM sodium hyroxide and 80 mM sodium chloride	43

TABLE 8.1 (continued)

Analytes	Separation conditions	Ref.
BODY FLUIDS:		
Purines in body fluids	borate/phosphate buffer containing 75 mM SDS (pH 9)	44
Biological thiols and their S-nitrosated derivatives	0.01 M sodium phosphate, 0.01 M HCl (pH 2.3)	45
PHARMACEUTICALS AND DRUGS:		
Hydroquinone and ether derivatives in skin-toning cream	0.01 M borate (pH 9.5), 0.075 M SDS, 10% (v/v) methanol	46
Sulfonamides in pork meat extract	0.02 M phosphate, 0.02 M borate (pH 7)	47
Trifluperidol (R2498), Domperidone (R33812) and Motillium$_{TM}$ formulation and major known impurities	(1) citrate-phosphate (pH 2) (2) citrate-phosphate (pH 4) (3) citrate-HCl (pH 4)	48
2-hydroxy-3,5-diiodobenzoic acid (T 926) isomers	Borate–KCl buffer (pH 11), 10 mM and tetrabutylammonium hydrogen sulfonate	48
Benzetimide (R4929)	Citrate–NaOH–HCl buffer (pH 4)	48
Cefixime	20 mM phosphate buffer (pH 6.8) containing 3-[(3-cholamindopropyl)-dimethylamino]-1-propanesulfonate (PDAPS) (0.15%, w/v) and methanol (20%, v/v)	49
Salbutamol, terbutaline sulfate and fenoterol hydrobromide in pharmaceutical dosage forms	(1) Tris–acetate buffer (pH 5), after rinsing capillary with 0.1 M KOH (2) Histidine-acetate (pH 5)	50
Antibacterial drugs used in aquaculture industry	TRIZMA buffer (pH 2)	51
INORGANIC IONS:		
Inorganic anions	Phosphate buffer at pH 7 with 0.2 mM, 1 mM or 25 mM CTAC	52
Inorganic cations	10 mM Waters UVCat-1, 4 mM α-hydroxyisobutyric acid (HIBA), adjusted to pH 4.4 with acetic acid	53
Calcium ions	CITP: Leading: 0.01 M potassium acetate-acetic acid (pH 5.4). Terminating: 0.01 M n-hexanoic acid in presence of collagen	54
CARBOHYDRATES:		
1-Phenyl-3-methyl-5-pyrazolone (PMP) derivatives of reducing carbohydrates	Buffer containing alkaline metal salts, e.g. 100 mM calcium acetate	55

TABLE 8.1 (continued)

Analytes	Separation conditions	Ref.
Oligosaccharides [from hydrolysis] of poly (galact uronic acid)	CGE: Gel: 18% T, 3% C Buffer: 0.1 M Tris, 0.25 M borate, 2 mM EDTA (pH 8.48)	56
FOOD ANALYSIS:		
Phytate in soybeans	50 mM benzoic acid titrates at pH 6.2 with L-histidine	57
Hop bitter acids	40 mM SDS, pH 8.55 in 25 mM Tris buffer	58
POLYMERS:		
Poly (styrenesulfonate)	25 mM KH_2 PO_4 (pH 5), with 5.23 mg/ml or 10.03 mg/ml of HEC	59
NATURAL PRODUCTS:		
Leucinostatins	0.1 M phosphate buffer (pH 2.5)	60
Flavonol aglycones in plant extracts	25 mM borate (pH 8.3), 30 mM SDS	61
ENVIRONMENTAL POLLUTANTS:		
Polycyclic aromatic hydrocarbons	10 mM SDS, 2 mM γ -CD in 0.05 M phosphate–0.1 M borate buffer (pH 7)	62
Pyrethroid insecticides: alphamethrine and cypermethrine in water in soils	CITP: Leading: 0.01 M HCl, with creatinine and 0.05% PVA, pH_L = 4.8 Terminating: 0.005 M MES	63
TOXINS:		
Paralytic shellfish poisoning toxins	TRIZMA buffer (pH 7.2)	51

8.8 FUTURE TRENDS

One of the most likely questions to be raised by newcomers to the field as well as experienced users of CE at this point of time may be "What is the future potential of CE?". Although it is impossible to predict with certainty possible future developments, the capabilities of CE have been proven, the potential of the technique is tremendous and the prospect for growth of this relatively new separation technique appears to be excellent.

Table 8.2 presents a summary of the capability and potential of CE relative to HPLC and slab-gel electrophoresis. For a separation technique to gain popularity, the understanding of the principles of separation, efficiency, reproducibility, sensitivity, affordability and automation must reach an acceptable level. In these

TABLE 8.2

CAPABILITY AND POTENTIAL OF CE RELATIVE TO HPLC AND SLAB-GEL ELECTROPHORESIS

Criteria	CE	HPLC	Slab-gel
1. Principles	proven	proven	proven
2. Efficiency	very high	mod. high	mod. high
3. Reproducibility	high	high	high
4. Sensitivity:			
Mass	excellent	good	good
Concentration	good	excellent	good
5. Affordability	economical	economical	economical
6. Automation	easy	easy	mod. easy
7. Speed	rapid	rapid	slow
8. Quantitation	mod. easy	easy	difficult
9. Sample requirement capacity	nl range	μl and above	μl and above
10. Methods to control separation mechanism	developing	developed	developed
11. Selectivity:			
Diversity	excellent	excellent	limited
Implementation	easy	mod. easy	tedious
12. Developed methods	few	many	many
13. Literature sources	mod. extensive	extensive	extensive
14. Orthogonality:			
To HPLC	orthogonal	–	orthogonal
To slab-gel	orthogonal	orthogonal	–
15. Safety precautions	high voltage	high pressure	mod. high voltage
16. Equipment maintainence	simple	mod. simple	simple
17. Environmental friendliness (solvent usage/waste)	little waste and solvent	solvent/ waste	some waste
18. Consumable costs:			
Column/plate	low	mod. low	mod. low
Solvent/buffer gel	low	mod. low	mod. low
Lamps/light sources, electrodes	low	low	low
19. Prospect for future growth	excellent	mature	mature

Mod. = moderately

respects, CE compares favorably with existing techniques. In terms of speed of analysis, selectivity, quantitation, methods to control separation mechanism, orthogonality, CE performs better than conventional electrophoresis and is at about the same level as that of HPLC. There are clear advantages in using CE when environmental friendliness, consumable costs and equipment maintainence are considered. As for sample capacity or requirement, CE has the advantage that only minute amounts of samples are required, although it is not suitable for the separation and collection of large amount of samples. The areas which CE do not compare well with HPLC and slab-gel electrophoresis are the relatively fewer developed methods and literature sources on CE, mainly due to the much shorter history of CE. However, these are not indications of weakness of CE. Instead, these factors are positive indicators of ample room for further development of the technique.

The growth rate of CE has been increasing rapidly. A statistical analysis of the historical development of CE has been performed [64], which shows that there has been an exponential growth in publications and citations on CE for the period 1980–1991. The trend is expected to continue or even accelerate. CE is clearly becoming one of the most powerful separation techniques ever developed.

8.9 REFERENCES

1 H. Carchon and E. Eggermont, Amer. Lab., 24 (1992) 67
2 C.C. Campos and F.F. Simpson, J. Chromatogr., 30 (1992) 53
3 R. Chen and D.S. Burgi, Anal. Chem., 64 (1992) 489
4 E.V. Dose and G. Guiochon, Anal. Chem., 64 (1992) 123
5 R. Dadoo, L.A. Colon and R.N. Zare, J. High Resolut. Chromatogr., 15 (1992) 134
6 T. Higashijima, T. Fuchigami, T. Imasaka and N. Ishibashi, Anal. Chem., 64 (1992) 711
7 X. Huang, M.A. Quesada and R.A. Mathies, Anal. Chem., 64 (1992) 967
8 L.N. Amankwa, M. Albin and W.G. Kuhr, Trends Anal. Chem., 11 (1992) 114
9 J. Wu and J. Pawliszyn, Anal. Chem., 64 (1992) 225
10 J. Wu and J. Pawliszyn, Anal. Chem., 64 (1992) 219
11 T.J. O'Shea, R.D. Greenhagen, S.M. Lunte, C.E. Lunte, M.R. Smyth, D.M. Radzik and N. Watanabe, J. Chromatogr., 593 (1992) 305
12 Y.F. Yik and S.F.Y. Li, Trends Anal. Chem., 11 (1992) 325
13 M.W.F. Nielen, J. Chromatogr., 588 (1992) 321
14 T. Wang and R.A. Hartwick, J. Chromatogr., 589 (1992) 307
15 R.D. Smith, J.A. Loo, R.R. Ogorzalek, M. Busman and H.R. Udseth, Mass Spectrom. Rev., 10 (1991) 359
16 G.J. Bruin, G. Stegeman, A.C. van Asten, X. Xu, J.C. Kraak and H. Poppe, J. Chromatogr., 559 (1992) 163
17 S. Mayer and V. Schurig, J. High Resolut. Chromatogr., 15 (1992) 129
18 M. Huang, W.P. Vorkink and M.L. Lee, J. Microcol, 4 (1992) 135
19 Z. El. Rassi, Symp. on Column Liquid Chromatogr., Basel, Switzerland, 3–7 June, 1991
20 Y. Baba and M. Tsuhako, Trends Anal. Chem., in press (1992)
21 E. Kenndler and C. Schwer, J. Chromatogr., 595 (1992) 313
22 Y. Baba, T. Matsuura, K. Wakamoto, Y. Morita, Y. Nishitsu and M. Tsuhako, Anal. Chem., (1992)
23 A. Guttman, R.J. Nelson and N. Cooke, J. Chromatogr., 595 (1992) 297
24 S. Terabe, in: Capillary Electrophoresis Technology, N. Guzman (ed.), Marcel Dekker, New

York, N.Y. (in press)
25 E.L. Little and J.P. Foley, J. Microcol. Sep., 4 (1992) 145
26 C. Whang and E.S. Yeung, Anal. Chem., 64 (1992) 502
27 T. Tsuda, Anal. Chem., 64 (1992) 387
28 M.A. Hayes and A.G. Ewing, Anal. Chem., 64 (1992) 512
29 C. Wu, T. Lopes, B. Patel and C.S. Lee, Anal. Chem., 64 (1992) 886
30 T. Demana, U. Guhathakurta and M.D. Morris, Anal. Chem., 64 (1992) 390
31 D.N. Heiger, S.M. Carson, A.S. Cohen and B.L. Karger, Anal. Chem., 64 (1992) 192
32 J. Cai, J.T. Smith and Z. El. Rassi, J. High Resolut. Chromatogr., 15 (1992) 30
33 T. Hirokawa, R. Sugino and Y. Kiso, J. Chromatogr., 585 (1991) 364
34 Y.F. Yik and S.F.Y. Li, J. Chromatogr. (submitted)
35 T. Ueda, R. Mitchell, F. Kitamura and T. Metcalf, J. Chromatogr., 593 (1992) 265
36 J. Lu, Y. Hsieh, D. Wiesler and M. Novotny, Anal. Chem., 63 (1991) 408
37 B. Wanders, Beckman Application Note, BBU19210B (1992)
38 S. Wu and N.J. Dovichi, Talanta, 39 (1992) 173
39 P.M. Young, N.E. Astephen and T.E. Wheat, LC-GC, 10 (1992) 26
40 K.A. Cobb and M.V. Novotny, Anal. Chem., 64 (1992) 879
41 T.T. Lee and E.S. Yeung, J. Chromatogr., 595 (1992) 319
42 K. Tsuji and R.J. Little, J. Chromatogr., 594 (1992) 317
43 C. Silverman, M. Komar, K. Shields, G. Diegnan and J. Adamovics, J. Liq. Chromatogr., 15 (1992) 207
44 W. Thormann, A. Minger, S. Molteni, J. Casllavska and P. Gebauer, J. Chromatogr., 593 (1992) 275
45 J.S. Stamler and J. Loscalzo, Anal. Chem., 64 (1992) 779
46 I.K. Sakodinskaya, C. Desiderio, A. Nardi and S. Fanali, J. Chromatogr., 596 (1992) 95
47 M.T. Ackermans, J.L. Beckers and F.M. Everaerts, J. Chromatogr., 596 (1992) 101
48 A. Pluym, W. Van. Ael and M. De. Smet, Trends Anal. Chem., 11 (1992) 27
49 S. Honda, J. Chromatogr., 590 (1992) 364
50 M.T. Ackermans, J.L. Beckers, F.M. Everaerts and I.G.J.A. Seelen, J. Chromatogr., 590 (1992) 341
51 S. Pleasance, P. Thibault and J. Kelly, J. Chromatogr., 591 (1992) 325
52 T. Kaneta, S. Tanaka, M. Taga, H. Yoshida, Anal. Chem., 64 (1992) 798
53 A. Weston, P.R. Brown, P. Jandik, W.R. Jones and A.L. Heckenberg, J. Chromatogr., 593 (1992) 289
54 S. Matsushita, M. Sugita, I. Motooka, Y. Kanaji, J. Chromatogr., 586 (1991) 363
55 S. Honda, K. Yamamoto, S. Suzuki, M. Ueda and K. Kakehi, J. Chromatogr., 588 (1991) 327
56 J. Liu, O. Shirota and M.V. Novotny, Anal. Chem., 64 (1992) 973
57 A. Nardi, M. Cristalli, C. Desiderio, L. Ossicini, S.K. Shukla and S. Fanali, J. Microcol., 4 (1992) 9
58 J. Vindevogel and P. Sandra, J. High Resolut. Chromatogr., 14 (1991) 796
59 J.B. Poli and M.R. Schure, Anal. Chem., 64 (1992) 896
60 M.G. Quaglia, S. Fanali, A. Nardi, C. Rossi and M. Ricci, J. Chromatogr., 593 (1992) 259
61 C. Wesen and H. Mu, J. High Resolut. Chromatogr., 15 (1992) 136
62 Y.F. Yik, C.P. Ong, S.B. Khoo, H.K. Lee and S.F.Y. Li, J. Chromatogr., 589 (1992) 333
63 V. Dombek and Z. Stransky, Anal. Chim. Acta., 256 (1992) 69
64 T. Braun and S. Nagydiósi-Rózsa, Trends Anal. Chem., 10 (1991) 266

Subject Index

JOURNAL OF CHROMATOGRAPHY LIBRARY

A Series of Books Devoted to Chromatographic and Electrophoretic Techniques and their Applications

Although complementary to the *Journal of Chromatography,* each volume in the Library Series is an important and independent contribution in the field of chromatography and electrophoresis. The Library contains no material reprinted from the journal itself.

Other volumes in this series

CPSIA information can be obtained at www.ICGtesting.com
Printed in the USA
LVOW07s2159100315

429997LV00004B/29/P

9 780444 815903